HAZARDOUS
WASTES owl

Sources, Pathways, Receptor

Richard J. Watts

Department of Civil & Environmental Engineering
Washington *ersity*

JOHN WILEY & SONS,
New York • Chichester • Weinheim • Brisbane • ngapore • Toronto

Acquisitions Editor: Wayne Anderson
Marketing Manager: Katherine Hepburn
Production Editor: Tony VenGraitis
Text and Cover Designer: Harry Nolan
Illustration Editor: Sigmund Malinowski
Electronic Illustrations: Radiant
Cover Photographs: Background: Biohazard cylinders © Richard Gaul 1996, FPG International
　　　　　　　　　　Upper right: Recycling hazardous chemicals, Casa Grande, AZ. © Jeff Heger, 1994, FPG International
　　　　　　　　　　Lower left: Asbestos removal, C.T. Tracy, 1994, FPG International.
Outside Production Coordination: York Production Services

To order books or for customer service please, call 1(800)-CALL-WILEY (225-5945).

Library of Congress Cataloging in Publication Data:
Watts, Richard J.
　　Hazardous wastes : sources, pathways, receptors / Richard J. Watts.

　　　　p.　　cm.
　　Includes index.
　　ISBN 0-471-00238-0 (cloth : alk. paper)
　　1. Hazardous wastes.　　I. Title.

　TD1030.W38　1997

　628.4'2—dc21

97-28360

CIP

10 9 8 7 6 5 4 3

For Maura, Leah, Emily, and Jackson

Preface

Hazardous waste courses have recently been established at many universities throughout the United States; they vary in focus from management (regulations, manifest forms, etc.) to the design of unit processes for treating wastes disposed of under the Resource Conservation and Recovery Act (RCRA). This book is based on a course developed at Washington State University to provide senior and M.S. students with the scientific principles of hazardous waste management and engineering. In developing the course, and subsequently this book, I considered fundamental concepts that should be presented in an introductory hazardous waste class. After determining the knowledge required of entry-level engineers and scientists by consulting firms, industry, and government, and assessing the knowledge needed by graduate students in advanced hazardous waste classes, I developed material that covers the following topics:

- terminology, nomenclature, and properties of hazardous wastes and materials;
- behavior of hazardous chemicals in surface impoundments, soils, groundwater, and treatment systems;
- assessment of the toxicity and risk associated with exposure to hazardous chemicals;
- strategies to find information on nomenclature, transport and behavior, and toxicity for hazardous compounds; and
- application of the scientific principles of hazardous wastes to their management, remediation, and treatment.

In selecting the material for the book, I made an effort to avoid duplication of topics presented in standard environmental engineering, environmental science, and hydrogeology courses currently offered by most institutions. I also tried to develop material that would be fundamental in nature and tried to design a text that would be an educational document rather than a training manual.

Organization

The book is divided into three major parts—"Sources," "Pathways," and "Receptors"—and a fourth part that extends the fundamental principles—"Management and Design Applications." After an introductory chapter that describes hazardous waste problems and hazardous waste legislation, Chapter 2 of "Sources" provides information on nomenclature and structure of common hazardous contaminants, what industrial operations have generated the different classes of waste materials, and the types of contamination that have resulted from their disposal. In Chapter 3, the basic properties of common contaminants, such as water solubility, density, and chemical incompatibility, are covered. Source analysis, focusing on waste audits in industrial facilities, assessment of contaminated sites, sampling, and chemical analysis, is the topic of Chapter 4.

Using quantitative problem solving, "Pathways" provides a conceptual basis for understanding the behavior of hazardous chemicals, whether they are present in soil and groundwater systems, in hazardous waste landfills, in storage tanks, or in treatment systems such as air stripping towers. Chapter 5 covers partitioning phenomena, including theory, isotherms, and estimating sorption in soil-water systems. In Chapter 6, material on the theory of volatilization is presented, including equations for estimating volatilization rates from surface impoundments and soils. Concepts of abiotic and biotic transformations as a basis for the natural attenuation of contaminants at hazardous wastes sites and the design of treatment systems are covered in Chapter 7. The material in Chapters 5 through 7 is integrated in Chapter 8, in which the atmospheric and subsurface transport of hazardous chemicals away from the source is presented.

If a hazardous contaminant moves in the environment by one of the routes described in "Pathways," receptors (e.g., humans or wildlife) may be affected—the basis for the "Receptors" part. Chapter 9 deals with fundamental human and mammalian toxicology and explains the ways in which chemicals may be toxic. Quantitative toxicology and industrial hygiene are covered in Chapter 10, which serves as a basis for assessing the toxicity of hazardous contaminants. Chapter 11 emphasizes all of the concepts from previous chapters by addressing risk assessment. Risk is a function of exposure (covered under "Pathways") and hazard (covered under "Receptors"). Using the material of Chapters 2 through 10, the student not only becomes capable of conceptualizing hazardous waste dynamics and exposure through quantitative problem solving, but also develops the ability to perform elementary risk assessments.

In the final part, "Management and Design Applications," the fundamental principles of *sources, pathways*, and *receptors* are applied to hazardous waste management, remediation, and treatment. Remediation and treatment designs may be considered applications of the pathways covered in Part Two. An overview of pollution prevention, remediation, treatment and disposal is presented in Chapter 12. The principles learned in "Pathways" are then applied to the design of selected hazardous waste treatment systems in Chapter 13, based on each of the four chemodynamic pathways:

Pathway	Treatment Application
Sorption	Granular activated carbon
Volatilization	Air stripping
Abiotic transformations	Advanced oxidation processes
Biotic transformations	Slurry bioreactors

Use of This Book

Hazardous Wastes: Sources, Pathways, Receptors contains enough material to allow flexibility in teaching a one-semester hazardous waste course. If the class is taught in an environmental science or hydrogeology program where students do not have a design emphasis, Chapters 1 through 11 will provide a science-based hazardous waste course. A civil engineering course with a 33% design content would include Part Four, "Management and Design Applications," but omit parts of Chapters 5, 9, 10, and 11. Another option is a two-semester sequence in hazardous waste engineering. The first semester would emphasize engineering science and use Chapters 1 through 11

ond semester would consist of the design of hazardous waste treatment systems and could use Chapters 12 and 13 along with selected design manuals or hazardous waste design texts as they become available.

Chapter 2, "Common Hazardous Wastes: Nomenclature, Industrial Uses, Disposal Histories," should no doubt be covered, at least in part, if students have not completed organic chemistry. If organic chemistry is a prerequisite for the class, covering Chapter 2 may not be necessary. One option for students with a background in organic chemistry may be to rapidly cover Chapter 2 so that they have familiarity with chemicals germane to hazardous waste management (e.g., chlorinated solvents, PCBs, dioxins). Another alternative would be to provide information on classes of chemicals at specific points of the text. For example, detailed information on solvents could be presented in Chapter 6, "Volatilization," because most solvents are volatile organic compounds.

Some sections of the text have less problem solving content, and may receive less emphasis in lectures and deferred to the student as reference material. Some of these sections include 2.6, "Explosives"; 2.10, "Metals and Inorganic Nonmetals"; 4.6, "Sampling away from the Source"; 11.3, "Ecological Risk Assessments"; and a number of topics in Chapters 9 and 10 on toxicology.

Based on the flexibility inherent in the text, potential emphases of a one-semester course include (1) hazardous wastes with a science emphasis, and (2) hazardous wastes with engineering science and design components.

Hazardous Wastes (Science Emphasis). A fundamental approach to the concepts of hazardous wastes, with the study of both currently generated hazardous wastes and the assessment and characterization of contaminated sites, would focus on the majority of Parts One, "Sources," Two, "Pathways," and Three, "Receptors":

Chapter	Topic
1	Introduction
2	Common Hazardous Wastes: Nomenclature, Industrial Uses, Disposal Histories
3	Common Hazardous Wastes: Properties and Classification
4	Source Analysis
5	Partitioning, Sorption, and Exchange at Surfaces
6	Volatilization
7	Abiotic and Biotic Transformations
8	Dynamics of Transport away from the Source
9	Concepts of Hazardous Waste Toxicology
10	Quantitative Toxicology
11	Hazardous Waste Risk Assessment

Hazardous Wastes (Engineering Science with Engineering Design Components). Because engineering science (covered in Part Two, "Pathways") serves as the basis for engineering design, a one-semester hazardous waste class with approximately 33%–50% engineering design content would focus primarily on Parts Two, "Pathways," and Four, "Management and Design Applications," with support from Part One, "Sources":

Chapter	Topic
1	Introduction
2	Common Hazardous Wastes: Nomenclature, Industrial Uses, Disposal Histories
3	Common Hazardous Wastes: Properties and Classification
4	Source Analysis
5	Partitioning, Sorption, and Exchange at Surfaces
6	Volatilization
7	Abiotic and Biotic Transformations
8	Dynamics of Transport away from the Source
12	Approaches to Hazardous Waste Minimization, Remediation, Treatment, and Disposal
13	Design of Selected Pathway Applications

Other emphases may also be created using appropriate sections of the text. Some other potential areas of emphasis include contaminated site management, RCRA hazardous waste management, contaminant fate and transport, and hazardous waste risk assessment.

Metric units are used in most cases throughout the book, with English units following parenthetically. The only cases in which English units receive primary emphasis are in the presentation of historical or anecdotal information, which occurs mostly in Chapters 1 and 2.

Acknowledgments

Completion of this text would have not been possible without the assistance of a group of gracious colleagues and students. My former graduate student Alex Jones served as an invaluable aide de camp during his time in Pullman by tracking down references, developing tables and figures, solving example and chapter problems, and proofreading the manuscript. My colleagues at Washington State University, particularly Drs. Wade Hathhorn, Jim Hoover, Brian Lamb, and Amy Teel, provided helpful suggestions on the manuscript. Dr. Hoover was instrumental in obtaining photographs of the Hanford nuclear reservation, and Dr. Teel's eye for detail was invaluable as she helped in proofreading the manuscript. Drs. Tom Hess of the University of Idaho and Dave Atkinson of the Idaho National Engineering and Environmental Laboratory also provided valuable comments. Numerous former students provided assistance, including Brett Bottenberg, Gerry Brown, Jay Bower, Brett Freeborn, Dan Haller, Michael Harrington, Jimmy Howsakeng, Pat McGuire, Rolf Parsloe, and Cindy Spencer. I would also like to thank Kathy Cox for typing sections of the manuscript.

The John Wiley & Sons review team provided thorough guidance on the pedagogical development, presentation, and accuracy of the manuscript. This excellent group of reviewers included Dr. Paul Bishop (University of Cincinnati), Dr. Peter Fox (Arizona State University), Dr. Mriganka Ghosh (University of Tennessee), Dr. John Harkin (University of Wisconsin), Dr. Steve Hrudey (University of Alberta), Dr. Glenn Miller (University of Nevada), Dr. Jeffrey Peirce (Duke University), and Dr. Jim Seiber (University of Nevada). I am also indebted to the editors at John Wiley & Sons, including the late Cliff Robichaud, Wayne Anderson, Catherine Beckham, and Sharon Smith. I would also like to thank Tony VenGraitis and Sigmund Malinowski at John Wiley & Sons and Kirsten Kauffman at York Production Services.

On a more personal note, I would like to thank my wife, Lennis Boyer, who tolerated long hours of writing while she was practicing medicine. Finally, the seeds for this project were no doubt planted decades ago by my mother and late father, who taught me the importance of environmental quality at an early age by inspiring me to stalk *Salmo aquabonita* through alpine "greensfields."

Richard J. Watts
Pullman, Washington

Acronyms and Abbreviations

AA - Atomic Absorption

ACGIH - American Conference of Governmental Industrial Hygienists

ACMA - Agricultural Chemicals Manufacturing Association

ADI - Acceptable Daily Intake

AOPs - Advanced Oxidation Processes

ARARs - Applicable or Relevant and Appropriate Requirements

BCF - Bioconcentration Factor

BDST - Bed Depth Service Time

BTEX - Benzene, Toluene, Ethylbenzene, and Xylenes

CAA - Clean Air Act

CAS - Chemical Abstract Service

CCA - Copper chrome arsenate

CDI - Chronic Daily Intake

CEC - Cation exchange capacity

CERCLA - Comprehensive Environmental Response, Compensation and Liability Act

CERCLIS - Comprehensive Environmental Response, Compensation and Liability Information System

CFR - Code of Federal Regulations

CFSTR - Continuous Flow Stirred Tank Reactor

CMA - Chemical Manufactures Association

CMC - Critical Micelle Concentration

CSI - Common Sense Initiative

CWA - Clean Water Act

2,4-D - 2,4-Dichlorophenoxyacetic acid

DBCP - 1,2-Dibromo-3-chloropropane

DCE - Dichloroethylene (various isomers)

DDT - Dichlorodiphenyltrichloroethane

DNAPL - Dense Nonaqueous Phase Liquid

DOT - Department of Transportation

DSMA - Disodium methyl arsenate

EBDC - Ethylene-*bis*-dithiocarbamate

ECD - Electron Capture Detector

ED - Effective dose

EDB - Ethylene dibromide

ELP - Environmental Leadership Program

EP - Extraction Procedure (toxicity test)

EPA - Environmental Protection Agency

EPCRA - Emergency Planning and Community Right-to-Know Act

FID - Flame Ionization Detector

GAC - Granular activated carbon

GC - Gas Chromatography

HAP - Hazardous Air Pollutant

HI - Hazard Index

HLW - High-level wastes (nuclear)

HPLC - High-Performance Liquid Chromatography

HRS - Hazard Ranking System

HSWA - Hazardous and Solid Waste Amendments of 1984

IARC - International Agency for Research on Cancer

ICP - Inductively Coupled Plasma

IUPAC - International Union of Pure and Applied Chemists

LD - Lethal dose

LEPC - Local Emergency Planning Committee
LFL - Lower Flammability Limit
LLW - Low-level wastes (nuclear)
LNAPL - Light Nonaqueous Phase Liquid
MCL - Maximum Contaminant Level
MEK - Methyl ethyl ketone
MIBK - Methyl isobutyl ketone
MS - Mass Spectrometer
MSDS - Material Safety Data Sheet
MSMA - Monosodium methyl arsenate
NAAQS - National Ambient Air Quality Standards
NCP - National Contingency Plan
NFPA - National Fire Protection Association
NOAEL - No Observed Adverse Effect Level
NPDES - National Pollutant Discharge Elimination System
NPL - National Priorities List
OCDD - Octachlorodibenzo-*p*-dioxin
OU - Operable Unit
OVA - Organic Vapor Analyzer
PA - Preliminary Assessment
PAHs - Polycyclic Aromatic Hydrocarbons
PCBs - Polychlorinated Biphenyls
PCDDs - Polychlorinated Dibenzo-*p*-dioxins
PCDFs - Polychlorinated Dibenzofurans
PCE - Perchloroethylene
PCP - Pentachlorophenol
PFR - Plug Flow Reactor
PID - Photoionization Detector
PPA - Federal Pollution Prevention Act (of 1990)
PRP - Potentially Responsible Party
QSARs - Quantitative Structural-Activity Relationships
RCRA - Resource Conservation and Recovery Act

RI/FS - Remedial Investigation/ Feasibility Study
RfD - Reference Dose
ROD - Record of Decision
SARA - Superfund Amendments and Reauthorization Act (of 1986)
SCAP - Superfund Comprehensive Accomplishments Plan
SDWA - Safe Drinking Water Act
SERC - State Emergency Response Commission
SF - Slope factor
SI - Site Inspection
SOC - Soil organic carbon
SOM - Soil organic matter
STEL - Short-Term Exposure Limit
SVE - Soil Vapor Extraction
2,4,5-T - 2,4,5-Trichlorophenoxyacetic acid
TCA - 1,1,1-Trichloroethane
TCDD - 2,3,7,8-Tetrachlorodibenzo-*p*-dioxin
TCE - Trichloroethylene
TEF - Toxicity Equivalent Factor
TLV - Threshold Limit Value
TNT - 2,4,6-Trinitrotoluene
TOC - Total Organic Carbon
TPH - Total Petroleum Hydrocarbons
TRI - Toxics Release Inventory
TCLP - Toxicity Characteristic Leaching Procedure
TSCA - Toxic Substances Control Act
TSD - Treatment, Storage, and Disposal (facilities)
TWA - Time-Weighted Average
UFL - Upper Flammability Limit
UN/NA - United Nations/North American
USTs - Underground Storage Tanks
VOA - Volatile Organic Analysis
VOCs - Volatile Organic Compounds
WHO - World Health Organization

Table of Contents

CHAPTER 3:   COMMON HAZARDOUS WASTES: PROPERTIES AND CLASSIFICATION   155

CHAPTER 7:   ABIOTIC AND BIOTIC TRANSFORMATIONS   333

CHAPTER 8:   CONTAMINANT RELEASE AND TRANSPORT FROM THE SOURCE   405

PART FOUR: MANAGEMENT AND DESIGN APPLICATIONS   545

CHAPTER 12:   APPROACHES TO HAZARDOUS WASTE MINIMIZATION, REMEDIATION, TREATMENT, AND DISPOSAL   546

CHAPTER 13:   DESIGN OF SELECTED PATHWAY APPLICATIONS   588

APPENDICES

Chapter 1

Introduction

Waste materials are a part of the high standard of living to which we have become accustomed in an industrialized society. The manufacture of products that we use in everyday life results in the generation of wastes, some of which may be persistent, toxic, flammable, corrosive, or explosive. For example, the production of computer and semiconductor components requires halogenated solvents. Aircraft construction and maintenance activities generate petroleum, solvent, and heavy metal wastes. The synthesis of plastics, paints, and pesticides produces organic solvents, by-products, and sludges. Every industry that has produced manufactured goods has also generated wastes.

Industries, municipalities, and government currently generate between 30 and 60 million tons of hazardous waste per year in the United States that are subject to federal regulations. Hazardous waste not covered by federal legislation but regulated by states is conservatively thought to add another 230 to 260 million tons per year, with some estimates as high as 750 million tons per year [1.1, 1.2]. Therefore, even conservative estimates put the annual rate of hazardous waste generated in the United States at 1 ton per person [1.3]. You may think, "I surely do not produce one ton of hazardous waste per year!" But if you eat food grown by agribusiness, drive a car, ride a bicycle, use a computer, read books or newspapers, own furniture, own anything coated with paint, get clothes laundered at a dry cleaner, buy household cleaning products, fly in airplanes, heat your residence with electricity or petroleum, write with a pen, . . . , you have contributed to the hazardous waste problem.

The quantity and diversity of hazardous wastes have grown with the progression of technology. Until the 1800s, most materials used in homes and industries were natural products such as lard or plant extracts. As the world entered the petroleum age in the nineteenth century, kerosene and other petroleum distillates were used as solvents and fuels. From the 1930s through the 1950s, chemists discovered that the industrial properties of petroleum products could be improved by a variety of synthetic techniques. One of the processes developed was halogenation (i.e., adding halogens—chlorine, fluorine, or bromine—to petroleum-based chemicals). Halogenated aliphatic solvents were found to be more effective degreasing agents than petroleum distillates and had an additional benefit of nonflammability [1.4]. Halogenated pesticides were found to have properties that enhance their toxicity and corresponding control of insects, plants, and fungi [1.5]. However, many of these halogenated organic compounds persist for

1

long periods of time and are harmful to public health and to the environment—characteristics that lead us to consider what constitutes a hazardous waste.

The term *hazardous waste* does not have an exact scientific definition because of the wide range of properties that can make a chemical a threat to public health or the environment. Some of the hazardous effects of chemicals include short-term toxicity to humans, long-term toxicity to humans, ecotoxicity, flammability, explosivity, and corrosivity. Furthermore, each of the thousands of chemicals used by industry is characterized by a different degree of hazard for any one of these characteristics. Consider one of these effects—short-term toxicity to humans. For the thousands of chemicals that are potential hazardous wastes, a range of short-term toxicities may be found. For example, exposure to a few milligrams of one chemical may cause death to the average adult. At the other end of the spectrum, adults may survive after ingesting kilograms of a different chemical. Between these two extremes lies a gradient of toxicities for each of the thousands of different chemicals. If the definition of a hazardous waste is based on acute toxicity, would a dose of 1 mg provide the definition of hazardous? Should 5 mg be the standard? Would 100 mg be sufficient? The same case may be made for other characteristics and effects, such as chronic toxicity or explosivity.

Due to the difficulty of quantifying the detrimental characteristics of chemicals, hazardous wastes are typically defined by government regulations—a procedure that is by no means perfect or without controversy. Nonetheless, a commonly used general definition of a hazardous waste is *a waste that, due to its chemical activity or flammable, explosive, toxic, or corrosive properties, is likely to result in danger to human health or the environment* [1.6]. Detailed regulatory definitions of hazardous wastes are presented in Section 1.2, but before examining the regulatory definitions of hazardous waste and the associated requirements for their safe treatment and disposal, a perspective of hazardous waste problems is appropriate. Hazardous waste management has an intriguing history, and some examples of recent hazardous waste episodes provide cultural and scientific perspectives for the forthcoming regulatory definitions of hazardous wastes.

1.1 PREREGULATORY DISPOSAL OF HAZARDOUS WASTES

1.1.1 Past Disposal Practices

The use of industrial chemicals has increased steadily over the past 50 years. Trends in the production of benzene, vinyl chloride, acrylonitrile, and polychlorinated biphenyls (PCBs), shown in Figure 1.1, document a general increase for each of the first three compounds with the exception of aberrations related to recessions and market fluctuations. Their rates of production are representative of the tens of thousands of chemicals in use throughout the United States and the world. The production of PCBs is characterized by a different trend: These persistent chemicals were banned in 1979 because trace residuals, which cause cancer, were detected throughout the United States population. A positive outcome of its discontinued use is the corresponding decrease in PCB human tissue concentrations [1.8].

Where were these and other chemicals disposed of before strict guidelines were outlined under the first federal hazardous waste mandate—the Resource Conservation and Recovery Act of 1976? Most disposal occurred through routes that were easiest for

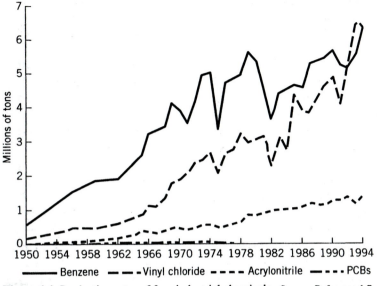

Figure 1.1 Production rates of four industrial chemicals. *Source:* Reference 1.7.

maintenance and industrial personnel. Workers would simply find the most simple and least-cost means to dispose of waste chemicals, which usually meant storing them indefinitely on the company's property, placing them in a shallow pit, or sending them off to a landfill with nonhazardous solid waste. Jerry K. was a plant superintendent during the 1950s and 1960s for a company that produced about 5000 tons of pesticides per year. Although the plant production rate was not high, the company still generated significant quantities (approximately three hundred 55-gallon drums per year) of concentrated waste chemical precursors such as hexachlorocyclopentadiene, chlorophenols, benzene, chlorobenzenes, and solvents (see Chapter 2). Jerry initially directed his employees to dump the drum contents onto the soil near the boundary of the plant property. Later, he had them dig a shallow pit in which to empty the drums. Jerry saw nothing wrong with the process—the waste simply disappeared and the problem was solved. In fact, the problem was just postponed and made worse. Like Jerry, most industrial workers decades ago did not realize that hazardous chemicals persist for decades or centuries, and that a 5-gallon bucket of solvent has the potential to contaminate an entire groundwater system. Furthermore, they had no comprehension of the potential adverse health effects of these chemicals. At thousands of sites throughout the United States and the world, petroleum derivatives, solvents, pesticides, transformer fluids, and electroplating wastes were disposed of improperly. The waste chemicals then migrated through surface soils into groundwater, where they were resistant to abiotic and biological degradation processes. Twenty years after Jerry began dumping the chemicals onto the soil, benzene was discovered in a water supply well 3 miles away. Another chemical, *p*-dichlorophenol, was found in the well 6 months later. An environmental consulting firm, after drilling and sampling 80 monitoring wells, found almost all of the chemicals that had been "disposed of." Every few weeks a new chemical would appear in the groundwater samples, suggesting that each waste compound traveled at a different rate in the subsurface.

Mike S., an electronic technician for a major utility company, was responsible for restoring power after thunderstorms. While his boss was answering telephone calls in attempts to calm the tempers of utility customers, Mike frantically washed electrical contacts with solvents while transformer oil leaked onto the ground. Spillage and disposal were minor concerns when the phone was ringing off the hook with angry customers. After a dozen such power outages the substation soil was contaminated with chlorinated solvents and PCB transformer oils.

Before strict regulatory measures were promulgated in 1980, numerous "quick and dirty" disposal techniques were used. Some of the more common disposal practices that contributed to the problems are described next.

Soil Spreading. Waste liquids, especially lubricating oils and other petroleum residues, were commonly spread on soils and unpaved roads. The practice not only provided a means of disposing of the waste, but also mitigated blowing dust. However, one of the problems with placing petroleum on soils is the presence of cancer-causing polycyclic aromatic hydrocarbons (PAHs). In addition, petroleum products were sometimes mixed with waste solvents and pesticides. The worst case of soil spreading with tainted waste oils occurred at Times Beach, Missouri (see Section 2.9).

Pesticide Rinse and Formulation Areas. Agricultural chemicals have been used widely in the United States and throughout the world. In almost all cases, pesticides have been received by farmers and aerial applicators in concentrated form and then diluted on site for application. The areas where the pesticides have been formulated, which may be next to a barn or near the landing strip used by aerial applicators, are usually open spaces containing little or no vegetation. Large polyethylene vats have been used to mix the concentrated pesticides with water. Because of the remote nature of many of these sites and the lack of regulatory control, spillage and improper disposal of the rinsate have become commonplace. These areas have become contaminated from both past and present disposal practices because many of the pesticides, especially those used in the 1940s through 1970s, are resistant to natural degradation processes.

Underground Storage Tanks. Throughout the twentieth century, approximately 5 million underground storage tanks (USTs) were installed to hold gasoline, jet fuel, solvents, heating oil, and other industrial compounds. Unfortunately, soil is a corrosive environment and, after a relatively short period of time, the tanks rust, corrode, and leak. Recent studies have shown that only 1.8% of tanks less than 5 years old leak, but the most common tank age range for leakage is 11 to 25 years. A few resistant tanks (11.8%) last longer than 25 years [1.9]. All of the tanks installed throughout the United States are expected to leak eventually, and some estimates have 10 to 25% of tanks in the United States leaking at any one time [1.10]. The obvious result of such leaks is the saturation of soil with the chemicals and eventual contamination of groundwater. A common sight, which may be witnessed in the smallest town to the largest city, is the excavation of a UST at a corner gasoline station (Figure 1.2). Such "tank-yanking" activities can be costly for independent station owners. Strict UST installation regulations have been incorporated into the Resource Conservation and Recovery Act, and a growing number of petroleum distributors have begun using aboveground tanks, such as those shown in Figure 1.3.

Figure 1.2 Excavation of a leaking underground storage tank (UST) at a closed gasoline station.

Figure 1.3 Aboveground storage tanks, an alternative to USTs.

Pits/Ponds/Lagoons. Before the regulated management of hazardous wastes, many industries disposed of their chemicals on site by placing them in unlined pits, ponds, and lagoons. Workers simply dug a pit into which wastes were poured; the wastes then disappeared by seeping through the soil. In less permeable soils, the wastes were held at the surface, so these areas were called ponds or lagoons (Figure 1.4).

Sanitary Landfills. Every sanitary landfill that was designed to accept newspapers, cans, bottles, and other household wastes also received waste petroleum products, solvents, pesticides, and transformer oils. Liquid hazardous wastes were often disposed of in drums, or in some cases were poured directly into the landfills. These sanitary landfills were unlined, so the wastes have often migrated to surface and groundwaters. Some regulators believe that all sanitary landfills should be considered hazardous waste sites because they received hazardous wastes, particularly before 1980.

Drum Storage Areas. Waste chemicals stored in 55-gallon drums were often placed on loading docks, concrete pads, or other temporary storage areas until they could be disposed of. Because of lack of money or administrative action, the drums often accumulated, sometimes to the point where thousands were stored and stacked. Drums stored in this manner eventually corroded and leaked, with chemical releases to the underlying soil and groundwater.

Unlined Hazardous Waste Landfills. Many large industries constructed their own landfills that were used primarily for the land disposal of industrial by-products, such as building materials or out-of-date equipment. Unfortunately, chemical wastes were

Figure 1.4 A pit-pond-lagoon system from the surface disposal of hazardous wastes.

Figure 1.5 A soil pit used for the improper disposal of drums of hazardous wastes. *Source:* U.S. Department of Energy, Richland, WA, office.

also disposed of in these landfills along with the nonhazardous solid materials. Leakage to soil and groundwater was commonplace because these systems were not designed to contain hazardous wastes.

Midnight Dumping. Although hazardous wastes were easily disposed of before comprehensive hazardous waste regulations were enacted, the wastes were sometimes surreptitiously transported and disposed of on others' property or in isolated locations, such as a wooded area of a desert canyon (Figure 1.5). These "midnight dumping" practices often resulted in pockets of contamination on private, federal, and state land. Many sites have gone undiscovered, and others have been found during excavation activities for new highways and buildings. Construction managers, who are often uneducated in hazardous waste management, have encountered buried drums and stained soils more often than what they would like. The cleanup of hazardous wastes at construction sites has delayed the completion of hundreds of new highways and buildings.

Uncontrolled Incineration. Burning and incinerating some hazardous wastes, such as chlorophenols and polychlorinated biphenyls, has sometimes resulted in incomplete combustion, with the formation of more toxic products in the ash and the emission of hazardous air pollutants. The safe incineration of many hazardous chemicals requires specific conditions of temperature, turbulence, and residence—conditions that are provided by the permitting process under the dominant hazardous waste management law, the Resource Conservation and Recovery Act.

1.1.2 Case Histories

The waste disposal practices just described were the most common cases of improper hazardous waste management. Most incidents involved a small number of drums disposed of in soil pits or spilled on soils, resulting in localized contamination. However, a number of large-scale, unsafe disposals occurred, with significant threats to public health. A few of these landmark cases are described next.

Love Canal. Lured by the availability of inexpensive and readily available electricity near Niagara Falls, New York, William T. Love devised a plan for an industrial park and accompanying housing development in the late 1800s. An integral part of the development was an 8-mile canal drawn off of the Niagara River. Unfortunately, as the canal was nearing completion in the early twentieth century, the geographic dependence of energy use shifted dramatically as the result of the development of alternating current. Canal construction was never completed; nonetheless, industries were drawn to the area by the availability of inexpensive electricity. Most of the energy-intensive industries that settled near the canal were involved in electrochemistry and chemical synthesis. For example, Hooker Chemical produced the insecticide DDT, the herbicide 2,4,5-T, and chlorinated solvents. In the 1930s, Hooker began using part of the unfinished canal for the disposal of chemical wastes. In 1952, waste disposal into the canal ceased, and it was covered. Hooker was pressured into selling the contaminated land, and a school was built with the full knowledge of the authorities that chemical wastes had been disposed of on the property. The availability of a school prompted the development of a typical subdivision around the abandoned Hooker dump site. However, no one anticipated any problems from chemicals buried under the property.

The horrors that followed, primarily in the 1970s, constituted one of the worst environmental disasters in United States history. Liquid and gaseous wastes migrated into the school and into the basements of houses. The residents of Love Canal were exposed to hazardous chemicals primarily through subsurface waste migration. The exposure was not by drinking contaminated groundwater, but through inhalation after the wastes had migrated into basements and homes in the area. Children even played in ditch water into which toxic organic chemicals had migrated. The incidence of cancer and birth defects soared, according to residents and the popular press. In addition, the lowest birth weights of children born to women of Love Canal occurred during the time of highest exposure [1.11].

Many subsequent hazard exposure investigations were performed on Love Canal residents. Paigen et al. [1.12], in a study of 523 Love Canal children, documented increased prevalence of seizures, learning problems, skin rashes, and abdominal pain relative to a control population. However, Janerich et al. [1.13] found no increase in liver cancer, lymphomas, or leukemias in a survey of 700 census tract residents relative to the New York State population. The National Research Council [1.14] cautioned that these results are difficult to interpret because other populations of New York state are also exposed to hazardous waste. The Council concluded that, with exposure of the Love Canal population to over 200 chemicals, the cause-effect relationships of the disaster are still not well understood. Nonetheless, the EPA probably overreacted to the Love Canal contamination. The problem may have been less serious than originally believed.

Stringfellow Acid Pits. Three thousand miles to the west of Love Canal, a hazardous waste landfill operated from 1956 through 1976 near the town of West Avon in Riverside County, California. Several hydrogeological studies concluded that this landfill east of Los Angeles, known as the Stringfellow site, would not pose a threat to the underlying Chino Basin aquifer, which is a primary water source for this region of Southern California. However, the subsurface investigation was inaccurate—fractured rock and subsurface intrusion of water complicated the analyses.

By the mid-1970s, the potential for off-site waste migration became apparent. Heavy rains flooded the site and toxic runoff flowed through the streets of West Avon. The potential for groundwater contamination was also evident. Baker et al. [1.15] documented an increase in bronchitis, asthma, and skin rashes in populations near the Stringfellow Acid Pits relative to a control population. In 1977, two cleanup options were evaluated: (1) total removal of the contamination, and (2) containment, on-site neutralization of contaminants, and capping. The second option was selected, but by this time, remedial costs had increased by over 1000% from an initial cost projection 5 years earlier.

Hardeman County, Tennessee. A 200-acre pesticide waste dump operated from 1964 through 1972 in Hardeman County, Tennessee. Approximately 300,000 drums of waste pesticides and solvents were buried at the site in shallow trenches. Five years after the site was closed, nearby residents noticed foul odors in their well water. In addition, some experienced nausea, cramping, shortness of breath, respiratory infections, and a number of other health problems. Because the occurrence of chemical odors correlated with the onset of medical problems, the well water in the area was analyzed for organic contaminants. Chlorinated solvents, such as carbon tetrachloride and perchloroethylene, and intermediates in pesticide synthesis, such as hexachlorocyclopentadiene, were found in alarmingly high concentrations. In addition, detectable concentrations of the same chemicals were documented in the indoor air of homes. Clark et al. [1.16] described the results of a toxicological study for the area. The contaminant found in the highest concentration in the well water was carbon tetrachloride, and its presence correlated with liver abnormalities and high levels of marker enzymes.

Numerous other landmark hazardous waste incidents have been reported. At Times Beach, Missouri, the waste herbicide 2,4,5-T containing the highly toxic impurity 2,3,7,8-tetrachlorodibenzo-*p*-dioxin was spread on roads and animal pens for dust control (see Section 2.9). The area was eventually evacuated and the property purchased under the United States EPA Superfund program. The Bunker Hill site in Kellogg, Idaho, is contaminated with arsenic, lead, and cadmium as a result of mining, ore concentration, and smelting. Hazardous waste problems are not limited to the United States. An explosion in a chemical plant in Seveso, Italy, where the industrial chemical 2,4,5-trichlorophenol was synthesized, produced a cloud of dioxin with widespread contamination and public exposure.

1.1.3 The Magnitude of the Problem

The hazards resulting from the disposal of the thousands of industrial compounds commonly used by industry were largely unknown until the 1970s. Drums containing waste petroleum, solvents, pesticides, industrial intermediates, and other compounds were commonly disposed of by unsafe methods for decades. These hazardous and persistent wastes migrated through soils and groundwater to water supply wells and began appearing in drinking water during the 1970s. Large hazardous waste problems were subsequently discovered with such disasters as Love Canal and the Stringfellow Acid Pits.

The National Academy of Sciences [1.17] reported that, depending on the characteristics of the site and the contaminants, up to 500 years will be required to see the effects of improper hazardous waste disposal because of the slow migration rate in some subsurface systems.

The Number of Hazardous Waste Sites. The EPA estimated a total of 50,000 hazardous waste sites in the United States, in which approximately 60 million tons of wastes had been disposed when landmark hazardous waste laws were passed in the late 1970s [1.1]. Similar estimates were detailed more than a decade later, as noted in Table 1.1. At least 30,000 other sites in different categories, such as those under the jurisdiction of the Department of Defense, are listed in Table 1.1. In a more formal listing procedure, over 1200 sites have been placed on the National Priorities List (NPL), a system for ranking the worst sites for clean up under the Comprehensive Environmental Response, Compensation and Liability Act (CERCLA), a law passed in 1980 to provide a mechanism for cleaning up past improper hazardous waste disposal [1.12]. The states with the largest number of NPL sites are New Jersey, Pennsylvania, California, Michigan, and New York, with a cumulative total of 464 sites. Other estimates of the number of potential NPL sites are even higher. The Office of Technology Assessment proposed as many as 439,000 candidate sites if mining, USTs, pesticide rinse areas, and radioactive sites are included [1.19].

Under CERCLA, more than 31,000 sites have been reported to the *CERCLA Information System* (CERCLIS), which is a more comprehensive listing of sites from which the NPL is developed. The EPA has completed more than 27,000 preliminary assessments and has conducted detailed investigations at over 9000 sites [1.19]. Not all sites are reported to the CERCLIS program. Other systems for site documentation include the Emergency Response Notification System, the Federal Facilities Hazardous Waste Compliance Docket, the Department of Defense, and the Department of the Interior's Abandoned Mine Lands Remediation Program. In addition, states track and regulate hazardous waste sites. A recent survey reported approximately 33,000 sites on state lists [1.12].

In summary, the multiple reporting and regulatory systems at federal, state, and local levels complicate attempts to quantify the absolute number of hazardous waste sites in the United States. Nonetheless, review of the available data confirms that there are thousands of serious sites, and the number of smaller sites may range in the hundreds of thousands.

Table 1.1 Potential Number of Hazardous Waste Sites and
Associated Cleanup Costs

	No. of Potential Sites	Estimated Cost (billions of dollars)
Superfund sites	60,000	50
Sites of RCRA cleanups	2,400	23
State-funded cleanup sites	22,000	45
Department of Defense sites	7,200	11–15
Department of Energy sites	Not reported	66–110

Source: Reference 1.18.

Hazards from Improper Disposal. Most hazardous wastes spilled or disposed of on land migrate through surface and subsurface soils (i.e., through the unsaturated zone) to groundwater. Therefore, subsurface migration is a primary pathway for reaching a place where the public is exposed (i.e., a receptor). The EPA [1.20] reported survey results of potential pathways for the release of chemicals from Superfund sites. Migration to groundwater was cited as the primary pathway of contaminants at these hazardous waste sites, a trend confirmed by the data in Table 1.2; 37% of sites involve releases to groundwater and 23% are characterized by releases to both groundwater and surface water. Other studies document the widespread nature of the problem. The EPA, in a survey of 466 public water supply wells, found that one or more volatile organic compounds (VOCs) were detected in 16.8% of small water systems and 28% of large water systems. The VOCs found most often in the survey were trichloroethylene and perchloroethylene [1.21].

A survey of 7000 wells conducted in California from 1984 through 1988 showed that approximately 1500 contained detectable concentrations of organic chemicals, although only about 400 had concentrations exceeding the state's regulatory requirement or the Maximum Contaminant Level (MCL) prescribed by the Safe Drinking Water Act [1.22]. The most common chemicals detected were perchloroethylene, trichloroethylene, chloroform, 1,1,1-trichloroethane, and carbon tetrachloride (see Chapter 2).

The importance of minimizing the effects of hazardous wastes is accentuated by the fact that 48% of the United States population receives its drinking water from groundwater [1.23]. A state-by-state accounting of the reliance on groundwater as a water source is noted in Figure 1.6. Approximately 33% of the United States cities use groundwater; however, 95% of the rural United States population relies on groundwater for domestic use [1.24].

More than 600 chemicals have been discovered at Superfund sites. The contaminants that are found most frequently at NPL sites are lead (43% of sites), trichloroethylene (42%), chromium (35%), benzene (34%), perchloroethylene (28%), arsenic (28%), and toluene (27%) [1.25]. Common concerns resulting from the improper disposal of these chemicals focus on chronic health effects such as cancer. More than 4 million people live within 1 mile of a Superfund site, and more than 40 million live within 4 miles of a site [1.25]. Of the 4 million that live within 1 mile, 1.9 million are women of child-

Table 1.2 Pathways of Releases of Hazardous Chemicals from National Priority List Landfills

Observed Releases from NPL Landfills to Water and Air	Percent
Groundwater only	37
Groundwater and surface water	23
None observed	15
Surface water only	9
Groundwater, surface water, air	8
Groundwater and air	3
Surface water and air	3
Air only	2

Source: Reference 1.20.

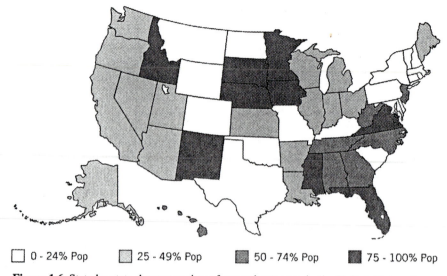

Figure 1.6 State-by-state documentation of groundwater use in the United States. *Source:* Reference 1.21.

bearing age, young children, or elderly—all of whom are at a greater risk from toxic exposure. The Agency for Toxic Substances and Disease Registry reported that, of the 1189 NPL sites in 1991, 109 sites were characterized by "urgent public health concern" or "public health concern" because of actual human exposures or probable exposures to hazardous chemicals. An estimated 715,000 people live within a 1-mile radius of these 109 problem sites.

A drawback in the early, improper disposal of hazardous wastes was a lack of understanding of the behavior of these chemicals in the environment (i.e., that they follow the laws of thermodynamics, and matter and energy are conservative). Waste chemicals have to go somewhere—they do not just vanish and become a nonproblem. They can volatilize into the air, partition onto soils, move through saturated and unsaturated subsurface soils, partition into biomass, and degrade. The concept that contaminants enter a series of pathways when released to the environment is the basis for a fundamental approach to hazardous waste assessment, management, and control.

Another, albeit less scientific, basis for managing hazardous wastes and contaminated sites is the passage of sweeping federal legislation, which occurred in the late 1970s in response to the media reports of Love Canal, the Stringfellow Acid Pits, and hundreds of other sites. Federal legislation has brought about widespread effects on industry, as well as the role of environmental engineers and scientists. The series of federal laws spawned state and local legislation, and industries that generate hazardous wastes have become highly regulated communities. The new legislation had an effect on the scientific and engineering professions as well. These regulations drove the development of a new hazardous waste industry: Engineers and scientists are now faced with challenges of minimizing and treating newly generated hazardous wastes as well as assessing and cleaning up contaminated sites that were created before the regulations were in place.

1.2   HAZARDOUS WASTE LEGISLATION

The improper preregulatory disposal of hazardous wastes (with the creation of the associated thousands of contaminated sites) triggered the passage of sweeping legislation in the late 1970s. The resulting hazardous waste regulations have driven the cleanup of widespread soil and groundwater contamination and ensure that hazardous wastes generated in the future will be minimized, treated, and disposed of safely. The following description provides an historical perspective that shows the importance of regulations in the evolution of hazardous waste management. Hazardous waste regulations are as complex as the scientific and engineering principles of hazardous waste management, and a comprehensive discussion of the regulations is beyond the scope of this text; furthermore, hazardous waste regulations are dynamic and may change dramatically over relatively short periods of time. Detailed summaries of the most current hazardous waste regulations may be found in the *Environmental Law Handbook* [1.26], which is revised every 2 years.

Many environmental laws are written in Congress and, because of their complexity, passed in general, nonspecific forms. Federal agencies (usually the EPA for environmental legislation) then provide detailed requirements (regulations), which are published in the *Code of Federal Regulations* (CFR) under Title 40, the section on *Protection of the Environment.*

Hazardous waste is a subcategory of solid waste. The first major federal legislation related to solid waste disposal was the Solid Waste Disposal Act of 1965, which was amended and expanded as the Resource Recovery Act of 1970. The 1965 and 1970 solid waste laws did not have a major impact on solving hazardous waste problems, although they did initiate a federal role in solid waste management. Hazardous waste problems were almost unheard of at the time; therefore, no hazardous wastes provisions were provided in the two early pieces of legislation. However, these laws created a series of guidelines that affected federal facilities and contractors in the areas of storage and collection of solid wastes, including the operation of sanitary landfills, incineration, and resource recovery.

1.2.1   The Resource Conservation and Recovery Act and the Hazardous and Solid Waste Amendments

The *Resource Conservation and Recovery Act* (RCRA)—legislation that requires total documentation of where a waste is generated and where it is disposed of—was passed in 1976 in response to widespread environmental contamination. Although RCRA, as written by Congress, was comprehensive, it was promulgated in the 1980s—a time when the EPA underwent significant budget cuts. The EPA was criticized for irresponsibly carrying out Congress' hazardous waste mandate under RCRA. As a result, Congress passed the *Hazardous and Solid Waste Amendments* of 1984 (HSWA), which contained significantly more technical details (and less latitude for the EPA) than the original RCRA of 1976. The 1984 amendments significantly strengthened the act and therefore have their own name.

To ensure that the EPA would carry out the provisions outlined within RCRA, *hammer provisions* were written into the legislation. If the EPA did not follow through on promulgating RCRA, the "hammer would fall" on the EPA, resulting in even more stringent regulations. Furthermore, RCRA has provisions for citizen actions in which the public can file a suit against an industry regulated under RCRA.

The primary goals of RCRA and HSWA are (1) to protect public health and the environment from hazardous and other solid wastes, and (2) to preserve natural resources through resource recovery and conservation. A large part of RCRA focuses on defining a hazardous waste. A series of lists of specific chemicals and industrial processes that generate wastes is one method by which hazardous wastes are defined. The other method is one of definition; specific criteria have been developed that define a waste as corrosive, ignitable, explosive, or toxic. The management goal of RCRA is to control hazardous wastes from "cradle to grave" by tracking their movement from the point of generation, through transit, and finally to treatment, storage, and disposal (TSD) facilities. Under the RCRA cradle-to-grave management of hazardous wastes, an industry that generates hazardous wastes is liable for its wastes. Other aspects of RCRA deal with leaking underground storage tanks and a ban on the land disposal of hazardous wastes.

Definition of a RCRA Hazardous Waste

Because RCRA had its origins in the Solid Waste Disposal Act of 1965, hazardous wastes are considered a subset of solid wastes. Therefore, before a waste can be considered hazardous, it must first be classified as a solid waste. Under RCRA, *a solid waste* is defined as

> *any garbage, refuse, sludge from a waste treatment plant, water supply treatment plant or air pollution control facility and other discarded material including solid, liquid, semisolid, or contained gaseous materials resulting from industrial, commercial, mining, and agricultural activities and from community activities but does not include solid or dissolved material in domestic sewage, or solid or dissolved materials in irrigation return flows or industrial charges which are point sources subject to permits under Section 402 of the Federal Water Pollution Control Act, as amended, or source, special nuclear, or byproduct material as defined by the Atomic Energy Act of 1954, as amended (68 Stat. 923).*

Note that RCRA defines wastes that are gaseous, liquid, and semisolid (as well as solid) as solid wastes. The rationale for this definition is that wastes, regardless of their physical state, have usually been disposed of in landfills and other solid waste repositories. In addition, wastes covered by other environmental regulations (e.g., the Clean Water Act) are not specifically covered under RCRA.

If a waste is classified as a solid waste, the next step is to determine if it is hazardous. Under RCRA, a *hazardous waste* is generally defined as

> *a solid waste, or combination of solid wastes, which because of its quantity, concentration, or physical, chemical, or infectious characteristics, may: 1) cause, or significantly contribute to an increase in mortality or an increase in serious irreversible, or incapacitating reversible illness, or 2) pose a substantial present or potential hazard to human health or the environment when improperly treated, stored, transported, or disposed of, or otherwise managed.*

The working definition of a RCRA hazardous waste is more involved than this general description. It consists of a series of exemptions, hazardous waste lists, hazardous waste characteristics, and delisting procedures. The definition of a RCRA hazardous waste is illustrated in Figure 1.7, a flowchart that provides a summary of the following discussion of the RCRA waste identification process.

Exempted Wastes. Some hazardous solid wastes may not be classified as hazardous because of an exemption rule. Nine categories of solid wastes (listed in Table 1.3) are exempt from regulatory control under RCRA, even if they are classified as hazardous by other criteria.

Hazardous Wastes Lists. If a waste material is a solid waste and is not exempt, a series of lists must be examined to assess if it is hazardous. The first of these is the *F list—hazardous wastes from nonspecific sources.* There are 20 F classifications listed from F001 to F029. A common F list waste is F001:

> *The following spent halogenated solvents used in degreasing: tetrachloroethylene, trichloroethylene, methylene chloride, 1,1,1-trichloroethane, carbon tetrachloride, and chlorinated fluorocarbons; all spent solvent mixtures/blends used*

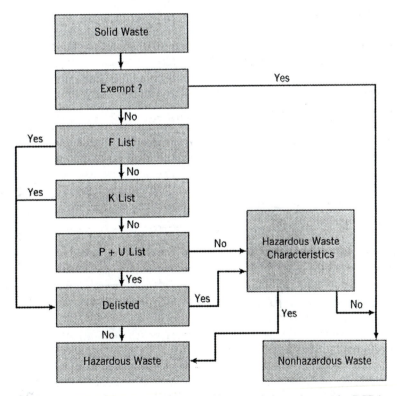

Figure 1.7 Critical path for determining if a waste is hazardous under RCRA.

Table 1.3 Solid Wastes Exempt from RCRA Hazardous Waste Management

1. Household waste
2. Agricultural waste returned to the ground
3. Mining overburden returned to the mine site
4. Utility wastes from coal combustion
5. Oil and gas exploration drilling waste
6. Wastes from the extraction and processing of ores and minerals
7. Cement kiln wastes
8. Arsenic-treated wood wastes generated by end users of such wood
9. Certain chromium-bearing wastes

*in degreasing containing, before use, a total of 10 percent or more (by volume)
of one or more of those solvents listed in F002, F004, and F005; and still bottoms from the recovery of these spent solvents and spent solvent mixtures.*

The second list is the K list—*wastes from specific sources.* If an industry generates any
of 87 source-specific wastes, the waste is considered hazardous under RCRA. A typical K waste is K001:

*Bottom sediment sludge from the treatment of wastewaters from wood preserv-
ing processes that use creosote and/or pentachlorophenol.*

The F and K lists represent waste from industrial processes. The third category, comprising the P and U lists, is not composed of by-products of industrial operations, but
instead commercial products that must be disposed of because they are off-specification or out-of-date. Containers with hazardous residues and spills of commercial products are also included in this category. Because of the presence of the P and U lists,
commercial products are classified as RCRA hazardous wastes if they are disposed of
in any manner. The *P list* targets chemical products considered *acutely hazardous*
when discarded; they are regulated more strictly than wastes on other RCRA lists. The
second category of commercial products is the U list, which are classified as toxic.
Unlike the products on the P list, U list chemicals are regulated in the same manner as
other hazardous wastes (e.g., F and K wastes). The entire F, K, P, and U lists can be
found in Appendix A.

Characteristic Wastes. If a waste material is not a listed waste, it may still be classified as a hazardous waste by four hazardous characteristics: ignitability, corrosivity, reactivity, and toxicity. Definitions of the four characteristics are listed in Table 1.4. Of
particular concern is the toxicity characteristic, which is evaluated by laboratory leaching procedures followed by chemical analysis of the leachate. The first toxicity test developed by the EPA was the *Extraction Procedure* (EP) toxicity test. The procedure involves grinding solid samples with subsequent extraction in dilute acetic acid for 24
hours. The extract is then filtered, analyzed, and compared to established values, which
are generally 100 times greater than Maximum Contaminant Levels (MCLs) established under the Safe Drinking Water Act. The EP toxicity test has been replaced by
the *Toxicity Characteristic Leaching Procedure* (TCLP), which is considered to be
more representative of landfill conditions and to have greater quality control (e.g., lab-

Table 1.4 Hazardous Waste Characteristics

Ignitability (D001)

The waste
1. is a liquid, other than an aqueous solution containing less than 24% alcohol (v/v) with a flash point of less than 60°C;
2. is a nonliquid which under normal conditions can cause fire through friction, absorption of moisture, or spontaneous chemical changes, and burns so vigorously when ignited that it creates a hazard;
3. is an ignitable compressed gas as defined by the DOT regulations 49 CFR 173.300, and;
4. exhibits ignitability as an oxidizer as defined by 40 CFR 173.151.

Corrosivity (D002)

The waste
1. is aqueous and has a pH ≤ 2.0 or ≥ 12.5, and
2. is a liquid and corrodes steel at a rate greater than 6.35 mm/year.

Reactivity (D003)

The waste
1. is normally unstable and readily undergoes violent change without detonation,
2. violently reacts with water,
3. forms potentially explosive mixtures with water,
4. generates toxic gases or fumes in a quantity sufficient to present a danger to public health or the environment when mixed with water,
5. is a cyanide- or sulfide-bearing waste that, when exposed to pH between 2 and 12.5, can generate toxic gases,
6. is capable of detonation or explosive reaction if it is subjected to a strong initiating source or it is heated under confinement,
7. is readily capable of detonation or explosive decomposition or reaction at standard temperature and pressure, and
8. is a forbidden explosive, as defined by DOT regulations (49 CFR 173.51, 173.53, and 173.88).

Toxicity (D004-D043)

The waste is pH adjusted to landfill conditions, and a leaching procedure is performed. This analysis, called the *Toxicity Characteristic Leaching Procedure* (TCLP), compares the leachate concentrations to a list of 31 organic chemicals and 8 inorganic chemicals. If the waste leachate concentrations are greater than the TCLP list shown below, the waste is characterized by toxicity.

(continued on next page)

oratory procedures that minimize volatilization). Wastes that are characterized as hazardous by the four characteristics are *D wastes*; for example, a corrosive waste is given the RCRA waste code D002.

Mixture Rule. Another criterion for defining a hazardous waste is the *mixture rule*, which states that if a listed hazardous waste is mixed with a nonhazardous solid waste,

Table 1.4 Continued

Toxicity Characteristic Constituents and Regulatory Levels

EPA HW No.[a]	Constituent	Regulatory Level (mg/L)
D004	Arsenic	5.0
D005	Barium	100.0
D018	Benzene	0.5
D006	Cadmium	1.0
D019	Carbon tetrachloride	0.5
D020	Chlordane	0.03
D021	Chlorobenzene	100.0
D022	Chloroform	6.0
D007	Chromium	5.0
D023	o-Cresol	200.0[b]
D024	m-Cresol	200.0[b]
D025	p-Cresol	200.0[b]
D026	Cresol	200.0[b]
D016	2,4-D	10.0
D027	1,4-Dichlorobenzene	7.5
D028	1,2-Dichloroethane	0.5
D029	1,1-Dichloroethylene	0.7
D030	2,4-Dinitrotoluene	0.13[c]
D012	Endrin	0.02
D031	Heptachlor (and its epoxide)	0.008
D032	Hexachlorobenzene	0.13[c]
D033	Hexachloro-1,3-butadiene	0.5
D034	Hexachloroethane	3.0
D008	Lead	5.0
D013	Lindane	0.4
D009	Mercury	0.2
D014	Methoxychlor	10.0
D035	Methyl ethyl ketone	200.0
D036	Nitrobenzene	2.0
D037	Pentachlorophenol	100.0
D038	Pyridine	5.0[c]
D010	Selenium	1.0
D011	Silver	5.0
D039	Tetrachloroethylene	0.7
D015	Toxaphene	0.5
D040	Trichloroethylene	0.5
D041	2,4,5-Trichlorophenol	400.0
D042	2,4,6-Trichlorophenol	2.0
D017	2,4,5-TP (Silvex)	1.0
D043	Vinyl chloride	0.2

[a]RCRA waste number
[b]If o-, m-, and p-cresol concentrations cannot be differentiated, the total cresol (D026) concentration is used. The regulatory level for total cresol is 200 mg/L.
[c]Quantitation limit is greater than the calculated regulatory level. The quantitation limit therefore becomes the regulatory level.

the mixture is still a hazardous waste regulated under RCRA. Therefore, industries cannot rid themselves of hazardous wastes by simply diluting them. Some of these mixtures, however, are no longer RCRA wastes if (1) the original hazardous waste is listed only because it exhibited a characteristic, and the mixture does not have the characteristic, (2) the mixture includes a specific group of hazardous chemicals mixed with wastewater, or (3) the mixture contains small quantities (termed *de minimis*) of discarded commercial products that result from regular losses during manufacturing.

Derived-From Hazardous Wastes. Under the *derived-from* rule, a waste generated from a TSD facility is also a hazardous waste, unless there is an exemption for it. Typical derived-from wastes include incineration ash and landfill leachate.

Delisting. A hazardous waste generator that has reasonable cause may petition the EPA to *delist* a waste so that it is no longer considered hazardous under RCRA. Delisting is usually pursued when waste characteristics are borderline, if the industry has changed unit processes to decrease the severity of the waste, or if companies are partially treating the waste.

EXAMPLE 1.1 *Evaluation of a Waste under RCRA*

A large painting operation generates 2500 kg of 60% spent methyl ethyl ketone by weight in water. Is this a hazardous waste under RCRA?

SOLUTION

First, the waste is not an exempted waste (Table 1.3) under RCRA. Second, this waste is listed under the F list (Appendix A) as F005. Therefore, it is a hazardous waste under RCRA.

Cradle-to-Grave Hazardous Waste Management

Three groups are involved in the management of hazardous wastes—generators, transporters, and TSD facilities. Under RCRA, guidelines and responsibilities are specified for each group so that hazardous wastes are accounted for and disposed of properly.

Generators. A *generator* is defined by the EPA as any person whose act or process produces hazardous waste or whose act causes hazardous waste to become subject to regulation. By person, the EPA is usually referring to a company or corporation that generates over 1000 kg of hazardous waste per month. Many small businesses do not generate enough hazardous waste to be regulated under RCRA.

The first duty of hazardous waste generators is to determine if the waste is hazardous using the procedure shown in Figure 1.7. If they are generating a hazardous waste, it is their responsibility to obtain an EPA identification number by filing the *Notification of Hazardous Waste Activity* form (Figure 1.8), which serves as a mechanism for track-

Please print or type ELITE type *(12 characters/inch)* in the unshaded areas only.

Form Approved OMB No. 2000-0098
GSA No. 0246-EPA-OT Expiration Date 12/31/86

U.S. ENVIRONMENTAL PROTECTION AGENCY
NOTIFICATION OF HAZARDOUS WASTE ACTIVITY

INSTALLA-TION's EPA I.D. NO.

I. NAME OF IN-STALLATION

II. INSTALLA-TION MAILING ADDRESS

III. LOCATION OF INSTAL-LATION

PLEASE PLACE LABEL IN THIS SPACE

INSTRUCTIONS: If you received a preprinted label, affix it in the space at left. If any of the information on the label is incorrect, draw a line through it and supply the correct information in the appropriate section below. If the label is complete and correct, leave items I, II, and III below blank. If you did not receive a preprinted label, complete all items. "Installation" means a single site where hazardous waste is generated, treated, stored and/or disposed of, or a transporter's principal place of business. Please refer to the INSTRUCTIONS FOR FILING NOTIFICATION, before completing this form. The information requested herein is required by law *(Section 3010 of the Resource Conservation and Recovery Act).*

FOR OFFICIAL USE ONLY

COMMENTS

INSTALLATION'S EPA I.D. NUMBER   APPROVED   DATE RECEIVED *(yr., mo., & day)*

I. NAME OF INSTALLATION

II. INSTALLATION MAILING ADDRESS

STREET OR P.O.BOX

CITY OR TOWN   ST.   ZIP CODE

III. LOCATION OF INSTALLATION

STREET OR ROUTE NUMBER

CITY OR TOWN   ST.   ZIP CODE

IV. INSTALLATION CONTACT

NAME AND TITLE *(last, first, & job title)*

V. OWNERSHIP

A. NAME OF INSTALLATION'S LEGAL OWNER

B. TYPE OF OWNERSHIP
(enter the appropriate letter into box)

F = FEDERAL
F = NON–FEDERAL

VII. TYPE OF HAZARDOUS WASTE ACTIVITY *(enter "X" in the appropriate box(es))*

☐ A. GENERATION   ☐ B. TRANSPORTATION *(complete item VII)*
☐ C. TREAT/STORE/DISPOSE   ☐ D. UNDERGROUND INJECTION

VII. MODE OF TRANSPORTATION *(transporters only – enter "X" in the appropriate box(es))*

☐ A. AIR   ☐ B. RAIL   ☐ C. HIGHWAY   ☐ D. WATER   ☐ E. OTHER *(specify):*

VIII. FIRST OR SUBSEQUENT NOTIFICATION

Mark "X" in the appropriate box to indicate whether this is your installation's first notification of hazardous waste activity or a subsequent notification. If this is not your first notification, enter your installation's EPA I.D. Number in the space provided below.

C. INSTALLATION'S EPA I.D. NO.

☐ A. FIRST NOTIFICATION   ☐ B. SUBSEQUENT NOTIFICATION *(complete item C)*

IX. DESCRIPTION OF HAZARDOUS WASTES

Please go to the reverse of this form and provide the requested information.

EPA Form 8700-12 (6-85)

CONTINUE ON REVERSE

Figure 1.8 A RCRA Notification of Hazardous Waste Activity form.

ing the waste by regulators until it is disposed of. Waste can be stored for up to 90 days before being transported off site for disposal. If the generators need a longer time to store their waste, a Treatment, Storage, and Disposal (TSD) permit must be obtained.

Generators have additional responsibilities when it comes time to dispose of their waste. First, a licensed hazardous waste transporter must be hired and shipment of the waste must be arranged. A *Uniform Hazardous Waste Manifest* (Figure 1.9) is prepared that details the classifications and quantities of the wastes. The manifest provides documentation and must accompany the waste until it is properly disposed of, and all parties must keep a copy for their records. Also, generators must prepare the shipment of hazardous waste in a safe manner using specific Department of Transportation regulations, which include safe packaging, labeling, marking, and placards (see Section 3.8). Furthermore, generators must provide personnel training and contingency plans.

Industries that generate less than 1000 kg per month were not regulated under RCRA when it was first promulgated in 1980, but with HSWA, the generation re-

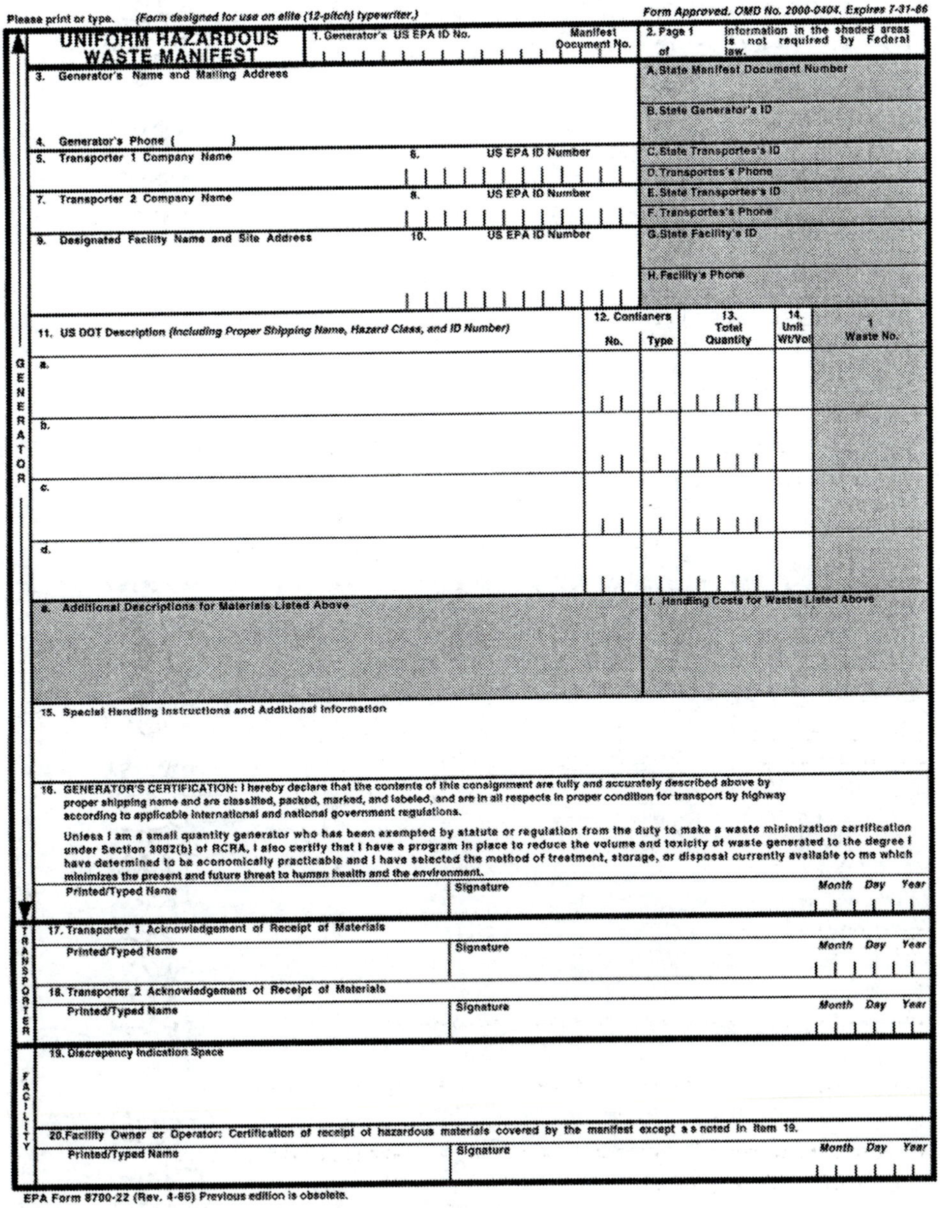
Please print or type. (Form designed for use on elite (12-pitch) typewriter.)
Form Approved. OMB No. 2000-0404. Expires 7-31-86

UNIFORM HAZARDOUS WASTE MANIFEST
1. Generator's US EPA ID No.
Manifest Document No.
2. Page 1 of
Information in the shaded areas is not required by Federal law.
3. Generator's Name and Mailing Address
A. State Manifest Document Number
B. State Generator's ID
4. Generator's Phone ( )
5. Transporter 1 Company Name
6. US EPA ID Number
C. State Transporter's ID
D. Transporter's Phone
7. Transporter 2 Company Name
8. US EPA ID Number
E. State Transporter's ID
F. Transporter's Phone
9. Designated Facility Name and Site Address
10. US EPA ID Number
G. State Facility's ID
H. Facility's Phone
11. US DOT Description (Including Proper Shipping Name, Hazard Class, and ID Number)
12. Contianers No. Type
13. Total Quantity
14. Unit Wt/Vol
I. Waste No.
a.
b.
c.
d.
GENERATOR
J. Additional Descriptions for Materials Listed Above
K. Handling Costs for Wastes Listed Above
15. Special Handling Instructions and Additional Information
16. GENERATOR'S CERTIFICATION: I hereby declare that the contents of this consignment are fully and accurately described above by proper shipping name and are classified, packed, marked, and labeled, and are in all respects in proper condition for transport by highway according to applicable international and national government regulations.
Unless I am a small quantity generator who has been exempted by statute or regulation from the duty to make a waste minimization certification under Section 3002(b) of RCRA, I also certify that I have a program in place to reduce the volume and toxicity of waste generated to the degree I have determined to be economically practicable and I have selected the method of treatment, storage, or disposal currently available to me which minimizes the present and future threat to human health and the environment.
Printed/Typed Name | Signature | Month Day Year
TRANSPORTER
17. Transporter 1 Acknowledgement of Receipt of Materials
Printed/Typed Name | Signature | Month Day Year
18. Transporter 2 Acknowledgement of Receipt of Materials
Printed/Typed Name | Signature | Month Day Year
FACILITY
19. Discrepancy Indication Space
20. Facility Owner or Operator: Certification of receipt of hazardous materials covered by the manifest except as noted in Item 19.
Printed/Typed Name | Signature | Month Day Year
EPA Form 8700-22 (Rev. 4-85) Previous edition is obsolete.

Figure 1.9 A RCRA Hazardous Waste Manifest form.

quirement was lowered to include a new class of businesses called *small quantity generators*. These industries produce hazardous wastes in the range of 100 to 1000 kg/month. More lenient rules are provided for small quantity generators; for example, they may keep up to 1000 kg of waste on their property for 180 days, and if the waste has to be shipped over 200 miles, it may be stored for up to 270 days.

Transporters. Any person engaging in the off-site transportation of hazardous waste by air, rail, highway, or water is defined as a *hazardous waste transporter.* Most transporters are truckers, and although minimal training is required, they must obtain an EPA transporter identification number. A commonly used reinforced cylindrical tank truck used for transporting hazardous wastes is shown in Figure 1.10. Transporters can accept only wastes accompanied by a manifest and must keep the manifest with them at all times during transport. They can hold the waste for up to 10 days without obtaining a TSD permit. Transporters are responsible for accidental and intentional releases when the waste is in their possession. If the waste is spilled, they must be prepared to clean it up after contacting local police and fire departments. When the waste is delivered to a TSD facility, the manifest must be signed by the TSD operator and copies must be retained by both parties.

Treatment, Storage, and Disposal Facilities. Common *Treatment, Storage, and Disposal* (TSD) facilities include landfills, incinerators, acid neutralization tanks, holding tanks, and lined surface impoundments. These facilities must be approved by the EPA and, especially during the first 10 years of RCRA, the permitting procedure was accomplished by a two-step process—the Part A and Part B permitting system. *Part A* permits were interim documents for which a TSD facility must have met three criteria: (1) been in operation on November 19, 1980, the day RCRA was promulgated, (2)

Figure 1.10 A typical truck used for hazardous waste transport.

notified the EPA of the facility's hazardous waste activities, and (3) filed for the interim Part A permit. The rationale for the two-phase permitting system was logistics. Part A permits were relatively simple documents whereas *Part B* permits were extensive and have required large time commitments for their preparation and review. Without the Part A process, it would have taken years to provide permits for TSD facilities operating in 1980. A few years after RCRA was promulgated, the backlog in the RCRA permitting process dissipated and Part A permits became an item of the past.

Current RCRA permits are extensive, time-consuming reports that provide a detailed description of the TSD operation. They are written in a narrative (report) format, and are often documents of 500 to 2000 pages. The detailed information of RCRA permits related to the facility include treatment system design and operating criteria, chemical analyses of incoming wastes, personnel training, fire and spill control plans, and record-keeping procedures.

Although TSD facilities include a range of operations, such as acid neutralization units, biological treatment systems, and long-term (greater than 90-day) storage areas, perhaps the most important are hazardous waste landfills and incinerators. Operators of landfills are required to provide covers and to cap the landfill at the end of its operation, and to provide long-term monitoring of groundwater. Under HSWA, minimum technology requirements have been established for landfills, including the installation of double liners and collection of leachate. Technology requirements for incinerators include 99.99% destruction for most hazardous organic compounds and 99.9999% for acutely hazardous wastes, such as those on the P list.

Minimum Technology Requirement and Restrictions on Land Disposal. Research over the past decade has documented that the integrity of even the most impervious landfill liners is impaired by the corrosive nature of hazardous wastes. Therefore, HSWA contains provisions for restrictions on the land disposal of hazardous wastes. The *land disposal restrictions* (LDRs) were accomplished in phases, with each phase successively more restrictive in the wastes that could be disposed of in landfills. The first wastes banned were dioxins and solvents, followed by the "California list" (a group of chemicals previously banned in California consisting of free cyanides greater than 100 mg/L, specified concentrations of heavy metals, PCB concentrations greater than 50 mg/kg, etc.). This "land ban" followed with restrictions on the disposal of uncontained liquid hazardous wastes. Contained wastes can be disposed of in landfills only if no alternative is available that is environmentally acceptable.

Corrective Actions. As the federal environmental regulatory agency, the EPA has the authority to enforce RCRA, although in specific instances, RCRA enforcement is delegated to states. This authority was extended under HSWA, in which *corrective actions* may be required of generators to clean up contamination for which they are responsible, both on and off of their property. Stiff penalties (in terms of tens of thousands of dollars) may be assessed to generators, transporters, or TSD facilities that do not carry out the proper manifesting or that violate other hazardous waste regulations.

1.2.2   Underground Storage Tank Legislation

Because of the hundreds of thousands of underground storage tanks that have the potential for leaking, HSWA included a new section (Subtitle I) to regulate the installation of new tanks and the mitigation of leaking tanks. Underground storage tanks (USTs), as defined by RCRA, are

> *any one or combination of tanks (including underground pipes connected thereto) which is used to contain an accumulation of regulated substances, and the volume of which (including the volume of the underground pipes connected thereto) is 10% or more beneath the surface of the ground.*

Under Subtitle I of HSWA, a *regulated substance* is (1) any substance defined in Section 101 (M) of the Comprehensive Environmental Response, Compensation and Liability Act (CERCLA) except for hazardous wastes described under RCRA Subtitle C, and (2) any fraction of petroleum that is a liquid under standard environmental conditions (15.6°C and 1 atm). Many tanks are exempted, including farm or residential tanks less than 1000 gallons (3790 L) used to store fuel, tanks for storing heating oil for use on the premises, septic tanks, pit-pond-lagoon systems, and a few others.

Under Section 9003(C) of RCRA, the EPA was required to establish standards for properly designed and constructed new tanks that will not leak in the future. If property owners plan to install a new UST, they must notify the state regulatory agency at least 30 days before the tank is put into use. Four types of tanks have been approved by the EPA, including those constructed of (1) fiberglass-reinforced plastic, (2) cathodically protected steel, (3) steel/fiberglass-reinforced plastic, and (4) metal tanks, which may be installed only in noncorrosive environments. In addition, new tanks must be fitted with spill and overfill protection. Piping systems must be constructed of the same materials as the new tanks.

Corrosion protection systems must be maintained and tested on all new steel UST systems under Subtitle I. For example, cathodic protection systems must be tested at least every 3 years.

The UST regulations went into effect on December 22, 1988. Within 10 years of promulgation, all USTs in existence must meet the standards for the new tanks or face being closed down.

The second aspect of the UST regulations is mitigation of contaminated soils and water resulting from the failure of old or new tanks. When a release occurs, mitigation must be initiated, including immediate assessment and corrective measures involving safety (e.g., noxious gases, flammability) and the treatment of contaminated soils and groundwater.

1.2.3 The Comprehensive Environmental Response, Compensation and Liability Act and the Superfund Amendments and Reauthorization Act

Superfund, known formally as the *Comprehensive Environmental Response, Compensation and Liability Act of 1980 (CERCLA)*, was enacted to respond to spills and other releases of hazardous substances to the environment, particularly those resulting from the past disposal of hazardous wastes. The law was passed in the final session of the 96th Congress in response to Love Canal, Times Beach, and other hazardous waste disasters. The purpose of the act is to provide a mechanism for the cleanup of hazardous waste contamination resulting from large-scale accidental spills or chronic environmental damage. While RCRA deals with the cradle-to-grave management of hazardous waste that is currently generated, CERCLA focuses primarily on contaminants at sites where disposal took place in the past (i.e., pre-1980). It was reauthorized as the *Superfund Amendments and Reauthorization Act of 1986* (SARA). The 1980 act was funded at $1.8 billion, but SARA created an $8.5 billion fund. The other significant provision of SARA is the *Community Right to Know* clause, which requires industries to inform the public of the quantities of hazardous substances used through documentation known as the *Toxics Release Inventory.*

CERCLA Hazardous Substances. Although RCRA has a very specific definition of hazardous wastes, hazardous substances are more broadly defined under CERCLA. A CERCLA hazardous substance does not need to be a waste or waste material. It can be a commercial formulation, a product, and so on. The definition of a CERCLA hazardous substance is more broad and less exact than that for RCRA wastes; in fact, the CERCLA definition is based on other environmental regulations. A CERCLA *hazardous substance* is defined as any chemical regulated under the Clean Water Act, the Clean Air Act, the Toxic Substances Control Act, or the Resource Conservation and Recovery Act. Two materials that are excluded from the hazardous substances list are petroleum and natural gas.

Another classification under CERCLA covers almost any chemical that EPA believes is a threat to human health. This second category, called *pollutants or contaminants*, is defined as any other chemical or agent that "will or may reasonably be anticipated to cause harmful effects to human or ecological health." In summary, these two groups—hazardous substances and pollutants or contaminants—encompass a broad group of chemicals.

Environmental Releases. The event that triggers an action under Superfund is termed an *environmental release*, which is defined as the spilling, leaking, or disposing of a hazardous substance into the environment. From a practical standpoint, two types of environmental releases are the focal points of CERCLA: (1) spills, and (2) releases from past hazardous waste disposal sites. The implementation of Superfund has focused more on responding to contaminated sites than to spills. Four types of environmental releases are exempt from CERCLA regulatory activity: (1) workplace exposures, (2) vehicle exhausts, (3) radioactive contamination covered by other statutes, and (4) the application of fertilizers. However, in defining the *environment* relative to environmental releases, CERCLA is wide ranging. *Environment* consists of surface waters, groundwater, soils, subsurface strata, ambient air, and just about any other part of the physical environment with the exception of indoor air.

The National Contingency Plan. The blueprint for the cleanup of contaminated sites is the *National Contingency Plan* (NCP), which includes guidelines for site ranking, site assessment, feasibility studies, and cleanup actions. The first step of the NCP mandated the EPA to develop and maintain criteria for ranking sites for cleanup, known as the *National Priority List* (NPL). Site ranking is based on a Preliminary Assessment and a Site Inspection. The *Preliminary Assessment* (PA) usually involves a paper study of information on site history and disposal records. The *Site Inspection* (SI) usually includes a walk around the site as a minimum. The PA and SI data are then used to calculate the *Hazard Ranking System* (HRS) score. The EPA developed the HRS based on the MITRE model (after the MITRE Corporation, which developed the model), in which sites are ranked based on the extent of contamination, toxicity of the contaminants, hydrogeological characteristics of the site, and distance to groundwater or a population. High-ranking sites are placed on the NPL and await cleanup in order of their score.

The NCP provides a critical pathway by which sites are assessed and by which remediation systems are designed, constructed, and operated. The cleanup of Superfund sites is a lengthy, complicated process in which years may be required to complete the site assessment and remediation. The process involves legal, managerial, scientific, and engineering personnel to accomplish the task. The EPA does not have the resources to manage more than a handful of cleanup efforts at one time, so it outlines a plan each fiscal year called the *Superfund Comprehensive Accomplishments Plan* (SCAP), which details goals for each site on the NPL. The major elements of cleanup include the Remedial Investigation/Feasibility Study (RI/FS), The Record of Decision (ROD), the determination of how clean is clean, and in some cases, the formation of smaller Operable Units (OUs) at some Superfund sites.

The Remedial Investigation/Feasibility Study (RI/FS) is two separate investigations, but are usually performed together by the same contractor. The *Remedial Investigation* is the first step in defining contamination at the site and focuses on documenting sources, degree of contamination, environmental pathways, and potential adverse effects on human health and the environment. The primary goal of the Remedial Investigation is to provide a risk assessment at the site in which a decimal value is determined that describes the probability of an undesirable event (e.g., cancer) as the result of human exposure to the contaminants. Although sampling data are used as a basis for risk assessment, the Remedial Investigation process relies primarily on pathway and exposure models. The *Feasibility Study* focuses on assessing remedial designs, and provides engineering criteria for feasibility, costs, operational control, and so on. The Feasibility Study may provide 2, 3, or perhaps up to 20 design options. The EPA selects the appropriate remedial design in the *Record of Decision* (ROD) process. The design options may range from no action to extensive treatment such as incineration, soil washing, and bioremediation.

Protection of public health is a primary directive of CERCLA; the potential human health effects are evaluated by quantitative risk assessment, in which Superfund sites are assessed on a site-by-site basis to determine the extent of cleanup that is necessary. Specific cleanup criteria have not been established for the hazardous substances regulated under CERCLA because of the unique hazards posed at each site. All Superfund sites are assessed on a site-by-site basis to determine *how clean is clean,* with the rationale that the hazard of a contaminant is a function of its potential to reach a receptor (e.g., groundwater, population) and its hazard *if* the receptor is exposed to the contaminant. The potential for a contaminant to migrate and degrade, as well as the distance to a receptor of concern (i.e., the risk), is site-specific. Only by assessing each site individually and determining the risk to receptors is it possible to provide efficient and cost-effective cleanup of the thousands of hazardous waste sites throughout the United States.

In 1986, SARA brought about *Applicable or Relevant and Appropriate Requirements* (ARARs), which aid in the assessment of how clean is clean. The ARARs are usually derived from other environmental laws, such as the Safe Drinking Water Act

(SDWA) or RCRA. One of the more common ARARs is the use of SDWA Maximum Contaminant Levels (MCLs) as a cleanup level for contaminated groundwater (see Section 1.2.6).

Most Superfund sites are large, difficult to manage, and complicated. Therefore, an increasing trend has been to divide large sites into small segments, which are called *Operable Units* (OUs).

Liability. Under Superfund, a liability scheme has been established that requires responsible parties to clean up the waste that was disposed of. *Potentially Responsible Parties* (PRPs) are individuals or companies whose wastes have been disposed of at a CERCLA-regulated site. Although a large fund has been established from taxes on industry (hence, the name Superfund), remuneration from PRPs is mandated to keep the Emergency Response Fund intact. Common PRPs include current and past owners of the contaminated site, operators of the facilities (e.g., a company that leased land and used it as a poorly designed industrial landfill), generators (which can number in the hundreds), and transporters of wastes to the site. In other words, liability under Superfund is highly comprehensive. The wide-ranging liability has resulted in numerous legal battles, because many companies do not agree that they are responsible or may be unwilling to pay the costs of cleanup. As a result, Superfund has been mired in legal battles. Of the 12 years required for cleanup of a typical Superfund site, 5 years of legal battles are usually involved. Future versions of Superfund will no doubt be streamlined with better-defined liability provisions.

1.2.4   The Emergency Planning and Community Right-to-Know Act

As a basis for preventing large-scale chemical disasters, the *Emergency Planning and Community Right-to-Know Act* (EPCRA) was passed in parallel to SARA in 1986. This disaster-prevention legislation focuses on the development of emergency plans and data reporting for the life-threatening chemicals used by industry. In other words, businesses that use a group of chemicals known as *extremely hazardous substances* are required to notify state and local agencies about how much and what kinds of these chemicals are used at their facility. Furthermore, industries must report any planned or unplanned hazardous substance releases to these local and state authorities. Related to potential chemical releases, EPCRA requires the formation of *State Emergency Response Commissions* (SERCs) and *Local Emergency Planning Committees* (LEPCs)— entities that are responsible for implementing emergency response plans at the regional and local levels.

Numerous reporting requirements have been implemented under EPCRH. First, industries that generate extremely hazardous substances must submit copies of *material safety data sheets* (MSDSs), which are summaries of chemical properties and hazards, to the SERCs and LEPCs. In addition, Emergency and Hazardous Chemical Inventory Forms, which provide data on the quantities and locations of the hazardous substances, are required as supplements to the MSDSs.

The promulgation of EPCRA has had a substantial impact as a mechanism for providing a database of toxic substance releases in the United States. The *Toxics Release Inventory* (TRI), which has been developed out of EPCRA, has been a closely watched indicator of public health exposures. These data have provided citizens' groups with information to assess the exposure of residents living near industrial facilities.

1.2.5   The Clean Water Act

Although hazardous wastes are covered primarily under solid waste regulations, the Clean Water Act also focuses on toxic chemicals as a result of a major legal action. Congress enacted the *Federal Water Pollution Control Act (the Clean Water Act)* in 1972. The law set national goals for providing swimable and fishable waters by July 1, 1983, and for eliminating the discharge of pollutants into United States waters by 1985. In developing standards for the Clean Water Act, the EPA focused almost entirely on nonspecific pollution measures (i.e., conventional pollutants) such as biochemical oxygen demand (BOD), suspended solids, and pH. The emphasis on conventional pollutants was considered a source of weakness because toxic pollutants found in some effluents pose a greater threat to public health and the environment than solids and oxygen-demanding materials.

The Natural Resources Defense Council (NRDC) filed a lawsuit in 1976 against the EPA that resulted in a mandate requiring regulation of what eventually became known as the *priority pollutants.* The EPA was required to regulate the discharge of these 65 pollutants (129 specific chemicals) from 21 industrial categories (see Section 3.10). Congress, in passing the 1977 Amendments to the Clean Water Act, adopted essentially all of the NRDC mandate. Therefore, with the addition of the 1977 amendments, the Clean Water Act regulated the discharge of toxic pollutants, as well as conventional pollutants, from end-of-pipe aqueous waste streams to surface waters.

1.2.6   The Toxic Substances Control Act

In 1976, Congress passed the *Toxic Substances Control Act* (TSCA), a law that authorizes the testing of chemicals to assess their potential threat to public health upon entering the environment. Under TSCA, the EPA has authority to require industry to provide data on the potential adverse effects of new chemicals. The TSCA has a unique niche in environmental legislation in that other laws regulate chemicals that become waste streams or are released to the environment; TSCA, on the other hand, regulates the manufacture of new or even old chemicals. This procedure, the *Premanufacture Notification*, provides EPA with the authority to identify possible problem chemicals and require their evaluation.

The most important aspect of TSCA in relation to hazardous wastes concerns polychlorinated biphenyls (PCBs). These toxic and persistent transformer oils are not regulated under RCRA or CERCLA; their regulation is only through TSCA, under which they were banned in 1979.

1.2.7   The Safe Drinking Water Act

Passed in 1974, the purpose of the *Safe Drinking Water Act* (SDWA) is to provide drinking water that meets strict safe chemical and microbiological requirements. The original act of 1974 regulated the *Maximum Contaminant Levels* (MCLs) of only 13 compounds—7 metals and 6 pesticides. However, when the law was reauthorized in 1986, the list jumped to 39, which included solvents and industrial compounds that have found their way to water supplies as a result of improper disposal. The importance of these MCLs in hazardous waste management is related to cleanup standards.

Many states have promoted MCLs as the cleanup standard for the remediation of contaminated groundwater. At the federal level, MCLs have been incorporated into CERCLA as Applicable or Relevant and Appropriate Requirements (ARARs).

The recently reauthorized SDWA was signed into law in August 1996, creating several new programs and providing $12 billion for a number of drinking water programs from 1997 through 2003. Among the many provisions of the act are (1) a comprehensive study of the health effects of arsenic, with a final National Drinking Water Regulation (NDWR) to be published by the EPA by January 1, 2001, (2) further work on contaminant selection and a drinking water standard for these contaminants, (3) coordination of groundwater protection programs, and (4) drinking water research. Numerous other programs of the 1996 amendments to the SDWA are outlined in Reference 1.27.

1.2.8   The Clean Air Act

The purpose of the *Clean Air Act* (CAA), passed in 1970, has been to achieve *National Ambient Air Quality Standards* (NAAQS)—concentrations that are not to be exceeded for "criteria" pollutants, which include ozone, carbon monoxide, and particulate matter.

Most relevant to hazardous waste management are the *Hazardous Air Pollutants* (HAPs). A risk-based approach used in the 1970 CAA provided only seven HAPs; however, 189 HAPs are listed under the 1990 version of the CAA. This list was developed from technology-based standards—sources that emit 10 tons/year of a single HAP or greater than 25 tons/year of a combination of HAPs are classified as major sources, which are most highly regulated.

1.2.9   The Federal Pollution Prevention Act of 1990

As stated by its title, the purpose of the *Federal Pollution Prevention Act* (PPA) of 1990 is to promote the minimization and source reduction of hazardous wastes. Recycling, treatment, and disposal (e.g., landfilling) receive lower priorities in the management of hazardous wastes than do waste minimization and pollution prevention. The basis for the PPA is the *Common Sense Initiative* (CSI), in which a preventative approach is taken to pollution control rather than a crisis approach (e.g., water pollution, hazardous waste sites).

The Common Sense Initiative has been hailed by the EPA as their new approach to protecting public health and the environment. Rather than passing laws on a crisis-by-crisis basis, pollution prevention through the Common Sense Initiative is meant to provide long-term maintenance of a quality environment.

The PPA is also related to the Emergency Planning and Community Right-to-Know Act (EPCRA) by the Toxic Release Inventory (TRI). Industries that file TRI forms must also provide information on pollution prevention efforts for each chemical on the TRI.

Many other programs are part of the PPA, including the *Environmental Leadership Program* (ELP) Pilot Project Proposals. With the ELP, companies are encouraged to develop pollution prevention programs, not based on regulatory driving forces, but for achieving environmental goals.

1.2.10   State and Local Legislation

Although RCRA, CERCLA, and other federal legislation dominate the regulatory framework, state and local regulations are often also important. Some state and local governments had existing hazardous waste regulations when the federal legislation was passed; other regulations were passed by states and local entities to provide complimentary or more stringent regulations relative to the federal laws. Some states, such as California and New Jersey, have passed hazardous waste regulations that are more stringent than the corresponding federal statutes. Furthermore, many states have received the authority to administer federal legislation, such as RCRA, if their enforcement is at least as stringent as the federal regulations.

1.2.11   Hazardous Wastes, Hazardous Materials, Hazardous Substances

The definition of hazardous chemicals is dictated more by regulation and administrative details than by chemical properties and classification. Specific definitions, which are detailed and may involve lists and testing procedures, are outlined under RCRA, CERCLA, and Department of Transportation regulations. Based on these regulations, the most common three working definitions of hazardous chemicals are listed below:

Hazardous wastes are chemicals that are disposed of under RCRA. Lists, characteristics, and laboratory testing procedures are used to define RCRA hazardous wastes.

Hazardous substances are chemicals regulated under the Comprehensive Environmental Response, Compensation and Liability Act (CERCLA); that is, they are hazardous chemicals found at Superfund sites. They are defined by a "list of lists" of chemicals listed under RCRA, the Clean Water Act, the Clean Air Act, and other environmental statutes.

Hazardous materials are chemicals that are transported by truck, rail, air, or pipeline under United States Department of Transportation regulations.

A working knowledge of these three definitions of hazardous chemicals is essential because they are used throughout the text and in the practice of hazardous waste management.

1.3   CURRENT HAZARDOUS WASTE GENERATION AND MANAGEMENT

The thousands of hazardous waste sites resulting from preregulatory environmental releases of hazardous wastes will require decades to assess and clean up. As remediation efforts proceed, we are faced with yet another problem—how to dispose of the millions of pounds of hazardous waste generated each year. What is the magnitude of the problem? How much hazardous waste is actually generated per year in the United States? The most comprehensive report of current hazardous waste generation was completed by Baker et al. [1.2, 1.28], who provided summaries of surveys conducted by the Research Triangle Institute for the EPA. The survey, which was initiated in 1987, used data for the calendar year 1986. A sample of 2600 industries was asked to mark questionnaires (consisting of 328 pages!). The surveys were completed in 1989, and the data were published in 1992. The results show that during 1986, 12,478 industries generated 747.4 million tons of hazardous waste. This quantity included 695.1

million tons regulated under RCRA and 52.3 million tons regulated under state or other non-RCRA regulations. These generation rates are obviously higher than previously reported rates of 250 million tons per year [1.3].

Detailed statistical analyses were conducted on the data collected by the Research Triangle Institute. The relative amounts of hazardous waste generated in each of the EPA's 10 regions are shown in Figure 1.11. As expected, the highest generation rates are in industrial regions such as the mid-Atlantic states (Regions II and III), the upper Midwest (Region V), and the South (Regions IV and VI). In contrast, the Rocky Mountain and Prairie States of Region VIII produce only 0.6% of the nation's hazardous waste.

The survey also divides hazardous waste generation into primary sources, which result from production-related activities, and secondary sources, which result from waste management activities. Primary sources resulted in the generation of 539.9 million tons, and secondary sources accounted for 131.9 million tons. Of the primary sources, the industries that generated the wastes are classified in Figure 1.12, which shows that, after a miscellaneous category, the processes that produce the largest quantities of waste are electroplating, hydrogenation, and distillation/fractionation. One note of interest provided by the authors is that 5% of the industries generated 90% of the waste.

The data of Baker et al. [1.28] for the physicochemical state of the waste (Figure 1.13) show that over 75% of the wastes are inorganic liquids. What are these inorganic liquids that comprise so much of the United States' hazardous waste? We can make some sense of this by viewing Figure 1.14, which classifies the quantity of hazardous

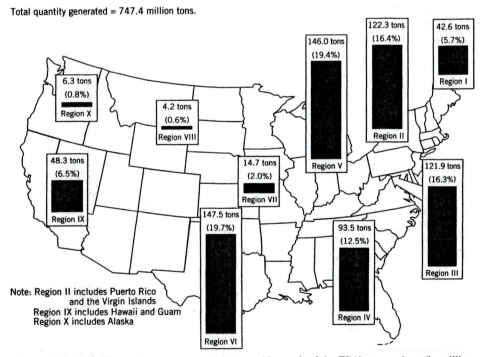

Figure 1.11 Quantities of hazardous waste generated by each of the EPA's ten regions (in million tons). *Source:* Reference 1.2.

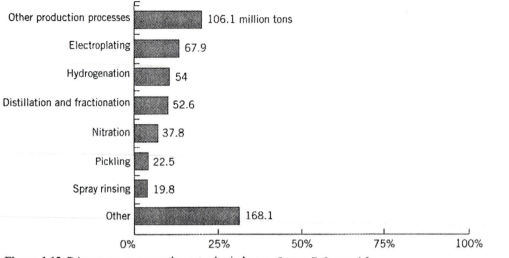

Figure 1.12 Primary waste generation rates by industry. *Source:* Reference 1.2.

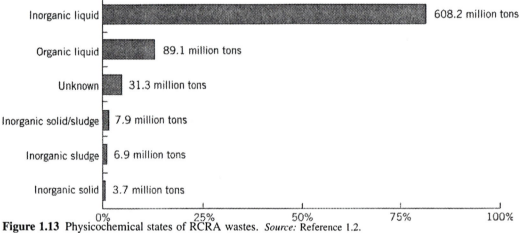

Figure 1.13 Physicochemical states of RCRA wastes. *Source:* Reference 1.2.

waste generated by RCRA waste code. This pie chart shows that acid waste streams (D002) and mixtures with D002 comprise 57.1% of the nation's hazardous waste.

The categories of industries that generated the hazardous wastes are listed in Figure 1.15. The chemical products industry produces over half of the waste at 383 million tons, with substantially lower hazardous waste generation rates for electronics, petroleum/coal products, primary metals, and transportation equipment. Baker et al. [1.28] also showed that more than 93% of the hazardous waste generators were private companies; the balance was accounted for by government and the military.

The ownership of hazardous waste management facilities, based on a survey of 2971 operations, is shown in Figure 1.16. As expected, the majority (89%) of waste management is under the auspices of the private sector. Through the 1980s, hazardous waste landfills regulated under RCRA were the primary method for the ultimate dis-

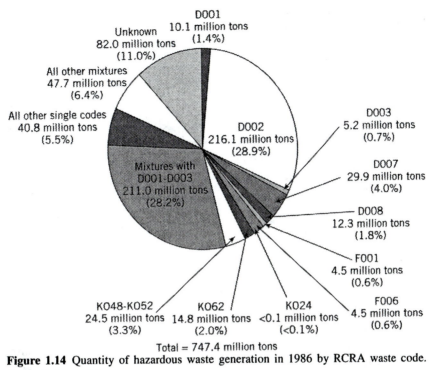

Figure 1.14 Quantity of hazardous waste generation in 1986 by RCRA waste code.
Source: Reference 1.2.

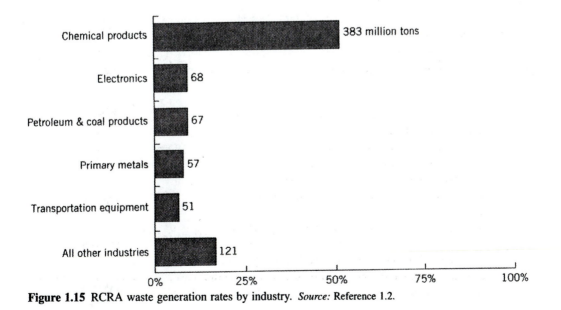

Figure 1.15 RCRA waste generation rates by industry. *Source:* Reference 1.2.

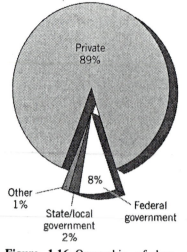

Figure 1.16 Ownership of hazardous waste management facilities.
Source: Reference 1.29.

posal of hazardous wastes. Although incineration produces air pollutants and, under improper operating conditions, may produce highly toxic chlorinated dioxins and furans [1.30], some expect incineration to have a larger share in the future management of hazardous wastes.

Ultimate disposal is not the only challenge associated with the current and future generation of hazardous wastes. Environmental releases of these chemicals have become a commonplace occurrence during their transportation and storage. Numerous spills occur during the transportation of hazardous materials; Ryckman and Ryckman [1.31] reported that 18,000 hazardous materials spills occurred during the year 1976. In 1983, spills from 4829 highway and 851 railroad incidences resulted in 8 deaths, 191 injuries, and damages exceeding $110 million.

Contamination from leaking underground storage tanks (USTs) may be considered a current and future hazardous waste problem, in addition to a past problem as discussed in Section 1.1. From the 1930s through the 1960s, approximately 2 million underground tanks were installed at gasoline stations, petroleum distribution points, industries, and residences (for home heating oil). Corrosion gradually weakened the tanks and, beginning in the mid-1980s, leaking UST incidents became commonplace in the United States. Recent studies estimate that 300,000 underground tanks will leak, with resulting soil and groundwater contamination, before the old tanks are replaced with corrosion-resistant new tanks [1.32]. Although many old tanks have been replaced with their modern counterparts, tank management will continue to be a significant part of hazardous waste management.

1.4 THE NATURE OF HAZARDOUS WASTE MANAGEMENT, ASSESSMENT, AND CONTROL

The hazardous waste field is interdisciplinary, and professionals from a number of environmental disciplines work together to solve complex hazardous waste problems. Consulting firms, industries, federal and state regulatory agencies, and the federal government (e.g., the United States Navy, the United States Geological Survey) employ professionals educated in environmental engineering, environmental chemistry, soil science, environmental microbiology, toxicology, and hydrogeology. What are some of the professional responsibilities of hazardous waste professionals? The nature of some typical hazardous waste projects are described next.

Site Assessment. When a contaminated site, leaking UST, or spill is reported or discovered, the extent of contamination must be assessed in order to determine how much, if any, of the soil and groundwater must be cleaned up. Depending on the size of the site, hazardous waste scientists and engineers spend weeks, months, or even years collecting data to quantify the spatial and temporal distribution of the contaminants.

Numerous procedures are used to assess contamination at hazardous waste sites. Calculations to quantify the rates of waste transformation and transport may be performed using hand calculators, chemodynamic models, and groundwater models. However, field data are usually required, which involves sampling of soils and groundwater followed by chemical analyses. Field techniques, such as ground-penetrating radar, soil-gas sampling, and portable gas chromatography are becoming more commonplace to determine the extent of contamination.

Risk Assessment. Once a site is assessed and the extent of contamination has been quantified, the data may be used to evaluate the potential threat to public health and the environment. Risk is a function of exposure and hazard; if the waste is far from a population and immobile or if it biodegrades rapidly, the risk from the contamination may be minimal. If the waste is deep and immobile, excavating the site may cause more problems than leaving the soil intact. For sites where contamination poses a low risk, cleanup may take a low priority relative to other sites, or it may not be cleaned up at all. Furthermore, cleaning up every contaminated site in the United States would cost at least $200 billion [1.12]. Risk assessment serves as an effective decision tool by which the sites that pose the greatest threat to public health and the environment are cleaned up in relation to their potential danger.

Emergency Response Assessment and Hazardous Materials Spill Control. Hazardous wastes and materials are commonly transported on highways, by rail, and on ships. Furthermore, releases of acutely toxic chemicals from industrial facilities may pose potential environmental disasters. Accidents and weather events often result in spills of hazardous wastes and materials from chemical plants or during transport, resulting in potential public health threats if the spilled material is acutely toxic. Emergency response personnel are involved in assessing the risk of hazardous material releases and mitigating any harmful effects. Teams of personnel evaluate contaminant concentrations, exposure pathways, and potential toxic effects on the exposed populations. In many cases, emergency response teams are on 24-hour call; if a spill occurs, they use source data (such as labels and placards on trucks and tanker cars), databases of chemical properties, and chemodynamic models to rapidly assess the movement of contaminant plumes and toxicity of the spilled chemicals. If rapid spill cleanup is required, emergency response personnel design and implement mitigation measures to protect the exposed populations from toxic effects.

Soil and Groundwater Remediation. If a contaminated site, after it is assessed and characterized, requires cleanup based on risk or standards, soil and groundwater treatment is implemented. In many cases, the most effective processes for the remediation of contaminated sites are based on natural pathways of chemicals in the environment. For example, contaminants that tend to volatilize from water and soils are also effectively treated by engineering processes that promote their volatilization. The remediation of contaminated sites can be implemented using in situ (in place) and ex situ processes. A commonly used in situ volatilization process is soil vapor extraction, in which vacuum is applied through a matrix of wells placed throughout the contaminated site. Pump-and-treat air stripping systems, in which contaminated groundwater is pumped to the surface, passed through an air stripper (Figure 1.17), and returned to the subsurface, are common examples of ex situ treatment systems. In the air stripper, volatilization is optimized by blowing air up through the stripper while the contaminated water cascades over high-area surfaces, resulting in volatilization of the contaminants and cleaner water at the bottom of the air stripper. Similar treatment processes are based on other naturally occurring phenomena, such as sorption, chemical oxidations, biodegradation, and thermal decomposition.

Brownsfield Development. Extensive land holdings in industrial areas have been contaminated with relatively low levels of industrial chemicals. In some cases, these "environmentally tainted" properties can be modified by simple cleanup methods or the installation of caps or barriers to provide an industrial property that poses minimal risk to the public or workers. This revitalizing of industrial properties, called *Brownsfield development*, has been a recent practice of the mid-1990s in response to the high cost of returning many contaminated sites to pristine conditions [1.33]. Brownsfields came

Figure 1.17 A low-profile air stripping system for the pump-and-treat remediation of groundwater. *Source:* Kleinfelder, Inc.

by the name because they lie somewhere between pristine green fields and the blackened soil of sites such as Love Canal and the Stringfellow Acid Pits.

Development of these "tainted" properties after cleanup to acceptable levels is a recent emphasis in contaminated site management and has become an economically attractive means of redeveloping these areas. The presence of many contaminated industrial properties has had other effects besides being a potential environmental and public health threat; these properties also represent underused land that negatively affects the economic and social development of cities. Some legislative considerations that are driving the Brownsfield movement include possible protections against liability and tax incentives for developers. Brownsfield development will no doubt be an integral part of future CERCLA and state legislation; incentives inherent in such legislation will likely include explicitly clarified liability and tax incentives for industries that develop Brownsfields.

Treatment, Storage, and Disposal Design and Permitting.   Under RCRA, TSD facilities include hazardous waste landfills, incinerators, heavy metal recycling operations, solvent recycling facilities, and traditional treatment processes such as neutralization, oxidation, and precipitation. Although RCRA landfills are receiving less emphasis as a hazardous waste management alternative because of the ban on the land disposal of hazardous wastes, other types of waste disposal facilities will be needed for the 250 to 750 million tons of hazardous waste generated per year.

Permits for TSD facilities are lengthy documents that provide detailed design criteria, groundwater protection systems, safety and health plans, and data on the contaminants treated at the facility. Hazardous waste disposal facilities usually contract with engineering consulting firms for TSD permit preparation. Engineers and scientists who prepare the permits characterize the site soils, geology, and hydrogeology, design monitoring well placement, and prepare sampling plans. Safety and emergency response plans are also part of the RCRA permitting process. As with many other aspects of hazardous waste management, an interdisciplinary team is usually required to accomplish the permitting process.

Waste Minimization and Pollution Prevention.   Reducing the mass and volume of hazardous wastes that are used in manufacturing and commercial operations is receiving greater emphasis from government and industry. Waste minimization may provide the key to ultimately decreasing the millions of tons of hazardous waste generated each year. Chemical engineers with experience in hazardous waste management are involved in altering process designs in plants where pesticides, solvents, and other petrochemicals are synthesized and manufactured. Environmental engineers and environmental chemists are often consulted by utilities, electroplating shops, dry cleaners, semiconductor manufacturers, and facilities that clean machinery to help in finding solvents, transformer oils, and degreasing agents that are less hazardous, easier to dispose of, and less costly.

Hazardous Waste Management.   Environmental engineers and scientists are employed by industries (or the industry will hire the professionals indirectly by contracting with a consulting firm) to supervise the cradle-to-grave management of hazardous wastes. Hazardous waste professionals manage RCRA compliance by ensuring that the paperwork required of the generator is submitted and that wastes are segregated, stored prop-

Figure 1.18 Protective clothing used in hazardous waste management. *Source:* Blasland, Bouck and Lee.

erly, and collected by a hazardous waste transporter within the 90-day holding time required by RCRA. They are responsible for health and safety measures (e.g., respirators, protective clothing) for workers who come into contact with hazardous materials, and they supervise the cleanup of spills and environmental releases (Figure 1.18).

Hazardous Waste Treatment. In order to solve the treatment and disposal problems associated with the generation of hazardous wastes, economical and environmentally sound treatment technologies will require development and design. New restrictions on the disposal of hazardous wastes are shifting the emphasis to destruction and treatment. Destruction processes that will see increased use include incineration, land treatment, and oxidative processes. Incineration, under properly controlled conditions of high temperature and oxygen, has the potential to destroy most hazardous organic residuals. Land treatment—the incorporation of organic wastes into soil containing acclimated microorganisms—is not as fully developed as incineration, but does have a high degree of potential for hazardous waste destruction. Advanced oxidation processes using ozone, ultraviolet light, and hydrogen peroxide have the potential to destroy compounds that are resistant to biodegradation.

1.5   SOURCE-PATHWAY-RECEPTOR ANALYSIS

Unlike the management and treatment of other wastes, such as municipal wastewater, hazardous wastes are difficult to define and assess before treatment systems can be designed or even considered. There are a number of reasons for this difficulty. First, unlike domestic wastewater, which is usually measured as the nonspecific parameter of biochemical oxygen demand, hazardous wastes may consist of complex mixtures of hundreds of different synthetic chemicals with wide-ranging physicochemical properties (e.g., water solubility, vapor pressure, biodegradability, toxicity, etc.). Hazardous waste problems are also multimedia in nature. Hazardous wastes that were disposed of on soils may evaporate into the atmosphere or migrate to surface waters and groundwaters. In many cases, hazardous waste problems are difficult to evaluate because they are hard to see. Suspended solids in wastewater are visible to the eye, but many of the hazardous organic chemicals found in contaminated soils and waters do not reflect visible light and, as a result, are colorless. In addition, wastewater influent is in plain view during end-of-pipe treatment evaluations. Standing on a hazardous waste site, it is often impossible to tell if there is any contamination at all, because the wastes are where they cannot be seen—in the soil, the unsaturated zone, and the groundwater.

How do we solve hazardous waste problems in which mixtures of complex wastes have been disposed of on land and may have migrated through different media? One approach, which has served as the basis for the risk assessment of contaminated sites, has been to divide the problem into three conceptual categories—*sources, pathways, and receptors* (Table 1.5). The first step in assessing a hazardous waste problem is defining the waste components at their *source*, including their concentrations and their properties such as density, water solubility, and flash point. A prerequisite for defining and understanding hazardous waste sources is knowing the nomenclature (i.e., the naming procedure) and structures of common hazardous chemicals. Such fundamental knowledge provides a basis for locating data on the chemicals and assessing their behavior in natural and engineered systems. After the source is characterized, *pathway* analysis focuses on quantifying rates at which the waste compounds volatilize, degrade, and migrate from the source. Pathway analysis is built on source information—it is impossible to quantify the potential for the waste to migrate, degrade, or be treated if the nomenclature, characteristics, and properties are not known or understood. Pathway analyses may show that the contaminant will biodegrade within weeks and cease to be a problem, or they may show that the compound persists in the environment for hundreds of years and will reach a receptor (e.g., a drinking water well) within 5 years. If the pathway analysis shows that exposure will result in an effect on *receptors* (humans, endangered species, etc.), the hazard must be assessed through toxicological and industrial hygiene data.

The importance of source-pathway-receptor analysis lies in its potential for providing a systematic analysis of hazardous waste site assessments. As an example of source-pathway-receptor analysis in assessing the hazard of contaminated sites, consider two sites, A and B. Site A is a pesticide formulation area that is characterized by 500 gallons of spillage of the insecticide malathion over 10 years. The soil is a silty loam high in natural organic matter, and the depth to groundwater is 45 ft. Use of the area for pesticide formulation ceased 10 ten years ago. Site B is a soil pit behind a dry cleaner that received 50 gallons of the waste solvent perchloroethylene over 2 years. The surface soil is a sandy loam containing low organic matter, and the depth to groundwater (a drinking water supply) is 45 ft.

Which site poses a greater threat to the groundwater supply? The answer is straightforward, based on source-pathway-receptor analysis. Although Site A received more waste, malathion degrades rapidly through hydrolysis and biodegradation, especially in soils containing high concentrations of organic matter. Moreover, it is also held in

Table 1.5 Summary of Source-Pathway-Receptor Analysis

	Variables	Hypotheses to be Tested
Sources	• Contaminants • Concentrations • Time • Locations	• Source exists • Source can be contained • Source can be removed and disposed • Source can be treated
Pathways	• Media • Rates of migration • Time • Loss and gain functions	• Pathway exists • Pathway can be interrupted • Pathway can be eliminated
Receptors	• Types • Sensitivities • Time • Concentrations • Numbers	• Receptor is not impacted by migration of contaminants • Receptor can be relocated • Institutional controls can be applied • Receptor can be protected

the upper soil horizons by the organic matter. Therefore, the malathion likely biodegraded years ago. Conversely, perchloroethylene is dense, highly mobile in soils, and slowly biodegraded. The soils at the site are permeable and contain low concentrations of organic matter. Therefore, Site B poses a greater threat to groundwater, even though the source volume is only one-tenth that of Site A.

The source-pathway-receptor concept was developed primarily for assessing the degree of hazard at contaminated sites. However, this conceptual theme also provides a fundamental basis for approaching other hazardous waste problems, such as the evaluation of process designs for soil and groundwater remediation systems, TSD design and permitting, waste minimization, and hazardous waste treatment.

As an example of how a fundamental understanding of source-pathway-receptor analysis can aid in designing groundwater treatment processes, consider two groundwater systems, one contaminated with carbon tetrachloride and the other with benzo[a]pyrene—chemicals that will be described in detail in Chapter 2. The most common groundwater remediation options are (1) air stripping (blowing air through a cascade of water to evaporate, or strip, the contaminant from the water), (2) sorption, in which the contaminated water is passed through a column of granular activated carbon, which removes the contaminant from the water by weak chemical bonds, and (3) biological treatment. Which design option would be your choice for each of the two contaminated groundwaters? Some of the important physical constants that are used in pathway analysis calculations, which will be covered in Chapters 5, 6, 7, and 8, are listed in Table 1.6.

The data show that both compounds are characterized by low rates of biodegradability, but carbon tetrachloride has a high Henry's Law constant (which means it has a tendency to escape from water) and benzo[a]pyrene has a high octanol–water partition coefficient (which means it is hydrophobic and will partition onto solids). Based on these data, sorption onto activated carbon would be a likely choice to treat

Table 1.6 Important Parameters for Two Groundwater Contaminants

	Carbon Tetrachloride	Benzo[a]pyrene
Log octanol–water partition coefficient	2.73	5.38
Henry's Law constant (atm-m^3 mole)	0.0302	2.4×10^{-6}
Aerobic first-order biodegradation rate constant (hr^{-1})	< 0.0096	< 0.0022

benzo[a]pyrene and air stripping is a logical process option for treating the carbon tetrachloride. Although this is a simple example, it shows that a fundamental understanding of source-pathway-receptor analysis provides a basis for process design.

The source-pathway-receptor theme will be used continuously throughout this text as a basis for approaching hazardous waste problems. Such a basis for understanding the complexity of hazardous wastes, ranging from site assessment to waste minimization, will provide fundamental knowledge and problem-solving skills far into the future.

1.6 SUMMARY OF IMPORTANT POINTS AND CONCEPTS

- Hazardous wastes have been generated from essentially all industrial activities. Prior to the passage and promulgation of federal legislation in the late 1970s, hazardous wastes were often disposed of improperly in pits, ponds, and lagoons, on surface soils, and in landfills.

- The Resource Conservation and Recovery Act (RCRA) was passed in 1976 to provide cradle-to-grave management of hazardous wastes; it was amended as the Hazardous and Solid Waste Amendments (HSWA) in 1984. Hazardous waste generators, transporters, and treatment/storage/disposal facility operators have responsibilities that provide safeguards against improper hazardous waste disposal.

- The Comprehensive Environmental Response, Compensation and Liability Act (CERCLA), or Superfund, was passed in 1980 to provide a mechanism for the mitigation of chronic environmental damage, particularly the cleanup of contaminated sites. Superfund was amended in 1986 as the Superfund Amendments and Reauthorization Act (SARA).

- Definitions of hazardous chemicals are based on regulatory and administrative criteria. Hazardous wastes are defined by RCRA, hazardous substances by CERCLA, and hazardous materials by Department of Transportation regulations.

- The current estimate of hazardous waste generation is approximately 750 million tons (680.4 million metric tons) per year in the United States. Most of the waste is classified as corrosive, and can be treated by neutralization.

- The hazardous waste field is multidisciplinary and requires the expertise of environmental engineers, environmental chemists and microbiologists, soil scientists, toxicologists, and hydrogeologists. Hazardous waste professionals have a number of responsibilities, including site assessments, risk assessment, soil and ground-water remediation, RCRA TSD permitting, hazardous waste management, and hazardous waste treatment.

• Hazardous waste problems can be approached using the conceptual theme of sources, pathways, and receptors. Although this analysis has been used primarily for risk assessment, it also aids in conceptualizing site assessments and evaluating process selection for the remediation and treatment of soils, groundwater, and other media.

PROBLEMS

1.1. Using your knowledge of introductory hydrology and chemistry, list (a) five environmental or site characteristics and (b) five contaminant properties that are likely to affect the movement of hazardous wastes disposed of at hazardous waste sites and facilities.

1.2. Each year the June issue of *Water Environment Research* is an extensive review of the scientific and engineering literature. Read the most recent review, "Hazardous Wastes: Assessment, Management, Minimization," and list all of the terms and acronyms with which you are not familiar.

1.3. Review the article "Hazardous Waste Control" [*Environmental Science & Technology,* 1983, **17**(7), 281A–285A], and describe the rationale used by the National Academy of Sciences for the estimated rates of hazardous waste generation.

1.4. Determine, using the procedures outlined by RCRA, whether each of the following is a hazardous waste. State the reason why it is or is not a hazardous waste. If it is a hazardous waste, list the RCRA waste category number. Assume that the industry producing the waste is a RCRA hazardous waste generator.
   a. Soil on an industrial property contaminated by the spill of a 5-gallon (18.9-L) container of endrin.
   b. A drum of off-specification ethyl acetate that is to be disposed of.
   c. Cement kiln waste.
   d. Sawdust in a warehouse contaminated by a spill of pentachlorobenzene.
   e. Sludge from the treatment of water from the chemical conversion coating of aluminum.
   f. An aqueous industrial waste stream from a plastic manufacturing plant that is discharged to a river.
   g. A drum of sulfuric acid at pH 1.2.
   h. A drum of waste 22% trichloroethylene.

1.5. Using the lists provided, determine whether each of the following is a hazardous waste by the RCRA listing procedure. If it is a listed waste, provide the waste number. If it is not a hazardous waste, state why it is not. Assume that the industry producing the waste is a RCRA hazardous waste generator.
   a. 2800 kg/month of acetone waste generated as the result of metal parts washing.

b. 5.7 million L/day of 150 mg/L trichloroethylene in an industrial waste stream that is being discharged to a municipal wastewater treatment plant.

c. 235 kg of endrin that is being disposed of because it is off-specification and out-of-date.

d. 185 kg/month of spent cyanide bath plating solutions from electroplating operations.

e. 4500 kg/month of wastewater treatment sludge from the production of iron blue pigments.

f. Sawdust contaminated by a 125-kg spill of commercially available methyl parathion.

1.6. A process in a manufacturing plant generates 1500 kg of nitrogen dioxide waste per month as a by-product of a manufacturing process. Nitrogen dioxide is listed on the P list. What responsibility does the generator have to dispose of this waste under RCRA?

1.7. Describe, using a flow chart or outline, how a waste is determined as hazardous under RCRA.

1.8. What characteristics are used to describe a
a. corrosive waste? c. reactive waste?
b. ignitable waste? d. TCLP toxic waste?

1.9. Under RCRA, the F and K lists are used to regulate wastes from specific and nonspecific sources. What general class of waste chemicals is regulated by the P and U lists, and what prompts the use of the P and U lists rather than the F and K lists?

1.10. You are the hazardous waste management supervisor for a large chemical company that manufactures pesticides. An employee calls you and reports that a 55-gallon drum containing parathion has fallen off a truck and has spilled its contents onto the pavement in a warehouse. Is there any action that needs to be taken based on federal regulations? Why or why not?

1.11. How are responsibilities different under RCRA for small quantity generators vs. large quantity hazardous waste generators?

1.12. As the new supervisor of hazardous waste management at a chemical manufacturing plant that generates 14,000 kg of hazardous waste per month, outline all of the steps you would take to characterize the waste, hire a transporter, and arrange disposal.

1.13. What is the difference between a RCRA Part A and Part B permit? What are some of the elements of a TSD facility that are addressed in a RCRA TSD permit?

1.14. List and briefly describe five types of sites that resulted from improper handling and disposal of hazardous wastes prior to the promulgation of RCRA.

1.15. What is the difference between a "hazardous substance" and "pollutants and contaminants" under CERCLA?

1.16. What is the difference between a hazardous waste (RCRA) and a hazardous substance (CERCLA)?

1.17. Define and describe each of the following:
a. NCP   b. NPL
c. ROD   d. Environmental release
e. "Environment"

1.18. The National Contingency Plan has been described as a "blueprint for cleanup" under Superfund. List and describe, using one or two sentences, three different programs under the NCP.

1.19. Describe the RI/FS process including the following:
a. What is involved in the RI?
b. What is involved in the FS?
c. What is the ROD?

1.20. Describe in detail what is meant by "how clean is clean?" and the rationale for this cleanup criterion.

1.21. Describe the regulatory requirements for the installation of new underground storage tanks.

1.22. Describe what is meant by source-pathway-receptor analysis and how it can be applied to
a. site assessment.
b. TSD permitting.
c. soil and groundwater remediation.

REFERENCES

1.1. *Technologies and Management Strategies for Hazardous Waste Control*, Office of Technology Assessment, U.S. Government Printing Office, Washington, DC, 1983.

1.2. Baker, R. D. and J. L. Warren, "Generation of hazardous waste in the United States," *Hazard Waste Hazard Mater.,* **9**, 19–35 (1992).

1.3. Hileman, B., "Hazardous waste control," *Environ. Sci. Technol.,* **17**, 281A–285A (1983).

1.4. Collins, A. J. and S. G. Luxon, *Safe Use of Solvents,* Academic Press, New York, 1982.

1.5. Matsumura, F., *Toxicology of Insecticides,* Plenum Press, New York, 1985.

1.6. Batstone, R. J., E. Smith, Jr., and D. Wilson, *The Safe Disposal of Hazardous Waste,* World Bank Technical Paper No. 983, Washington, DC, 1989.

1.7. *Synthetic Organic Chemicals,* U.S. International Trade Commission, Washington, DC, 1995.

1.8. *Environmental Trends,* Council on Environmental Quality, U.S. Government Printing Office, Washington, DC, 1989.

1.9. Shwendeman, T. D. and H. K. Wilcox, *Underground Storage Systems: Leak Detection and Monitoring,* Lewis Publishers, Chelsea, MI, 1987.

1.10. Kramer, W. H., "Groundwater pollution from gasoline," *Ground Water Monit. Rev.,* 2(2), 18–22 (1982).

1.11. Vianna, N. J. and A. K. Polan, "Incidence of low birth weight among Love Canal residents," *Science,* **226**, 1217–1219 (1984).

1.12. Paigen, B., L. R. Goldman, J. H. Highland, and A. T. Steegman, Jr., "Prevalence of health problems in children living near Love Canal," *Hazard. Waste Hazard. Mater.,* **2,** 23–43 (1985).

1.13. Janerich, D. T., W. S. Burnett, G. Feci, M. Hoff, P. Nasca, A. P. Polednak, P. Greenwald, and N. Vianna, "Cancer incidence in the Love Canal area," *Science,* **212,** 1404–1407 (1981).

1.14. *Environmental Epidemiology. Vol. 1. Public Health and Hazardous Wastes,* National Research Council, National Academy Press, Washington, DC, 1991.

1.15. Baker, D. B., S. Greenland, J. Mendlein, and P. Harmon, "A health study of two communities near the Stringfellow waste disposal site," *Arch. Environ. Health,* **43,** 325–334 (1988).

1.16. Clark, C. S., C. R. Meyer, P. S. Gartside, V. A. Majete, B. Specker, W. F. Balisteri, and V. J. Elia, "An environmental health survey of drinking water contamination by leachate from a pesticide waste dump in Hardeman County, Tennessee," *Arch. Environ. Health,* **37,** 9–18 (1982).

1.17. *Management of Hazardous Industrial Wastes: Research and Development Needs,* National Academy of Sciences National Academy Press, Washington, DC, 1983.

1.18. Rubin, D. K., J. J. Kosowatz, and P. Kemezis, "Cleanup dollars flow like water but industry is awash in problems," *Eng. News Rec.,* **222**(10), 33–43 (1989).

1.19. *Coming Clean: Superfund's Problem Can be Solved,* OTA-ITE-433, Office of Technology Assessment, U.S. Government Printing Office, Washington, DC, 1989.

1.20. *Report to Congress: Solid Waste Disposal in the United States, Vol. II,* EPA/530-SW-88-011B, U.S. Environmental Protection Agency, U.S. Government Printing Office, Washington, DC, 1988.

1.21. Westrick, J. J., J. W. Mills, and R. F. Thomas, *The Ground Water Supply Survey: Summary of Volatile Organic Contaminant Occurrence Data,* U.S. EPA, Office of Drinking Water, Cincinnati, OH, 1983.

1.22. MacKay, D. M., and L. A. Smith, "Agricultural chemicals in groundwater: Monitoring and management in California," *J. Soil Water Conserv.,* **45,** 253–255 (1990).

1.23. *The Quality of Our Nation's Water: A Summary of the 1988 National Water Quality Inventory,* EPA 440/4-90-005, U.S. Environmental Protection Agency, U.S. Government Printing Office, Washington, DC, 1990.

1.24. Patrick, R., *Groundwater Contamination in the United States,* National Academy Press, Washington, DC, 1983.

1.25. *ATSDR Biannual Report to Congress: October 17, 1986–September 30, 1988,* Agency for Toxic Substances and Disease Registry, U.S. Public Health Service, Atlanta, GA, 1989.

1.26. Arbuckle, J. G., M. E. Bosco, D. R. Case, E. P. Laws, J. C. Martin, M. L. Miller, R. D. Moran, R. V. Randle, D. M. Steinway, R. G. Stoll, T. F. P. Sullivan, T. A. Vanderver, and P. A. J. Wilson, *Environmental Law Handbook,* 14th Edition, Government Institutes, Inc., Rockville, MD, 1997.

1.27. Pontious, F. W., "Overview of the Safe Drinking Water Act Amendments of 1996," *J. Am. Water Works Assoc.,* **88**(10), 22–27 (1996).

1.28. Baker, R. D., J. L. Warren, N. Behmanesh and D. T. Allen, "Management of hazardous waste in the United States," *Hazard. Waste Hazard. Mater.,* **9,** 37–60 (1992).

1.29. Krieger, J., "Hazardous waste management database starts to take shape," *Chem. Eng. News,* Feb. 6, 1989, 19–21 (1989).

1.30. Markland, S., L. O. Kjeller, M. Hansson, M. Tsyklind, C. Rappe, C. Ryan, H. Cooazo, and R. Dougherty, "Determination of PCDDs and PCDFs in incineration samples and pyrolytic products," In Rappe, C., et al. (Eds.), *Chlorinated Dioxins and Dibenzofurans in Perspective,* Lewis Publishers, Chelsea, MI, 1986.

1.31. Ryckman, D. W. and M. D. Ryckman, "Organizing to cope with hazardous materials spills," *J. Am. Water Works Assoc.,* **72,** 196–200 (1980).

1.32. Bauman, B. J., "Soils contaminated by motor fuels: Research activities and perspectives of the American Petroleum Institute," In Kostecki, P. T. and E. J. Calabrese (Eds.), *Petroleum Contaminated Soils,* Lewis Publishers, Chelsea, MI, 1988.

1.33. Maldonado, M., "Brownsfield boom," *Civil Eng.,* **66**(5), 36–40 (1996).

Part One

SOURCES

The first step in the fundamental study of hazardous waste management and engineering focuses on the important hazardous chemicals including their structures, acronyms, and naming procedures (i.e., nomenclature). These topics, which are covered in Chapter 2, are based on qualitative environmental organic and inorganic chemistry. Chapter 2 also emphasizes how chemicals of concern are used and how they have been disposed of, which provides an appreciation of how they have become a problem.

After the nomenclature and structure of source materials have been introduced, their physical and chemical properties and classifications are presented in Chapter 3, including water solubility, density, flammability limits, flash point, chemical incompatibility, placards, and Chemical Abstract Services (CAS) registry numbers. The information in Chapter 3 provides the basis for beginning to recognize the behavior of hazardous chemicals and potential hazards of storage and transportation.

The analysis of hazardous waste sources, such as industrial facilities, treatment/storage/disposal facilities, and contaminated sites, is covered in Chapter 4. Topics include procedures for site assessments, industrial audits, sampling schemes, and methods of chemical analysis.

Completion of Part I provides a solid basis for study of the remainder of the material of the text. Knowledge of contaminant nomenclature, structure, and properties combined with source characterization is the foundation for quantifying chemodynamic pathways, assessing contaminant toxicity, defining risk, and designing remedial processes.

Chapter *2*

Common Hazardous Wastes: Nomenclature, Industrial Uses, Disposal Histories

Regardless of the regulatory scheme used to classify a waste or material as hazardous (e.g., RCRA, CERCLA, or state regulations), some chemicals are encountered repeatedly. In this chapter, the most common classes of hazardous compounds are introduced, including petroleum products, solvents, pesticides, polychlorinated biphenyls, dioxins, and metals. During most hazardous waste projects, the names of contaminants are a key part of a professional's conversation, and it is not uncommon for a handful of six-syllable chemical names or their acronyms to be used in a single sentence. Knowledge of hazardous compound nomenclature and structure is also necessary for accessing information to assess rates of transformation and transport, evaluate toxicity and risk, and design processes for hazardous waste treatment. In addition to providing a working knowledge of common hazardous compounds, this chapter includes information on the industrial processes that have produced the wastes and common disposal practices for each waste class.

Students who have not completed a course in organic chemistry should consider Section 2.1 a prerequisite before covering Section 2.2 on the composition of petroleum products. Readers with a working knowledge of organic chemistry will find this chapter beneficial for learning the hazardous compounds not covered in traditional organic chemistry classes. These chemicals will be encountered frequently in the text and, more important, in practice as a hazardous waste professional.

2.1 INTRODUCTION TO ORGANIC CHEMISTRY

Chemists define organic compounds as those that are carbon based. *Organic* molecules are composed of carbon and hydrogen atoms, but may also contain oxygen, nitrogen, phosphorus, sulfur, and halogens. The word *organic* is a holdover from the nineteenth century when chemists believed there were but two sources of materials—minerals and living sources. Biological sources produced organic compounds; chemicals derived from minerals were termed *inorganic* species. Compounds from biological sources had the common property of being carbon based, and although chemists have since found that carbon-based compounds can be synthesized in the laboratory, the name *organic* remains to describe these carbon compounds.

While most organic chemistry books emphasize the synthesis of large molecules, synthesis reactions are not included in this text because of their minimal importance in hazardous waste management. However, references are provided at the end of the chapter for those who desire further reading in organic chemistry [2.1–2.6].

2.1.1   Carbon Bonding

Two types of chemical bonds are usually covered in introductory chemistry—ionic and covalent. Although such a dichotomy is an oversimplification of the chemical bonding process, it does provide a working model. An *ionic bond* is characterized by the donation of a valence electron from an electropositive atom to an electronegative atom. Ionic bonding results from the stability associated with filled electron orbitals as defined by the octet rule; that is, atoms are thermodynamically stable when their outer s and p orbitals are filled. Sodium, with atomic number 11, is not stable with only one single unpaired electron in its $3s$ orbital:

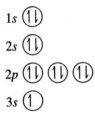

Chlorine, with atomic number 17, is not stable with seven electrons in its $3s$ and $3p$ orbitals:

Sodium donates its $3s$ electron to chlorine during bonding and, as a result, both atoms achieve thermodynamically stable ionic states with completely filled octets in their outermost shells:

Compounds with ionic bonds are usually composed of one element from the far left side of the periodic table and one from the far right side. Examples include LiF, NaCl, $MgCl_2$, and KCl.

Two atoms sharing valence electrons exhibit *covalent bonding,* which occurs when at least one atom lies in the central region of the periodic table (e.g., nitrogen, oxygen, silicon, carbon, and sulfur). As with an ionic bond, electrostatic forces hold the nuclei together. However, the electrons of a covalent bond exist between both nuclei:

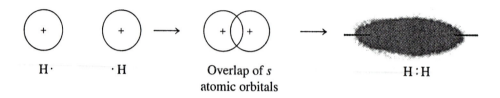

H·   ·H   Overlap of *s*   H : H
                                          atomic orbitals

Water, ammonia, and hydrogen sulfide are common examples of covalently bonded compounds.

The bonds of organic compounds are covalent. Although organic molecules may contain electronegative atoms, the bonds are covalent because carbon atoms are not capable of completely donating an electron to electronegative atoms such as halogens. The diverse group of compounds shown below has one commonality—their bonds are covalent:

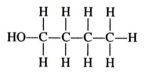

n-Butyl alcohol

Methane

Cl
|
Cl—C—Cl
|
Cl

Carbon tetrachloride

Acetaldehyde

Polarity of Bonds and Molecules

Many chemical bonds exhibit characteristics that are both covalent and ionic. In fact, there is a continuum between bonds that are totally ionic and totally covalent, a variation that exists because of the large number of differences in electronegativity between atoms. A bond between two identical atoms will show no evidence of polarity because the electron cloud is equally attracted to both nuclei. In molecules containing different atoms that are covalently bonded, the electron cloud surrounding the two different nuclei is not equally distributed between them, resulting in a charge separation or dipole:

The atom with higher electron density is termed the negative dipole, and the atom with fewer electrons is the positive dipole. Chemists designate polarity with the symbols δ^+ and δ^-, which show partial charges on the atoms.

Bond polarity and shifts of electron clouds affect the behavior of organic pollutants in natural and engineered systems. When electrons are drawn away from an electropositive atom by an adjacent electronegative atom, the low electron density allows attack by nucleophiles (species that are attracted to the atomic nucleus). For instance, the insecticide parathion is susceptible to nucleophilic attack at the phosphorus atom because the adjacent sulfur draws electrons, which allows nucleophiles (e.g., OH^-) to attack, resulting in its decomposition and detoxification:

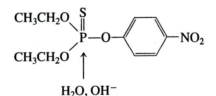

Parathion

In contrast, aerobic biological degradation rates are lowered by electron-poor areas of organic compounds; for example, organic contaminants containing a high degree of chlorine substitution biodegrade very slowly under aerobic conditions. Electron densities in relation to abiotic and biotic transformations of hazardous compounds are discussed in Chapter 7.

2.1.2 Nomenclature of Organic Compounds

To understand the chemical composition of petroleum and other organic hazardous wastes, familiarity with the names and structures of alkanes, alkenes, alkynes, aromatic hydrocarbons, and other organic contaminants is necessary. The following discussion provides those who do not have an organic chemistry background with a working knowledge of organic nomenclature and structures.

The diversity of carbon bonding results in a wide array of organic chemicals; two major divisions are aliphatic and aromatic compounds. *Aliphatic compounds* are composed of straight or branched chains of carbon atoms, and are classified as alkanes, alkenes, alkynes, alcohols, ethers, and so on. *Aromatic compounds* are characterized by carbon-based rings with resonating conjugated double bonds. Benzene, polycyclic aromatic hydrocarbons, and phenols are common aromatic chemicals of concern in hazardous waste management.

Isomerism

Organic compounds may be described in at least three ways, and each presents information about the chemical in a different manner. The three kinds of formulas that can be used for organic compounds are empirical, chemical, and structural. The *empirical formula* is no more than a ratio of elements in the molecule. The empirical formula of the sugar glucose is CH_2O, but its chemical formula is $C_6H_{12}O_6$; the empirical formula for the alkane pentane is the same as its chemical formula, C_5H_{12}. *Chemical formulas* describe the number of each atom in a molecule—a glucose molecule has 6 carbon atoms, 12 hydrogen atoms, and 6 oxygen atoms. However, more information is often needed to completely describe larger organic compounds.

Because of the different possible arrangements of atoms, organic compounds containing greater than three carbons may be found in more than one structural configuration. The different arrangements are called *isomers*. The greater the number of carbons, the greater the number of corresponding isomers, and they increase exponentially. The exponential increase in isomers with chain size for alkanes (straight and branched compounds of carbon and hydrogen) is listed in Table 2.1.

Table 2.1 Numbers of Possible Isomers
for Alkanes

Molecular Formula	Number of Isomers
C_4H_{10}	2
C_5H_{12}	3
C_6H_{14}	5
C_7H_{16}	9
C_8H_{18}	18
C_9H_{20}	35
$C_{10}H_{22}$	75
$C_{11}H_{24}$	159
$C_{13}H_{28}$	802
$C_{14}H_{30}$	1858
$C_{20}H_{42}$	366,319
$C_{30}H_{62}$	$> 4.1 \times 10^9$
$C_{40}H_{82}$	$> 6.2 \times 10^{12}$

As the result of isomerism, the chemical formula of an organic compound may not provide all of the information needed to describe its structure. For example, the compound with the chemical formula C_5H_{12} can exist as three possible isomers:

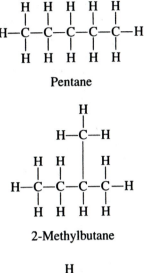

Pentane

2-Methylbutane

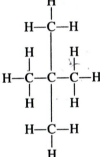

2,2-Dimethylpropane

Isomers can have widely varying chemical and physical properties that affect water solubilities, volatilities, environmental transport rates, degradation rates, and toxicity.

Organic Nomenclature Systems

Because larger organic molecules can exist in numerous isomeric forms, the chemical formula will not usually provide the necessary information to identify the compound. However, if the structural formula is given, the identity of the compound is evident. For example, it is impossible to recognize the molecular formula C_5H_{12} as the particular isomer called 2-methylbutane without listing the structural formula:

$$CH_3-CH_2-\overset{\overset{\displaystyle CH_3}{|}}{CH}-CH_3$$

Another way to describe a specific isomer associated with a chemical formula without drawing the compound's structure is by assigning a name to the compound—a name that specifies its structural formula. Three nomenclature systems are used for organic compounds: (1) the *International Union of Pure and Applied Chemists (IUPAC)* system, a formal process that accurately names all organic compounds; (2) *common,* or *trivial* names in which, for example, the carbon chain minus one hydrogen is termed a *radical* or *substituent,* and is bonded to some other molecular moiety, and; (3) *trade names.* Although the IUPAC system is universally accepted, it is useful only for smaller molecules; complicated molecules, such as pesticides, are more effectively described using common or trade names. Some of these names are acronyms, such as DDT. Because all three systems are used in the laboratory, in the field, and in design, familiarity with each is necessary.

Alkanes

Alkanes, such as methane, butane, and isooctane, are characterized by straight or branched chains of carbon and hydrogen.

Methane

Ethane

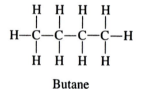

Butane

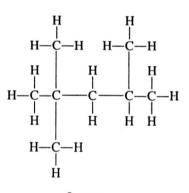

Isooctane

The straight lines between neighboring carbon atoms symbolize pairs of electrons shared in covalent bonds. The carbon-hydrogen bonds are usually omitted when organic structures are drawn:

$$CH_4$$

Methane

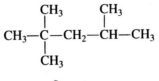

$$CH_3—CH_2—CH_2—CH_3$$

Butane

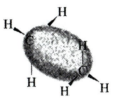

Isooctane

Abbreviated notations are less cumbersome, particularly when drawing the structures of large molecules.

The bonds of alkanes, which involve *hybrid sp³ orbitals*, are called *sigma bonds*. They have the general shape of a long tube in which the electrons are swirling about two nuclei:

Sigma bonds are not fixed; as a result, single carbon-carbon bonds are in constant rotation. One popular misconception about carbon bonding is related to the bond angles of the atoms. In two-dimensional chemical structures on a plane of paper, the bond angles appear to be 90°. Bonds of alkanes, however, point out at equal angles from each carbon atom toward the corners of a tetrahedron and have angles of 109.5°:

The IUPAC nomenclature of alkanes is based on the number of carbons in the longest chain of the molecule. The names of straight-chain alkanes up to 20 carbons are listed in Table 2.2. Nomenclature procedures for other groups of organic compounds are based on the names of the alkanes presented in Table 2.2.

Alkanes with four or more carbons can have isomeric forms; as a result, the names listed in Table 2.2 do not describe all of the isomers for a given number of carbons. The large number of isomers that can exist as the number of carbons in a molecule increases could result in a nomenclature nightmare; however, this naming problem can be solved by radical nomenclature.

Radical Groups and Alkyl Nomenclature

You may recall from basic chemistry that complex ions have names in and of themselves. The compound $MgSO_4$ is composed of Mg^{2+} and SO_4^{2-} ions, with the latter simply called sulfate ion; the compound name is magnesium sulfate. Similar complex ions are CO_3^{2-}, carbonate, and NO_3^-, nitrate.

Table 2.2 Names of Straight-Chain Alkanes
Containing up to 20 Carbons

Name of Compound	Number of Carbons
Methane	1
Ethane	2
Propane	3
Butane	4
Pentane	5
Hexane	6
Heptane	7
Octane	8
Nonane	9
Decane	10
Undecane	11
Dodecane	12
Tridecane	13
Tetradecane	14
Pentadecane	15
Hexadecane	16
Heptadecane	17
Octadecane	18
Nonadecane	19
Eicosane	20

Organic chemists have developed a similar system for segments of molecules called *radicals* or *substituent groups*. For alkanes, the radicals are called *alkyl groups* and lack one hydrogen atom relative to the parent molecule. Typical alkyl groups are —CH_3, methyl; —CH_2—CH_3, ethyl; and —CH_2—CH_2—CH_3, n-propyl. The root of an alkyl group is derived from the name of the alkane, but the suffix is simply changed from -*ane* to -*yl*. The most common alkyl groups are shown in Table 2.3. Abbrevia-

Table 2.3 Common Alkyl Groups

—CH_3	—CH_2—CH_3	—CH_2—CH_2—CH_3
Methyl (Me)	Ethyl (Et)	n-Propyl (n-Pr)
—CH(CH$_3$)(CH$_3$)	—CH_2—CH_2—CH_2—CH_3	—CH(CH$_3$)—CH_2—CH_3
Isopropyl (i-Pr)	n-Butyl (n-Bu)	sec-Butyl (s-Bu)
	—CH_2—CH(CH$_3$)—CH_3	—C(CH$_3$)(CH$_3$)—CH_3
	Isobutyl (i-Bu)	tert-Butyl (t-Bu)

tions for the alkyl radical groups used in *Chemical Abstracts* are given in parentheses.

One problem with alkyl nomenclature is isomerism, including two propyl isomers and four butyl isomers. The two propyl isomers are *n*-propyl (normal) and isopropyl. The four forms of the butyl group are *n*-butyl, *sec*-butyl (secondary), isobutyl, and *tert*-butyl (tertiary). The term *normal* means the group is a straight chain. In the *secondary* form, bonding to the parent compound occurs at the second carbon from the end of the group. In the *tertiary* structure, the attachment carbon is bonded to the three other carbons of the butyl group. *Iso-* has the characteristic of the two last carbons being branched.

IUPAC Nomenclature

The most systematic and accurate method for naming organic chemicals, the IUPAC system, is the only feasible procedure for linking the names of large organics to their structures. The IUPAC system uses very specific rules, shown in Table 2.4, for naming alkanes. The standard procedure under IUPAC for the placement of substituent groups is from smallest to largest. However, the Chemical Abstract Service (CAS) of the American Chemical Society uses the alphabetical method described in Table 2.4. Because of the straightforward nature of alphabetical ordering and its ubiquitous use, this method will be followed throughout the text.

Table 2.4   IUPAC and CAS Rules for Naming Alkanes

1. Select the carbon skeleton with the longest chain as the parent compound. Consider substitution groups (alkyl groups) to be replacements of hydrogen on the carbon skeleton.
2. Assign numbers to the carbons where alkyl groups are attached. Start at the end of the skeleton that allows use of the lowest possible numbers. For example, the structure

$$\text{H}_3\text{C}-\text{CH}_2-\text{CH}_2-\text{CH}_2-\overset{\displaystyle \overset{\text{CH}_3}{|}}{\text{CH}}-\text{CH}_2-\text{CH}_3$$

   would be named 3-methylheptane rather than 5-methylheptane.
3. If more than one of a given substituent group is attached to the skeleton, use a prefix (di-, tri-, tetra-, penta-, etc.) to show how many of these groups are on the molecule. Precede the group names with their numbered locations on the skeleton. With four methyl groups on the pentane skeleton, the compound

$$\text{H}_3\text{C}-\overset{\displaystyle \overset{\text{CH}_3}{|}}{\underset{\displaystyle \underset{\text{CH}_3}{|}}{\text{C}}}-\overset{\displaystyle \overset{\text{CH}_3}{|}}{\underset{\displaystyle \underset{\text{CH}_3}{|}}{\text{C}}}-\text{CH}_2-\text{CH}_3$$

   is named 2,2,3,3-tetramethylpentane.
4. Commas always separate numbers, and hyphens are placed between numbers and words: e.g., 4-ethyl-2,3,3-trimethylheptane.
5. For molecules that contain different substituents attached to the carbon skeleton, the order of the substituents used in the names is alphabetical; for example,

Table 2.4 (Continued)

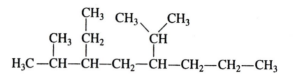

is named 3-ethyl-2-methyl-5-isopropyloctane. The sequence of the numbers indicating the positions of the substituents is irrelevant.

The prefixes di-, tri-, iso-, *n*-, *sec*-, *tert*-, etc. are not considered in the alphabetical ordering [2.3].

EXAMPLE 2.1 *Nomenclature of Alkanes*

Name the following alkane:

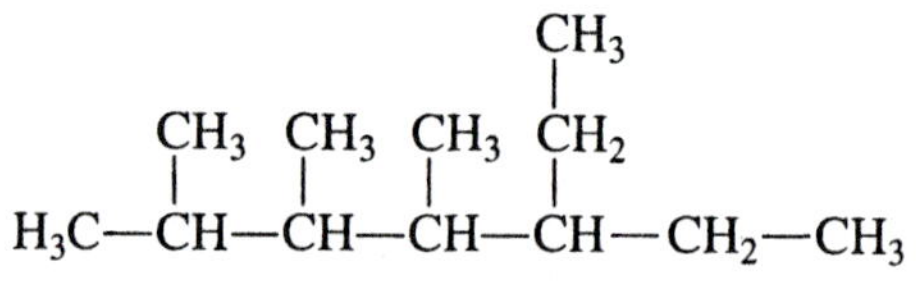

$$CH_3$$
$$|$$
$$CH_3 \quad CH_3 \quad CH_3 \quad CH_2$$
$$|\qquad|\qquad|\qquad|$$
$$H_3C-CH-CH-CH-CH-CH_2-CH_3$$

SOLUTION

1. The longest straight or parent chain has seven carbons and is derived from the alkane heptane in Table 2.2:

$$H_3C-CH_2-CH_2-CH_2-CH_2-CH_2-CH_3$$

2. Number the parent chain using the lowest possible numbers for the substituted alkyl groups, which include

$$-CH_3$$

2,3,4-Trimethyl

and

$$-CH_2-CH_3$$

5-Ethyl

3. Placing the substituted alkyl groups in alphabetical order and combining them with the parent chain yields the name 5-ethyl-2,3,4-trimethylheptane, which exactly describes the chemical structure.

EXAMPLE 2.2 *Drawing the Structure of Alkanes*

Draw the structural formula of 4-*tert*-butyl-3-ethyl-2,6-dimethyloctane.

SOLUTION

1. The longest chain in the molecule is octane:

$$CH_3-CH_2-CH_2-CH_2-CH_2-CH_2-CH_2-CH_3$$

2. Substituent groups include

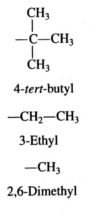

4-*tert*-butyl

$$-CH_2-CH_3$$

3-Ethyl

$$-CH_3$$

2,6-Dimethyl

3. Placing the alkyl groups on the parent chain in place of hydrogen atoms results in the structure

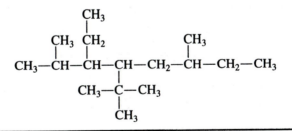

Common Names for Alkanes and Alkyl Derivatives

The prefix *iso-* is used extensively in the petroleum industry as a common, or trivial method for naming alkanes branched at one end. It is used primarily for smaller alkanes that are straight chains with branching of the last two carbons:

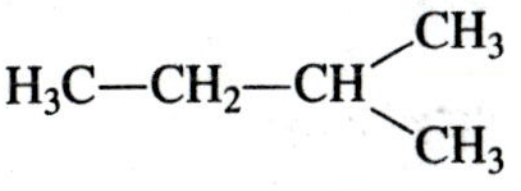

$$H_3C-CH_2-CH\begin{smallmatrix}CH_3\\\\CH_3\end{smallmatrix}$$

Isopentane

With this common naming procedure, the *pentane* root of isopentane describes the total number of carbons in the molecule (five), rather than the number in the parent chain. An anomaly to the general *iso* designation is the name given to the important gasoline constituent isooctane, which is the IUPAC compound 2,2,4-trimethylpentane, not 2-methylheptane:

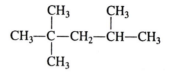

Isooctane

Isooctane is used as a standard in the rating of gasoline because it is the alkane exhibiting the greatest antiknock properties in internal combustion engines.

Common names for alkane derivatives are based on the nomenclature of alkyl groups. Common hazardous chemicals that are alkane derivatives include alkyl halides (or chloroalkanes), such as 2-chloropropane. Organic compounds substituted with halogens and nitrogen groups are important hazardous wastes and will be seen throughout this text. Some of the more important substituents are listed in Table 2.5. In industry, the less formal common naming procedure is used, in which the name of the alkyl group is followed by the name of the anion (e.g., bromide, chloride, fluoride, nitrate). In other words, an alkyl group bonded to a substituent listed in Table 2.5 is named "alkyl halide" or "alkyl nitrate." For example, the common name for 2-chloropropane is isopropyl chloride:

$$CH_3{-}CH{-}CH_3$$
$$|$$
$$Cl$$

Isopropyl chloride (common name)
2-Chloropropane (IUPAC name)

For amines (organic compounds containing an amino group, $-NH_2$), the alkyl name is followed by the word *amine;* for example, $CH_3{-}NH_2$ is methylamine, $CH_3{-}CH_2{-}NH_2$ is ethylamine, and so on.

Table 2.5 Prefixes for Substitution Groups

Substituent	Prefix
$-Br$	Bromo-
$-Cl$	Chloro-
$-F$	Fluoro-
$-I$	Iodo-
$-NH_2$	Amino-
$-NO_2$	Nitro-

Alkenes

Alkenes are characterized by a double bond somewhere in the molecule. The double bond results in unique chemistry, both in terms of biochemical and chemical reactivity. Some common alkenes are

Ethene or ethylene

Propene or propylene

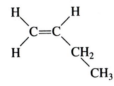

1-Butene or 1-butylene

The *-ylene* form is the older IUPAC suffix for *-ene*, and its use persists today. Perhaps the most widely used *ylene* name is trichloroethylene (TCE), a common solvent. The correct current IUPAC name for trichloroethylene is actually trichloroethene, but the *ylene* form is used almost exclusively in the hazardous waste literature.

Bond Angles and Molecular Structure of Alkenes. The carbon-carbon double bond of alkenes is planar, and other bonds radiate from the carbons almost symmetrically at angles near 120°. Slight deviations from the 120° bond occur as a result of atomic sizes and repulsion and attraction forces determined by the electronegativity of the atoms. The carbons of the double bond are fixed in space, so no rotation occurs around the double bond. The structure of the ethene molecule is shown in Figure 2.1, and the rules for naming alkenes are listed in Table 2.6.

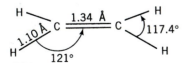

Figure 2.1 Structure of ethene.

Table 2.6 Rules for Naming Alkenes

1. The longest carbon skeleton containing the double bond serves as the basis for naming the compound. Use the alkane names listed in Table 2.2, but change the last syllable from *-ane* to *-ene*.
2. Number the first double-bonded carbon on the longest chain so that it has the lowest number possible.
3. Name alkyl substituents in the same manner as for alkanes.
4. Give a special prefix to an alkene that is unsymmetrical across the plane of the double bond. If the parent chain crosses the double bond and remains on the same side of the molecule, use the prefix *cis-*. If the parent chain continues on the opposite side of the double bond, use the prefix *trans-*.

EXAMPLE 2.3 *Nomenclature of Alkenes*

Name the following alkene:

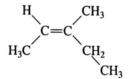

SOLUTION

1. The parent structure is the longest chain containing the double bond:

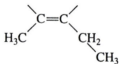

The double bond receives the lowest number possible:

2-pentene (not 3-pentene)

2. One substituent group is present on the number 3 carbon:

3-methyl

3. The continuation of the parent chain is on the same side of the double bond, so the *cis-* prefix must be noted.
4. The resulting IUPAC name is 3-methyl-*cis*-2-pentene.

Other Aliphatic Compounds

Other minor components of petroleum include alkynes (or acetylenes), organic acids, and cycloalkanes. *Alkynes* are characterized by a triple carbon-carbon bond:

$$H-C\equiv C-CH_3$$

Propyne

Alkynes are named in the same manner as alkenes, but *-ene* is replaced by *-yne*. Because alkynes are straight-chain compounds, they have no *cis-* or *trans-* designation.

Organic acids contain a carboxylic group (—COOH) in place of one terminal methyl group. In IUPAC nomenclature, the *-ane* of an alkane is replaced by *-anoic acid*; the compound commonly called butyric acid is therefore

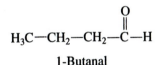

1-Butanoic acid

Aldehydes are structurally similar to organic acids but contain an —H instead of an —OH on the carbonyl, C=O, carbon. However, they do not behave as weak acids (see Chapter 3). To name an aldehyde using IUPAC nomenclature, the *-ane* for an alkane is replaced by *-al*:

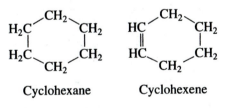

1-Butanal

Cyclohexane and cyclohexene are the most common cycloalkanes and cycloalkenes. Cyclohexane contains hydrogen-saturated carbon atoms in a six-carbon ring. Both compounds are members of the chemical class called alicyclic hydrocarbons:

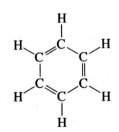

Cyclohexane   Cyclohexene

Aromatic Compounds

Aromatic compounds are often depicted by alternating single and double bonds between carbon atoms joined in a ring. The most common aromatic hydrocarbon is benzene:

A shorthand form of the benzene structure leaves out the carbon and hydrogen symbols:

Because the double bonds are resonating (i.e., not located between any particular carbons) and represent a cloud of electrons swirling around the six carbon nuclei, the benzene ring is often shown as:

Because the double bonds are resonating (i.e., not located between any particular carbons) and represent a cloud of electrons swirling around the six carbon nuclei, the benzene ring is often shown as:

The bonds of benzene and other aromatic compounds are called *pi bonds*—circular, electron-rich clouds of electrons that swirl around the ring of carbon nuclei:

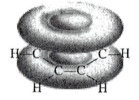

Benzene was discovered by Michael Faraday in 1825, and its commercial production began in 1849. Benzene production routinely ranks among the top 10 commodity chemicals [2.7]. It is highly flammable, has a characteristic aromatic odor, and has properties that make it a stable hazardous compound. With its resonating double bonds, benzene is more stable than alkenes. For example, cyclohexene is rapidly oxidized by dilute potassium permanganate ($KMnO_4$) or combines with hydriodic acid (HI), resulting in the addition of two OH groups or an H and I across its single double bond. However, benzene does not react with these oxidants because its double bonds are resonating, which enhances its stability [2.5]. Benzene also normally undergoes substitution reactions rather than addition reactions; that is, an atom or group is substituted for a hydrogen, and the aromatic ring is preserved. Because the benzene ring resists addition, it is more persistent in the environment than some other hydrocarbons such as alkenes.

Nomenclature of Benzene Derivatives

For most benzene derivatives, the compound is named by placing the substituent first followed by *benzene*. (See Table 2.5 for substituent prefixes). For example, some monosubstituted benzenes are

Chlorobenzene

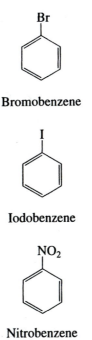

Bromobenzene

Iodobenzene

Nitrobenzene

By using the symbol Ph for a benzene ring, the CAS system uses abbreviations such as PhCl, PhBr, PhI, and $PhNO_2$ for these compounds. The symbol Ph is an abbreviation for *phenyl,* which is the radical name for benzene. Therefore, if a benzene ring is a substituent of a larger molecule, the term phenyl is used to describe its presence and location, just as methyl, ethyl, and so on are used to describe alkyl substituent groups. For example, the compound

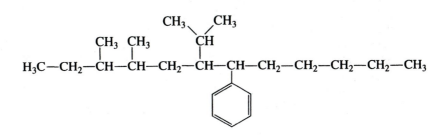

would be named 3,4-dimethyl-7-phenyl-6-isopropyldodecane.

Some monosubstituted benzenes that are common hazardous wastes have traditional names for which there is no systematic basis:

Phenol

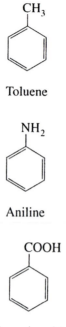

Toluene

Aniline

Benzoic acid

If two groups are substituted on the benzene ring, the groups on the ring must be named with their corresponding locations. One of the most common methods for describing their locations is through the use of the terms *ortho- (o-)*, *meta- (m-)*, or *para- (p-)*:

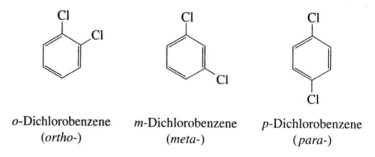

| *o*-Dichlorobenzene | *m*-Dichlorobenzene | *p*-Dichlorobenzene |
| (*ortho-*) | (*meta-*) | (*para-*) |

Using *ortho-*, *meta-*, and *para-* to designate positions of substituents on benzene rings is a common nomenclature procedure. The more formal IUPAC system is based on numerical positions of the substituents on the ring:

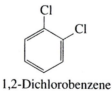

1,2-Dichlorobenzene

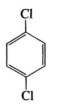

1,4-Dichlorobenzene

If the two substituents are different and no traditional name has been given to the compound, the groups are numbered and put in alphabetical order followed by *-ben-zene.* For derivatives of phenol, aniline, toluene, and benzoic acid, the name of the compound is based on its traditional name, and that substituent (OH, NH_2, CH_3, or COOH) is considered to be in the 1-position; the second substituent is given a number relative to it, for example, 2-chlorobenzoic acid or *o*-chlorobenzoic acid.

EXAMPLE 2.4 *Nomenclature of Benzene Derivatives*

Name the following benzene derivatives:

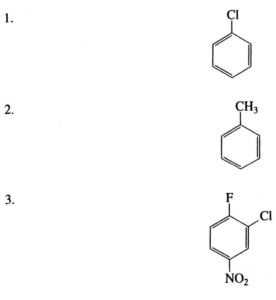

1.

2.

3.

SOLUTION

1. Based on the rule that substitution groups are treated as a prefix followed by *-ben-zene,* the compound is named chlorobenzene.
2. The compound is a monosubstituted benzene with a traditional name—toluene.

3. This is a trisubstituted benzene. Using the lowest sequence of numbers possible and placing the substituents in alphabetical order yields 2-chloro-1-fluoro-4-nitrobenzene.

Special Disubstituted Benzenes

Some benzene derivatives with two substitution groups are so common that they have been given traditional names. These include xylenes, cresols, and phthalic acids:

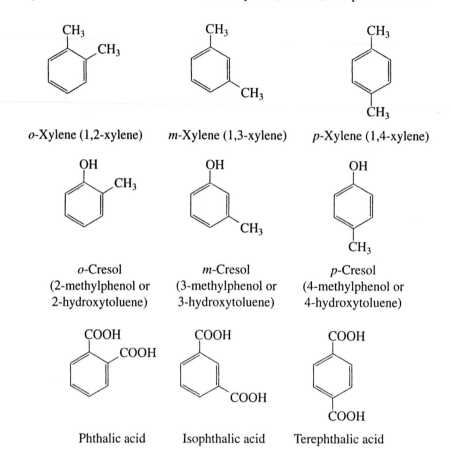

| o-Xylene (1,2-xylene) | m-Xylene (1,3-xylene) | p-Xylene (1,4-xylene) |

o-Cresol	m-Cresol	p-Cresol
(2-methylphenol or	(3-methylphenol or	(4-methylphenol or
2-hydroxytoluene)	3-hydroxytoluene)	4-hydroxytoluene)

| Phthalic acid | Isophthalic acid | Terephthalic acid |

These compounds can also be given systematic names (e.g., 2-methylphenol), which often results in some confusion in their nomenclature.

Polycyclic Aromatic Hydrocarbons

Two or more benzene rings can share pairs of carbon atoms, resulting in fused aromatic rings. Two names are given to this class of compounds: *polycyclic aromatic hydrocarbons (PAHs)* and *polynuclear aromatic compounds (PNAs)*. The PNA classification is more general and includes heterocyclic aromatic compounds, such as benzene-like rings containing an oxygen, nitrogen, or sulfur. The rings of PAHs contain only carbon. They are moderately stable materials that are formed as trace contaminants in the incomplete combustion of organic compounds and are found in the heavier fractions of petroleum products (e.g., lubricating oils, asphalt, and tarlike materials), as well as cigarette smoke and automobile exhaust. Because these compounds are formed during combustion processes, they are also found in trace concentrations in fly ash and the residues of incineration and other thermal processes. They occur as solids at room temperature and often partition onto the surfaces of soils and sludges. The first evidence of chemically induced cancer was linked to PAHs; during the eighteenth century, English chimney sweeps were found to have a high incidence of scrotum cancer, and the cancer was eventually connected with the PAH-laden soot to which the chimney sweeps were exposed. Two modern routes of personal exposure to PAHs are cigarette smoke (the tar component of the warning labels) and the blackened outside of barbecued food [2.8].

A major source of PAH contamination has been manufactured gas plants, which produced gas for residential and industrial uses from coal and oil. These facilities operated from the mid-1880s through about 1950, when their use was superseded by the installation of natural gas distribution networks. Over 1000 manufactured gas plants may have been in operation in the early 1900s, primarily in the midwestern and eastern United States. Soils and sludges around these areas are difficult to remediate because of their high concentrations of PAHs, long-chain aliphatics, and phenolics [2.9].

The smallest PAHs include naphthalene and anthracene, which are found in some fractions of petroleum:

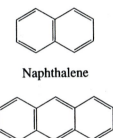

Naphthalene

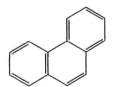

Anthracene

Some PAHs, such as the following three-, four- and five-ring structures, are larger:

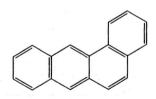

Phenanthrene

Benzo[*a*]anthracene

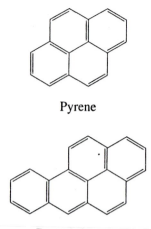

Pyrene

Benzo[*a*]pyrene

Polycyclic aromatic hydrocarbon nomenclature is based on 35 IUPAC-prescribed compounds (Table 2.7), which form the backbone of other more complex PAHs. The most commonly used nomenclature system, the fusion method, builds upon these prescribed structures for naming other more complicated PAHs. Before describing the fusion method, a summary of the IUPAC system for orientating and numbering PAHs is necessary, which is outlined in Table 2.8 and illustrated in Example 2.5.

Table 2.7 Prescribed Names of Polycyclic Aromatic Hydrocarbons

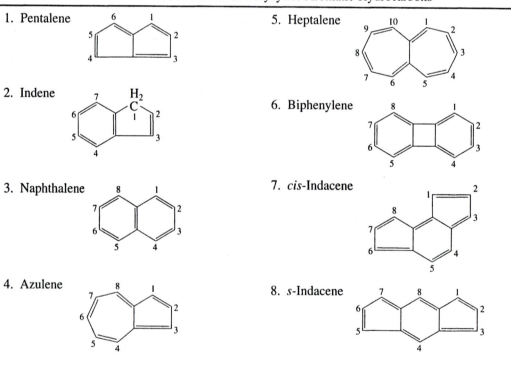

1. Pentalene

2. Indene

3. Naphthalene

4. Azulene

5. Heptalene

6. Biphenylene

7. *cis*-Indacene

8. *s*-Indacene

Table 2.7 (Continued)

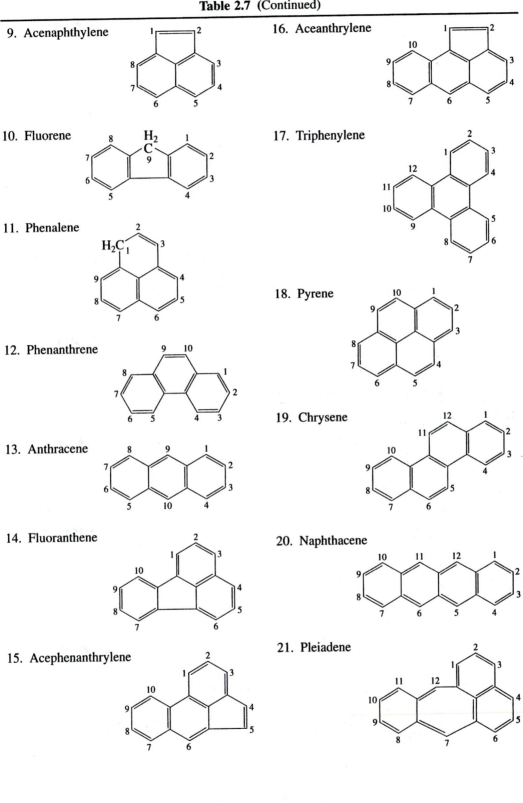

9. Acenaphthylene

10. Fluorene

11. Phenalene

12. Phenanthrene

13. Anthracene

14. Fluoranthene

15. Acephenanthrylene

16. Aceanthrylene

17. Triphenylene

18. Pyrene

19. Chrysene

20. Naphthacene

21. Pleiadene

Table 2.7 (Continued)

22. Picene

23. Perylene

24. Pentaphene

25. Pentacene

26. Tetraphenylene

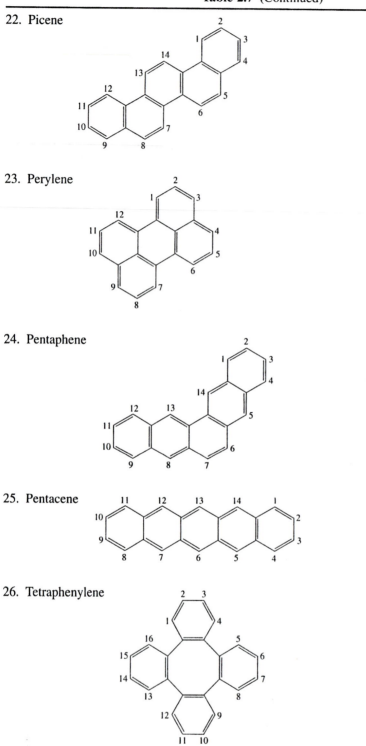

Table 2.7 (Continued)

27. Hexaphene

28. Hexacene

29. Rubicene

30. Coronene

31. Trinaphthylene

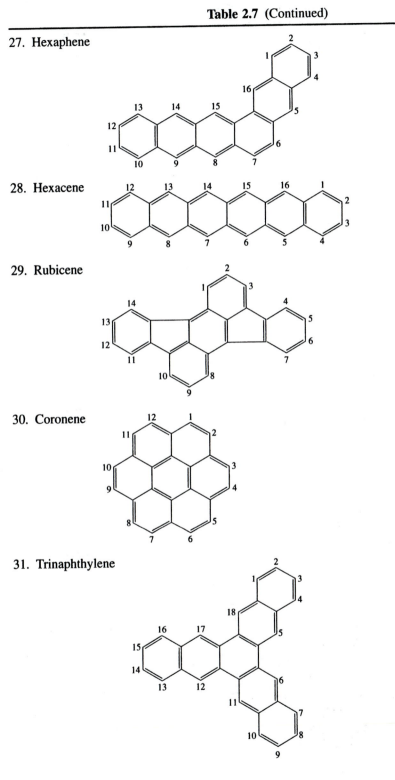

Table 2.7 (Continued)

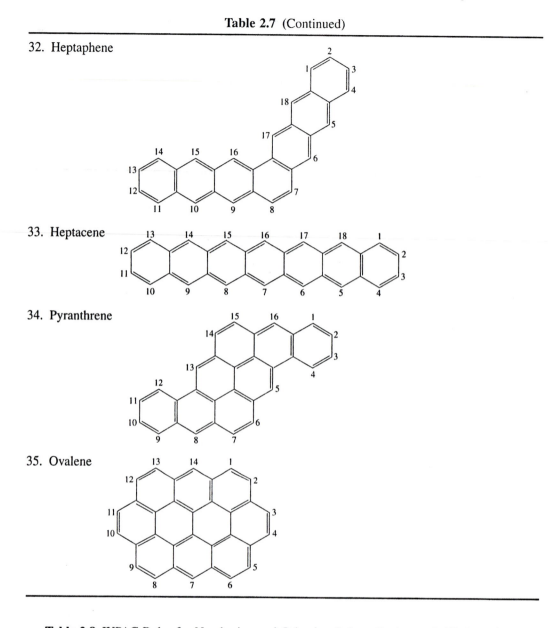

32. Heptaphene

33. Heptacene

34. Pyranthrene

35. Ovalene

Table 2.8 IUPAC Rules for Numbering and Orienting Polycyclic Aromatic Hydrocarbons

1. Draw the PAH rings so that two of the sides are vertical.
2. Draw as many rings as possible in a horizontal line. Consider the middle of the first row as the center of a circle about the molecule. Place as much of the remainder of the molecule as possible in the top right quadrant and as little as possible in the bottom left quadrant.
3. Number the periphery of the ring clockwise, starting at the first carbon atom not a part of ring fusion of the right hand ring of the top row. As numbering proceeds, the bonds potentially involved in ring fusion receive the small italicized letters *a*, *b*, *c*, etc. after the numbers of the preceding atoms.

EXAMPLE 2.5   *Orientation and Numbering of Polycyclic Aromatic Hydrocarbons*

Orient the following PAH in its proper position and number it using IUPAC rules:

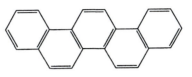

SOLUTION

1. Rotate the molecule using rules 1 and 2 from Table 2.8 so that (1) the longest sequence of rings is horizontal, and (2) the maximum amount of the molecule is in
   the top right quadrant, with as little as possible in the bottom left quadrant:

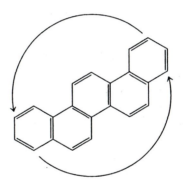

2. Beginning at the first carbon of the ring in the upper right corner of the molecule,
   number the carbons bearing hydrogens clockwise, and label the bonds with italicized letters (e.g., 1–2 = *a*, etc.):

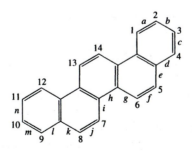

Note that this molecule corresponds to the structure of picene (number 22 of Table
2.7)

How are PAHs other than those prescribed in Table 2.7 named? The *fusion method* is the most common IUPAC nomenclature procedure for complex PAHs. The basis for the fusion method is adding a smaller aromatic to a prescribed PAH; that is, the two molecules are fused together to form a larger PAH. At the same time, numbers and letters corresponding to bonds and carbon atoms must be tracked. The rules for the fusion method are listed in Table 2.9, and their use is demonstrated in Example 2.6.

Table 2.9 Rules for Naming Polycyclic Aromatic Hydrocarbons by the IUPAC Fusion Method

1. Choose the largest possible prescribed structure from Table 2.7 that is part of the molecule. Label each of the sides with italic letters (i.e., *a* for side 1–2, *b* for side 3–4, etc.).
2. The aromatic that will be the new attachment is given an *-o* or *-a* designation (e.q., benzo for benzene, naphtho for naphthalene, anthra for anthracene, phenanthro for phenanthrene).
3. If the new attachment is a simple moiety (e.g., benzene, naphthalene), and the position on the attachment compound is not required, note only the position of fusion on the prescribed compound by italic letters (as always, make sure the point of fusion is the lowest letter/number possible on the prescribed PAH).
4. If the new attachment is a more complex moiety (e.g., anthracene, phenanthrene), the fusion position for both the attachment and the parent PAH must be documented. Numbers are used to denote the fusion position for the attachment compound; letters are used to show the point of fusion for the parent compound. Using the lowest numbers and letters possible for the fusion position (for the attachment compound and the parent compound, respectively), place the numbers followed by the italicized letters separated by a hyphen.
5. Using the procedure listed in Table 2.8, reorient the structure with the longest sequence of rings on the horizontal axis and the proper distribution in quadrants.

EXAMPLE 2.6   *Nomenclature of Polycyclic Aromatic Hydrocarbons*

Name the following PAH:

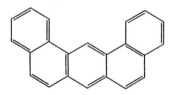

SOLUTION

1. Anthracene is the largest prescribed PAH on which this molecule is based. Therefore, the molecule may be considered as the fusion of anthracene and two benzene molecules:

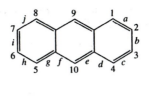

and

Note that anthracene has anomalous numbering.

2. Based on Rule 1 of Table 2.9, the fusion-attachment points are *a* and *j*. No attachment numbers are necessary for benzene. Therefore, the letters noted within the brackets are [*a,j*].
3. The resulting name is dibenzo[*a,j*]anthracene.
4. The final step is to reorient the fused molecule so that the longest sequence of rings is horizontal and as much of the molecule as possible is in the upper right quadrant, followed by renumbering of the molecule. The structure shown above is already found in this orientation, so no rotation is necessary.

2.2   PETROLEUM

Problems associated with the disposal and environmental release of petroleum products are widespread. One of the most common sources of pollution by petroleum has been leaking *underground storage tanks* (USTs). As a precaution against fire hazard, tanks containing liquid fuels have often been placed underground. Hundreds of thousands of these tanks are located at gasoline stations, and most of them have begun to leak in recent years [2.10]. Underground tanks that hold home heating fuel are another source of petroleum releases to the environment. Other problems include central storage tanks (fuel farms) in which large quantities of gasoline, jet fuel, or other petroleum products are stored prior to distribution to airports, military installations, and refineries [2.11] (Figure 2.2). Additional sources of petroleum contamination are unintentional releases including tanker truck accidents, pipeline ruptures, and oil rig blowouts at sea. Intentional contamination has resulted from disposal in pits, midnight dumping of waste oil, and disposal of waste oil in landfills.

The quantity of petroleum used in the United States has ranged from 1.51×10^9 to 2.43×10^9 barrels per year from 1960 through 1988 [2.12]. The volumes used in a given year are determined by weather patterns, demographics, fuel efficiency, and numerous other factors. Most of the petroleum is used in automobiles, heating of homes and buildings, and power generation. The quantity that enters the environment is not directly related to the amount used because most petroleum is not intentionally spilled. Most releases are from USTs or transportation accidents, and although they are related to the quantity of petroleum being consumed, environmental releases are determined more by tank corrosion, transportation safety, and other indirect cause-and-effect relationships.

Figure 2.2 A fuel farm used to store jet fuel.

To understand and assess the problems of petroleum releases, it is necessary to know what petroleum is, and what chemicals make up its different fractions that include gasoline, lubricating oil, diesel, and fuel oil. *Petroleum* is a naturally occurring, complex mixture of hydrocarbons and other organic compounds of nitrogen, sulfur, and oxygen that were formed by a complex series of chemical and biochemical reactions from organic material deposited over geological time. Petroleum can occur in gaseous (natural gas), liquid (crude oil), and solid (asphalt) phases [2.13].

Crude oil and the petroleum products that result from its refining consist of a highly variable mixture of hydrocarbons. Each source of petroleum is unique in composition, a phenomenon that may be explained by the differences in plant and animal life in regions of the earth at different eras of geologic time from which the petroleum originated [2.14]. However, despite such variability, many characteristics are common among the different crude oil sources.

Based on chemical analysis, petroleum consists primarily of aliphatic and aromatic hydrocarbons, with small quantities of organic sulfides and amines. Petroleum products are sometimes classified as four classes of hydrocarbons:

- *Paraffins*—alkanes
- *Olefins*—alkenes
- *Naphthenes*—cycloalkanes
- *Aromatics*—monocyclic aromatic hydrocarbons and PAHs

The proportions of these classes of hydrocarbons in a crude oil from Prudhoe Bay, Alaska, are listed in Table 2.10. These raw materials consist of a wide range of saturated aliphatic compounds up to C_{32} plus aromatic compounds up to six rings. Crude

Table 2.10 Characteristics of Prudhoe Bay,
Alaska, Crude Oil

Specific gravity	0.89
Sulfur content	1.0%
Volatile organic compounds	
$(C_1 - C_8)$	8%
$(C_9 - C_{15})$	15%
Major components of the crude oil	
Alkanes	33%
Alkenes	22%
Aromatics	30%
Residuum	15%
Alkane/Aromatic ratio	1.12

Source: Reference 2.15.

oil is rarely used in the form obtained from the source; instead, it is refined, reformed, and distilled into a number of petroleum products that provide hundreds of uses. The two major categories of petroleum products are

- fuels (e.g., gasoline, fuel oil, and diesel), and
- petrochemical feed stocks (e.g., solvents, plasticizers, industrial intermediates).

Refining processes can be grouped into three classes:

- *separation* (usually distillation) to isolate fractions of different volatility and boiling range,
- *conversion* (usually cracking) to change the molecular weight and boiling point, and
- *upgrading* (e.g., hydrocracking) to meet product quality specifications.

After crude oil is washed with water to remove salts and suspended particles, it is distilled using continuous-fractionation-plate towers at atmospheric pressure to separate fractions that boil below 370°C. Primary distillation products include naphtha, middle distillate (light and heavy gas oil), and the bottoms fraction. *Naphtha,* the lightest fraction of distillate, is used primarily for gasoline after octane improvement by catalytic reforming. Virgin naphtha is catalytically re-formed, usually over noble metal catalysts, in cyclic fixed-bed reactors to increase the aromatic content and the octane number.

Light virgin gas oil obtained after refining is used for diesel, jet fuel, and home heating oil. The *heavier gas oil fraction* is cracked catalytically (i.e., chemically converted to smaller molecules) and can then be used for high-octane gasoline. Bottoms from atmospheric distillation are used for two purposes—vacuum distillation and the production of long-chain hydrocarbon products. Vacuum distillation provides more distillate for catalytic cracking, which increases the yield of gasoline from crude oil.

Distillation ranges of petroleum products are shown in Table 2.11. Gasoline is the fraction with the lowest boiling point range (and the corresponding lowest carbon chain length) followed by kerosene, fuel oil, diesel, jet fuel, light gas oil, long-chain lubricating oils, and the nonvolatile residual fraction. Gasoline is the product in high-

Table 2.11 Characteristics of Common Petroleum Products

Use	Distillation Range	Chain Length of Hydrocarbons	General Composition
Gasoline	40–205°C	C_5–C_{10}	Alkanes, isoalkanes, cycloalkenes, aromatics, BTEX
Kerosene	175–325°C	C_{12}–C_{16}	Alkanes, isoalkenes, cycloalkenes, PAHs
Diesel	200–400°C	C_{15}–C_{25}	Alkanes, isoalkanes, cycloalkenes, PAHs
JP-4 jet fuel	40–250°C	C_5–C_{12}	65% gasoline; 35% light petroleum distillate
JP-5 jet fuel	175–300°C	C_{12}–C_{15}	Alkanes, cycloalkenes, BTEX
Fuel oil	140–400°C	C_{10}–C_{25}	Alkanes, isoalkanes, cycloalkenes, BTEX, PAHs
Lubricating oil	Residuum	C_{20}–C_{70}	Alkanes, PAHs

est demand, followed by the middle distillates. The exact chemical composition (i.e., the number of isomers for any C_n) of each of these products is highly variable, depending on the petroleum source and the cracking and refining processes. Theoretically, gasoline may contain more than 1500 different chemicals. Chemical analysis usually results in the identification of 150–180 compounds [2.16]. Detailed analyses have shown a range of normal and branched alkanes, cycloalkanes, and cycloalkenes in gasoline. The lower boiling fractions are mainly gases (propane, butane), which are liquefied by pressure and used as fuels, such as liquid propane. Small alkenes, such as ethylene and propylene, are also part of these fractions; they are used as chemical feedstocks for polymerization to low- and high-density polyethylene or polypropylene, or are chlorinated to trichloroethylene or similar solvents. One of the most important groups of compounds that is common to all gasoline products is benzene, toluene, and xylenes, often referred to as *BTX*. When ethylbenzene is included in this group, the acronym changes to *BTEX*. The ranges of *BTEX* found in gasoline are listed in Table 2.12. Benzene, the most toxic *BTEX* constituent, is found in concentrations up to 3.5%. Because *BTEX* comprises from 6 to 36% of gasoline (and less in diesel), more nonspecific chemical analyses are often used to quantify all of the hydrocarbons (i.e., *BTEX* plus alkylbenzenes, alkanes, cycloalkanes, and alkenes). To achieve this goal, gasoline, diesel, and other petroleum fractions are analyzed by infrared spectroscopy or gas chromatography (see Section 4.7), to obtain a value for the total concentration of compounds called *total petroleum hydrocarbons* (TPHs).

Lead, in concentrations of 400–800 mg/L, was added to gasoline formulations in the United States prior to 1985 to increase the octane rating and suppress preignition. Unleaded gasoline currently contains less than 20 mg/L of lead [2.18]. A relatively new class of gasoline additives are oxygenates, the most common being methyl *tert*-butyl ether. Oxygenates have been added at varying levels, and concentrations in the percent range are common [2.19].

Table 2.12 Concentration of Benzene, Ethylbenzene, Toluene,
and Xylenes (BTEX) in Gasoline

Compound	Weight Percent in Gasoline	Toxic Effects
Benzene	0.12–3.50	Leukemia
Toluene	2.73–21.80	Neurotoxicity
Ethylbenzene	0.36–2.86	Neurotoxicity
o-Xylene	0.68–2.86	Neurotoxicity
m-Xylene	1.77–3.87	Neurotoxicity
p-Xylene	0.77–1.58	Neurotoxicity
Total	6.43–36.47	

Source: Reference 2.17.

The information presented in Tables 2.10 through 2.12 shows that, although the exact compositions of petroleum and its products are highly variable, certain generalities can be made about its chemical composition. These general criteria for petroleum characteristics serve as an aid in assessing the pathways and toxicological risk of petroleum released to the environment.

2.3 NONHALOGENATED SOLVENTS

The most common nonhalogenated solvents are petroleum distillates, aliphatic and aromatic hydrocarbons, alcohols, ketones, esters, and ethers. Nonhalogenated solvents have been used in a number of industrial applications such as cold cleaning, which includes metal degreasing, parts cleaning, and paint stripping. Nonhalogenated solvents have also been used as carriers for paints, varnishes, and printing inks. After the solvent-pigment solution is applied, the solvent evaporates (i.e., the paint or ink dries), leaving the pigment in place. Prior to the promulgation of RCRA, waste solvents were disposed of in landfills, sewers, and soil pits. Now they are candidates for solvent recycling, fuel blending, and incineration.

2.3.1 Hydrocarbons

Just as petroleum fractions are used as different fuels and lubricants, several fractions are used as solvents. Common petroleum distillate solvents are listed in Table 2.13. One of the most common distillates, mineral spirits, has been used for dry cleaning and paint thinning, and its use may increase in the future because of the difficulty of disposing of chlorinated solvents.

Single-component aromatic and aliphatic hydrocarbons have also been used as industrial solvents. Common aromatic hydrocarbons have included benzene, toluene, xylenes, and a number of alkylbenzenes. Approximately 10% of the toluene used in

Table 2.13 Common Petroleum Distillates Used as Solvents

Distillate	Boiling Point Range	Common Compounds
Petroleum benzin	35–80°C	Primarily pentanes, hexanes
Ligroin	130–155°C	Alkanes
Petroleum ether/ high-flash V.M. & P. naphtha	141–160°C	
Stoddard solvent/ rule 66 Mineral Spirits	157–196°C	85% Nonane 15% Trimethylbenzenes
Turpentine	154–170°C	Distillate from *Pinus palustris*

the United States is for solvent and synthesis purposes; the remaining 90% is used as a component of gasoline [2.20]. Xylenes and ethylbenzene are used as solvents for a number of cleaning applications; they are also used as a starting material for the synthesis of polymers. Aliphatic hydrocarbons that have been commonly used include various isomers of C_5 through C_{12} alkanes (e.g., *n*-hexane, mixed hexanes, *n*-octane, *n*-decane), cycloalkanes, and alkylcycloalkanes.

2.3.2 Ketones

Low molecular weight *ketones,* which have the general form

are an important class of nonhalogenated solvents. The fundamental structure of ketones consists of R and R′ alkyl groups linked to a *keto-,* or *carbonyl,* group. The IUPAC nomenclature rules for ketones are listed in Table 2.14. In hazardous waste practice, however, common names are used almost exclusively for ketonic solvents. Common names for ketones are derived by following the two alkyl groups with *ketone.* Use of common and IUPAC nomenclature for ketones is illustrated in Example 2.7.

Table 2.14 IUPAC Nomenclature for Ketones

1. Select the longest chain possible that contains the keto group.
2. The parent chain is numbered so that the carbon with the keto group has the lowest number possible. It is named based on alkane nomenclature (Table 2.2), but the suffix *-ane* is replaced by *-anone.*
3. Substituent groups are numbered and ordered using rules defined previously (see Table 2.2).

EXAMPLE 2.7   *Nomenclature of Ketones*

Name the ketone

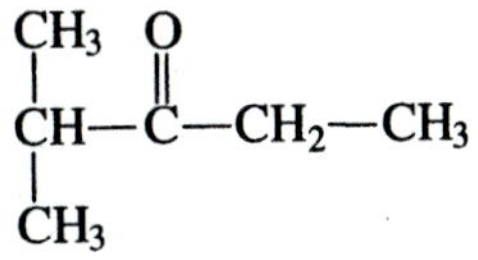

$$\begin{array}{c}
CH_3 \quad O \\
| \quad\quad || \\
CH-C-CH_2-CH_3 \\
| \\
CH_3
\end{array}$$

using (1) common nomenclature, and (2) IUPAC nomenclature.

SOLUTION

1. The two alkyl groups connected to the keto carbon are

$$-CH_2-CH_3$$

Ethyl

and

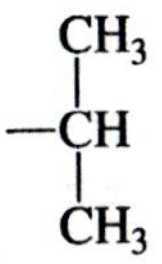

$$\begin{array}{c}
CH_3 \\
| \\
-CH \\
| \\
CH_3
\end{array}$$

Isopropyl

The alkyl groups are placed in order of size. The common name is therefore *ethyl isopropyl ketone.*

2. The parent carbon chain is five carbons long:

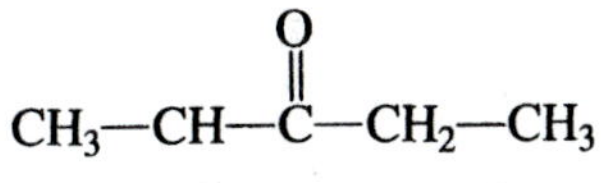

$$\begin{array}{c}
O \\
|| \\
CH_3-CH-C-CH_2-CH_3
\end{array}$$

with the keto group at the number three carbon. Therefore, the compound is a derivative of

3-pentanone

To obtain the lowest possible number for the substituted methyl group, it is labeled 2-methyl. The resulting IUPAC name for the ketone is *2-methyl-3-pentanone.*

The most common ketone solvents include:

$$H_3C-\overset{\overset{\displaystyle O}{\|}}{C}-CH_3$$

Acetone (common name)
2-Propanone (IUPAC name)

$$H_3C-\overset{\overset{\displaystyle O}{\|}}{C}-CH_2-CH_3$$

Methyl ethyl ketone—MEK (common name)
2-Butanone (IUPAC name)

$$H_3C-\overset{\overset{\displaystyle O}{\|}}{C}-CH_2-\overset{\overset{\displaystyle CH_3}{|}}{CH}-CH_3$$

Methyl isobutyl ketone—MIBK (common name)
4-Methyl-2-pentanone (IUPAC name)

2.3.3   Alcohols and Esters

Alcohols possess a hydroxyl (—OH) group and, in their IUPAC naming, the *-ane* of an alkane is replaced by *-anol*. In their common naming, the term *alcohol* is added to the alkyl radical to name compound:

$$H_3C-CH_2-CH_2-\overset{\overset{\displaystyle H}{|}}{\underset{\underset{\displaystyle H}{|}}{C}}-OH$$

n-Butyl alcohol (common name)
1-Butanol (IUPAC name)

Esters are derivatives of carboxylic acids, in which the acid hydrogen is replaced by a covalently bonded alkyl group:

$R-\overset{\overset{\displaystyle O}{\|}}{C}-OH$	$R-\overset{\overset{\displaystyle O}{\|}}{C}-O-R'$
Acid	Ester

Naming esters is based on alkyl nomenclature of the unprotonated salts of carboxylic acids. A carboxylic acid without the acidic hydrogen ion derives its name from alkanes, but *-ane* is replaced by *-anate*. For example, the salt of propionic acid is propanate:

$$H_3C-CH_2-\overset{\overset{\displaystyle O}{\|}}{C}-O^-$$

An anomaly in this naming system occurs with the two-carbon acid, which is commonly known as *acetate* rather than ethanate. The naming of esters proceeds by placing the name of the alkyl group first, followed by the name of the salt of the carboxylic acid, with no hyphen between them. The nomenclature for esters is demonstrated in Example 2.8.

EXAMPLE 2.8   *Nomenclature of Aliphatic Organic Solvents*

Name the following solvents:

1.
$$CH_3$$
$$HO—CH_2—CH—CH_3$$

2.
$$O$$
$$H_3C—C—O—CH_2—CH_3$$

SOLUTION

1. Using a 3-carbon skeleton with the —OH group at the number 1 carbon, the methyl group is then designated at the number 2 position, which yields

   2-methyl-1-propanol

   The common name is based on the —OH attached to the alkyl group (isobutyl), making the molecule *isobutyl alcohol* or *isobutanol.*

2. The parent structure of this ester is the salt of acetic acid, which is the anion acetate:

   The group that is bonded to the acetate oxygen is an ethyl group; therefore, the compound is *ethyl acetate.*

Several other nonhalogenated chemical classes are used as solvents. Their nomenclature is beyond the scope of this text, but a brief summary of each class is provided in Table 2.15.

2.4   HALOGENATED SOLVENTS

Halogenation of hydrocarbons yields compounds of lower flammability, higher density, higher viscosity, and improved solvent properties compared to nonhalogenated solvents. These *halogenated solvents* are an important category of hazardous wastes.

Table 2.15 Miscellaneous Nonhalogenated Solvents

Chemical Class	General Formula	Typical Compounds	Chemical Structure
Glycols (diols)	HO—R—R′—OH	Ethylene glycol	HO—CH_2—CH_2—OH
		Propylene glycol	HO—CH_2—$\overset{\displaystyle OH}{\overset{\displaystyle \vert}{CH}}$—$CH_3$
		1,3-Butanediol	HO—CH_2—CH_2—$\underset{\underset{\displaystyle OH}{\displaystyle \vert}}{CH}$—$CH_3$
Ethers	R—O—R′	Dimethyl ether	CH_3—O—CH_3
		Di-*n*-butyl ether	$CH_3(CH_2)_3$—O—$(CH_2)_3CH_3$
Amines	R—NH_2	Isopropylamine	H_2N—$\underset{\underset{\displaystyle CH_3}{\displaystyle \vert}}{CH}$—$CH_3$
		Dimethylamine	$\overset{\displaystyle CH_3}{\overset{\displaystyle \vert}{CH_3—NH}}$
Glycol ethers	H(R‴)—O—R—R′—O—R″	Ethylene glycol monomethyl ether	HO—CH_2—CH_2—O—CH_3
Nitroparaffins	R—NO_2	Nitroethane	NO_2—CH_2—CH_3
Furans		Furan	
		Furfuryl alcohol	
Nitriles	R—C≡N	Acetonitrile	H_3C—C≡N

Halogenation, particularly chlorination, of organics has become a widespread industrial practice; in the 1970s, 46.5% of the chlorine gas used in the United States was for the production of chlorinated organic compounds. Over 400,000 tons of halogenated solvents are used annually for metal cleaning, which is their largest application [2.21].

Halogenated solvents have been used for degreasing and cleaning a range of products, from machined parts to computer chips. To avoid exposure of personnel in the workplace, waste solvents have commonly been disposed of off-site, often in a casual manner; drums or smaller quantities have been disposed of in unlined pits, in landfills, or onto soil, with the conception that they would evaporate or degrade. Although these chemicals are volatile, they are also dense and migrate through the subsurface. They also sorb to soils, but are still soluble enough in water to move through the subsurface and contaminate groundwater systems. The properties of halogenated solvents that make them mobile in groundwater systems include high density, relatively high water solubility, and low degradability.

Halogenated solvents have become common groundwater contaminants in metropolitan areas such as Phoenix, Arizona, and San Jose, California, as well as other areas where they have been used for washing semiconductors and computer chips. When the solvents were disposed of by placing them in unlined pits or directly on soils, the underlying groundwater became contaminated. High-tech industries, originally thought of as nonpolluting, have contaminated millions of gallons of groundwater in metropolitan areas by improper disposal practices. The use of halogenated solvents in cleaning jet engines has caused similar problems at several U.S. military bases.

2.4.1   Nomenclature of Aliphatic Halocarbons

In hazardous waste practice, common names and abbreviations are used extensively for halogenated solvents, which are formally classified as *aliphatic halocarbons*. The common names of small halogenated alkanes are based on the nomenclature of the alkyl groups listed in Table 2.2. To name these compounds using common nomenclature, the name of the alkyl group is followed by chloride, bromide, iodide, fluoride, and so on.

Common names for halogenated C_2 alkenes are based on the same procedure as the IUPAC system, but *ethylene* replaces *ethene*. Ethylene is the older term for the IUPAC root ethene, and it is still used extensively in the common naming of halogenated solvents. The use of common and IUPAC nonenclature for halogenated solvents is illustrated in Example 2.9.

EXAMPLE 2.9   *Nomenclature of Halogenated Solvents*

Name the following compounds using common nomenclature:

1. $H_3C-CHCl-CH_3$

2. CH_2Cl_2

3.

SOLUTION

1. The group $CH_3-CH-CH_3$ is the isopropyl radical. The common name for the compound is therefore isopropyl chloride.

2. The H—C—H moiety is the methylene radical. The chlorinated methylene compound is therefore methylene chloride. The IUPAC name is dichloromethane.

3. The IUPAC name is trichloroethene. Numbering the locations of the chlorines provides no information over not having the numbers so there is no need for their numbering. The common name is simply trichloroethylene (TCE).

2.4.2   Uses and Disposal of Common Halogenated Solvents

What are the most common halogenated solvents? They are generally derivatives of small aliphatic hydrocarbons. Some are derivatives of methane:

Methylene chloride (common name)
Dichloromethane (IUPAC name)

Chloroform (common name)   Carbon tetrachloride (common name)
Trichloromethane (IUPAC name)   Tetrachloromethane (IUPAC name)

Chlorinated methane derivatives have been used extensively in a number of industrial processes. A major application of methylene chloride has been in paint stripping, where it has often been mixed with alcohols, acids, and amines. Methylene chloride has also been used for the extraction of caffeine from coffee and other beverage stock. The applications that have resulted in perhaps the greatest releases to the environment are the cleaning of metal parts, primarily from dip cleaning. Other methylene chloride uses include industrial intermediates, an additive in adhesives, and a solvent in the manufacture and analysis of pesticides.

Chloroform has been used most frequently as a dry cleaning spot remover and for cleaning machine and engine parts laden with oils and grease. Other applications have included industrial intermediates, insecticides, and the purification of alkaloids and vitamins. Some of the chemical derivatives similar to chloroform, trihalomethanes (e.g., bromo- and chloro-methanes), are also commonly found in chlorinated drinking water.

Carbon tetrachloride was a common commercial product and was widely used in the U.S. until 1970, when it was banned because of its toxicity. Its primary applications were in dry cleaning, metal degreasing, and fire extinguishers. Carbon tetrachloride is still used for some industrial purposes in the U.S., which include the production of chlorofluorocarbons (e.g., trichlorofluoromethane, Freon 11), grain fumigation, and as a reaction medium in the industrial synthesis of chemicals. It is found at contaminated sites decades after it was disposed of because of its long environmental persistence.

The other major categories of chlorinated solvents are derivatives of ethane and ethene (ethylene):

1,1,1-Trichloroethane (TCA)

Trichloroethylene—TCE (common name)
Trichloroethene (IUPAC name)

Perchloroethylene—PCE (common name)
Tetrachloroethene (IUPAC name)

1,1,1-Trichloroethane (TCA) has been used for hundreds of industrial solvent applications including the cleaning of electrical parts, electronic components, printed circuit boards, motors, and metal and plastic manufacturing components. It readily dissolves most greases, oils, tars, and other organic materials, and is used for both hot and cold cleaning applications. Advantages of TCA are its excellent solvent properties and its low toxicity relative to other chlorinated solvents.

Trichloroethylene (TCE) is an excellent solvent for a large number of natural and industrial materials. It is moderately toxic, is nonflammable, and is slowly oxidized. Trichloroethylene has had greatest use in the vapor degreasing of fabricated metal parts, where a reflux unit supplies fresh solvent to the part that is suspended in its vapors (Figure 2.3). The solvent vapors then degrease the part as it heats to the equilibrium temperature of the system. Vapor degreasing has been an effective method for cleaning metal parts, particularly in cases where solvent residues are undesirable. Approximately 80% of the TCE consumed in the United States has been used for vapor degreasing. The remaining 20% has been used for a variety of purposes—a component of adhesive and paint-stripping formulations, a heat-transfer medium, and a carrier solvent. Its use has decreased since 1970 as the result of environmental and health concerns, but production rates are currently in the order of 100,000–200,000 tons per year [2.22].

Perchloroethylene (PCE), sometimes called *perc*, is one of the most stable of the chlorinated aliphatic solvents due to the total chlorination (perchlorination) of the ethene molecule. Perchloroethylene has excellent solvent characteristics; its primary application has been in dry cleaning, and about 80% of the cleaning industry uses PCE. Other applications include vapor degreasing, cold cleaning of metals, and the manufacture of chlorofluorocarbons.

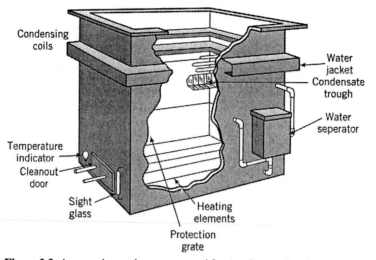

Figure 2.3 A vapor degreasing system used for metal parts cleaning. *Source:* Reference 2.21.

2.5   PESTICIDES

A *pesticide* is a chemical, physical, or biological agent that destroys or controls pest organisms including insects, plants, fungi, rodents, fish, and nematodes. The most common types of pesticides are *insecticides* (targeting insects), *herbicides* (vascular plants), and *fungicides* (fungi). Over 900 different pesticides are used in the United States under 25,000 trade names [2.23]. Approximately 500,000 tons of pesticides are applied to crops each year in the United States [2.24].

Pesticides have played an important role in improving public health through disease vector reduction and increased food production. Nearly one million species of insects are found in the world and some ten thousand are significant pests [2.25]. One widely used insecticide is DDT, which successfully controlled the insect vectors that caused the spread of malaria, typhus, yellow fever, and bubonic plague. Between 1942 and 1952, the use of DDT saved approximately 5 million lives and prevented 100 million illnesses [2.26]; it also enhanced crop yields through insect control.

The restriction (i.e., banning) of DDT occurred after long debates between those opposed to its use because of adverse environmental effects and public health proponents who believed that the environmental effects were outweighed by disease prevention. The adverse environmental effects of DDT became evident throughout the 1960s. Residue analysis results confirmed that DDT was present in wildlife, marine and freshwater sediments, and human lipids. Furthermore, DDT was found to affect calcium metabolism in birds, resulting in thin eggshells that cause premature hatching and loss of young birds.

Although the benefits of disease reduction and increased agricultural productivity are undisputed, pesticide use and disposal have resulted in undesirable releases of these toxic chemicals to the environment. Thousands of hazardous waste sites have been created as a result of inappropriate disposal practices. A range of pesticide management facilities may be found, depending on the sophistication of the operation and local regulations. Some large-scale commercial operations effectively manage their wastes;

Figure 2.4 A pesticide rinse and formulation area.

however, many on-site facilities are found in isolated locations with minimal regulation and improper disposal practices. Every landfill has also received waste pesticides, either as discarded products or container residues, and numerous abandoned pesticide manufacturing plants have become Superfund sites with associated large-scale cleanup efforts. Perhaps the most common hazardous waste sites contaminated with pesticides are mixing and loading areas. At these locations, found primarily in agricultural regions, farmers and aerial applicators have diluted and mixed pesticides for field application or rinsed equipment and containers. These sites are characterized by surface soil and groundwater contamination from deliberate and accidental spillage of pesticides and rinsate (Figure 2.4). Concentrations of pesticides in soil and groundwater resulting from improper disposal practices are listed in Table 2.16.

The IUPAC nomenclature for pesticides is beyond the scope of this text. Most pesticides are large molecules, and their complex IUPAC names are rarely used. Common names and trade names are employed most often, and if an IUPAC name is needed, it can be found in reference texts, such as *The Farm Chemicals Handbook* [2.23]. The use of reference material for finding information on pesticides and other hazardous chemicals is addressed in Section 2.12. The following brief survey serves as an introduction to some of the more common pesticides.

2.5.1 Insecticides

Numerous classes of insecticides have been developed through empirical data collection over the past hundred years and, in the last decade, by detailed structure-activity studies. Natural products (e.g., nicotine, rotenone) were used for centuries until, in the

Table 2.16 Pesticide Concentrations of Soils and Groundwater near Pesticide
Mixing and Loading Areas

Pesticide Category/Names	In Pools and Soils in Loading and Rinse Areas (μg/L)	Groundwater in Affected Wells and Seeps (μg/L)	Local Background Groundwater (μg/L)
Herbicides			
Atrazine	70,000	65.0	N.D.–0.65
Alachlor	270,000	145.0	N.D.–1.30
Cyanizine	225,000	36.0	N.D.–0.26
Metolachlor	270,000	50.0	N.D.–0.80
Metribuzin	52,000	8.0	N.D.
Trifluralin	1,000	0.20	N.D.
Insecticides			
Carbofuran	1,000	N.D.	N.D.
Fonofos	1,000	1.30	N.D.–0.30
Fumigants			
Ethylene Dibromide	10–100	1.00	N.D.

N.D. = Nondetectable
Source: Reference 2.27.

nineteenth century, the "first-generation" insecticides were developed. These were metals such as lead arsenate and methyl arsenate (see Section 2.10.1). The "second generation" consisted of synthetic organic insecticides, which first included organochlorine compounds; their use began during the 1940s after the insecticidal action of DDT was accidentally discovered in 1939. Organophosphorus ester insecticides followed as modifications of nerve gases synthesized for possible use in World War II. During the mid-twentieth century, thousands of organic insecticides were developed and tested. However, most of the organochlorine compounds and many other insecticides used from 1940 through 1970 are no longer available because they have been banned by the EPA. In addition, widespread use of insecticides since the 1940s has resulted in selective pressure on the gene pool, resulting in more than 400 insect species that are resistant to one or more classes of insecticides [2.28]. As a result, pesticide chemists are developing a third generation of insecticides, which include natural products and pheromones (sex-attractants). Use of these natural processes in conjunction with second-generation insecticides falls under the practice of *integrated pest management.*

Organochlorine Insecticides

Organochlorine insecticides are chlorinated aliphatic and aromatic compounds that may also contain oxygen and sulfur. They were used extensively from the mid-1940s through the early 1970s. These insecticides are highly lipophilic (fat-loving) compounds that readily penetrate the cuticles and cellular membranes of insects. The mechanism of toxicity involves disruption of the sodium-potassium balance in cellular membranes and disruption of neurological functions. Most uses of organochlorine insecticides have been restricted (banned) by the EPA because of their biorecalcitrance, tendency to bioconcentrate, chronic toxicity, and detrimental ecological effects. Although the majority of the organochlorine insecticides has been banned in the United States for at least a decade, residues are still found at many disposal sites because (1) material disposed of decades ago has still not degraded due to its persistent nature and (2) although not used in the United States, some of these compounds are still manufactured here for export. Common organochlorine insecticides include DDT, methoxychlor, lindane, aldrin, dieldrin, and endosulfan.

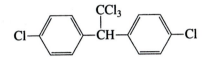

DDT—*D*ichloro*d*iphenyl*t*richloroethane (common name)
1,1,1-Trichloro-2,2-*bis*(4-chlorophenyl)ethane (IUPAC name)

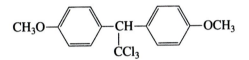

Methoxychlor (common name)
1,1,1-Trichloro-2,2-*bis*(4-methoxyphenyl)ethane (IUPAC name)

Although DDT was first synthesized in 1874, its insecticidal properties were not discovered until 1939. Until the registration of DDT was canceled in 1972 as a result of the ecological damage it caused, DDT had been used for the control of hundreds of species of insects. It was used in gardens, orchards, fields, and forests for agricultural and public health purposes, with peak production in 1963 at 88,200 tons. In 1985, after DDT was banned, 330 tons were exported from the United States. [2.29]

Methoxychlor was developed as a more biodegradable alternative to DDT. The methoxy (CH_3O) groups on the benzene rings enhance aerobic biodegradability. The concept of biodegradability in relation to chemical structure is covered in Chapter 7.

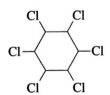

Lindane (common name)
γ-Benzenehexachloride (γ-BHC) (common name)
γ-1,2,3,4,5,6-Hexachlorocyclohexane (IUPAC name)

Lindane is the γ-isomer of 1,2,3,4,5,6-hexachlorocyclohexane; the γ indicates a specific orientation of the chlorines above and below the cyclohexane ring, which is in the form of a chair- or boat-shaped configuration. Lindane has been used as a soil sterilant and for the control of grasshoppers and insects that infest cotton. Hexachlorocyclohexane was first synthesized in 1825, but its use as an insecticide in the form of Lindane did not begin until 1942. Lindane was banned in 1976.

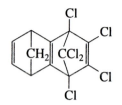

Aldrin (common name)
(1R,4S,5S,8R)-1,2,3,4,10,10 hexachloro-1,4,4a-,5,8,8a-
hexahydro-1,4:5,8-dimethanonaphthalene (IUPAC name)

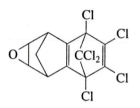

Dieldrin (common name)
1,2,3,4,10,10-hexachloro-1R,4S,4aS,5R,6R,7S,8S,8aR-octahydro-
6,7-epoxy-1,4:5,8-dimethanonaphthalen (IUPAC name)

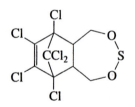

Endosulfan (common name)
6,7,8,9,10,10-Hexachloro-1,5,5a,6,9,9a-hexahydro-6,9-
methano-2,4,3-benzodioxathiepin-3-oxide (IUPAC name)

Aldrin, dieldrin, and endosulfan are chlorinated cyclodienes. Aldrin–named after Kurt Alder, a codiscoverer of the Diels-Alder condensation reaction used in its synthesis—was used as a soil insecticide and for insects that infest fruit trees and cotton. Dieldrin, named after Otto Diels, the other codiscoverer of the Diels-Alder reaction, has been used to control mosquitoes as vectors of malaria. Aldrin is rapidly converted to dieldrin in the environment. Endosulfan is a broad-spectrum insecticide that was used for decades on vegetables, fruits, and field crops. However, because of the persistence and chronic toxicity of these compounds, their registrations were canceled by the EPA in 1975.

Toxaphene is not a single compound, but a group of isomers with the empirical formula $C_{10}H_{10}Cl_8$. It is prepared by chlorination of the monoterpene camphene, which yields a mixture of 175 isomers containing 67–69% chlorine. One of the more potent isomers is 2,2,5-*endo*-6-*exo*-8,9,9,10-octachlorobornane.

Organophosphorus Esters

Organophosphorus esters are nonselective, broad-spectrum insecticides that inhibit acetyl cholinesterase, an important enzyme in neurological functions. Although organophosphorus esters were first synthesized in 1854, their use did not become widespread until the 1970s after the registrations of most of the organochlorine insecticides were canceled. One reason for the widespread use of organophosphorus esters is their low persistence. When released to the environment, organophosphates persist for only days to weeks compared to years for organochlorine compounds. Because of their relatively rapid degradation rates, the potential for bioaccumulation, chronic toxicity, and detrimental ecological effects is low relative to organochlorine insecticides. However, the acute toxicity of organophosphorus esters is much higher than that of the organochlorines—only three drops of parathion applied to the skin will kill the average adult [2.30].

Approximately 100 different organophosphorus insecticides are used in the United States. Their general structure is

where

    R = a methyl or ethyl group
    R′ = an alkoxy, alkyl, aryl, amino, or allylthio group
    X = a labile leaving group

The most common organophosphorus ester insecticides are

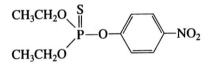

(Ethyl) Parathion (common name)
O,O-Diethyl *O*-4-nitrophenyl phosphorothioate (IUPAC name)

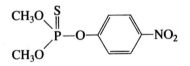

Methyl parathion (common name)
O,O-Dimethyl *O*-4-nitrophenyl phosphorothioate (IUPAC name)

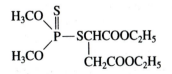

Malathion (common name)
S-1,2-*bis*(ethoxycarbonyl)ethyl
O,O-dimethyl phosphorodithioate (IUPAC name)

Parathion and methyl parathion have become the most widely used organophosphorus insecticides. Parathion is effective against a wider variety of insects than any other insecticide. However, it is also extremely toxic to humans, and a number of fatalities have resulted from its improper use. Methyl parathion is more effective than parathion on some insects (e.g., aphids) and is also less toxic to humans. Because of their high toxicities, both are classified as restricted-use pesticides; they can be applied only by personnel who have received pesticide application and safety training.

Malathion is a minimally persistent, general-use insecticide of relatively low toxicity to humans. It has been used extensively in homes, gardens, and orchards, on turf grasses, and for public health purposes. Because it is highly toxic to birds, its use on lawns has been discontinued.

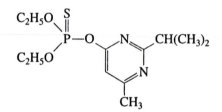

Diazinon (common name)
O,O-Diethyl *O*-2-isopropyl-6-methylpyrimidin-4-yl
phosphorothioate (IUPAC name)

Diazinon is a general-use insecticide used to control soil insects, and insects on vegetables, fruits, and field crops.

Carbamate Esters

Carbamate ester insecticides are *N*-methyl carbamates, which are modeled after the natural product physostigmine, a neurotoxic alkaloid isolated from the calabar bean. Physostigmine and its analogs have been used in the treatment of glaucoma and myasthenia gravis. The first successful carbamate insecticides were developed in 1954. By 1985, more than 25 carbamates had been developed or were in the development stage. Carbamates, along with organophosphorus esters, are currently the most widely used insecticides in the United States.

Carbamate insecticides are derivatives of carbamic acid:

$$HO-\overset{\overset{\displaystyle O}{\|}}{C}-NH_2$$

which is so unstable that it is nonexistent; however, substituting an alkyl or other leaving group for the hydrogen atom attached to the oxygen, combined with replacing one of the hydrogens of the amino moiety with a methyl group, provides a stable molecule that is an effective insecticide. Like organophosphorus esters, the toxicological mechanism of carbamates is the inactivation of acetyl cholinesterase. In addition to functioning as insecticides, carbamate esters have been used as fungicides, herbicides, and pharmaceuticals.

Carbamate ester insecticides are sometimes referred to as ideal pesticides because they are moderately labile in the environment, have low acute toxicity, and pose a minimal long-term threat to the environment. Carbaryl, carbofuran, and aldicarb are common carbamate insecticides:

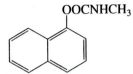

Carbaryl (common name)
Sevin (trade name)
1-Naphthyl methylcarbamate (IUPAC name)

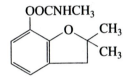

Carbofuran (common name)
Furadan (trade name)
2,3-Dihydro-2,2-dimethylbenzofuran-7-yl methylcarbamate (IUPAC name)

$$CH_3-S-\underset{\underset{CH_3}{|}}{\overset{\overset{CH_3}{|}}{C}}-CH=NO\overset{\overset{O}{\|}}{C}-NH-CH_3$$

Aldicarb (common name)
Temik (trade name)
2-Methyl-2-(methylthio)propionaldehyde *O*-methylcarbamoyloxime (IUPAC name)

Carbaryl is a broad-spectrum insecticide used on more than 100 crops. Carbofuran is also a broad-spectrum insecticide, but it is used primarily for soil treatment. Aldicarb is a systemic insecticide used for soil and seed treatments, mainly on citrus fruit and house plants. When systemic insecticides are applied to the soil, they are taken up by the roots and distributed throughout the plant, where they provide protection from insects.

2.5.2   Herbicides

Herbicides are agents that destroy vascular plants and are employed mainly as weed killers. They are the most widely used of all the classes of pesticides and are generally not as persistent or chronically toxic as many insecticides used over the past 40 years, especially organochlorines. Measurable herbicide residuals in topsoil last from a few days to over a year after application or release to the environment. Degradation rates are a function of the structure of the herbicide and environmental conditions.

Herbicides kill plant life by a variety of mechanisms including interference with DNA replication (uracil derivatives) and mimicking plant growth hormones (phenoxy derivatives). Approximately 120 different compounds with 242 registered trade names are used in the United States [2.31]. Most herbicides are classified into 20 families based on chemical structure. A survey of five common families follows.

Acid Amides

The *acid amide herbicides* are characterized by an amide (carboxamide) moiety:

Most acid amides are used for the selective control of weed seedlings by either pre-emergence or preplant application; they inhibit seedling growth by interfering with cell division. Although the exact mechanism of action is not known, they appear to interrupt protein synthesis and nucleic acid replication. Commonly used acid amide herbicides are alachlor (Lasso) and propanil (Stampede):

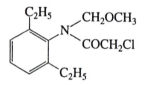

Alachlor (common name)
Lasso (trade name)
2-Chloro-2′,6′-diethyl-*N*-methoxymethylacetanalide (IUPAC name)

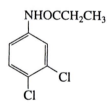

Propanil (common name)
Stampede (trade name)
N-(3,4-dichlorophenyl)propionamide (IUPAC name)

Aliphatics

The *aliphatic herbicides* contain no aromatic moieties; they are a family of miscellaneous compounds that include glyphosate (Roundup) and its analogs, chlorinated aliphatic acids, organic arsenicals, and methyl bromide:

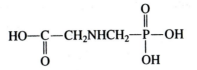

Glyphosate (common name)
Roundup (trade name)
N-(Phosphonomethyl)glycine (IUPAC name)

Methyl bromide (common name)
Bromomethane (IUPAC name)

The mode of herbicidal action for glyphosate is interference with amino acid synthesis and, therefore, plant protein and enzyme production. Methyl bromide is toxic to most organisms, including bacteria, fungi, and humans; because it is a gas at ambient temperatures, methyl bromide is applied as a fumigant (see Section 2.5.4).

Phenoxy Herbicides

Phenoxy herbicides are phenyl ethers with an oxygen linkage to the benzene ring. The most important phenoxy herbicides are 2,4-D, 2,4,5-T, and Silvex:

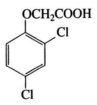

2,4-D (common name)
(2,4-Dichlorophenoxy)acetic acid (IUPAC name)

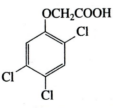

2,4,5-T (common name)
(2,4,5-Trichlorophenoxy)acetic acid (IUPAC name)

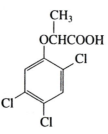

Silvex (common name)
2-(2,4,5-Trichlorophenoxy)propanoic acid (IUPAC name)

Over 27,000 tons of 2,4-D are used per year in the United States. Important analogs include the alkylamine salts and alkyl esters of the parent compounds. Agent Orange, used as a Vietnam War defoliant from 1962 to 1971, is a 50:50 mixture of the *n*-butyl esters of 2,4-D and 2,4,5-T. In 1979, 2,4,5-T was banned by the EPA.

The herbicidal actions of phenoxy herbicides occur by mimicking auxins, which are growth regulators produced by plants. Although phenoxys are transformed relatively rapidly in the environment, impurities formed during their synthesis, polychlorinated dibenzodioxins and polychlorinated dibenzofurans, are highly stable and toxic (see Section 2.9). As a result, the use of some phenoxys, particularly 2,4,5-T and Silvex, has been restricted by the EPA.

Substituted Ureas

Substituted urea herbicides are derivatives of urea, a bound form of nitrogen commonly excreted by higher animals:

$$H_2N-\overset{\overset{\textstyle O}{\|}}{C}-NH_2$$

Urea

Substituents on one nitrogen are indicated by the prefix *N*-, and substituents on the other by the prefix *N'*-. Some of the more common substituted ureas are

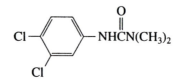

Diuron (common name)
3-(3,4-Dichlorophenyl)-1,1-dimethylurea (IUPAC name)

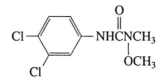

Linuron (common name)
3-(3,4-Dichlorophenyl)-1-methoxy-1-methylurea (IUPAC name)

Substituted urea herbicides are most effective against broadleaf plants and are used primarily as preemergent weed killers. Most of these compounds persist in soils for up to a year.

Triazines

The triazines are predominantly *N*-alkylated derivatives of 2-chloro-4,6-diamino-*s*-triazine a compound that was first used in the synthesis of dyes:

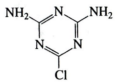

The most widely used *s*-triazines are atrazine (AAtrex) and cyanazine (Bladex). The *s*- used in these names indicates the 1,3,5-, or symmetrical distribution, of the three carbon and three nitrogen atoms in the benzene-like aromatic ring. A popular *asym*- or *as*- (for asymmetrical) triazine is metribuzin (Sencor). The primary route of toxicity to plants for the triazines is through interference with photosynthesis.

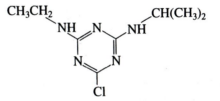

Atrazine (common name)
AAtrex (trade name)
2-Chloro-4-ethylamino-6-isopropylamino-1,3,5-triazine (IUPAC name)

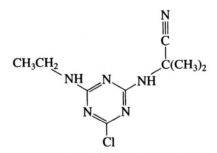

Cyanazine (common name)
Bladex (trade name)
2-Chloro-4-(1-cyano-1-methylethylamino)-6-ethylamino-1,3,5-trazine) (IUPAC name)

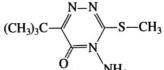

Metribuzin (common name)
Senecor (trade name)
4-Amino-6-*tert*-butyl-3-(methylthio-1,2,4-triazin-5(4H)-one (IUPAC name)

Since their introduction in the mid-1960s, triazines have been considered ideal herbicides because they (1) lose their phytotoxic activity within a normal growing season, and (2) are characterized by low acute and chronic human toxicity. Their main degradation products, which are hydroxylated species formed by hydrolysis reactions catalyzed on soil organic matter, are nontoxic to plants and animals. Moreover, these hydrolysis products are strongly sorbed to soils and therefore migrate slowly to groundwater. However, investigations of pesticide residues in groundwater, begun in the mid-1980s, have revealed a significant degree of contamination, both in the United States and Europe, by traces of unchanged atrazine, suggesting that the parent compound is more mobile than its hydroxylated metabolites. Triazine contamination of groundwater has been documented even in aquifers underlying soils of low permeability.

2.5.3 Fungicides

Fungicides are used against fungi and molds that may cause the destruction or deterioration of fruit, grain, and vegetables. Three groups of fungicides, the ethylene-*bis*-dithiocarbamates (EBDCs), captan, and the benzimidazoles, comprise 70% of the fungicides now used in the United States. Organomercury fungicides, widely used in the 1950s and 1960s, are now banned because of their bioaccumulation in fish.

Organic fungicides were introduced in the early 1930s and now number over 100. The ethylene-*bis*-dithiocarbamates (EBDCs) are metal salts of thiocarbamic acid and are the most widely used fungicides in the United States. Mancozeb, a common EBDC, contains both manganese and zinc in the proportions of x and y in the formula:

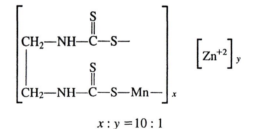

$$x : y = 10 : 1$$

Mancozeb (common name)
Manganese ethylene *bis*(dithiocarbamate) polymeric complex with zinc salt
(IUPAC name)

Captan, introduced in 1949, was the first of the class of phthalimide fungicides; together with its analogs, it ranks second in use to the EBDCs.

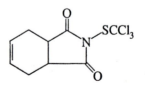

Captan (Common name)
1,2,3,6-tetrahydro-*N*-(trichloromethylthio) phthalimide (IUPAC name)

The benzimidazole fungicides were introduced in the 1960s. Benomyl is a widely used benzimidazole that has functional groups characteristic of carbamate and substituted ureas:

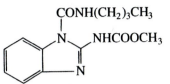

Benomyl (common name)
Methyl 1-(butylcarbamoyl)benzimidazol-2-ylcarbamate (IUPAC name)

An organochlorine compound that has been used as a fungicide, wood preservative, and insecticide is pentachlorophenol, which is also called *Penta* or *PCP*; it is applied mainly as its water-soluble sodium salt.

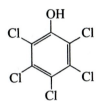

Pentachlorophenol

Penta has been used extensively for preserving railroad ties, telephone poles, fence posts, lumber, and other timber materials (Figure 2.5). It has also been used as an

Figure 2.5 Telephone poles in storage that have been treated with *Penta.*

insecticide, fungicide, aracacide, herbicide, disinfectant, and antifouling agent in paint, with 88% of its consumption as a wood preservative. Hundreds of hazardous waste sites, some of which are Superfund sites, are contaminated with pentachlorophenol. Although pentachlorophenol itself is only moderately biorefractory, its formulations usually contain dioxins, such as octachlorodibenzo-*p*-dioxin, which are toxic and biorefractory (see Section 2.9). Although still used for commercial applications, PCP is now classified as a restricted use chemical and is no longer available for home use.

Creosote, a dark, oily liquid produced from the distillation of coal tar, is a commonly used wood preservative. Over 100 compounds have been identified in creosote, and some of the most common constituents are listed in Table 2.17. The primary chemicals in creosote are naphthalene and its alkyl derivatives; in addition, creosote contains 15–21% PAHs.

2.5.4 Soil Fumigants

Small halogenated organic compounds have been used as soil fumigants to control a range of pests. Early fumigants, which have long been banned, included carbon tetrachloride, carbon disulfide, and hydrogen cyanide. Two commonly used fumigants that have resulted in significant groundwater contamination problems are 1,2-dibromo-3-chloropropane (DBCP) and ethylene dibromide (EDB):

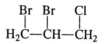

$$\underset{\text{H}_2\text{C}-\text{CH}-\text{CH}_2}{\overset{\overset{\displaystyle \text{Br}}{|}\ \overset{\displaystyle \text{Br}}{|}\ \overset{\displaystyle \text{Cl}}{|}}{}}$$

1,2-Dibromo-3-chloropropane (DBCP)

$$\underset{\text{H}_2\text{C}-\text{CH}_2}{\overset{\overset{\displaystyle \text{Br}}{|}\ \overset{\displaystyle \text{Br}}{|}}{}}$$

Ethylene dibromide-EDB (common name)
1,2-Dibromoethane (IUPAC name)

These compounds, which are injected into the soil at depths from 15 cm to 1 m (6 in. to 39 in.), have been used to control nematodes and fungi. A study of several rural central California counties in the early 1980s provided data showing detectable DBCP concentrations in 92 of 262 well samples [2.33]. The use of DBCP was subsequently banned by the state of California, and the EPA followed with a nationwide ban. Because of its structural similarity, studies of EDB followed. In 1983, EDB was detected in 15 of 137 California wells and, like DBCP, was banned soon thereafter [2.34].

2.6 EXPLOSIVES

An *explosive* is a quasi-stable chemical that rapidly changes from a solid or liquid to a gas following activation or detonation. An extreme increase in pressure by gas evolution results in the explosion, with the energy liberated a function of the explosive's

Table 2.17 Partial List of Compounds
Identified in Creosote

Benzene
Toluene
Xylene
Trimethylbenzene
Methylethylbenzenes
Styrene
Phenol
Benzofuran and dibenzofuran
Benzonitrile and methylbenzonitrile
Methylstyrene
Cresols
Indenes and methylindenes
Xylenols
Naphthalene and methylnaphthalene
Benzothiophenes
Quinoline and isoquinoline
Diphenyl
Dimethylnaphthalene
Ethylnapthalene
Acenaphthene
Fluorene
Dibenzothiophene
Acenaphthylene
Benzo[g,h,i]perylene
Anthracene
Pyrene
Penanthrene
Crysene
Benzo[e]pyrene
Dibenzo[a,h]anthracene
Benzo[k]fluoranthrene
Benzo[a]fluorene

Source: Reference 2.32.

thermodynamic properties. Most high-explosive molecules are characterized by an aliphatic or aromatic structure with substituted nitro groups. These compounds have regions of greatly differing oxidation-reduction potentials so that, when detonated, they produce evolution of gases through rapid intra- and intermolecular oxidation reactions. The activation energy necessary to detonate explosives can be supplied by high-pressure shock waves, friction, electricity, or heat, such as from the electrical discharge of unstable compounds known as primary explosives.

Explosives are used primarily in mining, quarrying, highway construction, demolition, and military applications. Most military hazardous waste problems involve explosives that are manufactured or stored at a site; releases of the explosives to the en-

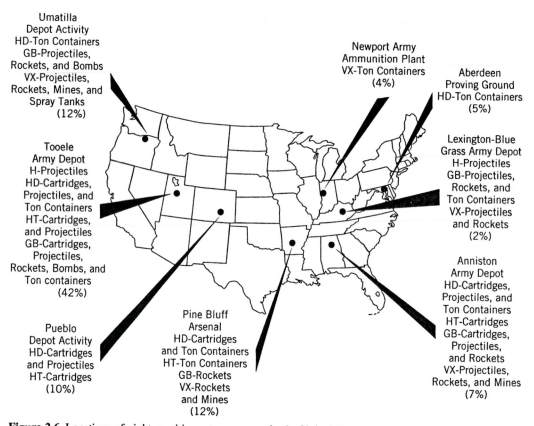

Figure 2.6 Locations of eight munitions storage areas in the United States. *Source:* Reference 2.35.

vironment have contaminated soils, water, and structures. In addition, large quantities of off-specification explosives must be disposed of under RCRA guidelines. In the United States, private corporations operate many of the government plants that manufacture military explosives. These operations, termed GOCOs (Government-Owned, Contractor-Operated facilities), are often investigated through Department of Defense hazardous waste programs. The locations of the eight largest munitions storage sites are shown in Figure 2.6.

Explosives pose hazardous waste problems because they are persistent and toxic, even at relatively low concentrations. Most organic nitro compounds are inhibitors of oxidative phosphorylation, the respiratory process that produces adenosine triphosphate (ATP), the primary energy source for lower and higher organisms. At high concentrations, contaminated sludges and soils may actually be explosive. Pit-pond-lagoon systems and surface waters contaminated with some explosives (e.g., 2,4,6-trinitrotoluene or TNT) and their degradation products reflect red light and have been labeled *pink water.*

Explosives are classified based on their velocity of detonation as low (or propellant) and high (or detonating) explosives. *Low explosives* react slowly and burn rapidly instead of detonating. Black powder, nitrocellulose, smokeless gunpowder, solid rocket fuels, and common fireworks are examples of low explosives. *High explosives,* which produce extreme pressures by rapid detonation, are usually more stable to detonation than low explosives; their detonation can be initiated only by exceeding a higher activation energy through shock. High explosives are characterized by detonation rates as high as 6400 m/sec (21,100 ft/sec): low explosives are in the range of 80 m/sec (260 ft/sec) [2.36]. High explosives are divided into three groups based on chemical structure—aliphatic nitrate esters (i.e., esters of nitric acid), nitramines, and nitroaromatics.

2.6.1 Aliphatic Nitrate Esters

Aliphatic nitrate esters are among the most powerful explosives. They are produced by treating polyhydric alcohols, such as glycerol or pentaerythritol, with concentrated nitric acid and include

Nitroglycerine

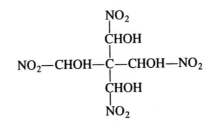

Pentaerythritol tetranitrate

Nitroglycerin is an unstable component of dynamite; therefore, it is often soaked into an absorbent for stability during transport. As implied by Western movies in which miners gingerly transport a wagon load of nitroglycerin, its detonation is highly sensitive to shock and impact. Dynamite (developed by Alfred Nobel) is made by mixing an inert substance, such as Fuller's earth or pulpwood, with nitroglycerin to reduce its sensitivity. Pentaerythritol tetranitrate is used in blasting caps, detonating fuses (Primacord), and sheet explosives. It is sometimes mixed with TNT (2,4,6-trinitrotoluene) and RDX (cyclotrimethylene-nitramine) to form less sensitive explosives.

2.6.2 Nitramines

Nitramines contain a nitro ($-NO_2$) group attached to a nitrogen atom. The three most common nitramines are

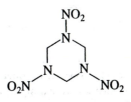

RDX (military code name)
Cyclotrimethylenetrinitramine

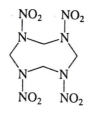

HMX (military code name)
Cyclotetramethylenetetranitramine

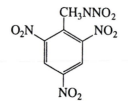

Tetryl (common name)
2,4,6-Trinitrophenylmethylnitramine (IUPAC name)

Both RDX and HMX are used as military explosives and as rocket propellants. Tetryl has been used most commonly as a base charge in detonators and blasting caps.

Nitroaromatics

The most common nitroaromatics are

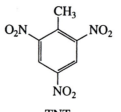

TNT
2,4,6-Trinitrotoluene

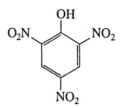

Picric acid (common name)
2,4,6-Trinitrophenol (IUPAC name)

Discovered in 1902, TNT is one of the most toxic and persistent explosives, making it a common hazardous waste. It is usually biodegraded only under anaerobic conditions and is currently the focus of intense bioremediation research. Picric acid was one of the first high explosives developed, but it is no longer used in military applications. It was once used for treating burns and preparing crystalline complexes to help identify aromatic hydrocarbons, but it has been replaced for both of these purposes. Removal of stores of picric acid formerly used for these applications has resulted in its disposal in a number of municipal landfills. Furthermore, some commercial and academic laboratories have old supplies of picric acid on their shelves—a practice that poses a severe explosion hazard.

2.6.3 Primary Explosives

Primary, or *initiator explosives* (detonators), are most sensitive to heat, friction, impact, shock, and electrostatic energy. Primary explosives burn rapidly but do not detonate upon ignition. Examples of primary explosives are mercury fulminate, diazodinitrophenol, and tetrazene:

$$Hg(ONC)_2$$

Mercury fulminate

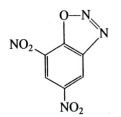

2-Diazo-4,6-dinitrophenol

Tetrazene

2.7 INDUSTRIAL INTERMEDIATES

Numerous aliphatic and aromatic compounds are used in the synthesis of pesticides, plastics, artificial fibers (e.g., nylon, polyester, acrylics), dyestuffs, paint, and a range of other useful consumer products. The most common of these *industrial intermediates* are benzene and its derivatives, phthalate esters, chlorobenzenes, chlorophenols, and chlorotoluenes. Varying amounts of these compounds are often found as contaminants in aqueous industrial waste streams and as more concentrated wastes, such as drummed liquids and sludges, which are defined as hazardous under RCRA.

Aromatic hydrocarbons have been commonly used as building blocks for larger molecules. Benzene is a precursor in the production of styrene, cyclohexane, phenolics, and chlorinated benzenes; xylenes are used in the synthesis of manufacturing polymers. Phenol is the starting material for the production of chlorophenols, and is also a precursor in the production of phenolic resins, which are used as adhesives for plywood, construction, and in automobiles.

Phthalate esters are used primarily as plasticizers (or softeners) in the manufacture of plastics; all of the important phthalate esters are derivatives of *o*-phthalic acid:

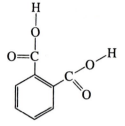

o-Phthalic acid

Alkyl groups may be linked to the carboxylic groups to form esters:

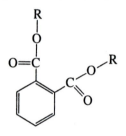

R = alkyl groups

Esters of *o*-phthalic acid are important materials in hazardous waste management. Among the most significant are dimethyl phthalate and di-*n*-octyl phthalate:

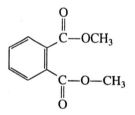

Dimethyl phthalate

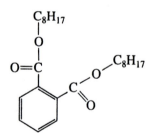

Di-*n*-octyl phthalate

Chlorobenzenes and chlorophenols are used as intermediates in the synthesis of pesticides, pharmaceuticals, antiseptics, plastics, and other products. The naming of chlorobenzenes is based on the previous nomenclature principles outlined in Section

2.1.2. The chlorines are numbered (using the lowest possible sequence) and the number of chloro constituents (1, 2, 3 or more) is preceded by di, tri, tetra, penta, or hexa

1,4-Dichlorobenzene or *p*-dichlorobenzene

1,2,3,5-Tetrachlorobenzene

Chlorobenzenes have been produced in the United States since 1915. Twelve chlorinated benzenes can be synthesized, ranging from chlorobenzene to hexachlorobenzene. Most are produced by chlorinating benzene with chlorine gas in the presence of an acidic catalyst such as ferric chloride or aluminum chloride. Historically, one of the largest uses of chlorobenzene has been in the manufacture of DDT before it was banned in the early 1970s (although smaller quantities are still used in its production for export). The current primary use of chlorobenzene is in the production of chloronitrobenzenes as intermediates in the synthesis of aniline dyes and a variety of pesticides. 1,2-Dichlorobenzene is an intermediate in the synthesis of 3,4-dichloroaniline, and 1,4-dichlorobenzene has limited functions as an intermediate but is used extensively in mothballs. 1,2,4-Trichlorobenzene is used as a dye carrier in the textile industry. 1,2,4,5-Tetrachlorobenzene is important as the precursor of 2,4,5-trichlorophenol, used to make 2,4,5-T, Silvex, and the antiseptic hexachlorophene, which is still used in medicinal soaps.

Chlorophenols are named using the same procedure as chlorobenzenes, but the hydroxyl group of the molecule is always designated as the number one position:

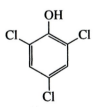

2,4,6-Trichlorophenol

Nineteen different chlorinated phenols are commercially available. The most important are *o*-chlorophenol, *p*-chlorophenol, 2,4-dichlorophenol, and 2,4,5-trichlorophenol. The two monochlorophenols are used as intermediates in dyestuffs, as preservatives, and in the manufacture of disinfectants. The primary use of 2,4-dichlorophenol is in the manufacture of 2,4-D, and 2,4,5-trichlorophenol is used as an intermediate in the manufacture of 2,4,5-T and Silvex.

EXAMPLE 2.10 *Nomenclature of Industrial Intermediates*

Name the following industrial intermediates.

1.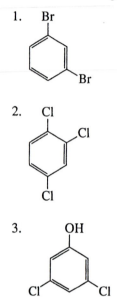

2.

3.

SOLUTION

1. The IUPAC name is 1,3-dibromobenzene; less formally, it is *m*-dibromobenzene.
2. This trichlorobenzene is named using the lowest possible sequence of numbers:

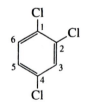

Hence, it is named 1,2,4-trichlorobenzene.
3. Chlorophenols are named based on the —OH in the number one position:

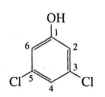

Therefore, this chlorophenol is 3,5-dichlorophenol.

Chlorinated toluenes include mono- and dichloro–species:

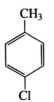

4-Chlorotoluene or *p*-chlorotoluene

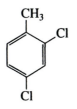

2,4-Dichlorotoluene

These compounds are used in the production of pesticides, dyes, pharmaceuticals, and peroxides.

Anilines are aminobenzenes. The greatest industrial use of analine derivatives is in the synthesis of inks, dyes, drugs, and photographic developers. The simplest compound is aniline itself:

Aniline

Some amino aromatics, such as β-naphthylamine, are known to be strong human carcinogens.

An industrial intermediate used in the manufacture of numerous pesticides, polymers, and flame retardants is hexachlorocyclopentadiene:

Hexachlorocyclopentadiene

Insecticides synthesized from this intermediate include aldrin, dieldrin, chlordane, endosulfan, and mirex. Its use in industrial synthesis is related to its reactivity as a chlorinated diene—it reacts with a number of reagents by substitution and addition reactions to produce multi-ring, chlorinated compounds.

2.8   POLYCHLORINATED BIPHENYLS

Polychlorinated biphenyls (PCBs) are heat-stable oils once used extensively as transformer and hydraulic fluids. The production of PCBs began at Monsanto in 1929 in response to the need for a stable, nonflammable transformer dielectric cooling oil. Mineral oil had been used previously, but PCBs were more stable and less flammable for transformer applications, in which power surges can cause arcs and ignite the oil. A group of such transformers in a storage yard is shown in Figure 2.7. Prior to 1971, mixtures of PCBs containing up to 68% chlorine were also used as plasticizers (e.g., for vinyl upholstery in automobiles), heat transfer fluids, hydraulic fluids, and lubricants. Under the Toxic Substances Control Act, Congress banned the manufacture and limited the distribution of PCBs after 1979; however, these stable compounds are still present in thousands of transformers, capacitors, and other electrical devices such as motors in refrigerators and freezers.

The basic unit of PCBs is the aromatic hydrocarbon biphenyl:

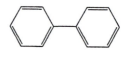

The biphenyl molecule has ten sites where chlorine can be substituted for hydrogen atoms; five positions in one ring are numbered 2 through 6, and five positions in the

Figure 2.7 Transformers in a maintenance storage yard.

second ring are numbered 2′ through 6′:

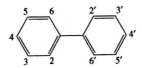

In assigning location numbers to the chloro substituents, lower numbers (primed or un-primed) take precedence over higher numbers, and unprimed numbers (e.g., 2, 3) are listed over equivalent primed (e.g., 2′, 3′) numbers. Chlorination yields up to 207 products (called *congeners* because they are all formed together by the same basic process of chlorination) with the substitution of one to ten chlorine atoms on the biphenyl molecule. Polychlorinated biphenyls are therefore given the general formula

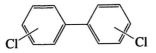

This structure represents all of the possible PCB congeners such as

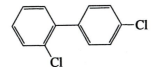

2,4′-Dichlorobiphenyl

and

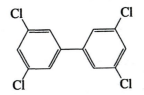

3,3′,5,5′-Tetrachlorobiphenyl

Congeners with equal numbers of chlorine atoms are, by definition, isomers.

Commercial polychlorinated biphenyl formulations never contain one isomer or congener by itself because, during synthesis, control of the number of substitutions and the positions of the chlorines on the biphenyl rings is impossible. In the United States, PCBs have been marketed by Monsanto under the trademark Aroclor. International trade names include Fenchlor (Italy), Kaneclor (Japan), Pyralene (France), and Clophen (West Germany). The different degrees of chlorination are noted by four-digit numbers after the trade name *Aroclor,* such as Aroclor 1221 or Aroclor 1232. The first two digits refer to the 12 carbons in biphenyl. For all Aroclors but one, the last two digits of the four-digit term represent the percentage of chlorination by mass of the PCB mixtures (e.g., Aroclor 1248 is a mixture containing 48% by weight of chlorine). The one exception is Aroclor 1016, which is 41% chlorinated.

Polychlorinated biphenyls containing five or more chlorine atoms are considered higher chlorobiphenyls, a distinction related not only to their higher chlorine content, but also to their longer environmental persistence and higher toxicity.

EXAMPLE 2.11   *Nomenclature of PCBs*

Name the following PCB.

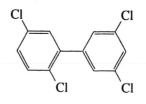

SOLUTION

Using the PCB numbering scheme:

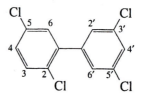

and ordering the chlorines numerically, the structure is named 2,3′,5,5′-tetrachloro-biphenyl.

2.9   POLYCHLORINATED DIBENZODIOXINS AND DIBENZOFURANS

Unlike almost all other organic compounds, such as those discussed previously in this chapter, polychlorinated dibenzodioxins (PCDDs) and dibenzofurans (PCDFs) are not synthesized or formulated for an industrial or a domestic use. Petroleum products, solvents, pesticides, and explosives are refined or manufactured for beneficial uses, but polychlorinated dibenzodioxins and dibenzofurans are trace impurities formed during the manufacture, chlorination, or combustion of other organic compounds. Even trace amounts of these compounds are important because of their extreme toxicity.

These chemicals are aromatic, nearly planar, and characterized by similar physical and chemical properties. Chlorinated dioxins are derivatives of dibenzo-*p*-dioxin:

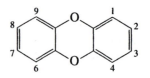

in which the locations of chlorine substitution are designated by the numbers shown, keeping the numbers as low as possible. Depending on the precursors and the conditions of formation, between one and eight chlorine atoms may be substituted on the dioxin molecule, resulting in 75 possible congeners. Two examples of PCDD congeners are

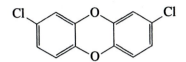

2,8-Dichlorodibenzo-*p*-dioxin

and

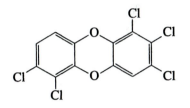

1,2,3,6,7-Pentachlorodibenzo-*p*-dioxin

The naming of chlorinated dioxins is demonstrated in Example 2.12.

Polychlorinated dibenzofurans are chlorine-substituted derivatives of dibenzofuran that have biological and physiological properties similar to those of PCDDs:

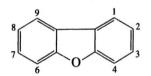

As with PCDDs, ring numbers designate locations of chlorine substitution and different congeners are found, resulting from one to eight chlorine substitutions. One possible PCDF isomer is

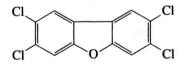

2,3,7,8-Tetrachlorodibenzofuran

The formation of chlorinated dioxins and furans occurs primarily during industrial synthesis and combustion processes. For example, during the alkali-catalyzed synthesis of 2,4,5-T from trichlorophenol and chloroacetic acid, two of the trichlorophenol molecules may dimerize to form 2,3,7,8-tetrachlorodibenzo-*p*-dioxin (2,3,7,8-TCDD):

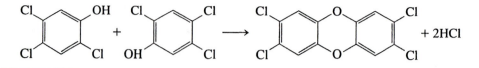

Polychlorinated dibenzodioxins and polychlorinated dibenzofurans are trace contaminants found in many commercial products. The most commonly contaminated materials are those in which chlorophenols are used as a precursor or are the final product [2.37]. Important PCDD- and PCDF-contaminated products include the herbicides 2,4,5-T and 2,4-D and the wood preservative pentachlorophenol. Another source of PCDD and PCDF formation is the incineration of nonchlorinated or chlorinated organic chemicals in the presence of chlorinated compounds. Just as polycyclic aromatic hydrocarbons are formed during the combustion of hydrocarbons, chlorinated dioxins and furans are produced as stable products during combustion at low-to-moderate temperatures, even of such relatively innocuous materials as firewood. The presence of dioxins appears to be ubiquitous; they have been found in chlorine-bleached paper and numerous commercial products [2.38].

A heightened awareness of chlorinated dioxins and furans resulted from the teratogenicity (the potential for birth defects) found with the use of 2,4,5-T. In one sampling scheme, concentrations of the most toxic dioxin congener, 2,3,7,8-TCDD, in drums of 2,4,5-T were 0.02–54 mg/kg, and Rappe [2.38] suggested that TCDD levels could be as high as 100 mg/kg in Agent Orange.

One of the largest hazardous waste problems involving dioxins occurred at Times Beach, Missouri. After 2,3,7,8-TCDD was recognized as a major impurity in 2,4,5-T, measures were taken to remove TCDD from the product. Because TCDD is a neutral molecule and 2,4,5-trichlorophenol exists as a phenoxide ion in alkaline solution during manufacture, TCDD could be isolated from the reaction product by extraction with toluene. Evaporation of the extract to recover the toluene left concentrated dioxin in still bottoms (the residue from distillation), which a waste oil hauler (ironically named Bliss) mixed with waste oils and spread on unpaved roads and a horse stable near Times Beach for dust control. Acutely toxic concentrations of dioxins in surface soils through such improper disposal of dioxin-contaminated waste oils resulted in evacuation of the town and purchase of the homeowners' property by the EPA under Superfund.

EXAMPLE 2.12   ***NOMENCLATURE OF CHLORINATED DIOXINS***

Name the following chlorinated dioxin.

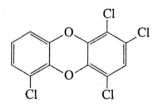

SOLUTION

Using the standard dibenzo-*p*-dioxin numbering system, the structure

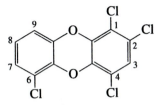

is named 1,2,4,6-tetrachlorodibenzo-*p*-dioxin.

The results of animal studies have demonstrated that significant differences are found in the toxic effects of PCDD and PCDF isomers. For example, toxicities can vary by 1000 to 10,000 times for such closely related congeners as 2,3,7,8-TCDD and 1,2,3,8-TCDD. However, some general trends in toxicities have been established. Congeners containing four to six chlorines with substitution of all of the lateral carbons (i.e., the 2, 3, 7, and 8 positions) are the most toxic. The toxic response of PCDD and PCDF congeners are most often compared to the most toxic PCDD congener, 2,3,7,8-TCDD, as *Toxicity Equivalent Factors* (TEFs). These equivalents are most commonly used to convert analytical data for specific isomers into TCDD equivalents. Based on a reference TEF of 1.0 for 2,3,7,8-TCDD, some of the more toxic PCDD and PCDF congeners include 1,2,3,7,8-PentaCDD (TEF = 0.5) and 2,3,4,7,8-PentaCFD (TEF = 0.5). Three hexaCDDs and four hexaCDF congeners are characterized by TEFs of 0.1.

2.10   METALS AND INORGANIC NONMETALS

Over 40,000 organic chemicals are used by industry, and their disposal has resulted in a broad range of hazardous waste problems. Because of the large number of organic compounds used in commerce and their wide-ranging characteristics and properties, the chemodynamic pathways of organic wastes in the environment and in treatment systems are complex. Consequently, most hazardous waste projects have a strong emphasis on organic waste components. Metals and their derivatives, on the other hand, are a relatively small group of elements. Although 65 elements listed in the periodic table are classified as metals, only about 30 have been widely used by industry, either in elemental form or as their salts or organometallic compounds. Because fewer metals are used by industry relative to organic compounds, and metals nomenclature is more straightforward, they have received less emphasis in this text.

A *metal* is defined as an element that

1. conducts electricity,
2. has high thermal conductivity,
3. has a high density, and
4. is characterized by malleability and ductility.

The dark line running diagonally from the upper left to the lower right of the periodic table (inside the front cover) divides metals from nonmetals—in general, metals lie below and nonmetals above the line. The periodic table is arranged into 7 horizontal rows (periods) and 16 vertical columns (families). The 8 main families are designated as 1A to VIIA and 0, which correlate with the thermodynamically stable oxidation state of the element. Beginning in the third column from the left are the groups IIA and IIIA, which are called the *transition elements.* Because of the ease of interchange of electrons in their higher *s* and *p* orbitals, these metals exist in a number of possible oxidation states. *Heavy metals,* which have been defined as (1) elements with atomic numbers greater than iron, and (2) metals with densities greater than 5.0 g/cm^3, are of high environmental concern because of their toxicity [2.39].

The speciation of metals has significant effects on their water solubility, transport, toxicity, and treatment. In this section, the industrial uses and speciation of common metals are introduced. The transport of metals in soils and groundwater will be covered in Sections 5.7–5.9, and the toxicity of metals will be the focus of Section 9.5. In addition, a primer on the treatment of metals in hazardous waste streams will be covered in Section 12.4.

The most common sources of waste metals have historically been spent electroplating solutions and sludges. A typical electroplating system, shown in Figure 2.8, consists of a bath containing a metal salt into which parts that are to be plated are lowered. Current is applied to the bath, and the metal is reduced and deposited at the part, which serves as the cathode. Electroplating operations use metals such as chromium, copper, nickel, and cadmium in aqueous solutions into which metal parts are dipped. The solutions eventually need replacement, and prior to the promulgation of RCRA, the metal-bearing liquids were disposed of in landfills, pit-pond-lagoons, or other land disposal areas. Sometimes the metals were precipitated into a sludge so that the clarified water could be discharged to a sanitary sewer. The sludge was then disposed of in landfills, soil pits, and so on.

Under RCRA, plating sludges must be disposed of in secure landfills. In the future, metals waste management will focus on selective precipitation and complexation of specific metals so that they can be recycled and reused in a relatively pure form.

2.10.1 Arsenic (As)

Arsenic, in the VA family of the periodic table, has five electrons in its outermost shell and can exist in $5+$, $3+$, 0, and $3-$ oxidation states. Elemental arsenic is a gray, crystalline material with a density of 5.727 g/cm^3. Arsenic is often classified as a metal, but in reality it is a nonmetal or metalloid. It occurs naturally in over 150 minerals, including loellingite ($FeAs_2$), sulforlite ($CoAs$), realgar (AsS), and arsenopyrite ($FeAsS$).

The chemistry of arsenic is complex; the element is subject to chemical, biochemical, and geochemical reactions. Arsenic can exist in the solid, dissolved, and gaseous states as inorganic or organometallic species. Metallic arsenic, As(0), is thermodynamically stable only at low oxidation-reduction potentials. Under very strong reducing conditions, arsine (AsH_3) gas may be formed. Arsenic can be converted to organic arsenic acids ($3+$ and $5+$) and to methylated arsines ($3-$ valence state) under anaerobic conditions by bacteria, fungi, and yeasts, which use these reactions as a mechanism of detoxification. The oxyacid anion arsenite AsO_3^- ($3+$) may be present if the

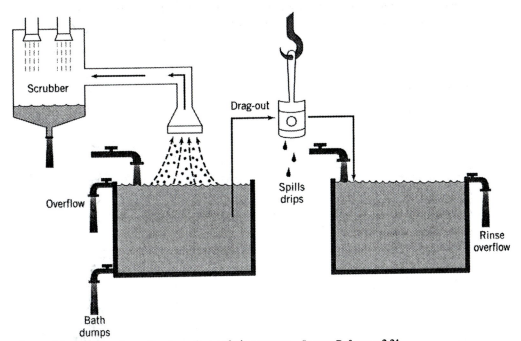

Figure 2.8 Process schematic of an electroplating system. *Source:* Reference 2.21.

$E_h < 0.1$ V, where E_h is the oxidation-reduction potential. Under aerobic conditions, arsenate AsO_4^{3-} (5+) is the most common species. Arsenate exhibits low mobility in the environment because, like its analog phosphate, it precipitates easily and is strongly sorbed to soils and sediments.

The most toxic form of arsenic is AsH_3, arsine gas. Arsenite (As^{3+}) is characterized by intermediate toxicity, and arsenate (As^{5+}) is the least toxic form of arsenic (but is still toxic). One of the largest public health catastrophes involving arsenic was the poisoning of over 12,000 Japanese children who drank milk contaminated with arsenic, resulting in 130 deaths [2.40].

The largest traditional use of arsenic has been in agriculture. The herbicides monosodium methyl arsenate (MSMA), disodium methyl arsenate (DSMA), and cacodylic acid (dimethylarsenic acid) were used extensively, particularly to control weeds in cotton fields. Sodium arsenite has been the most commonly used arsenic-containing veterinary product as an insecticidal ingredient of cattle and sheep dips. It has also been used for debarking and killing nuisance trees (e.g., cottonwoods), and for controlling aquatic weeds. Copper chrome arsenate (CCA) has been used as a wood preservative. Some organoarsenic compounds are still used for postemergent control of crabgrass on lawns.

Industrial uses for arsenic include the manufacture of glass, growth stimulants for plants and animals, and fungicides. Arsenic trioxide has been used to remove the color from glass during its production. A more recent use of arsenic has been in the semiconductor industry. Gallium arsenide has been used in photovoltaic cells and is replacing silicon in computer chips for some applications. Copper smelting releases significant quantities of arsenic; one of the largest hazardous waste sites resulting from smelting may be found at Bunker Hill, Idaho (see Section 1.1).

2.10.2   Cadmium (Cd)

Cadmium is a member of the IIB group of the periodic table; it is almost always found in the 2+ valence state. Cadmium has properties similar to zinc and forms complexes with cyanide, amines, and halides. In pure form, it is a silver-white malleable material. The most common naturally occurring form of cadmium, CdS (greenockite), is almost always associated with sulfides of zinc, lead, and copper. As a result, cadmium compounds are usually prepared from metallic cadmium obtained during lead-zinc production. Commercially important forms are made from the oxide, sulfide, selenite, chloride, sulfate, nitrate, hydroxide, and organic salts of cadmium.

Cadmium is a highly toxic metal. It is cumulative and is excreted with a half-life of 20 to 30 years. Cadmium exposure results in a range of health effects from hypertension to cancer.

The distribution of cadmium in products is shown in Figure 2.9. Its most important use has been in metal finishing. A cyanide bath, usually containing 35 g/L of cadmium oxide and 75 g/L of sodium cyanide, is used in plating operations. Cadmium is a superior plating metal because of its high rate of deposition, good resistance to alkali, and high ductility. Cadmium sulfide has been used in paint pigments, providing coloration for yellow, orange, red, and maroon. In electrochemical applications, $Cd(OH)_2$ serves as anode material in silver-cadmium and nickel-cadmium batteries. Organocadmium salts are also used as heat and light stabilizers for plastics and as catalysts in polymerization reactions.

2.10.3   Chromium (Cr)

Chromium, obtained by mining chromite ($FeO \cdot Cr_2O_3$), is in Group VIB of the periodic table. It is a highly acid-resistant material attacked only by HCl, HF, and H_2SO_4.

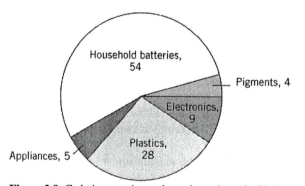

Figure 2.9 Cadmium use in products throughout the United States, based on weight percent. *Source:* Reference 2.41.

The United States is completely dependent on foreign sources of chromium ores; most of it is imported from South Africa.

Chromium is a transition element exhibiting oxidation states from $2-$ to $6+$. The $1-$ and $2-$ states are of minor significance; the $2+$, $3+$, and the $6+$ states are the most common, with $6+$ having the greatest commercial importance because of its high oxidation potential. The $6+$ state is thermodynamically stable as the anions chromate $(CrO_4{}^{2-})$ and dichromate $(Cr_2O_7{}^{2-})$, which are soluble in aqueous solutions. The $3+$ state, however, is insoluble in water. Therefore, treatment of aqueous waste streams usually involves reduction of Cr^{6+} to Cr^{3+} prior to precipitation and sedimentation.

The most important industrial uses of chromium are in electroplating (25% of its use) and chromium pigments (e.g., chrome yellow, chrome orange, chrome oxide green), which, at 33%, provide the major source of chromium consumption. Leather tanning is another significant use, followed by wood preservation, catalysts, and commercial electronic equipment.

2.10.4 Lead (Pb)

Lead is a soft, gray, moldable metal that is found in three oxidation states—0, $2+$, and $4+$—with the $2+$ state the most common. In general, lead (II) compounds exhibit ionic bonds and lead (IV) chemicals (e.g., tetraethyl-lead) are characterized by covalent bonds. Lead is a constituent of over 200 minerals, most of which are rare. Only three lead minerals are common: galena (PbS), angelsite ($PbSO_4$), and cerrusite ($PbCO_3$). The annual world production of lead is approximately 9 million tons. In the United States, about 5 million tons are mined each year and about 6 million tons are recovered, primarily through the recycling of lead storage batteries [2.42].

Several of its properties make lead a valuable industrial material, including a low melting point, high density, acid resistance, chemical stability, and ease of fabrication. Current uses, which are detailed in Figure 2.10, include storage batteries, pigments, electroplating, metallurgy, radiation protection equipment, electronics equipment, and

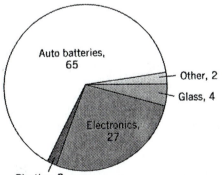

Figure 2.10 Lead use in products throughout the United States, based on weight percent.
Source: Reference 2.41.

plastics. The use of lead pigments in residential paint was banned in 1977. However, lead pigments remain in use for outdoor applications because of their weather-resistant properties. The most widespread use of lead has been as a gasoline additive to produce tetraethyl-lead. Now approximately 50% of lead is used for batteries.

As a result of its use in gasoline, paints, and water supply systems, lead is now ubiquitous in the environment. It is present in rain and snow, soils and dust, and industrial and municipal discharges. Fortunately, lead is characterized by low mobility in most soil-water systems owing to its low water solubility and strong tendency to sorb and exchange on solids.

2.10.5   Nickel (Ni)

Nickel occurs as a transition element in group VIIIB of the periodic table and exists in valences of 0, 1+, 2+, 3+, and 4+ (which is rarely encountered). The 2+ state is most commonly found in aquatic and soil environments. The primary sources of nickel are the minerals copyrite, pyrrhotite, and pentalandite. Use in the United States has been relatively constant at 240,000 tons per year [2.43].

Nickel is a high-melting element with a ductile crystal structure and is used extensively in alloys. Forty-four percent of nickel use in the United States is in stainless steel and alloy steels, 33% in iron ferrous alloys, 17% in electroplating, and 6% in catalysts, ceramics, and salts.

2.10.6   Mercury (Hg)

Mercury, a silver-white liquid at ambient temperatures, exists in valence states of 0, 1+, and 2+. The chemical symbol, Hg, comes from the Greek word *hydrargyros* (silver water). Mercury occurs naturally in the geosphere in concentrations ranging from 10 to 1000 μg/kg, primarily in association with sulfur in minerals such as cinnebar (HgS). Approximately 2000 tons of mercury are released to the environment each year, primarily from mining and smelting [2.44]. Mercury is found in three forms: (1) elemental, (2) inorganic, and (3) organic. Organic mercury may occur as long-chain alkyl or aryl mercury and as the more familiar short-chain (methyl and ethyl) mercury forms produced as a result of microbial alkylation.

Mercury is different from most metals in that it is a liquid at room temperature and has a uniform coefficient of expansion over its liquid range. These properties and its high density (13.6 g/cm^3) make it ideal for use in thermometers, barometers, and manometers. The distribution of mercury in products is shown in Figure 2.11. The largest specific use of mercury is in miniature batteries, which are employed in hearing aids, calculators, radios, cameras, and so on. These small batteries are often disposed of in municipal landfills as household solid waste. Another electrical use is the mercury vapor lamp, which uses an electrical discharge tube containing mercury vapor. Mercury is also used as an industrial catalyst in the production of caustic soda, chlorine, and chlorinated compounds. Overall, mercury has also been used widely in the pharmaceutical industry and as a fungicide and an insecticide. Mercury and its derivatives mercuric oxide, mercuric chloride, mercuric cyanide, mercuric iodide, and others are not so widely used as they once were.

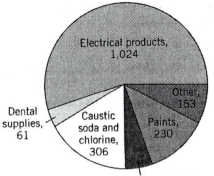

Figure 2.11 Mercury use in products throughout the United States, based on weight percent. *Source:* Reference 2.41.

2.10.7 Cyanides

Cyanide is an inorganic nonmetal anion with the structure

$$C \equiv N^-$$

It is the conjugate base of the weak acid hydrogen cyanide (HCN):

$$HCN \rightleftharpoons CN^- + H^+ \qquad K_a = 7.94 \times 10^{-10} \qquad (2.1)$$

Hydrogen cyanide is a gas that is somewhat soluble in water and has the odor of bitter almonds, which can be sensed by only 50% of the population [2.45]. It is also highly toxic and an acute poison; the mechanism for toxicity involves binding to cytochromes (enzymes that transfer electrons in respiration); in sufficiently high concentrations, cyanide causes death by asphyxia.

Hydrogen cyanide is produced by combining ammonia and methane under controlled conditions. The most common metal cyanide salts are those of sodium, potassium, and calcium. They are produced by combining sodium or potassium carbonate with carbon and ammonia or hydrogen cyanide with sodium or potassium hydroxide. Sodium cyanide and potassium cyanide are sold as powders, flakes, granules, and blocks.

One of the most common uses of cyanides is in electroplating. Hydrogen cyanide is also used as an industrial intermediate in the production of methyl methacrylate and cyanuric chloride, and as the chelating agent and detergent builder nitrilotriacetic acid (NTA). Sodium cyanide and potassium cyanide are white, crystalline solids, and both are used extensively in electroplating baths to maintain the solubility of metal ions. They are also used in the recovery of metals (e.g., silver and gold) during ore refining. Advances in mining engineering have provided processes by which gold and silver can be extracted from low-grade ores using large quantities of cyanide salts by heap leaching. Most of the cyanide is recovered and recycled; unrecovered cyanide can be degraded to cyanates (which are less toxic) by reactions with chlorine or other oxidants. However, in the western United States, large ponds have been constructed to hold the waste cyanide, and a series of environmental problems, ranging from the killing of water fowl to groundwater contamination, has resulted from poor pond design.

Two common hazardous consequences occur in the management of cyanide solutions. Sometimes acids are accidentally added to cyanide plating baths, resulting in the evolution of hydrogen cyanide gas. An analogous case is the inadvertent mixing of waste acid and waste cyanide salts, which has the same result—release of toxic cyanide gas.

2.10.8   Asbestos

Asbestos is a generic label applied to a variety of naturally occurring fibrous metal silicates of sodium, magnesium, calcium, and iron. The term *asbestos* is also a commercial label given to a group of six mineral fibers, all of which are classified as hydrated magnesium silicates. Asbestos is obtained by mining rocks rich in the magnesium silicates. The rocks are crushed and milled to separate the asbestos fibers. Worldwide production of asbestos peaked in the early 1970s at 5.4 million tons per year. Asbestos is not used nearly as much today; however, over 27.2 million tons are currently in use [2.46].

There are two major groups of asbestos materials based on their resistance to acids—serpentine and amphibole. The most common forms of asbestos are listed in Table 2.18. Although each class of fibers has an empirical formula and generic structure, the exact chemical composition varies because different concentrations of iron, chromium, cobalt, and nickel may be part of the crystalline structure. The chemical composition of a given species of asbestos can vary; for example, anthophyllite can have different ratios of magnesium and iron, which is reflected in its chemical composition by the use of brackets and a comma $(Mg,Fe)_5$. Chrysotile, in the serpentine group, is the form used in most manufactured products. The chemical formula of chrysotile is similar to that of the clay mineral kaolinite, but this asbestos mineral is different because of its fibrous structure. Chrysotile is a curly fiber found commonly in intertwined bundles. In contrast, all of the five amphibole fibers are composed of straight needles. Approximately 95% of the asbestos used in the United States is chrysotile [2.46].

The physical properties of asbestos—chemical and physical resilience, strength, and fibrous nature—have made it a desirable building and product material [2.47]. It is nonflammable, strong, and resistant to acids; over 3000 uses of asbestos have been developed to date. The most common are listed in Table 2.19.

Table 2.18 Common Forms of Asbestos

Species	Variety	Approximate Chemical Composition
Chrysotile	Serpentine	$Mg_3Si_2O_5(OH)_4$
Anthophyllite	Amphibole	$Mg_2(Mg,Fe)_5Si_2O_{22}(OH)_2$
Amosite	Amphibole	$Fe_5Mg_2Si_2O_{22}(OH)_2$
Actinolite	Amphibole	$Ca_2(Mg,Fe)_5SigO_{22}(OH)_2$
Tremolite	Amphibole	$Ca_2Mg_5Si_2O_{22}(OH)_2$
Crocidolite	Amphibole	$Na_2Fe_2Mg_3Si_2O_{22}(OH)_2$

Table 2.19 Distribution of Asbestos
Applications for the Year 1988

Product	Percentage of Use
Roofing products	28
Friction products	26
Asbestos-cement pipe	14
Packing and gaskets	13
Paper	6
Other	13

Source: Reference 2.48.

In addition to the applications listed in Table 2.19, asbestos has been used in floor tiles, insulation, brake linings, rope, countertops, and fireproof clothing for firefighters. In the last 20–25 years, medical research has documented that inhaled asbestos fibers cause chronic lung damage, although ingested asbestos fibers are apparently harmless. The major threat to health by asbestos is now indoor exposure in buildings. As a result, federal and state regulations now require that asbestos insulation be removed from public buildings. Over the past several years, thousands of schools, hospitals, and office buildings have been closed temporarily so that asbestos could be removed by specially trained and outfitted contractors. Epidemiological studies have shown crocidolite as the most toxic; amosite displays intermediate toxicity; and chrysotile is the least toxic [2.49]. However, in real-world situations, asbestos is usually composed of a mix of fibers, as well as impurities such as heavy metals, which confounds generalizations of toxicity.

2.11 NUCLEAR WASTES

Although nuclear waste sites represent a small proportion of the number of hazardous waste sites and facilities, nuclear waste management and site remediation are extremely important in some regions of the United States. Radioactive waste management has traditionally been regulated by the Nuclear Regulatory Commission under the Atomic Energy Act of 1954. Because RCRA and CERCLA are the primary hazardous waste management laws (and are enforced by the EPA rather than the Nuclear Regulatory Commission), hazardous waste management has focused primarily on nonradioactive wastes. Although radioactive wastes are not part of RCRA lists or waste characteristics, mixed hazardous and nuclear wastes are regulated under RCRA and CERCLA, not the Atomic Energy Act. Therefore numerous areas contaminated with mixed wastes have become Superfund sites, including Rocky Flats, Colorado; Hanford, Washington; and Maxey Flats, Kentucky.

Radioactive wastes, like nonradioactive hazardous wastes, can be organic or inorganic. Nonradioactive hazardous wastes (e.g., solvents, pesticides, heavy metals) represent a health hazard because of the toxicity they induce by interacting with biological molecules in the body. For example, organophosphate insecticides interfere with nerve transmission by inhibiting the enzyme acetylcholinesterase. Radioactive wastes, on the other hand, pose a hazard by emitting ionizing radiation, which is a function of the element's nuclear properties. Therefore, an introduction to radioactivity and ionizing radiation is a prerequisite to understanding the hazards of radioactive wastes.

Radioactivity and Ionizing Radiation

Radioactivity, which results from the instability of the nuclei in some isotopes of atoms, is the release of particles or radiation from the nucleus of an atom. Nuclear chemistry provides a basis for the fundamental understanding of radioactivity. Rutherford proposed the nuclear structure of the atom in 1911, a concept still held, although the existence of smaller subatomic particles has been established. The fundamental model of atomic structure is based on a dense, positively charged nucleus with negatively charged electrons of negligible mass surrounding the nucleus. The nucleus is composed of protons and neutrons, which are usually (but not always) found in approximately equal numbers. The number of protons for a given element defines its elemental status (i.e., 6 protons makes carbon the element it is) and is also the element's *atomic number* (Z). If an element does not carry a charge, the number of electrons equals the number of protons. The number of neutrons (N) provides the remaining mass; therefore,

$$A = Z + N \tag{2.2}$$

where

A = the atomic weight of the element.

Because the weight of an electron is only 1/1836 that of a proton or neutron, the mass of electrons can be neglected in estimating the atomic weight, even for heavier elements in the periodic table. Species with the same number of protons but different numbers of neutrons are called *isotopes.* To illustrate the relationship between protons, neutrons, and electrons for the simplest element, the atomic structures of three hydrogen isotopes are shown in Figure 2.12.

The International Union of Pure and Applied Chemistry (IUPAC) has the following standard notation for elements:

$$^{\text{Atomic Weight}(A)}_{\text{Number of Protons}(Z)}\text{Element Symbol}$$

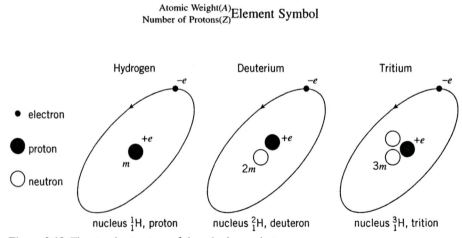

Figure 2.12 The atomic structure of three hydrogen isotopes.

For example, one isotope of lead is

$$^{204}_{82}Pb$$

which denotes that it has 82 protons (Z) and an atomic weight (A) of 204. By difference, this isotope of lead has $204 - 82 = 122$ neutrons (N).

The ratio of the number of neutrons (N) to the number of protons (Z) for an element is a measure of its stability. For stable light nuclei, $N/Z \approx 1$; for stable heavy nuclei, ratios of N/Z approach 2. Nuclei that are unstable based on deviations from these neutron-to-proton ratios may go through a natural process of decay by emission of radiation or particles that ultimately result in their stability. The natural decay process results in radioactivity, and the chemicals that are radioactive are known as *radioisotopes* or *radionuclides*. Alternatively, some isotopes are stable and, as a result, do not emit radiation or particles. Some of the more common stable and radioactive isotopes for elements with low atomic numbers are listed in Table 2.20. Many of these isotopes are not only used in industry but also in research as tracers. For example, ^{14}C is used to track the transformation of organic compounds when they degrade chemically and biochemically. Stable isotopes, such as ^3H and ^{13}C, can be differentiated from ^1H and ^{12}C by mass spectrometry. Innovative uses of carbon labeling have been used to evaluate the behavior and transformation of xenobiotic chemicals.

Three types of radiation are emitted from nuclei; the kind that emanates from a particular radioisotope is based on its nuclear characteristics. High atomic weight compounds [those with Z greater than 82 or atomic weight (A) greater than 200] are characteristically unstable and naturally decompose to lower molecular weight elements (with corresponding lower Z). An example of a complex decay process, the transition of ^{238}U to stable ^{206}Pb, is shown in Figure 2.13. This series of decay events documents that radioactivity is a complex process that may produce a series of progeny elements, possibly over thousands of years.

Table 2.20 Stable and Radioactive Isotopes of Selected Elements

Element	Isotopes					
Hydrogen	^1_1H	^2_1H	^3_1H			
	Stable	Stable	Stable			
Carbon	$^{11}_6C$	$^{12}_6C$	$^{13}_6C$	$^{14}_6C$		
	Radioactive	Stable	Stable	Radioactive		
Nitrogen	$^{12}_7N$	$^{13}_7N$	$^{14}_7N$	$^{15}_7N$	$^{16}_7N$	
	Radioactive	Stable	Stable	Stable	Radioactive	
Oxygen	$^{14}_8O$	$^{15}_8O$	$^{16}_8O$	$^{17}_8O$	$^{18}_8O$	$^{19}_8O$
	Radioactive	Radioactive	Stable	Stable	Stable	Radioactive
Phosphorus	$^{29}_{15}P$	$^{30}_{15}P$	$^{31}_{15}P$	$^{32}_{15}P$	$^{33}_{15}P$	
	Radioactive	Radioactive	Stable	Radioactive	Radioactive	
Chlorine	$^{34}_{17}Cl$	$^{35}_{17}Cl$	$^{36}_{17}Cl$	$^{37}_{17}Cl$	$^{38}_{17}Cl$	
	Radioactive	Stable	Radioactive	Stable	Radioactive	

Source: Reference 2.50.

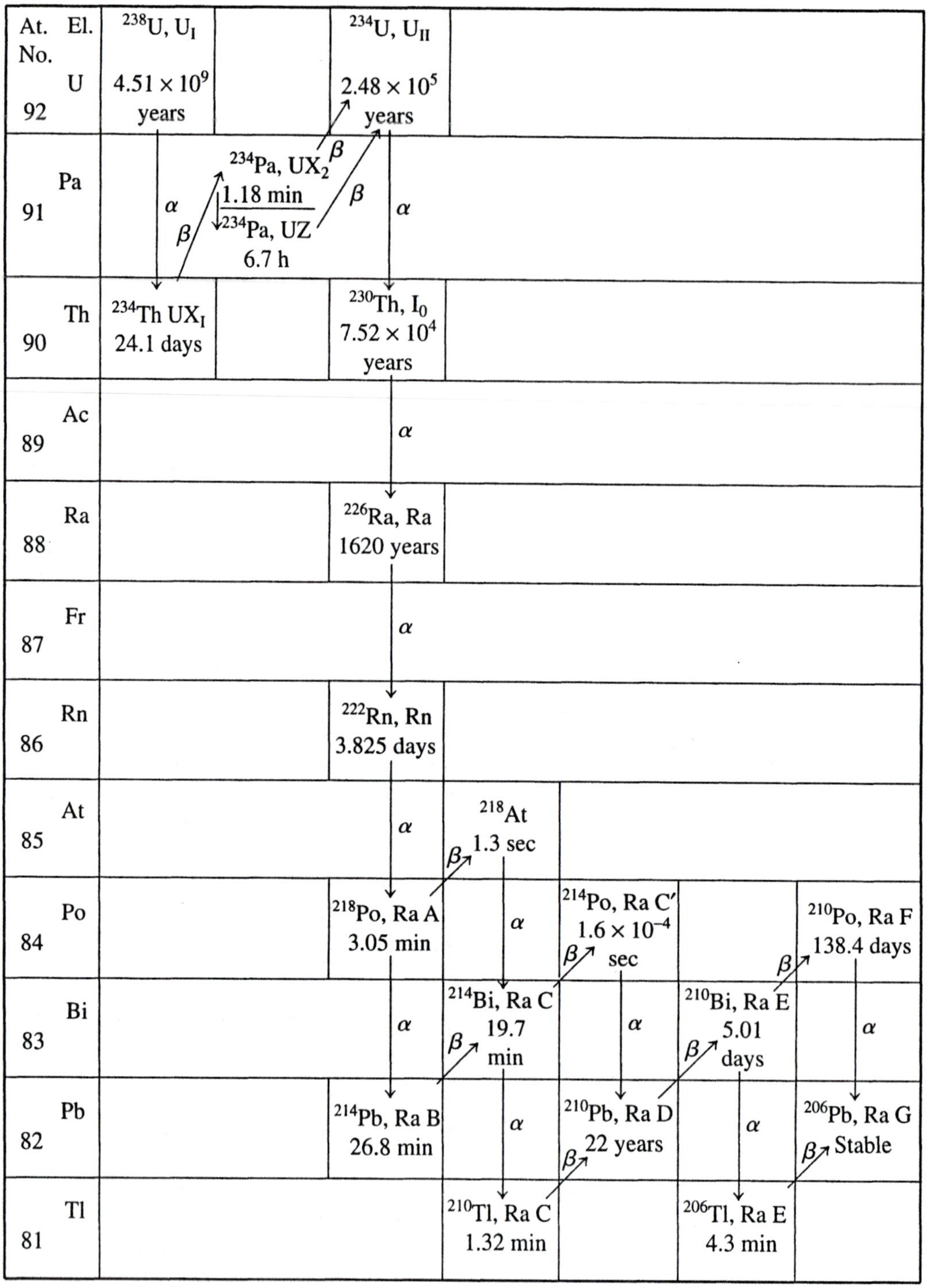

At. No. El.							
U 92	^{238}U, U$_I$ 4.51×10^9 years		^{234}U, U$_{II}$ 2.48×10^5 years				
Pa 91	α β	^{234}Pa, UX$_2$ β 1.18 min ^{234}Pa, UZ 6.7 h	β α				
Th 90	^{234}Th UX$_I$ 24.1 days		^{230}Th, I$_0$ 7.52×10^4 years				
Ac 89			α				
Ra 88			^{226}Ra, Ra 1620 years				
Fr 87			α				
Rn 86			^{222}Rn, Rn 3.825 days				
At 85			α	^{218}At 1.3 sec β			
Po 84			^{218}Po, Ra A 3.05 min	α	^{214}Po, Ra C$'$ 1.6×10^{-4} sec β		^{210}Po, Ra F 138.4 days β
Bi 83			α	^{214}Bi, Ra C 19.7 min β	α	^{210}Bi, Ra E 5.01 days β	α
Pb 82			^{214}Pb, Ra B 26.8 min	α	^{210}Pb, Ra D 22 years β	α	^{206}Pb, Ra G Stable β
Tl 81				^{210}Tl, Ra C 1.32 min		^{206}Tl, Ra E 4.3 min	

Figure 2.13 Decay process for ^{238}U to the stable element ^{206}Pb; individual steps involved in the release of α or β particles. *Source:* Reference 2.50.

One of the natural decay processes that commonly occurs is the loss of a nuclear fragment containing two protons and two neutrons, which results in transformation of the parent element to a new element of lower atomic weight (by 4 units) and atomic number (by 2 units):

$$_{Z}^{A}X \rightarrow _{Z-2}^{A-4}X + _{2}^{4}He \qquad (2.3)$$

The two proton–two neutron entity is called an *alpha* (α) *particle,* and is equivalent to the nucleus of a helium atom. An emission process that involves the release of an alpha particle is the decay of $_{92}^{234}U$ to $_{90}^{230}Th$:

$$_{92}^{234}U \rightarrow _{90}^{230}Th + _{2}^{4}He \qquad (2.4)$$

Alpha particles are characterized by a high degree of ionizing action but do not travel more than 10 cm in air and are stopped by thin barriers such as a sheet of paper.

A second mechanism by which neutron-to-proton ratios are lowered to achieve stability is through the decay of a neutron into a proton and an electron. The electron, known as a *beta* (β) *particle,* is emitted and the proton stays in the nucleus to increase the stability of the radionuclide, raising its atomic number by one and creating a new element:

$$_{Z}^{A}X \rightarrow _{Z+1}^{A}X + \beta^{-} \qquad (2.5)$$

Beta particles are negatively charged and travel at 30 to 99% of the speed of light. Their velocity varies with the element from which they are emitted. Although they travel hundreds of meters in air, they are stopped by a moderate barrier such as a sheet of aluminum foil.

The emission of α or β particles brings about a change in the nuclear structure and composition of the parent element, resulting in the formation of new progeny elements. The emission of *gamma* (γ) *radiation,* however, is synomomous with the release of energy. Gamma radiation is emitted when an unstable nucleus decays to a more stable state. The nucleus maintains its same proton-neutron composition, but the excess energy radiates from the nucleus as gamma rays. Gamma radiation is true electromagnetic radiation that travels at the speed of light. It readily penetrates most solids (including humans) and can be stopped only by thick inorganic solids such as a meter of concrete or several centimeters of lead. Gamma radiation often accompanies the emission of α and β particles.

Similar to γ radiation, *x-rays* are high-energy electromagnetic radiation. Gamma rays result from the transition from one nuclear exited state to another within the nucleus. X-rays, on the other hand, result from the transition of electrons from a higher to a lower orbital or energy state outside the nucleus. Because penetrating power increases as a function of frequency, x-rays penetrate far less than most γ-rays.

Lifetimes for the decay process of different radionuclides range from 10 seconds to 10^{18} years [2.51]. Half-lives of common radioactive elements and the accompanying decay species are listed in Table 2.21. The transformation of an unstable radioactive element by the emission of an α particle and a β particle is illustrated in Example 2.13.

Table 2.21 Half-Lives of Common Radioactive Isotopes

Atomic Number	Nuclide	Half-Life	Nature of Radiation
1	^{3}H	12.3 yr	β
6	^{14}C	5730 yr	β
11	^{24}Na	15.0 h	β, γ
15	^{32}P	14.3 days	β
16	^{35}S	88 days	β
19	^{40}K	1.28×10^9 yr	β
27	^{60}Co	5.3 yr	β, γ
35	^{78}Br	6.4 min	β, γ
38	^{90}Sr	28.1 yr	β
53	^{131}I	8.0 days	β, γ
55	^{137}Cs	30 yr	β
88	^{226}Ra	1600 yr	α, γ
92	^{238}U	4.51×10^9 yr	α

Source: Reference 2.50.

EXAMPLE 2.13   *Decay of a Radioisotope by Emission of α and β Particles*

An element $^{232}_{104}$X decays by the emission of an α particle followed by loss of a β particle. Describe the reactions that takes place in terms of IUPAC element notation.

SOLUTION

Element $^{232}_{104}$X first emits an α particle (2 protons and 2 neutrons = atomic mass of 4). Therefore, the first decay proceeds as

$$^{232}_{104}X \rightarrow ^{228}_{102}X + ^4_2He$$

The second decay occurs through the emission of a β particle:

$$^{228}_{102}X \rightarrow ^{228}_{103}X + \beta^-$$

Radioactive Waste Problems and Management

Nuclear wastes are usually segregated based on their potential hazard, but no strict quantitative standards have been developed to classify the wastes that are generated from their wide range of applications. The most common scheme used to classify nuclear wastes is as high-level, transuranic, and low-level. *High-level wastes* (HLWs), which have a high degree of activity measured in curies and also generate large amounts of heat, result primarily from fuel-rod reprocessing and the spent fuel from nuclear reactors. *Transuranic* wastes contain isotopes with atomic numbers higher than uranium. These elements, which include plutonium and americium, are characterized by relatively low radioactivity but long half-lives and must be stored and isolated for thousands of years. Transuranic wastes are produced primarily from nuclear weapons production, warhead destruction, and nuclear power generation. *Low-level wastes* (LLWs) are generated in large volumes by hospitals, universities, and industry, and are characterized by minimal radioactivity.

Radioactive wastes originate from a variety of sources, including (1) mining and milling activities, (2) commercial sources (hospitals, universities, etc.), (3) nuclear reactors, and (4) nuclear weapons facilities. In terms of volume, the largest production of radioactive wastes have originated from mining and milling operations, with 60 m^3 of tailings produced per megawatt of power generated. These ores (which are not regulated under RCRA) release ^{222}Rn into the atmosphere for thousands of years after the tailings are stockpiled.

Many radioisotopes, such as ^{14}C, ^{32}P, and ^{131}I, have been used extensively in industry and research as tracers. Applications have included using ^{14}C to track carbon pathways in industrial processes such as petroleum refining and in biochemical transformations and chemical synthesis. In hazardous waste remediation research, biological or physicochemical transformations are often confirmed using ^{14}C-labeled hazardous compounds. The radioisotope ^{32}P has been used extensively in biochemical and medical research because phosphorus is a component of certain classes of biological molecules, such as the nucleic acids in chromosomes and energy-rich intermediates in cells. Research on thyroid metabolism has used ^{131}I to track iodine uptake and excretion. The wastes that have been generated from commercial and academic activities are generally diluted with water and are therefore classified as low-level. They are usually disposed of in the low-level nuclear waste landfills listed in Table 2.22. Burial of drums containing low-level nuclear wastes at Hanford, Washington, is illustrated in Figure 2.14.

Another problem in nuclear waste management is mixed wastes. Nuclear weapons facilities, such as those at Hanford, Washington, and Rocky Flats, Colorado, as well as private and municipal radioactive waste generators, have produced wastes that contain

Table 2.22 Disposal Sites for Commercial Low-Level Radioactive Wastes

Disposal Site	Years in Operation	Cumulative Amounts in 10^3 m^3	Approximate Percentage of Total Production
Barnwell, SC	1971–present	583	45
Hanford, WA	1965–present	306	24
Maxey Flats, KY	1963–1977	136	10
Beatty, NV	1962–present	113	9
Sheffield, IL	1967–1978	87.8	7
West Valley, NY	1963–1975	70.8	5
Total		1296.6	100

Source: Reference 2.52.

Figure 2.14 Burial of low-level nuclear waste at the Hanford nuclear facility. *Source:* U.S. Department of Energy, Richland, WA Office.

radioisotopes mixed with solvents, aromatic compounds, metals, and other nonradioactive but hazardous chemicals. These wastes represent difficult disposal challenges. In general, mixed wastes are being stored while long-term radioactive disposal facilities are being developed.

Nuclear power plants employ uranium and plutonium (^{233}U, ^{235}U, and ^{239}Pu) as fuel for nuclear fission. After sufficient fuel is used to achieve a critical mass (i.e., enough fuel to initiate a nuclear reaction), the rate of fission is controlled by absorbing some of the neutrons emitted with neutron absorbers such as cadmium, graphite, or D_2O (heavy water).

Over 100 nuclear reactors are currently in operation in the United States, and each produces about 10 m^3 of spent fuel annually. High-level radioactive wastes from nuclear reactors are usually generated by the reprocessing of commercial used fuel. The nuclear waste is aqueous and results from an extraction system for reprocessing irradiated reactor fuels. The waste may also include sludge, precipitates, or other solid products resulting from the concentration processes. These materials are characterized by high decay energy and require correspondingly high levels of shielding.

Although the high-level wastes produced from nuclear power plants are a serious concern, the materials that have been generated from nuclear weapons production in the United States have resulted in the most serious radioactive contamination outside of the Soviet Union. These nuclear weapons disposal problems are as old as the Manhattan Project of the early 1940s. The production of nuclear warheads required enormous industrial plants that used or produced uranium, plutonium, and tritium. Fourteen weapons-related plants in 12 states were constructed in the 1940s and 1950s, particularly after the Soviet Union detonated a nuclear warhead in 1949, initiating a 40-year Cold War arms race. The locations of these sites are shown in Figure 2.15, and a summary of their functions in weapons production and their waste management problems is listed in Table 2.23.

The production of plutonium has been an inefficient process. One kilogram of weapons-grade plutonium generates 1420 L (375 gal) of high-level radioactive waste and 230,000 L (60,700 gal) of low-level waste. Most of the high-level nuclear waste (96% by volume and 92% by radioactivity) in the United States is located at the Hanford and Savannah River weapons production sites (Figure 2.16).

The Hanford site, located near Richland in south-central Washington, is a 200-square-mile desert reservation that originally produced the plutonium for the 1945 Nagasaki atomic bomb. Plutonium production stopped in 1987 because of safety concerns about the site's N reactor. Hanford's nuclear and hazardous waste problems are varied and widespread, and waste management was often handled on a "fire drill" basis (Figure 2.17). These tanks were managed by cascading the wastes from tank to tank. If a tank was filled, the waste would be transferred to another tank in order to provide a free one. Radioactive wastes were also discharged directly to soils. Scavenging of wastes provided other mixed waste problems. The wastes were mined (or scavenged) for uranium, and the highly radioactive waste left was sent to soil pits, or cribs. The EPA estimated that 127 million gallons of radioactive wastes were discharged to Hanford soils [2.53].

Soil and groundwater at Hanford have also been contaminated by plutonium, tritium, carbon tetrachloride, TCE, cyanide, mercury, and lead. In addition, wastes in

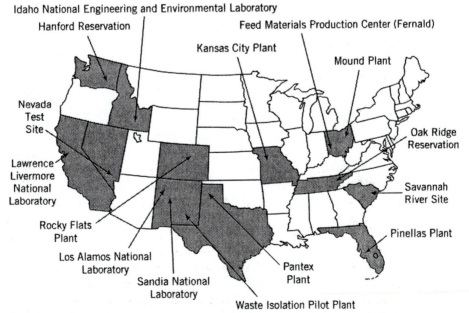

Figure 2.15 Locations of nuclear weapons facilities throughout the United States. *Source:* Reference 2.51.

Table 2.23 Summary of Contamination at U.S. Nuclear Weapons Sites

Site and Location	Function	Site Contamination
Hanford (near Richland, Washington)	Production of plutonium	Soil and groundwater contamination; leaking tanks containing high-level liquid and semisolid wastes
Savannah River (near Aikin, South Carolina)	Production of tritium and plutonium	Aging production reactors; leaking steel tanks; groundwater contamination with chlorinated solvents
Idaho National Engineering Laboratory (near Idaho Falls, Idaho)	Reprocessing of naval reactor fuel to recover uranium	Soil and groundwater contamination with carbon tetrachloride, TCE, and tritium
Fernald (near Cincinnati, Ohio)	Manufacture of uranium ingots, rods and tubes for other weapons production facilities	Soil and groundwater contamination by radioactive and nonradioactive waste chemicals; uranium dust has blown off-site
Rocky Flats (near Denver, Colorado)	Production of plutonium and beryllium weapons components	Soil and groundwater contamination from spills and accidents; potential contamination of Denver drinking water supply

Mound Plant (Miamisburg, Ohio)	Manufacture of small weapons components and radioisotopic batteries	Contaminated soils, groundwater, and surface waters; potential contamination of a sole-source aquifer
Pinellas Plant (St. Petersburg, Florida)	Manufacture of weapons neutron triggers using tritium	Discharges of tritium, lead, and chromium; potential groundwater contamination
Kansas City Plant	Manufacture of electronic and mechanical weapons components	Releases of uranium, cadmium, and lead; soil and groundwater contamination
Pantex Plant (Amarillo, Texas)	Final assembly of nuclear weapons, recycling old warheads	Radioactive and nonradioactive contamination of lagoons, landfills, storage areas, and buildings
Oak Ridge National Laboratory (Oak Ridge, Tennessee)	Manufacture of weapons components, recycling obsolete weapons	Releases of low-level and mixed wastes
Lawrence Livermore Laboratory (Livermore, California)	Weapons research and development	Soil and groundwater contamination by hydrocarbons, lead, and chromium
Los Alamos National Laboratory (Los Alamos, New Mexico)	Weapons research and development	Groundwater wells have shown low levels of contamination
Nevada Test Site (near Las Vegas, Nevada)	Above- and below-ground nuclear weapons testing	Widespread radioactive soil contamination from decades of weapons testing
Sandia Waste Isolation Pilot Plant	Produced plutonium and beryllium components	Soil and groundwater contamination resulting from accidents and spills

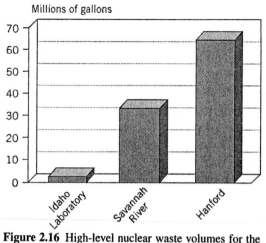

Figure 2.16 High-level nuclear waste volumes for the three largest sites. *Source:* Reference 2.51.

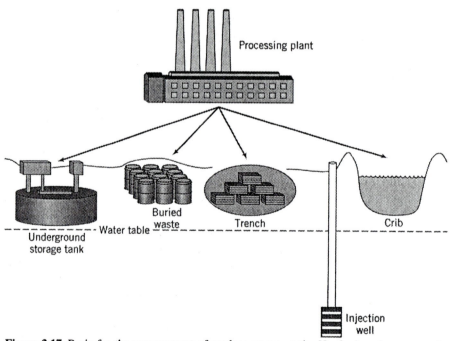

Figure 2.17 Basis for the management of nuclear wastes at the Hanford nuclear reservation. *Source:* Reference 2.52.

large tanks have been a chronic problem at Hanford. A total of 149 older single-walled tanks, which have developed numerous leaks, have been replaced by new tanks; 28 new double-walled tanks have recently been installed. In May 1989, the EPA, the State of Washington, and the Department of Energy signed an agreement on the cleanup of Hanford. The Hanford reservation is now a Superfund site.

The Savannah River nuclear materials production site is located on 192,000 acres near Aiken, South Carolina. Weapons-grade plutonium and tritium were produced there since the 1940s; however, the reactors that generate plutonium were shut down in 1988 due to safety concerns. The Department of Energy is keeping the K Reactor in operation to produce tritium—a more short-lived isotope.

The contamination by both nuclear and hazardous wastes at the Savannah River site is widespread, and the cleanup efforts represent complex challenges. Ten of the fifty-one large steel tanks used to store high-level nuclear wastes have developed leaks, resulting in the contamination of thousands of cubic meters of soil. Chlorinated solvents, which have been disposed of on soils or which leaked from the tanks, have migrated to the Tuscaloosa aquifer. Contaminated groundwater is being treated using pump-and-treat technologies (see Chapter 13).

Over 2.64×10^8 gallons of high-level nuclear wastes, which are currently stored in underground tanks, will be vitrified (solidified into a glasslike material by passing a strong electrical current through the wastes). To accomplish this complex task, a new Defense Waste Processing Facility is under construction for waste vitrification. The glass cylinders will eventually be permanently stored when a nuclear waste repository (e.g., Yucca Mountain, Nevada) has been constructed.

2.12   INFORMATION SOURCES ON CONTAMINANT NOMENCLATURE, STRUCTURE, AND PROPERTIES

Although the compounds covered in Sections 2.3 through 2.11 are those most commonly encountered in hazardous waste management, tens of thousands of other hazardous chemicals exist for which the hazardous waste professional may need to obtain information. The first step in obtaining information on contaminants is to determine the chemical structure and other relevant nomenclature information.

The most widely accepted first-cut reference on chemical structure, nomenclature, industrial uses, and other general properties is the *Merck Index,* a chemical encyclopedia [2.54]. The *Merck Index* lists compounds alphabetically, usually by common name, trade name, or acronym. Each entry is given a number in recent editions, and a comprehensive cross-reference index at the back of the *Merck Index* provides an extensive system for determining its location in the book, regardless of the nomenclature system. For instance, the herbicide bromacil may be found under the trade name bromacil or under the IUPAC name 2-bromo-3-*sec*-butyl-6-methyluracil. The insecticide DDT may be found in the cross-reference list under the common name DDT or the IUPAC name 1,1,1-trichloro-2,2-*bis*-(*p*-chlorophenyl)ethane. Both indicate the number in the body of the book under which information on DDT can be found.

The entry for DDT from the *Merck Index* is shown in Table 2.24, and lists information for its IUPAC nomenclature, boiling point, melting point, density, refractory index, and route of chemical synthesis. The entry also provides brief information on industrial, commercial, and agricultural uses. The *Merck Index* therefore provides a good

Table 2.24 Entry for DDT from the *Merck Index*

2898. DDT. *1,1'-(2,2,2-Trichloroethylidene)bis[4-chlorobenzene]: 1,1,1-trichloro-2,2-bis(p-chlorophenyl)ethane; α,a-bis(p-chlorophenyl)-β,β,β-trichlorethane;* dichlorodiphenyltrichloroethane; chlorophenothane; clofenotane; dicophane; pentachlorin; *p,p'*-DDT; Agritan; Gesapon; Gesarex; Gesarol; Guesapon; Neocid. $C_{14}H_9Cl_5$, mol wt 354.49. C 47.44%, H 2.56%, Cl 50.01%. Polychlorinated nondegradable pesticide. Prepn: Zeidler, *Ber.* **7**, 1180 (1874); P. Müller, **U.S. pat. 2,329,074** (1944 to Geigy); Rueggeberg, Torrans, *Ind. Eng. Chem.* **38**, 211 (1946); Cook *et al., ibid.* **39**, 868, 1683 (1947). Convenient lab procedures: Bailes, *J. Chem. Ed.* **22**, 122 (1945); Ginsburg, *Science* **108**, 339 (1948). Large scale production: Mosher *et al. Ind. Eng. Chem.* **38**, 916 (1946). Chemical composition of technical grade: H. L. Haller *et al., J. Am. Chem. Soc.* **67**, 1591 (1945). Activity: P. Müller, *Helv. Chim. Acta* **29**, 1560 (1946). Comprehensive monograph (in English and German): *DDT.* P. Müller, Ed., 3 vols (Birkhäuser Verlag, Basel and Stuttgart, 1955). Toxicity data: Gaines, *Toxicol. Appl. Pharmacol.* **2**, 88 (1960); **14**, 515 (1969). Review of toxicology and human exposure: *Toxicological Profile for 4.4'-DDT, 4,4'-DDE, 4,4'-DDD* (PB95-100137, 1994) 192 p.

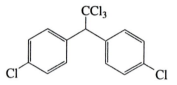

Biaxial elongated tablets, needles from 95% alc. mp 108.5–109°, uv max (95% alc): 236 nm. Vapor pressure at 20° = 1.5×10^{-7} mm Hg. Practically insol in water, dil acids, alkalies. Soly (g/100 ml): acetone 58; benzene 78; benzyl benzoate 42; carbon tetrachloride 45; chlorobenzene 74; cyclohexanone 116; 95% alc 2; ethyl ether 28; gasoline 10; isopropanol 3; kerosene 8–10; morpholine 75; peanut oil 11; pine oil 10–16; tetralin 61; tributyl phosphate 50; freely sol in pyridine, dioxane. The soly in organic solvents increases sharply with a rise in temp. Resistant to destruction by light and oxidation. Its unusual stability has resulted in difficulties in residue removal from water, soil and foodstuffs. Should not be kept in iron containers and should not be mixed with iron and aluminum salts nor with alkaline substances. High storage temps should also be avoided. Setting point of technical grade: 88.6–91.4°. LD_{50} in male, female rats (mg/kg): 113, 118 orally (Gaines, 1960).

Caution: Poisoning may occur by ingestion or by absorption through skin or respiratory tract. Potential symptoms of overexposure are numbness, paresthesias, malaise, headache, sore throat, fatigue, weakness, coarse tremors, convulsions and coma; death due to respiratory failure. *See Clinical Toxicology of Commercial Products,* R. E. Gosselin *et al.,* Eds. (Williams & Wilkins, Baltimore, 5th ed., 1984) Section III, pp. 134–138. This substance may reasonably be anticipated to be a carcinogen: *Seventh Annual Report on Carcinogens* (PB95-109781, 1994) p. 139.

USE: Contact insecticide.

THERAP CAT: Ectoparasiticide; pediculicide.

Source: Reference 2.54.

introductory source of information for nearly all environmental contaminants. However, the *Merck Index* is somewhat lacking in detailed quantitative information for physical and chemical properties (e.g., water solubility); other references with more extensive information on contaminant properties are discussed in Chapter 3.

2.13 SUMMARY OF IMPORTANT POINTS AND CONCEPTS

- Because of the thousands of organic chemicals classified as a hazardous waste or hazardous substance, knowledge of organic nomenclature is essential to understanding hazardous waste management.

- Two major divisions of organic compounds are aliphatic and aromatic chemicals. Important organic chemical classes that are ubiquitous hazardous wastes include

Aliphatics	Esters
Alkanes	Cycloalkanes
Alkenes	Monocyclic Aromatics
Alcohols	Polycyclic Aromatic Hydrocarbons (PAHs)

- Petroleum and its refined products contain hundreds of different organic compounds classed as alkanes, alkenes, cycloalkanes, and aromatics. One of the most common sources of petroleum contamination has been leaking underground storage tanks (USTs).

- A wide range of chemicals has been used as solvents, including petroleum distillates, benzene derivatives, alcohols, esters, ethers, ketones, chlorinated alkanes, and chlorinated alkenes. Halogenated solvents have become common groundwater pollutants due to their high densities, relatively high water solubilities, and long environmental peristence.

- Pesticides are agents that destroy insects, plants, fungi, and other pest organisms, and include a wide range of toxic and biorefractory organic compounds. They have been disposed of at large chemical manufacturing sites where they are synthesized and, more commonly, at small pesticide mixing and formulation areas.

- Polychlorinated biphenyls have been used primarily as transformer oils. They are persistent compounds that are highly hydrophobic.

- Polychlorinated dibenzodioxins and dibenzofurans are impurities formed during industrial synthesis and thermal processes. They are highly hydrophobic and lipophilic, and some exhibit serious acute and chronic toxicity.

- Heavy metals commonly present as hazardous wastes include arsenic, cadmium, chromium, lead, mercury, and nickel. Most of these metals, as well as cyanide, are used in electroplating, and plating wastes are a major disposal problem.

- Although nuclear wastes are not regulated under RCRA, weapons disposal sites contaminated with mixed hazardous and radioactive wastes are being regulated increasingly under CERCLA. Hanford, Washington, and Savannah River, South Carolina, are the most contaminated of the 14 nuclear weapons facilities in the United States.

- The best starting source of information on the nomenclature, structure, and industrial uses of most chemicals found in hazardous wastes is the *Merck Index*.

PROBLEMS

2.1. Name the following hydrocarbons using IUPAC rules.

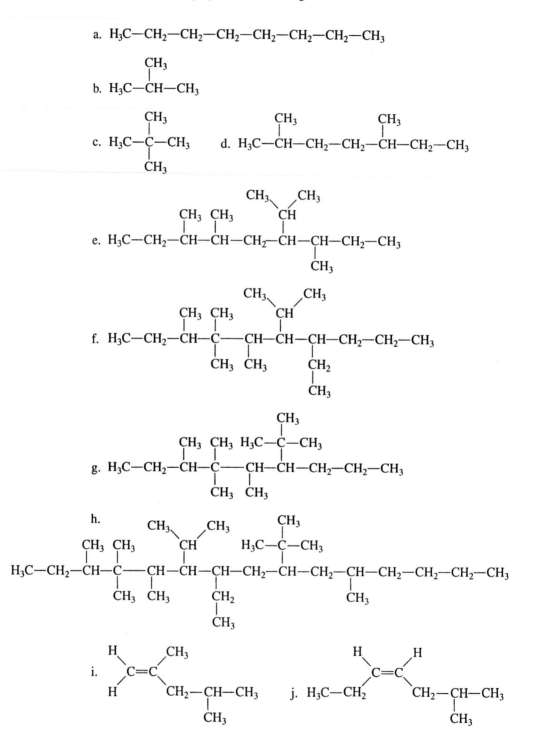

k.

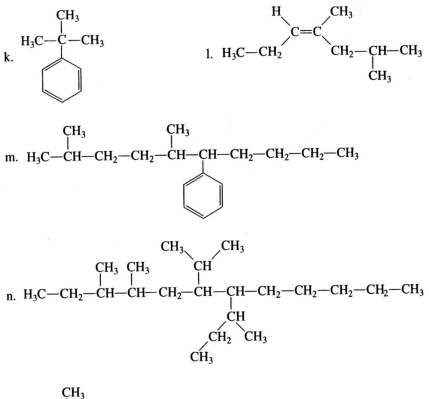

l. H_3C-CH_2 ... $CH_2-CH-CH_3$

m. $H_3C-CH-CH_2-CH_2-CH-CH-CH_2-CH_2-CH_2-CH_3$

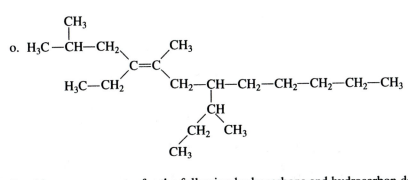

n. $H_3C-CH_2-CH-CH-CH_2-CH-CH-CH_2-CH_2-CH_2-CH_2-CH_3$

o. $H_3C-CH-CH_2$ ...

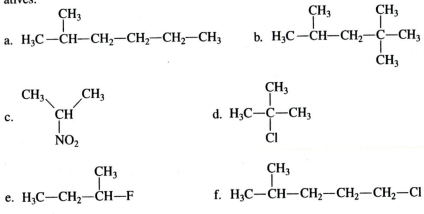

2.2. Provide common names for the following hydrocarbons and hydrocarbon derivatives.

a. $H_3C-CH-CH_2-CH_2-CH_2-CH_3$

b. $H_3C-CH-CH_2-C-CH_3$

c.

d. $H_3C-C-CH_3$

e. H_3C-CH_2-CH-F

f. $H_3C-CH-CH_2-CH_2-CH_2-Cl$

2.3. Name the following aromatic compounds.

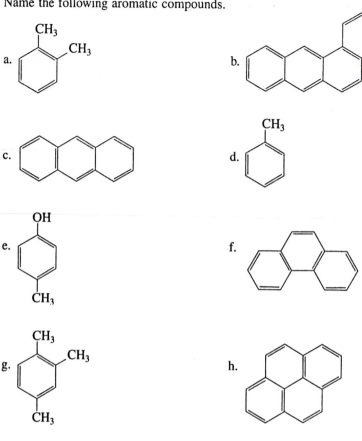

2.4. Draw the structures of the following hydrocarbons and hydrocarbon derivatives.
a. Benzene
b. *n*-Butylbenzene
c. Hexadecane
d. 2,2-Dimethylpentane
e. Isopentane
f. Isooctane
g. 3-Ethyl-2-methylheptane
h. 4-Isopropyl-3,5-dimethylnonane
i. 6-*n*-Butyl-6-ethyl-7-isopropyldodecane
j. 5,7-Di-*sec*-butyl-4-ethyl-3,4,6-trimethylundecane
k. 4-Methyl-1-pentene
l. 4-Ethyl-3,5-dimethyl-*cis*-3-heptene
m. *o*-Cresol
n. *sec*-Butylbenzene
o. *p*-Xylene
p. 1,2,4-Trimethylbenzene
q. 5-Isobutyl-5-phenyl-2,4,6-trimethylundecane
r. Anthracene
s. Chrysene
t. Benzo[*a*]pyrene
u. Picene
v. 1-Aminonaphthalene
w. 2,4,7-Trimethylnaphthalene
x. 3,5,10,12-Tetramethylchrysene
y. Benzo[*g,h,i*]perylene
z. Indeno[1,2,3-*c,d*]pyrene

2.5. Name and number the following polycyclic aromatic hydrocarbons using IUPAC rules.

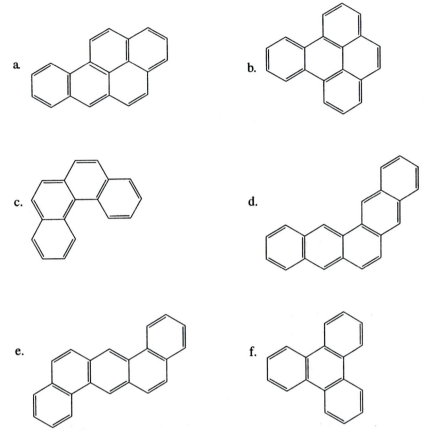

a.

b.

c.

d.

e.

f.

2.6. Describe the processes that convert crude oil into petroleum products.

2.7. Describe the differences between crude oil, gasoline, diesel, and lubricating oils.

2.8. A hazardous waste site has received gasoline contamination. A new person in your office looks at the BTEX and TPH analyses that have come back from the laboratory and asks three questions:
   a. "What is in gasoline?"
   b. "What do BTEX and TPH stand for?"
   c. "Does BTEX measure everything in gasoline?"
   Provide answers to your colleague's questions.

2.9. During many site assessment and remediation efforts involving gasoline, BTEX is used to monitor the gasoline rather than quantifying every chemical. Name and draw the chemical structures of the six chemicals that make up BTEX.

2.10. For the compounds shown in Problem 2.1, list the petroleum fractions to which each potentially belongs.

2.11. For the compounds shown in Problem 2.2, list the petroleum fractions to which each potentially belongs.

2.12. The following compounds have been identified as biological metabolites of *n*-hexane:

a. *n*-Hexanol
b. 1-Hexanal
c. Hexanoic acid
d. 2-Hexanol
e. Methyl-*n*-butyl ketone
f. 2,5-Hexanediol
g. 5-Hydroxy-2-hexanone
h. 2,5-Hexanedione

Draw the chemical structure of each of the metabolites.

2.13. Draw the structures of

a. 4-*sec*-Butyl-2,3,5-trimethyl-4-isopropylnonane
b. *n*-Hexyl chloride
c. Isopropyl phenyl ketone
d. 2,4,6-Trinitrophenol
e. *trans*-2-Pentene
f. Isopropyl acetate
g. Methyl *n*-butonate
h. Ethyl isopentanoate

2.14. Name the following chlorinated aliphatic compounds using IUPAC nomenclature.

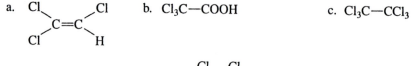

a. (structure) b. $Cl_3C\!-\!COOH$ c. $Cl_3C\!-\!CCl_3$

d. (structure) e. $H_3C\!-\!\overset{\displaystyle Cl}{\underset{\displaystyle |}{CH}}\!-\!\overset{\displaystyle Cl}{\underset{\displaystyle |}{CH}}\!-\!CH_2Cl$ f. $Cl_3C\!-\!CH_3$

2.15. Draw the structure of the following chlorinated aliphatic compounds.

a. *trans*-1,2-Dichloroethylene
b. PCE
c. TCA
d. TCE
e. 1,1,2,3-Tetrachlorobutane
f. Pentachloroethane

2.16. Name the following chlorinated aromatic compounds.

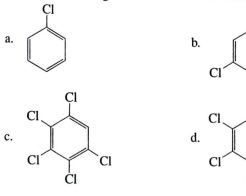

a.

b.

c.

d.

2.17. Draw the structure of the following chlorinated aromatic compounds.

a. Chlorobenzene
b. *p*-Dichlorobenzene
c. *m*-Dichlorobenzene
d. 2-Chlorophenol
e. 1,2,4,5-Tetrachlorobenzene
f. 2,4,6-Trichlorophenol
g. Hexachlorobenzene

2.18. List the family to which each of the following insecticides belongs.

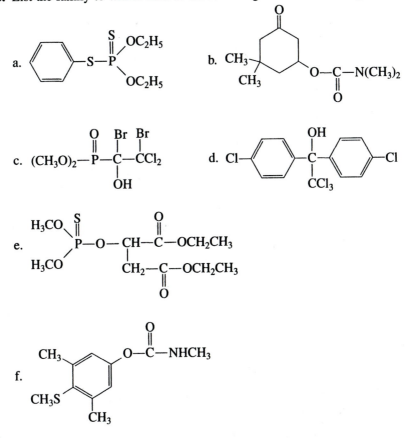

2.19. List the families to which each of the following herbicides belong.

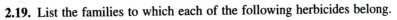

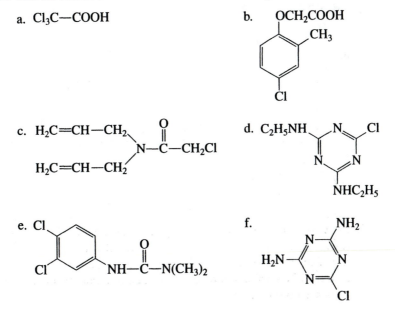

2.20. Draw the structures of the compounds that constitute Agent Orange.

2.21. A pesticide formulation and rinse site operated from 1946 through 1968. There were reports that approximately 200 kg (550 pounds) of insecticides may have been spilled at the site. Based on the types of insecticides that were used at that time:

a. Provide the common names for three possible insecticides that were used.

b. Would you expect residues of these chemicals to persist in soil today?

2.22. Describe, using chemical structures, the distinguishing characteristics that cause these pesticides to be classified into the families in which they are found.

a. DDT   b. Methoxychlor

c. Endosulfan   d. Dieldrin

e. Malathion   f. Carbaryl

g. Atrazine   h. Glyphosate

i. Diuron   j. Pentachlorophenol

k. Mancozeb

2.23. Name three uses of benzene.

2.24. Name the following industrial intermediates.

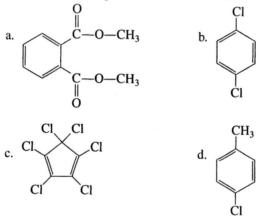

2.25. The following aromatics have been used as precursors in the industrial synthesis of dyes, pharmaceuticals, perfumes, and other chemicals. Draw the structures of each.

a. Isobutylbenzene   b. *p*-Ethyltoluene

c. 1,4-Benzenediol   d. 2-Phenyl-*cis*-2-butene

e. *n*-Propylbenzene   f. 2-Methyl-3-phenylpentane

g. *p-tert*-Butyltoluene

2.26. A medium-sized city is rezoning an industrial area for potential residential development. The light industries in the area, which have been in operation since 1924, have been known to dispose of their waste casually prior to the promulgation of RCRA. As a consulting environmental scientist/engineer, you have been called upon in a city council meeting (cold turkey—no preparation) to provide the city with potential hazardous waste problems from each industry. You get a few minutes to collect your thoughts because the mayor has called for a 5-

minute break just before you are to speak. During the break, you have a chance to jot down some information to tell the council. For each of the following industries, list a potential hazardous waste and a possible way it was released to the environment.
a. Semiconductor manufacturing
b. Dry cleaner
c. Fuel farm
d. Crop dusting operation
e. Paint stripping and furniture refinishing

2.27. A large dry cleaning operation disposed of its waste in soil pits. What hazardous chemicals would you expect to encounter at the site?

2.28. By what Aroclor number would each of the following PCBs be characterized?
a. 2,4′-Dichlorobiphenyl
b. 2,3,3′,4′-Tetrachlorobiphenyl
c. 2,3,3′,4,4′,5-Hexachlorobiphenyl
d. 2,2′,3,4,4′,5,6,6′-Octachlorobiphenyl

2.29. A series of three USTs, just a few feet apart, have each leaked benzene, chlorobenzene, and *p*-dichlorobenzene. Which compound would persist the longest and why?

2.30. You are the team leader for a preliminary site assessment of an industry that manufactures chlorinated cyclodiene insecticides. What chemicals would you expect to find at the site?

2.31. Name the following PCBs.

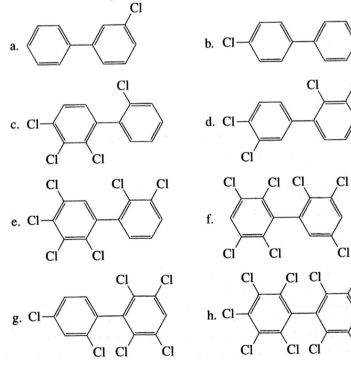

2.32. For the PCBs of Problem 2.31, what would be the Aroclor number if each were present as a single-component material (which is not common)?

2.33. Why is a commercial formulation of PCBs unlikely to contain only one PCB congener?

2.34. Name the following chlorinated dioxins and furans.

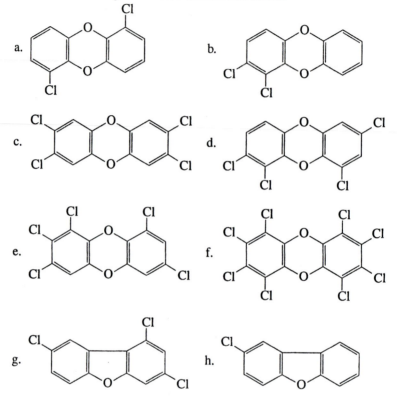

2.35. Which chlorinated dioxin congeners are most likely formed with the following compounds as precursors?
a. 2,4-Dichlorophenol and 2,4,6-trichlorophenol
b. 2,4,6-Trichlorophenol and 2,3,4,6-tetrachlorophenol
c. Two molecules of 2,4,6-trichlorophenol
d. Two molecules of pentachlorophenol

2.36. You are working on a project to evaluate the liability associated with wastes disposed of at a Superfund site. In such a forensic hazardous waste engineering study, you must locate potential sources of the hazardous wastes. Briefly describe the industries that could have disposed of each of the following inorganic species.
a. Arsenic   b. Cadmium
c. Chromium   d. Lead
e. Cyanide

2.37. Explain the most common method of heavy metals disposal prior to promulgation of RCRA.

2.38. An abandoned landfill received solid waste containing asbestos. What was the most likely source of this contamination?

2.39. List the particles emitted in each of the following decay steps.

$$^{218}_{86}A \rightarrow {}^{218}_{87}B \rightarrow {}^{218}_{88}C \rightarrow {}^{214}_{86}A \rightarrow {}^{214}_{87}B \rightarrow {}^{214}_{88}C \rightarrow {}^{210}_{86}A$$

2.40. For each step in the following decay chain, list the particles emitted.

$$^{214}_{84}Po \rightarrow {}^{210}_{82}Pb \rightarrow {}^{210}_{83}Bi \rightarrow {}^{210}_{84}Po \rightarrow {}^{206}_{82}Pb$$

2.41. What elements are described by the following nuclear states?
   a. $^{214}_{82}X$ b. $^{14}_{6}X$
   c. $^{90}_{38}X$ d. $^{226}_{88}X$
   e. $^{131}_{53}X$ f. $^{230}_{94}X$
   g. $^{18}_{8}X$ h. $^{45}_{20}X$

2.42. What species is formed from each of the following?
   a. $^{226}_{88}Ra$ emission of an α particle
   b. $^{210}_{82}Pb$ emission of a β particle
   c. $^{230}_{90}Th$ emission of an α particle
   d. $^{14}_{6}C$ emission of a β particle
   e. $^{210}_{83}Bi$ emission of a β particle

2.43. Write the balanced nuclear equation for the decay of each of the following species.
   a. ^{32}P emission of a β particle
   b. ^{222}Ru emission of an α particle
   c. ^{17}O emission of an α particle
   d. ^{14}C emission of a β particle

2.44. Use the *Merck Index* to draw the chemical structure of each of the following compounds and list its use.
   a. *p,p'*-Diaminodiphenylmethane
   b. Glyodin   c. *p*-Phenetidine
   d. Triglyme   e. Acetophenone
   f. Titanium tetrachloride   g. Selenious acid

2.45. The following compounds are usually used as industrial solvents and for other purposes. Using the *Merck Index,* draw the structure of each, and list all of its synonyms and industrial uses.
   a. Dimethyl sulfoxide   b. Pyridine
   c. Dioxane   d. Isophorone
   e. Dimethyl formamide   f. Isobutyl mercaptan
   g. Carbon disulfide

2.46. Hexachlorocyclohexane has been used as an organochlorine insecticide. Unlike benzene, which lies flat in a plane, hexachlorocyclohexane has bends and twists that result in four primary isomers (α, β, γ, and δ). The γ isomer is labeled lindane and has the highest insecticidal properties. Using the pesticide chemistry literature, draw the chemical structures of the four isomers and explain their structural differences.

2.47. Endrin, introduced in 1951, was one of the most toxic of the cyclodiene insecticides. Using the *Merck Index,* draw the chemical structure of endrin and list its uses.

2.48. Using the *Merck Index,* draw the structure of the following polycyclic aromatic hydrocarbon derivatives.
   a. Acenaphthene   b. 9H-Fluorene-2,7-diamine
   c. Anthranol   d. Benzanthrone

REFERENCES

2.1. Carey, F. A., *Organic Chemistry,* McGraw-Hill, New York, 1987.

2.2. Fresenden, R. and J. Fresenden, *Organic Chemistry,* 4th Edition, Brooks-Cole, 1990.

2.3. Neckers, D. C. and M. P. Doyle, *Organic Chemistry,* John Wiley & Sons, New York, 1977.

2.4. Pine, S. H, *Organic Chemistry,* 5th Edition, McGraw-Hill, New York, 1987.

2.5. Morrison, R. T. and R. N. Boyd, *Organic Chemistry,* 5th Edition, Allyn & Bacon, Boston, MA, 1987.

2.6. Solomons, T. W, *Organic Chemistry,* 4th Edition, John Wiley & Sons, New York, 1988.

2.7. Irons, R. D., "Benzene and other hemotoxins," In Sullivan, J. B., Jr. and G. R. Krieger (Eds.), *Hazardous Materials Toxicology,* Williams & Wilkins, Baltimore, MD, 1992.

2.8. Mazumdar, S., C. Redmond, W. Solecito, and N. Sussman, "An epidemiological study to exposure on coal tar pitch volatiles among coke oven workers," *J. Air Pollut. Control Assoc.,* **25,** 382–389 (1975).

2.9. Michelcic, J. R. and R. C. Luthy, "Degradation of polycyclic aromatic hydrocarbon compounds under various redox conditions in soil-water systems," *Appl. Environ. Microbiol.,* **54,** 1182–1187 (1988).

2.10. Schwendeman, T. D. and H. K. Wilcox, *Underground Storage Systems: Leak Detection and Monitoring,* Lewis Publishers, Chelsea, MI, 1987.

2.11. *Leaking Underground Fuel Tank Field Manual: Guidelines for Site Assessment, Cleanup, and Underground Storage Tank Closure,* Leaking Underground Fuel Tank Task Force, California Department of Health Services, Sacramento, CA, 1989.

2.12. *Reserves of Crude Oil, Natural Gas Liquids, and Natural Gas in the United States and Canada and United States Productive Capacity,* American Petroleum Institute and American Gas Association, Arlington, VA, 1990.

2.13. Hunt, J. M., "Distribution of carbon in the crust of the earth," *Am. Assoc. Pet. Geol. Bull.,* **56,** 2273–2277 (1972).

2.14. Ronov, A. B., "Global carbon geochemistry, volcanism, carbonate accumulation, and life," *Geochem. Int.,* **13,** 172–195 (1976).

2.15. *The Fate and Effect of Crude Oil Spilled on Subarctic Permafrost Terrain in Interior Alaska, EPA-600/3-80-040,* U.S. Environmental Protection Agency, Corvallis Environmental Research Laboratory, Corvallis, OR, 1980.

2.16. Maynard, J. B. and W. N. Sanders, "Determination of the detailed hydrocarbon composition and potential reactivity of full range motor gasolines," *J. Air Pollut. Control Assoc.,* **19,** 505–527 (1969).

2.17. *Leaking Underground Fuel Tank (LUFT) Manual,* California Department of Health Services and California State Water Resources Control Board, Sacramento, CA, 1988.

2.18. Kehoe, R. A., "Toxicological appraisal of lead in relation to the tolerable concentration in the ambient air," *J. Air Pollut. Control Assoc.,* **19,** 690–700 (1969).

2.19. Yeh, C. K. and J. T. Novak, "The effect of hydrogen peroxide on the degradation of methyl and ethyl *tert*-butyl ether in soils," *Water Environ. Res.,* **67,** 828–834 (1995).

2.20. *Toxicological Profile for Toluene,* Agency for Toxic Substances and Disease Registry, Contract No. 205-88-0608, Clement Associates, Atlanta, GA, 1989.

2.21. Higgins, T. E., *Hazardous Waste Minimization Handbook,* Lewis Publishers, Chelsea, MI, 1989.

2.22. Barcleoux, D. G., "Halogenated solvents," In Sullivan, J. B. and G. R. Krieger (Eds.), *Hazardous Materials Toxicology,* Williams & Wilkins, Baltimore, MD, 1992.

2.23. *The Farm Chemicals Handbook,* 82nd Edition, Meister Publishing, Willoughby, OH, 1996.

2.24. *Statistical Abstract of the United States, 1990,* 110th Edition, U.S. Department of Commerce, Superintendent of Documents, Washington, DC, 1990.

2.25. Chapman, R. F., *The Insects: Structure and Function,* Howard University Press, Cambridge, MA, 1982.

2.26. Ware, G. W., *Pesticides: Theory and Application,* W. H. Freeman, San Francisco, CA, 1983.

2.27. *Beneath the Bottom Line: Agricultural Approaches to Reduce Agrichemical Contamination of Groundwater,* Office of Technology Assessment, U.S. Government Printing Office, Washington, DC, 1990.

2.28. Jensen, A. A., *Residues of Pesticides and Other Contaminants in the Total Environment,* Springer-Verlag, NY, 1983.

2.29. *Toxicological Profile for p,p′-DDT, p,p′-DDE, p,p′-DDD,* Agency for Toxic Substances and Disease Registry, U.S. Public Health Service, U.S. Government Printing Office, Washington, DC, 1989.

2.30. Fest, C. and K. J. Schmidt, *The Chemistry of Organophosphorus Pesticides,* Springer-Verlag, NY, 1982.

2.31. Buchel, K. H., *Chemistry of Pesticides,* John Wiley & Sons, New York, 1983.

2.32. Heikkila, P. R., M. Hameila, L. Pyy, and P. Raunu, "Exposure to creosote in the impregnation and handling of impregnated wood," *Scand. J. Work Environ. Health,* **13,** 431–437 (1987).

2.33 Peoples, S. A., K. T. Maddy, W. Cusicle, T. Jackson, C. Cooper, and A. S. Fredrickson, "A study of samples of well water collected from selected areas in California to determine the presence of DBCP and certain other pesticide residues," *Bull. Environ. Contam. Toxicol.,* **24,** 611–618 (1980).

2.34. MacKay, D. M. and L. A. Smith, "Agricultural chemicals in groundwater: Monitoring and management in California," *J. Soil Water Conserv.,* **45,** 253–255 (1990).

2.35. Carnes, S. and P. Watson, "Disposing of U.S. chemical weapons stockpile," *J. Am. Med. Assoc.,* **262,** 653–659 (1989).

2.36. Myer, E., *Chemistry of Hazardous Material—Twelve Chemical Explosives,* Prentice-Hall, Englewood Cliffs, NJ, 1977.

2.37. Miller, G. C. and R. G. Zepp, "2,3,7,8-Tetrachlorodibenzo-*p*-dioxin: Environmental chemistry," In Exner, J. H. (Ed.), *Solving Hazardous Waste Problems. Learning from Dioxins,* American Chemical Society, Washington, DC, 1987.

2.38. Rappe, C., H. R. Buser, and H. P. Bosshardt, "Dioxins, dibenzofurans, and other poly-halogenated aromatics: Production, use, formation, and destruction," *Ann. N.Y. Acad. Sci.,* **320,** 1–18 (1979).

2.39. Förstner, U. and G. T. W. Wittmann, *Metal Pollution in the Aquatic Environment,* Springer-Verlag, New York, 1981.

2.40. Yamashita, N., M. Doi, and M. Nishio, "Recent observations of Kyoto children poisoned by arsenic tainted dry milk," *Jpn. J. Hyg.,* **27,** 364 (1972).

2.41. *Facing America's Track: What Next for Municipal Solid Waste,* Office of Technology Assessment, U.S. Government Printing Office, Washington, DC, 1989.

2.42. Woodbury, D. C., "Lead," In *Minerals Yearbook,* Bureau of Mines, United States Department of Commerce, Washington, DC, 1987.

2.43. Nriago, J. O., *Nickel in the Environment,* John Wiley & Sons, New York, 1980.

2.44. "Potential for human exposure," Chapter 5 in *Toxicologic Profile for Mercury,* Agency for Toxic Substances and Disease Registry, U.S. Public Health Service, Atlanta, GA, 1989.

2.45. Gonzales, E. R., "Cyanide evades some noses, overpowers others," *J. Am. Med. Assoc.,* **248,** 2211 (1982).

2.46. Selikoff, I. J. and D. H. Lee, *Asbestos and Disease,* Academic Press, New York, 1978.

2.47. Pooley, F. D., "Asbestos mineralogy," in Antman K. and J. Aisner (Eds.), *Asbestos-Related Malignancy,* Grune & Stratton, Orlando, FL, 1987.

2.48. *Mineral Commodity Summaries,* U.S. Bureau of Mines, U.S. Government Printing Office, Washington, DC, 1989.

2.49. Mossman, B. T. and J. B. Gee, "Asbestos-related diseases," *N. Engl. J. Med.,* **320,** 1721–1730 (1989).

2.50. Sawyer, C. N., P. L. McCarty, and G. F. Parkin, *Chemistry for Environmental Engineering,* McGraw-Hill, New York, 1994.

2.51. Lau, F. S., *Radioactivity and Nuclear Waste Disposal,* John Wiley & Sons, New York, 1987.

2.52. *Integrated Data Base for 1989: Spent Fuel and Radioactive Waste Investories, Projections, and Characteristics,* DOE/RW-0006, U.S. Department of Energy, U.S. Government Printing Office, Washington, DC, 1989.

2.53. *Complex Cleanup: The Environmental Legacy of Nuclear Weapons Production,* Office of Technology Assessment, Office of Government Printing, Washington, DC, 1991.

2.54. Budavari, S. (Ed.), *The Merck Index,* 12th Edition, Merck and Co., Whitehouse Station, NY, 1996.

Chapter 3

Common Hazardous Wastes: Properties and Classification

The information contained in Chapter 2 provides fundamental knowledge about common hazardous wastes, including their nomenclature and structure, their industrial uses, and where and how they have been treated and disposed of. The next step in understanding hazardous waste sources is the study of the physical and chemical properties of hazardous compounds. In this chapter, the properties of water solubility, density, flammability, and chemical incompatibility are the primary focus. In addition, classification systems are covered, including the hazardous property labeling systems used by the National Fire Protection Association and the U.S. Department of Transportation, Chemical Abstract Service (CAS) registry numbers, and the list of priority pollutants developed under the Clean Water Act.

Toxicity is also an important hazardous waste property. Under RCRA, toxicity is defined by the Toxicity Characteristic Leaching Procedure (TCLP)—a physicochemical property determined by extracting a solid waste under acidic conditions and analyzing the extract for selected toxic species. The fundamental evaluation of contaminant toxicity, covered under the subject of toxicology, is an extension of biochemistry and physiology, and requires years of study to master. An introduction to applied toxicology as a basis for a conceptual understanding of the toxicity of hazardous wastes is presented in Chapters 9 and 10.

3.1 COMMON CONCENTRATION UNITS

Different units are used to quantify the concentration of hazardous chemicals depending on whether they are present in water, solids, or air. The most common concentration unit for contaminants dissolved in water is mg/L, although in some systems where concentrations are very low (e.g., groundwater), the values may be reported in μg/L. Aqueous concentrations are sometimes reported in mmoles/L (mM) or μmoles/L (μM), but these units are usually limited to the research literature. Expressions that are commonly, but often incorrectly, used for aqueous concentrations are the mass/mass units of part per million (ppm) and part per billion (ppb). The definition of ppm is

$$\frac{\text{mass of compound}}{\text{mass of } 10^6 \text{ units of medium}} \tag{3.1}$$

Typical units of ppm concentration include:

$$\frac{\text{mg of compound}}{\text{kg of medium}} = \frac{10^{-6} \text{ g}}{1 \text{ g}} = \frac{\mu\text{g of compound}}{\text{g of medium}} \tag{3.2}$$

In a similar sense, ppb is defined as

$$\frac{\text{mass of compound}}{10^9 \text{ units of medium}} \tag{3.3}$$

A typical expression of ppb units is

$$\frac{\text{g of compound}}{10^9 \text{ g of medium}} = \frac{10^{-6} \text{ g}}{10^3 \text{ g}} = \frac{\mu\text{g of compound}}{\text{kg of medium}} \tag{3.4}$$

The use of ppm and ppb units for aqueous solutions has been perpetuated because 1 L of water has a mass of approximately 1 kg. Therefore,

$$\frac{\text{mg}}{\text{L}} \approx \frac{\text{mg}}{\text{kg}} \tag{3.5}$$

If the temperature of the water is 4°C, then the specific gravity of water = 1.00 and mg/L is equal to mg/kg. However, at 20°C, the density of water = 0.998 g/cm^3, and a 1.000 mg/L solution does not equal 1.000 ppm:

$$\frac{1.000 \text{ mg}}{\text{L}} \times \frac{\text{L}}{0.998 \text{ kg}} = \frac{1.002 \text{ mg}}{\text{kg}} \tag{3.6}$$

Contaminant concentrations in soils and sludges, however, are based on mass/mass relationships. The most common expression is mg contaminant/kg of dry soil (which is, by definition, ppm). In addition, μg/kg of solid (ppb) is used for very low concentrations. Although the expressions ppm and ppb are technically correct as units of mass/mass concentrations, some research journals will not accept them and prefer the more formal expressions of mg/kg or μg/kg.

Two common concentration units used to quantify the concentration of hazardous chemicals in air are

$$\frac{\mu\text{g}}{\text{m}^3} = \frac{\mu\text{g contaminant}}{\text{m}^3 \text{ of air}} \tag{3.7}$$

and

$$\text{ppm} = \frac{1 \text{ part contaminant (by volume)}}{10^6 \text{ parts air (by volume)}} \tag{3.8}$$

The mg/m^3 term may be used to quantify the concentration of both gaseous and particulate pollutants; however, ppm is used only for gaseous pollutants (gases and vapors of volatile liquids). The use of ppm for air measurements is fundamentally different from its use for solids; ppm concentrations in air are volume/volume, and ppm values in solids are mass/mass.

The units of mg/m^3 and ppm may be easily interconverted if the temperature and pressure at which the measurements were made are known. The basis for the conversion is the Ideal Gas Law:

$$PV = nRT \tag{3.9}$$

where

P = pressure (atm)
V = volume (L)
n = moles of gas
T = temperature (°K)
R = the universal gas constant (0.082 L-atm/mole-°K)

From Equation 3.9, the volume occupied by one mole of gas at standard temperature and pressure (0°C or 273°K and 760 mm Hg or 1 atm) is 22.4 L, or 0.0224 m^3 (1000 L = 1 m^3). Using this information, the volume of the contaminant may be calculated at standard temperature and pressure:

$$\text{volume}_{\text{contaminant, STP}}(m^3) = \frac{\text{mass contaminant (g)}}{\text{molecular weight} \dfrac{g}{\text{mole}}} \times 0.0224 \, \frac{m^3}{\text{mole}} \tag{3.10}$$

If the mg/m^3 data are collected at a temperature and pressure other than standard conditions, the Ideal Gas Law may be normalized for these atmospheric conditions. Because the volume of a gas is directly proportional to the absolute temperature and indirectly proportional to the pressure (in mm Hg), the contaminant volume can be normalized to standard temperature and pressure based on ambient data:

$$\text{volume}_{\text{contaminant, STP}} = \frac{T}{273°K} \times \frac{760 \text{ mm Hg}}{P} \times \text{volume}_{\text{contaminant, T,P}} \tag{3.11}$$

where

T = ambient temperature (°K) of the gas sample
P = atmospheric pressure (mm Hg) of the gas sample

Equations 3.10 and 3.11 may be combined; the volume of the contaminant normalized for standard temperature and pressure is then

$$\text{volume}_{\text{contaminant, STP}}(m^3) = \frac{\text{mass contaminant (g)}}{\text{molecular weight } \dfrac{g}{\text{mole}}} \times$$

$$0.0224 \frac{m^3}{\text{mole}} \times \frac{T}{273°K} \times \frac{760 \text{ mm Hg}}{P} \qquad (3.12)$$

The final step is to relate the volume of the contaminant to the volume of air and incorporate a factor of 10^6 based on the definition of ppm.

$$\text{ppm} = \frac{\text{volume}_{\text{contaminant}} (m^3)}{\text{volume}_{\text{air}}(m^3)} \times 10^6 \qquad (3.13)$$

Based on the fundamental volume-mass relationships of Equations 3.12 and 3.13, g/m^3 concentrations may also be converted to ppm using the simplified equation

$$C_{\text{ppm}} = C\left(\frac{RT}{PM}\right) \times 10^6 \qquad (3.14)$$

where

C_{ppm} = the contaminant concentration in ppm
C = the contaminant concentration in g/m^3
R = the ideal gas constant (8.21×10^{-5} m^3-atm/mole-°K)
T = temperature (°K)
P = absolute pressure (atm)
M = molecular weight of the contaminant (g/mole)

The use of Equation 3.14 is illustrated in Example 3.1.

EXAMPLE 3.1 *Conversion of Air Concentrations from mg/m^3 to ppm*

During routine air sampling, the concentration of perchloroethylene (PCE) near a hazardous waste site was 120 mg/m^3. The temperature was 16°C, and the atmospheric pressure was 730 mm Hg. What was the PCE concentration in ppm?

SOLUTION

First, calculate the molecular weight of PCE ($Cl_2C = CCl_2$):
   The atomic weights of the substituent atoms are C = 12 g/mole and Cl = 35.5 g/mole.

$$MW_{\text{PCE}} = 2(12) + 4(35.5) = 166 \frac{g}{\text{mole}}$$

Next, convert the temperature (°C) to °K:

$$273 + 16°C = 289°K$$

Convert the pressure to atmospheres

$$730 \text{ mm Hg} \times \frac{1 \text{ atm}}{760 \text{ mm Hg}} = 0.961 \text{ atm}$$

and change the concentration from $\frac{mg}{m^3}$ to $\frac{g}{m^3}$:

$$\frac{120 \text{ mg}}{m^3} = \frac{0.120 \text{ g}}{m^3}$$

Finally, use Equation 3.14 to convert the g/m^3 value to ppm:

$$C_{ppm} = 0.120 \frac{g}{m^3} \times \frac{8.21 \times 10^{-5} \frac{m^3\text{-atm}}{\text{mole-°K}} \times 289°K}{0.961 \text{ atm} \times 166 \frac{g}{\text{mole}}} \times 10^6$$

$$C_{ppm} = 17.8 \text{ ppm}$$

Radioactivity is not measured as mass/volume, but as the rate of decay of nuclear particles. Radioactive elements, which are characteristically unstable, eventually reach stability by emitting α or β particles, a process known as decay (see Section 2.11).

The activity of radioactive materials is defined by the number of unstable nuclei present, which is measured indirectly by the rate of radioactive decay. The primary unit of radioactive decay is the *curie*, which is the amount of radioactive material containing unstable atoms characterized by 3.70×10^{10} disintegrations per second (dps). The specific activity of a radioactive compound is the activity per mass of material. Because the curie is a large unit, mCi and μCi are used more commonly. The determination of specific activity is demonstrated in Example 3.2.

EXAMPLE 3.2 *Determination of Specific Activity of a Radioisotope*

A 10-g sample of dry sludge contains mixed hazardous wastes (radionuclides as well as hazardous inorganic and organic chemicals) in which β particles are emitted at a rate of 3×10^8 dps. Determine the specific activity of this sample.

SOLUTION

Convert the emission rate of β particles to Ci:

$$3 \times 10^8 \text{ dps} \times \frac{1 \text{ Ci}}{3.7 \times 10^{10} \text{ dps}} = 8.11 \times 10^{-3} \text{ Ci} = 8.11 \text{ mCi}$$

Using the mass of dry sludge, calculate the specific activity:

$$\text{Specific activity} = \frac{8.11 \text{ mCi}}{10\text{g}} = 0.811 \frac{\text{mCi}}{\text{g}}$$

3.2 WATER SOLUBILITY

The extent to which a hazardous compound dissolves in water—its aqueous or *water solubility*—is one of its most important chemical properties. Water solubility often controls the environmental fate of waste compounds and the handling of concentrated hazardous materials, such as those in drums awaiting disposal or treatment. As an example of the importance of water solubility, consider a groundwater system contaminated with a large volume of gasoline. The undissolved *free product* floating above the water table exists because it is sparingly soluble in the groundwater and its density is lower than that of water. Therefore, the water solubility of a contaminant controls not only its concentration in groundwater but also the proportion that exists in the floating layer of free product. Compounds that are miscible (i.e., completely soluble) in water, such as methanol, acetone, or formaldehyde, would not form a floating layer, regardless of their densities.

Water solubility is also important in the treatment of soil and groundwater contaminants. Compounds that are *hydrophobic* (i.e., water-hating—a characteristic measured indirectly by low water solubility) partition onto solids, form emulsions, and float at the air-water interface or sink to the bottom of aquifers. These partitioning mechanisms decrease the effectiveness of biological and physicochemical treatment processes. In general, water solubility is inversely proportional to sorptivity, bioaccumulation, and volatilization from aqueous solutions [3.1]. Most data suggest that water solubility may also influence biodegradation, photolysis, chemical oxidation, and other pathways.

Water solubility is defined as the maximum (or saturation) concentration of a substance that will dissolve in water at a given temperature [3.2]. However, some substances are so finely divided that they do not settle out; these species may remain in colloidal suspension for weeks but are not truly dissolved. Therefore, the *practical* analytical definition of a dissolved chemical is based on chemical analysis; it is the material that passes through a GFC glass fiber filter. The pore size of GFC filters is approximately 1.2 μm [3.3].

Because temperature significantly affects water solubility, most data are obtained and reported at 25°C; this temperature is slightly above room temperature and can be easily kept constant to provide accurate measurements. The most common water solubility unit in environmental applications is mg/L, although some sources of theoretical information report solubility in mmoles/L or μmoles/L. Water solubilities of hazardous compounds typically range from 1 mg/L to over 100,000 mg/L [3.4]. Some compounds, such as methanol, ethanol, acetone, acetic acid, ethylene glycol, and formaldehyde, are miscible in water. Similarly, inorganic cations and anions exhibit wide-ranging water solubilities, which often depend on the nature and concentrations of other ions in the solution. However, complexes between metals and some organic molecules (e.g., hydroxy aromatic acids) may increase the water solubility of the metals [3.5].

The water solubility of both inorganic ions and organic compounds is related to their structure and size. The concept of "like dissolves like" is the best general rule for predicting water solubility. Water is a highly polar species; therefore, charged and very polar organic and inorganic species exhibit high water solubilities. For example, the nonpolar compound hexane has a low water solubility of 13 mg/L. The addition of a polar hydroxyl group (—OH) to hexane to form hexanol increases the water solubility to 5900 mg/L [3.6].

The solubility of a hazardous chemical in any solvent, which may be water or an organic liquid such as benzene, ethanol, or trichloroethylene (TCE), is based on attractive forces between the solute and solvent molecules. The most important attractive forces include (1) Van der Waals forces (the interaction of electron orbitals between adjacent molecules), (2) hydrogen bonding, which occurs primarily with molecules containing hydroxyl (—OH) and amino (—NH_2) groups, and (3) dipole-dipole interactions [3.7].

The size and shape of a molecule also affect its water solubility. The logarithm of water solubility for a homologous series of hydrocarbons was found to be inversely proportional to their molar volumes [3.8]. The effect of the number of aromatic rings of polycyclic aromatic hydrocarbons (PAHs) on water solubility is illustrated in Table 3.1. The larger PAHs are characterized by lower water solubility due largely to the inability of these large, nonpolar planar structures to fit between small, polar water molecules.

The presence of some functional groups (e.g., —OH, —COOH, —CO, —NH_2, —NO_2, —Cl, —Br, —CHO, and —CN) may or may not affect water solubility. For example, double bonds do not significantly affect the water solubility of hydrocarbons with equal numbers of carbons. Halogens decrease the water solubility of organic compounds by increasing their molecular volume [3.10]. Organic acids, amines, alcohols,

Table 3.1 Water Solubility of Polycyclic Aromatic Hydrocarbons

Number of Rings	Compound	Water Solubility (mg/L)
2	Naphthalene	32
3	Anthracene	0.059
4	Chrysene	0.0033
4	Pyrene	0.13
4	Benzo[a]anthracene	0.0012
5	Benzo[a]pyrene	0.0038
5	Dibenzo[a,h]anthracene	0.0005
6	Benzo[g,h,i]perylene	0.00026
6	Indeno[1,2,3-c,d]pyrene	0.00053

Source: Reference 3.9.

ethers, and ketones are more soluble in water than unsubstituted hydrocarbons. The water solubility of compounds within any of these chemical classes decreases as the size of the molecule increases. For the smaller polar organics, such as acetic acid, the polar section of the molecule ($-COOH$) has a significant effect in controlling its solubility. A large organic acid, such as hexadecanoic acid ($C_{15}H_{31}-COOH$), is dominated by the nonpolar $C_{15}H_{31}$ group; hence, it has much lower water solubility than acetic acid. Even the ionized salts of these higher organic acids are relatively insoluble in water [3.11].

Water Solubilities of Weak Acids and Bases. Organic compounds that are weak acids and bases have differing water solubilities in the protonated (acidic) and unprotonated (basic) forms. The general equilibrium expression for the dissociation of a weak acid in aqueous solution is:

$$R-COOH \rightleftharpoons R-COO^- + H^+ \tag{3.15}$$

(acidic form)   (basic form)

The equilibrium expression for the reaction of Equation 3.15 is

$$K_a = \frac{[\text{base}][H^+]}{[\text{acid}]} = \frac{[RCOO^-][H^+]}{[RCOOH]} \tag{3.16}$$

where K_a is the acid dissociation constant and the brackets indicate the concentration (in moles/L) of each species. The most common hazardous compounds that exhibit this behavior are chlorophenols. For example, pentachlorophenol acts as a weak acid by losing a proton:

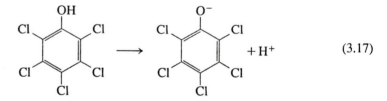

Pentachlorophenol   Pentachlorophenate

The equilibrium expression for the dissociation of pentachlorophenol is

$$K_a = \frac{[\text{pentachlorophenate}]\ [H^+]}{[\text{pentachlorophenol}]} \tag{3.18}$$

The equilibrium coefficient for Equation 3.18 has been determined experimentally

$$K_a = 1.78 \times 10^{-5} \tag{3.19}$$

The expression for K_a may be easily converted to pK_a, because $-\log X$ is defined as pX:

$$pK_a = 4.75 \tag{3.20}$$

Values of pK_a for chlorophenols are listed in Table 3.2; pK_a values for other common weak organic acids are found in Table 3.3.

The relative proportions of the acid form and the ionized salt of pentachlorophenol as a function of pH are shown in Figure 3.1. The concentration of the acidic (un-ionized) form decreases with increased pH, and the concentration of the pentachlorophenate salt (ionized) form increases as the pH is raised. The two forms exist at equal concentrations at the pK_a of 4.75. The basic, ionized form of weak organic acids ($RCOO^-$) is usually several orders of magnitude more water soluble than the neutral un-ionized form. For example, the water solubility of pentachlorophenate is approximately 100 times greater than the solubility of pentachlorophenol, which is 14 mg/L [3.12]. The water solubilities of most ionized salts of weak organic acids have not been measured, but these species are usually considered to be nearly miscible in water. Inspection of Figure 3.1 also shows that the net water solubility of a contaminant increases as a function of the proportion of its ionized form and that the net wa-

Table 3.2 Values of pK_a for Chlorophenols

Compound	pK_a
Phenol	9.98
2-Chlorophenol	8.53
3-Chlorophenol	9.13
4-Chlorophenol	9.43
2,4-Dichlorophenol	7.85
3,4-Dichlorophenol	8.63
3,5-Dichlorophenol	8.18
2,4,6-Trichlorophenol	6.15
2,4,5,6-Tetrachlorophenol	5.16
Pentachlorophenol	4.75

Source: Reference 3.13.

Table 3.3 Values of pK_a for Organic Acids

Compound	pK_a
Amiben	3.4
Benzoic acid	4.2
Bromoacetic acid	2.9
Chloroacetic acid	2.85
Cyanoacetic acid	2.47
Dicambia	1.93
Dichloroacetic acid	2.90
o-Nitrophenol	7.17
Trichloroacetic acid	0.70
2,4,6-Trinitrophenol	0.38

Source: Reference 3.13.

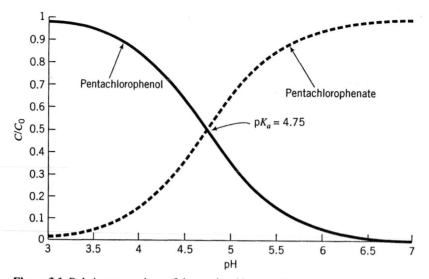

Figure 3.1 Relative proportions of the weak acid pentachlorophenol and its salt pentachlorophenate as a function of pH.

ter solubility is based on the proportionate contribution of the ionized and un-ionized forms.

The ratio of the concentration of a weak acid to its ionized form can be determined by relating the pK_a to the pH of the solution containing the contaminant. The first step in such an approach is to isolate the [H$^+$] into a separate term and then take the $-\log$ of both sides of Equation 3.21:

$$K_a = [H^+] \cdot \frac{[\text{base}]}{[\text{acid}]} \tag{3.21}$$

or

$$-\log K_a = -\log H^+ + -\log \frac{[\text{base}]}{[\text{acid}]} \tag{3.22}$$

Because $-\log X$ is defined as pX, Equation 3.22 can be simplified:

$$pK_a = pH - \log \frac{[\text{base}]}{[\text{acid}]} \tag{3.23}$$

Rearranging yields

$$pH = pK_a + \log \frac{[\text{base}]}{[\text{acid}]} \tag{3.24}$$

Equation 3.24, known as the *Henderson-Hasselbalch equation*, provides a straightforward method for calculating the degree of acidic dissociation at a given pH. Its use is demonstrated in Example 3.3.

By analogy, a weak organic base is characterized by accepting a proton:

$$R + H_2O \rightleftharpoons RH^+ + OH^- \qquad (3.25)$$

The base association constant K_b is defined as

$$K_b = \frac{[RH^+][OH^-]}{[R]} \qquad (3.26)$$

or

$$K_b = \frac{[RH^+]\, K_w}{[R][H^+]} \qquad (3.27)$$

where

$$K_w = [H^+][OH^-] = 10^{-14} \qquad (3.28)$$

Values of pK_a for selected weak organic bases are listed in Table 3.4.

A measure of the dissociation of weak acids is the parameter α, which is the fraction of un-ionized acid:

$$\alpha = \left(1 + \frac{K_a}{[H^+]}\right)^{-1} \qquad (3.29)$$

The fraction of un-ionized base, β, is

$$\beta = \left(\frac{1 + K_b[H^+]}{K_w}\right)^{-1} \qquad (3.30)$$

Because of the large difference in water solubility (and, therefore, in transport rates) between ionized and un-ionized forms of weak acids and bases, the terms α and β are often used to correct for contaminant dissociation in groundwater transport models (see Section 5.7).

Table 3.4 Values of pK_a for Organic Bases

Compound	pK_a
Aniline	9.37
Atrazine	12.3
p-Nitroaniline	13.0
Propazine	12.2
Pyridine	8.8
Simazine	12.4

Source: Reference 3.13.

EXAMPLE 3.3 *Use of the Henderson-Hasselbalch Equation*

What are the concentrations of 2,4,6-trichlorophenol and 2,4,6-trichlorophenate if a 0.1 mM 2,4,6-trichlorophenol solution is adjusted to a pH of 7.0? Also, determine α for these conditions.

SOLUTION

The pK_a of trichlorophenol = 6.15. At pH 7.0, the Henderson-Hasselbalch equation (Equation 3.24) is

$$7.0 = 6.15 + \log \frac{[\text{trichlorophenate}]}{[\text{trichlorophenol}]}$$

Solving for the ratio $\dfrac{[\text{trichlorophenate}]}{[\text{trichlorophenol}]}$ gives

$$\frac{[\text{trichlorophenate}]}{[\text{trichlorophenol}]} = 7.08,$$

and

$$[\text{trichlorophenate}] + [\text{trichlorophenol}] = 0.1 \text{ mM}$$

Solving these equations, the concentration of trichlorophenate = 0.0876 mM and the trichlorophenol concentration = 0.0124 mM.

$$\alpha = \frac{0.0124 \text{ M un-ionized trichlorophenol}}{0.1 \text{ M total}} = 0.124$$

α may also be calculated using Equation 3.28:

$$\alpha = \left[1 + \frac{K_a}{[H^+]}\right]^{-1} = \left[1 + \frac{7.08 \times 10^{-7}}{1 \times 10^{-7}}\right]^{-1} = 0.124$$

The acidity of chlorophenols increases with higher chlorine substitution, a trend that is illustrated by the pK_a values of Table 3.2. Chlorine is an electronegative element; therefore, the more chlorines on the ring, the more electrons are drawn away from electron-rich phenoxide (PhO⁻):

Withdrawing electrons from the phenoxide lowers its negative charge and its corresponding potential to hold an oppositely charged species, H^+. Therefore, the more chlorines on the ring, the more easily a proton is released and the greater the acidity.

Water solubility data are available for most compounds of environmental significance, and data for some of the most common hazardous compounds are shown in Table 3.5. More extensive tables of water solubility data are available in Appendix B.

Inspection of the data of Table 3.5 and Appendix B reveals a wide range of values reported for some compounds, which is sometimes the result of poor experimental methods, particularly when some values deviate substantially from others. Although most of these experimental values are published in the peer-reviewed literature, some may not have been obtained under the experimental conditions necessary for the accurate determination of the data. Often, however, the range is simply due to experimental variability.

3.3   DENSITY AND SPECIFIC GRAVITY

The level (in elevation, not concentration) at which an insoluble waste equilibrates in a natural system, a waste drum, or a solvent recovery system is dictated by its density. Density is an important property that aids in assessing the behavior of nonaqueous phases in soil and groundwater systems and in the sampling of drum and storage tank contents. *Density* is defined as the ratio of mass to volume and is usually expressed as g/mL or kg/m³ (SI units) for liquids and in g/cm³ for solids. *Specific gravity* is a dimensionless number analogous to density, derived by relating the compound's density to that of water, which has a density of 1.0 g/mL at 4°C, the temperature of its highest density:

$$\text{specific gravity} = \frac{\text{mass of a volume of compound}}{\text{mass of the same volume of water}} \tag{3.31}$$

or

$$\text{density}_{\text{compound}} = \text{specific gravity}_{\text{compound}} \times \text{density of water at 4°C (1.0 g/mL)} \tag{3.32}$$

Specific gravities range from 0.6 to 2.9 for organic compounds and are usually related to the atomic numbers of their constituent atoms. In general, the density of a non-metallic element increases with increasing atomic number [3.14]. For example, the density of hydrogen is less than that of carbon, which in turn is less than the density of chlorine. Consequently, hydrocarbons are lighter than water and their density increases as a function of molecular weight. Specific gravities of aliphatic hydrocarbons range from 0.65 to 0.85. Substitution of elements with atomic numbers higher than carbon ($n = 6$) results in a more dense compound. For example, substitution of a chlorine atom on a hydrocarbon increases its density. As expected with the general rule of density being proportional to atomic number, the relative densities of monosubstituted hydrocarbons are RI > RBr > RCl > RF. Oxygen-substituted organics may or may not be more dense than water [3.15]. The density of oxygen-containing molecules is a function of the percentage of atomic oxygen relative to carbon and the extent of double bonding. Ethers, alcohols, and esters are all less dense than water; however, phenol is more dense.

Table 3.5 Water Solubilities of Common Hazardous Compounds

Compound	Mean Water Solubility (mg/L) @ 25°C	Range
Aliphatic Hydrocarbons		
Cyclohexane	59.3	58.4–66.5
Cyclohexene	213	
Isooctane	1.25	0.56–2.46
n-Decane	0.021	0.020–0.022
n-Heptane	2.91	2.24–3.57
n-Octane	1.39	0.431–3.37
Monocyclic Aromatic Hydrocarbons		
Benzene	1770	1740–1800
Ethylbenzene	181	152–208
m-Xylene	163	146–173
o-Xylene	221	171–297
p-Xylene	189	156–214
Toluene	546	515–627
Polycyclic Aromatic Hydrocarbons		
Anthracene	0.059	0.045–0.073
Benzo[a]anthracene	0.012	0.0094–0.014
Benzo[a]pyrene	0.0038	
Chrysene	0.0033	0.0018–0.006
Fluorene	1.84	1.69–1.98
Naphthalene	31.9	30.0–34.4
Phenanthrene	1.09	0.994–1.29
Pyrene	0.134	0.132–0.135
Nonhalogenated Solvents		
Acetone	miscible	
Methyl butyl ketone	35,000	
Methyl ethyl ketone	256,000	
Methyl isobutyl ketone	19,100	
Halogenated Solvents		
Carbon tetrachloride	911	757–1160
Chloroform	7870	7100–9300
Methylene chloride	18,400	16,700–19,400
Perchloroethylene	275	150–400
1,1,1-Trichloroethane	657	
Trichloroethylene	1310	1100–1470
Insecticides		
Aldrin	0.011	
Carbaryl	76.5	70–83

Table 3.5 (Continued)

Compound	Mean Water Solubility (mg/L) @ 25°C	Range
Insecticides (cont.)		
Dieldrin	0.198	0.195–0.200
DDT	0.0024	0.0012–0.020
Hexachlorobenzene	0.005	
Lindane	7.30	6.80–7.80
Malathion	145	
Methyl parathion	57	
Parathion	24.0	
Herbicides		
Atrazine	33.0	
2,4-D	697	522–890
2,4,5-T	273	268–278
Trifluralin	1.0	
Fungicides		
Pentachlorophenol	14.0 @20°C	
Industrial Intermediates		
Chlorobenzene	460	295–503
2,4-Dichlorophenol	9750	4500–15,000
Dimethyl phthalate	4160	4000–4320
Di-*n*-octyl phthalate	3	
Hexachlorocyclopentadiene	1.80	
Phenol	77,900	67,000–84,700
2,4,6-Trichlorophenol	800	
Explosives		
Picric acid	14,000 @20°C	
2,4,6-Trinitrotoluene	130 @20°C	
Polychlorinated Biphenyls		
PCB 1016	0.42	
PCB 1232	1.45	
PCB 1248	0.036	0.017–0.054
PCB 1254	0.011	0.010–0.012
PCB 1260	0.0027	
Chlorinated Dioxins		
2,3,7,8-Tetrachlorodibenzo-*p*-dioxin	0.00032	

Source: Appendix B

Metals are more dense than water, and the hazardous metals of greatest concern are at least four times more dense. As noted in Chapter 2, the most problematic metals—those defined as heavy metals—have densities greater than 5.0 g/cm^3 [3.5]. Most metals rarely exist in pure form in environmental systems; instead, they are found as metal oxides, carbonates, sulfates, chlorides, and sulfides, or sometimes in chelated form attached to soil organic matter. These metal salts are less dense than the pure metal—usually in the range of 2.5 to 4.5 g/cm^3 [3.16]. Specific gravities of some common hazardous organic compounds are listed in Table 3.6, and additional specific gravity data are available in Appendix C.

Vapor density is analogous to specific gravity. It is defined as the density of a compound in the vapor phase (with no air present) to the density of air. By definition, air has a relative vapor density of 1.0. Vapor density is important in assessing the fate of a gas after its release. For example, chlorine gas, if allowed to escape from a tank, stays close to the ground because its vapor density is 3.8; that is, it is 3.8 times more dense than air. Vapor densities for common hazardous compounds are listed in Table 3.7.

3.4 LIGHT AND DENSE NONAQUEOUS PHASE LIQUIDS

The presence of water-insoluble hazardous chemicals has become so commonplace, and their problems so complex, that they have been given a generic name—*nonaqueous phase liquids* (NAPLs). Two other terms based on density have recently been established to describe the behavior of water-insoluble organic compounds—*LNAPLs* (light nonaqueous phase liquids) and *DNAPLs* (dense nonaqueous phase liquids). The stratification of contaminants relative to water is only important if they are in a NAPL phase. Chemicals that are dissolved in water are dispersed by Brownian motion, resulting in homogeneous distribution throughout the solution.

The presence of NAPLs results in increased complexity in the assessment and remediation of subsurface systems. The problems associated with NAPLs that have migrated to groundwater are depicted in Figure 3.2. The LNAPLs are more easily detected because they float on the surface of the groundwater and are often observed by visual inspection of well samples, soil cores, or well cuttings. However, LNAPLs may also be present even if they are not visible; they may be trapped by capillarity and surface tension in soil pores or may be present in locations not sampled [3.18]. If DNAPLs have migrated to an aquifer, the situation is significantly worse because they sink to the bottom of the subsurface system. There they form pools and lenses in pockets and depressions, where they are difficult to detect and remove [3.19].

Nonaqueous phase liquids may consist of a single compound or a multicomponent mixture. For example, disposal of PCE in a soil pit by a dry cleaner would result in a single component NAPL problem. A soil pit or industrial landfill in which hundreds of solvents and industrial compounds were disposed of would provide a source for a multicomponent mixture of NAPLs.

Depending on the mass of contaminant in contact with an aqueous system (e.g., a surface water, groundwater, or water storage system), the volume of the aqueous system, and the compound water solubility, a range of proportions of the contaminant will be found in the nonaqueous phase and the dissolved phase, as illustrated in Example 3.4.

Table 3.6 Specific Gravities of Common
Hazardous Compounds

Compound	Specific Gravity
Aliphatic Hydrocarbons	
Cyclohexane	0.779
Cyclohexene	0.810
n-Heptane	0.684
Isooctane	0.692
n-Octane	0.703
n-Decane	0.730
Monocylic Aromatic Hydrocarbons	
Benzene	0.877
Ethylbenzene	0.867
Toluene	0.867
o-Xylene	0.880
m-Xylene	0.864
p-Xylene	0.881
Polycylic Aromatic Hydrocarbons	
Anthracene	1.283
Benzo[a]anthracene	1.274
Benzo[a]pyrene	1.351
Chrysene	1.274
Fluorene	1.203
Naphthalene	1.145
Phenanthrene	0.980
Pyrene	1.271
Nonhalogenated Solvents	
Acetone	0.790
Methyl butyl ketone	0.811
Methyl ethyl ketone	0.805
Methyl isobutyl ketone	0.798
Halogenated Solvents	
Carbon tetrachloride	1.594
Chloroform	1.483
Methylene chloride	1.327
Perchloroethylene	1.623
1,1,1-Trichloroethane	1.339
Trichloroethylene	1.464

Table 3.6 (Continued)

Compound	Specific Gravity
Insecticides	
Aldrin	1.70
Carbaryl	1.232
Dieldrin	1.75
DDT	1.56
Hexachlorobenzene	1.569
Lindane	1.891
Malathion	1.23
Methyl parathion	1.352
Parathion	1.26
Herbicides	
Atrazine	1.20
2,4-D	1.416
2,4,5-T	1.80
Fungicides	
Pentachlorophenol	1.978
Industrial Intermediates	
Chlorobenzene	1.106
2,4-Dichlorophenol	1.40
Dimethyl phthalate	1.191
Di-*n*-octyl phthalate	0.990
Hexachlorocyclopentadiene	1.702
Phenol	1.058
2,4,6-Trichlorophenol	1.490
Explosives	
Picric acid	1.763
2,4,6-Trinitrotolulene	1.654
Polychlorinated Biphenyls	
PCB 1016	1.33
PCB 1232	1.24
PCB 1248	1.41
PCB 1254	1.51
PCB 1260	1.57
Chlorinated Dioxins	
2,3,7,8-Tetrachlorodibenzo-*p*-dioxin	1.827

Source: Appendix C

Table 3.7 Vapor Density of Several Common Hazardous Compounds

Compound	Vapor Density
Aliphatic Hydrocarbons	
Cyclohexane	2.9
Isopentane	2.5
n-Heptane	3.5
n-Hexane	3.0
n-Octane	3.9
Pentane	2.5
Aromatic Hydrocarbons	
Benzene	2.8
Ethylbenzene	3.7
m-Xylene	3.7
o-Xylene	3.7
p-Xylene	3.7
Toluene	3.1
Polycyclic Aromatic Hydrocarbons	
Naphthalene	4.4
Fuels	
Gasoline	3.0 to 4.0
No. 1 fuel oil	4.5
No. 2 fuel oil	2.0
No. 4 fuel oil	2.0
JP-1 jet fuel	3.0
JP-3 jet fuel	3.0
JP-5 jet fuel	3.0
Kerosene	4.1
Nonhalogenated Solvents	
Acetone	2.0
Ethanol	1.6
Methyl ethyl ketone	2.5
Methyl isobutyl ketone	3.5
Cyclohexanol	3.5
Turpentine	4.8
Halogenated Solvents	
Carbon tetrachloride	8.2
Chloroform	4.1
Methylene chloride	2.9

Table 3.7 (Continued)

Compound	Vapor Density
Halogenated Solvents (cont.)	
Perchloroethylene	5.8
Trichloroethylene	4.5
Herbicides	
2,4-D	7.6
Insecticides	
Aldrin	12.0
Dieldrin	13.2
DDT	12.0
Endrin	13.0
Parathion	10.0
Industrial Intermediates	
Chlorobenzene	3.1
Dichlorophenol	5.6
Hexachlorocyclopentadiene	9.4
Phenol	3.2
2,4,6-Trichlorophenol	6.8
Explosives	
2,4-Dinitrotoluene	6.3

Source: Reference 3.17.

EXAMPLE 3.4 *Proportions of NAPL and Dissolved Phases*

A 208-L (55-gal) drum of TCE has spilled into a 10,000-L (2640-gal) covered water storage reservoir. If volatilization losses are negligible and the TCE dissolves to the level of its experimentally determined water solubility, estimate the proportion of TCE in both the NAPL and the dissolved phases. Where will the NAPL phase be found— at the top or the bottom of the reservoir?

SOLUTION

The total mass of TCE is a function of the density of TCE and the volume spilled.

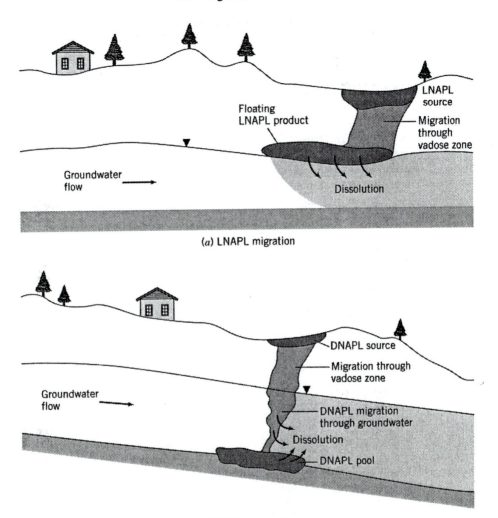

(a) LNAPL migration

(b) DNAPL migration

Figure 3.2 Migration and fate of (a) LNAPLs and (b) DNAPLs in the subsurface.

The density of TCE (Table 3.6) = 1.464 g/mL = 1.464 kg/L

$$208 \text{ L} \times 1.464 \text{ kg/L} = 305 \text{ kg TCE}$$

The mass of TCE dissolved (based on the TCE water solubility—Table 3.5) is:

$$10,000 \text{ L water} \times 1310 \text{ mg TCE/L water} \times \text{kg}/10^6 \text{ mg} = 13.1 \text{ kg dissolved}$$

Therefore the proportion of TCE dissolved is:

$$\frac{13.1 \text{ kg aqueous phase TCE}}{305 \text{ kg total TCE}} \times 100\% = 4.30\% \text{ dissolved}$$

$$100\% - 4.30\% = 95.7\% \text{ in the NAPL phase}$$

The NAPL phase would be found at the bottom of the reservoir because the specific gravity (1.464) of TCE is greater than the specific gravity of water (1.0).

Dissolution, the transfer of a compound from its insoluble phase into the water, is the mechanism that controls the rate at which contaminants move into aqueous systems; this process is also often the rate-limiting step in the remediation of contaminated groundwater [3.20]. Dissolution occurs when the chemicals within an overlying or underlying pool of NAPL move across the NAPL-water phase boundary. The potential for dissolution is proportional to the difference between the water solubility and the actual concentration in the aqueous phase:

$$\frac{dC}{dt} \, \alpha \, (C_s - C) \tag{3.33}$$

where

$\dfrac{dC}{dt}$ = rate of change of dissolution $\left(\dfrac{mg}{L/min}\right)$

C_s = contaminant water solubility (mg/L)

C = residual contaminant concentration (mg/L)

Incorporation of a mass-transfer coefficient, K, that describes the rate of transfer of the contaminant from the nonaqueous phase to the aqueous phase results in an equality:

$$\frac{dC}{dt} = K \, (C_s - C) \tag{3.34}$$

where K = a mass-transfer coefficient (min^{-1}). Higher mass-transfer coefficients are associated with increased mixing and surface-to-volume ratios of NAPL lenses [3.21].

Experimental water solubility data presented in Table 3.5 and Appendix B provide a basis for estimating the theoretical solubility of a NAPL in water at equilibrium. However, the equilibrium concentration in actual groundwater systems (i.e., the actual water solubility) is often less than the value measured in the laboratory. For example, Anderson et al. [3.22] reported that the maximum concentration of PCE actually found in groundwater is 5 mg/L, approximately 3% of its theoretical water solubility. The mechanisms for this reduced solubility have not been elucidated, but they may be related to reduced mixing of the NAPL with water, resulting in insufficient contact to reach C_s. Other possible explanations include water chemistry factors, such as the ionic composition of the groundwater.

The solubility behavior of mixtures of NAPLs is characteristically more complex than that of single-component systems. With such mixtures, the *effective solubility* is the true solubility under real-world conditions, which is often less than the water solubility of a single compound in the laboratory. The effective solubility of one compound, S^e, is estimated by normalizing its single-phase water solubility for its mole fraction in the mixture:

$$S_i^e = X_i S_i \gamma_i \tag{3.35}$$

where

S_i^e = the effective solubility of compound i (mg/L)
X_i = mole fraction of compound i in the NAPL mixture
S_i = experimental water solubility of i in a single component system (mg/L)
γ_i = correction factor that normalizes solubility based on field conditions and water chemistry

Therefore, the steady-state contaminant concentrations rarely reach the experimental water solubilities reported in Table 3.5 and Appendix B; this observation serves as the basis for the correction factor (γ) listed in Equation 3.35. The use of Equation 3.35 is demonstrated in Example 3.5.

EXAMPLE 3.5   *Estimation of Effective Solubility of a NAPL*

A release of gasoline from an underground storage tank (UST) has resulted in the presence of a large lens of an LNAPL on the surface of an aquifer. Estimate the effective solubility of benzene in the aquifer if $X_{benzene} = 0.06$ and $\gamma_{benzene} = 0.5$.

SOLUTION

The water solubility of benzene = 1770 mg/L.
Therefore,

$$S^e_{benzene} = X_{benzene} \cdot S_{benzene} \cdot \gamma_{benzene}$$

$$= (0.06)\ (1770\ \text{mg/L})\ (0.5)$$

$$= 53\ \text{mg/L}$$

In the remediation of subsurface systems, the removal of NAPLs is a prerequisite to treating the aqueous phase; otherwise, the NAPL will serve as a continuous source input to the groundwater. The removal of free product LNAPLs is relatively straightforward (see Section 12.2); however, the pockets of DNAPLs found at the bottom of aquifers present a formidable task. To date, no full-scale DNAPL-contaminated aquifers have been cleaned up, and although some have proposed giving up on these systems as "groundwater management zones," others hope that new and emerging engineering processes will hold the key to their remediation [3.23].

3.5 FLAMMABILITY LIMITS

Not all chemicals are classified as hazardous because of their acute or chronic toxicity. Hazardous wastes, as defined by RCRA, may also be corrosive, explosive, or flammable. Some chemicals volatilize if stored in open drums or vats; the vapors may then ignite in the presence of a spark or open flame if they are within a specific range of concentrations in the air.

Fire is defined as the rapid, exothermic oxidation of fuel. The three elements of fire are the fuel, an oxidizer, and ignition [3.24]. For chemicals that volatilize and are ignitable, there is a range of their concentrations in air for which flammable conditions exist. At low vapor concentrations, insufficient mass is available for flame propagation (i.e., the mixture is "too lean" to burn or explode). At high concentrations, there is a threshold concentration that limits combustion (i.e., the mixture is "too rich"). These concentrations, usually expressed in percent, are the *lower and upper flammability limits* (LFLs and UFLs). Lower and upper flammability limits for selected hazardous compounds are listed in Table 3.8.

Table 3.8 Upper and Lower Flammability Limits of Common Hazardous Compounds

Compound	LFL (%)	UFL (%)
Aliphatic Hydrocarbons		
Cyclohexane	1.3	8.0
n-Decane	0.8	9.2
Isopentane	1.4	8.3
n-Heptane	1.1	6.7
n-Hexane	1.1	7.5
n-Octane	1.0	5.6
Pentane	1.5	7.8
Aromatic Hydrocarbons		
Benzene	1.3	7.9
Ethylbenzene	1.0	6.6
m-Xylene	1.1	7.0
o-Xylene	1.0	7.0
p-Xylene	1.1	7.0
Toluene	1.2	7.1
Polycyclic Aromatic Hydrocarbons		
Naphthalene	0.9	5.9
Fuels		
Gasoline	1.4	7.8
No. 1 fuel oil	0.7	5.0
No. 2 fuel oil	1.3	6.0
No. 4 fuel oil	1.0	5.0
JP-1 jet fuel	0.7	5.0
JP-4 jet fuel	1.3	8.0
JP-5 jet fuel	0.6	4.6
Kerosene	0.7	5.0
Nonhalogenated Solvents		
Acetone	2.5	13.0
Ethanol	3.3	19.0
Methyl ethyl ketone	1.4	11.4
Methyl isobutyl ketone	1.2	8.0

Table 3.8 (Continued)

Halogenated Solvents		
Methylene chloride	12.0	19.0
Trichloroethylene	8.0	10.5
Industrial Intermediates		
Chlorobenzene	1.3	9.6
Phenol	1.8	8.6

Source: Reference 3.17.

Flammability Limits of Mixtures. Although LFLs and UFLs have been determined for hundreds of flammable compounds, most hazardous waste disposal and storage areas contain mixtures of chemicals that, not surprisingly, have flammability limits different from single component systems. The LFLs and UFLs for mixtures were developed by LeChatelier [3.25]:

$$LFL_{mixture} = \frac{1}{\sum\limits_{i=1}^{n} \dfrac{y_i}{LFL_i}} \tag{3.36}$$

where

LFL_i = the LFL for compound i
y_i = mole fraction of compound i in the mixture
n = the number of compounds in the mixture

and

$$UFL_{mixture} = \frac{1}{\sum\limits_{i=1}^{n} \dfrac{y_i}{UFL_i}} \tag{3.37}$$

where UFL_i = the UFL for compound i. The use of Equations 3.36 and 3.37 is demonstrated in Example 3.6.

EXAMPLE 3.6 *Flammability Limits of a Gaseous Mixture*

Determine the LFL and UFL of a gaseous mixture (v/v) of 0.65% acetone, 0.2% decane, and 0.3% hexane.

SOLUTION

1. Calculate the mole fraction of each species; based on the Ideal Gas Law, the number of moles of each gas is directly proportional to its volume:

$$X_{\text{acetone}} = \frac{0.65}{0.65 + 0.2 + 0.3} = 0.565$$

$$X_{\text{decane}} = \frac{0.2}{0.65 + 0.2 + 0.3} = 0.174$$

$$X_{\text{hexane}} = \frac{0.3}{0.65 + 0.2 + 0.3} = 0.261$$

2. Using data from Table 3.8 and Equations 3.36 and 3.37, determine the LFL and UFL for the mixture:

$$LFL_{\text{mixture}} = \frac{1}{\dfrac{0.565}{2.6} + \dfrac{0.174}{0.8} + \dfrac{0.261}{1.1}} = 1.49\%$$

$$UFL_{\text{mixture}} = \frac{1}{\dfrac{0.565}{12.8} + \dfrac{0.174}{9.2} + \dfrac{0.261}{7.5}} = 10.2\%$$

Estimation Methods for Flammability Limits. Although LFLs and UFLs are available for hundreds of hazardous chemicals, the assessment of facilities that contain chemicals for which data are not available is often necessary. In such cases, empirical relationships may be used, which are based on the stoichiometric requirements for combustion between the compound and air [3.26]:

$$LFL = 0.55 \, C_{\text{st}} \tag{3.38}$$

$$UFL = 3.50 \, C_{\text{st}} \tag{3.39}$$

where C_{st} = the % fuel in air, which is defined by combustion stoichiometry. The combustion of most organic compounds may be described by

$$C_mH_xO_y + zO_2 \rightarrow mCO_2 + \frac{x}{2} H_2O \tag{3.40}$$

From stoichiometry:

$$z = m + \frac{x}{4} - \frac{y}{2} \tag{3.41}$$

where z = moles O_2/moles organic compound burned.

Based on the combustion stoichiometry, the following relationships may be developed:

$$C_{\text{st}} = \frac{\text{moles compound}}{\text{moles compound} + \text{moles air}} \times 100\% \tag{3.42}$$

$$C_{st} = \frac{100}{1 + \left(\dfrac{z}{100}\right)} \tag{3.43}$$

Solving for z in Equation 3.43 and combining with Equations 3.38 and 3.39 yields

$$LFL = \frac{0.55(100)}{4.76m + 1.19x - 2.38y + 1} \tag{3.44}$$

$$UFL = \frac{3.50(100)}{4.76m + 1.19x - 2.38y + 1} \tag{3.45}$$

The use of Equations 3.44 and 3.45 in estimating flammability limits is illustrated in Example 3.7.

EXAMPLE 3.7   *Estimation of Flammability Limits*

Estimate the LFL and UFL for acetone. Compare the estimated values to the tabulated values.

SOLUTION

Acetone has the molecular formula:

$$H_3C—CO—CH_3 = C_3H_6O$$

The stoichiometric combustion of acetone may be described by:

$$C_3H_6O + 4O_2 \rightarrow 3CO_2 + 3H_2O$$

Based on the balanced equation

$$m = 3$$
$$x = 6$$
$$y = 1$$
$$z = 4$$

Estimate the LFL and UFL:

$$LFL = \frac{0.55(100)}{4.76(3) + 1.19(6) - 2.38(1) + 1} = 2.74\%$$

The tabulated value = 2.6%

$$UFL = \frac{3.50(100)}{4.76(3) + 1.19(6) - 2.38(1) + 1} = 17.5\%$$

The tabulated value = 12.8%

Therefore, the stoichiometric estimation procedure provides a close approximation to the experimental values.

3.6 FLASH POINT AND IGNITION TEMPERATURE

A measure of ignitability for organic materials based on flammability limits, *flash point* is defined as the minimum temperature at which a compound emits sufficient vapor to form an ignitable mixture with air. An *ignitable mixture* is a vapor-air combination within the flammability limits capable of propagating a flame away from a source of ignition.

Numerous systems are used to experimentally determine flash points. The EPA Method 1010 uses the *Pensky-Martens* closed-cup testing system to determine the flash point for hydrocarbon mixtures, solids suspensions, and other liquids (Figure 3.3). A stirred sample is heated at a slow and constant rate; at periodic intervals when the stirring is stopped, a small flame is directed into the cup. The lowest temperature at which the flame ignites the vapor in the cup is the flash point of the sample.

Two other closed-system devices are commonly used to determine flash points. The *Seta Flash Closed Tester,* the basis for EPA Method 1020, is used for jet fuels, paints, enamels, lacquers, and similar products with flash points between 0 and 145°C. The

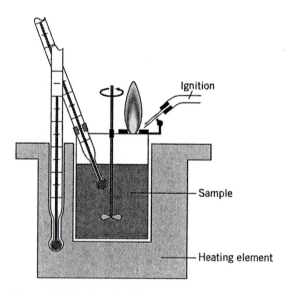

Figure 3.3 The Pensky-Martens closed-cup system used for determining the flash points of hydrocarbon mixtures and other liquids. *Source:* Reference 3.27.

sample is injected into the system with a syringe. The temperature of the system is increased through the anticipated range, and a test flame is applied at intervals until a flash is observed. The *Tag Closed Tester* is used for evaluating liquids with flash points lower than 93°C. The *Cleveland Open Cup Tester* is used for petroleum products, except fuel having a flash point less than 79°C, and the *Tag Open Cup* System is used for liquids with low flash points to simulate open tanks.

Flash points obtained from closed systems are usually 10 to 20% lower than flash points measured in open systems. Therefore, open-cup flash-point data are usually identified with the letters OC.

A property analogous to flash point is *ignition temperature*, which is the minimum temperature required to initiate direct combustion of a material, whether it is a solid, liquid, or gas. Because of the difficulty of measurement, ignition temperatures are considered approximations. Flash points and ignition temperatures for selected hazardous compounds are listed in Table 3.9.

Table 3.9 Flash Points and Ignition Temperatures of Several Common Hazardous Compounds

Compound	Flash Point (°C)	Ignition Temperature (°C)
Aliphatic Hydrocarbons		
Cyclohexane	−20.0	245.0
Isopentane	−56.6	426.6
n-Heptane	−3.8	203.8
n-Hexane	−21.6	225.0
n-Octane	13.3	206.1
Pentane	−40.0	260.0
Aromatic Hydrocarbons		
Benzene	−11.1	497.7
Ethylbenzene	−6.6	432.2
m-Xylene	27.2	527.7
o-Xylene	32.2	463.8
p-Xylene	27.2	528.8
Toluene	4.4	480.0
Polycyclic Aromatic Hydrocarbons		
Naphthalene	78.8	526.1
Fuels		
Gasoline	−42.7	280.0
No. 1 fuel oil	37.0 to 72.0	210.0 to 228.0
No. 2 fuel oil	52.0 to 95.0	256.6
No. 5 fuel oil	68.0 to 168.0	407.2
JP-1 jet fuel	37.7	228.8

Table 3.9 (Continued)

Compound	Flash Point (°C)	Ignition Temperature (°C)
Fuels (cont.)		
JP-3 jet fuel	43.0 to 65.0	N/A
JP-4 jet fuel	−23.0 to −1.0	240.0
JP-5 jet fuel	60.0	246.1
Nonhalogenated Solvents		
Acetone	−17.7	465.0
Ethanol	12.7	362.7
Methyl ethyl ketone	−8.8	403.8
Methyl isobutyl ketone	17.7	448.8
Cyclohexanol	71.0	300.0
Turpentine	35.0	253.3
Halogenated Solvents		
Methylene chloride	N/A	556.1
Trichloroethylene	32.2	420.0
Herbicides		
2,4-D	79.4	N/A
Trifluralin	85.0	N/A
Insecticides		
DDT	72.2	N/A
Endrin	26.6	N/A
Malathion	163.0	N/A
Industrial Intermediates		
Chlorobenzene	28.9	640.0
Dichlorophenol	93.3	N/A
Phenol	79.4	715.0
Explosives		
2,4-Dinitrotoluene	206.6	N/A

Source: Reference 3.28.

3.7 CHEMICAL INCOMPATIBILITY

Hazardous wastes and materials are often collected and stored for a number of purposes. Hazardous wastes generated by an industry are usually stored under RCRA guidelines while awaiting treatment or disposal, and hazardous materials and commercial products are stored in preparation for sale or industrial use. Contaminated soils, wastes collected from tank bottoms, and other materials from contaminated sites are often placed in containers before they are treated or sent to a landfill. These materials must be combined and stored in a rational manner so that mixing does not result in chemical reactions that cause immediate safety and health hazards.

A range of undesirable reactions can result from chemical incompatibility. The most common consequences are fires, explosions, and the evolution of heat and toxic gases.

Fires are propagated only if there is sufficient fuel, an oxidizer in sufficient concentrations, and an ignition source to initiate combustion. The list of common fuels and oxidizers, shown in Table 3.10, shows that oxygen is not the only oxidant in fires; chemical fires may also be promoted by a number of oxidation-reduction reactions. Fires can result from chemical incompatibility; for instance, the mixing of hydrogen sulfide and calcium hypochlorite produces a strongly exothermic redox reaction that results in a chemically generated fire.

Chemical reactivity (i.e., explosivity) is analogous to flammability; the primary difference is that an explosion propagates more rapidly than a fire. An explosion is characterized by a substantial energy release in a short time within a small space or volume. Explosions occur most commonly in hazardous waste management when a combustible mixture is ignited in a confined space, such as a waste storage drum, a railroad tank car, or a building. After ignition, a flame front propagates away from the ignition source; the pressure at which the front propagates determines the nature and damage of the explosion.

Explosions are classified as detonations or deflagrations, divisions that are based on the characteristics of the shock wave produced by the expansion of gases, which is the result of either a change in the number of moles of gases or the effect of thermal expansion. *Deflagrations* are defined by a pressure wave slower than the speed of sound. Increases in pressure of several atmospheres are common for deflagrations. *Detonations* have pressure increases at least an order of magnitude greater than deflagrations and have shock waves that propagate faster than the speed of sound.

Chemical incompatibility can also result from the heat evolved by mixing two incompatible solutions. Under conditions of ideal mixing, such as combining infinitely dilute acids and bases, the enthalpy change of mixing is zero:

$$\Delta H_{mix} = 0 \qquad (3.46)$$

However, concentrated solutions of acids and bases do not behave ideally, and heat is evolved when a concentrated acid is mixed with a base, or when a concentrated acid or base is diluted with water. Excessive heat evolution is commonplace during such

Table 3.10 Common Components of Fires and Explosions

	Liquids	Solids	Gases
Fuels	Petroleum Nonhalogenated solvents Pesticide concentrates	Wood products Plastics	Acetylene Propane Butane
Oxidizers	Hydrogen peroxide Nitric acid Perchloric acid	Metal peroxides Ammonium nitrate Sodium nitrate	Oxygen Chlorine Fluorine

Table 3.11 General Guidelines on Chemical Incompatibilities

Chemical Class	Incompatible Class of Chemicals	Consequences
Alkaline materials (Examples: spent caustics, lime wastewaters, acetylene sludge)	Acidic materials (Examples: battery acid, acid sludges, pickling liquor)	Heat generation or violent reaction
Zero-valence alkali and alkaline earth metals [Examples: Na(0), K(0), Al(0), Ca(0)]	Acids, bases, and water	Fire, explosion, or generation of flammable hydrogen gas
Alcohols Water	Water-reactive chemicals [Examples: acids, bases, K(0), Li(O), SOCl$_2$, PCl$_3$]	Fire, explosion, heat generation, or toxic/flammable gas generation
Reactive solvents and organic compounds (Examples: alcohols, aldehydes, halogenated compounds, nitroorganics, alkenes)	Acids, bases, zero-valence metals	Fire, explosion, or violent reaction
Weak acids of toxic, volatile gases (Examples, cyanide, sulfides)	Acids	Generation of toxic gases
Strong oxidizers (Examples: chlorine, chromic acid, perchlorates, permanganates, peroxides)	Organic and mineral acids, zero-valence metals, solvents and organics	Fire, explosion, or violent reaction

mixing procedures, and the heat generated is proportional to the concentrations of the two species being mixed. For example, mixing equal volumes of concentrated (96%) sulfuric acid and water evolves sufficient heat to make the mixture boil.

The evolution of toxic gases usually results from mixing an acid, such as hydrochloric or sulfuric acid, with the ionized salt of a weak acid to produce an unionized volatile and toxic species. Alternatively, adding a caustic solution to a weak base may result in the evolution of a toxic gas:

$$A^- + H^+ \longrightarrow AH_{(g)}(\text{toxic}) \tag{3.47}$$

or

$$BH^+ + OH^- \longrightarrow B_{(g)}(\text{toxic}) \tag{3.48}$$

The two common examples of this incompatibility include the addition of acid to cyanide and sulfide:

$$CN^- + H^+ \longrightarrow HCN \qquad pK_a = 9.1 \tag{3.49}$$

$$S^{2-} + 2H^+ \longrightarrow H_2S \qquad pK_a = 7.1 \tag{3.50}$$

Both hydrogen cyanide and hydrogen sulfide are volatile gases that rapidly escape from aqueous solutions. After volatilizing, they pose significant toxicity problems to humans and animals.

The examples cited here—chemical fires, heat production, chemical explosions, and toxic gas generation—represent the most common mechanisms of chemical incompatibility. Predicting chemical incompatibilities is based on a range of chemical, physical, and toxicological properties such as oxidation-reduction potential, enthalpies of mixing, and toxicity of gaseous products. Although chemical incompatibilities can be evaluated based on theoretical concepts, the use of lists is often more effective in their prediction. General guidelines for chemical incompatibilities, listed in Table 3.11 on page 186, provide rules for a wide range of chemicals, such as not mixing acids with bases, not mixing metals with reducing agents, and not combining cyanides and other potentially toxic ionized compounds with acids. To complement these general statements, a detailed list of chemical incompatibilities, including explosive or flammable combinations and toxic products that are produced from mixing various chemicals, is available in Appendix D. In addition, more specific information related to the incompatibilities of the most common hazardous chemicals is provided in Table 3.12. The use of the tabular data to assess chemical incompatibilities is demonstrated in Example 3.7.

EXAMPLE 3.7 *Evaluation of Chemical Incompatibility*

A proposed waste plan is to mix waste trichloroethylene with sulfuric acid. Are these chemicals compatible?

SOLUTION

Chemicals that are incompatible with halogenated organics, such as TCE, are listed in Table 3.12. Some of the incompatible chemicals include acids, amines, azo compounds, caustics, and cyanides. Mixing TCE with sulfuric acid will result in heat generation and the evolution of toxic gas; therefore, these two chemicals are incompatible.

Table 3.12 Incompatible Mixtures of Common Hazardous Chemicals

Chemical Class	Incompatible Chemicals	Hazard
Hydrocarbons		
Aliphatic saturated hydrocarbons	Oxidizing mineral acids	Heat, fire
Aliphatic unsaturated hydrocarbons	Nonoxidizing mineral acids	Heat
	Oxidizing mineral acids	Heat, fire
	Aldehydes	Heat
	Elemental and alloy metals	Heat, fire
Aromatic hydrocarbons	Oxidizing mineral acids	Heat, fire
Nonhalogenated Solvents		
Ketones	Nonoxidizing mineral acids	Heat
	Oxidizing mineral acids	Heat, fire
	Azo compounds and hydrazines	Heat, nontoxic/nonflammable gas
	Cyanides	Heat
	Strong bases	Heat
Aldehydes	Nonoxidizing mineral acids	Heat, violent polymerization
	Oxidizing mineral acids	Heat, violent polymerization
	Organic acids	Heat, violent polymerization
Organic acids	Oxidizing mineral acids	Heat, nontoxic/nonflammable gas
Alcohols and glycols	Nonoxidizing mineral acids	Heat, nontoxic/nonflammable gas
	Oxidizing mineral acids	Heat
	Organic acids	Heat, fire
		Heat, violent polymerization
Aliphatic and aromatic amines	Nonoxidizing mineral acids	Heat
	Oxidizing mineral acids	Heat, toxic gas
	Organic acids	Heat
	Aldehydes	Heat
Esters	Nonoxidizing mineral acids	Heat
	Oxidizing mineral acids	Heat, fire
	Azo compounds and hydrazines	Heat, nontoxic/nonflammable gas
	Caustics	Heat

Table 3.12 (Continued)

Chemical Class	Incompatible Chemicals	Hazard
Nonhalogenated Solvents (cont.)		
Ethers	Nonoxidizing mineral acids	Heat
	Oxidizing mineral acids	Heat, fire
Nitroorganics	Oxidizing mineral acids	Heat, fire, toxic gas
	Aldehydes	Heat
	Caustics	Heat, explosion
	Alkalai and alkaline earth metals	Heat, flammable gas, explosion
	Nitrites	Heat, flammable gas, explosion
Halogenated Solvents/	Nonoxidizing mineral acids	Heat, toxic gas
Halogenated Organics	Oxidizing mineral acids	Heat, fire, toxic gas
	Amines	Heat, toxic gas
	Azo compounds and hydrazines	Heat, nontoxic/nonflammable gas
	Caustics	Heat, flammable gas
	Cyanides	Heat
Pesticides		
Organophosphates	Nonoxidizing mineral acids	Heat, toxic gas
	Oxidizing mineral acids	Heat, toxic gas
	Caustics	Heat, explosion
	Alkali and alkaline earth metals	Heat
Carbamates	Nonoxidizing mineral acids	Heat, nontoxic/nonflammable gas
	Oxidizing mineral acids	Heat, toxic gas
	Azo compounds and hydrazines	Heat, nontoxic/nonflammable gas
Amides	Nonoxidizing mineral acids	Heat
	Oxidizing mineral acids	Heat, toxic gas
Explosives		
	Oxidizing mineral acids	Heat, explosion
	Nonoxidizing mineral acids	Heat, explosion
	Organic acids	Heat, explosion
	Azo compounds and hydrazines	Heat, explosion
	Caustics	Heat, explosion
	Esters	Heat, explosion
	Alkali and alkaline earth metals	Heat, explosion
	Elemental and alloy metals (e.g., powders, vapors)	Heat, explosion

Table 3.12 (Continued)

Chemical Class	Incompatible Chemicals	Hazard
Explosives (cont.)		
	Elemental and alloy metals (e.g., sheets, rods)	Heat, explosion
	Toxic/heavy metals	Heat, explosion
	Peroxides	Heat, explosion
	Phenols and cresols	Heat, explosion
	Sulfides	Heat, polymerization
Industrial Intermediates		
Phenols and cresols	Nonoxidizing mineral acids	Heat
	Oxidizing mineral acids	Heat, fire
	Azo compounds and hydrazines	Heat, nontoxic/nonflammable gas
	Isocyanates	Heat, violent polymerzation
	Alkalai and alkaline earth metals	Heat, flammable gas
	Nitriles	Heat, flammable gas
	Peroxides	Heat
Alkali and Alkaline Earth Elemental Metals		
	Nonoxidizing mineral acids	Flammable gas, heat, fire
	Oxidizing mineral acids	Flammable gas, heat, fire
	Organic acids	
	Alcohols and glycols	
	Aldehydes	
	Amides	Flammable gas, heat
	Amines	Flammable gas, heat
	Azo compounds and hydrazines	Flammable gas, heat
	Carbamates	Flammable gas, heat
	Caustics	Flammable gas, heat
	Cyanides	Flammable gas, heat
	Esters	Flammable gas, heat
	Halogenated organics	Heat, explosion
	Isocyanates	Flammable gas, heat
	Ketones	Flammable gas, heat
	Mercaptans and organic sulfides	Flammable gas, heat
Toxic/Heavy Metals		
	Nonoxidizing mineral acids	Release of toxic metals
	Oxidizing mineral acids	Release of toxic metals

Table 3.12 (Continued)

Chemical Class	Incompatible Chemicals	Hazard
Toxic/Heavy Metals (cont.)		
	Organic acids	Release of toxic metals
	Amides	Release of toxic metals
	Amines	Release of toxic metals
	Caustics	Release of toxic metals
Cyanides		
	Nonoxidizing mineral acids	Toxic gases, flammable gas
	Oxidizing mineral acids	Toxic gases, flammable gas
	Organic acids	Toxic gases, flammable gas
	Azo compounds and hydrazines	Nontoxic/nonflammable gas

3.8 LABELS AND PLACARDS

Placards and labeling systems are used to identify potential dangers during the transportation and storage of hazardous wastes and hazardous materials. Two systems are used for identifying hazardous materials and wastes—one for transportation and the other for storage. The U.S. Department of Transportation (DOT) requires the placement of placards on the outside of vehicles transporting hazardous materials to aid in assessing the potential hazard of a spill. Hazardous wastes and materials under stationary conditions (e.g., storage, manufacturing) require labels under a system developed by the National Fire Protection Association (NFPA).

The DOT requires that trucks and rail cars carrying hazardous materials display *placards*, which must be placed on all four sides of the vehicle. Typical placards used in transportation are shown in Figure 3.4. Fire personnel, emergency medical personnel, and others who come under the category of *first responders* to traffic accidents are often faced with a tank truck that has spilled its load, which may involve chemicals ranging from hydrochloric acid to methyl parathion. Placards provide the most effective means of identifying hazards under such emergency response conditions.

The DOT requires placards when 1000 pounds (454 kg) or more of hazardous materials are transported by rail or truck unless the materials are classified as explosives, poisons, radioactive, or flammable solids. In these cases, placarding is required, regardless of the quantity.

The DOT classifies materials based on nine hazard classes, which are listed in Table 3.13. Each hazardous material is placed in one class based on its most significant hazards, even though it may be characterized by more than one hazardous property. For example, chlorine gas is both a poisonous gas and an oxidizer, but DOT classifies it as a poisonous gas—its most significant hazard.

Two variations of a typical placard are shown in Figure 3.5. These signs are diamond shaped with sides of 10.75 inches (27.3 cm) and have a colored background, with colors

Figure 3.4 Use of placards in transportation.

Table 3.13 The Nine Classes of Hazardous Materials Used by the DOT

DOT Hazard Code Number	DOT Symbol	Hazard Class
1		Explosives
2		Gases
3		Flammable liquids
4		Flammable solids
5		Spontaneously combustible materials
		Materials dangerous when wet
		Oxidizers
		Organic peroxides
6		Poisonous materials
		Biohazard
7		Radioactive materials
8		Corrosives
9		Other regulated materials

Source: Reference 3.29.

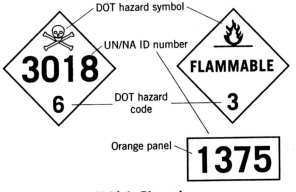

Vehicle Placards

Figure 3.5 Placards with locations of DOT hazard information. *Source:* Reference 3.29.

specific to each hazard class. Three pieces of information are displayed on a placard: (1) the DOT hazard symbol, (2) the corresponding DOT hazard code (Table 3.13), and (3) a United Nations/North American (UN/NA) identification number. Hundreds of UN/NA identification numbers are listed in the *DOT Emergency Response Guidebook* [3.30]. A brief sample of the UN/NA numbers and hazard classes for common hazardous chemicals are listed in Table 3.14. A symbol at the top of the placard specifies the DOT hazard class (e.g., biohazard, corrosive, explosive, flammable liquid, flammable solid, gas, oxidizer, poison, or radioactive material), and the corresponding DOT code for the hazard class is located at the bottom of the placard (Table 3.13). The UN/NA identification numbers, which are required on shipments in the United States and Canada, are located in the center of a placard or beside the placard on an orange rectangular sign in blue letters. If the UN/NA number is placed on a placard, it is located in the center in place of the hazard class (Figure 3.5). For hazardous materials classified as poison gas, radioactive, or explosive, DOT regulations require that the UN/NA identification number be placed on the separate orange rectangular sign. The orange rectangular sign must be positioned adjacent to the diamond-shaped placard. Examples of various placards used in the transportation industry are illustrated in Figure 3.6.

The NFPA system is used for labeling materials in storage areas, such as drums awaiting treatment or disposal. The system is based on a simple and easily understood procedure that provides information on the immediate hazards of short-term health effects, flammability, and explosivity. The labeling system was developed to provide a mechanism to protect the health of firefighters, plant operation engineers, and safety personnel in the event of a spill or fire.

The system focuses on three categories in defining hazards: health, flammability, and reactivity. Five levels of hazard have been defined for each category, ranging from 4 (for severe hazard) to 0 (for no hazard). Qualitative descriptions of the five degrees of hazard for the three categories are listed in Table 3.15, and a listing of hazards for common chemicals is presented in Table 3.16. The health hazard rating is applicable more to fire fighting than to occupational or public health. Firefighters are exposed to hazardous materials for a few seconds to an hour, and the NFPA system classifies

Table 3.14 DOT Hazard Classifications and United Nations/North American Identification Numbers of Several Common Hazardous Compounds

Compound	DOT Hazard Class	UN/NA Number
Aliphatic Hydrocarbons		
Cyclohexane	Flammable Liquid	UN 1145
Hexane	Flammable Liquid	UN 1208
Isopentane	Flammable Liquid	UN 1265
n-Heptane	Flammable Liquid	UN 1206
n-Octane	Flammable Liquid	UN 1262
Pentane	Flammable Liquid	UN 1265
Aromatic Hydrocarbons		
Benzene	Flammable Liquid	UN 1114
Ethylbenzene	Flammable Liquid	UN 1175
m-Xylene	Flammable Liquid	UN 1307
o-Xylene	Flammable Liquid	UN 1307
p-Xylene	Flammable Liquid	UN 1307
Toluene	Flammable Liquid	UN 1294
Polycyclic Aromatic Hydrocarbons		
Naphthalene	N/A	NA 2304
Fuels		
Gasoline	Flammable Liquid	UN 1203
No. 1 fuel oil	Combustible Liquid	NA 1993
No. 5 fuel oil	Combustible Liquid	NA 1993
JP-1 jet fuel	Flammable Liquid	UN 1223
JP-3 jet fuel	Flammable Liquid	UN 1223
JP-4 jet fuel	Flammable Liquid	UN 1863
JP-5 jet fuel	Flammable Liquid	UN 2761
Kerosene	Combustible Liquid	UN 1223
Nonhalogenated Solvents		
Acetone	Flammable Liquid	UN 1090
Ethanol	Flammable Liquid	UN 1170
Methyl ethyl ketone	Flammable Liquid	UN 1193
Methyl isobutyl ketone	Flammable Liquid	UN 1245
Cyclohexanol	Combustible Liquid	NA 1993
Turpentine	Flammable Liquid	UN 1299

Table 3.14 (Continued)

Compound	DOT Hazard Class	UN/NA Number
Halogenated Solvents		
Carbon tetrachloride	ORM-A	UN 1846
Chloroform	ORM-A	UN 1888
Methylene chloride	ORM-A	UN 1593
Perchloroethylene	ORM-A	UN 1897
Trichloroethylene	ORM-A	UN 1710
Herbicides		
2,4-D Esters	ORM-E	UN 2765
Trifluralin	Poison	UN 1609
Fungicides		
Pentachlorophenol	ORM-E	NA 2020
Insecticides		
Aldrin	Poison B	NA 2761
Dieldrin	ORM-A	NA 2761
DDT	ORM-A	NA 2761
Endrin	Poison B	NA 2761
Lindane	ORM-A	NA 2761
Malathion	ORM-A	NA 2783
Parathion	Poison B	NA 2783
Industrial Intermediates		
Chlorobenzene	Flammable Liquid	UN 1134
2,4-Dichlorophenol	Poison A	UN 2020
Hexachlorocyclopentadiene	Corrosive	UN 2646
Phenol	Poison B	UN 1671
2,4,6-Trichlorophenol	ORM-A	NA 2020
Explosives		
2,4-Dinitrotoluene	ORM-E	NA 1600
Gelatine dynamite	Explosives A	N/A
Polychlorinated Biphenyls		
PCBs	ORM-E	UN 2315

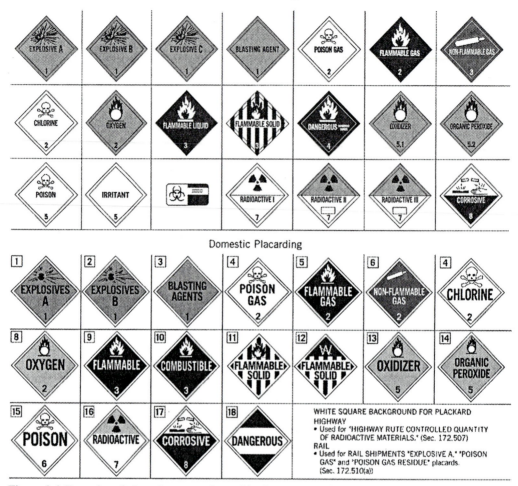

Figure 3.6 Examples of placards commonly used in the transportation industry.

health hazards based solely on such short-term exposures. Occupational and public health concerns are based on exposure over years and decades, and use an entirely different system for assessing toxicity and hazard (see Chapter 10).

The numbers associated with the three categories—fire, health, and reactivity—are displayed on hazardous commercial products, drums, tanks, and storage areas with a diamond-shaped label on the container or displayed on a wall or door of the storage area (Figure 3.7). The placement of the hazard numbers is shown in Figure 3.8. The health hazard is displayed on the left quarter-diamond, fire is at the top, and reactivity is on the right. If the material is incompatible with water, a W is listed in the bottom field.

Table 3.15   Hazard Ranking of the National Fire Protection Association

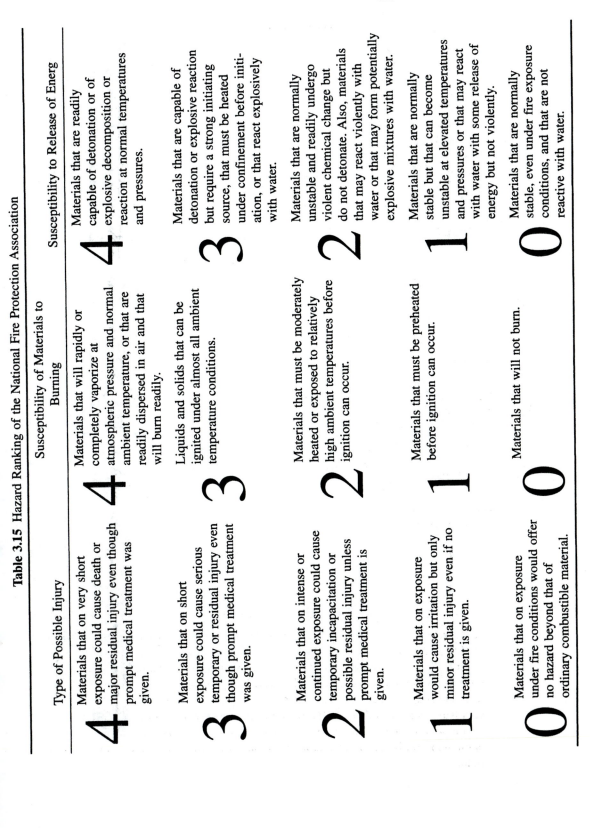

Type of Possible Injury	Susceptibility of Materials to Burning	Susceptibility to Release of Energ
4 Materials that on very short exposure could cause death or major residual injury even though prompt medical treatment was given.	**4** Materials that will rapidly or completely vaporize at atmospheric pressure and normal ambient temperature, or that are readily dispersed in air and that will burn readily.	**4** Materials that are readily capable of detonation or of explosive decomposition or reaction at normal temperatures and pressures.
3 Materials that on short exposure could cause serious temporary or residual injury even though prompt medical treatment was given.	**3** Liquids and solids that can be ignited under almost all ambient temperature conditions.	**3** Materials that are capable of detonation or explosive reaction but require a strong initiating source, that must be heated under confinement before initiation, or that react explosively with water.
2 Materials that on intense or continued exposure could cause temporary incapacitation or possible residual injury unless prompt medical treatment is given.	**2** Materials that must be moderately heated or exposed to relatively high ambient temperatures before ignition can occur.	**2** Materials that are normally unstable and readily undergo violent chemical change but do not detonate. Also, materials that may react violently with water or that may form potentially explosive mixtures with water.
1 Materials that on exposure would cause irritation but only minor residual injury even if no treatment is given.	**1** Materials that must be preheated before ignition can occur.	**1** Materials that are normally stable but that can become unstable at elevated temperatures and pressures or that may react with water with some release of energy but not violently.
0 Materials that on exposure under fire conditions would offer no hazard beyond that of ordinary combustible material.	**0** Materials that will not burn.	**0** Materials that are normally stable, even under fire exposure conditions, and that are not reactive with water.

Table 3.16 Flammability and Hazard Data for Common Hazardous Chemicals

Chemical	Health	Hazard Identification Flammability	Reactivity
Acetone	1	3	0
Acetonitrile	3	3	3
Acetophenone	1	1	0
Acrylonitrile	4	3	2
Aniline	3	2	0
Benzene	2	3	0
Chlorobenzene	2	3	0
o-Chlorophenol	3	2	0
o-Cresol	2	2	0
m-Cresol	2	1	0
Cyclohexane	1	3	0
Decane	0	2	0
o-Dichlorobenzene	2	2	0
1,4-Dioxane	2	3	1
Dodecane	0	2	0
Ethylbenzene	2	3	0
Heptane	1	3	0
Hexadecane	0	1	0
Hexane	1	3	0
Hydrogen cyanide	4	4	2
Methylene chloride	2	0	0
Methyl ethyl ketone	1	3	0
Naphthalene	2	2	0
Nonane	0	3	0
Octane	0	3	0
Phenol	2	1	0
Toluene	2	3	0
o-Xylene	2	3	0
m-Xylene	2	3	0
p-Xylene	2	3	0

3.9   CHEMICAL ABSTRACT SERVICE REGISTRY NUMBERS

One problem with the IUPAC and common nomenclature systems described in Chapter 2 is the large number of synonymous names that can be given to a specific chemical. As a result, a significant amount of ambiguity is found in chemical nomenclature. An alternative to the numerous IUPAC, common, and trade names that provides a unique numerical identity to a chemical is the *Chemical Abstract Service (CAS) registry number.* This classification is based on multidigit numbers derived from an unambiguous computer language description of a chemical's molecular structure. Even molecular stereochemistry is included in the numerical code. Registry numbers are assigned continually to new chemicals, so the system is analogous to social security numbers for chemicals. These registry numbers have been developed for all of the

Figure 3.7 A National Fire Protection Association placard for identifying hazards in an industrial storage area.

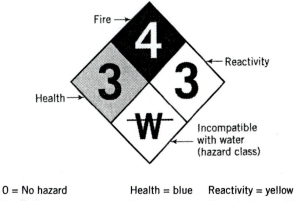

0 = No hazard	Health = blue	Reactivity = yellow
4 = Extremely hazardous	Fire = red	Hazard class = white

Figure 3.8 A typical NFPA placard with locations of hazard numbers. *Source:* Reference 3.29.

compounds indexed in *Chemical Abstracts* since 1965. In addition, publications and references are surveyed continually for novel chemicals that are added to the registry.

A short list of CAS registry numbers for common hazardous compounds is shown in Table 3.17, and registry numbers for the RCRA P and U lists are provided in Appendix A. The most exhaustive cross-reference between CAS registry numbers, IUPAC

Table 3.17 Chemical Abstracts Service (CAS) Registry Numbers
for Common Hazardous Chemicals

Compound	CAS Registry Number
Aliphatic Hydrocarbons	
Cyclohexane	110-82-7
Cyclohexene	110-83-8
n-Decane	124-18-5
n-Dodecane	112-40-3
n-Heptane	142-82-5
Isooctane	540-84-1
n-Octane	110-65-9
Monocyclic Aromatic Hydrocarbons	
Benzene	71-43-2
Ethylbenzene	100-41-4
Toluene	108-88-3
o-Xylene	95-47-6
m-Xylene	108-38-3
p-Xylene	106-42-3
Polycyclic Aromatic Hydrocarbons	
Anthracene	120-12-7
Benzo[a]anthracene	56-55-3
Benzo[a]pyrene	50-32-8
Chrysene	218-01-9
Fluorene	86-73-7
Naphthalene	91-20-3
Phenanthrene	85-01-8
Pyrene	129-00-0
Unchlorinated Solvents	
Acetone	67-64-1
Methyl butyl ketone	591-78-6
Methyl ethyl ketone	78-93-3
Methyl isobutyl ketone	108-10-1
Halogenated Solvents	
Carbon tetrachloride	56-23-5
Chloroform	67-66-3
Methylene chloride	75-09-2
Perchloroethylene	127-18-4
1,1,1-Trichloroethane	71-55-6
Trichloroethylene	79-01-6

Table 3.17 (Continued)

Compound	CAS Registry Number
Insecticides	
Aldrin	309-00-2
Carbaryl	63-25-2
Dieldrin	60-57-1
DDT	50-29-3
Hexachlorobenzene	118-74-1
Lindane	58-89-9
Malathion	121-75-5
Methoxychlor	72-43-5
Parathion	56-38-2
Herbicides	
2,4-D	94-75-7
2,4,5-T	93-76-5
Trifluralin	1528-09-8
Fungicides	
Pentachlorophenol	87-86-5
Industrial Intermediates	
Chlorobenzene	108-90-7
2,4-Dichlorophenol	120-83-2
Dimethyl phthalate	131-11-3
Di-*n*-octyl phthalate	117-84-0
Hexachlorocyclopentadiene	77-47-4
Phenol	108-95-2
2,4,6-Trichlorophenol	88-06-2
Explosives	
Picric acid	88-89-1
2,4,6-Trinitrotoluene	118-96-7
Polychlorinated Biphenyls	
PCB-1016	12674-11-2
PCB-1232	11141-16-5
PCB-1248	12672-29-6
PCB-1254	11097-69-1
PCB-1260	11096-82-5
Chlorinated Dioxins	
2,3,7,8-Tetrachlorodibenzo-*p*-dioxin	1746-01-6

names, common names, and trade names is the *Cross Reference Index of Hazardous Chemicals, Synonyms, and CAS Registry Numbers* [3.31]. One of the primary uses of CAS registry numbers is in searching computerized databases for information on individual elements or compounds.

3.10   PRIORITY POLLUTANTS

As a result of a law suit filed by the Natural Resources Defense Council (NRDC) against the EPA, the 1977 Amendments to the Clean Water Act (Section 1.2.5) included an emphasis on toxic discharges. Besides identifying the industries that would be regulated and the methods of controlling the discharge of toxic chemicals, a primary emphasis of the settlement was identification of the chemicals that would be regulated under the Clean Water Act.

The compounds identified under this consent degree are called *priority pollutants* (Table 3.18). The EPA was charged to develop a program to regulate the discharge of 65 categories of priority pollutants (129 specific chemicals) by 34 industry categories. As the result of the regulation of these chemicals, more than 70% of U.S. industries were affected by the 1977 Clean Water Act amendments.

Although the priority pollutants were developed for end-of-pipe discharges under the Clean Water Act, their use has been adopted in hazardous waste management. The priority pollutants are part of the CERCLA "list of lists"; more importantly, these chemicals are widely recognized industrial pollutants that are commonly found at thousands of contaminated sites.

3.11   SUPPLEMENTAL DATA ON CONTAMINANT PROPERTIES

The information in this chapter provides descriptions of important hazardous waste properties and classification schemes. Tables 3.1 through 3.18 include data on water solubility, density, ignition temperature and flash point, NFPA hazard labeling, CAS registry numbers, and priority pollutant classification for the most common hazardous compounds likely to be encountered in hazardous waste practice. However, information for any one of the thousands of hazardous chemicals once used or currently in use may be needed in a hazardous waste source assessment.

There are several sources of data available for the contaminant properties described in this chapter. For water solubility and density, a useful source of information is the *Handbook of Environmental Data on Organic Chemicals* [3.9], which contains data on water solubility, vapor pressure, biodegradation rate, and a number of other properties. The *CRC Handbook of Chemistry and Physics* [3.13] is an additional resource for water solubility and density. For data on flash points, ignition temperature, and hazard labeling, the *Fire Protection Guide on Hazardous Materials,* published by the National Fire Protection Association [3.28], is the most comprehensive source. Another good source for the same type of data is the *Firefighter's Hazardous Materials Reference Book* [3.17]. A good, widely available reference for CAS registry numbers is the *Merck Index* [3.32]. Much of this information is also provided on Material Safety Data Sheets (MSDSs) which all chemical manufacturers and suppliers must provide to customers on delivery of chemical shipments in accordance with the right-to-know provisions of HSWA. The Chemical Manufacturers Association has compiled and continues to update collections of MSDSs. The Agricultural Chemicals Manufacturing Association

Table 3.18 U.S. EPA Priority Pollutants

1. Acenaphthene
2. Acrolein
3. Acrylonitrile
4. Benzene
5. Benzidine
6. Carbon tetrachloride

Chlorinated Benzenes
(other than dichlorobenzenes)
7. Chlorobenzene
8. 1,2,4-Trichlorobenzene
9. Hexachlorobenzene

Chlorinated Ethanes
10. 1,2-Dichloroethane
11. 1,1,1-Trichloroethane
12. Hexachloroethane
13. 1,1-Dichloroethane
14. 1,1,2-Trichloroethane
15. 1,1,2,2-Tetrachloroethane
16. Chloroethane

Chloroalkyl Ethers
17. Bis(chloromethyl) ether
18. Bis(2-chloroethyl) ether
19. 2-Chloroethyl vinyl ether

Chlorinated Naphthalene
20. 2-Chloronaphthalene

Chlorinated Phenols
21. 2,4,6-Trichlorophenol
22. *p*–Chloro-*m*-cresol
23. Chloroform (trichloromethane)
24. 2-Chlorophenol

Dichlorobenzenes
25. 1,2-Dichlorobenzene
26. 1,3-Dichlorobenzene
27. 1,4-Dichlorobenzene

Dichlorobenzidine
28. 3,3-Dichlorobenzidine

Dichloroethylenes
29. 1,1-Dichloroethylene
30. 1,2-*trans*-Dichloroethylene
31. 2,4-Dichlorophenol

Dichloropropane and Dichloropropylene
32. 1,2-Dichloropropane

33. 1,2-Dichloropropylene (1,3-Dichloropropene)
34. 2,4-Dimethylphenol

Dinitrotoluene
35. 2,4-Dinitrotoluene
36. 2,6-Dinitrotoluene
37. 1,2-Diphenylhydrazine
38. Ethylbenzene
39. Fluoranthene

Haloethers (other than those listed elsewhere)
40. 4-Chlorophenyl phenyl ether
41. 4-Bromophenyl phenyl ether
42. Bis(2-chloroisopropyl) ether
43. Bis(2-chloroethoxy) methane

Halomethanes (other than those listed elsewhere)
44. Methylene chloride
45. Methyl chloride
46. Methyl bromide
47. Bromoform
48. Dichlorobromomethane
49. Trichlorofluoromethane
50. Dichlorodifluoromethane
51. Chlorodibromomethane
52. Hexachlorobutadiene
53. Hexachlorocyclopentadiene
54. Isophorone
55. Naphthalene
56. Nitrobenzene

Nitrophenols
57. 2-Nitrophenol
58. 4-Nitrophenol
59. 2,4-Dinitrophenol
60. 4,6-Dinitro-*o*-cresol

Nitrosamines
61. *N*-Nitrosodimethylamine
62. *N*-Nitrosodiphenylamine
63. *N*-Nitrosodi-*n*-propyl-amine

64. Pentachlorophenol
65. Phenol

Phthalate Esters
66. Bis(2-ethylhexyl) phthalate
67. Butyl benzyl phthalate
68. Di-*n*-butyl phthalate
69. Di-*n*-octyl phthalate
70. Diethyl phthalate
71. Dimethyl phthalate

Polynuclear Aromatic Hydrocarbons
72. Benzo[*a*]anthracene
73. Benzo[*a*]pyrene
74. 3,4-Benzofluoranthene
75. Benz[*k*]fluoranthane
76. Chrysene
77. Acenaphthylene
78. Anthracene
79. Benzo[*g,h,i*]perylene
80. Fluorene
81. Phenanthrene
82. Dibenzo[*a,h*]anthracene
83. Indeno[1,2,3-*c,d*]pyrene
84. Pyrene
85. Tetrachloroethylene
86. Toluene
87. Trichloroethylene
88. Vinyl chloride

Pesticides and Metabolites
89. Aldrin
90. Dieldrin
91. Chlordane
92. 4,4'-DDT
93. 4,4'-DDE
94. 4,4'-DDD

Endosulfan and Metabolites
95. α-Endosulfan
96. β-Endosulfan
97. Endosulfan sulfate

Endosulfan and Metabolites
98. Endrin
99. Endrin aldehyde

Heptachlor and Metabolites
100. Heptachlor
101. Heptachlor epoxide

Table 3.18 (Continued)

Hexachlorocyclohexane	109. PCB-1232 (Aroclor	119. Chromium (total)
(all isomers)	1232)	120. Copper (total)
102. α-BHC	110. PCB-1248 (Aroclor	121. Cyanide (total)
103. β-BHC	1248)	122. Lead (total)
104. γ-BHC (Lindane)	111. PCB-1260 (Aroclor	123. Mercury (total)
105. δ-BHC	1260)	124. Nickel (total)
Polychlorinated	112. PCB-1016 (Aroclor	125. Selenium (total)
Biphenyls (PCBs)	1016)	126. Silver (total)
106. PCB-1242 (Aroclor	113. Toxaphene	127. Thallium (total)
1242)	114. Antimony (total)	128. Zinc (total)
107. PCB-1254 (Aroclor	115. Arsenic (total)	129. 2,3,7,8-Tetra-
1254)	116. Asbestos (fibrous)	chlorodibenzo-
108. PCB-1221 (Aroclor	117. Beryllium (total)	p-dioxin (TCDD)
1221)	118. Cadmium (total)	

(ACMA) has a similar collection for pesticide information. *The Farm Chemicals Handbook* [3.33] summarizes information on all pesticides manufactured worldwide.

3.12 SUMMARY OF IMPORTANT POINTS AND CONCEPTS

- Water solubility significantly influences environmental transport and transformations. The water solubility of organic hazardous chemicals is a function of polarity, molar volume, and degree of ionization. The water solubility of metal cations is complicated by their reactions with organic and inorganic ligands.

- Hazardous contaminants not dissolved in water form another phase on top of or below the aqueous phase; the location of this phase is dictated by the chemical's density or specific gravity. Two classifications have been given to these systems: LNAPLS (light nonaqueous phase liquids) and DNAPLs (dense nonaqueous phase liquids).

- A number of parameters aids in assessing the explosivity and flammability of hazardous materials and hazardous wastes. Flammability limits describe the range of flammable concentrations in air. Flash point is the temperature that results in sufficient volatilization for the compound to reach the lower flammability limit in the air above a liquid. Ignition temperature is the minimum temperature that, when applied to a material, results in its combustion.

- Toxic gas emission and/or excessive heat, fire, or explosion results when two or more incompatible compounds are mixed. Although chemical incompatibility can sometimes be predicted based on thermodynamic properties, empirical data from tables are most often used for its assessment.

- Two label and placard systems are used to display potential hazards of hazardous wastes and materials. In the transportation industry, the Department of Transportation requires the use of placards containing the DOT hazard code and the

United Nations/North American identification number. For storage and other stationary systems, the National Fire Protection Association system is used, which consists of relative hazards rated from 0 through 4 for health, fire, and reactivity.

- Chemical Abstracts Service (CAS) numbers provide a definitive means of identifying hazardous chemicals, and are used in electronic search systems.

- The EPA priority pollutants are 129 of the most common chemicals in industrial wastewater discharges that pose a threat to public health and the environment. The priority pollutants are also commonly found in hazardous waste sites, so site and facility assessments often target the presence of these compounds.

PROBLEMS

3.1. A water sample at a temperature of 34°C contains 344 ppb of dissolved mercury ion. What is the Hg concentration in μg/L?

3.2. Show that 5.0 μg/L of PCE in water at 30°C is not equal to 5.0 ppb.

3.3. A 500-L (132 gal) tank contains 255 mg/L 2,4,6-trichlorophenol dissolved in hexane. What is the trichlorophenol concentration in ppm? Assume the temperature of the system is 20°C.

3.4. Repeat Problem 3.3, but substitute trichloroethylene for hexane as the system solvent.

3.5. Convert each of the following air concentrations to ppm if the temperature at the time of sampling is 28°C and the atmospheric pressure is 720 mm Hg.
   a. 28 mg/m^3 toluene
   b. 12 mg/m^3 hexachloroethane
   c. 21 mg/m^3 anthracene

3.6. A gas sampling device containing 200 g of the sorbent Tenax is used to collect organic impurities from air that may be contaminated with 1,1,2-trichloro-1,2,2-trifluoroethane. At the end of sampling, 240 L of air had been drawn through the sampling system, and the concentration of the chlorofluorocarbon in the Tenax was 811 mg/kg. Assuming 100% trapping efficiency because no weight gain was observed in a second Tenax tube placed in series after the first, what was the contaminant concentration in the air in ppm? The ambient temperature was 18°C and pressure was 750 mm Hg.

3.7. The concentration of TCE in an air sample near a hazardous waste landfill is 34 ppm, the ambient air temperature is 31°C, and the atmospheric pressure is 0.94 atm. What is the TCE concentration in mg/m^3?

3.8. A sample of air has been collected from above an air stripping tower to evaluate its performance and to determine if toxic air emission standards are being violated. The temperature at the time of collection was 18°C, and the atmospheric pressure was 740 mm Hg. The concentrations in the sample were 4.8 mg/m^3 of benzene, 1.7 mg/m^3 of PCE, and 2.0 mg/m^3 of methylene chloride. What are the concentrations in ppm?

3.9. Determine the specific activity of 100 g of a dry contaminated soil containing mixed hazardous waste that emits α radiation at a rate of 6.7×10^9 dps.

3.10. A 10-g sample of ^{14}C-atrazine is characterized by 2.5×10^7 dps. What is the specific activity of this sample?

3.11. Describe the relationship between molecular structure and water solubility.

3.12. A drum contains an aqueous solution of 2,4-dichlorophenol at pH 8.2. Determine the value of α.

3.13. An aqueous solution in a 55-gallon (208-L) drum, characterized by minimal buffering capacity, contains 4 kg of phenol and 1.5 kg of sodium phenate. What is the pH of the solution?

3.14. The pH of an aqueous solution of aniline is 11.1. What are the values of α and β?

3.15. A waste drum has received approximately half acetone and half water. In sampling the drum, where can you expect to find the acetone and the water?

3.16. A drum of waste chemicals contains benzene, aniline, and toluene. Would you expect the presence of different phases? If so, where in the drum?

3.17. List the specific gravity and water solubility of the following compounds:
a. 1,2,3,5-Tetrachlorobenzene
b. Mevinphos
c. Pentachloroethane
d. Dieldrin
e. Dodecane
f. Isopropyl chloride
g. Ronnel

3.18. Using the *Merck Index*, find the specific gravities of each of the following forms of cadmium.
a. Cd metal
b. CdS
c. Cd(OH)$_2$
d. CdCO$_3$

3.19. From their known water solubilities and densities, determine the maximum equilibrium aqueous phase concentration and the position in a waste drum relative to water if the nonaqueous phase is
a. TCE
b. Naphthalene
c. Methylene chloride
d. Acetone
e. Toluene

3.20. Hydrogeologists classify contaminants as "floaters" and "sinkers" based on where the nonaqueous phase ("free product" or undissolved organic contaminants) reaches equilibrium in groundwater. Classify each of the following based on their density.
a. Isooctane
b. PCE
c. Anthracene
d. Carbon tetrachloride
e. Aniline
f. Pentachlorophenol
g. Sulfotepp
h. Methyl isobutyl ketone (MIBK)

3.21. A common adage is "A small bucket of PCE can contaminate an entire aquifer." If a 18.9-L (5-gal) bucket is released (and equally distributed) into a 3.79×10^8-L (100 million-gallon) aquifer, will the PCE standard of 5 μg/L under the Safe Drinking Water Act be exceeded?

3.22. A 208-L (55-gal) drum of benzene is spilled into a 37,900-L (10,000-gal) water storage tank. What percentage of the spilled benzene is present in the aqueous phase? Where would you find the benzene that is not dissolved?

3.23. Repeat Problem 3.22, but with the drum contents consisting of hexachloroethane.

3.24. A 37,900-L (10,000-gal) UST containing toluene has leaked and the toluene is floating at the top of a shallow aquifer as free product. Its water solubility is controlling the concentration of toluene in the groundwater. If 37.9×10^6 L (10 million gallons) of groundwater have been polluted with toluene at a concentration equal to its water solubility, what volume of toluene remains as floating product?

3.25. Determine the effective solubility of TCE in groundwater if it is present in a mixture at a mole fraction = 0.47 and $\gamma = 0.2$.

3.26. Gasoline, containing 2% (w/w) benzene, has migrated to a shallow aquifer. Determine the effective benzene concentration in the groundwater ($\gamma = 0.5$). Assume that the average molecular weight of the gasoline components is 120 g/mole.

3.27. A hydrocarbon mixture containing essentially all BTEX is in contact with a groundwater system. If the six BTEX compounds are present in equal mass proportions, determine the effective solubility for each. Assume $\gamma = 0.5$.

3.28. Determine the LFL and UFL for each of the following mixtures of chemicals.
   a. 44% benzene and 56% n-octane
   b. 37% MEK and 63% TCE
   c. 17% chlorobenzene, 45% gasoline, and 38% methylene chloride.
   d. 39% naphthalene, 41% MIBK, and 20% phenol

3.29. Estimate the LFL and UFL for
   a. Hexane   b. Dodecane
   c. MEK   d. Naphthalene
   e. Phenol
Compare the estimated values to the tabulated values.

3.30. Estimate the LFL and UFL for
   a. Tetradecane   b. Dimethyl phthalate
   c. MIBK   d. Diethyl ether
   e. Anthracene

3.31. Based on the ignition temperatures of chlorinated ethanes and ethenes, what conclusion can be made about the effect of chlorination on flammability?

3.32. A steel storage shed containing an open drum of trichloroethylene has reached a temperature of 138°F (59°C). Is there a fire hazard in the building, and why?

3.33. Using the information provided in Table 3.16, list the labeling categories of fire, reactivity, and health for the following chemicals.
   a. Hexane   b. Dioxane
   c. Chlorobenzene   d. Phenol

3.34. Why are each of the following mixtures potentially dangerous?
   a. Sodium sulfide and battery acid

b. Sodium metal and aqueous sodium hydroxide

c. A strong solution of sodium carbonate and concentrated hydrochloric acid

d. Sodium cyanide and acetic acid

e. Bleach (calcium hypochlorite) and ammonia

3.35. Determine if the following wastes are incompatible. If they are incompatible, state the hazardous conditions that would result from mixing the wastes.

a. Crystalline sodium chloride and water

b. Waste pentaldehyde and a lime slurry

c. Plating bath waste containing cyanide and pickling liquor

d. Waste calcium metal and water

e. Battery acid and NaOH pellets

f. $KMnO_4$ crystals and waste sulfuric acid

g. $KMnO_4$ crystals and waste muriatic (technical-grade hydrochloric) acid

h. Concentrated ethanol and water

3.36. Describe the physical or chemical reaction that takes place when each of the following are mixed.

a. Hydrochloric acid and cyanide

b. Sodium metal and water

c. Battery acid and caustic lime sludge

3.37. Sketch a DOT placard, including the UN/NA identification number, for tanker trucks containing each of the following hazardous materials.

a. Acetone   b. TCE

c. Parathion   d. Benzene

e. Phenol

3.38. A 90-day hazardous waste storage area is to be designed to receive 208-L (55-gal) drums containing the following waste chemicals at the following general rates:

Battery acid (100 L/mo)

Hydrated lime (400 L/mo)

10,000 mg/L Aqueous sodium sulfide (500 L/mo)

n-Butyl nitrate (50 L/mo)

Pentaldehyde (50 L/mo)

Design a 90-day storage facility with the following features:

a. Waste segregation based on chemical compatibility

b. NFPA placards

c. Area requirements for the storage and movement of drums

d. Waste containment in the event of a spill

3.39. The primary activities at an industrial facility include electroplating, metal parts cleaning, and paint stripping. The usual monthly disposal of the wastes include

Acetone	Ten 208-L (55-gal) drums (222,000 mg/L in water)
MEK	Eighteen 208-L (55-gal) drums (50,000 mg/L in water)
Acrylonitrile	Two 208-L (55-gal) drums (95% v/v)
TCE	Two 208-L (55-gal) drums (90% v/v)

Sodium cyanide	Five 208-L (55-gal) drums (174,000 mg/L)
Potassium dichromate	Electroplating sludge (1020 kg solids at 2.16 kg/L and 12,600 L of water)
Sulfuric acid	80% of a 208-L (55-gal) drum

The drums are stored for 90 days before they are transported off site by a RCRA-permitted transporter. They are currently stored casually in a corner of the facility, which is a safety hazard. The upper management wants you to design a floor plan for a small building to safely store the waste drums. Some key features of the plan include the number and area requirements of segregated sections of the building to negate chemical incompatibility. A 208-L (55-gal) drum is 0.6 m in diameter. (Feel free to use a safety factor to provide more area for increased future waste generation rates). Each section will be isolated using a 0.5-m-high concrete berm, which would contain spilled materials from any drum ruptures. Construction of the berm is expensive, so the best design would use the minimum number of segregation cells but still provide safety in the building if different chemicals within one cell come into contact. Design a storage area for the wastes, and sketch a plan view showing the location of the waste drums and the berms.

3.40. Using data from the *Fire Protection Guide on Hazardous Materials*, determine the LFL and UFL for each of the following mixtures of chemicals.
a. 28% acrylonitrile and 72% decane
b. 40% *n*-heptane and 60% hydrogen cyanide
c. 24% acetone, 38% cyclohexane, and 38% dodecane.
d. 10% acrylonitrile, 39% toluene, and 51% *p*-xylene

3.41. Using reference materials, find the water solubility and specific gravity of the following hazardous compounds:
a. Acrolein
b. Dimethoate
c. 2,4-Pentanedione
d. Dibrom
e. Methomyl
f. Neburon
g. Vacron

REFERENCES

3.1. Lyman, W. J., "Solubility in water," in Lyman, W. J., et al. (Eds.), *Handbook of Chemical Property Estimation Methods*, McGraw-Hill, New York, 1982.

3.2. *Standard Methods for the Examination of Water and Wastewater*, American Public Health Association, American Water Works Association, Water Environment Federation, Washington, DC, 1992.

3.3. Tchobanoglous, G. and E. D. Schroeder, *Water Quality*, Addison-Wesley, Reading, MA, 1985.

3.4. Montgomery, J. H. and L. M Welkom, *Groundwater Chemicals Desk Reference*, Lewis Publishers, Chelsea, MI, 1991.

3.5. Snoeyink, V. A. and D. Jenkins, *Water Chemistry*, John Wiley & Sons, New York, 1981.

3.6. Stephenson, R., J. Stuart, and M. Tabak, "Mutual solubility of water and aliphatic alcohols," *J. Chem. Eng. Data*, **29**, 287–290 (1984).

3.7. Nirmalakhandan, N. N. and R. E. Speece, "Prediction of aqueous solubility of organic compounds based on molecular structure," *Environ. Sci. Technol.,* **22,** 328–338 (1988).

3.8. Yalkowsky, S. H., R. J. Orr, and S. C. Valvani, "Solubility and partitioning. 3. The solubility of halobenzenes in water," *Ind. Eng. Chem. Fundam.,* **18,** 351–353 (1979).

3.9. Verschueren, K., *Handbook of Environmental Data on Organic Chemicals,* Van Nostrand Reinhold, New York, 1983.

3.10. Yalkowsky, S. H. and S. C. Valvani, "Solubilities and partitioning. 2. Relationships between aqueous solubilities, partition coefficients, and molecular surface areas of rigid aromatic hydrocarbons," *J. Chem. Eng. Data,* **24,** 127–129 (1979).

3.11. Bell, G. H., "Solubilities of normal aliphatic acids, alcohols and alkanes in water," *Chem. Phys. Lipids,* **10,** 1–10 (1973).

3.12. Jafvert, C. T., J. C. Westall, E. Grieder, and R. P. Schwartzenbach, "Distribution of hydrophobic ionogenic organic compounds between octanol and water: Organic acids," *Environ. Sci. Technol.,* **24,** 1975–1803 (1990).

3.13. Lide, D. R. (Ed.), *CRC Handbook of Chemistry and Physics,* 77th Edition, CRC Press, Boca Raton, FL, 1996.

3.14. Leonard, J., B. Lygo, and G. Procter, *Advanced Practical Organic Chemistry,* 2nd Edition, Chapman & Hall, New York, 1995.

3.15. Kier, L. B., W. J. Murray, M. Randie, and L. H. Hall, "Molecular connectivity: V. Connectivity series concept applied to density," *J. Pharm. Sci.,* **65,** 1226–1230 (1976).

3.16. Martell, A. E. and R. D. Hancock, *Metal Complexes in Aqueous Solutions,* Plenum Press, New York, 1996.

3.17. Davis, D. J. and G. T. Christianson, *Firefighter's Hazardous Materials Reference Book,* Van Nostrand Reinhold, New York, 1991.

3.18. *Light Nonaqueous Phase Liquids,* EPA/540/5-95/500, Robert S. Kerr Environmental Research Laboratory, U.S. Environmental Protection Agency, Ada, OK, 1995.

3.19. Hunt, J. R., N. Sitar, and K. D. Udell, "Nonaqueous phase liquid transport and cleanup, analysis of mechanisms," *Water Resour. Res.,* **24,** 1247–1258 (1991).

3.20. Mercer, J. W. and R. M. Cohen, "A review of immiscible fluids in the subsurface: Properties, models, characterization and remediation," *J. Contam. Hydrol.,* **6,** 107–163 (1990).

3.21. Miller, C. T., M. M. Poirier-McNeill, and A. S. Mayer, "Dissolution of trapped nonaqueous phase liquids: Mass transfer characteristics," *Water Resour. Res.,* **26,** 2783–2796 (1990).

3.22. Anderson, M. R., R. L. Johnson, and J. F. Pankow, "Dissolution of dense chlorinated solvents into groundwater. 1. Dissolution from a well-defined residual source," *Ground Water,* **30,** 250–256 (1992).

3.23. Apgar, M. A., P. J. Cherry, J. H. Jordan, and S. N. Williams, "The ground water management zone—An alternative to costly remediation," *Ground Water Monit. Rev.,* **9,** 70–73 (1989).

3.24. Crowl, D. A. and J. F. Louvar, *Chemical Process Safety: Fundamentals with Applications,* Prentice-Hall, Englewood Cliffs, NJ, 1990.

3.25. LeChatelier, H., "Estimation of firedamp by flammability limits," *Ann. Mines.,* **19,** 388–395 (1891).

3.26. Jones, G. W., "Inflammation limits and their practical application in hazardous industrial operations," *Chem. Rev.,* **22,** 1–26 (1938).

3.27. Whiston, J. (Ed.), *Safety in Chemical Production*, Proceedings of the First IUPAC Workshop on Safety in Chemical Production, International Union of Pure and Applied Chemistry, Basle, Switzerland, Sept. 9–13, 1990.

3.28. *Fire Protection Guide on Hazardous Materials*, 6th Ed., National Fire Protection Association, Boston, 1975.

3.29. Sullivan, J. B., Jr. and G. R. Krieger (Eds.), *Hazardous Materials Toxicology: Clinical Principles of Environmental Health*, Williams & Wilkins, Baltimore, MD, 1992.

3.30. *Emergency Response Guidebook*, DOT 5800.4, U.S. Department of Transportation, Office of Hazardous Materials Transportation, Washington, DC, 1987.

3.31. *Cross Reference Index of Hazardous Chemicals, Synonyms, and CAS Registry Numbers*, The Forum for Scientific Excellence, Inc., J.B. Lippincott, Philadelphia, 1990.

3.32. Budavari, S. (Ed.), *The Merck Index*, 12th Edition, Merck and Co., Whitehouse Station, NY, 1996.

3.33. *The Farm Chemicals Handbook*, 82nd Edition, Meister Publishing, Willoughby, OH, 1996.

Chapter 4

Source Analysis

Tens of thousands of chemicals are used by industry in manufacturing, petrochemical refining, cleaning, degreasing, and hundreds of other processes. For example, Boeing personnel report that over 11,000 chemicals are needed to manufacture a commercial airliner [4.1]. Industries in the United States and other countries are faced with increased regulatory pressure to minimize their hazardous wastes and to manage them properly. Large manufacturing companies can no longer dispose of their wastes in a casual fashion but must track each waste component and manifest its disposal under RCRA. In addition, the HSWA amendments to RCRA require industries to reduce their rate of hazardous waste generation. The management of hazardous wastes has become a time-intensive and costly necessity for business. It is not surprising that many industries are voluntarily reducing their waste volumes and finding substitute source materials that are not classified as hazardous under RCRA [4.2]. These hazardous waste management and minimization efforts require analysis of the source before plans can be established and implemented.

The thousands of contaminated sites throughout the United States and the world also require characterization before remediation systems are designed and constructed. In some cases, no cleanup may be the best alternative to protect public health and the environment. Nonetheless, the complexity of contaminated sites requires a multiphase assessment process in which the sites are identified, screened, and characterized. The procedure for assessing hazardous waste sites is based on a phase-by-phase procedure of increasing complexity and cost. These hazardous waste site assessment procedures have evolved over the past 10 years and now usually include a Phase I assessment, which involves a records search and cursory site inspection, followed by a Phase II evaluation, which focuses on sampling and chemical analysis.

In Chapter 4, procedures for describing, assessing, and sampling hazardous wastes at industrial facilities and at contaminated sites are presented. Specific procedures addressed include waste audits, fundamentals in the assessment of contaminated sites, estimation of mass' concentrations of spills, sampling fundamentals, and chemical analyses.

4.1 MATERIALS BALANCES AND WASTE AUDITS

Hazardous waste management has become a complicated and expensive undertaking for companies that handle large volumes of chemicals. As shown in Figure 4.1, hazardous wastes must be organized, stored, and disposed of in a safe and efficient man-

Figure 4.1 Management of drums containing hazardous wastes at a 90-day RCRA storage facility. *Source:* U.S. Department of Energy Richland, WA office.

ner. The tracking of hazardous wastes within an industry, a waste transfer station, or after excavation from a contaminated site is called a *hazardous waste audit*. The first step for an industry in assessing, organizing, tracking, and minimizing hazardous wastes is to quantify all of the source materials that it receives (inputs) and balance these with the wastes leaving the facility, which usually takes place through waste disposal routes. Such an assessment provides the generator with a means of bookkeeping and a procedure for the management, disposal, and manifesting of hazardous wastes. Small quantity generators often track their source materials using database spreadsheets; more sophisticated software may be necessary for industries that use hundreds or thousands of chemicals [4.3].

The basis for a hazardous waste audit is a *source materials balance*, which is the same as a reactor balance, but without a reaction term (Figure 4.2). The rates at which the materials enter and exit the system are multiplied by their corresponding concentrations to provide a mass change per time. For example, if twenty 208-L (55-gal) drums are generated per month that contain 800 mg/L trichloroethylene (TCE), the mass of TCE in the drums is

$$20 \times 208 \text{ L} = 4160 \text{ L} \tag{4.1}$$

$$800 \text{ mg/L} \times 4160 \text{ L} = 3,330,000 \text{ mg TCE} = 3.33 \text{ kg TCE} \tag{4.2}$$

The most common source inputs for materials that enter a hazardous waste inventory are purchases; invoices and other purchase records serve as an appropriate source for the input term on the materials balance. Other inputs, which are usually minor, include the acquisition of hazardous materials through property acquisitions and by "donations" from other companies or subsidiaries.

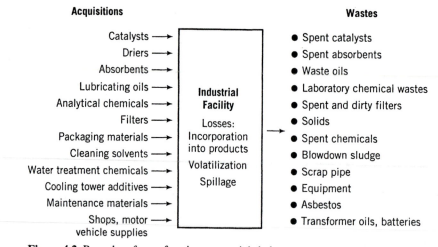

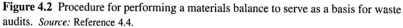

Figure 4.2 Procedure for performing a materials balance to serve as a basis for waste audits. *Source:* Reference 4.4.

For industrial synthesis processes (e.g., the production of pesticides, dyes, etc. from chlorobenzenes, chlorophenols, and anilines), most of the source material is converted to the products. One of the goals of pollution prevention programs (see Section 12.1) is the optimization of process conditions (e.g., reactor temperature and pressure) to improve yields of the products while decreasing the volume of side streams that contain unreacted materials and waste products [4.5].

The primary routes by which unaccounted wastes leave an industrial facility are (1) industrial wastewater discharges, (2) RCRA-related waste management activities, (3) volatilization (e.g., fugitive emissions), and (4) spillage and other unaccounted losses [4.6]. Such unaccounted wastes include solvents, paint removers, and other chemicals that are not consumed in chemical reactions. Chlorophenols, hexachlorocyclopentadiene, and other industrial intermediates (Section 2.7) are usually consumed in chemical reactions during industrial synthesis processes, and their output is usually minimal [4.7]. In fact, the goal in the chemical processing of reactive materials is to incorporate these intermediates into products. They are the building blocks of the products of synthesis, and the presence of an unreacted intermediate results in extra cost. The use of a hazardous waste audit for tracking wastes in industrial facilities is demonstrated in Example 4.1.

EXAMPLE 4.1 *Hazardous Waste Audit of a Waste Transfer Facility*

The state regulatory agency has accused your company, a waste transfer facility, of massive spillage due to improper housekeeping practices over the past year. The agency alleges that the upper 5 cm of the soils on your property (approximately 140,000 kg of soil) are contaminated with polychlorinated biphenyls (PCBs) above the regulatory action level of 50 ppm. You can sample your property and analyze the soils for PCBs to refute the agency's claims, but the cost is estimated at $60,000. As an alternative, perform a waste audit on the transformer oils that were exchanged at the waste transfer facility over the past year. Based on your company's records given here, can you show that PCBs were not spilled on the property at a level greater than 50 ppm? Because PCBs are characterized by low volatility and water solubility, you can assume that volatilization and leaching losses were negligible.

Industrial Records

Incoming shipments of transformer oils:

	Volume		
Invoice No.	(L)	(gal)	PCB Concentration (mg/L)
01137	208	55	42,000
01138	1740	460	1,300
01139	795	210	22,000
01140	3260	860	8,200
01141	1440	380	38,000

Industrial wastewater:

Flow: 1,890,000 L/day
Mean concentration: 0.002 mg/L

RCRA disposal manifests:

	Volume		
Manifest No.	(L)	(gal)	PCB Concentration (mg/L)
0124	500	1890	22,100
0125	500	1890	18,400
0126	500	1890	9,550
0127	500	1890	6,420

SOLUTION

1. Determine the incoming mass of PCBs to the waste transfer facility.

	Volume			
Invoice No.	(L)	(gal)	PCB Concentration (mg/L)	Mass (kg)*
01137	208	55	42,000	8.75
01138	1740	460	1,300	2.27
01139	795	210	22,000	17.5
01140	3260	860	8,200	26.7
01141	1440	380	38,000	54.6
Total				109.8 kg

*Total volume (L) × PCB concentration (mg/L) × kg/1,000,000 mg

2. Determine the mass leaving the facility.

Wastewater:

$$1,890,000 \text{ L/day} \times 365 \text{ d} \times 0.002 \text{ mg/L} = 1.38 \text{ kg}$$

RCRA Manifests:

| Manifest No. | Volume | | PCB Concentration (mg/L) | Mass (kg) |
	(L)	(gal)		
0124	1890	500	22,100	41.8
0125	1890	500	18,400	34.8
0126	1890	500	9,550	18.0
0127	1890	500	6,420	12.1

Total: RCRA Manifests   106.7 kg
Total: Wastewater and RCRA Manifests   108.1 kg
Total mass unaccounted for: 109.8 kg − 108.1 kg = 1.7 kg
Total mass to trigger regulatory action: 50 mg/kg × 140,000 kg soil =
                                          7,000,000 mg = 7.0 kg

Because the total mass unaccounted for in waste disposal practices is less than the regulatory action level, a case could be presented to the state regulatory agency that the losses do not provide sufficient PCB concentrations in the soil to exceed the regulatory action level.

4.2 HAZARDOUS WASTE SITE ASSESSMENTS

Abandoned landfills, surface impoundments, and other hazardous waste sites must be assessed to determine the extent of contamination before cleanup is initiated. Unlike municipal wastewater or industrial waste streams, hazardous wastes disposed of improperly at waste sites are not usually evident to the naked eye because most of the contamination lies below the soil surface. Therefore, a series of site assessment procedures must be conducted, ranging from record searches to detailed subsurface sampling.

Site assessments follow a common procedure that is divided into three phases, each of which is increasingly complex. Site assessments are initiated whenever the existence of a hazardous waste site is suspected, which may include a leaking underground storage tank (UST), the discovery of hazardous chemicals in a drinking water supply, a high incidence of localized illness, or a routine property transfer [4.8]. The pattern is much the same for small sites as for large ones. In fact, the Preliminary Assessment and Remedial Investigation/Feasibility Study (RI/FS) programs of CERCLA follow the same format as the general site assessment phases described below.

The three site assessment phases are appropriately named Phase I, Phase II, and Phase III, and each builds upon the previous phase with more extensive information [4.9]. A *Phase I* assessment, which may also be called a Level 1 assessment, preliminary assessment, or initial assessment study, is begun in an attempt to confirm the suspicions of the presence of hazardous wastes. The procedure focuses on "soft," or non-scientific, information and is analogous to a background search. Phase I studies involve paper research including a chemical inventory evaluation, interviews with current and former personnel and neighbors, and regulatory agency record searches and interviews. Title searches and reviews of historical ownership are also necessary under CERCLA liability clauses. Other records that are often reviewed include aerial photographs, National Pollutant Discharge Elimination System (NPDES) permits and violations, zoning maps, tax records, fire records, and newspaper articles.

The Phase I process usually starts with a search of historical documents including newspaper reports of spills and official records of any safety violations that may have taken place. More detailed records analysis may include searches of orders, invoices, and the inventory of hazardous chemicals. The assumption in these investigations is that, if a large volume of hazardous materials has passed through the property, there is a high probability that some of it was disposed of improperly. Although many Phase I assessments focus on information from the past few decades, some waste disposal sites in the eastern United States date back to the mid-1800s [4.10]. Therefore, records searches in some regions should investigate disposal practices for at least 100 years.

The on-site inspection and personnel interviews provide information and perspectives that cannot be obtained through records searches. Often retirees, neighbors, and employees of 20+ years are able to relay information on past disposal practices and the location of buried wastes. A "walk-through" of the site can often provide clues of improper waste disposal. Signs such as stained soil, an unlined pit, or concrete pads associated with an abandoned fuel farm, solvent storage area, pesticide mixing/loading zone, or gasoline station may provide the evidence necessary to initiate a Phase II study (Figure 4.3). Phase I reports do not certify that the site is free of contamination, but they provide a basis for further study or investigation.

If the suspicions that were first raised to initiate the site assessment are confirmed in the Phase I evaluation, a *Phase II* study is warranted to confirm or deny the presence of hazardous wastes at the site. A Phase II study includes finalizing any record searches that were not completed in the Phase I assessment. A detailed evaluation of pathways and potential receptors is begun, which may include an analysis of the subsurface by a hydrogeologist to assess groundwater flow directions and travel times to drinking water wells or other receptors. If ecological damage is evident, a biologist may assess critical habitat or the need for ecological risk assessment.

Phase II assessments often involve an increased sampling effort. Based on where the contamination is expected, surface soil "grab samples" (random samples collected without any guidance from prior knowledge), soil cores, surface water samples, and groundwater samples are collected after the installation of monitoring wells. There are no firm rules on the degree to which sampling is conducted during Phase II assessments; sampling intensity evolves on an ad hoc basis through negotiations among site owners, their consultants, and state regulators [4.11].

If Phase II studies show that the site is contaminated, a *Phase III* study is initiated. The purpose of a Phase III investigation is to detail the extent of contamination in terms of the area, volume, and contaminant concentrations. Depending on the source characteristics, age of the site, and predominant pathways, the source and adjacent areas (soils, subsurface, and/or groundwater) may be sampled extensively. With appropriate sampling designs, contaminant concentration data over depth and area provide sufficient information to assess the site hazard (i.e., the need for site cleanup) and provide criteria for the design of remedial processes.

Figure 4.3 Irresponsible management of containers and an aging tank provide evidence of possible environmental releases at a pesticide rinse and formulation area.

4.3 ESTIMATION OF SOURCE CONCENTRATIONS FOR HAZARDOUS MATERIAL SPILLS

The rapid assessment of hazardous material spills is often necessary to evaluate the potential pathways of acutely toxic chemicals and the potential health effects to nearby populations. Under the short time constraints of such *emergency response assessments*, sampling and chemical analyses are often impractical [4.12]. Therefore, source contaminant concentrations may be evaluated for hazardous material spills if some of the characteristics of the hazardous chemicals and the site are known or can be estimated. The basis for the method is estimating (1) the mass of the material spilled and (2) the mass of the soil or solids onto which the chemical has been spilled in order to compute a mass/mass concentration. This procedure is not accurate for determining hazardous waste concentrations in soils and sludges for detailed site assessments. Rather, it is a first-cut calculation that can be used for emergency response planning after a spill of hazardous wastes or materials. Furthermore, the procedure is only applicable to recent spills where loss and transformation processes have not occurred.

The first step in the procedure is determination of the total mass of contaminant spilled. Some chemical formulations, particularly pesticides, are diluted in water, a factor that must be taken into account in mass calculations. The volume of material is usually available and is converted to a mass value using the contaminant density. In the second step, the volume of contaminated soil is estimated, a process usually accomplished in the field by visual inspection. The mass of contaminated soil may then be estimated using typical values for the soil bulk density. The procedure for estimating source contaminant concentrations is demonstrated in Example 4.2.

EXAMPLE 4.2   *Estimation of a Source Concentration after a Chemical Spill*

A 1890-L (500-gal) tanker containing a 40% formulation of 2,4-D in acetone has spilled, contaminating an area of soil approximately 200 m² and 50 cm deep. If the soil bulk density is 1800 kg/m³, estimate the 2,4-D concentration in the soil.

SOLUTION

$$\text{The volume of 2,4-D spilled} = 1890 \text{ L} \times 0.40 = 756 \text{ L}$$

Convert the 2,4-D volume to mass using its specific gravity (Table 3.6).

$$756 \text{ L} \times \frac{1000 \text{ mL}}{\text{L}} \times \frac{1.416 \text{ g}}{\text{mL}} = 1.07 \times 10^6 \text{ g} = 1.07 \times 10^9 \text{ mg}$$

Determine the volume and mass of soil.

$$200 \text{ m}^2 \times 0.50 \text{ m} = 100 \text{ m}^3$$

$$100 \text{ m}^3 \times 1800 \frac{\text{kg}}{\text{m}^3} = 180{,}000 \text{ kg}$$

$$\text{The estimated concentration} = \frac{1.07 \times 10^9 \text{ mg 2,4-D}}{180{,}000 \text{ kg soil}} = 5960 \frac{\text{mg}}{\text{kg}} \text{ 2,4-D}$$

4.4   SOURCE SAMPLING

Sampling of source materials, including contaminated soils, sludges, lagoons, and drum contents, is necessary to assess their degree of hazard and to comply with regulations. Numerous sampling schemes have been developed based on the physical characteristics of the source and its degree of heterogeneity. If the sample is totally homogeneous, such as an aqueous waste drum with no concentration gradients, one sample from anywhere would characterize the source. However, almost all hazardous waste sources, especially soils and sludges, are heterogeneous, and their characterization is relatively complicated.

A pile of contaminated soil that requires sampling to quantify the level of contamination is shown in Figure 4.4. Such a source characterization is often a costly procedure for potentially responsible parties (PRPs), treatment, storage, and disposal (TSD)

Figure 4.4 A pile of contaminated soil awaiting sampling to assess the degree of contamination.

facilities, and industries, because it involves sampling, reliable analyses with sophisticated instrumentation, quality assurance, and report preparation. Although the analytical laboratory procedures are well defined, the results are only as good as the weakest link in the chain of characterization, which is often sampling. Sampling errors are characteristically greater than analytical errors for heterogeneous media such as soils and sludges, but hazardous waste personnel often do not design adequate sampling schemes to ensure that sources are sampled using statistically valid procedures [4.13]. Therefore, an effective sampling scheme that fully characterizes the site or facility in an efficient and economical manner is necessary to ensure the quality of the analytical results.

Statistical Fundamentals for Sampling

A basic knowledge of statistics is necessary to assess sampling requirements. Two fundamental statistical concepts that are important in developing sampling plans are accuracy and precision. *Accuracy* is defined by how close a measured value is to the true value. Assuming that the field sampling and analytical laboratory procedures provide the necessary accuracy, results close to the true value can be obtained; however, in theory, a true value of contaminant concentration can be achieved only by taking an infinite number of samples. *Precision* is a measure of the variability between samples. Soils and sludges are characteristically heterogeneous, and the concentration of hazardous chemicals in each subsample are likely to be quite different; as a result, hazardous waste samples are often characterized by low precision and high variability.

Statistical analyses are concerned with two aspects related to accuracy and precision. First, the level of accuracy must be specified. The parameter α is the error allowed and $(1 - \alpha)$ is the corresponding confidence level. For example, a specified error of 5% corresponds to a 95% confidence that the sample value is an accurate estimate of the true value. Secondly, a *precision requirement* (D) is the deviation from the true value; in hazardous waste management, this true value is often represented by a regulatory standard, such as TCLP value [4.14].

Two other fundamental concepts in statistical analysis are populations and samples. A *population* represents the true values of the global data set for the system under consideration; that is, the population may be thought of as the universe of data within the system boundaries. The true value of a population variable is called a *parameter*. The population may be described by such parameters as a mean (μ), with a variance (σ^2) and standard deviation (σ).

A *sample* is a data set, regardless of size, collected from the population. Sample variables include data points (X_i) and the number of data points in the sample (n). The *sample mean* ($\overline{X}$) is

$$\overline{X} = \frac{\Sigma X_i}{n} \tag{4.3}$$

and the *sample variance*, a measure of the spread of data about the mean, is

$$S^2 = \frac{\Sigma (X_i - \overline{X})^2}{n - 1} \tag{4.4}$$

The sample standard deviation, S, is simply $\sqrt{S^2}$.

Degrees of freedom (df), a parameter used in several statistical distributions, is an integer equal to the sample size (n) minus the number of population parameters being evaluated through analysis of the sample. In most cases, one parameter (e.g., variance or differences in population means through the t-distribution) is being evaluated, so df $= n - 1$.

Sample precision is promoted by collecting an adequate number of samples, which is apparent from inspection of Equation 4.4; the variance, which is an indirect measure of precision because it describes the spread of data about the mean, decreases as the number of samples (n) increases.

A hazardous waste source from which samples are drawn (e.g., a sludge drying bed) may be considered a subpopulation composed of many individuals. A group of these individuals, such as a core or a slice of given dimensions, is defined as a *sampling unit* [4.15], which may be conceptualized as soil cores of specific dimensions, spade loads, or squares of contaminated concrete. Sampling units must be separate and distinct, and their total number must comprise the entire population; that is, if all of the sampling units of the source could be tested, the entire population would be described. However, such a procedure is not practical or economical. Rather, the population parameters (i.e., the true values) are estimated within a specified confidence interval by collecting data from a portion of the sampling units to obtain a set of numbers called *statistics*.

To visualize sampling a hazardous waste source, think of the system (which could be a bed of plating sludge or a contaminated surface soil) divided into a large number of separate and discrete blocks, or sampling units, perhaps 10 cm square and 20 cm deep. One goal of the sampling plan is to designate which sampling units will make up the sample. In other words, the fundamental question in sampling a source is *how many and which of these blocks need to be sampled in order to achieve a given level of precision?*

Sampling Procedures

Many procedures have been developed for selecting sampling units, some of which have statistical validity and some of which do not. Sampling can be divided into a number of general categories; the most common include haphazard sampling, search sampling, judgment sampling, and probability sampling. Only probability sampling is based on fundamental statistical techniques in which the concentration mean and its precision can be determined based on a specified confidence level [4.16].

Haphazard sampling, which is the least valid method, involves collecting samples with no emphasis on randomness and often is based on convenience. In many cases, samples are collected at a location because it involves minimal walking over contaminated soil or at a location that will keep the person who is sampling out of the rain. Such "haphazardness" may result in biased sample data; therefore, haphazard sampling is fundamentally flawed as a method for hazardous waste source characterization.

Search sampling uses historical information and prior knowledge of the site to evaluate areas of elevated contamination. For example, a surface soil contaminated with motor oil will often have stained regions; these darkened areas would be the sites of highest contamination and therefore the places where search samples would be taken.

In *judgment, or biased, sampling,* the selection of sampling units is random, but they are not chosen based on statistical procedures. Rather, sampling units are chosen based on what appears to be representative samples of the site. Judgment samples often tend to be more representative than probability sampling for very low ($n = 1$–3) sample sizes [4.17]. However, the accuracy of true random sampling (i.e., probability sampling) increases rapidly as n increases, and the accuracy of probability sampling usually becomes greater than judgment samples with just a few more sampling points.

Probability sampling is designed so that samples, which have an equal chance of being chosen, are collected randomly in order to provide maximum precision. Under the category of probability sampling are simple random sampling and stratified random sampling. *Simple random sampling* is two-dimensional and does not focus on changes in sample values with depth. In simple random sampling, sampling units are selected separately, randomly, and independently of others. For sources in which three dimensional heterogeneity is random, simple random sampling provides a statistically valid means of determining the dimensions of the sampling units and the number of samples collected. However, if a waste source is characterized by distinct layers or stratification, stratified random sampling is necessary.

In *stratified random sampling*, the site is divided into a number of nonoverlapping layers called *strata* in order to obtain a better estimate of the mean for each stratum or for all of the strata. Stratified random sampling may be thought of as a series of simple random sampling systems stacked on top of one another.

Other probability sampling schemes include multistage sampling, cluster sampling, and systematic sampling. Detailed discussions of these procedures are available in References 4.18–4.20.

Probability sampling is based on the *t distribution*—a probability density function that is used to evaluate sample means when the population variance (σ^2) is not known but can be estimated by S^2. The *t* distribution is defined as

$$t = \frac{\overline{X} - \mu}{S/\sqrt{n}} \tag{4.5}$$

The term $S/\sqrt{n}$ is also called the *standard deviation of the mean*, $S_{\overline{X}}$. Values of the *t* distribution, which will be used to determine the required numbers of samples for a given confidence level, are listed in Appendix E.

Random sampling is usually achieved by dividing the source area into a grid of sampling units. Each of the squares on the grid is assigned a number, and a sample is collected from each sampling unit designated by a random number. The most reliable procedure for selecting a random sample of *n* measurements is to use a computer-based random number generator or a table of random numbers (Appendix F). In using such a table, the numbers are random regardless of the point where selection begins. For example, to choose a random sample of $n = 8$ from a sampling grid of 100 sampling units, 8 two-digit numbers would be selected starting at any point on the table. Moving vertically down the columns of Appendix F has also been recommended as a method for selecting random numbers [4.21].

Initial or previous data are usually necessary to determine the number of samples required to achieve a given level of precision. The initial data provide preliminary estimates of $\overline{X}$ and S^2 before the random sampling design is finalized. The number of samples required for a designated level of precision may be calculated by rearranging Equation 4.5:

$$n = \frac{t^2 S^2}{D^2} \tag{4.6}$$

where

t = Student's two-sided *t* with $n - 1$ degrees of freedom for a confidence level of $(1 - \alpha)\%$
S^2 = the sample variance for the initial data
D = a specified limit relative to the sample mean

The value of D is derived from the term $\overline{X} - \mu$; in hazardous waste source assessments, a regulatory threshold or specified level of contaminant concentration based on risk is used in place of μ. For example, the use of D in sampling RCRA waste sources is $D = \overline{X} - $ RT where RT = a regulatory threshold (e.g., a toxicity characteristic leaching procedure (TCLP) value of 5.0 mg/L for lead).

An additional parameter needed for a specified simple random sampling procedure under RCRA (Table 4.1) is the *confidence interval* (CI) around the sample mean:

$$\text{CI} = \overline{X} \pm t_\alpha (S^2/n)^{1/2} \tag{4.7}$$

Equation 4.7 is used for collecting a minimum number of randomly collected samples to document that the upper bound of the CI for μ is less than the regulatory concentration.

Table 4.1 Procedures for the Simple Randomized Sampling of RCRA Hazardous Waste Sources

1. Collect a few (3–6) random samples to obtain preliminary estimates of $\overline{X}$ and S^2.
2. Using Equation 4.6, estimate the minimum number of samples (n) using a specified confidence level, e.g., 95% or 99.9% (1 in 20 or 1 in 1,000).
3. Using a sampling grid and random number assignments, collect and analyze at least the minimum number of samples from the waste source.
4. Determine the values of $\overline{X}$ and S^2 for the detailed sampling plan.
5. If the sample mean, $\overline{X}$, is $\geq$ the regulatory threshold (RT), the compound is present in hazardous concentrations. However, if $\overline{X}$ < RT, determine the confidence interval using Equation 4.7. If the upper CI < RT, the compound is not present at a hazardous level and the sampling and analysis are finished.

Source: Adapted from Reference 4.14.

A specific application of Equation 4.5 is the strategy for sampling RCRA waste sources, that is, determining the number of samples that needs to be collected, a procedure that is outlined in Table 4.1. A more general application of a sampling design is demonstrated in Example 4.3.

EXAMPLE 4.3 *Source Sampling Design*

A drying bed holding sludge from an electroplating process is to be sampled for cadmium content. The dimensions of the drying bed are 6 m × 8 m, and the sample volume will require an area 40 cm × 40 cm. Five preliminary samples were collected randomly with the following results: 25, 36, 49, 28, and 48 mg/kg Cd. Based on this information, develop a simple randomized sampling scheme. Determine (1) the number of samples required for 95% confidence limits within 5 mg/kg of the sample mean, and (2) the location of the samples in the sludge bed.

SOLUTION

For the five preliminary samples

$$\overline{X} = \frac{25 + 36 + 49 + 28 + 48}{5} = 37\frac{\text{mg}}{\text{kg}} \text{ Cd}$$

$$S^2 = \frac{\Sigma(X_i - \overline{X})^2}{n - 1}: (25 - 37)^2 = 144$$

$$(36 - 37)^2 = 1$$

$$(49 - 37)^2 = 144$$

$$(28 - 37)^2 = 81$$

$$\underline{(48 - 37)^2 = 121}$$

$$\Sigma(X_i - \overline{X})^2 = 491$$

$$S^2 = \frac{491}{4} = 123$$

$$S = \sqrt{123} = 11$$

From the Student's t table (Appendix E), $t_{95\%} = 2.776$ (*Note:* Use the $\alpha = 0.025$ column, because $2 \times 0.025 = 0.05$ [5%], which is the error level for the 95% confidence level). The number of subsequent samples (n) is

$$n = \frac{t^2_{95\%}S^2}{D^2} = \frac{(2.776)^2(123)}{5^2} = 37.9 \text{ samples; round up to 38 samples.}$$

The number of sampling units in one direction is 8 m/0.4 m = 20 units; the number in the other direction is 6 m/0.4 m = 15 units. The $15 \times 20 = 300$ sampling units (Figure 4.5) are numbered consecutively, and then 38 random numbers are selected from Appendix F. Random number selection may be started at any point on the table; for this example, begin selection at line 1, column 5 and move down the columns. Three-digit numbers within the window 1 to 300 are taken from each group. During random number selection, only numbers within the range of 001 through 300 are selected (i.e., numbers outside of this range are disregarded). If a number comes up twice, it is ignored the second time, because it is assumed that the sampling unit does not contain sufficient medium to be sampled twice. Based on these procedures, the 38 locations of the random samples are

078	151	047	150
061	186	263	125
277	185	287	179
188	141	153	208
174	248	253	082
232	187	081	221
091	058	300	066
133	176	015	069
197	298	011	
279	136	234	

The locations of the sampling units are shaded in Figure 4.5.

Stratified Random Sampling

The basis for stratified random sampling is dividing the site into layers, or strata, each of which is then sampled using randomized sampling techniques. There are two rea-

1	21	41	61	81	101	121	141	161	181	201	221	241	261	281
2	22	42	62	82	102	122	142	162	182	202	222	242	262	282
3	23	43	63	83	103	123	143	163	183	203	223	243	263	283
4	24	44	64	84	104	124	144	164	184	204	224	244	264	284
5	25	45	65	85	105	125	145	165	185	205	225	245	265	285
6	26	46	66	86	106	126	146	166	186	206	226	246	266	286
7	27	47	67	87	107	127	147	167	187	207	227	247	267	287
8	28	48	68	88	108	128	148	168	188	208	228	248	268	288
9	29	49	69	89	109	129	149	169	189	209	229	249	269	289
10	30	50	70	90	110	130	150	170	190	210	230	250	270	290
11	31	51	71	91	111	131	151	171	191	211	231	251	271	291
12	32	52	72	92	112	132	152	172	192	212	232	252	272	292
13	33	53	73	93	113	133	153	173	193	213	233	253	273	293
14	34	54	74	94	114	134	154	174	194	214	234	254	274	294
15	35	55	75	95	115	135	155	175	195	215	235	255	275	295
16	36	56	76	96	116	136	156	176	196	216	236	256	276	296
17	37	57	77	97	117	137	157	177	197	217	237	257	277	297
18	38	58	78	98	118	138	158	178	198	218	238	258	278	298
19	39	59	79	99	119	139	159	179	199	219	239	259	279	299
20	40	60	80	100	120	140	160	180	200	220	240	260	280	300

Figure 4.5 Sampling grid developed for Example 4.3. The shaded sampling units have been selected using random numbers.

sons for using stratified random sampling to characterize a source: (1) to evaluate each strata separately, and (2) to increase the sampling precision for the entire population.

The fundamental approach used in stratified random sampling is breaking the population down into a number of subpopulations. In selecting the most common subpopulations of depth in natural soils, soil horizons (layers designated A, B, C, etc. according to their origin and characteristics) are usually considered as subpopulations and are sampled separately. Poor results have been obtained by placing more than one horizon in a subpopulation [4.22].

After sampling and analyses have been completed, the estimate of the mean in one stratum may be determined by

$$\bar{y} = \left(\sum_{i=1}^{n} y_i \right) \Big/ n \tag{4.8}$$

where

$\bar{y}$ = the estimate of the mean in one stratum
y_i = the value observed for the ith sample collected
n = the number of samples collected

The mean over all strata is

$$\bar{\bar{y}} = \left(\sum_{h=1}^{L} N_h \bar{y}_h \right) \Big/ N \tag{4.9}$$

where

$\bar{\bar{y}}$ = the estimate of the mean over all strata
L = the total number of strata
N_h = the total number of sampling units in the hth stratum
$\bar{y}_h$ = the mean value in the hth stratum
N = the total number of sampling units in all strata

The sample variance within each stratum may also be calculated:

$$V(\bar{y}_h) = \frac{S_h^2}{n_h} = \frac{1}{n_h \cdot (n_h - 1)} \sum_{i=1}^{n_h} (y_i - \bar{y}_h)^2 \tag{4.10}$$

where

$$V(\bar{y}_h) = \frac{S_h^2}{n_h} = \text{the sample variance within each stratum}$$

n_h = the number of sampling units measured in each stratum
y_i = the value observed for the ith sampling unit

The variance of the sample mean over all strata is then

$$V(\bar{\bar{y}}) = \frac{1}{N^2} \sum_{h=1}^{L} N_h^2 \left(\frac{S_h^2}{n_h} \right) \tag{4.11}$$

where $V(\bar{\bar{y}})$ = the variance of the sample mean over all strata.

Analysis of data resulting from stratified random sampling is illustrated in Example 4.4.

EXAMPLE 4.4 *Stratified Random Sampling Analysis*

Stratified random sampling for di-*n*-octyl phthalate was conducted at four depths below a soil pit with the following results.

Depth	[Di-*n*-octyl phthalate], mg/kg				
2 cm	24.6	38.1	26.7	50.6	48.1
4 cm	21.1	30.2	18.9	38.7	31.0
6 cm	32.4	10.7	17.8	20.1	15.5
8 cm	10.8	33.4	23.5	16.1	11.9

Determine the mean and variance for each of the strata and for the plot as a whole.

SOLUTION

The mean di-*n*-octyl phthalate concentration, $\bar{\bar{y}}$, may be computed:

$$\bar{\bar{y}} = (24.6 + 38.1 + 26.7 + 50.6 + 48.1 + 21.1 + 30.2 + 18.9 + \cdots)/20 =$$
$$520.2/20 = 26.0$$

The sample variance may then be calculated using Equation 4.11. The first step in calculating the sample variance is to determine the variance for each depth h using Equation 4.10.

For the 2 cm depth

$$\bar{y}_h = 24.6 + 38.1 + 26.7 + 50.6 + 48.1/5 = 188.1/5 = 37.6$$

$$\frac{S_h^2}{n_h} = \frac{1}{n_h \cdot (n_h - 1)} \sum_{i=1}^{n_h} (y_i - \bar{y}_h)^2$$

$$\frac{S_h^2}{n_h} = \frac{1}{5 \cdot (5 - 1)} \cdot [(24.6 - 37.6)^2 + (38.1 - 37.6)^2 + \cdots + (48.1 - 37.6)^2]$$

$$\frac{S_h^2}{n_h} = 28.4$$

Repeat the procedure for the remaining three strata. The mean and variance within each stratum for the succeeding depths are

$$\bar{y}_{4\ cm} = 28.0 \qquad V(\bar{y}_{4\ cm}) = 12.9$$

$$\bar{y}_{6\ cm} = 19.3 \qquad V(\bar{y}_{6\ cm}) = 15.2$$

$$\bar{y}_{8\ cm} = 19.1 \qquad V(\bar{y}_{8\ cm}) = 17.7$$

The variance for all of the strata combined may be calculated using Equation 4.11:

$$V(\bar{y}) = \frac{1}{N^2} \sum_{h=1}^{L} N_h^2 \left(\frac{S_h^2}{n_h}\right) = \frac{1}{20^2}[5^2(28.4) + 5^2(12.9) + 5^2(15.2) + 5^2(17.7)]$$

$$V(\bar{y}) = 4.6$$

The confidence interval may also be determined using Equation 4.7.

$$CI = \bar{\bar{y}} \pm t_\alpha \left(\frac{S^2}{n}\right)^{1/2}$$

$$= 26.0 \pm 2.093(4.6)^{1/2}$$

$$= 26.0 \pm 4.5$$

In summary, the true population mean for the concentration of di-*n*-octyl phthalate in all strata lies in the range of 21.5 to 30.5 mg/kg with a probability of 0.95 (a 95% confidence interval).

4.5   SOURCE SAMPLING PROCEDURES AND STRATEGIES

The practical aspects of source sampling involve a knowledge of sampling devices and the strategies for their use.

4.5.1   Sampling Devices

Instruments ranging from simple to complex have been used to sample hazardous wastes contained in drums and tanks. The *coliwasa* (composite liquid waste sampler), used to sample liquids, is a glass, plastic, or metal tube fitted with an end closure that can be closed when the tube is submerged in the sample. *Weighted bottles* are also used to sample liquids and nonviscous, free-flowing slurries. These containers are simple glass or plastic bottles fitted with a sinker, a stopper, a line or rod to lower the bottle into the liquid, and a device to open the bottle. A *dipper* is simply a glass or plastic beaker clamped to the end of a telescoping aluminum or fiberglass pole. A *thief* is used to sample dry granules or powders. It is constructed of two slotted, concentric steel or brass tubes. The outer tube has a pointed tip; the inner tube is rotated to open and close the sampler. Diagrams of these sampling devices are shown in Figure 4.6.

Surface soils and sludges may be sampled using a number of devices. Depending on the magnitude of the site or facility assessment, mechanical drill rigs or hand-driven soil samplers can be used to sample surface soils and sludges; drill rigs are used to sample soils down to 30 m (98 ft.). Augers commonly used with drill rigs include

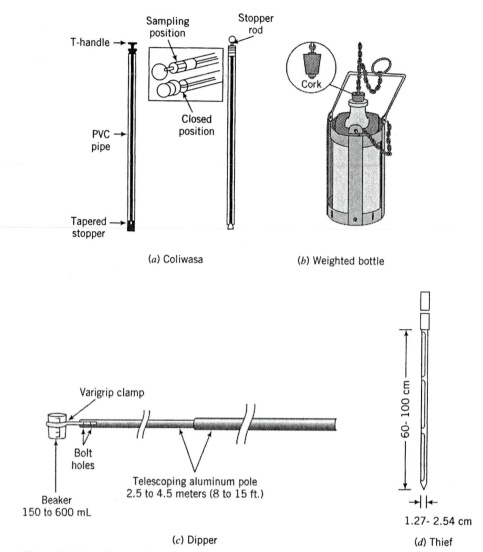

Figure 4.6 Sampling devices for tanks and drums. *Source:* Reference 4.14.

hollow-stem augers, Shelby tubes (thin-walled volumetric samplers), and split-spoon or split-barrel drive samplers. *Hollow-stem augers* are connected in 1.5-m (5-ft.) sections, and provide an efficient method of sampling deep into the vadose zone. *Shelby tubes* come in models of 5 cm (2 in.), 8 cm (3 in.), and 13 cm (5 in.) outside diameters, and are usually 76 cm (30 in.) in length. They are pushed into the soil by a drill rig and, because of compaction, the sample recovered is shorter than the distance packed. *Split-barrel drive samplers* consist of two split-barrel halves, a drive shoe, and a sampler head containing a ball check-valve. These systems have traditionally been used for geotechnical engineering sampling and evaluations.

Manual, hand-powered systems have been developed for sampling surface soils, and provide low-tech methods for sampling. *Screw-type augers* consist of a screw or flight auger of ~4 cm (~1.5 in.) diameter. The upper end has a threaded assembly to add a crossbar or extensions; the auger essentially screws into the soil. *Barrel or bucket augers* (e.g., Soil Conservation Service augers), which comprise a pair of penetrating stainless steel bits attached to a barrel (i.e., a short cylinder that catches the soil) of slightly smaller diameter, may also be used for sampling shallow contamination. Using a T-handle, the sampler is turned, or screwed, into the ground until the barrel is filled. Common *post hole augers* are simple and readily available. A tapered barrel helps to hold the soil once it is sampled. Some of these soil sampling devices are illustrated in Figure 4.7.

4.5.2 Sampling Strategies

Weighted bottles, coliwasas, thieves, and dippers are used to sample drums. Drum sampling strategies vary depending on (1) the number of containers requiring sampling, (2) the heterogeneity within the drums, and (3) sampling access. If a set of drums has all received the same waste and each is homogeneous in physical characteristics, each drum could, in theory, be sampled to characterize the population. However, if hundreds of drums are staged for disposal, a one-dimensional sampling scheme is usually used. The drums are numbered and n samples are selected (based on Equation 4.6) by first numbering the drums consecutively and selecting n samples using random number generation [4.23].

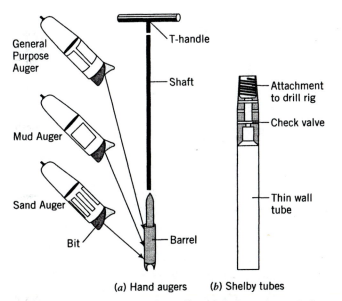

(a) Hand augers   (b) Shelby tubes

Figure 4.7 Devices for sampling soils and sludges. *Source:* Reference 4.14.

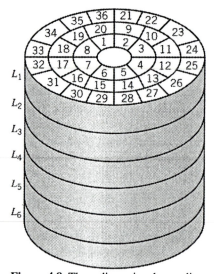

Figure 4.8 Three-dimensional sampling grid used for drums and tanks. *Source:* Reference 4.14.

If no information is known about the waste contents of drums, a three-dimensional sampling grid is often used. The reason for such detailed sampling is the tendency of drummed wastes to be heterogeneous in the vertical direction because of (1) settling of drum contents, (2) the formation of distinct phases, and (3) the order in which the contents were placed in the drum. The EPA [4.14] recommends the use of a three-dimensional stratified random sampling grid (Figure 4.8). The height of the vertical sampling unit, L, should be at least as high as the sampling device. An alternative two-dimensional sampling scheme has been used when the vertical contents are known to be homogeneous. In this case, a two-dimensional, simple random sampling grid (using principles of sampling units and random number generation) coupled with collecting a composite sample with depth is used.

4.6 SAMPLING AWAY FROM THE SOURCE

Part II of this book, "Pathways," focuses on the movement and transformation of hazardous chemicals at the source and after they are released from the source. The approach of Part II provides a basis for learning about the behavior of hazardous chemicals in the environment and in engineered systems through the use of calculations and models for estimating contaminant concentrations away from the source. However, sampling away from the source, whether it is a RCRA TSD facility or a contaminated site, is usually necessary to determine the concentrations of hazardous chemicals. Furthermore, sampling away from the source is often a prerequisite before designing remediation systems and documenting their effectiveness during operation. The most common media that are sampled are air, surface water, and groundwater. The sampling of volatile emissions and the installation of monitoring wells are field techniques that are beyond the scope of this text. A number of professional references are available that dedicate hundreds of pages to these techniques [4.24–4.27]. The following brief discussion serves only as an introduction to sampling away from the source.

4.6.1 Sampling and Monitoring of Air and Volatile Emissions

Air sampling data are needed for a number of reasons, including (1) complying with toxic air emission standards (e.g., Hazardous Air Pollutants under the Clean Air Act), (2) maintaining worker safety by monitoring and comparing actual values in the air to guidelines used by industrial hygienists—Threshold Limit Values (see Section 10.4), and (3) evaluating the treatment of volatile compounds. For example, volatile emissions released from drums, industrial processes, and contaminated sites require monitoring to assess hazard, protect public health, and provide data for the design of treatment and remediation systems (Figure 4.9). Monitoring of emissions from soil treatment processes such as soil vapor extraction systems (see Section 12.4.2) is also commonplace.

Two approaches are used in air monitoring—direct (i.e., real-time) measurements in which data are obtained instantaneously, and indirect measurements. The primary advantage of direct measuring devices is the immediate availability of the results. However, these instruments are often incapable of measuring low concentrations and are only available for selected hazardous compounds. Indirect measurements, which involve collecting the hazardous compounds over a longer time period coupled with sub-

Figure 4.9 Sampling a potentially dangerous drum with an organic vapor analyzer. *Source:* U.S. Department of Energy, Richland, WA office.

sequent laboratory analysis, are more sensitive and can measure a wide range of chemicals; however, the data may take days to obtain.

Real-Time Air Monitoring

The use of a direct-reading air monitoring device must be matched to the chemicals that are likely emitted from the source. The most common real-time air sampling instruments are electrochemical cells, photoionization detectors, organic vapor analyzers, and gas detector tubes.

Electrochemical cells are based on the principle of membrane electrolysis. Gases present in the atmosphere pass through a membrane where they react with an electrolyte, which then causes a change in standing current proportional to the concentration of the gas in the air. Because of the specific nature of the reactions in electrochemical cells, most of these are single-compound analytical systems.

Photoionization detector (PID) monitors are more nonspecific systems relative to electrochemical cells. Molecules that are drawn through the detector are excited by a source of ultraviolet light, and the excited species produce a change in standing current proportional to the concentration of the contaminants.

Organic vapor analyzers (OVAs) measure a spectrum of volatile organic compounds in the air using a nonselective detector, such as a flame ionization detector (FID). The FID has a flame, and as the sample passes through the detector, the standing current rises in proportion to the increase in the flame from combustion of the organic compounds.

Gas detector tubes are fitted with colorimetric detectors, and the contaminant in a standard volume of air reacts with a specific reagent designed to produce a colorimetric response to the compound. The response is produced in proportion to its concentration and is quantified on a calibrated meter. Gas detector tubes have been developed for total hydrocarbons, benzene, hydrogen cyanide, phosgene, arsine, carbon monoxide, and hydrogen sulfide.

Air Sample Collection and Concentration

Many systems have been developed to concentrate and collect contaminants in air for subsequent analysis by gas chromatography or atomic absorption spectrophotometry (see Section 4.7). The most common sample concentration/collection systems are solid sorbents, passive diffusion monitors, and impingers.

Solid sorbent tubes are used to sample gases and vapors by drawing air through a tube of solid material onto which the contaminants partition by weak chemical bonds (e.g., hydrogen bonding, van der Waals forces). Common sorbents include activated carbon, silica gel, and proprietary materials such as Tenax and XAD-4. Contaminants collected on these sorbent tubes are then removed by exhaustive extraction procedures (e.g., a recirculating solvent flush known as a Soxhlet extraction) with subsequent analysis of the extract by gas chromatography.

Passive diffusion monitors are similar to sorbent tubes, but air is not actively drawn through the tube; instead, it is collected by passive diffusion. The solid sorbents are then analyzed by extraction and gas chromatography.

In collecting the contaminant in *impingers*, air is passed through a sparger with a solution in which the contaminant is soluble. The solution is then analyzed by laboratory methods.

4.6.2 Groundwater Sampling

Sampling of groundwater away from contaminated sites and RCRA facilities often requires installing monitoring wells followed by sampling to obtain groundwater samples at periodic times. Well installation involves drilling a borehole followed by placement of a casing in the vadose zone and screens in the saturated zone, and sealing the well with an expanding clay, usually bentonite. After the well is fit with a secure well head, it is developed, that is, purged to remove particulates and contamination that may have resulted from construction. A typical monitoring well is shown in Figure 4.10.

The placement of monitoring wells is an intricate task, and is sometimes considered more an art than a science. A minimum of one monitoring well upgradient and three wells downgradient is required for RCRA facilities. In assessing and monitoring contaminated sites, a range of a few to hundreds of monitoring wells may be installed. Therefore, a fundamental understanding of the types of monitoring wells, their applications, and their limitations is essential for assessing hazardous waste sources.

Borehole Drilling

The first step in well installation is drilling the borehole, which is no more than a deep, narrow hole in the ground. Numerous borehole drilling methods are available, and se-

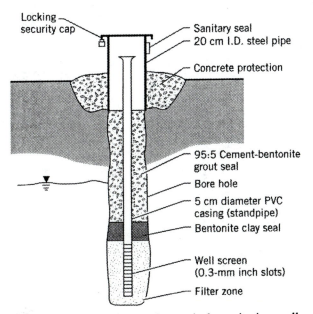

Figure 4.10 Components of a typical monitoring well.
Source: Reference 4.24.

Table 4.2 Summary of Methods for Drilling Monitoring Wells

Method	Subsurface Material	Maximum Depth	Diameters	Limitations
Auger	Unconsolidated materials: clay, silt, sand, gravel (<5 cm diameter)	25 m (82 ft)	15–90 cm (6–36 in.)	Difficult to obtain surface material samples. Difficult to keep hole open under certain conditions (sidewall collapse).
Mud rotary	Silt, sand, gravel (<2 cm) Silt-hard consolidated rock	450 m (1500 ft)	8–45 cm (3–18 in.)	Water source required for mud. Contaminants may be added by the addition of mud.
Air rotary	Silt, sand, gravel (<5 cm) Soft-hard consolidated rock	600 m (2000 ft)	30–50 cm (12–20 in.)	Air source may add contaminants to borehole. May strip volatile contaminants.
Cable tool	Almost all surface materials (unconsolidated, fractured rock, boulders, etc.)	600 m (2000 ft)	10–75 cm (4–30 in.)	Drilling is slow. Steel pipe may be subject to corrosion.
Driven wells	Unconsolidated material containing no rock or boulders	15 m (50 ft)	3–10 cm (1–4 in.)	No subsurface solid samples can be collected.
Dual-wall reverse circulation rotary	Silt, sand, gravel (<10 cm)	60 m (200 ft)	40–120 cm (16–50 in.)	Air source may contaminate bore hole. Expensive.
Jetted	Unconsolidated	15 m (50 ft.)	3–10 cm (1–4 in.)	Wash water may dilute groundwater. Slow.

lection is based primarily on the type of subsurface material (e.g., unconsolidated solids vs. rock), the required drilling depth, and the need to minimize contamination from drilling materials. Several drilling methods are described here, and more specific information on each method is listed in Table 4.2. The process of drilling a borehole is illustrated in Figure 4.11.

Auger Method. *Augers* are commonly used devices that can be hand-operated for drilling shallow wells in loose soils or power-driven for deeper wells. Of the power-driven augers, hollow-stem models are most commonly used for drilling monitoring wells. Other auger systems include bucket augers for large-diameter wells and solid-stem augers for drilling through consolidated subsurface materials.

Hollow-stem augers consist of sections of steel tubes with spiral flights and a cutter head at the tip (Figure 4.12). The best use of these systems is in drilling through unconsolidated material such as sands, silts, and gravels.

Mud Rotary Drilling. Adopted from the oil industry, *mud rotary drilling* (which has also been called *direct rotary, hydraulic rotary,* and *rotary drilling*) uses a hollow bit through which drilling mud or a clay-water slurry is forced during drilling (Figure 4.13). As the mud is pumped through the drill rods and the bit, it flows upward through the annular space (the area between the borehole wall and the drill rods). The injection of the mud serves three purposes: (1) it carries the drilling debris back to the sur-

Figure 4.11 A portable rig drilling a monitoring well. *Source:* U.S. Department of Energy, Richland, WA office.

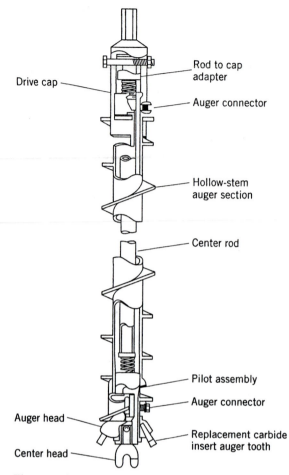

Figure 4.12 A hollow-stem auger drilling system.
Source: Reference 4.25.

face, (2) it holds the borehole open, and (3) it cools the drill bit. The source of the drilling mud is a simple pit or trough. As the mud exits the borehole, it is pumped back to the pit or trough, where the heavy cuttings settle to the bottom and the lighter mud is recycled to the borehole.

Reverse Circulation Rotary Drilling. A method similar to mud rotary drilling, *reverse circulation rotary drilling* is based on pumping the mud up the drilling pipes rather then down. Larger mud pumps are required compared to the standard mud drilling practice because the boreholes are usually larger. The primary advantage of this method is the minimal disturbance of the borehole and rapid removal of the cuttings.

Air Rotary Drilling. Also termed *direct rotary drilling,* the *air rotary process* is similar to the mud rotary process, but compressed air is delivered through the hollow stem and bit. As in rotary mud drilling, the air cools the drill bit and blows the cuttings to the surface, where they accumulate around the borehole. Two methods are used to keep

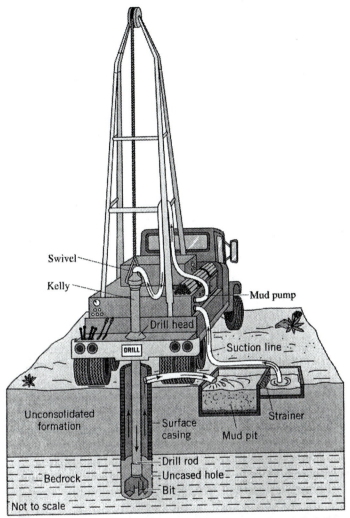

Figure 4.13 A mud rotary borehole drilling system. *Source:* Reference 4.27.

the borehole open when drilling in unconsolidated material. One method involves the use of drilling foam to fill the well as it is drilled. In the second procedure, the well casing is inserted just behind the drilling zone.

Cable Tool Drilling. By continuously lifting and dropping a string of tools suspended on a cable using a rotating pulley, the *cable tool drilling method* is effective in penetrating nearly every type of subsurface strata including unconsolidated formations, consolidated rock, fractured rock, broken cavernous rock, and boulders. At the end of the string is a percussion drill bit that has the capability of penetrating both unconsolidated and consolidated materials. As the well casing is driven down simultaneously within the outer borehole casing, the cuttings accumulate and mix with the groundwater (or water added down the borehole) to form a slurry that is pumped to the surface.

Jetted Well. For drilling shallow wells in unconsolidated media, the jetted system is inexpensive and uses light equipment that can be operated by personnel with minimal training. The *jetted well* is installed by the penetration of a pointed bit containing jets through which water is forced. The drilling process is accomplished by a combination of the jetting action of the water and the mechanical action of the bit. As the subsurface solids are pushed aside by the process, the well casing slides down the hole under gravity. Water exiting the system through the annular space carries the loosened solids to the surface.

Drive-Point Wells. By using a series of connected pipes driven into the soil with a sledgehammer, posthole driver, drive weight, or pneumatic hammer, a quick and inexpensive method of drilling is provided by *drive-point wells.* The drilling system consists of threaded interlocking pipes or pipes connected by threaded couplings atop a steel driving cone. Drive-point wells are usually installed only in unconsolidated rock without rocks or large gravel.

Well Casings and Screens

Well casings are placed in the borehole to keep the hole open and minimize dissociation of soils and gravel. *Well screens* replace the casings in the saturated zone and allow collection of the groundwater. Installation of the well casings and screens is often completed simultaneously with drilling or may occur shortly thereafter. The five materials commonly used for these internal well components include mild steel, stainless steel, polyvinyl chloride (PVC), Teflon, and polypropylene. Selection of the casing material is based on strength considerations for deep and wide wells, which, because of their configurations, are characterized by more forces on the well casings. If strength and structural integrity are the primary factors in casing and screening selection, then mild steel and stainless steel are the materials of choice. Teflon, PVC, and polypropylene have lower tensile strength and less rigidity than steel materials.

A primary concern in the construction of monitoring wells is chemical incompatibility and the potential for sample contamination. Sampling for organic contaminants is a common goal for groundwater monitoring well installation, and in such cases, PVC and polypropylene are not used. These materials contain plasticizers (e.g., phthalate esters) and solvents from adhesives used to join pipe sections, which leach out during sampling with subsequent contamination of the samples. Furthermore, PVC may sorb some groundwater contaminants and is reactive with ketones, esters, and BTEX compounds. Teflon is the most inert of the casing and screening materials; it is resistant to attack from almost all chemicals and does not leach organic compounds, but it is expensive and not very strong.

Mild steel and even stainless steel, although structurally sound, may be chemically reactive. Both types of steel may leach traces of heavy metals into groundwater samples and may also fix metals by ion exchange. Stainless steel, under corrosive conditions, especially in the presence of H_2S in anaerobic aquifers, may bleed chromium into groundwater samples and may also act as a catalyst that promotes the transformation of some organic contaminants.

Besides the materials of which they are constructed, the most important criteria for well screens are (1) slot sizing to minimize the entry of aquifer solids and (2) sufficient structural support. Of the most common well screens that have been used for monitoring wells, wire-wound continuous-slot screens are the most expensive, but are durable, exclude aquifer solids, and provide good yield into the well. Wire-wound perforated pipe is effective where coarse aquifer material is present; it also has high structural integrity. Louvre or bridge-slot screens can be purchased in a wide range of sizes but clog easily.

Filter Packs. To create a more permeable zone surrounding the well screen and to increase the effective diameter of the well, a porous material called *filter packs* is added around the well screen at a thickness of 8–15 cm (3–6 in.) in the annular space of the borehole. Filter packs are composed of clear, uniform, smooth, well-rounded sand and gravel. The EPA recommends the use of washed filter pack material for all monitoring wells. Filter packs offer a number of benefits, including stabilizing the aquifer material and reducing head loss to the well. Bentonite or concrete is placed above the filter pack within the annular space as grouting. The purpose of the grouting material is to prevent water and other extraneous materials from entering the well from the surface, and to prevent vertical water movement from different layers in the aquifer by short-circuiting down the borehole alongside the well casing.

Monitoring Well Development and Sampling. After a monitoring well has been constructed, it must be developed before sampling of the groundwater is begun. *Well development* is the process of pumping or bailing water from freshly installed wells until the turbidity associated with suspended fines from the drilling and well installation has been removed, and the clear water from the well achieves a constant composition reflecting that of the groundwater. If the well is shallow and sampling will be performed infrequently, the well can be sampled manually with a bailer or siphon. Deeper wells are sampled with submersible pumps, centrifugal pumps, or surging with compressed air.

4.7   PRIORITY POLLUTANT AND SAMPLE ANALYSES

As part of the 1977 Amendments to the Clean Water Act, the EPA was charged with developing a list of common industrial pollutants that pose an imminent threat to public health and the environment. The list consists of 129 compounds: 114 organics, 13 metals, 1 mineral, and 1 inorganic nonmetal. These *priority pollutants* (see Section 3.10) are divided into categories based on their method of analysis.

Most organic contaminants found in water and solid samples are analyzed after they have been extracted into an organic solvent such as hexane, ethyl acetate, or a cosolvent system (e.g., hexane-acetonitrile). Extraction procedures usually involve shaking the water or solid with the solvent, which results in transfer of the contaminant from the environmental media to the solvent. Due to the difficulty of extracting soils and solids because of strong sorptive forces (see Section 5.1), these media are often flushed continuously with solvent using a device called a *Soxhlet tube*. In a similar sense, most solid media require digestion with a strong acid solution before the analysis of metal contaminants.

The organic priority pollutants are usually analyzed using a *gas chromatograph* (GC). A microliter quantity of sample extract is injected into a long, narrow heated column packed with a solid sorbent. Modern columns are made of glass capillary tubing usually tens of meters long coated on the inside with a waxy liquid such as methyl phenyl silicones which serves as the sorbent. The chemicals are swept through the heated column by an inert carrier gas such as nitrogen or helium, and are separated based on their volatility and polarity (Figure 4.14). A detector at the end of the column then provides a signal to an integrator. The most common detectors are the nonselective *flame ionization detector* (FID), which burns the sample components and sends a response to the integrator in proportion to the flame, and the *electron capture detector* (ECD), in which an electron source (usually ^{63}Ni) is present in the detector through which the sample passes, producing an electron flow. Compounds with electrophilic moieties (e.g., —Cl, —NO_2) temporarily capture the electrons and provide a change in the baseline current, which is measured at the recorder. More recently, gas chromatographs have been interfaced with quadrapole or high-resolution *mass spectrometers* (MSs), which act as detectors for analyzing trace organic compounds, while providing a high degree of proof of their presence in samples. When a large number of samples are to be analyzed, GCs are fitted with automatic injectors, which sample and inject test solutions from sequences of sealed vials, with intermediate rinses of pure solvent.

Low-volatility organic pollutants (e.g., triazine herbicides) are analyzed by *high-performance liquid chromatography* (HPLC), which uses a solvent or solvent mixture under high pressure as a mobile phase, flowing through a packed column of solid absorbent, e.g., silica gel or alkylated (C_{18}) silica gel as a stationary phase. Numerous detectors (UV absorption, conductivity, mass spectrometers) are used with HPLCs.

The most common method for metals analysis is *atomic absorption* (AA) *spectrophotometry,* which is based on aspiration (spraying through an atomizer) of an aqueous sample (usually dissolved in nitric acid) into a flame; absorption of a light beam passing through the flame is measured and is proportional to the metal concentration in the sample (Figure 4.15). The light beam emits the characteristic wavelength for the element being analyzed. To enhance the presence of atoms in a dispersed quasi-gaseous state suitable for absorbing this light, some instruments use an electrically heated carbon furnace instead of a flame. Another effective method of elemental analysis is *inductively coupled argon plasma* (ICP), which can be used for simultaneous multielement analysis and can be coupled with mass spectrometers for supersensitive or single isotope analyses.

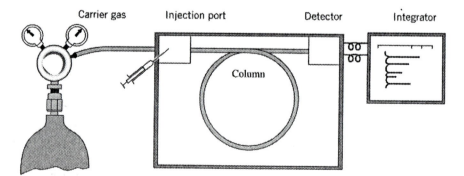

Figure 4.14 Schematic of a gas chromatograph. *Source:* Reference 4.28.

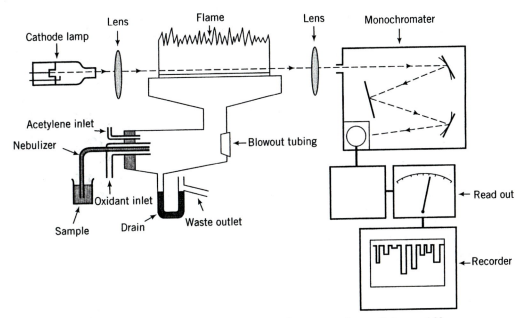

Figure 4.15 Schematic of an atomic absorption spectrophotometer. *Source:* Reference 4.28.

Organic compounds with similar physical and chemical properties are often analyzed together following well established, standardized analytical methods. For example, 21 of the purgeable priority pollutants are analyzed simultaneously using EPA Method 601. ("Purgeable" implies that the compounds are volatile and can be stripped from aqueous solution by a stream of air). The methods, which are given the categories 601 through 625, are listed in Table 4.3.

The 600 series analyses are EPA-approved procedures for aqueous samples [4.29]. The 8000 series has been developed for soils and sludges [4.14]. The numbering for most of the 8000 series analysis is analogous to the 600 series. For example, Method 8010 is for volatile organics in soils and sludges, just as Method 601 is for volatiles in water.

4.8   SUMMARY OF IMPORTANT POINTS AND CONCEPTS

- Waste audits, which are based on materials balances around industrial facilities, can be used to determine losses from volatilization, spillage, authorized removals to secure landfills or incineration, or improper management. Waste audits also serve as the basis for waste minimization and pollution prevention plans.

- The assessment of contaminated sites is based on a multistep approach, with each phase characterized by succeeding complexity. Phase I assessments focus on the review of historical data combined with a cursory site visit. Phase II assessments emphasize source sampling and sampling away from the source (e.g., groundwater, air).

Table 4.3 U.S. Environmental Protection Agency 600 Series for Priority Pollutant Analyses

Method 601—Purgeable Halocarbons

Bromodichloromethane
Bromoform
Bromomethane
Carbon tetrachloride
Chlorobenzene
Chloroethane
2-Chloroethylvinyl ether
Chloroform
Chloromethane
Dibromochloromethane
1,2-Dichlorobenzene
1,3-Dichlorobenzene
1,4-Dichlorobenzene
Dichlorodifluoromethane
1,1-Dichloroethane
1,2-Dichloroethane
1,1-Dichloroethene
trans-1,2-Dichloroethene
1,2-Dichloropropane
cis-1,3-Dichloropropene
trans-1,3-Dichloropropene
Methylene chloride
1,1,2,2-Tetrachloroethane
Tetrachloroethene
1,1,1-Trichloroethane (TCA)
1,1,2-Trichloroethane
Trichloroethene (TCE)
Trichlorofluoromethane
Vinyl chloride

Method 602—Purgeable Aromatics

Benzene
Chlorobenzene
1,2-Dichlorobenzene
1,3-Dichlorobenzene
1,4-Dichlorobenzene
Ethylbenzene
Toluene

Method 603—Acrolein and Acylonitrile

Acrolein
Acrylonitrile

Method 604—Phenols

4-Chloro-3-methylphenol
2-Chlorophenol
2,4-Dichlorophenol
2,4-Dimethylphenol
2,4-Dinitrophenol
2-Methyl-4,6-dinitrophenol
2-Nitrophenol
4-Nitrophenol
Pentachlorophenol
Phenol
2,4,6-Trichlorophenol

Method 605—Benzidines

Benzidine
3,3′-Dichlorobenzidine

Method 606—Phthalate Esters

Bis(2-ethylhexyl) phthalate
Butyl benzyl phthalate
Di-n-butyl phthalate
Diethyl phthalate
Dimethyl phthalate
Di-n-octyl phthalate

Method 607—Nitrosamines

N-Nitrosodimethylamine
N-Nitrosodiphenylamine
N-Nitrosodi-n-propylamine

Method 608—Pesticides and PCBs

Aldrin
α-BHC
β-BHC
δ-BHC
γ-BHC
Chlordane
4,4′-DDD
4,4′-DDE
4,4′-DDT
Dieldrin
Endosulfan I
Endosulfan II
Endosulfan sulfate
Endrin
Endrin aldehyde
Heptachlor

Method 608—Continued

Heptachlor epoxide
Toxaphene
PCB-1016
PCB-1221
PCB-1232
PCB-1242
PCB-1248
PCB-1254
PCB-1260

Method 609—Nitroaromatics and Isophorone

2,4-Dinitrotoluene
2,6-Dinitrotoluene
Isophorone
Nitrobenzene

Method 610—Polynuclear Aromatic Hydrocarbons

Acenaphthene
Acenaphthylene
Anthracene
Benzo[a]anthracene
Benzo[a]pyrene
Benzo[b]fluoranthene
Benzo[g,h,i]perylene
Benzo[k]fluoranthene
Chrysene
Dibenzo[a,h]anthracene
Fluoranthene
Fluorene
Indeno(1,2,3-c,d)pyrene

Naphthalene
Phenanthrene
Pyrene

Method 611—Haloethers

Bis(2-chloroisopropyl)ether
Bis(2-chloroethoxy)methane
Bis(2-chloroethyl)ether
4-Chlorophenyl phenyl ether
4-Bromophenyl phenyl ether

Method 612—Chlorinated Hydrocarbons

2-Chloronaphthalene
1,2-Dichlorobenzene
1,3-Dichlorobenzene
1,4-Dichlorobenzene
Hexachlorobenzene
Hexachlorobutadiene
Hexachlorocyclopentadiene
Hexachloroethane
1,2,4-Trichlorobenzene

Method 613—2,3,7,8-Tetrachlorodibenzo-p-dioxin

2,3,7,8-Tetrachlorodibenzo-p-dioxin

Method 624—Purgeables

Benzene
Bromodichloromethane
Bromoform
Bromomethane
Carbon tetrachloride

Chlorobenzene
Chloroethane
2-Chloroethylvinyl ether
Chloroform
Chloromethane
Dibromochloromethane
1,2-Dichlorobenzene
1,3-Dichlorobenzene
1,4-Dichlorobenzene
1,1-Dichloroethane
1,2-Dichloroethane
1,1-Dichloroethene
trans-1,2-Dichloroethene
1,2-Dichloropropane
cis-1,3-Dichloropropene
trans-1,3-Dichloropropene
Ethyl benzene
Methylene chloride
1,1,2,2-Tetrachloroethane
Tetrachloroethene
Toluene
1,1,1-Trichloroethane
1,1,2-Trichloroethane
Trichloroethene
Trichlorofluoromethane
Vinyl chloride

Method 625—Base/neutrals, Acids and Pesticides (includes compounds listed under Methods 604, 605, 606, 607, 608, 609, 610, 611, and 612)

- For the purpose of rapid assessment under emergency response conditions, source concentrations may be estimated based on the mass of contaminant spilled, the volume of soil contaminated, and the bulk density of the soil.

- Source materials, such as soils and sludges, are most effectively sampled using probability sampling. Simplified random sampling is used in systems in which there is no gradient in contaminant concentration with depth. Stratified random sampling is used to sample systems where different strata contain a range of concentrations.

- Contaminants are often sampled at the source and as they migrate from the source after release. The most common sources that require sampling are drums, tanks, soils, and sludges. Although all environmental compartments often require sampling as contaminants migrate from the source, air and groundwater are the media that receive the most focus. Air is sampled by real-time instruments or after collection on solid media followed by laboratory analysis. Groundwater is sampled after the installation of monitoring wells.

- The EPA priority pollutants are analyzed by methodology in which the chemicals are grouped by physical properties (e.g., volatility) or similar functional groups (e.g., nitroaromatics). The EPA 600 series are used for water samples and the 8000 series have been developed for solid media.

PROBLEMS

4.1. A solvent distilling and recycling facility, which primarily recycles TCE, is under investigation for the release of toxic air emissions. As the hazardous waste manager at the site, your goal is to account for the solvents coming in and out of the facility using a waste audit based on a mass balance. The wastes coming into the facility over a 1-year time period include

Invoice Number	Volume (L)	TCE Concentration (mg/L)
204	10,000,000	130
205	5,000,000	210
206	5,000,000	200

These aqueous solutions were distilled to concentrate the TCE. The concentrated product was then shipped (over the same 1-year time period) from the facility in tanker trucks as 90% (900 g/L) formulations (the other 10% were various impurities):

Shipping No.	Volume (L)	TCE Concentration (g/L)
1021	2200	900
1022	1400	900

The state regulatory agency allows 200 kg/year TCE emitted into the air from your facility. Using a materials balance, show that the plant is meeting this regulatory criterion.

4.2. The following records are available for a small company that manufactures semi-conductors. The chemical used most at the plant is TCA, and the primary loss mechanism is volatilization. Using the following data, estimate the loss of TCA from the plant by volatilization.

Acquisitions:

Date	Volume (L)	Purity (%)
1/8	208	92
2/6	833	95
2/26	416	88
3/15	208	97
4/11	208	90
5/2	625	95

Discharges:

Wastewater characteristics: 18.9×10^6 L/day (5 MGD) at 0.5 mg/L TCA.

RCRA manifests:

Date	RCRA Waste No.	Volume (L)	Concentration (%)
2/28	F001	1140	22
4/20	F001	386	13
5/21	F001	462	45
5/30	F001	1287	36

Based on these data, estimate the maximum volatilization rate.

4.3. A plastic formulation facility receives one of its synthetic chemicals, dimethyl phthalate, by steady flow through a supply pipe at a rate of 200 L/day. The dimethyl phthalate is used in synthesis reactors for making a plasticized polymer at a rate of 0.8 mole/min. Partially reacted polymers are precipitated from specific reactions containing fluids with 1200 mg/kg (dry wt.) dimethyl phthalate. If 30,000 kg of the sludge (with an average water content of 85%) are generated each day, how much dimethyl phthalate is unaccounted for?

4.4. You are the hazardous waste manager for an industrial shop. The primary activities at the facility include electroplating, metal parts cleaning, and paint stripping. The monthly input of hazardous chemicals to the plant is as follows:

Acetone	Five 208-L (55-gal) drums
MEK	Eight 208-L (55-gal) drums
Acrylonitrile	Three 208-L (55-gal) drums
TCE	Five 208-L (55-gal) drums
Sodium cyanide	227 kg (500 lb)
Potassium dichromate	22.7 kg (50 lb)
Sulfuric acid	One 208-L (55-gal) drum

All of the chemicals purchased are at > 99% purity.

The usual monthly disposal of the wastes includes:

Acetone	Ten 208-L (55-gal) drums (222,000 mg/L in water)
MEK	Eighteen 208-L (55-gal) drums (50,000 mg/L in water)
Acrylonitrile	Two 208-L (55-gal) drums (95% v/v)
TCE	Two 208-L (55-gal) drums (90% v/v)
Sodium cyanide	Five 208-L (55-gal) drums (174,000 mg/L)
Potassium dichromate	Electroplating sludge (1020 kg solids at 216 ppm Cr and 12,600 L of water)
Sulfuric acid	80% of a 208-L (55-gal) drum

The state regulatory agency has required you to perform a waste audit on the facility to document losses. Provide a listing of losses of chemicals in the facility between procurement and disposal. Based on the physical/chemical properties of these wastes, what are some possible routes of these losses for each chemical?

4.5. A UST soil remediation operation is under the scrutiny of the state regulatory agency for possible emissions of hydrocarbons during excavation and handling activities. The soil is being excavated in lots of approximately 1000 kg, stockpiled for about 2 weeks, and then transported to a RCRA landfill. The state regulatory agency is concerned about potential volatilization of BTEX during the stockpiling. The following listings are lots of soil and their documented BTEX concentrations.

Excavation records:

Excavation Lot	BTEX Concentration (mg/kg)	Soil Mass (kg)
1	2340	1030
2	1880	1320
3	3870	1000
4	3140	1220
5	2850	1460
6	4110	1190

Disposal records at RCRA landfill:

RCRA Manifest No.	BTEX Concentration (mg/kg)	Soil Mass (kg)
4763	2460	3400
4764	1840	2010
4765	2510	1810

The state regulatory agency is concerned about volatilization during the waste handling activities and is enforcing a maximum emission standard of 1 kg/day. If volatilization is the only significant loss mechanism, document that your remedial operation is in compliance.

4.6. Describe how you would approach the Phase I assessment of an abandoned petroleum distributorship that is being sold to a real estate developer who plans to construct a housing project.

4.7. A spill of a radioactive waste tanker truck containing 1000 L (264 gal.) of low-level ^{14}C-labeled compounds (12.5 mCi/L) has occurred. Estimate the mean specific activity of the ^{14}C-labeled compounds in the soil if it has been spilled on to a circular area 11 m in diameter and has penetrated 6 cm deep. The soil bulk density = 1600 kg/m^3.

4.8. A 1000-L (264 gal.) tanker truck containing an 80% (v/v) formulation of carbofuran has spilled onto a hardpan soil (a soil with a compacted subsurface layer). Emergency response personnel estimate that the spill is approximately 20 m × 40 m × 40 cm deep, and that the bulk density of the soil is about 2800 kg/m^3. Estimate its concentration in the soil in ppm.

4.9. An aqueous solution of 500 mg/L $Hg(NO_3)_2$ has leaked from a ruptured 208-L (55 gal.) drum. Visual inspection of the area around the drum suggests that the compound is confined to an area approximately 4 m in diameter and 0.2 m deep. Estimate the concentration of mercury in the soil if the soil bulk density = 1800 kg/m^3.

4.10. A transformer containing 80 kg of 0.5% (w/w) Aroclor 1260 has ruptured, resulting in a spill approximately 5 cm deep in an area 12 m × 10 m. If the soil bulk density is approximately 1800 kg/m^3, estimate the PCB soil concentration.

4.11. A pesticide application truck containing 1900 L (500 gal.) of 1800 mg/L butacarb tipped over in an accident. The butacarb formulation spilled and flowed down-slope over an area approximately 200 m^2 and 0.5 m deep. If the soil bulk density is approximately 1.5 g/cm^3, estimate the mean concentration of butacarb in the soil.

4.12. The following data were collected from a sampling grid for octachlorodibenzo-p-dioxin (OCDD) in the upper layer of a surface soil.

[Octachlorodibenzo-p-dioxin], mg/kg:

23.4	22.2
13.1	2.9
30.7	11.6
8.8	34.7
16.2	19.0

Determine the mean, variance, standard deviation, and 95% confidence interval for these data.

4.13. Four samples have been collected randomly from a soil pit containing arsenic. These initial TCLP concentrations are 6.1, 4.2, 5.8, and 6.5 mg/L. Determine the number of samples required to estimate the true population mean at the 95% confidence level based on the difference between the initial mean and the TCLP value.

4.14. An area 10 m × 10 m at a rural service station received waste oil, and is being assessed for total petroleum hydrocarbon (TPH) contamination. Simple random sampling is to be used with sampling units of 0.5 m × 0.5 m × 0.75 m deep. Four initial samples were collected randomly from the site, and their TPH concentrations were 18,100 mg/kg, 17,900 mg/kg, 18,400 mg/kg, and 17,500 mg/kg. If the maximum tolerance from the sample mean is 200 mg/kg, determine the number of samples required for a confidence level of 95%. Plot the location of the samples on a grid.

4.15. An electroplating facility generates a chromium sludge. The facility management would like to petition the State Department of Environmental Protection (SDEP—which has EPA authority to regulate RCRA in the state) to delist the plating sludge as a RCRA hazardous waste. Before it is delisted, SDEP must be reasonably certain that the sludge is not a RCRA waste based on characteristics. The SDEP requires that the population mean for the Cr TCLP be estimated at the 95% confidence level.

a. The plating sludge is pumped into a drying bed 20 m × 20 m. Five samples are collected initially, and the Cr TCLP values are 4.2 mg/L, 5.8 mg/L, 2.9 mg/L, 5.2 mg/L, and 3.9 mg/L. Using these data, verify that 28 samples are required to estimate the population mean at the 95% confidence level.

b. The following data have been provided from Cr TCLP analysis for the sludge.

Cr TCLP (mg/L)

4.8	3.6	5.8	4.3	2.5
4.9	6.1	4.1	5.8	3.4
4.0	5.2	3.7	6.0	4.2
2.9	4.4	6.1	4.8	5.3
4.3	4.1	3.9	4.0	4.2
3.9	4.6	5.1		

Do the data show that the Cr TCLP concentration is less than the TCLP regulatory threshold for Cr at the 95% confidence level?

4.16. A subsurface system was sampled for ^{90}Sr using stratified random sampling. The following data were obtained for the concentrations of ^{90}Sr in three strata of a subsurface system.

[^{90}Sr], mCi
Depth = 0–0.25 m

130	243	135	304
111	188	219	107
214	202	164	207
144	121	198	186

Depth = 0.26–0.5 m

244	267	311	188
323	289	344	199
333	259	306	168
266	403	284	268

Depth = 0.51–0.75 m

344	198	278	397
326	363	388	290
307	278	272	415
326	282	336	432

Using these sampling results, determine (a) the total mean and variance, and (b) the mean for each stratum and the variance between strata.

4.17. An abandoned hazardous waste site received contamination from improper disposal of nitroaromatic explosives. What 600 series EPA analysis would you recommend on groundwater samples collected at the site?

4.18. Groundwater near a former unregulated dump site has been contaminated by wastes from a dry cleaner, a circuit-board preparing operation, a pesticide manufacturing operation, and a wood preservation company. What 600 series analyses would you recommend?

4.19. A plastic manufacturing firm has over 100 drums containing an aqueous waste that must be analyzed for the most likely contaminants. To save your client money, your supervisor would like you to use only one EPA 600 series analysis. What analysis would you choose?

REFERENCES

4.1. Wirth, D., Personal communication, Boeing Corporation, Seattle, WA, 1994.

4.2. Freeman, H. M., "Pollution prevention: The U.S. experience," *Environ. Prog*, **14,** 214–218 (1995).

4.3. Mullins, M. L., "Pollution prevention progress," *Water Environ. Technol.,* **5,** 90–94 (1993).

4.4. Freeman, H. M. and J. Lounsbury, "Waste minimization as a waste management strategy in the United States," in Freeman, H. M. (Ed.), *Hazardous Waste Minimization,* McGraw-Hill, New York, 1990.

4.5. Hopper, J. R., C. L. Yaws, M. Vichailak, and T. C. Ho, "Pollution prevention by process modification: Reactions and separations," *Waste Manage.,* **14,** 187–202 (1994).

4.6. Filice, F. and G. Brosseau, "For pollution prevention: Consider the source," *Water Environ. Technol.,* **5,** 84–89 (1993).

4.7. Roberge, H. D. and B. W. Baetz, "Optimizing modeling for industrial waste reduction planning," *Waste Manage.,* **14,** 35–48 (1994).

4.8. Ünlü, K., "Assessing risk of ground water pollution from land-disposed wastes," *J. Environ. Eng.,* **120,** 1578–1597 (1994).

4.9. Bernath, T., "Environmental audit and property liability assessment," *Pollut. Eng.,* **20,** 110–115 (1988).

4.10. Hess, K., *Environmental Site Assessments. Phase I. A Basic Guide,* CRC Press, Boca Raton, FL, 1993.

4.11. Goodrich, M. T. and J. T. McCord, "Quantification of uncertainty in exposure assessments at hazardous waste sites," *Ground Water,* **33,** 727–732 (1995).

4.12. *Rapid Assessment of Potential Ground-Water Contamination Under Emergency Response Conditions,* EPA-600/8-83-030, U.S. Environmental Protection Agency, Office of Health and Environmental Assessment, Washingon, DC, 1983.

4.13. Flatman, G. T. and A. A. Yfantis, "Geostatistical strategy for soil sampling: The survey and the census," *Environ. Monit. Assess.,* **4,** 335–350 (1984).

4.14. *Test Methods for the Evaluation of Solid and Hazardous Wastes,* SW-846, EPA-361-082/315, U.S. Environmental Protection Agency, U.S. Government Printing Office, Washington, DC, 1982.

4.15. Cline, M. D., "Principles of soil sampling," *Soil Sci.,* **58,** 275–288 (1944).

4.16. Nelson, J. D. and R. C. Ward, "Statistical considerations and sampling techniques for ground-water quality monitoring," *Ground Water,* **19,** 617–625 (1981).

4.17. Schweitzer, G. E. and S. C. Black, "Monitoring statistics," *Environ. Sci. Technol.,* **19,** 1026–1030 (1985).

4.18. Gilbert, R. O., *Statistical Methods for Environmental Pollution Monitoring,* Van Nostrand Rheinhold, New York, 1987.

4.19. Gilbert, R. O. and P. G. Doctor, "Determining the number and size of soil aliquots for assessing particulate contaminant concentrations," *J. Environ. Qual.,* **14,** 286–292 (1985).

4.20. Snedecor, G. W. and W. G. Cochran, *Statistical Methods,* 7th Edition, Iowa State University Press, Ames, IA, 1980.

4.21. Ott, L., *An Introduction to Statistical Methods and Data Analysis,* Duxbury Press, North Scituate, MA, 1977.

4.22. Petersen, R. G. and L. D. Calvin, "Sampling," in Klute A. (Ed.), *Methods of Soil Analysis. Part I. Physical and Mineralogical Methods,* American Society of Agronomy, Madison, WI, 1986.

4.23. Provost, L. P., "Statistical methods in environmental sampling," in Schweitzer, G. E. and J. A. Santolucito (Eds.), *Environmental Sampling for Hazardous Wastes,* ACS Symposium Series 267, American Chemical Society, Washington, DC, 1984.

4.24. LaGrega, M. D., P. L. Buckingham, and J. C. Evans. *Hazardous Waste Management,* McGraw-Hill, New York, 1994.

4.25. Devinny, J. S., L. G. Everett, J. C. S. Lu, and R. L. Stollar, *Subsurface Migration of Hazardous Wastes,* Van Nostrand Rheinhold, New York, 1990.

4.26. Smith, N., *Monitoring and Remediation Wells,* CRC Press, Boca Raton, FL, 1995.

4.27. Fetter, C. W., *Contaminant Hydrogeology,* Macmillan, New York, 1993.

4.28. Marr, I L. and M. S. Cresser, *Environmental Chemical Analysis,* Chapman & Hall, New York, 1983.

4.29. *Standard Methods for the Examination of Water and Wastewater,* American Public Health Association, New York, 1993.

Part Two

PATHWAYS

Hazardous chemicals found in a storage tanks, surface impoundments, landfills, soils, groundwaters, and treatment systems behave through a series of dynamic processes. In the environment, these pathways affect the release of contaminants from the source and, subsequently, their ultimate fate, which governs whether they pose a hazard to public health or the environment. For engineered systems, the rates at which contaminants move between phases or are transformed control the effectiveness of remediation systems and treatment processes.

Just as water moves in the environment through the hydrologic cycle, so do hazardous chemicals, although they do so at different rates. In Part II, the primary pathways of hazardous compounds are covered, including sorption (Chapter 5), volatilization (Chapter 6), and abiotic and biotic transformations (Chapter 7). The principles of Chapters 5 through 7 are then extended further in Chapter 8—release and transport from the source in air and groundwater. Knowledge of the processes presented in Part II is important in a number of areas of hazardous waste management by providing (1) a conceptual understanding of contaminant behavior for Phase I and II site assessments, Remedial Investigation/Feasibility Studies, and audits at RCRA hazardous waste facilities; (2) a basis for designing sampling schemes; (3) quantitative problem solving skills related to exposure evaluations during risk assessments; and (4) a fundamental background for the treatment of RCRA hazardous wastes and the remediation of contaminated sites including process selection and system design.

Chapter 5

Partitioning, Sorption, and Exchange at Surfaces

In a two-phase system containing solids and water, hazardous chemicals usually partition between the two phases. This partitioning controls the extent to which contaminants exist in the aqueous phase of groundwater systems, in the vadose or unsaturated zone of soils, and in treatment systems such as granular activated carbon (GAC) contactors and slurry bioreactors. This chapter focuses on assessing the equilibrium concentration of contaminants at surfaces based on partitioning, sorption, and exchange. *Partitioning* is a general phenomenon that describes the tendency of a contaminant to exist at equilibrium between two phases. The term *partitioning* is used to describe not only contaminants distributed between solids and water, but also air-water and water-biotic equilibria. *Sorption* is a specific partitioning and exchange phenomenon in which contaminants accumulate at the solid surface from the surrounding solution. *Exchange* is most commonly defined as the displacement of charged species, such as metal ions, by another species that has a higher affinity for the same site on the solid surface. However, the term *exchange* has also been used in another context: Because sorbed compounds are in dynamic equilibrium with the surrounding solution and sorbed contaminants are continuously released and sorbed again, sorption has been called solid-liquid exchange.

Paterson and Mackay [5.1] developed an approach to partitioning of contaminants in the environment based on *fugacity*, which is a molecule's tendency or driving force to escape the system it is in. By quantifying the relative tendencies of all of the molecules within the most important environmental compartments (water, air, solids, biota) to escape, then the direction and magnitude of contaminant transfer and transformation may be evaluated. In this approach, the potential for contaminants to move from one compartment to another (such as the partitioning between the aqueous phase and the sorbed phase or from water to air) is based on the gradients of fugacity between the two media. The concept of fugacity for pathway analysis has been the basis for a number of environmental fate models (see Section 11.3.2), and this conceptual approach of contaminants moving from compartment to compartment provides an appropriate model as hazardous waste pathways are considered in Chapters 5, 6, 7, and 8.

A fundamental understanding of partitioning and sorption theory and calculations is necessary in hazardous waste management for several reasons. First, sorption is one of the most important mechanisms controlling the rate at which contaminants move in soil and subsurface systems. Sorption isotherms and their extension, the soil distribution coefficient, serve as the basis for calculating the retardation factor, an important parameter that provides the basis for the determination of subsurface contaminant transport rates.

Second, sorption affects a number of other pathways such as volatilization, hydrolysis, oxidation processes, and biodegradation. Most hazardous waste treatment processes (e.g., bioremediation, oxidation processes) occur in the aqueous phase of solid-water systems. If a contaminant is sorbed, it is often considered unavailable for transformation by biochemical or physicochemical processes [5.2–5.3]. As a result, sorption can dramatically reduce the effectiveness of many hazardous waste treatment processes. Therefore, hazardous waste destruction technologies such as bioremediation or chemical oxidation are often coupled with processes that enhance desorption.

Third, sorption processes are commonly used to remove contaminants from aqueous and gaseous waste streams. The selection and design of treatment systems for soils, groundwater, and RCRA wastes can be highly complicated. Process designs based on partitioning and sorption mechanisms (e.g., soil washing, granular activated carbon and resin sorption, ion exchange) are effective systems for treating some contaminants. A fundamental grasp of which compounds sorb and the matrix effects involved (e.g., the effects of impermeable soils or competition for sorption sites by nonhazardous organic compounds) is essential in the design of effective sorption treatment systems. Therefore, a conceptual and quantitative understanding of sorption phenomena provides the basis for describing, assessing, and treating hazardous wastes in the natural environment and engineered systems.

5.1 SORPTION THEORY

Sorption is an equilibrium phenomenon in which hydrophobic compounds partition onto surfaces in a two-phase system containing a liquid, such as water, and a solid phase such as soil, humus, or a treatment matrix such as activated carbon. The physical process that occurs when a contaminant partitions onto a solid usually involves weak, reversible bonds between the contaminant and the solid [5.4]. Hazardous compounds in a two-phase system can attach (sorb) and detach (desorb) repeatedly; with this reversible process taking place, the contaminants move through a groundwater system or an activated carbon column, but they do so at a slower rate than the water.

By definition, *adsorption* is the separation of a compound from solution and its deposition on a solid-liquid interface. *Desorption* is the opposite of adsorption and involves the disassociation or release of the compound from the solid. A *sorbent* is the solid phase to which the chemical is sorbed. In hazardous waste management, sorbents include soils, sludges, the subsurface strata of a groundwater system, hazardous waste containment material (e.g., clay liners of landfills), and materials used for the treatment of organic contaminants in aqueous systems (e.g., granular activated carbon). The *sorbate* is the compound in the liquid phase that partitions onto the sorbent. Sorbates are often organic contaminants but may include inorganic species, such as those described in Chapter 2. The liquid phase in almost all hazardous waste systems is water.

In some specialized cases sorption from an organic solvent may be encountered, such as in the design of a waste minimization system to remove a nonpolar organic solvent from a more polar one such as acetone. However, sorption phenomena between a sorbent and an aqueous solution of the contaminant (or sorbate) are most commonly encountered. Examples of these systems include groundwater, surface waters, aqueous wastes from industries, and aqueous wastes in storage tanks and drums.

Sorption is an equilibrium process in which three classes of sorption mechanisms have been described—electrostatic, physical, and chemical [5.5]. Some of the mechanisms are illustrated in Figure 5.1. Electrostatic mechanisms occur as the result of charged attractive forces associated with the sorbent and the sorbate. An example of electrostatic sorption is the partitioning of heavy metals onto negatively charged hydrous iron oxides (e.g., $\alpha = FeOOH$) in soils. The most important sorbate property related to electrostatic sorption is the charge of the species [5.7]. For example, Fe^{3+} has

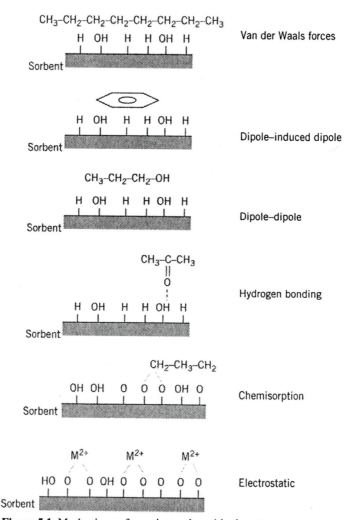

Figure 5.1 Mechanisms of sorption and partitioning. *Source:* Reference 5.6.

a higher potential for anion attraction than Na^+. For ions of equal charge, smaller ions exhibit greater sorptivity because of their higher charge-to-mass ratio.

Physical mechanisms are probably the most important for the sorption of organic contaminants to soils, subsurface strata, and granular activated carbon. The general mechanism attributed to physical attraction is weak bonding processes caused by the attractions of intermolecular forces, such as hydrogen bonding and Van der Waals forces [5.8]. Israelachvili [5.9] described physical sorption in detail, classifying physical mechanisms by four types of dipole interactions. *Dipole moments* are shifts in electron density within molecules that produce a partial separation of charges. The dipole moments can be permanent, as in a polar molecule such as acetic acid. Polar molecules can interact with other polar molecules and sorption can occur through *dipole-dipole* interactions. In addition, interactions can occur between polar molecules and nonpolar molecules (*dipole-induced dipole* interactions). Still another possibility is the *instantaneous dipole-induced dipole* interaction, in which a fluctuating electron cloud results in short-term dipole moments that serve as the basis for weak bonding between the sorbent and sorbate.

Chemical sorption, or *chemisorption*, is characterized by a stronger, more permanent bond relative to physical sorption mechanisms. Chemisorption has all of the characteristics of a true chemical bond. Because of its high enthalpy (heat) of sorption, chemisorption is favored by high temperatures.

Sorption phenomena are complex and, to some degree, each of the three classes of sorption—electrostatic, physical, and chemical—occurs simultaneously in a given soil-water system or treatment process.

Many compounds, when sorbed to soils, become strongly or irreversibly bound, which can ultimately affect their degradability and transport. Examples of such chemicals (manifested by their resistance to solvent extraction during chemical analyses) include chloroalkanes, chlorophenols [5.10], and chloroalkenes [5.11]. The mechanisms include irreversible binding to organic or inorganic soil fractions [5.12] and trapping in micropores [5.13]. Desorption measurements have often displayed a two-stage process—the first stage consisting of a more weakly bound contaminant fraction and the second stage considered strongly or irreversibly bound.

Related to the concept of strongly sorbed contaminants is *aging* in soils and solids. Over time periods of months-to-years, some contaminants become more strongly sorbed and more unavailable to solvent extraction during chemical analysis. Numerous mechanisms have been proposed for aging, including migration to micropores and the formation of bound residues (i.e., the reaction with the contaminant, often enzyme-catalyzed, to form longer chains). For example, Bollag et al. [5.14] found that anilines polymerized into lignin-like polymers and Hatcher et al. [5.15] documented the covalent bonding of 2,4-dichlorophenol to humic acids in the presence of the enzyme peroxidase.

The presence of strongly sorbed contaminants in soils can cause significant problems in hazardous waste assessment and remediation. These fractions can drastically affect the sampling and analysis of soils and subsurface solids. Sometimes the standard extraction procedures (e.g., the EPA 8000 series) will provide low extraction efficiencies resulting in false negative results. In some cases, an opposite phenomenon will occur. The chemical analyses are accurate but natural or engineered pathways are affected; that is, some strongly bonded fractions are more persistent under natural conditions or are not responsive to remediation processes such as thermal desorption, soil washing, or bioremediation.

5.2   THE GOVERNING VARIABLES: SORBENT CHARACTERISTICS, CONTAMINANT HYDROPHOBICITY, AND THE SOLVENT

The tendency for a compound to sorb is a function of the properties of the sorbent, the sorbate, and the liquid medium [5.4]. Sorbent characteristics that promote sorption of organic compounds include its hydrophobicity (i.e., its nonpolar nature) and its specific surface area. For example, granular activated carbon and soil organic matter are nonpolar sorbents and hydrophobic organic compounds tend to sorb to their surfaces. Clays and smaller soil particles have surface charges, which increases their importance as sorbents for charged species [5.16].

The sorbate characteristic that correlates most with adsorptivity of organic contaminants is hydrophobicity. Compounds that are hydrophobic tend to partition onto sorbents rather than remaining dissolved in the aqueous phase. It is not surprising that low water solubility has been correlated with the tendency of a compound to sorb. Another measure of hydrophobicity, the *octanol-water partition coefficient* (K_{ow}), is an even more effective and commonly used predictor of sorptivity.

Another factor that affects sorption is the solvent. Although the liquid phase in hazardous waste systems is almost always water, recent concerns about emissions of hazardous air pollutants (HAPs) have resulted in increased interest in another two-phase system involving a stationary and a fluid medium—solids and air. Since the 1989 Amendments to the Clean Air Act regulating the emissions of HAPs, granular activated carbon is being used increasingly to remove volatile organic compounds (VOCs) from contaminated air. The sorption of volatile contaminants as they diffuse through unsaturated soils is also a topic of widespread current research, especially in relation to petroleum spills and leaking underground storage tanks.

5.3   PROPERTIES OF SOILS AND OTHER SORBENTS

Preregulatory hazardous waste disposal occurred primarily on soils; therefore, a fundamental knowledge of soil characteristics is necessary to evaluate and quantify the extent of contaminant sorption as a basis for site characterization and facility assessments. The physical structure and chemical composition of surface and subsurface soils are highly variable. Compared to aqueous systems, soils are complex media composed of solid, liquid, and gaseous phases. Soils, whether saturated or unsaturated, are not completely mixed as most aquatic systems are, so soil and groundwater contamination tends to be heterogeneous. Because of such "patchiness," it is not uncommon to find contaminated sections of soil adjacent to nearly clean areas. Important soil properties that relate to hazardous waste management include the soil textural class, soil mineralogy, bulk density, porosity, and organic carbon content.

Soils are composed of four major components: an inorganic (or mineral) fraction, organic matter, water, and air. The proportion of each of the four components, with some typical ranges, is shown in Figure 5.2. Approximately half of soils is pore space, which consists of air and water. The remaining 50% is soil-solids—the mineral fraction (45% to 49.9%) and organic matter (0.1% to 5%). These numbers are highly variable and depend on site-specific conditions. For example, water-saturated soils of wet-

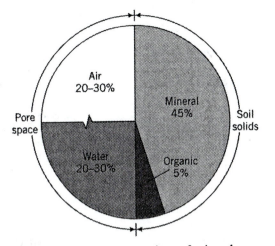

Figure 5.2 Relative proportions of minerals, organic matter, water, and air in a silt loam soil. *Source:* Reference 5.17.

lands contain less air but more water, and forest soils contain higher levels of organic matter compared to desert soils.

Particle Size Distribution. Systems developed by the U.S. Department of Agriculture (USDA), the British Standards Institution, and the International Society of Soil Science are used to classify soils by *particle size*. The method most commonly used in hazardous waste management is the USDA system. The size fractions classified as gravel, sand, silt, and clay, which are called *soil separates*, are listed in Table 5.1.

Classification of a natural material by such a simple size fractionation is impossible because almost all natural soils contain a percentage of each of the size fractions listed in Table 5.1. Therefore, three general *textural classes* have been developed—sands, clays, and loams. *Sands* are soils that contain at least 70% of the sand separate. Soils classified as *clays* consist of at least 35% clay. A *loam* soil is a general classification that contains nearly equal weights of sand, silt, and clay separates. The U.S. Department of Agriculture has developed a relationship between the percentage of soil separates and its textural classes, which is shown in Figure 5.3. Classification of soil textures is illustrated in Example 5.1 on the following page.

Soil Minerals. *Primary minerals* are small pieces of parent material (i.e., rocks) that have formed through physical soil genesis. Typical primary minerals include quartz

Table 5.1 USDA Soil Particle Size Separates

Soil Type	Particle Size Range
Gravel	2.0–15 mm
Sand	0.075–2.0 mm
Silt	0.002–0.075 mm
Clay	< 0.002 mm

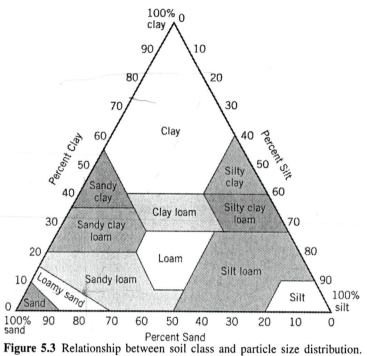

Figure 5.3 Relationship between soil class and particle size distribution.
Source: Reference 5.17.

(SiO_2) and feldspar $(CaAl_2O_4[SiO_2]_2)$. Primary minerals contain the relatively strong Si—O covalent bonds, which makes them more resistant to chemical weathering than ionically bonded species.

EXAMPLE 5.1 *Classification of Soil Texture*

What is the textural class of a soil containing 74% sand, 20% silt, and 6% clay?

SOLUTION

Using Figure 5.3, locate the percentage of silt and project a line parallel to the clay side of the triangle. Next, find the percentage of clay on Figure 5.3 and project a line parallel to the sand side of the triangle. The location of the intersection of the two lines is the region of the textural class for the soil. Therefore, the textural class of this soil is a *sandy loam*.

Secondary minerals are usually formed through chemical transformations (termed weathering). *Clays,* such as kaolinite $[Si_4Al_4O_{10}(OH)_8]$ and smectite

[$M_x(Si,Al)_8(Al,Fe,Mg)_4O_{20}(OH)_4$ where M = an interlayer cation], are products from the weathering of primary minerals.

The influence of clay minerals on the interaction of hazardous chemicals with soils is related to their high surface areas and negative surface charges. Clays also interact physically with other soil components. They are usually linked to humic materials and iron or manganese oxides; these complexes have a significant role in retarding the movement of hazardous organic and inorganic chemicals in soils and groundwater. *Metal oxide* secondary minerals [e.g., goethite (α-FeOOH) and hematite (α-Fe$_2$O$_3$)] can originate from the hydrolysis and displacement of silica from such secondary minerals as smectite and kaolinite. Some metal oxides, such as goethite, are the thermodynamically stable forms of weathering and represent the end points of chemical transformations in soils. Iron and manganese oxides occur in soils as (1) coatings on soil particles often mixed with clays; (2) fillings in voids; and (3) concentric nodules. Important clay mineral groups are listed in Table 5.2, and important metal oxides and oxyhydroxide minerals are shown in Table 5.3. Some structures of typical secondary minerals are illustrated in Figure 5.4. Because of their high surface areas and unique chemical structures, clay minerals and metal oxides can dramatically influence the transport and transformation of organic and inorganic hazardous chemicals in soils.

Cation Exchange Capacity. The surface density of exchange sites on a soil is measured by the *cation exchange capacity* (CEC), which is defined as the quantity of cations sorbed per mass of soil and is expressed as milliequivalents of positive charge per 100 grams of soil (meq/100 g). It is most commonly determined by saturating the soil with a buffered cation (e.g., sodium acetate) and measuring the amount of sodium exchanged. Cation exchange capacity is a characteristic of the soil, not the cations, because it is measured as the total positive charges per mass of soil that can be ex-

Table 5.2 Clay Mineral Groups

Group	Typical Chemical Formula[a]
Kaolinite	$[Si_4]Al_4O_{10}(OH)_8 \cdot nH_2O$ ($n = 0$ or 4)
Illite	$M_x[Si_{6.8}Al_{1.2}]Al_3Fe_{0.25}Mg_{0.75}O_{20}(OH)_4$
Vermiculite	$M_x[Si_7Al]Al_3Fe_{0.5}Mg_{0.5}O_{20}(OH)_4$
Smectite[b]	$M_x[Si_8]Al_{3.2}Fe_{0.2}Mg_{0.6}O_{20}(OH)_4$
Chlorite	$Al(OH_{2.55})_4 \cdot [Si_{6.8}Al_{1.2}]Al_{3.4}Mg_{0.6}O_{20}(OH)_4$

[a]$n = 0$ is kaolinite and $n = 4$ is halloysite; M = monovalent interlayer cations.
[b]Principally montmorillonite in soils.

Table 5.3 Metal Oxides, Oxyhydroxides, and Hydroxides Commonly Found in Soils

Name	Chemical Formula	Name	Chemical Formula
Anatase	TiO_2	Hematite	α-Fe$_2$O$_3$
Birnessite	$Na_{0.7}Ca_{0.3}Mn_7O_{14} \cdot 2H_2O$	Ilmenite	$FeTiO_3$
Boehmite	γ-AlOOH	Lepidocrocite	γ-FeOOH
Ferrihydrite	$Fe_{10}O_{15} \cdot 9H_2O$	Lithiophorite	$(Al,Li)MnO_2(OH)_2$
Gibbsite	γ-Al(OH)$_3$	Maghemite	γ-Fe$_2$O$_3$
Goethite	α-FeOOH	Magnetite	Fe_3O_4

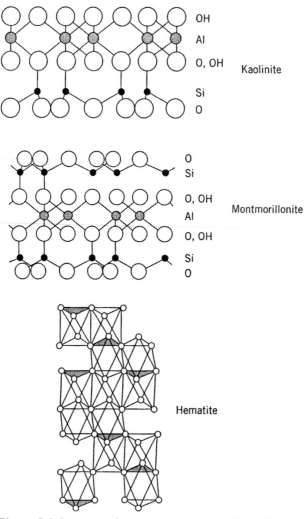

Figure 5.4 Structure of some common secondary minerals.
Source: Reference 5.18.

changed. Soils with higher clay and organic matter contents are characterized by higher CECs. Cation exchange capacities range from 2 to 7 meq/100 g for sandy soils, 9 to 27 meq/100 g for silt loams, and 5 to 60 meq/100 g for clay soils [5.17]. Cation exchange capacity calculations are demonstrated in Example 5.2.

EXAMPLE 5.3 *Cation Exchange Capacity*

Using a solution of sodium acetate as an exchange medium, a sandy soil was found to have a CEC of 0.14 meq/100 g. Additional analyses using atomic adsorption spectrophotometry showed that 40% of the sites contained sodium, 50% of the sites contained calcium, and 10% of the sites exchanged with chromium (III). Determine the concentrations of the three exchanged species in mg/kg.

SOLUTION

The mass concentration of each cation may be determined by first converting its concentration to moles/g of soil, followed by conversion to its mass equivalent based on its molecular weight.

For sodium (Na^+):

$$(0.40) \times \left(\frac{0.14 \text{ meq}}{100 \text{ g soil}} \right) = 0.00056 \frac{\text{meq}}{\text{g soil}}$$

$$\left(0.00056 \frac{\text{meq}}{\text{g soil}} \right) \times \left(\frac{1 \text{ mmole}}{1 \text{ meq}} \right) = 5.60 \times 10^{-4} \frac{\text{mmoles}}{\text{g soil}}$$

$$\left(5.60 \times 10^{-4} \frac{\text{mmole}}{\text{g soil}} \right) \times \left(23.0 \frac{\text{mg } Na^+}{\text{mmole}} \right) \times \frac{1000 \text{ g}}{\text{kg}} = 12.9 \frac{\text{mg}}{\text{kg}} Na^+$$

For calcium (Ca^{2+}):

$$(0.50) \times \left(\frac{0.14 \text{ meq}}{100 \text{ g soil}} \right) = 0.00070 \frac{\text{meq}}{\text{g soil}}$$

$$\left(0.00070 \frac{\text{meq}}{\text{g soil}} \right) \times \left(\frac{1 \text{ mmole}}{2 \text{ meq}} \right) = 3.50 \times 10^{-4} \frac{\text{mmole}}{\text{g soil}}$$

$$\left(3.50 \times 10^{-4} \frac{\text{mmole}}{\text{g soil}} \right) \times \left(40.1 \frac{\text{mg } Ca^{2+}}{\text{mmole}} \right) \times \frac{1000 \text{ g}}{\text{kg}} = 14.0 \frac{\text{mg}}{\text{kg}} Ca^{2+}$$

For chromium (III) (Cr^{3+}):

$$(0.10) \times \left(\frac{0.14 \text{ meq}}{100 \text{ g soil}} \right) = 0.00014 \frac{\text{meq}}{\text{g soil}}$$

$$\left(0.00014 \frac{\text{meq}}{\text{g soil}} \right) \times \left(\frac{1 \text{ mmole}}{3 \text{ meq}} \right) = 4.67 \times 10^{-5} \frac{\text{mmoles}}{\text{g soil}}$$

$$\left(4.67 \times 10^{-5} \frac{\text{mmoles}}{\text{g soil}} \right) \times \left(52.0 \text{ mg} \frac{Cr^{3+}}{\text{mmole}} \right) \times \frac{1000 \text{ g}}{\text{kg}} = 2.43 \frac{\text{mg}}{\text{kg}} Cr^{3+}$$

Bulk Density. The density of a soil while in place (i.e., not compacted or disturbed) is defined as its *bulk density*, which has SI units of kg/m^3. Typical bulk densities are listed in Table 5.4. Fine-grained soils (silts and clays) have characteristically lower bulk densities than sands and gravels.

Table 5.4 Typical Soil Bulk Densities

Soil Textural Class	Bulk Density (kg/m^3)
Organic soils	200–300
Cultivated surface mineral soils	1250–1450
Clays, clay loams, silt loams	1000–1600
Sands and sandy loams	1200–1800

Porosity. The *total porosity* of a soil, which is defined as the fractional or percent content of the soil occupied by the pore space, is an important parameter related to groundwater transport, retardation, and mass transfer during soil remediation. Typical soil porosity values, listed in Table 5.5 as percentages, are in the range of 30 to 50%. Total porosity is used in calculating the retardation factor (Section 5.7). *Effective porosity* does not include the space of dead-end pores, and is used for calculating the pore water velocity (the flow velocity within the voids) based on Darcy's Law.

Volumetric Water Content. The *volumetric water content* of a soil is the fraction of the soil pores filled with water. Therefore, if the soil is saturated, as groundwater systems are, the volumetric water content is 100%. The volumetric water content for most surface soils ranges from 5 to 50%, but can approach saturation under some instances. Volumetric water content is often used for estimating the pore water velocity of the unsaturated zone [5.19].

Soil Organic Matter. One of the most important soil properties that influences the transport of hazardous compounds is naturally occurring *organic matter*. Soil organic matter may be divided into two categories—humic and nonhumic materials. Nonhumic chemicals are unaltered amino acids, carbohydrates, fats, and other biochemicals

Table 5.5 Representative Values of Soil Porosity

Material	Porosity (%)	Material	Porosity (%)
Gravel, coarse	28	Loess	49
Gravel, medium	32	Peat	92
Gravel, fine	34	Schist	38
Sand, coarse	39	Siltstone	35
Sand, medium	39	Claystone	43
Sand, fine	43	Shale	6
Silt	46	Till, predominantly silt	34
Clay	42	Till, predominantly sand	31
Sandstone, fine-grained	33	Tuff	41
Sandstone, medium-grained	37	Basalt	17
Limestone	30	Gabbro, weathered	43
Dolomite	26	Granite, weathered	45
Dune sand	45	Granite, weathered	45

Source: Reference 5.17.

that occur in the soil as a result of the presence of living organisms. Humic materials are yellow to dark brown polymers formed by microbial mediated reactions. This material, also called *humus*, is nearly thermodynamically stable and originates from the carbons of plants, animals, and microorganisms. Humus is produced when microbial metabolism decomposes soil biomass to reactive intermediates, thought to be primarily quinones, which rapidly polymerize to large, biochemically inert polymers. Humus typically contains polymerized phenols with accompanying carboxylic, carbonyl, ester, and methoxy groups with 44–53% carbon, 3.6–5.4% hydrogen, 40–47% oxygen, and 1.8–3.6% nitrogen [5.20].

Humus is often classified based on chemical analyses, in which three fractions are found: (1) humin, insoluble at high pH, (2) humic acids, soluble in acid and insoluble at high pH, and (3) fulvic acids, soluble at both high and low pH. Although the chemical composition of these classes of organic matter is highly variable, the generic empirical formula is $C_{187}H_{186}O_{89}N_9S$ for humic acids and $C_{135}H_{182}O_{95}N_5S_2$ for fulvic acids [5.19]. A generic structure of humic acids is shown in Figure 5.5. Humus is often strongly sorbed to the inorganic fraction (e.g., secondary iron minerals fixed on the surfaces of sand particles) providing a mineral–organic matter complex that often serves as the primary sorbent in soils [5.21].

Because of its variable composition, organic matter in soils and sludges is usually measured as organic carbon. The soil or solid is digested using an oxidant, such as dichromate, and the organic carbon is measured as evolved carbon dioxide or an equivalent reduction of dichromate. Based on average values of hydrogen and oxygen, the relationship between soil organic carbon (SOC) and organic matter (SOM) is

$$SOM = 1.724 \cdot SOC \qquad (5.1)$$

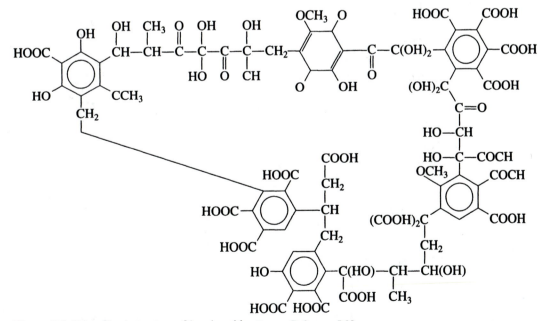

Figure 5.5 Generalized structure of humic acids. *Source:* Reference 5.20.

Low Organic Carbon (e.g., desert) Soil

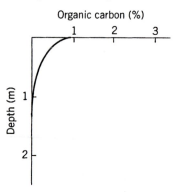

High Organic Carbon (e.g., forest) Soil

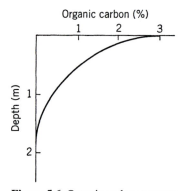

Figure 5.6 Organic carbon content as a function of soil depth.

Table 5.6 Average Organic Matter Contents and Ranges of Mineral Surface Soils in Several Areas of the United States

Soils	Organic Matter (%)	
	Range	Average
240 West Virginia soils	0.74–15.1	2.88
15 Pennsylvania soils	1.70–9.9	3.60
117 Kansas soils	0.11–3.62	3.38
30 Nebraska soils	2.43–5.29	3.83
9 Minnesota prairie soils	3.45–7.41	5.15
21 Southern Great Plains soils	1.16–2.16	1.55
21 Utah soils	1.54–4.93	2.69

Source: Reference 5.19.

The input of organic carbon in soils is from the soil surface; therefore, an exponential decrease in organic carbon is usually found as a function of soil depth (Figure 5.6). Typical organic carbon concentrations below the influence of the surface [> 1 m (3.3 ft) deep] are typically < 0.1%. As noted in Table 5.6, surface concentrations for organic matter are usually in the range of 1 to 5%. Concentrations of organic matter in this range can have a significant influence on the sorption of organic and inorganic hazardous chemicals.

5.4   SORPTION ISOTHERMS

Numerous quantitative empirical mathematical expressions, called isotherms, have been developed to describe sorption. An *isotherm* is a plot of the extent to which sorption occurs at successively higher concentrations of sorbate at constant temperature. Isotherm plots provide a graphical representation of material sorbed (mass of sorbate/mass of sorbent) as a function of the equilibrium concentration of sorbate. Depending on the sorbent and sorbate, one of the sorption models used may describe the system better than others; however, characteristics of more than one model are usually found for any system.

The two most common sorption models are the Langmuir and Freundlich isotherms. The *Langmuir model* is based on the assumption that a single monolayer of sorbate accumulates at the solid surface. A typical Langmuir isotherm plot (Figure 5.7) shows that, as the concentration of the sorbate is increased in the liquid phase, proportionately more of the sorbent surface is covered with the sorbate. At higher concentrations of the material in the solution phase the sorbent is completely saturated (i.e., all sorption sites are occupied), and above that concentration no more of the material is sorbed. In Figure 5.7, if the concentration of the sorbate in the aqueous phase is raised above 75 mg/L, no increase in sorption occurs.

The equation that describes the Langmuir system is

$$C_s = \frac{x}{m} = \frac{ab\,C_e}{1 + bC_e} \tag{5.2}$$

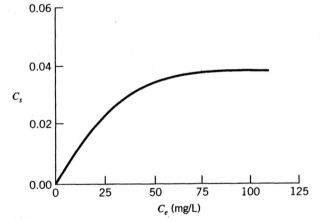

Figure 5.7 A typical Langmuir isotherm plot.

where

C_s = contaminant concentration sorbed on the solid (dimensionless)
C_e = concentration of contaminant remaining in solution at equilibrium (g/m³)
a = empirical constant
b = saturation coefficient (m³/g)
x = mass of material sorbed (sorbate) on the solid phase (g)
m = mass of sorbent (g)

The Langmuir equation may be transformed to a linear expression by inverting Equation 5.2 and separating variables:

$$\frac{C_e}{C_s} = \frac{1}{ab} + \frac{C_e}{a} \tag{5.3}$$

The empirical coefficients a and b may be obtained by plotting C_e/C_s as a function of C_e. Using linear regression, the slope = $1/a$ and the y-intercept = $1/(a \cdot b)$.

The *Freundlich model* is characterized by sorption that continues as the concentration of sorbate increases in the aqueous phase. A typical Freundlich isotherm plot (Figure 5.8) shows that the mass of material sorbed is proportional to the aqueous phase concentration at low sorbate concentrations and decreases as the sorbate accumulates on the sorbent surface. Sorption then continues with increasing aqueous phase sorbate concentrations, but to a diminishing degree. The Freundlich isotherm is quantified by

$$C_s = \frac{x}{m} = K_F \, C_e^{1/n} \tag{5.4}$$

where

K_F = Freundlich sorption coefficient
n = an empirical coefficient

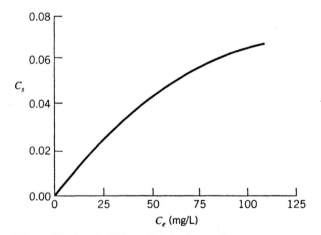

Figure 5.8 A typical Freundlich isotherm plot.

The coefficients K_F and n may be determined by plotting experimental data on log-log paper. Alternatively, Equation 5.4 can be transformed logarithmically:

$$\ln C_s = \ln K_F + \frac{1}{n} \ln C_e \qquad (5.5)$$

Logarithmic transformations of experimental data may then be plotted on arithmetic paper as $\ln C_s$ as a function of $\ln C_e$; through linear regression, $K_F = e^{y\text{-intercept}}$ and $1/n = $ slope.

Several approaches are used to determine the sorption characteristics of a system. Although theoretical attempts have been made to quantify sorption, the most common method involves laboratory studies followed by best-fit analysis of the sorption isotherm equations. Empirical determination of the best-fit isotherm model and its coefficients can be accomplished by two methods. The first uses an equal volume of liquid with varying concentrations of contaminant placed in separate beakers. An equal mass of sorbent (e.g., soil, granular activated carbon) is added to each beaker and stirred until equilibrium is established. The residual concentration of the material in the liquid phase is then analyzed. The second method uses an equal volume of the aqueous phase with the same contaminant concentration in each beaker, but a different mass of sorbent is added to each. The use of isotherms in assessing sorbent characteristics is demonstrated in Example 5.3.

In most sources of soil and groundwater contamination the contaminant concentrations in the aqueous phase are low, which results in the equivalent of a *linear isotherm*. Furthermore, the aqueous phase concentrations in treatment reactors, such as activated carbon and ion exchange contactors, are also often low, which also results in conditions that approximate linear isotherm conditions. These low contaminant concentrations provide the basis for a number of fundamental sorbent-sorbate relationships which will be addressed in Section 5.6.

EXAMPLE 5.4 *Use of Sorption Isotherms*

Five beakers are filled with one liter of 400 mg/L 2,4-D. Two grams of soil are added to the first beaker, 3 g to the second, 4 g to the third, and 5 g to the last beaker. The fifth beaker serves as a control and receives no addition of soil. The beakers are then stirred and allowed to reach equilibrium. The following aqueous concentrations of 2,4-D result:

Beaker	C_e 2,4-D (mg/L)
1	289
2	234
3	179
4	124

No 2,4-D loss was found in the control beaker, which confirmed that losses due to volatilization or sorption to the beaker were negligible. Fit these data to the Langmuir and Freundlich isotherms and determine the empirical constants for the best fit isotherm.

SOLUTION

1. Arrange the data to find C_e/C_s and to generate a Langmuir plot:

	C_e (mg/L)	x (mg 2,4-D sorbed)	C_s (mg/mg)	C_e/C_s (mg/L)
1	289	400–289 = 111*	111/2000 = 0.0555	289/0.0555 = 5210
2	234	400–234 = 166	166/3000 = 0.0553	234/0.0553 = 4230
3	179	400–179 = 221	221/4000 = 0.05525	179/0.05525 = 3240
4	124	400–124 = 276	276/5000 = 0.0552	124/0.0552 = 2250

*(400 mg/L −289 mg/L) · 1L = 111 mg

2. Convert the data to the ln of C_s and C_e to generate data for a Freundlich plot:

	ln C_s	ln C_e
1	−2.89	5.67
2	−2.89	5.46
3	−2.90	5.19
4	−2.90	4.82

3. Plot C_e/C_s as a function of C_e (mg/L) (Langmuir plot) and ln C_s vs. ln C_e (Freundlich plot):

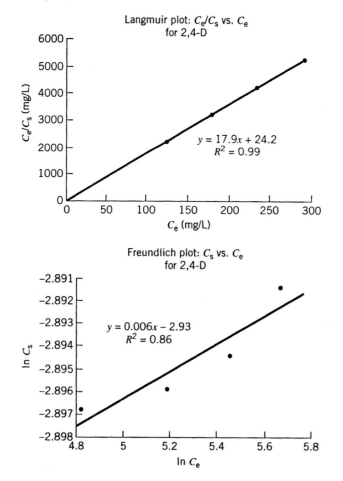

Langmuir plot: C_e/C_s vs. C_e for 2,4-D

$y = 17.9x + 24.2$
$R^2 = 0.99$

Freundlich plot: C_s vs. C_e for 2,4-D

$y = 0.006x - 2.93$
$R^2 = 0.86$

4. The Langmuir plot is characterized by a greater R^2 value, 0.99; therefore, the Langmuir equation is used to find the coefficients a and b. Using linear regression, the slope is 17.9 and the y-intercept 24.2. Therefore,

$$1/a = 17.9, \qquad a = 0.0559$$

$$1/(a \cdot b) = 24.2, \qquad b = 0.739$$

and the Langmuir equation is

$$C_s = \frac{ab\, C_e}{1 + bC_e} = \frac{0.0413 C_e}{1 + 0.739 C_e}$$

5.5   THE OCTANOL-WATER PARTITION COEFFICIENT

Hydrophobicity is a primary driving force for sorption. An effective model of hydrophobic partitioning, the *octanol-water partition coefficient* (K_{ow}), was developed in the pharmaceutical sciences and its use has been adopted widely in the environmental sciences. The K_{ow} is a parameter determined experimentally by measuring the distribution of a compound in a flask containing deionized water and n-octanol, known as a shake-flask study. A flask containing n-octanol and deionized water is shaken until each of the two solvents is saturated in the other. A trace concentration of the compound to be measured is then added to the aqueous phase and the flask is shaken again; the phases are allowed to separate and each phase is analyzed for the concentration of the compound [5.22]. The K_{ow} is then defined as

$$K_{ow} = \frac{\text{concentration in water-saturated } n\text{-octanol (mg/L)}}{\text{concentration in } n\text{-octanol-saturated water (mg/L)}} \tag{5.6}$$

Values for K_{ow} range from 0.001 to over 100,000,000. Because of the wide range, the data are usually reported as log K_{ow}.

A limitation of the shake-flask method for determining octanol-water partition coefficients of high-K_{ow} compounds ($>10^5$) is the extremely low concentrations that are found in the aqueous phase. Therefore, surrogate methods, such as correlations between K_{ow} and high-performance liquid chromatography (HPLC) retention times have been developed [5.23] for such cases.

The K_{ow} measurement system is unique in that some of the octanol is soluble in water (5.5×10^{-3} mole/L) and some of the water is soluble in octanol (2.3 mole/L). Therefore, K_{ow} is not the same as the ratio of the solubility of the compound in octanol to the solubility in water. The K_{ow} has also been weakly correlated with water solubility, but obviously the two measurements are not the same (Figure 5.9).

The K_{ow} is an important predictor of the behavior of environmental contaminants. It has been correlated with aquatic bioconcentration factors [5.24], toxicity [5.25], and sorption to soils and sediments through their organic matter or organic carbon contents [5.26].

Octanol-water partition coefficients are now widely available for nearly all hazardous chemicals. During most hazardous waste investigations, K_{ow} values are ob-

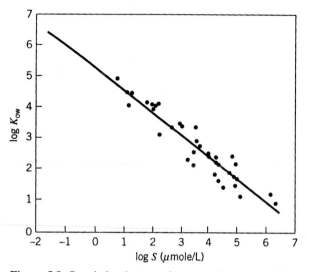

Figure 5.9 Correlation between log octanol-water partition coefficient and log water solubility. *Source:* Reference 5.28.

tained from tabular data. A listing of K_{ow} for the most common hazardous compounds is shown in Table 5.7, and data for over 300 chemicals are available in Appendix G. Nonetheless, chemicals for which octanol-water partition coefficients are not available will sometimes be encountered during hazardous waste investigations. Therefore, estimation methods have been developed; the most common procedure, Hansch and Leo's [5.27] fragment constants, is detailed in Appendix H. Other estimation methods include correlations with water solubility and chromatographic retention times; these methods are outlined in References 5.27 and 5.28.

5.6 THE SOIL ADSORPTION COEFFICIENT AND THE SOIL DISTRIBUTION COEFFICIENT

When a hazardous compound is released to the environment, it will partition onto soils or subsurface solids as water moves through the system. The degree to which the contaminant is distributed between the sorbed and aqueous phases is described by the appropriate sorption isotherm for the compound and its concentration in the system. A more general expression for contaminant sorption is the *soil distribution coefficient* (K_d), which is defined as

$$K_d \text{ (mL/g)} = \frac{\text{mass of contaminant sorbed (mg/g)}}{\text{mass of contaminant dissolved in the aqueous phase (mg/mL)}} \quad (5.7)$$

The K_d can be determined experimentally, which provides the most accurate measure of sorption in a solid-water system [5.29]. The experimental procedure involves adding the contaminant to a flask containing the soil or subsurface solids and water and shaking until equilibrium is achieved, which is usually within 24 hours. An inhibitor of microbial metabolism, such as azide or mercury, is often added to ensure

Table 5.7 Octanol-Water Partition Coefficients for Common Hazardous Compounds

Compound	Mean Log K_{ow}	Range
Aliphatic Hydrocarbons		
Cyclohexane	3.44	
Cyclohexene	2.86	
n-Decane	6.69	
n-Heptane	4.66	
Isooctane	5.83	
n-Octane	5.18	
Monocyclic Aromatic Hydrocarbons		
Benzene	2.05	1.95–2.15
Toluene	2.58	2.21–2.79
Ethylbenzene	3.11	3.05–3.15
o-Xylene	3.11	2.77–3.13
m-Xylene	3.20	
p-Xylene	3.17	3.15–3.18
Polycyclic Aromatic Hydrocarbons		
Anthracene	4.34	
Benzo[a]anthracene	5.91	5.90–5.91
Benzo[a]pyrene	6.06	
Chrysene	5.71	5.60–5.91
Fluorene	4.38	
Naphthalene	3.51	3.01–4.70
Phenanthrene	4.52	
Pyrene	5.32	
Nonhalogenated Solvents		
Acetone	−0.24	
Methyl butyl ketone	1.38	
Methyl ethyl ketone	0.28	0.26–0.29
Methyl isobutyl ketone	1.09	
Halogenated Solvents		
Carbon tetrachloride	2.73	
Chloroform	1.94	1.90–1.97
Methylene chloride	1.28	1.25–1.30
Perchloroethylene	2.79	2.53–2.88
1,1,1-Trichloroethane	2.33	2.18–2.47
Trichloroethylene	2.33	2.29–2.42
Insecticides		
Aldrin	5.17	
Carbaryl	2.38	2.31–2.56

Table 5.7 (Continued)

Compound	Mean Log K_{ow}	Range
Insecticides (cont.)		
Dieldrin	5.16	
DDT	6.11	5.76–6.36
Endrin	5.02	4.56–5.34
Hexachlorobenzene	5.65	5.45–6.18
Lindane	3.76	3.66–3.85
Malathion	2.84	
Parathion	3.43	2.15–3.93
Herbicides		
Atrazine	2.68	
2,4-D	2.94	1.57–4.88
2,4,5-T	3.40	
Trifluralin	5.31	5.28–5.34
Fungicides		
Pentachlorophenol	4.41	3.81–5.01
Industrial Intermediates		
Chlorobenzene	2.84	2.71–2.98
2,4-Dichlorophenol	3.08	
Dimethyl phthalate	1.70	1.47–2.00
Di-*n*-octyl phthalate	9.54	9.20–9.87
Hexachlorocyclopentadiene	4.52	4.00–5.04
Phenol	1.47	1.46–1.48
2,4,6-Trichlorophenol	3.24	
Explosives		
Picric acid	1.69	1.34–2.03
2,4,6-Trinitrotoluene	2.25	
Polychlorinated Biphenyls		
Aroclor 1016	5.58	
Aroclor 1232	3.87	3.20–4.54
Aroclor 1248	6.11	
Aroclor 1254	6.31	
Aroclor 1260	6.91	
Chlorinated Dioxins		
2,3,7,8-Tetrachlorodibenzo-*p*-dioxin	5.77	5.38–6.15

Source: Appendix G.

that biodegradation does not interfere with the K_d measurement.   The aqueous and solid phases are then analyzed for the contaminant. Metals are analyzed by atomic absorption spectrophotometry. The most accurate measure of organic contaminant concentrations in experimental K_d determinations is through the use of ^{14}C-labeled material, which is easily quantified through liquid scintillation counting to measure the emission of β-particles. If extraction efficiencies are sufficiently high, the contaminant can be analyzed after extraction from each phase by gas chromatography or high-performance liquid chromatography. However, many hazardous compounds, such as triazine pesticides, are irreversibly sorbed to soils in the form of bound residues (in humic-like polymers), and this immobilization can bias the distribution results [5.30].

In many cases, the experimental determination of K_d may not be practical or economical. In addition, subsurface strata are highly heterogeneous and, if only a small number of K_d determinations are made, the results may be biased. Therefore, estimation techniques are often used to determine K_d. The estimation methods are based on the contaminant properties (e.g., the octanol-water partition coefficient or water solubility) and the soil property of organic carbon content.

The basis for K_d estimation is the Freundlich isotherm. As shown in Figure 5.8, the lower region of the Freundlich isotherm is nearly linear; that is, the mass of contaminant sorbed is directly proportional to its mass in the aqueous phase. In this region of the curve, n is equal to 1 in the Freundlich isotherm (Equation 5.4):

$$C_s = \frac{x}{m} = K_F C_e \qquad (5.8)$$

By convention, K_F may be replaced by K_d; rearranging Equation 5.8 yields

$$K_d = \frac{C_s}{C_e} \qquad (5.9)$$

Inspection of Equation 5.9 shows that the slope of the isotherm is equal to the soil distribution coefficient. This simplification of the Freundlich isotherm, which equates with the definition of K_d (Equation 5.7), is based on two assumptions. First, the concentration of the contaminant in the system must be low. This is a valid assumption for many subsurface systems. Second, equilibrium must be achieved, which is a relatively valid assumption for most environmental contaminants, particularly those that are hydrophobic.

Numerous site and contaminant characteristics have been studied to predict or estimate values for K_d. Soil properties that have been related to soil sorptivity include particle size distribution, soil organic carbon content, and soil surface area. Of all the factors investigated, the highest correlation with sorptivity was soil organic carbon content [5.31]. At first, this correlation may not seem logical because contaminants are often thought to sorb to clays and smaller particles. However, clays, silts, and other fine particles have negative surface charges and are not characterized by a nonpolar surface. Soil organic matter, although it also has negatively charged ionic groupings, has a nonpolar core material—an amorphous polymer formed mainly by the degradation of lignin from woody plants—to which hydrophobic contaminants tend to sorb. The organic matter in soils is usually strongly bound as a clay-organic complex. In soils with significant amounts of soil organic matter, the clay-organic matter matrix acts as the primary sorbent. Therefore, although other soil properties are sometimes correlated with sorption (e.g., pH, surface area, iron oxide content, and cation exchange capacity), the soil property that usually influences sorption the most is the soil organic matter content.

The results described above [5.31] do not completely describe the importance of soil organic matter as the only sorbent of importance in soils. When soils or subsurface solids contain low organic carbon contents, no correlation has been established between sorption and the organic carbon [5.32]. Therefore, when soil organic matter concentrations are low, the inorganic soil fraction is the primary sorbent. The concentration at which organic carbon no longer provides the role of primary sorbent, f_{oc}^*, has most commonly been 0.1%, although organic carbon levels of 1% or even higher have been reported as the concentration at which it becomes the primary sorbent [5.33]. The role of soil organic matter as the primary sorbent is a complex phenomenon that is likely a function of the organic matter, the soil characteristics, contaminant properties, and other factors. Dragun [5.34] provided general guidelines to assess the role of soil organic matter as the primary sorbent. He proposed that, if the clay/organic carbon ratio (%/%) is less than 25 for organic compounds with polar functional groups or is less than 60 for nonpolar organic contaminants, then the soil organic matter serves as the primary sorbent.

McCarty et al. [5.35] provided a more quantitative basis for estimating f_{oc}^*; they proposed that f_{oc}^* is related to the surface area of the soil and the K_{ow}:

$$f_{oc}^* = \frac{S_a}{200(K_{ow})^{0.84}}$$

(5.10)

where

f_{oc}^* = the minimum SOC content in which organic matter serves as the primary sorbent (g/g)
S_a = the soil surface area (m^2/g)

Typical surface areas of soils and soil minerals range from 1 to 1100 m^2/g [5.33]. For example, surface areas of sandy loam soils = 10–40 m^2/g, loam soils = 50–100 m^2/g, silty loam soils = 50–150 m^2/g, and clays = 150–250 m^2/g. Inspection of Equation 5.10 reveals that f_{oc}^* increases as a function of S_a and decreases with K_{ow}. The effect of K_{ow} on f_{oc}^* is demonstrated in Example 5.4.

EXAMPLE 5.5 *Estimation of f_{oc}^**

For a soil with a surface area of 8.2 m^2/g, determine f_{oc}^* for (1) MEK, (2) 1,2,3-trichlorobenzene, and (3) benzo[a]pyrene.

SOLUTION

1. First, obtain the K_{ow} for MEK. From Table 5.7, log K_{ow} = 0.28, so K_{ow} = 1.91. Use Equation 5.10, to determine f_{oc}^*.

$$f_{oc}^* = 8.2/(200 \cdot 1.91^{0.84}) = 0.0239 = 2.39\%$$

2. Obtain the K_{ow} for 1,2,3-trichlorobenzene. From Appendix G, log K_{ow} = 4.07, so K_{ow} = 11,700. Using equation 5.10 gives

$$f_{oc}^* = 8.2/(200 \cdot 11,700^{0.84}) = 1.6 \times 10^{-5} = 0.0016\%$$

3. The log K_{ow} for benzo[a]pyrene = 6.06, so K_{ow} = 1.15 × 10⁶. Using Equation 5.10 gives

$$f_{oc}^* = 8.2/(200 \cdot 1,150,000^{0.84}) = 3.33 \times 10^{-7} = 0.000033\%$$

These calculations demonstrate that f_{oc}^* decreases significantly with higher K_{ow}s.

Because soil organic matter is the most common sorbent in soils, the estimation of K_d may be based on the organic carbon content of the soil and a more specific distribution, the *soil adsorption coefficient*, K_{oc}:

$$K_{oc} \text{ (mL/g)} = \frac{\text{mass of contaminant sorbed to the soil organic carbon (mg/g)}}{\text{mass of contaminant in the aqueous phase (mg/mL)}} \quad (5.11)$$

The K_{oc} is based on the premise that the contaminant sorbs primarily to the soil organic matter (the primary sorbent), not to the soil as a whole. Values for K_{oc} have been determined experimentally for a small number of hazardous compounds (Table 5.8 and Appendix J). Use of K_{oc} is preferred when the information is available; however, K_{oc}

Table 5.8 Measured K_{oc} Values for Selected Hazardous Compounds

Compound	K_{oc} (mL/g)
Aldrin	410
Anthracene	26,000
Atrazine	175*
Benzene	83
Carbaryl	270*
2,4-D	39*
DDT	130,000*
1,2-Dichlorobenzene	347
2,2',4,4',5,5'-Hexachlorobiphenyl	416,869
2,2',4,4',6,6'-Hexachlorobiphenyl	1,200,000
Malathion	1,778
Naphthalene	1,300
Parathion	7,751*
Pyrene	73,350*
2,3,7,8-Tetrachlorodibenzo-p-dioxin (TCDD)	481,340
Tetrachloroethylene (PCE)	363
1,1,1-Trichloroethane (TCA)	178

Source: Reference 5.32.
*Mean values

Table 5.9 Regression Equations for Estimating K_{oc}

Regression Equation	Equation No.	Organic Carbon	Applicable Compounds	Reference
$\log K_{oc} = 1.029 \log K_{ow} - 0.18$	(5.14)	Wide range	Broad range of herbicides and insecticides (atrazine, bromacil, carbofuran, 2,4-D, DDT, diuron, malathion, methyl parathion, simazine)	5.36
$\log K_{oc} = 0.94 \log K_{ow} + 0.22$	(5.15)	Wide range	s-Triazine herbicides	5.36
$\log K_{oc} = 0.524 \log K_{ow} + 0.855$	(5.16)	1.0–4.0%	Phenyl urea herbicides [e.g., 3-(3-chlorophenyl) urea, 1,1-dimethyl-(3,3,4-dichlorophenyl) urea]	5.37
$\log K_{om} = 0.52 \log K_{ow} + 0.64$	(5.17)	1.09–5.92% (organic matter)	Phenyl ureas, carbamates, organochlorine insecticides, 4-bromophenol, captan, bromo and chloroanilines, chloro and methyl anilines, bromo- and chloro-nitrobenzenes, diphenyl amino folpet, hexachlorobenzene, naphthalene, phenol, simazine	5.32
$\log K_{oc} = 0.544 \log K_{ow} + 1.377$	(5.18)	Wide range	PAHs, organochlorine insecticides, benzene, uracil herbicides, carbamates, acid amides, phenoxy herbicides, pentachlorophenol, DBCP, organophosphate esters, PCBs, ethylene dibromide	5.39
$\log K_{oc} = 0.72 \log K_{ow} + 0.49$	(5.19)	<0.01–33%	Chlorobenzene, chlorobenzenes, methylbenzenes, toluene, PCE, n-butylbenzene	5.40
$\log K_{oc} = 0.989 \log K_{ow} - 0.346$	(5.20)	0.66–2.38%	Benzene, PAHs	5.41
$\log K_{oc} = \log K_{ow} - 0.21$	(5.21)	0.09–3.29%	PAHs	5.42
$\log K_{oc} = -0.82 \log S + 4.07$ S = water solubility (mg/L)	(5.22)	0.11–2.38%	PAHs	5.43
$\log K_{om} = -0.62 \log K_{ow} + 2.04$	(5.23)	3.51%	Triazines, pyrimidines, pyridazines	5.44
$\log K_{oc} = 0.601 \log K_{ow} + 1.991$	(5.24)	Wide range	Phenylsulfinyl and phenylsulfonyl acetates	5.45

data are not as widely available as values for K_{ow}s. Therefore, a series of correlation equations has been developed that relate K_{oc} to K_{ow} for a variety of compound classes; some are listed in Table 5.9. Guidelines for using the regression equations are provided in Table 5.10.

Table 5.10 Guidelines for Using Correlation Equations to Obtain K_{oc}

1. Determine if the equations listed in Table 5.9 are valid using the following criteria.
   a. Organic matter must serve as the primary sorbent. Using Equation 5.10, the organic carbon content of the soil is the primary sorbent if $f_{oc} > f^*_{oc}$.
   b. The compound's molecular weight must not be greater than 400 (which negates the potential of Van der Waals forces as the primary sorption mechanism).
   c. The compound must not contain functional groups that promote other sorption mechanisms such as ion exchange, interaction of molecular fragments, or coordination.
2. Select the appropriate equation for characterizing the chemical property (S or K_{ow}) with K_{oc} or K_{om}. Use the equation that has been developed with compounds of closely related chemical structures. If two or more equations were developed with similar chemicals, use all applicable equations and average the resulting K_{oc} or K_{om}.

After K_{oc} is selected from Table 5.8 or estimated using the appropriate regression equation, the soil distribution coefficient, K_d, may be calculated by normalizing K_{oc} for the fraction of organic carbon in the soil (f_{oc}):

$$K_d = K_{oc} \cdot f_{oc} \qquad (5.12)$$

The use of the equations in Table 5.9 to estimate K_d for a contaminant is illustrated in Example 5.5.

When the soil organic matter is considered the primary sorbent (i.e., when K_{oc} is used as the model for sorption), the highly variable nature of the organic matter between, and even within, soils requires consideration. For example, Garbarini and Lion [5.46] documented that the distribution of TCE and toluene in soil-water slurries was a function of the size distribution and structure of the soil humus.

The soil distribution coefficient, K_d, is a measure of the ratio of equilibrium contaminant concentrations in the sorbed and aqueous phases. It provides a fundamental parameter for calculating the rate of contaminant transport relative to the rate of groundwater movement in subsurface systems. These calculations, in which K_d is used to calculate the retardation factor, are described next in Section 5.7 and used in subsurface transport calculations in Chapter 8.

EXAMPLE 5.6   *Estimation of K_{oc} and K_d*

Estimate K_{oc} and K_d for PCE in a soil-water system with a soil organic carbon content of 0.2% and a surface area of 12 m²/g.

SOLUTION

1. Verify that organic matter serves as the primary sorbent using Equation 5.10: From Table 5.7, the log K_{ow} for PCE = 2.79, so K_{ow} = 617

$$f_{oc}^* = \frac{S_a}{200(K_{ow})^{0.84}}$$

where

$$S_a = 12 \text{ m}^2/\text{g}$$
$$K_{ow} = 617$$

$$f_{oc}^* = \frac{12}{200(617)^{0.84}}$$

$$f_{oc}^* = 0.027\%$$

2. The f_{oc}^* of 0.027% is less than 0.2%; therefore, organic matter is the primary sorbent. Next, use Equation 5.19 to obtain K_{oc} and K_d:

$$\log K_{oc} = 0.72 \ (2.79) + 0.49$$

$$\log K_{oc} = 2.50$$

$$K_{oc} = 316$$

3. Determine the K_d by normalizing K_{oc} for the fraction of organic carbon in the soil:

$$K_d = K_{oc} \cdot f_{oc}$$

$$K_d = (316) \cdot (0.002)$$

$$K_d = 0.63$$

5.7 THE RETARDATION FACTOR

The soil distribution coefficient, K_d, describes the partitioning of a contaminant between the sorbed and bulk soil-water phases. Because a contaminant must be in the bulk phase to be transported, the K_d serves as a basis for the relative contaminant velocity, which is quantitatively described by the *retardation factor*:

$$R = 1 + \frac{\rho_B}{n} K_d \qquad (5.13)$$

where

$$R = \text{retardation factor} \left(\frac{\text{groundwater velocity}}{\text{contaminant velocity}} \right)$$

ρ_B = soil bulk density (g/cm^3)
n = effective porosity
K_d = soil distribution coefficient

The retardation factor is related directly to contaminant hydrophobicity through correlations with K_d, K_{oc}, and K_{ow}. For contaminants that are weak acids, such as chlorophenols, R is often normalized for α, the fraction of the contaminant in the un-ionized form (see Section 3.2). The assumption in this calculation is the ionized contaminant is essentially miscible in water and, therefore, not retarded in solid-water systems.

The retardation factor provides an estimate of the velocity of groundwater relative to the velocity of the contaminant. However, the use of the retardation factor itself is limited because it does not take into account dispersion or transformation reactions. These processes may be considered through the use of groundwater transport equations in which the retardation factor and a term for contaminant transformation are incorporated. Determination of the retardation factor is demonstrated in Example 5.6.

EXAMPLE 5.6 *Determination of the Retardation Factor*

Find the retardation factor for chlorobenzene in a groundwater system containing 0.2% organic carbon with bulk density of 1.4 g/cm^3 and porosity of 0.4 (assume S_a = 12 m^2/g).

SOLUTION

1. Verify that organic carbon is the primary sorbent:
   From Table 5.7 (or Appendix G), the log K_{ow} for chlorobenzene = 2.84.

$$f^*_{oc} = \frac{S_a}{200(K_{ow})^{0.84}}$$

where

S_a = 12 m^2/g
K_{ow} = 692

f^*_{oc} = 0.025% which is less than the soil organic carbon content of 0.2%; therefore, organic carbon is the primary sorbent.

2. Estimate K_d using Equation 5.17,

$$\log K_{oc} = 0.72 (2.84) + 0.49$$

$$\log K_{oc} = 2.54$$

$$K_{oc} = 10^{2.54} = 347$$

$$K_d = K_{oc} \cdot f_{oc}$$

$$K_d = 347 \cdot 0.002 = 0.69$$

3. Determine R using Equation 5.13:

$$R = 1 + \frac{\rho_B}{n} \cdot K_d = 1 + \frac{1.4}{0.4} \cdot 0.69$$

$$R = 3.42$$

Therefore, the relative velocity of the groundwater is 3.42 times greater than the rate of transport of chlorobenzene.

5.8 REACTIONS OF METALS IN SOILS AND SOLIDS

Just as organic contaminants are sorbed and retarded in soils, the rate of metals migration is usually slower than the rate of water movement due to fixation and partitioning processes on organic matter and the surfaces of secondary minerals. Fixation may also involve incorporation of the metal within the inner structure of the minerals, resulting in more permanent immobilization [5.47]. The partitioning of metals is highly complex and depends on a number of environmental and sorbent properties such as the cation exchange capacity, pH, iron oxide content, redox potential, and so on. Unlike the modeling of organic hazardous compounds, reliable compound-specific distribution coefficients have not been developed for metals partitioning and attenuation. Even the eloquent equilibrium models that are used for the thermodynamic characterization of metals in aqueous systems sometimes do not provide accurate predictions of metals partitioning in soils and the subsurface.

As with organic contaminants, sorption of metals can occur by physical and chemical mechanisms. Physical sorption is through long-range Van der Waals forces, a mechanism that does not involve direct contact between the sorbent and sorbate. In contrast, chemical sorption of metals often involves valency forces such as ion exchange. In the ion exchange mechanism, an ion displaces another that was originally present on the exchange surface. Both anions and cations are involved in ion exchange, although in hazardous waste systems, cations (usually heavy metals) are the more common concern. An important predictor of the amount of cation exchange that can take place in a soil or subsurface is the cation exchange capacity (CEC). As with organic contaminants, partitioning of metals on soil surfaces decreases the aqueous concentration and, correspondingly, their ability to move in the subsurface. Relative to organic contaminants with low to moderate K_{ow}, most metals are characterized by low rates of migration, especially in oxic soils at neutral pH ranges [5.48].

The difficulty in predicting the partitioning of metals may be appreciated by viewing Table 5.11, which lists ranges of experimentally determined K_ds for metals. The ranges of K_d listed in Table 5.11 are extremely wide, suggesting that numerous variables govern the partitioning of metals in solid-water systems [5.49].

The governing variables that control the soluble concentration of metals in soil-water are shown in Figure 5.10. Any one of these parameters can significantly affect the soluble metal concentration. Furthermore, the variables are interactive; for example, pH may affect redox potential, ion exchange, and complexation. Oxidation-reduction is also interactive with precipitation; most oxidized metals (e.g., Cd^{2+}, Fe^{3+}, Pb^{2+}) are characterized by significantly lower solubility product (K_{sp}) values relative

Table 5.11 Ranges of K_d (mL/g) for Selected Elements in Soils and Clays of pH 4.5 to 9.0

Element	Observed Range of K_d	No. Observed	μ^a	σ^b
Ag	10–1,000	16	4.7	1.3
Am	1.0–47,230	46	6.7	3.0
As(III)	1.0–8.3	19	1.2	0.61
As(V)	1.9–18	37	1.9	0.52
Ca	1.2–9.8	10	1.4	0.78
Cd	1.26–26.8	28	1.9	0.86
Ce	58–6,000	16	7.0	1.3
Cm	93.3–51,900	31	8.1	1.9
Co	0.2–3,800	57	4.0	2.3
Cr(II)	470–150,000	15	7.7	1.2
Cr(VI)	1.2–1,800	18	3.6	2.2
Cs	10–52,000	135	7.0	1.9
Cu	1.4–333	55	3.1	1.1
Fe	1.4–1,000	30	4.0	1.7
K	2.0–9.0	10	1.7	0.49
Mg	1.6–13.5	58	1.7	0.52
Mn	0.2–10,000	45	5.0	2.7
Mo	0.37–400	17	3.0	2.1
Np	0.16–929	44	2.4	2.3
Pb	4.5–7,640	125	4.6	1.7
Po	196–1,063	6	6.3	0.65
Pu	11–300,000	40	7.5	2.3
Ru	48–1,000	17	5.4	1.0
Se(IV)	1.2–8.6	19	1.0	0.65
Sr	0.15–3,300	218	3.3	2.0
Tc	0.0029–0.28	24	−3.4	1.1
Th	2000–510,000	17	11	1.5
U	10.5–4,400	24	3.8	1.3
Zn	0.1–8,000	146	2.8	1.9

[a]The mean of the logarithms of the observed values.
[b]The standard deviation of the logarithms of the observed values.

Source: Reference 5.48.

to their lower oxidation states (e.g., Cd^+). Qualitative descriptions of each of the processes shown in Figure 5.10 are provided next. More detailed discussions of these variables may be found in References 5.33 and 5.51.

Oxidation-Reduction. Redox conditions in soils, sludges, groundwater, and hazardous waste treatment systems can have a significant effect on the mobility of metals. The oxidation-reduction state is based on the presence of aqueous electrons (i.e., the free electron activity). Two conventions used to quantify the system redox potential are (1) the pE, the negative log of the free aqueous electron activity, and (2) the E_h, the difference in potential (expressed in millivolts) between a platinum electrode and a stan-

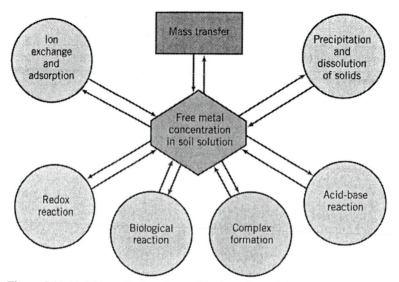

Figure 5.10 Variables affecting the partitioning of metals in soils. *Source:* Reference 5.50.

dard hydrogen electrode. Positive values of pE or E_h correlate with oxidizing conditions and negative values are associated with reducing conditions. In general, reduced forms of metals (e.g., Cd^+ vs. Cd^{2+}, Cu^+ vs. Cu^{2+}) are more mobile than the oxidized species. However, two anomalies occur to this rule—Cr^{6+} is more soluble than Cr^{3+} and Se^{6+} is more soluble than Se^{4+}.

pH. Low pH conditions also correlate with increased metal mobility. A number of interactive processes occur at low pH levels (i.e., pH 2–5), which may involve metals dissolution, ion exchange, and oxidation-reduction. Acidic pH often correlates with reducing conditions, which provide an environment that promotes metal dissolution. For example, a metal hydroxide, such as cadmium (II) hydroxide, may be released into the aqueous phase of a soil, groundwater, or sludge by the presence of hydrogen ions:

$$Cd(OH)_2 + 2H^+ \longrightarrow Cd^{2+} + 2H_2O \tag{5.25}$$

Furthermore, because of the interactive relationship between pH and E_h, metals may become reduced to lower oxidation states and, as a result, become more mobile. Although it is beyond the scope of this text, the interactive relationship between pH and E_h may be derived for any metal [5.51]. An example of an E_h-pH diagram for cadmium is shown in Figure 5.11. Diagrams of E_h-pH relationships for most hazardous metals are provided by Dragun [5.33].

Ion Exchange. Most heavy metals are present as cations in soils and groundwater, and their affinity for the surfaces of soil solids is a function of the nature of the negative charges inherent in soils. Clay surfaces have negative charges that result from isomorphic substitution (the replacement of metal ions within the clay lattice by other metals

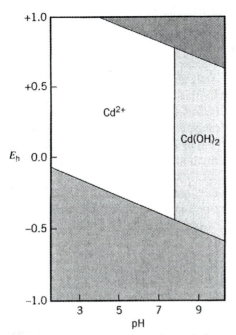

Figure 5.11 E_h-pH diagram for cadmium.
Source: Reference 5.33.

of higher positive charge, resulting in a corresponding negative charge on the clay surface). These negative surfaces are the sites of cation exchange. Humus (humic and fulvic acids) and metal oxide minerals also play a significant role in ion exchange.

Negative charges associated with clays, iron and manganese oxides, and humus are ubiquitous in soils, and must be balanced by cations to maintain electrochemical neutrality. However, the cations that are present at any given time may be displaced by another species that has a higher affinity for the soil surface:

$$
\begin{array}{c}
\rangle\text{O---Na} \\
\qquad\qquad\qquad + \text{Cd}^{2+} \longrightarrow \\
\rangle\text{O---Na}
\end{array}
\begin{array}{c}
\rangle\text{O} \\
\qquad\quad\text{Cd} + 2\text{Na}^+ \\
\rangle\text{O}
\end{array}
\qquad (5.26)
$$

In the ion exchange process, one metal displaces another on the inorganic or humus exchange sites, but the exchange sites on the soil surface are always filled to maintain electroneutrality. A hierarchy exists for the exchange of these metals based on their charge and charge-to-molecular-weight ratio. Trivalent cations are exchanged over divalent cations, which in turn are exchanged over the monovalent species. Within each group of valences there is a hierarchy of exchange, the *lyotrophic series*, for metals:

$$\text{Ba}^{2+} > \text{Sr}^{2+} > \text{Ca}^{2+} > \text{Mg}^{2+} \qquad (5.27)$$

$$\text{Cs}^+ > \text{Rb}^+ > \text{K}^+ > \text{Na}^+ > \text{Li}^+ \qquad (5.28)$$

The law of mass action, in which a high concentration of a less exchangeable species displaces a more exchangeable ion, is applicable to ion exchange just as it is throughout equilibrium chemistry. For example, a soil or ion exchange resin flooded with brine (sodium) will displace a more favorable species, such as Cd^{2+}. Hydrogen ions displace most exchanged metals; therefore, soils and sludges that receive acidic wastes will release the metals that were previously fixed.

The exchange of metals occurs because of their electrostatic attraction to a sorbent, some of which have more significance than others. Both organic and inorganic sorbents (including metal oxides and oxyhydroxides) are important in ion exchange, although the role of the inorganic fraction usually outweighs that of organic matter. Clays are characterized by significantly higher CECs relative to silt and sand, with CECs often three orders of magnitude greater than sand [5.16]. Ion exchange is also used in hazardous waste treatment processes; synthetic ion exchange resins, such as zeolites, effectively remove metals from electroplating and other metal-bearing waste streams.

Selectivity. Not only do metals have different affinities for solids; sorbents also have different affinities for metals—a phenomenon known as sorbent *selectivity*. The different selectivities are a function of the negative charges and surface characteristics of the sorbents and charge-to-atomic-radius relationships for the metals. Alloway [5.52] described selectivity in terms of the Lewis hard-soft acid-base (HSAB) principle, in which a hard Lewis acid (i.e., a species with high electronegativity and small ionic radius) is attracted to a Lewis base (which has low electronegativity). Some orders of selectivity for a number of sorbents are listed in Table 5.12. Because varying degrees of hardness and softness between both acids (metals) and bases (sorbents) are found, selectivity is sorbent-specific. Nonetheless, some trends predominate. Lead and copper are fixed strongly to most soil components; cadmium, nickel, and mercury are among the most mobile compounds of those listed for most sorbents.

Complexation. Hazardous metals may form complexes with organic or inorganic anions, or ligands. The anions that complex with metals include hydroxide, carbonate, sulfate, salts of organic acids, polypeptides, and salts of humic acids. The reactions of

Table 5.12 Sorbent Selectivity for Divalent Metals

Sorbent	Selectivity Order	Reference
Montmorillonite (Na)	Ca > Pb > Cu > Mg > Cd > Zn	5.52
	Cd = Zn > Ni	5.53
Illite (Na)	Pb > Cu > Zn > Ca > Cd > Mg	5.52
Kaolinite (Na)	Pb > Ca > Cu > Mg > Zn > Cd	5.52
	Cd > Zn > Ni	5.53
Ferrihydrite	Pb > Cu > Zn > Ni > Cd > Co > Sr > Mg	5.54
Hematite	Pb > Cu > Zn > Co > Ni	5.55
Goethite	Cu > Pb > Zn > Co > Cd	5.56
Peat	Pb > Cu > Cd = Zn > Ca	5.57

hazardous metals with these ligands (L) are typically:

$$M_i + L_j \rightleftharpoons M_iL_j \qquad (5.29)$$

where L_j is the ligand and M_iL_j is the coupled metal-ligand. The ligand often carries a positive or negative charge and may not be highly hydrophobic. A range of sorption effects may be seen with complexes. The complex may be sorbed more than the free metal, or the complex may be more soluble than the metal. Therefore, complexation complicates the partitioning of metals in soils, and makes their quantification difficult.

Precipitation. Another significant mechanism for the loss of metals from solution is precipitation, in which the solubility product of the metal with carbonates, hydroxides, or other species is exceeded. For example, most metals exhibit low solubility at high pH regimes due to their extremely low solubility products with carbonates and hydroxides:

$$Cd^{2+} + 2OH^- \rightleftharpoons Cd(OH)_2 \qquad K_{sp} = 10^{-14.30} \qquad (5.30)$$

$$Cd^{2+} + CO_3^{2-} \rightleftharpoons CdCO_3 \qquad K_{sp} = 10^{-11.60} \qquad (5.31)$$

A process similar to precipitation is *coprecipitation,* in which a metal is enmeshed or sorbed to another species that is undergoing precipitation.

5.9 SYNOPSIS OF THE PARTITIONING BEHAVIOR OF IMPORTANT HAZARDOUS METALS

Although pH, redox conditions, ion exchange, and other factors result in significant variability in the partitioning behavior of metals in soil and subsurface systems, some general trends are found for each metal. The following brief qualitative descriptions provide a synopsis of the behavior of some of the more important metals in soils.

Arsenic. A metalloid element, arsenic is often classified as a nonmetal; it is found most commonly as an anionic species. Arsenic exists in the $5+$ oxidation state in oxic environments, whereas arsenic (III) is most common under reducing conditions. Arsenic acid (H_3AsO_4) is the most common form of arsenic (V). It is a weak acid with pK_as of 2.2, 7.0, and 11.5. Arsenic (III) is commonly found as arsenous acid (H_3AsO_3) with pK_as of 9.2, 12.1, and 13.4. The ratio of As(V) to As(III) is usually a function of E_h/pH relationships. Attenuation of arsenic in soils has been shown to follow both Langmuir and Freundlich isotherms, and is complicated by changes in redox states and microbial processes; nonetheless, arsenic (V) species tend to be more mobile in soils than arsenic (III). Like mercury, arsenic can be transformed microbially through methylation reactions. However, unlike mercury, the organoarsenic compounds are less toxic than the inorganic forms of arsenic.

Cadmium. Cadmium is relatively mobile in soils and groundwater compared to other metals such as lead and copper. Sorption, rather than precipitation, often controls the distribution of cadmium in soils. As with all metal species in soils, complex formation complicates its partitioning phenomena. Alloway [5.52] reported that Cd^{2+} predominated in soils, with concentrations of the neutral species $CdSO_4$ and $CdCl_2$ increasing as a function of pH. The key factors that control the sorption of cadmium include pH, organic matter, and hydrous oxide content. For example, a threefold increase in cadmium sorption was noted for every pH unit. The increased sorption at higher pH regimes resulted from increased hydrolysis, increased sorption capacity, and accompanying increased pH-dependent negative charges [5.58].

Chromium. The most stable oxidation states for chromium are Cr(III) and Cr(VI). Chromium, unlike most other hazardous metals, is most commonly found as an anion. Its two most common valence states are Cr(III), which exists as chromate (CrO_4^-), and chromium (VI), which is found as dichromate ($Cr_2O_7^{2-}$). The most toxic form of chromium is dichromate. In contrast, chromium (III), is less mobile as well as less toxic than Cr(VI).

Lead. As with most metals, pH and CEC are the primary factors affecting lead mobility. The solubility of lead is regulated by $Pb(OH)_2$, $Pb_3(PO_4)_2$, and $Pb_5(PO_4)_3OH$ in noncalcareous soils and $PbCO_3$ in calcareous soils. In general, lead is considered to be one of the more immobile metals. However, it is also an insidious toxin to the central nervous system (see Chapter 9).

Mercury. Three different valences are important in the dynamics of mercury in two-phase systems—Hg^0, Hg^{1+}, and Hg^{2+}. The most important chemical parameters that affect the speciation, and consequently the mobility, of mercury are pH and chloride. Mercury complexes strongly with sulfide, although sulfide exists only under reducing conditions. Mercury reacts with chloride at acidic pH above a redox potential of 0.4 V to form $HgCl^0$. However, at pH greater than 7, the most common complex is $Hg(OH)_2$. Even with the strong tendency of mercury to form complexes (e.g., $HgCl_2$), partitioning of the complexes is usually not significant. The ionic species Hg^{2+}, however, is strongly bound to soil minerals or sorbed to inorganic surfaces and organic ligands. The tendency of soils to fix Hg^{2+} results in minimal migration of mercury. For example, Hogg et al. [5.59] found that inorganic and organic mercury did not leach below 20 cm (0.66 ft) in soil columns.

Nickel. The fate of nickel in solid-water systems is governed by its 2+ valence state. Although nickel ferrite ($NiFe_2O_4$) is, thermodynamically, the most probable precipitate in soils, experimental data have shown that nickel oxides and nickel carbonates dominate at high pH and nickel (II), nickel sulfates, and nickel phosphates are the primary species at neutral pH. The most important factor in nickel mobility is pH, followed by the clay content of the soil and the presence of hydrous iron and manganese oxides [5.60].

To summarize the relative mobility of metals, some partitioning behavior of different metals on a number of soils is illustrated in Figure 5.12. The trends of Figure 5.12 show that copper and lead exhibit the lowest mobility and arsenic, chromium, and mercury are characterized by the highest mobility. The trends shown in Figure 5.12 correlate with the sorbent selectivity data listed in Table 5.12. The degree of attenuation is not the same for all soils, which is expected based on differing sorbent selectivities.

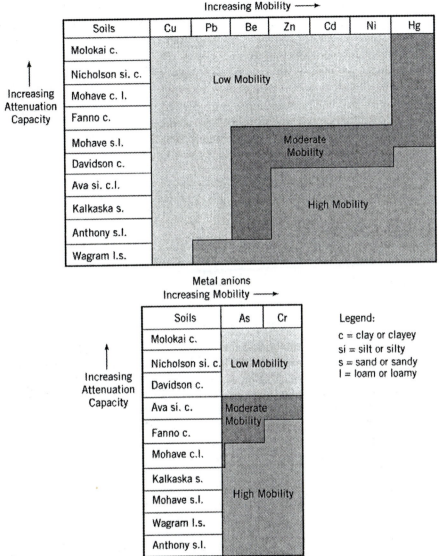

Figure 5.12 Relative mobility of metals in soils. *Source:* Reference 5.48.

5.10   ESTIMATION OF PARTITIONING AND POTENTIAL MOBILITY OF METALS

The variables described in Sections 5.7 and 5.8 provide a common theme for all metals—their partitioning behavior is highly complex and difficult to quantify. If metals partitioning and attenuation is so variable, how is it possible to quantify metals migration rates in soils and the subsurface? As discussed in Section 5.6, the most accurate measure of metals partitioning is through laboratory K_d measurements. If more detailed partitioning-concentration relationships are needed, isotherm analyses may be performed. Although the use of tabulated metal K_d values is not accurate (Table 5.11), a technique based on column studies, developed by Amoozegar-Fard et al. [5.61] provides a reliable algorithm for the migration of three metals in soils. In their work, the relative rates of cadmium, nickel, and zinc migration in landfill leachate were quantified based on collecting surface soil or subsurface samples and conducting column studies by loading the systems with aqueous metals solutions. Through multiple regression analyses, the relative rates of metals transport were correlated with soil properties and empirical equations were derived. Amoozegar-Fard et al. [5.61] found that the soil properties that were the best predictors of metals transport were the organic carbon content, the iron oxide content, and the percentage of clay in the soil, which was described quantitatively by

$$V_{C/C_0} = \left(\frac{V}{25}\right)(A_1X_1 + A_2X_2 + \cdots + A_nX_n + K) \tag{5.32}$$

where

V_{C/C_0} = velocity of the relative metal concentration being evaluated (cm/day)
V = the pore water velocity (cm/day)
A_i = regression coefficients for each parameter (listed in Tables 5.13–5.15)
X_i = the concentration of the parameters listed in Tables 5.13–5.15
K = regression constant (listed in Tables 5.13–5.15)

The relative concentration coefficients for Equation 5.32 are listed in Tables 5.13–5.15. The basis for using Equation 5.32 and Tables 5.13–5.15 is first determining the relative concentration of the metal of interest and then selecting the appropriate coefficients, which, when entered into Equation 5.32, provide a relative rate of metals transport. The use of Equation 5.32 is illustrated in Example 5.6.

EXAMPLE 5.7 *Estimation of Relative Metals Migration Rates*

A landfill leachate contains 10 mg/L Ni and migrates with a pore water velocity of 4 cm/day. The leachate contains 0.08% total suspended solids (TSS) and 0.20% total organic carbon (TOC). The soil through which the leachate is migrating contains 2% free iron oxides, 8% clay, 22% sand, and 70% silt. Determine the time required for a concentration of 5 mg/L Ni to migrate 50 m.

SOLUTION

1. Determine the relative concentration of Ni to be evaluated.

$$\frac{C}{C_0} = \frac{5 \text{ mg/L Ni}}{10 \text{ mg/L Ni}} = 0.5$$

Table 5.13 Regression Coefficients for the Propagation Velocities of C/C_0 = 0.1, 0.2, . . . , 0.9 for Cd in Eq. [5.32]

Variables†	Relative Concentration, C/C_0								
	0.1	0.2	0.3	0.4	0.5	0.6	0.7	0.8	0.9
$(\% \text{ Clay})^{-1}$	32.02	31.07	30.65	29.54	29.91	29.52	29.18	28.65	28.10
% Sand	−0.233	−0.227	−0.223	−0.217	−0.217	−0.214	−0.212	−0.209	−0.204
$(\% \text{ Sand})^2$	0.00117	0.00114	0.00112	0.00109	0.00108	0.00107	0.00106	0.00105	0.00102
$(\% \text{ FeO})^2$	0.0116	0.0111	0.0106	0.00916	0.0101	0.00996	0.00970	0.00970	0.00939
$(\% \text{ FeO})^{-1}$	9.846	9.385	9.057	8.817	8.532	8.293	8.048	7.761	7.349
% TSS	90.83	89.19	87.32	92.20	84.13	83.06	81.63	79.10	77.81
$(\% \text{ TSS})^2$	−218.6	−216.4	−212.3	−234.1	−205.1	−203.0	−199.8	−194.9	−193.9
% TOC	0.322	0.370	0.339	−0.148	0.442	0.494	0.511	0.559	0.426
Constant	−2.090	−2.015	−1.923	−2.069	−1.760	−1.702	−1.585	−1.388	−1.307
r^2	0.844	0.843	0.844	0.845	0.844	0.842	0.840	0.838	0.838

†For clay, sand, and FeO, 1% is equivalent to 10 g kg^{-1}; for TSS and TOC, 1% is equivalent to 10 g L^{-1}.
Source: Reference 5.61.

Table 5.14 Regression Coefficients for the Propagation Velocities of $C/C_0 = 0.1, 0.2, \ldots, 0.9$ for Ni in Eq. [5.32]

Variables[†]	Relative Concentration, C/C_0								
	0.1	0.2	0.3	0.4	0.5	0.6	0.7	0.08	0.9
$(\% \text{ Clay})^{-1}$	17.18	15.71	174.72	14.04	13.460	13.09	12.70	12.62	12.85
$(\% \text{ Sand})^{-1}$	29.29	27.37	26.28	24.78	23.85	23.06	21.49	20.89	19.00
% FeO	0.350	0.333	0.323	0.312	0.304	0.297	0.290	0.280	0.263
$(\% \text{ FeO})^{-1}$	7.541	7.026	6.775	6.413	6.200	5.972	5.734	5.448	5.013
% TSS	82.19	76.34	73.00	71.06	69.32	67.59	65.56	66.12	65.15
$(\% \text{ TSS})^2$	−266.9	−246.5	−234.4	−228.6	−222.9	−217.3	−210.0	−212.7	−210.5
% TOC	−40.13	−37.44	−36.28	−35.63	−34.71	−34.20	−32.33	−32.19	−32.57
$(\% \text{ TOC})^2$	132.2	122.9	120.5	117.5	114.8	113.6	106.4	112.7	116.6
Constant	−6.791	−6.298	−6.052	−5.711	−5.534	−5.350	−5.195	−5.130	−4.927
r^2	0.908	0.905	0.907	0.892	0.885	0.874	0.861	0.839	0.795

[†]For clay, sand, and FeO, 1% is equivalent to 10 g kg^{-1}; for TSS and TOC, 1% is equivalent to 10 g L^{-1}.

Source: Reference 5.61.

Table 5.15 Regression Coefficients for the Propagation Velocities of $C/C_0 = 0.1, 0.2, \ldots, 0.9$ for Zn in Eq. 5.32

Variables[†]	Relative Concentration, C/C_0								
	0.1	0.2	0.3	0.4	0.5	0.6	0.7	0.8	0.9
% Sand	0.0210	0.0173	0.0161	0.0149	0.0136	0.0111	0.00954	0.00803	0.00487
$(\% \text{ Silt})^2$	0.000657	0.000591	0.000570	0.000550	0.000526	0.000482	0.000466	0.000443	0.000398
$(\% \text{ FeO})^{-1}$	2.784	2.747	2.697	2.660	2.637	2.638	2.660	2.648	2.698
$(\% \text{ FeO})^2$	0.00373	0.00336	0.00324	0.00310	0.00288	0.00255	0.00254	0.00236	0.00211
% TOC	−15.38	−14.92	−14.86	−14.72	−14.75	−14.57	−14.76	−14.96	−15.82
$(\% \text{ TOC})^2$	82.37	80.25	79.18	78.42	77.79	77.12	76.71	76.57	78.93
% TSS	67.11	60.90	58.93	56.97	54.66	51.12	51.03	49.37	46.16
$(\% \text{ TSS})^2$	−188.2	−167.5	−162.1	−156.3	−148.2	−135.9	−137.1	−132.3	−122.0
Constant	−6.44	−5.88	−5.68	−5.48	−5.27	−4.92	−4.82	−4.61	−4.22
r^2	0.841	0.844	0.846	0.848	0.847	0.848	0.848	0.848	0.846

[†]For sand, silt, and FeO, 1% is equivalent to 10 g kg^{-1}; for TSS and TOC, 1% is equivalent to 10 g L^{-1}.

Source: Reference 5.61.

293

2. Use Equation 5.32 to determine the velocity of Ni through the aquifer:

$$V_{C/C_0} = \left(\frac{V}{25}\right)(A_1X_1 + A_2X_2 + \cdots + A_nX_n + K)$$

where

$V = 4$ cm/day
% TSS $= 0.08$
% TOC $= 0.2$

and the aquifer characteristics are:

Clay $= 8\%$
Sand $= 22\%$
FeO $= 2\%$
Silt $= 70\%$ (Ni migration is independent of silt content)

3. Regression coefficients and the regression constant are obtained from Table 5.14. Substituting the appropriate values into Equation 5.32 yields

$$V_{0.5} = (4/25)[(8)^{-1}(13.46) + (22)^{-1}(23.85) + (2)(0.304) + (2)^{-1}(6.20) +$$
$$(0.08)(69.32) + (0.08)^2(-222.9) + (0.2)(-34.71) + (0.2)^2(114.8) - 5.534]$$

$$= 0.434 \text{ cm/day}$$

4. Determine the time to migrate 50 m:

$$t = (5000 \text{ cm})/(0.434 \text{ cm/day}) = 11,520 \text{ days} = 31.6 \text{ years}$$

5.11   SUMMARY OF IMPORTANT POINTS AND CONCEPTS

- Electrostatic, physical, and chemical mechanisms contribute to sorption, which is defined as the accumulation of chemicals onto surfaces from the surrounding solution.

- The two most common models for sorption onto soils and solids are the Langmuir and Freundlich isotherms. The Langmuir isotherm is based on a sorbed monolayer on the solid surface. The Freundlich model is characterized by a nearly linear relationship between sorbate and sorbent, with decreased sorption as the contaminant concentration increases in the aqueous phase.

- The octanol-water partition coefficient is the most commonly used contaminant physicochemical property that correlates with potential for sorption. Sorptivity is also inversely related to water solubility.

- The soil distribution coefficient, K_d, is defined as the ratio of the mass of sorbed contaminant to the mass in the soil water. It may be determined experimentally by measuring the contaminant concentrations in the solid and liquid phases.

- In soil-water systems containing significant concentrations of organic carbon (i.e., $> f_{oc}^*$), the organic carbon is the primary sorbent and the soil adsorption coefficient, K_{oc}, may be used to describe the partitioning of contaminants between the water and the organic carbon. A series of correlation equations has been developed between K_{ow} and K_{oc}; in addition, K_d, the soil distribution coefficient, may be related to K_{oc} by normalizing K_{oc} for the organic carbon content.

• Metals partitioning is complicated by a number of competing reactions including ion exchange, oxidation-reduction, pH, complexation, and precipitation. Although some K_d values have been developed for metals partitioning, they are soil-specific because of their dependence on CEC, hydrous oxide contents, and clay contents. Experimental determination of K_ds or isotherm coefficients is the best method for quantifying metals partitioning. Regression equations have been developed for the rates of migration of cadmium, nickel, and zinc in soils as a function of the soil properties that most affect their transport.

PROBLEMS

5.1. Describe the three mechanisms of sorption.

5.2. Describe the contaminant and sorbent characteristics that influence sorption.

5.3. What is the difference between a soil particle size and a soil textural class?

5.4. Explain the differences between primary and secondary minerals. Provide examples of each.

5.5. Using Figure 5.3, provide the soil textural class for each of the following particle size distributions.
  a. 20% sand, 30% silt, 50% clay   b. 88% sand, 6% silt, 6% clay
  c. 48% sand, 23% silt, 29% clay   d. 63% sand, 32% silt, 5% clay
  e. 30% sand, 17% silt, 53% clay

5.6. Using tables, find the K_{ow} for
  a. 4,6-Dinitro-*o*-cresol   b. DDT
  c. TCE   d. PCE
  e. Hexachlorobenzene   f. Naphthalene
  g. 1,1,1-Trichloroethane

5.7. The regression equation

$$\log S = -1.37 \log K_{ow} + 7.37$$

was derived to estimate K_{ow} from water solubility (S, in μmoles/L). Using the water solubility data of Table 3.5 for benzene, anthracene, and 2,4,6-trichlorophenol, use this equation to calculate K_{ow}. Compare the calculated values with the experimental values of Table 5.7.

5.8. A Langmuir isotherm analysis was conducted, and the results show that the slope $= 5.8$ and the y-intercept $= 2.2$. From these results, determine a and b for the Langmuir equation.

5.9. A soil evaluated for malathion sorption has been found to follow the Langmuir isotherm with $a = 0.01$ and $b = 4.5$. A remediation team has proposed adding

200 kg (441 lb) of the soil to a 500,000-L (132,100-gal) pond as an emergency response measure against a spill with a malathion concentration of 6 mg/L. If the aqueous concentration needs to be less than 0.01 mg/L, will the process work?

5.10. Batch isotherm analysis of a soil has shown that sorption of dimethyl phthalate follows the Freundlich isotherm with $n = 0.8$ and $K_F = 3.8$. If 1800 L (476 gal) of a 43 mg/L solution of dimethyl phthalate floods a 400-m³ (523-yd³) section of soil with a bulk density of 1800 kg/m³, what will its equilibrium concentration be?

5.11. A dry well receives 50,000 L (13,200 gal) of aqueous waste containing 125 mg/L MIBK, which sorbs to the soil in a manner characterized by the Langmuir isotherm with $a = 0.14$ and $b = 0.088$. The dry well is 1.5 m (4.9 ft) in diameter, and the soil has a bulk density of 2100 kg/m³. If the mean equilibrium concentration of the MIBK in the soil is 0.1 mg/L, how deep will the MIBK sorption be?

5.12. An isotherm analysis was performed on a surface soil and an aqueous solution of toluene. Six 1-L beakers were used, each containing 10 g of soil and varying concentrations of toluene. After allowing the system to equilibrate for 24 hours, analysis of the aqueous phase showed the following results:

Initial Aqueous Toluene Concentration (mg/L)	Aqueous Toluene Concentration after Sorption (mg/L)
215	112
180	87.6
140	48.5
114	4.8
86	2.5
48	1.8

Fit the data to Langmuir and Freundlich isotherms. Determine the empirical coefficients for the isotherm equation which best fits the data.

5.13. Using a groundwater contaminated with 74 mg/L dichromate, an isotherm analysis was performed using 500-mL aliquots and varying concentrations of subsurface solids. The groundwater was allowed to equilibrate with the sorbent, and the aqueous phase was analyzed for residual dichromate. The following aqueous phase concentrations were found:

Subsurface Solids Concentration (g)	Aqueous Dichromate Concentration after Sorption (mg/L)
0	73.8
2	54.3
4	39.8
6	26.5
8	18.1
10	10.7
12	5.9

Fit these data to the Langmuir and Freundlich isotherms and determine the empirical constants for the best fit isotherm.

5.14. What are the assumptions and limitations of using K_d as a basis for sorption?

5.15. To determine the soil distribution coefficient (K_d), 10 g of dry soil is spiked with 100 mg of ring-labeled ^{14}C-atrazine, 10 mL of deionized water is added, and the system is shaken for 24 h. The aqueous and solid phases are separated, and the solid phase is extracted with 50 mL of methylene chloride while the aqueous phase is extracted with three successive 5-mL portions of methylene chloride. The atrazine concentration in the 50 mL of soil extract is 10 mg/L, and the concentration in the combined 15 mL from the aqueous extractions is 0.4 mg/L. Based on this information, calculate K_d.

5.16. Describe how the soil distribution coefficient, K_d, has been developed from the Freundlich isotherm.

5.17. Describe the relationship between K_d, K_{oc}, soil organic matter, and why K_{ow} can be used as a predictor of adsorption.

5.18. A Freundlich plot developed from well cuttings was performed for the solvent MIBK. The slope in the linear portion was equal to 410. At low MIBK concentrations (in the linear region of the Freundlich curve), what percentage of the MIBK is in the aqueous phase?

5.19. A Freundlich plot developed from well cuttings was performed for the herbicide chlorosulfuron. The slope in the linear portion was equal to 34.
   a. At low chlorosulfuron concentrations (in the linear region of the Freundlich curve), what percentage of the chlorosulfuron is in the aqueous phase?
   b. What are K_d and K_{oc} for chlorosulfuron in this system? The organic carbon content of the well cuttings is 0.15%.

5.20. The following data have been collected for batch isotherm analysis of hexachlorocyclopentadiene in well cuttings from a contaminated groundwater system.

Volume of aqueous solution used: 500 mL
Mass of well cuttings: 50 g
Organic carbon content of well cuttings: 0.2%

C_0 (mg/L)	C_e (mg/L)
0.4	0.02
1.8	0.04
3.7	0.05
5.1	0.1
7.6	1.2
9.2	3.8
11.0	4.9
12.4	5.3

Based on these data, estimate K_d and K_{oc} for the groundwater system.

5.21. If the surface area of a soil is 42 m^2/g, determine the minimum soil organic carbon content at which the SOC is the primary sorbent for

    a. Toluene   b. Naphthalene

    c. TCA   d. Endrin

    e. Di-*n*-octyl phthalate

5.22. A soil with organic carbon content of 2% is contaminated with *p*-xylene. Estimate the K_d for *p*-xylene in the soil using the appropriate regression equation from Table 5.9. The soil S_a is 24 m^2/g.

5.23. A soil with organic carbon content of 0.75% and S_a of 16 m^2/g is contaminated with chrysene. Estimate the K_d for chrysene in the soil using the appropriate regression equation from Table 5.9.

5.24. A soil with organic carbon content of 4% and S_a of 12 m^2/g is contaminated with 2,4-dinitrotoluene. Estimate the K_d for 2,4-dinitrotoluene in the soil using the appropriate regression equation from Table 5.9.

5.25. A soil with organic carbon content of 3.5% is contaminated with carbofuran. Estimate the K_d for carbofuran in the soil using the appropriate regression equation from Table 5.9. The soil S_a is 24 m^2/g.

5.26. A soil with organic carbon content of 0.3% is contaminated with Aroclor 1242. Estimate the K_d for Aroclor 1242 in the soil using the appropriate regression equation from Table 5.9.

5.27. The following empirical relationship has been developed for PAHs:

$$\log K_{oc} = 14,300 \cdot S + 0.27$$

where S = water solubility in moles/L. For the PAH naphthalene, determine the K_d in a groundwater system with organic carbon content of 0.046%. Assume that organic carbon is the primary sorbent.

5.28. Determine K_d for the following compounds in a soil with 0.48% organic carbon. Assume that organic carbon is the primary sorbent.

    a. Benzene   b. TCE

    c. PCE   d. Anthracene

    e. Pyrene   f. DDT

    g. Pentachlorophenol

5.29. Repeat Problem 5.28, but calculate K_d for a soil with 4% organic carbon.

5.30. A soil profile contains organic carbon in the following concentrations.

A horizon (0–15 cm): 3.6% organic carbon
B horizon (15–50 cm): 1.2% organic carbon
C horizon (50–100 cm): 0.8% organic carbon
Subsurface: 0.3% organic carbon

Estimate the diuron K_d for each of the horizons of the soil profile. Assume that organic carbon is the primary sorbent in all of the soil horizons.

5.31. For a surface soil with an organic carbon content of 1.6%, estimate K_d for benzene, chlorobenzene, 1,4-dichlorobenzene, 1,3,5-trichlorobenzene, 1,2,3,5-tetrachlorobenzene, pentachlorobenzene, and hexachlorobenzene. What can be concluded about the effect of chlorination on K_d? Assume that the soil organic carbon is the primary sorbent.

5.32. Determine the soil distribution coefficient for perchloroethylene as a function of soil organic carbon from 0.2% to 2.0% at 0.2% intervals and plot the results. What can be concluded about the effect of soil organic carbon on the soil distribution coefficient?

5.33. Polychlorinated biphenyls and polychlorinated dibenzo-*p*-dioxins usually remain in the top 1 meter of soil when disposed of or spilled onto soil. Explain this phenomenon based on the soil and contaminant characteristics.

5.34. A transformer ruptured 10 years ago, releasing Aroclor 1260 onto the soil. Assuming degradation is negligible, at what depth in the soil profile would you expect to find the PCBs?

5.35. Spillage of organochlorine and organophosphate insecticides occurred at a pesticide rinse and formulation area. Which class of chemicals would migrate faster through the soil?

5.36. Based on the data of Appendix G, what conclusion(s) can you draw between PAH ring size and K_{ow}? What is the effect of chlorination on K_{ow}?

5.37. A groundwater with a pH of 7.5 containing acetone and trichloroacetic acid is to be treated. Granular activated carbon has been proposed to sorb the contaminants. Do you agree with this proposal? Why or why not?

5.38. Of gasoline, diesel, and motor oil, which would you expect to migrate the fastest in soils and groundwater? Which would be the slowest? Justify your answer using K_{ow}.

5.39. Determine the retardation factor for
   a. MEK   b. Naphthalene
   c. Hexachlorobenzene
for a soil-water system in which the bulk density = 1.8 g/cm^3, the porosity = 0.36, and the soil organic carbon content = 2.0%.

5.40. Repeat Problem 5.39 using a soil organic carbon content of 0.05%. Assume that the soil organic carbon is the primary sorbent.

5.41. Using vinyl chloride (low K_{ow}) and hexachlorobenzene (high K_{ow}), and the two organic carbon contents of the soils (2.0% and 0.05%), make a quantitative judgment of which variable (K_{ow} or f_{oc}) asserts the greatest influence on the retardation factor.

5.42. Using the retardation factor, estimate the time for DDT to reach a well 2 km (1.24 miles) away if the pore-water velocity = 0.1 m/day, the soil bulk density = 1.2 g/cm^3, $n = 0.45$, and the average soil organic carbon content = 0.01%. Assume that soil organic carbon is the primary sorbent and that degradation and dispersion are negligible.

5.43. A groundwater contaminated with 16 mg/L nickel also contains 240 mg/L suspended solids and 8.1 mg/L total organic carbon. The pore-water velocity is 55 cm/day, and the subsurface solids consist of 60% sand, 20% clay, and 1.4% free iron oxides. Based on the information given, develop a plot of nickel concentration as a function of time at a depth of 5 m (16.4 ft).

5.44. A groundwater system contains 10 mg/L Zn, and the pore-water velocity is 30 cm/day. The waste also contains 1200 mg/L suspended solids and 110 mg/L total organic carbon. The aquifer solids consist of 80% sand, 10% silt, and 2% free iron oxides. Using the procedure of Amoozegar-Fard et al., determine the time required for the aquifer solution of 1 mg/L to reach a well 1000 m (0.6 miles) downgradient.

5.45. A landfill leachate that has migrated to groundwater contains 5 mg/L Cd, 300 mg/L total organic carbon, and 550 mg/L suspended solids. The aquifer solids consist of 60% sand, 20% clay, and 1.5% free iron oxides. The pore-water velocity is 12 cm/day. How long will it take for a 3 mg/L concentration of Cd to reach a well 200 m (656 ft) away?

5.46. Read Reference 5.5. Write a two-page summary of the mechanistic models used to describe sorption.

5.47. Review Reference 5.61. Propose how you would design a laboratory study to quantify the migration rate of lead at a specific hazardous waste site.

REFERENCES

5.1. Paterson, S. and D. Mackay, "The fugacity concept in environmental modelling," in Hutzinger, O. (Ed.), *The Handbook of Environmental Chemistry: Reactions and Processes*, Springer-Verlag, Berlin, 1985.

5.2. Ogram, A V., R. E. Jessup, L. T. Ou, and P. S. C. Rao, "Effects of sorption on biological degradation of (2,4-dichlorophenoxy) acetic acid in soils," *Appl. Environ. Microbiol.*, **49**, 582–587 (1985).

5.3. Sedlak, D. L. and A. W. Andren, "Aqueous-phase oxidation of polychlorinated biphenyls by hydroxyl radicals," *Environ. Sci. Technol.*, **25**, 1419–1427 (1990).

5.4. Weber, J. B., "Properties and behavior of pesticides in soil," in Honeycutt, R. and D. Schabaker (Eds.), *Advances in Understanding the Mechanisms of Movement of Pesticides into Groundwater*, Lewis Publishers, Chelsea, MI, 1993.

5.5. Weber, W. J., Jr., P. M. McGinley, and L. E. Katz, "Sorption phenomena in subsurface systems: Concepts, models and effects on contaminant fate and transport," *Water Res.*, **25**, 499–528 (1991).

5.6. Schwarzenbach, R. P., P. M. Gschwend, and D. M. Imboden, *Environmental Organic Chemistry*, John Wiley & Sons, New York, 1993.

5.7. Chiou, C. T., L. J. Peters, and V. H. Freed, "A physical concept of soil-water equilibria for nonionic compounds," *Science (Wash. DC)*, **206**, 831–832 (1979).

5.8. Van Oss, C. J., D. R. Alsolom, and A. W. Neumann, "The hydrophobic effect: Essentially a Van der Waal's interaction," *Colliod. & Polym. Sci.*, **258**, 424–427 (1980).

5.9. Israelachvili, J. N., *Intermolecular and Surface Forces*, Academic Press, London, 1985.

5.10. Isaacson, P. J. and C. R. Frink, "Nonreversible sorption of phenolic compounds by sediment fractions: the role of sediment organic matter," *Environ. Sci. Technol.*, **18**, 43–48 (1984).

5.11. Pignatello, J. J., "Slowly reversible sorption of aliphatic halocarbons in soils. I. Formation of residual fractions," *Environ. Toxicol. Chem.*, **9**, 1107–1115 (1990).

5.12. Smith, A. E., "Persistence and transformation of the herbicides [14]C Fenoxaprop-ethyl and [14]C Fenthiaprop-ethyl in two prairie soils under laboratory and field conditions," *J. Agric. Food Chem.*, **33**, 483–488 (1985).

5.13. Steinberg, S. M., J. J. Pignatello, and B. L. Sawhney, "Persistence of 1,2-dibromoethane in soils: Entrapment in intraparticle micropores," *Environ. Sci. Technol.*, **21**, 1201–1208 (1987).

5.14. Bollag, J.-M., R. D. Minard, and S.-Y. Liu, "Cross-linkage between anilines and phenolic humus constituents," *Environ. Sci. Technol.*, **17**, 72–80 (1983).

5.15. Hatcher, P. G., J. M. Bortiatynski, R. D. Minard, J. Dec, and J.-M. Bollag, "Use of high resolution ^{13}C NMR to examine the enzymatic covalent binding of ^{13}C-labeled 2,4-dichlorophenol to humic substances," *Environ. Sci. Tech.*, **27**, 2098–2103 (1993).

5.16. Parks, G. A, "The isoelectric points of solid oxides, solid hydroxides, and aqueous hydroxo complex systems," *Chem. Rev.*, **65**, 177–198 (1965).

5.17. Brady, N. C., *The Nature and Properties of Soils*, 8th Edition, Macmillan, New York, 1974.

5.18. Sposito, G., *The Surface Chemistry of Soils*, Oxford University Press, New York, 1984.

5.19. Pettyjohn, W. A., D. C. Kent, T. A. Prickett, H. E. LeGrand, and F. E. Witz, *Methods for the Prediction of Leachate Plume Migration and Mixing*, U.S. EPA Municipal Environmental Research Laboratory, Cincinnati, OH, 1982.

5.20. Schnitzer, M. and S. U. Kahn, *Humic Substances in the Environment*, Marcel Dekker, New York, 1972.

5.21. Lyon, T. L., H. O. Buckman, and N. C. Brady, *The Nature and Properties of Soils*, Macmillan, New York, 1952.

5.22. Banerjee, S., "Solubility of organic mixtures in water," *Environ. Sci. Technol.*, **18**, 587–591 (1984).

5.23. Veith, G. D., N. M. Austin, and R. T. Morris, "A rapid method for estimating log P for organic chemicals," *Water Res.*, **13**, 43–47 (1979).

5.24. Veith, G. D., D. L. DeFoe, and B. V. Bergstedt, "Measuring and estimating the bioconcentration factor of chemicals in fish," *J. Fish Res. Board Can.*, **36**, 1040–1048 (1976).

5.25. Leo, A., C. Hansch, and D. Elkins, "Partition coefficients and their uses," *Chem. Rev.*, **71**, 525–553 (1971).

5.26. Karickhoff, S. W., D. S. Brown, and T. A. Scott, "Sorption of hydrophobic pollutants on natural sediments," *Water Res.*, **13**, 241–248 (1979).

5.27. Hansch, C. and A. Leo, *Substituent Constants for Correlation Analysis in Chemistry and Biology*, John Wiley & Sons, New York, 1979.

5.28. Lyman, W. J., W. F. Rheel, and D. H. Rosenblatt, *Handbook of Chemical Property Estimation Methods*, McGraw-Hill, New York, 1982.

5.29. Lion, L. W., T. B. Stauffer, and W. G. MacIntyre, "Sorption of hydrophobic compounds on aquifer materials: Analysis methods and the effect of organic carbon," *J. Contam. Hydrol.*, **5**, 215–234 (1990).

5.30. Winkelmann, D. A. and S. J. Klaine, "Degradation and bound residue formation of atrazine in a western Tennessee soil," *Environ. Toxicol. Chem.*, **10**, 335–345 (1991).

5.31. Karickhoff, S. W., "Organic pollutant sorption in aquatic system," *J. Hydraulic Eng.*, **110**, 707–735 (1984).

5.32. Stauffer, T. B., W. C. MacIntyre, and D. C. Wickman, "Sorption of nonpolar organic chemicals on low-carbon-content aquifer materials," *Environ. Toxicol. Chem.*, **8**, 845–852 (1989).

5.33. Lagrega, M. D., P. L. Buckingham, and J. C. Evans, *Hazardous Waste Management*, McGraw-Hill, New York, 1994.

5.34. Dragun, J. D., *The Soil Chemistry of Hazardous Materials*, Hazardous Materials Control Research Institute, Silver Spring, MD, 1988.

5.35. McCarty, P. L., M. Reinhard, and B. E. Rittmann, "Trace organics in groundwater," *Environ. Sci. Technol.*, **15**, 40–51 (1981).

5.36. Rao, P. S. C. and J. M. Davidson, "Estimation of pesticide retention and transformation parameters required in nonpoint source pollution models," in Overcash, M. R., and J. M. Davidson (Eds.), *Environmental Impact of Nonpoint Source Pollution*, Ann Arbor Science Publishers, Ann Arbor, MI, 1980.

5.37. Briggs, G. G., "A simple relationship between soil adsorption of organic chemicals and their octanol/water partition coefficients," *Proc. 7th British Insecticide and Fungicide Conf.*, Vol. 1, The Boots Company Ltd., Nottingham, UK (1973).

5.38. Briggs, G. G., "Adsorption of pesticides by some Australian soils," *Aust. J. Soil Res.*, **19**, 61–68 (1981).

5.39. Kenaga, E. E. and C. A. I. Goring, "Relationship between water solubility, soil-sorption, octanol-water partitioning, and bioconcentration of chemicals in biota," in *Aquatic Toxicology*, ASTM STP 707, American Society for Testing and Materials, Philadelphia, PA, 1980.

5.40. Schwartzenbach, R. P. and J. Westall, "Transport of nonpolar organic compounds from surface water to groundwater," *Environ. Sci. Technol.*, **15**, 1360–1367 (1981).

5.41. Karickhoff, S. W., "Semi-empirical estimation of sorption of hydrophobic pollutants on natural sediments and soils," *Chemosphere*, **10**, 833–846 (1981).

5.42. Karickhoff, S. W., D. S. Brown, and T. A. Scott, "Sorption of hydrophobic pollutants on natural sediments," *Water Res.*, **13**, 241–248 (1979).

5.43. Means, J. C., S. G. Wood, J. J. Hassett, and W. L. Banwart, "Sorption of polynuclear aromatic hydrocarbons by sediments and soils," *Environ. Sci. & Technol.*, **14**, 1524–1528 (1980).

5.44. Liao, Y.-Y., Z.-T. Wang, J.-W. Chen, S.-K. Han, L.-S. Wang, G.-Y. Lu, and T.-N. Zhao, "The prediction of soil sorption coefficients of heterocyclic nitrogen compounds by octanol/water partition coefficient, water solubility, and by molecular connectivity indices," *Bull. Environ. Contam. Toxicol.*, **56**, 711–716 (1996).

5.45. Hong, H., L. Wang, S. Han, Z. Zhang, and G. Zou, "Prediction of soil adsorption coefficient K_{oc} for phenylthio, phenylsulfinyl, and phenylsulfonyl acetates," *Chemosphere*, **34**, 827–834 (1997).

5.46. Garbarini, D. R. and L. W. Lion, "The influence of the nature of soil organics on the sorption of toluene and trichloroethylene," *Environ. Sci. Technol.*, **20**, 1263–1269 (1986).

5.47. Santillan-Medrano, J. and J. J. Jurinak, "The chemistry of lead and cadmium in soil: Solid phase formation," *Soil Sci. Soc. Am. Proc.*, **39**, 851–856 (1975).

5.48. Fuller, W. H., *Investigation of Landfill Leachate Pollutant Attenuation by Soils*, U.S. EPA Municipal Environmental Research Laboratoy, Cincinnati, OH, 1978.

5.49. Baes, C. F., III and R. D. Sharp, "A proposal for estimation of soil leaching and leaching constants for use in assessment models," *J. Environ. Qual.*, **12**, 17–28 (1983).

5.50. Mattigod, S. V., G. Sposito, and A. L. Page, "Factors affecting the solubilities of trace metals in soils," in *Chemistry in the Environment*, ASA Special Publication No. 40, American Society of Agronomy, Soil Science Society of America (1981).

5.51. Stumm, W. and J. J. Morgan, *Aquatic Chemistry: An Introduction Emphasizing Chemical Equilibria in Natural Waters*, John Wiley & Sons, New York, 1981.

5.52. Alloway, B. J., *Heavy Metals in Soils*, John Wiley & Sons, New York, 1990.

5.53. Bittel, J. E. and R. J. Miller, "Lead, cadmium, and calcium selectivity coefficients on a montmorillonite, illite, and kaolinite," *J. Environ. Qual.*, **3**, 250–253 (1974).

5.54. Puls, R. W. and H. L. Bohn, "Sorption of cadmium, nickel, and zinc by kaolinite and montmorillonite suspensions," *Soil Sci. Soc. Am. J.*, **52**, 1289–1292 (1988).

5.55. Kinniburgh, D. G., M. L. Jackson, and J. K. Syers, "Adsorption of alkaline-earth, transition, and heavy metal cations by hydrous oxide gels of iron and aluminum," *Soil Sci. Soc. Am. J.*, **40**, 769–799 (1976).

5.56. McKenzie, R. M., "Adsorption of lead and other heavy metals on oxides of manganese and iron," *Aust. J. Soil Res.*, **18**, 61–73 (1980).

5.57. Forbes, E. A., A. M. Posner, and J. P. Quirk, "Specific adsorption of divalent Cd, Co, Cu, Pb, and Zn on goethite," *J. Soil Sci.*, **27**, 154–166 (1976).

5.58. Bunzl, K., W. Schmidt, and B. Sansoni, "Kinetics of ion exchange in soil organic matter. 4. Adsorption and desorption of Pb^{2+}, Cu^{2+}, Cd^{2+}, Zn^{2+}, and Ca^{2+} by peat," *J. Soil Sci.*, **27**, 32–41 (1976).

5.59. Hogg, T. J., J. W. B. Stewart, and J. R. Bettany, "Influence of chemical form of mercury on its adsorption and ability to leach through soils," *J. Environ. Qual.*, **7**, 440–445 (1978).

5.60. Bowman, R. S., M. E. Essington, and G. A. O'Connor, "Soil sorption of nickel: Influence of solution composition," *Soil Sci. Soc. Am. J.*, **45**, 860–865 (1981).

5.61. Amoozegar-Fard, A., W. H. Fuller, and A. W. Warrick, "An approach to predicting the movement of selected polluting metals in soils," *J. Environ. Qual.*, **13**, 290–297 (1984).

Chapter **6**

Volatilization

Volatilization, or evaporation, is the transfer of chemicals from solids or liquids to the gaseous phase. Depending on the contaminant and site characteristics, volatilization can be an important mechanism for the loss of hazardous compounds from soils and liquid waste systems. Volatilization may decrease the concentration of the wastes at a facility or site; however, because of the law of mass conservation, contaminants will subsequently be found in the atmosphere. Therefore, air emissions from contaminated sites and RCRA hazardous waste management facilities, which are regulated under the Clean Air Act, may become hazardous air pollutants (HAPs) and result in short- or long-term health effects. Volatilization can also be applied as a hazardous waste treatment process to clean up contaminated groundwater through air stripping and to remediate soils by soil vapor extraction. The fundamental principles covered in this chapter serve as a basis for process selection and the design of treatment systems that promote volatilization. These design applications will be introduced in Chapters 12 and 13.

If volatilization occurs, it will do so primarily as a function of the contaminant's vapor pressure (if the compound is in relatively pure form) and the Henry's Law constant (if it is in aqueous solution). In this chapter, these two physical properties will first be described before three source release volatilization calculations are presented: (1) volatilization from an open container; (2) volatilization from a soil surface, and (3) volatilization from deep soil contamination. These calculations serve as a basis for the models covered in Chapter 8 that describe the path of volatile chemicals in the atmosphere after they are released from the source.

6.1   THE GOVERNING VARIABLES: VAPOR PRESSURE AND HENRY'S LAW

Vapor Pressure. Volatility of a pure compound is a function of its vapor pressure; conceptually, *vapor pressure* may be thought of as the pressure exerted by a chemical on the atmosphere. Compounds with higher vapor pressures exert more pressure against the atmosphere and, as a result, an increased driving force for volatilization. Vapor pressures of organic compounds range from 10^{-10} mm Hg to 760 mm Hg at 20°C.

Vapor pressure increases with temperature; the temperature at which a compound's vapor pressure reaches atmospheric pressure (760 mm Hg) is the boiling point of the compound. The relationship between vapor pressure and temperature is described by the Clausius-Claperon equation:

$$\ln VP = C - \frac{\Delta H_v}{R} \cdot \frac{1}{T} \tag{6.1}$$

where

   VP = vapor pressure (atm)

   C = constant

   ΔH_v = heat of vaporization $\left(\dfrac{J}{mole}\right)$

   R = gas constant $\left(8.314 \dfrac{J}{mole\text{-}°K}\right)$

   T = temperature (°K)

Equation 6.1 represents fundamental thermodynamic relationships related to vapor pressure but has limited practical significance. Instead, vapor pressures used in hazardous waste applications are usually obtained from tabulated values at specific temperatures.

Henry's Law. In a closed aqueous system containing a dilute contaminant, an equilibrium exists between the concentration in solution and the concentration in the overlying gaseous phase. *Henry's Law* states that the concentration of a compound in the aqueous phase is directly proportional to its partial pressure in the gaseous phase:

$$P = H \cdot X \tag{6.2}$$

where

   P = partial pressure (atm)
   H = Henry's Law constant (atm-m^3/mole)
   X = concentration of the compound in water (mole/m^3)

   The *Henry's Law constant* may also be considered a partition coefficient between air and water, just as the octanol-water partition coefficient describes the partitioning of a chemical between aqueous and nonpolar phases. As a rational basis for this partitioning, the Henry's Law constant may be written

$$H = \frac{VP}{S} \tag{6.3}$$

where

   VP = vapor pressure (atm)
   S = water solubility (mole/m^3)
   H = Henry's Law constant (atm-m^3/mole)

High water solubilities and low vapor pressures tend to decrease the potential for volatilization of dilute species from water. However, compounds with low vapor pressure may also have a high tendency to escape if their water solubility is low. For example, the insecticide DDT, which has a low vapor pressure of 1.9×10^{-7} mm Hg, has a moderate tendency to volatilize because of its low water solubility of 0.0017 mg/L. Conversely, methanol, which is highly volatile as a pure compound, has a lower tendency to evaporate from an aqueous solution because of its infinite water solubility.

The Henry's Law constant is the best indicator of the tendency of a chemical to volatilize from water. The relative volatilities of Henry's Law constants within different ranges of values are shown in Figure 6.1. A Henry's Law constant $\leq 10^{-7}$ indicates that a compound is essentially nonvolatile. In fact, some compounds with $H < 10^{-7}$ (e.g., dieldrin) will become more concentrated because the water evaporates more rapidly than the contaminant.

Henry's Law constants are influenced significantly by temperature. Therefore, empirical relationships have been developed to estimate Henry's Law constants as a function of temperature:

$$H = \exp\left(A - \frac{B}{T}\right) \qquad (6.4)$$

where

H = Henry's Law constant (atm-m^3/mol)
T = temperature (°K)
A, B = empirical constants

The empirical constants A and B describe the slope of the exponential increase in H as a function of temperature.

Henry's Law constants may also be expressed as a dimensionless quantity, which is used in a number of applications, such as two-film calculations and air stripper designs. The dimensionless Henry's Law constant, H', may be calculated as follows:

$$H' = \frac{H}{RT} \qquad (6.5)$$

where

H' = dimensionless Henry's Law constant
R = universal gas constant (8.2×10^{-5} m^3-atm/mole-°K)
T = temperature (°K)

Vapor pressures and Henry's Law constants for common hazardous chemicals are listed in Table 6.1 and more extensive listings are provided in Appendix J. Values of A and B for common hazardous compounds are listed in Table 6.2.

6.2 VOLATILIZATION FROM OPEN CONTAINERS

Volatilization is an important pathway for some hazardous chemicals stored in drums, vats, surface impoundments, and other liquid systems open to the atmosphere. Open receptacles are sometimes used to store spent solvents, pesticides, and industrial intermediates prior to their disposal or recycling under RCRA. A series of such receptacles

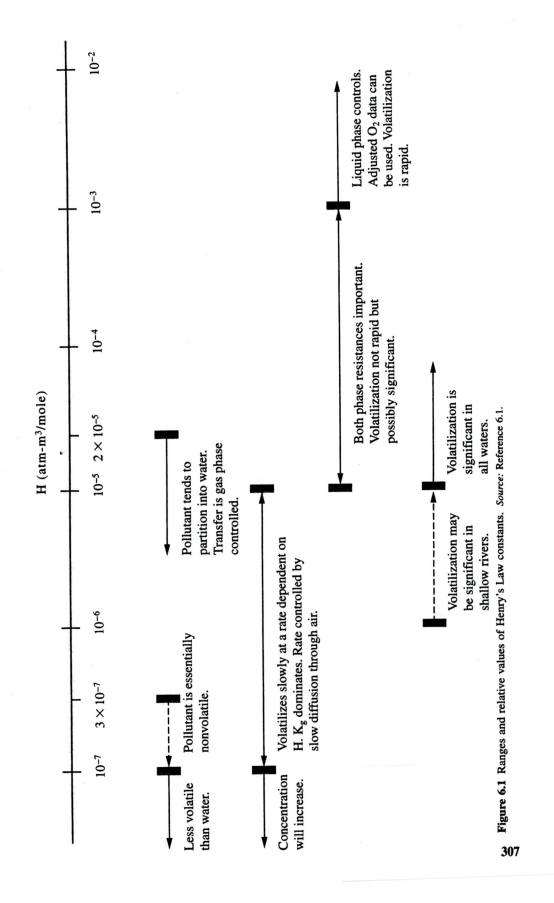

Figure 6.1 Ranges and relative values of Henry's Law constants. *Source:* Reference 6.1.

Table 6.1 Vapor Pressures and Henry's Law Constants for Common Hazardous Chemicals

Compound	Vapor Pressure (mm Hg) @ 20°C	Henry's Law Constant (atm-m^3/mole) @ 25°C
Aliphatic Hydrocarbons		
Cyclohexane	95	0.194
Cyclohexene	67	0.046
n-Decane	2.7	0.187
n-Heptane	40	2.04
Isooctane	40	3.01
n-Octane	10.4	3.23
Monocyclic Aromatic Hydrocarbons		
Benzene	76	5.48×10^{-3}
Ethylbenzene	7.08	8.68×10^{-3}
Toluene	22	6.74×10^{-3}
Polycyclic Aromatic Hydrocarbons		
Anthracene	1.7×10^{-5}	1.77×10^{-5}
Benzo[a]anthracene	2.2×10^{-8}	6.6×10^{-7}
Benzo[a]pyrene	5.0×10^{-7}	$< 2.4 \times 10^{-6}$
Chrysene	6.3×10^{-7}	7.26×10^{-20}
Fluorene	0.005	2.1×10^{-4}
Naphthalene	0.054	4.6×10^{-4}
Phenanthrene	2.1×10^{-4}	2.56×10^{-4}
Pyrene	2.5×10^{-6} @ 25°C	1.87×10^{-5}
Chlorinated Solvents		
Carbon tetrachloride	90	0.0302
Chloroform	160	3.2×10^{-3}
trans-1,2-Dichloroethene	265	0.0067
Hexachloroethane	0.18	2.5×10^{-3}
Methylene chloride	455 @ 25°C	2.69×10^{-3}
Perchloroethylene (PCE)	14	0.0153
1,1,1-Trichloroethane (TCA)	96	0.018
Trichloroethylene (TCE)	58	9.1×10^{-3}
Insecticides		
Aldrin	6×10^{-6}	1.4×10^{-6}
DDT	7.26×10^{-7} @ 30°C	5.20×10^{-5}
Dieldrin	1.78×10^{-7}	5.8×10^{-5}

308

Table 6.1 (Continued)

Compound	Vapor Pressure (mm Hg) @ 20°C	Henry's Law Constant (atm-m^3/mole) @ 25°C
Insecticides (cont.)		
4,6-Dinitro-o-cresol	3.2×10^{-4}	4.3×10^{-4}
Endrin	7×10^{-7}	5.0×10^{-7}
Hexachlorobenzene	1.09×10^{-5}	1.7×10^{-3}
Lindane	9.4×10^{-6}	3.25×10^{-6}
Herbicides		
2,4-D	4.7×10^{-3}	1.95×10^{-2} @20°C
Fungicides		
Pentachlorophenol	1.4×10^{-4}	3.4×10^{-6}
Industrial Intermediates		
Acrylonitrile	83	1.1×10^{-4}
Chlorobenzene	9.0	3.7×10^{-3}
o-Chlorophenol	1.42 @ 25°C	8.28×10^{-6}
Dibutyl phthalate	1×10^{-5} @ 25°C	6.3×10^{-5}
1,2-Dichlorobenzene	1	1.9×10^{-3}
1,3-Dichlorobenzene	2.30 @ 25°C	3.60×10^{-3}
Diethyl phthalate	0.05 @ 70°C	8.46×10^{-7}
Di-n-octyl phthalate	1.4×10^{-4} @ 25°C	1.41×10^{-12}
Hexachlorocyclopentadiene	0.081 @ 25°C	0.016
p-Nitrophenol	1×10^{-4}	3×10^{-5} @ 20°C
Phenol	0.2	3.97×10^{-7}
1,2,4-Trichlorobenzene	0.29 @ 25°C	2.32×10^{-3}
2,4,6-Trichlorophenol	0.017 @ 25°C	9.07×10^{-8}
Polychlorinated Biphenyls		
Aroclor 1016	4×10^{-4}	3.3×10^{-4}
Aroclor 1221	6.7×10^{-3}	3.24×10^{-4}
Aroclor 1232	4.60×10^{-3}	8.64×10^{-4}
Aroclor 1248	4.9×10^{-4} @ 25°C	3.5×10^{-3}
Aroclor 1260	4.1×10^{-5}	7.1×10^{-3}
Chlorinated Dioxins		
2,3,7,8-Tetrachlorodibenzo-p-dioxin (TCDD)	6.4×10^{-10}	5.40×10^{-23} @ 18–22°C

Source: Appendix J.

Table 6.2 Tabulated Empirical Constants for Use in Determining Henry's Law Constant for Selected Hazardous Chemicals (Equation 6.4)

Compound	A	B
Benzene	5.53	3190
Carbon tetrachloride	11.3	4410
Chlorobenzene	3.47	2690
Chlorodibromomethane	5.97	3120
Chloroform	9.84	4610
Chloromethane	9.36	4210
Cyclohexane	9.14	3240
Decalin	11.8	4130
Dibromochloromethane	14.6	6370
1,3-Dichlorobenzene	2.88	2560
1,4-Dichlorobenzene	3.37	2720
1,1-Dichloroethane	8.64	4130
1,2-Dichloroethane	−1.37	1520
cis-1,2-Dichloroethylene	8.48	4190
trans-1,1-Dichloroethylene	9.34	4180
Dichloromethane	6.65	3820
1,2-Dichloropropane	9.48	4710
Ethylbenzene	6.10	3240
Ethylene dibromide	5.70	3880
Hexachloroethane	3.74	2550
Hexane	2.53	7530
Methyl chloride	5.56	3180
Methylene chloride	8.48	4270
Methyl ethyl ketone	−48.4	2310
Phenol	11.3	4660
Propyl benzene	7.84	3680
Perchloroethylene (PCE)	12.4	4920
Toluene	5.13	3020
1,2,4-Trichlorobenzene	7.36	4030
1,1,1-Trichloroethane (TCA)	9.78	4130
Trichloroethylene (TCE)	11.4	4780
Trichlorofluoromethane	9.48	3510
1,1,2-Trichloro-1,2,2-trifluoroethane	9.65	3240
1,3,5-Trimethylbenzene	7.24	3630
Vinyl chloride	7.39	3290
m-Xylene	6.28	3340
o-Xylene	5.54	3220
p-Xylene	6.93	3520

Source: References 6.2–6.4.

is shown in Figure 6.2. Emissions from these open systems can be important because the resulting concentrations in air may pose a threat to public health; if concentrations in the air reach the percent range, fire and explosion may also occur.

Because most liquid organic hazardous wastes managed under RCRA are high strength, the focus of this section is volatile fluxes across the surfaces of concentrated waste solutions. Volatilization of chemicals from dilute aqueous solutions, such as surface waters, is evaluated using two-film theory and has been described by Mackay and Leinonen [6.5] and Liss and Slater [6.6].

6.2.1   Volatilization Fluxes across Liquid Surfaces

Volatilization rates of high-strength wastes out of an open container are proportional to their vapor pressures; however, the rate of vaporization is more accurately correlated with the difference between the chemical's vapor pressure and its steady state partial pressure; that is, the pressure deficit provides a physical driving force for volatilization:

$$Q \propto (P_{sat} - P) \qquad (6.6)$$

where

Q = the evaporation rate (mass/time)
P_{sat} = VP = the saturation vapor pressure of the liquid at the temperature of the system; i.e., the VP (atm)
P = the partial pressure of the compound above the liquid (atm)

Figure 6.2 Open drums in a storage area—a potential source of toxic air emissions.

Based on Equation 6.6, Hanna and Drivas [6.7] provided a quantitative description of volatilization:

$$Q = \frac{MKA(VP - P)}{RT} \qquad (6.7)$$

where

Q = the evaporation rate (g/sec)
M = contaminant molecular weight (g/mole)
K = a mass transfer coefficient per area A (m/sec)
R = the ideal gas constant (8.21×10^{-5} m^3-atm/mol-°K)
T = temperature (°K)

For most cases, $P = 0$ because the container is open; therefore

$$Q = \frac{MKA \cdot VP}{RT} \qquad (6.8)$$

Equation 6.8 is the working equation for estimating volatilization rates into the open atmosphere.

The mass transfer rate for a given compound may be estimated by

$$K = aD^{2/3} \qquad (6.9)$$

where

a = an empirical constant
D = the gas phase diffusion coefficient (cm^2/sec)

The ratio of two mass transfer coefficients may then be related to their gas diffusion coefficients:

$$\frac{K_1}{K_2} = \left(\frac{D_1}{D_2}\right)^{2/3} \qquad (6.10)$$

where

K_1, K_2 = mass transfer coefficients for compounds 1 and 2
D_1, D_2 = corresponding gas diffusion coefficients

A well-documented method for estimating gas diffusion coefficients uses the following equation

$$\frac{D_1}{D_2} = \sqrt{\frac{M_2}{M_1}} \qquad (6.11)$$

where

M_1, M_2 = corresponding molecular weights

Combining equations 6.10 and 6.11 yields

$$K_1 = K_2\left(\frac{M_2}{M_1}\right)^{1/3} \tag{6.12}$$

Matthiessen [6.8] reported that water ($K = 0.83$ cm/sec) is used most often as a mass transfer coefficient reference.

The use of Equations 6.8 and 6.12 in estimating the flux across the surface of a liquid in an open container is illustrated in Example 6.1.

EXAMPLE 6.1   *Estimation of Volatilization Flux from an Open Container*

An open vat 1.25 m × 0.75 m × 0.3 m deep (4.1 ft × 2.5 ft × 1 ft) is used to store spent MIBK prior to distilling it for reuse. The temperature is 20°C. Estimate the rate of volatilization across the surface of the vat.

SOLUTION

1. Estimate the mass transfer coefficient using $K = 0.83$ cm/sec for water as a reference.
   Equation 6.12 is used to estimate the mass transfer coefficient for MIBK.

$$M \text{ for water} = 18 \text{ g/mole}$$

$$M \text{ for MIBK} = 100 \text{ g/mole}$$

$$K_1 = K_2\left(\frac{M_2}{M_1}\right)^{1/3} = 0.83\left(\frac{18}{100}\right)^{1/3} = 0.47 \text{ cm/sec}$$

$$K_1 = 0.0047 \text{ m/sec}$$

2. Determine the surface area of the vat. The depth of the vat is not important.

$$A = (1.25 \text{ m})(0.75 \text{ m}) = 0.94 \text{ m}^2 \ (10.1 \text{ ft}^2)$$

3. Obtain the vapor pressure from Appendix J. The VP = 15 mm Hg × $\dfrac{\text{atm}}{760 \text{ mm Hg}}$ = 0.020 atm.

4. Find Q using Equation 6.8.

$$Q = \frac{MKA \cdot VP}{RT} = \frac{\left(100 \ \dfrac{\text{g}}{\text{mole}}\right)\left(0.0047 \ \dfrac{\text{m}}{\text{sec}}\right)(0.94 \text{ m}^2)(0.020 \text{ atm})}{\left(8.21 \times 10^{-5} \ \dfrac{\text{m}^3 \text{ atm}}{\text{mole-°K}}\right)(293°K)} = 0.37 \text{ g/sec}$$

6.2.2   Saturation Concentration in an Enclosed Area

In the event of a chemical spill or the improper management of volatile hazardous materials in enclosed areas, such as waste transfer facilities or drum storage areas, estimating the saturated vapor concentration may be necessary to assess potential toxicity or explosivity of the vapor.

The calculation of the steady-state saturation concentration resulting from volatilization is based on a mass balance on the enclosed space (Figure 6.3). The rate of change of the hazardous compound in the system is

$$\frac{VdC}{dt} = Q_m - kQ_vC \tag{6.13}$$

where

C = the vapor phase concentration of the compound (g/m^3)
V = volume of the enclosed area (m^3)
Q_m = volatilization rate of the compound (g/sec)
Q_v = ventilation rate for the enclosed area (m^3/sec)
k = a factor that accounts for incomplete mixing (0.1 to 0.5)

Assuming steady state,

$$\frac{VdC}{dt} = 0 \tag{6.14}$$

Therefore,

$$C = \frac{Q_m}{kQ_v} \tag{6.15}$$

Because C is more commonly expressed as ppm, it may be converted using the methods outlined in Section 3.1:

$$C_{ppm} = C\left(\frac{RT}{PM}\right) \times 10^6 \tag{6.16}$$

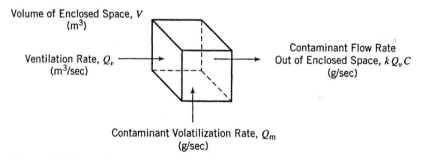

Volume of Enclosed Space, V (m^3)

Ventilation Rate, Q_v (m^3/sec)

Contaminant Flow Rate Out of Enclosed Space, kQ_vC (g/sec)

Contaminant Volatilization Rate, Q_m (g/sec)

Figure 6.3 Mass balance on an area receiving volatile emissions. *Source:* Reference 6.9.

where

R = the ideal gas constant (8.21×10^{-5} m³-atm/mole-°K)
T = temperature (°K)
P = absolute pressure (atm)
M = molecular weight (g/mole)

Combining Equations 6.15 and 6.16 yields

$$C_{ppm} = \frac{Q_m RT}{kQ_v PM} \times 10^6 \tag{6.17}$$

The use of Equation 6.17 in estimating the steady-state waste concentration in the air of an indoor facility is demonstrated in Example 6.2.

EXAMPLE 6.2 *Steady-State Concentration of Volatile Contaminants in an Enclosed Area*

A storage tank 1.5 m (4.9 ft) in diameter containing waste toluene has been left open in a small RCRA 90-day drum storage area 220 m³ (7770 ft³) in volume. The temperature is 20°C and the atmospheric pressure is 1 atm. Ventilation in the structure provides 12 changes of air per hour. Determine the steady-state toluene concentration in the structure. Assume $k = 0.2$.

SOLUTION

1. Find the volatilization rate of toluene.
   Using $K = 0.83$ cm/sec for water, estimate the mass transfer coefficient for toluene with Equation 6.12:

   M for water = 18 g/mole

   M for toluene = 92 g/mole

   $$K_1 = K_2 \left(\frac{M_2}{M_1}\right)^{1/3} = 0.83 \left(\frac{18}{92}\right)^{1/3} = 0.48 \text{ cm/sec}$$

   $K_1 = 0.0048$ m/sec

Determine the surface area of the tank.

$$A = \pi r^2 = \pi (0.75 \text{ m})^2 = 1.77 \text{ m}^2 \ (19.1 \text{ ft}^2)$$

Obtain the toluene vapor pressure from Table 6.1: VP = 22 mm Hg = 0.029 atm. Find Q_m using Equation 6.8:

$$Q_m = \frac{MKA \cdot VP}{RT} = \frac{(92 \text{ g/mole})(0.0048 \text{ m/s})(1.77 \text{ m}^2)(0.029 \text{ atm})}{\left(8.21 \times 10^{-5} \dfrac{\text{m}^3 \cdot \text{atm}}{\text{mole-}^\circ \text{K}}\right)(293^\circ \text{K})} = 0.94 \text{ g/sec}$$

2. Calculate the ventilation rate.

$$Q_v = (12 \text{ changes of air/hr}) \ (220 \text{ m}^3/\text{change of air}) \ (1 \text{ hr}/3600 \text{ sec}) = 0.73 \text{ m}^3/\text{sec}$$

3. Determine steady-state toluene concentration using Equation 6.17.

$$C_{ppm} = \frac{Q_m RT}{kQ_v PM} \times 10^6 = \frac{(0.94 \text{ g/sec})\left(8.21 \times 10^{-5} \dfrac{\text{m}^3 \cdot \text{atm}}{\text{mole-}^\circ \text{K}}\right)(293^\circ \text{K})}{(0.2)(0.73 \text{ m}^3/\text{sec})(1 \text{ atm})(92 \text{ g/mole})} \times 10^6 = 1680 \text{ ppm}$$

6.3   VOLATILIZATION FROM SOILS

Volatilization from soils is a complicated process, and a number of environmental factors and chemical properties influences the rate at which it occurs. Several equations and models have been developed to quantify volatilization from soils, some of which are relatively simple and some complex. The accuracy of these models is often low, which led Neely [6.10] to comment that attempting to predict volatile emissions from soils is futile. Furthermore, there appears to be no correlation between the complexity of the model and its accuracy. Some of the more eloquent models developed from partial differential equations are sometimes no more accurate than simple equations. The numerous variables that influence volatilization are, no doubt, a primary reason for the lack of accuracy. Even though soil volatilization calculations are inherently prone to error and in many cases yield only order-of-magnitude estimations of volatilization rates, they can still (1) provide a rational basis for site assessment and exposure decisions, (2) provide a conceptual basis for the importance of volatilization in site and facility assessments, and (3) promote a fundamental understanding of volatilization to aid in the design of processes such as soil vapor extraction (SVE) systems.

Volatilization from soils is classified into two areas: (1) from surface contamination and (2) from deeply contaminated soils. Obviously, contaminants that volatilize from the soil surface are not affected by diffusion through what can be tens of meters of soil pore space; as a result, volatilization occurs rapidly from the surface. Diffusion controls the volatilization of contaminants deep below the surface; therefore, its rate can be orders of magnitude lower than the volatilization of contaminants on the surface.

6.3.1   Environmental and Chemical Properties Affecting Volatilization from Soils

A knowledge of the distribution of contaminants in soils serves as the basis for understanding volatilization. As shown in Figure 6.4, contaminants in soils can partition between the soil-air, soil-water, and soil-solids (e.g., organic matter). (Note that the soil-air—the air in the soil voids—represents a microclimate within the soil and is very different from the ambient atmosphere above the soil). Because most contaminants are sorbed to the soil-solids, desorption is often the first mechanism in a series of processes that take place before contaminants volatilize to the ambient air. Most contaminants

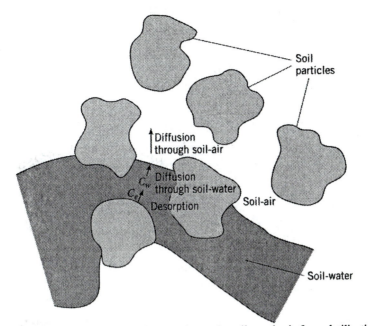

Figure 6.4 Distribution of contaminants in soils—a basis for volatilization.

desorb into the soil-water as the first step in volatilization (Figure 6.4). Subsequently, they diffuse from the soil-water to the soil-air and then diffuse through the soil-air to the ambient atmosphere. From this conceptual model of soil volatilization, it is obvious that the rate of desorption and the soil volumetric water content have a significant effect on the dynamics of volatilization of contaminants.

Variables inherent in soil volatilization include sorption, the soil-water content, diffusion, temperature, and wind. A brief description of each of these variables is provided next.

Sorption. As described in Chapter 5, sorption often controls a number of other processes such as subsurface transport, abiotic transformations, biotic transformations, and toxicity. In a manner similar to its influences on these other pathways, a model for the effects of soil sorption on volatilization may also be established. Based on the principles of Chapter 5, a sorbed compound cannot migrate to the atmosphere. Only through desorption can a contaminant migrate into the soil-water, then the soil-air, and eventually to the atmosphere. Not surprisingly, contaminants that are strongly sorbed have minimal potential for volatilization. As a result, vapor phase contaminant concentrations have been related to the surface area and organic carbon content of soils—two soil characteristics that correlate with sorptivity [6.11].

Soil Water Content. The volumetric water content of a soil can have a significant effect on contaminant volatilization. Most nonpolar and slightly polar contaminants can be displaced from the soil because water will partition preferentially to sorption sites. Therefore, contaminant volatilization is enhanced by the presence of water. Moreover, the presence of water provides a medium into which contaminants can desorb.

A common incorrect conceptualization of water evaporation is that when the water evaporates, it carries the contaminants with it. Instead, as the water evaporates it leaves the soil drier with more sorption sites available for contaminants.

Diffusion. Diffusion through the soil pore space is usually the rate-controlling process for contaminant volatilization [6.12]. Although volatilization of contaminants of even relatively low vapor pressures may occur rapidly at the soil surface, volatilization from deeply contaminated soils occurs slowly due to the low rates of diffusion. A common occurrence in the field is to scrape a few centimeters of soil from a contaminated area to find that it is dry but that staining is present below the surface layer because diffusion limits contaminant volatilization from these deeper sections (Figure 6.5).

Diffusion in soils can be nonvapor phase (i.e., liquid diffusion) or vapor phase (gaseous diffusion). *Liquid phase diffusion* is predominant in moist soils because contaminants must move through the water layers before they enter the air within the voids; this movement through the soil-water is often the rate-limiting process in contaminant volatilization.

Liquid phase diffusion coefficients are significantly lower than their gaseous phase counterparts. Liquid phase diffusion occurs at a rate of a few cm/month, whereas gaseous diffusion coefficients are approximately 10^4 times greater. Obviously, the volumetric water content of the soil is a controlling variable in determining whether liquid or gas phase diffusion predominates. The second variable of importance is bulk

Figure 6.5 Contaminant staining approximately 20 cm below the soil surface, a sign that diffusion limits volatilization in deeper soil horizons.

density (and the related parameter of porosity): Gaseous diffusion through tightly packed soils becomes more important when more pore space is available.

Although *vapor phase diffusion* rates are significantly greater than those in the liquid phase, diffusion via the vapor phase does not always predominate in soils. Depending on the volumetric water content and porosity of the soil, either liquid phase or gaseous phase diffusion may predominate. Because vapor phase diffusion rates are about 10^4 greater than those in the liquid phase, rates of liquid and gaseous diffusion are approximately equal when the contaminant partitions between the liquid and gaseous phases at a ratio of $10^4 : 1$.

Also related to the effect of water on volatilization rates of organic contaminants is a mechanism known as the *wick effect* or *wick evaporation*, in which the contaminant is transported to the surface by capillary action. In many cases, contaminants are carried along with the water, which increases their rate of volatilization. The concentration gradient resulting from volatilization then promotes continued migration to the surface. Note that the wick effect is not the same as evaporating out of solution with the water. In the wicking process, water only enhances the transport of the contaminants in the soil column; it does not enhance their volatilization rates out of the soil.

Temperature. When considering wide temperature ranges (i.e., 0–100°C), hazardous contaminants tend to volatilize more rapidly at higher temperatures. However, subsurface soils are not characterized by wide temperature differences, and the high temperatures of summer have little effect on rates of subsurface contaminant volatilization. Increasing the ambient temperature of soils to desorb and volatilize contaminants is the basis for the soil remediation process known as *thermal desorption* (see Chapter 12). From a more fundamental view, the mechanism of temperature effects is an increase in the vapor density, which then increases the potential flux out of the soil. In some cases, however, the variables affecting volatilization are interactive, resulting in changes that may be difficult to predict over narrow temperature changes [6.13].

Wind and Atmospheric Turbulence. High air flows at the soil surface increase diffusion in the ambient atmosphere, creating higher concentration gradients and more rapid rates of volatilization from the soil. However, atmospheric conditions affect only gradients near the soil surface; deeply contaminated soils and groundwater are not significantly affected by atmospheric conditions.

6.3.2   Volatilization from Soil Surfaces

Hazardous material spills or the spreading of hydrophobic hazardous chemicals on soils or surfaces is a problem commonly encountered in hazardous waste management. The rate of volatilization of contaminants from soil surfaces, quantified by researchers at Dow Chemical and described by Thomas [6.1], was found to follow first-order kinetics in which the rate of change of the compound is proportional to its concentration at time *t*:

$$-\frac{dC}{dt} \propto C \tag{6.18}$$

where C = the contaminant concentration.

Equation 6.18 can be converted to an equality by adding a proportionality constant k:

$$-\frac{dC}{dt} = kC \qquad (6.19)$$

After separating variables, the expression may be integrated:

$$-\int_{C_0}^{C_t} \frac{dC}{C} = k\int_0^t dt \qquad (6.20)$$

which yields

$$\ln\frac{C_t}{C_0} = -kt \qquad (6.21)$$

or

$$C_t = C_0 \cdot e^{-kt} \qquad (6.22)$$

Because ln $1/2 = 0.693$ the half life of a compound lost through first-order kinetics is

$$t_{\frac{1}{2}} = \frac{0.693}{k} \qquad (6.23)$$

The Dow equation for the volatilization of surface contaminants is based on their first-order loss from soils. In studying numerous environmental and chemical properties that may affect volatilization in a surface layer, Dow researchers found that volatilization rates increased with vapor pressure and decreased as a function of K_{oc} and water solubility. By empirical curve fitting, the first-order rate constant for contaminant loss from a surface spill was described by

$$k_v = 4.4 \times 10^7 \frac{\text{VP}}{K_{oc} \cdot S} \qquad (6.24)$$

where

k_v = volatilization rate constant (day^{-1})
VP = contaminant vapor pressure (mm Hg)
K_{oc} = soil adsorption coefficient (mL/g)
S = contaminant water solubility (mg/L)

Inspection of Equation 6.24 shows that the first-order volatilization rate is inversely proportional to K_{oc}, which confirms the theme emphasized in Chapter 5—that sorption controls many other dynamic processes in hazardous waste systems, such as volatilization as well as transformations reactions (see Chapter 7). A comparison of results between Equation 6.24 and laboratory data showed good agreement [6.1]. For example, the calculated volatilization half-life for carbofuran was 33 days compared to 24 days measured in the laboratory. The laboratory volatilization half-life for atrazine was 45 days compared to 69 days based on Equation 6.24. The use of Equation 6.24 is illustrated in Example 6.3.

EXAMPLE 6.3   *Estimation of Contaminant Volatilization Rates from a Soil Surface*

A surface spill of endrin has occurred on a soil. Estimate the volatilization half-life for endrin in the soil and the time required for 99.9% volatilization.

SOLUTION

1. The data needed for use of the Dow equation include

$$VP = 7 \times 10^{-7} \text{ mm Hg (Table 6.1)}$$
$$S = 0.24 \text{ mg/L (Table 3.1)}$$
$$\log K_{ow} = 5.02 \text{ (Appendix G)}$$

The mean K_{oc} is obtained by using Equations 5.14 and 5.18, (both of which were developed using organochlorine insecticides):

Equation 5.14: $\log K_{oc} = 1.029 (5.02) - 0.18$
$\log K_{oc} = 4.99$

Equation 5.18: $\log K_{oc} = 0.544 (5.02) + 1.377$
$\log K_{oc} = 4.11$
Average $\log K_{oc} = 4.55$
$K_{oc} = 10^{4.55} = 35,500$

2. Determine the first-order rate constant for endrin volatilization using Equation 6.24:

$$k_v = 4.4 \times 10^7 \frac{7 \times 10^{-7}}{35,500 \cdot 0.24}$$

$$k_v = 0.0036 \text{ day}^{-1}$$

3. Calculate the half-life for endrin volatilization:

$$t_{1/2} = \frac{0.693}{0.0036 \text{ day}^{-1}} = 193 \text{ days}$$

4. Determine the time required for 99.9% loss by volatilization:

$$\ln \frac{1}{1000} = -kt$$

$$\ln 0.001 = -0.0036 \, t$$

$$t = 1920 \text{ days} = 5.26 \text{ years}$$

6.3.3   Volatilization from Deep Soil Contamination

Numerous models have been developed to estimate the rate of volatilization from deeply contaminated soils. Computationally straightforward equations have been developed by Hartley [6.14] and Hamaker [6.15], and more complex models have been provided by Mayer et al. [6.16] and Jury et al. [6.17]. None of the equations accounts for all of the soil or contaminant variables; therefore, use of the equations is most accurate for the compounds from which the equations were developed. Because the accuracy of each of the equations is limited, the use of the more complex equations may provide no distinct advantage, particularly for compounds for which the model was not developed [6.1].

Hartley [6.14] developed an equation to estimate contaminant volatilization in soils where the wick effect and water flux are not important. The Hartley equation was derived with contaminant volatilization correlated to heat flux from the soil:

$$J = \frac{A_{\text{sat}} \dfrac{(1-h)}{\delta}}{\dfrac{1}{D} + \dfrac{\lambda^2 (A_{\text{sat}}) M}{kRT^2}} \tag{6.25}$$

where

J = contaminant flux (g/cm^2-sec)
A_{sat} = contaminant saturation concentration in air (at the temperature of outer air) (g/cm^3)
h = relative humidity ($0 < h < 1$)
δ = thickness of stagnant boundary layer (cm)
D = contaminant diffusion coefficient in air (cm^2/sec)
λ = latent heat of vaporization (cal/g)
M = contaminant molecular weight (g/mole)
k = thermal conductivity of air (61×10^{-6} cal/sec-cm-°K)
R = universal gas constant (1.987 cal/mol-°K)
T = temperature (°K)

Data for A_{sat} and λ for some common hazardous compounds are listed in Table 6.3. The second term in the denominator describes the resistance to volatilization due to thermal considerations and may be neglected for compounds that are significantly less volatile than water. The diffusion coefficient in air (D) may be estimated using Equation 6.11:

$$\frac{D_1}{D_2} = \sqrt{\frac{M_2}{M_1}}$$

D = diffusion coefficient in air (cm^2/sec)   (6.11)

M = molecular weight (g/mole)

Table 6.3 Saturation Concentrations in Air and Heats of Vaporization for Some Common Hazardous Compounds

Compound	A_{sat} (g/m^3)		λ (kJ/mole)
	@ 20°C	@ 30°C	
Acetone	553	825	29.1
Aniline	1.5	3.4	42.4
Benzene	319	485	30.7
Carbon tetrachloride	754	1109	29.8
p-Chloroaniline	0.01	0.34	44.4
Chlorobenzene	54	89	35.2
Chloroform	1027	1540	29.2
p-Chlorophenol	0.70	1.39	
o-Chlorotoluene	18.6	33.3	
o-Cresol	1.2	2.8	42.7
m-Cresol	0.24	0.68	61.7
p-Cresol	0.24	0.74	43.2
Cyclohexane	357	532	30.0
1,2-Dichlorobenzene	8.0	15	40.6
1,4-Dichlorobenzene	4.8	14	38.8
Ethyl acetate	336	533	31.9
Ethylbenzene	40	67	35.6
n-Heptane	196	306	31.8
n-Hexane	564	862	28.9
Methyl isobutyl ketone	27	53	34.5
Nitrobenzene	1.0	2.3	40.8
Perchloroethylene (PCE)	126	210	34.7
1,1,1-Trichloroethane (TCA)	726	1088	29.9
1,1,2-Trichloroethane	136	225	34.8
Trichloroethylene (TCE)	415	643	31.4
o-Xylene	29	50	36.2
m-Xylene	35	61	36.7
p-Xylene	38	67	35.7

Source: Reference 6.18.

Two established values for diffusion coefficients in air are 0.102 cm^2/sec for carbon disulfide and 0.089 cm^2/sec for diethyl ether [6.1].

The use of the Hartley equation in estimating the rate of contaminant volatilization from soils is illustrated in Example 6.4.

EXAMPLE 6.4   *Estimation of Soil Volatilization Flux Using the Hartley Equation*

A 10-m × 5-m (32.8-ft × 16.4-ft) soil pit is contaminated with chlorobenzene. Estimate the daily volatilization flux from the site if the temperature = 20°C, the relative humidity = 80%, and the stagnant boundary layer = 0.4 cm.

SOLUTION

1. Obtain the data needed for chlorobenzene.

$M = 112.6$ g/mol

$A_{sat} = 54$ g/m^3 = 5.4×10^{-5} g/cm^3 (Table 6.3)

$$\lambda = 35.7 \text{ kJ/mole} \times \frac{\text{Kcal}}{4.2 \text{ kJ}} \times \frac{1000 \text{ cal}}{\text{Kcal}} \times \frac{\text{mole}}{112.6 \text{ g}} = 75.5 \frac{\text{cal}}{\text{g}} \text{ (Table 6.3)}$$

2. Estimate the chlorobenzene diffusion coefficient through soils:

Compare D$_{chlorobenzene}$ to known values:

D based on carbon disulfide:

$$\frac{D_{chlorobenzene}}{0.102} = \sqrt{\frac{76}{112.6}}$$

$$D_{chlorobenzene} = 0.084 \frac{\text{cm}^2}{\text{sec}}$$

D based on diethyl ether:

$$\frac{D_{chlorobenzene}}{0.089} = \sqrt{\frac{74}{112.6}}$$

$$D_{chlorobenzene} = 0.072 \text{ cm}^2/\text{sec}$$

Mean D$_{chlorobenzene}$ = 0.078 cm^2/sec

3. Estimate the flux per cm^2 using Equation 6.25.

$$J = \frac{A_{sat}(1 - h)}{\delta} \bigg/ \left[\frac{1}{D} + \frac{\lambda^2 A_{sat} M}{kRT^2} \right]$$

$k = 61 \times 10^{-6}$ cal/cm-sec-°K

$R = 1.987$ cal/mol-K

$$J = \frac{(5.4 \times 10^{-5} \text{ g/cm}^3)\,(1 - 0.80)}{0.4 \text{ cm}} \bigg/$$

$$\left[\frac{1}{0.078} \text{ cm}^2/\text{sec} + \frac{\left(77.6 \dfrac{\text{cal}}{\text{g}}\right)^2 (5.4 \times 10^{-5} \text{ g/cm}^3) \left(112.6 \dfrac{\text{g}}{\text{mole}}\right)}{\left(61 \times 10^{-6} \dfrac{\text{cal}}{\text{sec-cm-}^\circ\text{K}}\right)\left(1.987 \dfrac{\text{cal}}{\text{mole-}^\circ\text{K}}\right)(293^\circ\text{K})^2} \right]$$

$$J = 1.60 \times 10^{-6} \frac{\text{g}}{\text{cm}^2 \cdot \text{sec}}$$

4. Estimate the daily volatilization flux from the soil.

$$J = (1.60 \times 10^{-6} \text{ g/sec-cm}^2)\,(50 \text{ m}^2)\,(10{,}000 \text{ cm}^2/\text{m}^2)\,(86{,}400 \text{ sec/day})$$

$$= 69{,}100 \text{ g/day}$$

Volatilization from deeply contaminated soils may also be calculated using the Hamaker Equation [6.13]. The assumption in its use is that the contamination is semi-infinite; that is, the contamination is essentially infinitely deep with a source of material sufficiently large so that the concentration is not significantly depleted for the purposes of the calculation. The Hamaker equation expresses loss of a chemical as a flux through an area of surface soil:

$$Q_t = 2C_0 \sqrt{\frac{Dt}{\pi}} \tag{6.26}$$

where

Q_t = volatilization of compound per unit surface area (g/cm^2)
C_0 = initial concentration in the soil (g/cm^3)
D = diffusion coefficient of vapor through soil (cm^2/sec)
π = 3.14
t = time (sec)

Use of the Hamaker equation is demonstrated in Example 6.5.

An equation subsequently developed [6.15] provides a method for estimating contaminant volatilization coupled to water flux (i.e., the wick effect) in the soil column:

$$Q_t = \frac{VP}{VP_{H_2O}} \frac{D_v}{D_{H_2O}} (f_w)_V + c(f_w)_L \qquad (6.27)$$

where

f_w = loss of water per unit area (g/cm^2)
VP = contaminant vapor pressure (mm Hg)
VP_{H_2O} = vapor pressure of water (mm Hg)
D_V = diffusion coefficient of the contaminant in air (cm^2/sec)
D_{H_2O} = diffusion coefficient of water vapor in air (cm^2/sec)
V = loss of vapor
L = loss of liquid
c = concentration of contaminant in the soil solution (mg/kg)

Very few data are available for diffusion coefficients of hazardous chemicals through soils. Two known values are 0.010 cm^2/sec for ethylene dibromide and 0.042 cm^2/sec for ethanol [6.1]. In a manner similar to finding diffusion coefficients for volatilization from an open container, vapor diffusion coefficients for other compounds may be estimated by

$$\frac{D_1}{D_2} = \sqrt{\frac{M_2}{M_1}} \qquad (6.11)$$

where

D = the soil diffusion coefficient (cm^2/sec)
M = molecular weight (g/mole)

EXAMPLE 6.5 *Contaminant Flux across a Soil Surface*

Estimate the flux of TCE from a deeply contaminated soil over 1 day with a concentration of 215 ppm and bulk density = 1.52 g/cm^3.

SOLUTION

1. Determine the initial concentration of TCE on a volumetric basis.

$$215 \text{ ppm} = 215 \text{ mg/kg} = 0.000215 \frac{g}{g} \times \frac{1.52 \text{ g}}{cm^3}$$

$$C_0 = 3.27 \times 10^{-4} \text{ g/cm}^3 \text{ TCE}$$

2. Estimate the TCE diffusion coefficient through soils by comparing D to known values.

$$D \text{ based on ethanol:} \quad \frac{D_{TCE}}{0.042} = \sqrt{\frac{46}{131.5}}$$

$$D_{TCE} = 0.025 \frac{cm^2}{sec}$$

$$D_{TCE} \text{ based on ethylene dibromide:} \quad \frac{D_{TCE}}{0.010} = \sqrt{\frac{188}{131.5}}$$

$$D_{TCE} = 0.012 \frac{cm^2}{sec}$$

$$\text{Mean } D_{TCE} = 0.0185 \frac{cm^2}{sec}$$

3. Determine the mass flux of TCE from the soil over one day using Equation 6.26:

$$Q = 2 \cdot 0.000327 \text{ g/cm}^3 \sqrt{\frac{0.0185 \text{ cm}^2/\text{sec } (86{,}400 \text{ sec})}{\pi}}$$

$$Q = 0.0148 \text{ g/cm}^2$$

One of the most common engineering applications of volatilization for soil remediation, soil vapor extraction (SVE), is based on placing vapor extraction wells into the vadose zone (see Section 12.4.2). The Hartley and Hamaker equations may be used to evaluate the potential for SVE as an in situ treatment process.

6.4 VOLATILIZATION OF METALS AND INORGANIC NONMETALS

When hazardous waste volatilization is considered, the chemicals known as volatile organic compounds (VOCs) are often the most significant class of contaminants. These VOCs, the most common of which are halogenated and nonhalogenated solvents, are the hazardous chemicals that provide the most significant volatilization problems. Heavy metals and inorganic nonmetals generally exhibit low volatilities. Organometallic compounds, however, behave much like organic compounds, and some exhibit significant volatility. For example, the microbial transformation of arsenic may result in volatile methylated arsenic compounds. However, these are oxidized rapidly upon exposure to aerobic conditions [6.18]. Dimethyl mercury is also highly volatile. Mercury is also the most volatile metal in its metallic state. Metallic liquid mercury is characterized by a vapor pressure of 0.0018 mm Hg at 25°C [6.20].

Of the inorganic nonmetals, hydrogen cyanide is the most volatile, exhibiting a vapor pressure of 741 mm Hg at 25°C [6.21]. However, hydrogen cyanide is a weak acid with $pK_a = 9.22$, so rapid volatilization would occur only from neutral and acidic solutions.

6.5 SUMMARY OF IMPORTANT POINTS AND CONCEPTS

- Vapor pressure, the pressure exerted on the atmosphere by a chemical, is an important variable in assessing the volatility of relatively pure solutions of contaminants. Vapor pressures of hazardous chemicals vary substantially, from those that are essentially nonvolatile to some that rapidly escape to the atmosphere.

- Henry's Law is traditionally defined as the partial pressure of a compound that is proportional to its concentration in the aqueous phase. The proportionality constant, H, the Henry's Law constant, is analogous to a partition coefficient between the aqueous phase and the air. Henry's Law is a useful predictor of volatilization from aqueous solutions.

- Vapor pressure is a primary variable in conjunction with the mass transfer coefficient, K, in calculating the flux across the surface of a concentrated waste liquid.

- Using a mass balance on an enclosed area with an input of hazardous compounds through volatilization from an open container, the steady-state contaminant concentration in the enclosed space may be evaluated as a basis for assessing toxic, flammable, or explosive concentrations of hazardous chemicals in air.

- Equations that estimate volatilization of hazardous contaminants from soils are subject to the high complexity and variability characteristic of soils including sorption, water content, diffusion, and temperature.

- An empirical equation developed by investigators at Dow Chemical may be used to estimate the first-order volatilization rate from surface soil contamination.

- Volatilization of hazardous chemicals from deeply contaminated soils may be estimated using the Hartley and Hamaker equations. The Hartley equation provides a flux across the soil surface, whereas the output of the Hamaker equation is a mass of contaminant volatilized over a specific time.

PROBLEMS

6.1. What is the difference between vapor pressure and Henry's Law constant? What is an application of each?

6.2. For each of the following compounds, estimate the Henry's Law constant from its water solubility and vapor pressure using the temperatures-dependent values listed in Appendices B and J. Compare the estimated values to the tabulated values.

    a. TCE   b. Benzene

    c. 2,4,6-Trichlorophenol   d. Dieldrin

    e. Aroclor 1248   f. Acrylonitrile

    g. 2,3,7,8-TCDD   h. Carbon tetrachloride

    i. Malathion

6.3. A pit-pond-lagoon system at 20°C contains 48 mg/L methanol, a volatile organic compound with a vapor pressure of 97.6 mm Hg. Would you expect a high methanol volatilization rate from the lagoon? Provide a conceptual basis for your answer using the Henry's Law constant, vapor pressure, and water solubility of methanol.

6.4. In a property transfer site assessment, a records search confirmed that the insecticide dieldrin dissolved in acetone was disposed of on a low-permeability soil 10 years ago. Would you expect either of the chemicals to be present 10 years later based on their volatilities?

6.5. Air stripping—the process of dropping water down a packed tower with a countercurrent of air blown in from the bottom—is the most commonly used design for the removal of volatile organics from groundwater after it is pumped to the surface (called pump-and-treat remediation). Based on the concepts that you have learned in Chapter 6, which of the following would you expect to be effectively and economically treated using air stripping?
  a. Pentachlorophenol   b. TCE
  c. PCE   d. Vinyl chloride
  e. MEK   f. Chloroform
  g. Dieldrin   h. BTEX
Justify your answer based on the chemical properties that control volatilization.

6.6. A small storage cabinet 2 m × 2 m × 3 m high (6.6 ft × 6.6 ft × 9.8 ft) contains an open 208-L (55-gal) drum of aqueous TCE at a concentration of 120 mg/L. The cabinet has been closed for over 2 months, so equilibrium conditions have been established. What is the TCE concentration in the water and vapor phases at this time? Assume that the temperature is 25°C.

6.7. A 208-L (55-gal) drum 0.4 m (1.3 ft) in diameter containing PCE is open in a small drum storage area 5 m × 5 m × 2.5 m high (16.4 ft × 16.4 ft × 8.2 ft). The temperature is 25°C and the atmospheric pressure is 1 atm. Ventilation in the structure provides four changes of air per hour. Determine (a) the volatilization rate of PCE from the drum, and (b) the steady-state PCE concentration in ppm in the structure. Assume $k = 0.2$.

6.8. A waste receptacle 3 m long × 1.5 m wide × 20 cm deep (9.8 ft × 4.9 ft × 0.66 ft) in a building that houses a metal parts cleaning operation contains a layer of TCA 8 cm (0.26 ft) deep. The temperature is 25°C and the atmospheric pressure is 1 atm. The ventilation rate in the building is 200 L/h. Determine the TCA volatilization rate from the waste receptacle. Assume $k = 0.4$.

6.9. Repeat problem 6.8 with (a) pentane and (b) dodecane as the compounds in the waste receptacle.

6.10. A large spill of PCE occurred in a storage building 10 m × 8 m × 6 m high (32.8 ft × 26.2 ft × 19.7 ft), resulting in a PCE depth of 0.1 m (0.33 ft) (release from the building was protected by a containment berm). The temperature in the building is 25°C. If $k = 0.3$ and the ventilation system provides 15 changes of air per hour, determine the steady-state concentration in the building.

6.11. A pesticide truck spills 91 kg (120 lb) of 80% DDT on a dirt road over 85 m² (914 ft²). The average depth for the DDT is 4 cm (1.6 in). The bulk density of the soil is 1800 kg/m^3. How long will it take for the DDT to volatilize to a level of 0.01 ppm in the soil? Assume negligible effects of temperature differences at which water solubility and vapor pressure are reported.

6.12. Two hundred kilograms (440 lb) of 4,6-dinitro-*o*-cresol (DNOC) have been spilled on an asphalt surface and held there at an average depth of 2 cm (0.8 in). The surface area of the spill is 100 m^2 (1080 ft^2), and the bulk density of the asphalt is 2100 kg/m^3. If volatilization is the only significant chemodynamic pathway, how long will it be until the DNOC level drops to 10 ppm? Assume negligible effects of temperature differences at which water solubility and vapor pressure are reported.

6.13. One hundred kilograms (220 lb) of 60% solution of nirvanum, an experimental new insecticide, is spilled over a 28,000 m^2 (301,000 ft^2) area of hardpan soil in a layer 79 mm thick. The log K_{oc} for nirvanum = 2.7, the vapor pressure = 2.3 × 10^{-5} mm Hg, the water solubility = 12 mg/L, and the soil bulk density = 2600 kg/m^3. Estimate the time required for the nirvanum soil concentration to drop to 1 mg/kg by volatilization.

6.14. A pesticide truck spills 55 kg (121 lb) of 80% endrin on a dirt road over 85 m^2 (914 ft^2). The average spill depth is 4 cm (1.6 in.). The bulk density of the soil is 1800 kg/m^3. How long will it take for the endrin to volatilize to a level of 1 ppb in the soil? Assume negligible effects of temperature differences at which water solubility and vapor pressure are reported.

6.15. Determine the half-life of hexachlorobenzene by volatilization if the contamination is on the soil surface. Assume negligible temperature effects.

6.16. One hundred cubic meters (131 yd^3) of gasoline-contaminated soil have been laid out on a concrete pad at a depth of 0.5 m (1.64 ft). As environmental facilities manager at the petrochemical facility where the waste has been placed, you have been approached by state regulatory personnel to control volatile gasoline emissions. Based on the physical properties, are the concerns of the regulators justified? List a possible measure that could be used to control the emissions.

6.17. The concentration of PCE in a soil with infinitely deep contamination is 1270 ppm. Determine the volatilization flux over 1 week using the Hamaker equation if the soil bulk density = 1800 kg/m^3 and D = 0.142 cm^2/sec.

6.18. Soil under a dried-up hazardous waste disposal pit contains DDT at an average soil concentration of 3,800 ppm. Using the soil-air D for ethylene dibromide and ethanol to estimate the soil-air D for DDT, estimate the time required for the DDT to drop to 0.01 ppm in the soil. Assume the depth of contamination is 10 m (32.8 ft), the surface area = 100 m^2 (1080 ft^2), and ρ_B = 1600 kg/m^3. Ignore the wick effect.

6.19. A soil disposal area is characterized by deep (semi-infinite) contamination of anthracene with a mean concentration of 240 mg/kg. If the bulk density of the soil is 1700 kg/m^3, determine the flux of anthracene out of the soil over a time period of 10 days. Ignore the wick effect.

6.20. Repeat Problem 6.19 accounting for the wick effect. The loss of water vapor from the soil over the 10-day period is 8.2 g/cm^2 and the loss of liquid water is 1.1 g/cm^2. Assume that 10% of the anthracene is present in the soil water.

6.21. A soil (ρ_B = 1700 kg/m^3) is deeply contaminated with 800 mg/kg anthracene. Estimate the volatilization flux at 1-day intervals up to 10 days using the Hamaker equation.

6.22. A hazardous waste landfill is contaminated with 1,1,2,2-tetrachloroethane. Assume that the relative humidity = 48%, the temperature = 20°C, and the thickness of the stagnant boundary layer = 0.5 cm. Estimate the flux of the 1,1,2,2-tetrachloroethane out of the soil.

6.23. A 208-L (55-gal) drum of toluene has ruptured and contaminated a section of soil 5 m (16.4 ft) in diameter and 1.5 m (4.9 ft) deep. Estimate the time required for the toluene to volatilize if the relative humidity = 26%, the temperature = 14°C, and the thickness of the stagnant boundary layer = 0.5 cm.

6.24. A soil disposal pit (ρ_B = 1700 kg/m^3) is contaminated with benzene at an average concentration of 285 mg/kg. The temperature = 22°C, the relative humidity = 70%, and the thickness of the stagnant boundary layer = 0.8 cm. Estimate and compare the flux of benzene over 1 day from the soil using both the Hartley and Hamaker equations.

6.25. Estimate the flux of carbon tetrachloride from a soil based on total water loss if the carbon tetrachloride concentration in the soil water is 0.0084%. The flux of water vapor is 0.048 g/cm^2-day, and the flux of liquid water is 0.013 g/cm^2-day.

6.26. Estimate the flux of chloroform from a soil based on total water loss if the flux of water vapor is 0.0095 g/cm^2-day. Assume that the flux of liquid water is negligible.

REFERENCES

6.1. Thomas, R. G., "Volatilization from soil," Chapter 16 in Lyman, W. J., W. F. Rheel, and D. H. Rosenblatt (Eds.), *Handbook of Chemical Property Estimation Methods*, McGraw-Hill, New York, 1982.

6.2. Howe, G. B., M. E. Mullins, and T. N. Rogers, "Evaluation and production of Henry's Law constants and aqueous solubilities for solvents and hydrocarbon fuel components. Volume I: Technical discussion," Final Report, No. ELS-86-66, Engineering and Services Lab, Tyndall Air Force Base, 1987.

6.3. Gossett, J. M., "Measurement of Henry's Law constants for C1 and C2 chlorinated hydrocarbons," *Environ. Sci. Technol.*, **21**, 202–208 (1987).

6.4. Howard, P. H., *Handbook of Environmental Fate and Exposure Data for Organic Chemicals*, Lewis Publishers, Boca Raton, FL, 1990.

6.5. Mackay, D. and P. J. Leinonen, "Rate of evaporation of low solubility contaminants from water bodies to atmosphere," *Environ. Sci. Technol.*, **9**, 1178–1180 (1975).

6.6. Liss, P. S. and P. G. Slater, "Flux of gases across the air-sea interface," *Nature (Lond.)*, **247**, 181–184 (1974).

6.7. Hanna, S. R. and P. J. Drivas, *Vapor Cloud Dispersion Models*, American Institute of Chemical Engineers, New York, 1988.

6.8. Matthiessen, R. C., "Estimating exposure levels in the workplace," *Chem. Eng. Prog.*, (April 30 1986).

6.9. Crowl, D. A. and J. F. Louvar, *Chemical Process Safety: Fundamentals with Applications*, Prentice-Hall, Englewood Cliffs, NJ, 1990.

6.10. Neely, W. B., *Chemicals in the Environment: Distribution, Transport, Fate, Analysis*, Marcel Dekker, New York, 1980.

6.11. Spencer, W. F. and M. M. Cliath, "Desorption of lindane from soil as related to vapor density," *Soil Sci. Soc. Am. Proc.*, **34**, 574–578 (1970).

6.12. Shearer, R. C., J. Letey, W. J. Farmer, and A. Klute, "Lindane diffusion in soil," *Soil Sci. Soc. Am. Proc.*, **37**, 189–193 (1973).

6.13. Guenzi, W. D. and W. E. Beard, "Volatilization of pesticides," Chapter 6 in Guenzi, W. D. (Ed.), *Pesticides in Soil and Water*, Soil Science Society of America, Madison, WI, 1974.

6.14. Hartley, G. S., "Evaporation of pesticides," Chapter II in "Pesticide formulations research, physical and colloidal chemical aspects," *Advances in Chemistry Series*, **86**, American Chemical Society, Washington, DC, 1969.

6.15. Hamaker, J. W., "Diffusion and volatilization," Chapter 5 in Goring, C. A. I. and J. W. Hamaker (Eds.), *Organic Chemicals in the Soil Environment*, Vol. 1, Marcel Dekker, New York, 1972.

6.16. Mayer, R., J. Letey, and W. J. Farmer, "Models for predicting volatilization of soil-incorporated pesticides," *Soil Sci. Soc. Am. Proc.*, **38**, 563–568 (1974).

6.17. Jury, W. A., R. Grover, W. F. Spencer, and W. J. Farmer, "Modeling vapor losses of soil-incorporated triallate," *Soil Sci. Soc. Am. J.*, **44**, 445–450 (1980).

6.18. Verschueren, K., *Handbook of Environmental Data on Organic Chemicals*, 2nd Edition, Van Nostrand Reinhold, New York, 1983.

6.19. Parris, G. E. and F. E. Brinckman, "Reactions which relate to the environmental mobility of arsenic and antimony. II. Oxidation of trimethylarsine and trimethylstibine," *Environ. Sci. Technol.*, **10**, 1128–1134 (1976).

6.20. Sienko, M. J. and R. A. Plane, *Chemistry: Principles and Properties*, McGraw-Hill, New York, 1966.

6.21. Dean, J. A. (Ed.), *Lange's Handbook of Chemistry*, 12th Edition, McGraw-Hill, New York, 1979.

Chapter 7

Abiotic and Biotic Transformations

Organic hazardous wastes managed under RCRA or disposed of on soils or in water can be transformed by abiotic (chemical or physical) and biotic (biological) processes. Transformation reactions of organic contaminants involve the breaking of chemical bonds and the subsequent formation of new ones. Metals, although not capable of being transformed to other species, can change oxidation states, which then significantly influences their transport in the environment or their behavior in treatment systems such as precipitation reactors or ion exchange contactors. A few metals, such as mercury, may be transformed to organometallic compounds under specific conditions. The formation of organomercury compounds has been an important process in anaerobic sediments of lakes and oceans, but it is usually not a significant pathway in soil and groundwater systems.

The abiotic transformation of organic hazardous contaminants, which does not usually play a significant role in the mineralization of the compounds (i.e., their conversion to carbon dioxide and water), can often initiate the transformation of biorefractory chemicals that would otherwise remain for long periods of time in the environment or biological treatment systems [7.1]. Although abiotic transformations do not result in the mineralization of contaminants, coupled abiotic-biotic processes are thought to be an important mechanism for the degradation and eventual mineralization of organic chemicals. An exception to this generalization is the use of advanced oxidation processes (AOPs), which have been shown to mineralize some organic contaminants [7.2–7.4]. On a global scale, biotic processes are the only significant pathway for the mineralization of both natural and anthropogenic organic compounds. Biotic processes also represent the most economical process design for the treatment of biodegradable contaminants.

Both abiotic and biotic transformations are covered in this chapter—an approach that parts with tradition. In environmental sciences and engineering, physicochemical and biological processes have long been considered separate subdisciplines; however, many similarities have recently been realized (and emphasized) between abiotic and biotic processes. This chapter focuses on both transformation routes because (1) abiotic oxidation-reduction reactions serve as a conceptual basis for understanding biochemical transformations, (2) contaminants often undergo similar oxidation-reduction reactions by either abiotic or biotic transformations, and (3) solving hazardous waste problems with engineered treatment systems often involves the use of integrated abiotic and biotic process designs.

A range of transformation processes is detailed in this chapter, some of which predominate in the natural environment (such as contaminated soils and groundwater). Others almost never occur naturally in the environment but serve as the basis for remediation and treatment processes. A perspective on the roles of each of the processes in the natural environment and in engineering process design will be outlined in Section 7.5.

The most important abiotic reactions include hydrolysis, oxidation-reduction (redox), and photolysis. Hydrolysis is the addition of water to a molecule, which usually occurs at specific functional groups on organic molecules and is often enhanced by the presence of acids or bases. Redox reactions are promoted in environments in which the chemical potential (i.e., a driving force for a reaction) favors (1) the addition of oxygen or the loss of electrons or a hydrogen atom (oxidation) or (2) the addition of electrons (reduction). Redox reactions are promoted by low concentrations of oxygen radicals, by coupled reactions on the surfaces of soil minerals, or by a number of treatment processes such as ultraviolet light–ozone, Fenton's reagent, titanium dioxide–mediated photocatalysis, and electron beams. These reactions are governed by the contaminant structure as well as the redox potential of the system (e.g., the pH, the dissolved oxygen concentration, and the presence of oxidizing and reducing agents). Photolysis is a light-induced redox reaction that has limited importance in the transformation of hazardous chemicals. Because the chemical must be exposed to light for the reaction to proceed, photolytic transformations of hazardous chemicals occur most often in surface impoundments and on soil surfaces. However, photochemical processes have been the focus of recent intensive research, and their use holds promise for the treatment of dilute aqueous hazardous wastes.

In biotic transformations, organic contaminants are metabolized by microorganisms in surface impoundments, surface waters, soils, groundwater, and treatment systems through biochemical redox reactions. Microorganisms often use the contaminant as a carbon source (to make more cellular material) and as an energy source (the energy contained in the chemical bonds is used for growth, reproduction, and motility). The importance of biodegradation as a pathway for organic hazardous chemicals cannot be overstated. Of the common physicochemical and biochemical degradation mechanisms (chemical oxidation, hydrolysis, photolysis, and biodegradation), biodegradation is by far the most significant process for the decomposition of contaminants in the natural environment [7.5].

Just as sorption and volatilization serve as the basis for treatment processes, abiotic and biotic processes can be promoted and optimized to clean up contaminated soils, groundwater, and RCRA wastes. Hydrolysis reactions are somewhat limited as a hazardous waste treatment process, but redox reactions represent important engineering systems for soil and groundwater remediation. The most important application of strong oxidants and reductants in the treatment and remediation of refractory organics—the use of zero-valent metals and AOPs, which generate hydroxyl radicals (OH·)—is providing new ways of destroying otherwise stable hazardous compounds. Enhanced biodegradation (i.e., bioremediation) is one of the most important engineering processes used for the remediation and treatment of contaminated soils, groundwater, and RCRA wastes. As with all engineering systems, an understanding of fundamental mechanisms serves as the basis for technology selection and design. Therefore, the material of this chapter provides fundamental concepts for understanding site characterization, facility assessments, and abiotic-biotic treatment processes.

7.1 THE GOVERNING VARIABLES: CHEMICAL STRUCTURE, PRESENCE OF REACTIVE SPECIES, AND AVAILABILITY

Some processes, such as biological growth, have been characterized by Liebig's Law of the Minimum, in which one essential factor, such as nitrogen concentration, limits growth of the organism. In reality, tens, or perhaps hundreds, of variables affect the transformation of hazardous contaminants in the environment and hazardous waste systems. However, there are usually no more than a few important variables at one time. Three variables have a significant effect on contaminant transformations and persistence in the environment, and on the process design of hazardous waste treatment systems. First, the contaminant must have a chemical structure that is reactive with a transforming species. Second, an *appropriate* transforming species must be available. Third, the contaminant must be physically available (i.e., not sorbed or in NAPL phase). These variables strongly interact with each other, as well as with other variables such as pH and redox potential (see Section 7.1.4). For example, an oxidized organic contaminant (e.g., carbon tetrachloride) requires an appropriate transforming species—one that will reduce the carbon tetrachloride. The presence of a reducing species (e.g., a reducing metal or a consortium of microorganisms that metabolize through reductive pathways) is often dictated by the redox potential and pH of the system. Similar trends are found for other abiotic and biotic processes in this chapter. The transformation mechanisms that occur on a molecular level continue to be the focus of intensive research, and the current state of knowledge on transformation processes will likely progress substantially over the next decade.

7.1.1 Chemical Structure: The Basis for Reactivity in Abiotic and Biotic Transformations

One of the problems of hazardous wastes, their environmental persistence, is related directly to chemical structure. Most of the common hazardous chemicals (see Chapter 2) are characterized by structures that affect their rates of abiotic and biotic transformations. Halogenated solvents, pesticides, explosives, PCBs, and PCDDs have been called *xenobiotics*, or chemicals that are foreign to native biological metabolism. Other descriptions of these persistent environmental chemicals are *biorefractory* or *biologically recalcitrant* compounds. Numerous chemical structure–biodegradability relationships have been reported, and Scow [7.6] has provided the most extensive review of the effect of chemical structure on biodegradability. Contaminant characteristics that have been reported to decrease rates of biodegradation include (1) branching of hydrocarbons, (2) high hydrophobicity, (3) hydrocarbon saturation (relative to unsaturated alkanes), and (4) increased halogenation (for aerobic processes). Another general rule is that compounds that have chemical structures similar to biochemical intermediates (e.g., keto acids) are more readily biodegradable. Although these are general relationships, the more detailed mechanisms of biodegradation discussed in this chapter provide a greater understanding of biological transformations.

Chemical structure affects all transformations, including hydrolysis and abiotic and biotic redox reactions. In the introduction to the bonding of organic compounds of Section 2.1, the polarity of carbon bonds was introduced. Covalent bonds involving carbon and oxygen, halogens, or nitro groups have varying degrees of polarity. The electron cloud is drawn to the more electronegative element (the oxygen, halogen, or nitro group). As a result, the carbon becomes a site of attack for nucleophilic (i.e., a nucleus- or positive-liking) species.

Hydrolysis is one of the most common nucleophilic reactions, in which an electron-rich species, such as water, attacks an electron-poor area of an organic contaminant. A paucity of electrons is maintained by the presence of electron-withdrawing groups such as halogens or carboxylic acid esters (C—O—CH_3):
$$\overset{\|}{\underset{O}{}}$$

$$CH_3\text{—}CH_2\text{—}CH_2\text{—}CH_2\overset{\rightarrow}{\text{—}}Cl$$

n-Butyl chloride

$$CH_3\text{—}CH_2\text{—}CH_2\text{—}\overset{\overset{O}{\|}\uparrow}{C}\underset{\rightarrow}{\text{—}}O\text{—}CH_3$$

Methyl butyrate

In the alkyl halide, *n*-butyl chloride, electrons are pulled toward the chlorine, leaving an electron-poor area at the first (α) carbon. Esters of carboxylic acids, such as methyl butyrate, are characterized by a carbonyl carbon that is electron poor and an area for electrophilic attack.

Another property, the contaminant *oxidation state,* is an important predictor of potential abiotic and biotic redox reactions in natural and engineered systems. An understanding of the oxidation state of contaminants is fundamental to understanding abiotic and biotic redox transformations. Elements and elemental compounds, such as O_2, Cl_2, or Fe metal, exhibit a zero oxidation state, and the oxidation state of anions and cations is the same as their charge. For example, H^+ is characterized by a $+1$ oxidation state. The carbon oxidation state of organic compounds is variable and is dictated by the chemical structure; that is, the carbon oxidation state is a function of the elements to which the carbon is bonded. The rules for characterizing carbon oxidation states are listed in Table 7.1, and the use of the rules is demonstrated in Example 7.1.

Table 7.1 Rules for the Determination of Oxidation State

1. Report the oxidation state by the atomic symbol followed by a Roman numeral in parentheses [e.g., Fe(III) or Cl($-$I)].
2. Elements and elemental compounds (e.g., Fe metal, O_2, Cl_2) by definition, have a zero oxidation state.
3. The oxidation states of anions and cations are equal to their charge.
4. When bonded to carbon in organic compounds, the oxidation state of hydrogen is ($+$I), the oxidation state of oxygen is ($-$II), and the oxidation state of halogens is ($-$I).
5. The oxidation state of nitrogen or carbon in a compound is based on the fixed oxidation states of hydrogen, oxygen, and halogens.

EXAMPLE 7.1 *Calculation of Oxidation States*

Determine the oxidation state of the carbon in (1) methane and (2) carbon tetrachloride.

SOLUTION

1. In methane, CH_4, the oxidation state of hydrogen is $4(+I) = (+IV)$. Because methane does not carry a charge, the carbon oxidation state must be $(-IV)$.
2. Carbon tetrachloride, CCl_4, has four chlorine atoms, represented by $4(-I) = (-IV)$. To maintain the neutrality of the compound, the oxidation state of the carbon is $(+IV)$.

Just as halogens draw electrons and increase the carbon oxidation state, other electron-withdrawing groups are common among hazardous chemicals; in contrast, other substituents donate electrons to carbon atoms. A common transformation reaction is electrophilic substitution of aromatic compounds, the attack on the electron-rich pi bonds by an electron-seeking species. Electron-withdrawing groups deactivate the ring toward such electrophilic attacks because they make the ring more electropositive and lower the driving force for attack by electrophiles. In contrast, other species donate electrons to the ring and make it more reactive to electrophilic attack. Some common ring *activating* and *deactivating* groups are listed in Table 7.2. The most important ring-deactivating groups on hazardous chemicals are halogens and nitro groups. Based on induced changes in the aromatic electron cloud, the electrophilic attack may be directed at either the *ortho* and *para* positions of the ring or the *meta* position. Electron-withdrawing groups also have an effect on similar reactions of aliphatic compounds. For example, the electrons of the carbon atoms of carbon tetrachloride and PCE are drawn toward chlorine, so the carbons of these compounds are highly oxidized. The importance of the contaminant oxidation state is straightforward—reduced organic contaminants, such as benzene, tend to degrade by oxidation reactions and oxidized organics, such as carbon tetrachloride, degrade by reacting with reductants.

7.1.2 Presence of Transforming Species

The transformation of hazardous contaminants will not occur unless another reactive species, such as a nucleophile, an oxidant, or a microorganism, is present. Furthermore, the *appropriate* transforming species is required; for example, the presence of an oxidant, such as hydroxyl radicals, is often necessary to transform a reduced contaminant (e.g., phenol). In contrast, an oxidized compound, such as carbon tetrachloride, reacts most effectively with reducing species, such as pyrite (FeS) or hydrated electrons, but is unreactive with hydroxyl radicals. Therefore, the presence of a thermodynamically (and kinetically) favorable reacting species is necessary for a transformation reaction to take place.

Table 7.2 Effect of Substituent Groups on Reactivity
Toward Electrophilic Attack

Deactivating	Activating (all are ortho/para directors)
$\longrightarrow$ **Ortho/Para Directors**	**Strongly Activating**
—F	—NH$_2$
—Cl	—OH
—Br	
—I	**Moderately Activating**
	—OR
$\longrightarrow$ **Meta Directors**	$\overset{\displaystyle O}{\overset{\displaystyle \|}{-NHC-CH_3}}$
—NO$_2$	
—N(CH$_3$)$_3^+$	
—CN	**Weakly Activating**
—COOH	
—SO$_3$H	—C$_6$H$_5$
—COR	—CH$_3$

In biotic transformations, the requirements for transforming species are more complex and depend on the genetic complement and cellular physiology of the microbial consortium. Many species of bacteria do not possess the enzymes required to metabolize specific compounds. For example, *Arthrobacter globiformis* transforms 4-chlorobenzoate as a sole carbon and energy source and *Xanthobacter autotrophicus* utilizes 1,2-dichloroethane as a carbon source, but they cannot metabolize many other compounds [7.7]. Biotic transformations for some contaminants may also be promoted by specific processes such as cometabolism and the presence of certain electron acceptors—mechanisms that will be described in Section 7.4. An overlying concept related to the presence and activity of transforming species is their dependence upon and interactions with other variables, such as pH, redox potential, and temperature.

7.1.3   Sorption and NAPLs: Inaccessibility to Transformation Processes

For hazardous compounds to degrade, they must have contact with a species that promotes their transformation. One of the best examples of inaccessibility to degradation is the large pools of petroleum found deep below the earth's surface. These huge reservoirs are in an area where microorganisms do not tend to grow; furthermore, conditions in the interior regions of these hydrocarbon systems are not conducive to microbial growth. If a microorganism found its way to the center of a large pool of oil, it would have more than enough carbon but would be lacking water, nutrients, and the other environmental factors necessary for it to grow and reproduce.

Nonaqueous phase liquids (NAPLs) behave much the same way as the ancient petroleum reserves. These pools of relatively pure materials are not a suitable environment for water-borne reactants such as chemical oxidants or microorganisms. Therefore, minimal transformation of contaminants in pools and lenses of NAPLs may be expected because of their immiscibility with water.

Sorption, the accumulation of contaminants at surfaces, provides an inaccessibility mechanism analogous to the presence of an NAPL phase. Sorption not only retards the rate of migration of hazardous contaminants and lowers the rate of phase transfer to the atmosphere, it also significantly affects rates of abiotic and biotic transformations. The conceptual basis for the effect of sorption is that the agents that transform organic contaminants (e.g., bacteria, oxidants, and reductants) exist in the aqueous phase. In other words, if a contaminant partitions onto soil organic matter or a mineral surface, it is not in direct contact with the agents responsible for its transformation.

In most cases, desorption is believed to control the transformation of hazardous organic contaminants in soils, groundwater, sludge, and treatment systems. For example, Sedlak and Andren [7.8] found that PCBs sorbed to diatomaceous earth were unavailable for oxidative attack by hydroxyl radicals. Contaminant accessibility also has an effect on biological processes. Ogram et al. [7.9] documented that sorption controlled the biological degradation of 2,4-D; similar results were reported for PAHs [7.10] and PCBs [7.11]. In contrast, other studies have found that, although sorption reduces the rate of biodegradation (and often controls it through desorption), a number of compounds can be biodegraded even when all the chemical is sorbed or when desorption is insignificant over the time biodegradation takes place [7.12–7.14].

Other studies have shown that some mechanisms can enhance desorption and subsequent transformation. For example, Watts et al. [7.15] showed that aggressive Fenton's reactions (hydrogen peroxide and catalysts that generate hydroxyl radicals) can oxidize hexachlorobenzene more rapidly than it is normally desorbed. In biological processes, naturally occurring surfactants, such as rhamnolipids, solubilize sorbed contaminants, resulting in increased rates of biodegradation [7.16].

7.1.4   Other Variables Affecting Transformation Processes

Dozens of variables influence abiotic and biotic transformations. A few of the most important are discussed next.

Dissolved Oxygen and Oxidation-Reduction Potential. The presence of oxygen and the redox potential of the system have a profound effect on the transformation pathway—they often dictate whether (1) abiotic oxidations or reductions will occur, or (2) aerobic or anaerobic biodegradation dominates.

Free electrons in abiotic redox processes in the presence of molecular oxygen are rapidly scavenged by oxygen resulting in the formation of superoxide radical anion:

$$O_2 + e_{aq}^- \longrightarrow O_2^{\cdot -} \tag{7.1}$$

Superoxide is a relatively unreactive species in most soil, groundwater, and treatment systems. Although it can act as a reductant, superoxide is not as effective as free electrons.

The presence of oxygen also governs the consortium of organisms, their metabolism, and the transformation reactions that occur. Some microorganisms are obligate *anaerobes* (i.e., organisms to which oxygen is toxic). On the other hand, most biodegradation occurs under *aerobic* conditions (i.e., with microorganisms that live in the presence of oxygen). Bacteria that can metabolize through either aerobic respiration or anaerobic fermentation depending on the presence of oxygen are *facultative anaerobes*. Aerobic microorganisms require a minimum dissolved oxygen level for maximum growth and metabolism. Although a paucity of information exists on the minimum dissolved oxygen concentration needed to promote the biodegradation of hazardous compounds, trends similar to biological wastewater treatment systems have been suggested. Activated sludge systems are characterized by negligible aerobic metabolism at undetectable dissolved oxygen concentrations, and the rate increases up to approximately 1 to 2 mg/L of dissolved oxygen; above these concentrations, the rate of microbial metabolism is relatively constant.

For biological processes in the presence of oxygen (aerobic metabolism), the reducing potential of the contaminant serves as the source for biochemical energy. However, in the absence of oxygen, biological metabolism proceeds through anaerobic metabolism, in which the activity of a cascade of *terminal electron acceptors* is dictated by the system thermodynamics (see Section 7.4):

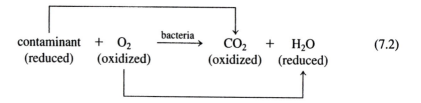

$$\text{contaminant} + \text{O}_2 \xrightarrow{\text{bacteria}} \text{CO}_2 + \text{H}_2\text{O} \qquad (7.2)$$
$$\text{(reduced)} \quad \text{(oxidized)} \qquad \text{(oxidized)} \quad \text{(reduced)}$$

Contaminant electrons (which serve as a source of energy) flow through a series of biochemical intermediates, including flavin nucleotides and cytochromes, where coupled reactions capture their energy in the high-energy phosphate bonds of adenosine triphosphate (ATP). The final fate of the electrons is a two-electron transfer with one-half mole of molecular oxygen and two moles of protons resulting in the formation of water:

$$2e^- \text{ (from cytochromes)} + \frac{1}{2}\text{O}_2 + 2\text{H}^+ \longrightarrow \text{H}_2\text{O} \qquad (7.3)$$

Another factor related to dissolved oxygen is its interrelationship with contaminant structure. Highly reduced organics, such as hydrocarbons, are degraded most rapidly in an oxidizing environment. On the other hand, highly oxidized organics, such as PCE, are more efficiently degraded under anaerobic conditions. Trichloroethylene, which is a less oxidized compound, is biodegraded under both aerobic and anaerobic conditions.

The foregoing discussion shows that biodegradation rates are influenced significantly by the concentration of dissolved oxygen. The variability in biodegradation rates is related not only to the gradient in concentrations of dissolved oxygen found in different environmental systems but also to the types of microorganisms that grow in the different dissolved oxygen regimes. Furthermore, the molecular structure and degree of oxidation of the organic compound (such as the degree of halogenation) affect the class of microorganisms that is capable of degrading the compound and the dissolved oxygen conditions under which that compound will be most efficiently biodegraded.

Temperature. Chemical and biochemical reactions usually obey the *Van't Hoff Rule,* which describes the effect of temperature on reaction rates:

$$Y = Ae^{-E_a/RT} \qquad (7.4)$$

where

Y, A = Arrhenius constants
E_a = activation energy (kcal/mole)
R = universal gas constant [0.00199 kcal/(mole-°K)]
T = absolute temperature (°K)

A common practice in biological treatment engineering is to group the term containing the activation energy of Equation 7.4 into one factor θ, which results in the following empirical equation:

$$k_{T_2} = k_{T_1} \theta^{T_2 - T_1} \qquad (7.5)$$

where

k_{T_1} = rate of reaction at temperature 1
k_{T_2} = rate of reaction at temperature 2

Values for θ range from 1.01 to 1.11 depending on the system. A θ value of 1.088 has been used to describe the effect of temperature on the metabolism of hydrocarbon-degrading bacteria [7.17]. The lower limit of microbial activity is the freezing point of water. In general, a twofold increase in reaction rate is found for each rise of 10°C in temperature up to approximately 40°C. Above 40°C, biodegradation rates usually decline due to denaturation of enzymes. Exceptions to this are the group of organisms called thermophiles, which have specially adapted enzymatic systems that allow them to metabolize at high temperatures—even above 100°C. Thermophiles have been found in hot springs and other areas where temperatures approach boiling.

Abiotic rate processes often also follow the Van't Hoff Law but with different coefficients, the result of which is often different temperature-rate relationships. For example, the temperature dependence of hydrolysis rate constants is greater than that of many other biotic and abiotic processes (a 2.5-fold increase in k for a 10°C rise in temperature) [7.18]. In addition, temperature can have interactive effects with catalyst solubilities in some abiotic reactions. Therefore, although reaction rates increase as a function of temperature, exact temperature-rate relationships are system-specific.

pH. Microorganisms are sensitive to pH extremes; in fact, acids and bases have been used to disinfect water and wastewater. The pH range generally accepted for optimum microbial growth is from pH 6 to pH 9, with most effective metabolism occurring between pH 6.5 and 7.5. Fungi, on the other hand, grow most effectively at the acidic pH regime between 4 and 6. Abiotic processes may or may not be sensitive to pH. One of the most important pH effects is in the application of Fenton's reagent in which iron(II) catalyzes the decomposition of hydrogen peroxide to hydroxyl radicals:

$$H_2O_2 + Fe^{2+} \longrightarrow OH\cdot + OH^- + Fe^{3+} \qquad (7.6)$$

The Fenton's reaction proceeds most effectively at pH 2–4—the acidic conditions maintain iron solubility and lower the redox potential of the system, which promotes the most efficient generation of hydroxyl radicals [7.4].

Other Variables. Abiotic and biotic degradation rates vary substantially, perhaps more than any other pathway of hazardous compounds in the environment or hazardous waste treatment systems. Tens of variables in soils, groundwater, and treatment systems affect the rates at which transformations occur. The variability may be attributed to environmental conditions, the species of microorganisms or reactants present, and the nature of the contaminant itself. A number of other factors may influence abiotic and biotic transformations, including the moisture content (for soils and sludges) and ionic strength. Under certain conditions, transformation reactions can be influenced by toxicity and quenching. Some contaminants, such as chlorophenols and nitrophenols, are toxic to microorganisms and may inhibit microbial growth [7.19]. Abiotic reactions can also be negated; for example, reactions of the strong oxidant hydroxyl radical are quenched by bicarbonate, chloride, and a number of other inorganic anions [7.20]. More detailed information on other variables that affect transformation reactions may be found in References 7.6 and 7.21.

7.2 RATES OF TRANSFORMATION

The kinetics of transformation reactions are more difficult to quantify than their thermodynamics. Furthermore, soil, groundwater, and hazardous waste treatment systems are so complex that the exact transformation pathway sometimes cannot be elucidated, which makes the quantitation of reaction rates difficult. Nonetheless, the determination of transformation rates is necessary in order to perform site characterizations and facility assessments, and to design hazardous waste treatment systems.

The loss of a compound through abiotic and biological transformation processes has been described by zero-, first-, second-, and mixed-order rate expressions. The definition of rate order is based on the empirical observation of contaminant loss:

$$\frac{-d[C]}{dt} = k[C]^n \tag{7.7}$$

where

$[C]$ = contaminant concentration (mg/L)
k = proportionality constant (units dependent on reaction order)
n = reaction order

The reaction order is based on the value of the exponent n in Equation 7.7 that provides the best fit of empirical data. A *zero-order reaction* is independent of the contaminant concentration and is defined by

$$\frac{-d[C]}{dt} = k \tag{7.8}$$

The integrated form of the equation is

$$C_t = C_0 - kt \tag{7.9}$$

A reaction that fits zero-order kinetics is characterized by a linear contaminant transformation rate.

The first-order rate equation was derived in Chapter 6:

$$C_t = C_0 \, e^{-kt} \qquad (6.22)$$

where k = the first-order rate constant (day^{-1}). Under many observed conditions, transformation reactions have been found to follow first-order kinetics. In true first-order reactions, the rate constant is independent of the contaminant concentration.

The rate of contaminant degradation in the environment or in engineered processes is usually more complex than first-order kinetics. The rate is often *second order* (in that the rate of change is proportional to two species—the contaminant and another reactant). For example,

$$\frac{-d[C]}{dt} = k[C][\text{enzyme}] \qquad (7.10)$$

or

$$\frac{-d[C]}{dt} = k[C][\text{OH}\cdot] \qquad (7.11)$$

often describes biological- and radical-mediated degradations, where [enzyme] is the concentration of enzyme and [OH·] is the concentration of the hydroxyl radicals that promote contaminant degradation. Often the reactant concentration is at steady state; that is, it remains constant over time:

$$\frac{-d[\text{OH}\cdot]}{dt} = 0 \qquad (7.12)$$

$$\frac{-d[\text{enzyme}]}{dt} = 0 \qquad (7.13)$$

and only the contaminant concentration then affects the rate of reaction. Equations 7.10 and 7.11 may then be simplified to an observed first-order rate:

$$\frac{-d[C]}{dt} = k[C] \qquad (6.19)$$

This phenomenon is termed *pseudo first-order kinetics.* The transformation rate is first order, but the first-order behavior is really a simplification because of the steady-state assumption.

EXAMPLE 7.2   *Steady-State Simplification of Second-Order Kinetics*

If the second-order rate constant for the biodegradation of chrysene is 1×10^{-13} L/(cell-h), determine its half-life in an oily sludge with a steady-state chrysene-degrading biomass of 10^5 cells/mL.

SOLUTION

The concentration of cells in the system is

$$10^5 \frac{\text{cells}}{\text{mL}} \cdot \frac{1000 \text{ mL}}{\text{L}} = \frac{10^8 \text{ cells}}{\text{L}}$$

The pseudo first-order rate constant is then

$$1 \times 10^{-13} \frac{\text{L}}{\text{cell-h}} \cdot \frac{10^8 \text{ cells}}{\text{L}} = 1 \times 10^{-5} \text{ h}^{-1}$$

The half-life is the time required for 50% biodegradation:

$$\ln \frac{50\%}{100\%} = 0.693 = -k(t)$$

$$t_{1/2} = \frac{0.693}{k} = \frac{0.693}{1 \times 10^{-5} \text{ h}^{-1}}$$

$$t_{1/2} = 69{,}300 \text{ h} = 2890 \text{ days} = 7.9 \text{ years}$$

Saturation kinetics are used to describe reaction rates catalyzed by a fixed mass of catalyst, which is usually an enzyme associated with a bacterial population. Alternatively, inorganic species, such as the surfaces of iron minerals, can be important catalysts in soils, groundwaters, and sludges. In a system with a fixed mass of catalyst, reaction rates increase nearly linearly as a function of substrate concentration, but then the rate of change decreases asymptotically to a maximum level termed $V_{\max}$ (Figure 7.1).

This saturation phenomenon, known as *Michaelis-Menton kinetics,* is described by

$$V = V_{\max} \frac{C}{C + K_m} \tag{7.14}$$

where

V = rate of transformation [mg/(L-h)]
$V_{\max}$ = maximum rate of transformation [mg/(L-h)]
C = contaminant concentration (mg/L)
K_m = half-saturation constant (mg/L)

The half-saturation constant, K_m, is the contaminant concentration that corresponds to one half of $V_{\max}$. Changes in $V_{\max}$ and K_m as a function of contaminant concentration (e.g., competitive and noncompetitive inhibition) have been used to describe microbial toxicity [7.23]. One of the most common procedures for assessing Michaelis-Menton kinetics is the evaluation of initial rates, which are then plotted as a function of the reactant concentration. Such a procedure is shown in Example 7.3.

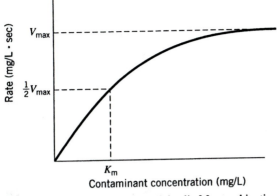

Figure 7.1 Curve describing Michaelis-Menton kinetics.

EXAMPLE 7.3 *Evaluation of Michaelis-Menton Kinetics*

The following data were obtained in a series of batch reactors containing 10^6 colony forming units (CFU)/mL of benzene-degrading bacteria.

Reactor	Initial [benzene] (mg/L)	Initial Rate [mg/(L-min)]
1	8	1.2
2	14	1.6
3	23	2.4
4	32	2.7
5	47	2.8
6	55	2.8
7	65	2.8

SOLUTION

A plot of the absolute initial rate of disappearance as a function of initial benzene concentration documents Michaelis-Menton kinetics.

Graphically illustrate.

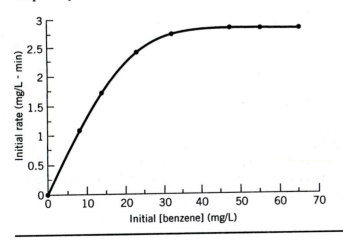

7.3 ABIOTIC TRANSFORMATIONS

Although not as important as biotic processes in the transformation of organic carbon on a global basis, abiotic processes are often significant in the transformation of biorefractory contaminants in the environment. However, one of the most important applications of abiotic processes is their use as physicochemical engineering processes, particularly advanced oxidation processes. Abiotic reactions include substitution, hydrolysis, elimination, and oxidation-reduction reactions. Of these, hydrolysis and redox reactions are the most important in the transformation of organic hazardous wastes in the environment and engineered systems.

7.3.1 Nucleophilic Substitution and Hydrolysis

Substitution reactions can be important transformation pathways for chemicals subject to nucleophilic attack. Compounds susceptible to substitution reactions have an electron-poor area that is open to attack by nucleophiles such as water, S^{2-}, Cl^-, or NO_3^-. These nucleophilic species are usually characterized by a partial or full negative charge and, as electron-rich species, are reactive with an electron-poor nucleophilic site on certain molecules. Although it is considered a simple conceptual model, the nucleophile may be seen as pushing a leaving group off of the molecule. No net change in contaminant electrons occurs during this reaction—two electrons of the new bond are brought by the nucleophile and, in turn, the leaving group takes two electrons as it is displaced. Two reaction mechanisms—S_N1 (first order) and S_N2 (second order)—provide a fundamental basis for nucleophilic attack. Information on these mechanisms is available in References 7.1 and 7.20. Nucleophilic attack can be promoted by a number of species (OH^-, S^{2-}, Cl^-, NO_3^-); however, because water is present in a much higher concentration (55.5 M vs. mM concentrations of other species), its contribution in nucleophilic attacks usually outweighs other species [7.22]. Therefore, hydrolysis is the most important substitution reaction in most hazardous waste systems.

A number of hazardous organic compounds present in aqueous media (e.g., soil-water, groundwater, aqueous RCRA wastes) may degrade by hydrolysis. The fundamental definition of hydrolysis is the breaking of a water molecule. However, during hydrolytic transformation of an organic compound, both the water molecule and the organic compound are decomposed.

The hydrolysis mechanism is based on nucleophilic attack on the organic compound. As with general substitution reactions, the nucleophile, water, attacks a region of the molecule from which electrons have been pulled, which is an area of the molecule with a partial net positive charge. When the nucleophilic atom is attacked, a leaving group, such as a halogen or —OR—, is displaced when water is added across the nucleus.

Some metal oxide surfaces have been shown to catalyze the hydrolysis of esters, a process that may be due to the reaction with surface hydroxyl groups [7.23] or the presence of higher concentrations of hydroxide ions in the microlayer near the mineral surface [7.24]. Further reading on hydrolysis mechanisms may be found in References 7.1 and 7.25.

Which organic compounds undergo hydrolysis reactions? Two classes of contaminants commonly undergo rapid rates of hydrolysis—alkyl halides and esters. Alkyl halides are characterized by a region on the molecule where the halogen pulls electrons and opens up the first carbon (the α-carbon) for nucleophilic attack:

$$R-CH_2-CH_2-Cl + H_2O \longrightarrow R-CH_2-CH_2-OH + Cl^- + H^+ \qquad (7.15)$$

Esters, which are cleaved into an organic acid and an alcohol, are the second class of compounds that react through hydrolysis:

$$R-CH_2-CH_2-\overset{\overset{\textstyle O}{\|}}{C}-O-R' + H_2O \longrightarrow R-CH_2-CH_2-\overset{\overset{\textstyle O}{\|}}{C}-OH + R'-OH \qquad (7.16)$$

Other esterlike compounds also undergo hydrolysis reactions including organophosphate esters, carbamate esters, and urea derivatives. Some typical hydrolysis reactions include attacks on the ester bond of parathion and the carbamate ester linkage of carbaryl:

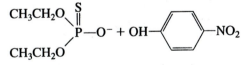

(7.17)

Parathion

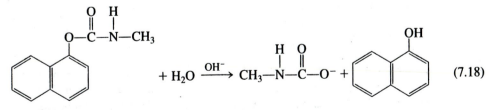

(7.18)

Carbaryl

Amides are also transformed through hydrolysis mechanisms but are less reactive than esters:

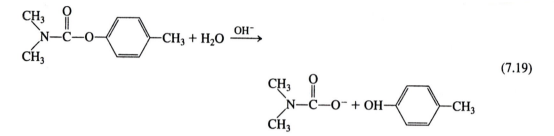

$$(7.19)$$

Some hydrolysis reactions are acid or base catalyzed; the presence of H^+ or OH^- results in a shift of electrons at the nucleophilic site, which then aids in the attack on the water-labile bond. The pH dependence of hydrolysis is described graphically by Figure 7.2. The trends in this figure depict pH effects as either U-shaped or V-shaped, which is a function of the ratio of neutral-pH rate constants to those at the pH extremes.

Because hydrolysis may be catalyzed by H^+ and OH^-, rate expressions contain additional second-order terms for acid-catalyzed and base-catalyzed hydrolysis as well as a first-order term for uncatalyzed hydrolysis:

$$-\frac{dC}{dt} = k_0[C] + k_{H+}[C][H^+] + k_{OH-}[C][OH^-] \qquad (7.20)$$

where

[C] = contaminant concentration (M)
k_0 = first-order hydrolysis rate constant (sec^{-1})
k_{H+} = second-order rate constant for acid-catalyzed hydrolysis ($M^{-1} sec^{-1}$)
k_{OH-} = second-order rate constant for base-catalyzed hydrolysis ($M^{-1} sec^{-1}$)

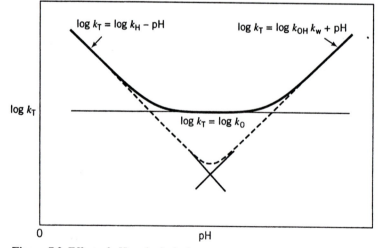

Figure 7.2 Effect of pH on hydrolysis rates. *Source:* Reference 7.22.

Some trends of acid- and base-catalyzed hydrolysis have been established. The hydrolysis of aliphatic halides is catalyzed by bases but not by acids. Therefore, their hydrolysis rates are generally independent of pH in most environments; however, in sodic saline (high-pH) soils, pH can be important. Neutral hydrolysis of amides is essentially unimportant compared to acid- and base-catalyzed hydrolysis. The hydrolysis rate of carboxylic acid esters is pH dependent and often catalyzed by both acids and bases. Based on these trends, adjusting soils contaminated with compounds subject to hydrolysis to the most reactive pH regime has been proposed for their remediation.

Common hydrolysis rate constants are listed in Table 7.3, and the calculation of hydrolysis rates is demonstrated in Example 7.4.

Table 7.3 Hydrolysis Rate Data for Selected Compounds

Compound	k_{H^+} (M^{-1} sec^{-1})	k_0 (sec^{-1})	k_{OH^-} (M^{-1} sec^{-1})	T (°C)
Organohalides				
Allyl iodide	—	4.1×10^{-6}	—	25
Benzyl chloride	—	7.0×10^{-6}	—	20
Chloroform	—	1.6×10^{-8}	6.0×10^{-5}	25
		7.3×10^{-8}		100
Fluoromethane	—	4.4×10^{-6}		100
			9.0×10^{-4}	95
Isopropyl chloride	—	1.0×10^{-3}	—	98
Methyl chloride	—	5.6×10^{-5}		90
			2.4×10^{-2}	95
Methyl bromide	—	1.1×10^{-4}		71
			3.5×10^{-1}	95
Methyl iodide	—	8.2×10^{-5}		80
			1.2×10^{-1}	95
n-Propyl bromide	—	1.6×10^{-4}	—	80
Epoxides				
Ethylene oxide	9.3×10^{-3}	6.8×10^{-7}	1.0×10^{-4}	25
Esters				
Benzyl acetate	1.1×10^{-4}	—	2.0×10^{-1}	25
Benzyl benzoate	—	—	5.9×10^{-3}	30
2,2-Dichloropropanoic acid	2.3×10^{-4}	1.5×10^{-5}	2.8×10^{3}	25
Dimethyl phthalate	—	—	6.9×10^{-2}	30
Ethyl acetate	1.1×10^{-4}	1.5×10^{-10}	1.1×10^{-1}	25
Methyl benzoate	1.7×10^{-4}	—		100
			1.9×10^{-3}	25
			9.0×10^{-3}	25

Table 7.3 (Continued)

Compound	k_{H^+} (M^{-1} sec^{-1})	k_0 (sec^{-1})	k_{OH^-} (M^{-1} sec^{-1})	T (°C)
Esters (cont.)				
Methyl formate	—	—	3.7×10^1	25
Phenyl acetate	—	6.6×10^{-8}	—	25
Amides				
Acetamide	1.0×10^{-3}	—	1.4×10^{-3}	75
Chloroacetamide	1.2×10^{-3}	—	1.4×10^{-1}	75
α-Phenylacetamide	5.2×10^{-4}	—	1.8×10^{-3}	75
Propionamide	1.2×10^{-3}	—	1.3×10^{-3}	75
Carbamates				
Baygon			5.0×10^{-1}	20
			4.6×10^{-1}	20
Dimetilan			5.7×10^{-5}	20
p-Nitrophenyl-N-methyl carbamate		$< 4 \times 10^{-5}$	3.0×10^3	25
Pyrolam			1.1×10^{-2}	20
Sevin			7.7×10^0	20
			3.4×10^0	23
Organophosphorus Esters				
Chlorpyrifos	—	1×10^{-7}	1×10^{-1}	25
Diazinon	2.1×10^{-2}	4.3×10^{-8}	5.3×10^{-3}	20
Malathion	4.8×10^{-5}	7.7×10^{-9}	5.5×10^0	27
Paraoxon	—	4.1×10^{-8}	1.3×10^{-1}	20
Parathion	—	4.5×10^{-8}	2.3×10^{-2}	20
Triphenyl phosphate	—	3×10^{-9}		82
			1.2×10^{-2}	25
2,4-D Esters				
n-Butoxyethyl	2.0×10^{-5}	2.0×10^{-5}	3.02×10^1	28
n-Butyl	6.6×10^{-4}	2.7×10^{-7}	5.0×10^3	67
Methyl			1.7×10^1	28
Miscellaneous				
DDT	—	1.9×10^{-9}	9.99×10^{-3}	27
Methoxychlor	—	2.2×10^{-8}	3.8×10^{-4}	27
Benzoyl chloride	—	4.2×10^{-2}	—	25
Ethylenimine	5.2×10^{-7}	—	—	25
β-Propiolactone	—	3.3×10^{-3}	—	25

Source: Reference 7.25.

EXAMPLE 7.6 *Calculation of Hydrolysis Half-Lives*

Determine the hydrolysis half-life of malathion at (1) pH 3, (2) pH 7, and (3) pH 10.

SOLUTION

From Table 7.3

$$k_{H+} = 4.8 \times 10^{-5} \text{ M}^{-1} \text{ sec}^{-1}$$

$$k_O = 7.7 \times 10^{-9} \text{ sec}^{-1}$$

$$k_{OH-} = 5.5 \text{ M}^{-1} \text{ sec}^{-1}$$

1. At pH 3, the concentrations of hydrogen ions and hydroxide ions are

$$[H^+] = 1 \times 10^{-3} \text{ M}$$

$$[OH^-] = 1 \times 10^{-11} \text{ M}$$

Therefore, the overall rate constant is

$$k_T = k_{H+}[H^+] + k_0 + k_{OH-}[OH^-] = (4.8 \times 10^{-5})(1 \times 10^{-3})$$
$$+ 7.7 \times 10^{-9} + (5.5)(1 \times 10^{-11})$$
$$k_T = 5.58 \times 10^{-8} \text{ sec}^{-1}$$

The half-life for malathion is

$$t_{1/2} = \frac{0.693}{5.58 \times 10^{-8} \text{ sec}^{-1}} = 12,400,000 \text{ sec} = 144 \text{ days}$$

2. At pH 7, the concentrations of hydrogen ions and hydroxide ions are

$$[H^+] = 1 \times 10^{-7} \text{ M}$$

$$[OH^-] = 1 \times 10^{-7} \text{ M}$$

Therefore, the overall rate constant is

$$k_T = k_{H+}[H^+] + k_0 + k_{OH-}[OH^-] = (4.8 \times 10^{-5})(1 \times 10^{-7})$$
$$+ 7.7 \times 10^{-9} + (5.5)(1 \times 10^{-7})$$
$$k_T = 5.58 \times 10^{-7} \text{ sec}^{-1}$$

The half-life for malathion is

$$t_{1/2} = \frac{0.693}{5.58 \times 10^{-7} \text{ sec}^{-1}} = 1,240,000 \text{ sec} = 14.4 \text{ days}$$

3. At pH 10, the concentrations of hydrogen ions and hydroxide ions are

$$[H^+] = 1 \times 10^{-10} \text{ M}$$

$$[OH^-] = 1 \times 10^{-4} \text{ M}$$

Therefore, the overall rate constant is

$$k_T = k_{H^+}[H^+] + k_O + k_{OH^-}[OH^-] = (4.8 \times 10^{-5})(1 \times 10^{-10})$$

$$+ 7.7 \times 10^{-9} + (5.5)(1 \times 10^{-4})$$

$$k_T = 5.5 \times 10^{-4} \text{ sec}^{-1}$$

The half-life for malathion is

$$t_{1/2} = \frac{0.693}{5.5 \times 10^{-4} \text{ sec}^{-1}} = 1260 \text{ sec} = 21.0 \text{ min}$$

7.3.2 Elimination Reactions

Aliphatic contaminants containing more than one halogen hydrolyze slowly under most conditions. However, ethane and propane derivatives containing a mixture of hydrogens and halogens may proceed through another pathway called *β-elimination*, which is also known as a *dehydrohalogenation* reaction:

$$(7.21)$$

Dehydrohalogenation rates are enhanced as the number of chlorines are increased on a single carbon but require the presence of hydrogens on a bordering carbon. These reactions are not as common as many other abiotic processes at hazardous waste sites and facilities. They are most important for compounds that do not rapidly degrade through hydrolysis reactions. In any case, half-lives in the range of years to decades are most common for these reactions.

7.3.3 Oxidation

Organic compounds are most commonly transformed through oxidation and reduction reactions—pathways that involve the transfer of electrons, such as reactions on mineral surfaces and with free radicals. Oxidative hazardous waste treatment processes often involve the generation of hydroxyl radicals, which are strong, nonspecific reactants. However, the most common redox reactions in the natural environment and treatment systems are biological; therefore, abiotic redox reactions (covered in this section) form a conceptual basis for understanding biochemical redox reactions, which are presented in Section 7.4.

Because an oxidation reaction must be coupled with a reduction, redox reactions are usually divided into *half-reactions*. For example, the oxidation of iron (II) by copper (II)

$$Cu^{2+} + Fe^{2+} \longrightarrow Cu^+ + Fe^{3+} \tag{7.22}$$

can be divided into half-reactions that include

$$Cu^{2+} + e^- \longrightarrow Cu^+ \tag{7.23}$$

$$Fe^{2+} \longrightarrow Fe^{3+} + e^- \tag{7.24}$$

Some common half-reactions, including their corresponding *electromotive forces* (E^0), are listed in Table 7.4. The reduction of protons (H^+) to hydrogen (H_2) is the reference reaction, with $E^0 = 0$. Half-reactions above the hydrogen reaction are oxidizing reactions; those below are reducing reactions. The E^0 for a combined redox reaction is obtained by balancing and combining the two half-reactions. The procedure for balancing half-reactions is listed in Table 7.5.

The potential for the flow of electrons provides a basis for chemical equilibrium. The electromotive force, E^0, is related to ΔG by

$$\Delta G^0 = -nFE^0 \tag{7.25}$$

where

ΔG = the Gibbs energy of the reaction at 1 atm and 25°C (kcal/mole)

n = the number of electrons in the reaction

F = caloric equivalent of the faraday = $23.06 \dfrac{\text{kcal}}{\text{volt-mole}}$

E^0 = electromotive force at 1 atm and 25°C (volts)

In addition, E^0 is related to the equilibrium constant, K, by

$$E^0 = \frac{RT}{nF} \ln K \tag{7.26}$$

where

R = the universal gas constant = 0.00199 kcal/mol-°K
T = temperature (°K)
K = the equilibrium constant

Recall from elementary chemistry that a reaction will proceed if ΔG is negative and K is positive. Balancing redox reactions and the use of Equations 7.25 and 7.26 are demonstrated in Example 7.5.

Table 7.4 Selected Half-Reactions and Standard Reduction
Potentials at 25°C and pH 7

Reaction	E^0 (V)
$Ag^{2+} + e^- \Leftrightarrow Ag^+$	+2.0
$Mn^{4+} + e^- \Leftrightarrow Mn^{3+}$	+1.65
$MnO_4^- + 8H^+ + 5e^- \Leftrightarrow Mn^{2+} + 4H_2O$	+1.51
$Cr_2O_7^{2-} + 14H^+ + 6e^- \Leftrightarrow 2Cr^{3+} + 7H_2O$	+1.33
$MnO_{2(s)} + 4H^+ + 2e^- \Leftrightarrow Mn^{2+} + 2H_2O$	+1.23
$CCl_3{-}CCl_3 + 2e^- \Leftrightarrow CCl_2{=}CCl_2 + 2Cl^-$	+1.13
$Fe(OH)_{3(s)} + 3H^+ + e^- \Leftrightarrow Fe^{2+} + 3H_2O$	+1.06
$2Hg^{2+} + 2e^- \Leftrightarrow Hg_2^+$	+0.91
$CBr_4 + H^+ + 2e^- \Leftrightarrow CHBr_3 + Br^-$	+0.83
$Ag^+ + e^- \Leftrightarrow Ag_{(s)}$	+0.80
$Fe^{3+} + e^- \Leftrightarrow Fe^{2+}$	+0.77
$CCl_4 + H^+ + 2e^- \Leftrightarrow CHCl_3 + Cl^-$	+0.67
$CHBr_3 + H^+ + 2e^- \Leftrightarrow CH_2Br_2 + Br^-$	+0.61
$MnO_4^- + 2H_2O + 3e^- \Leftrightarrow MnO_{2(s)} + 4OH^-$	+0.59
$CHCl_3 + H^+ + 2e^- \Leftrightarrow CH_2Cl_2 + Cl^-$	+0.56
$\text{⬡}{-}NO_2 + 6H^+ + 6e^- \rightleftharpoons \text{⬡}{-}NH_2 + 2H_2O$	+0.42
$Cu^{2+} + 2e^- \Leftrightarrow Cu_{(s)}$	+0.34
$Hg_2Cl_{2(s)} + 2e^- \Leftrightarrow 2Hg_{(l)} + 2Cl^-$	+0.27
$Cu^{2+} + e^- \Leftrightarrow Cu^+$	+0.16
$H^+ + e^- \Leftrightarrow \frac{1}{2}H_{2(g)}$	0
$Pb^{2+} + 2e^- \Leftrightarrow Pb_{(s)}$	−0.13
$Cd^{2+} + 2e^- \Leftrightarrow Cd_{(s)}$	−0.40
$Cr^{3+} + e^- \Leftrightarrow Cr^{2+}$	−0.41
$Fe^{2+} + 2e^- \Leftrightarrow Fe_{(s)}$	−0.44
$Zn^{2+} + 2e^- \Leftrightarrow Zn_{(s)}$	−0.76
$Mg^{2+} + 2e^- \Leftrightarrow Mg_{(s)}$	−2.37
$Na^+ + e^- \Leftrightarrow Na_{(s)}$	−2.72

Source: References 7.1, 7.21.

Table 7.5 Procedure for Balancing Redox Reactions

1. Identify each half-reaction, and write each half-reaction based on the total reaction.
2. Balance the atoms other than hydrogen and oxygen by multiplying the reactants and products by integers as necessary.
3. Balance oxygen by adding H_2O.
4. Balance hydrogen by adding H^+.
5. Balance the charge by adding electrons.
6. Multiply the reactions as necessary so that the number of electrons in each half-reaction are equal.
7. Add the two half-reactions.

EXAMPLE 7.5 *Balancing Redox Reactions*

Balance the reaction that describes the reduction of nitrobenzene by iron (II) and determine the equilibrium constant for the reaction.

SOLUTION

1. The unbalanced reaction for nitrobenzene reduction by iron (II) is

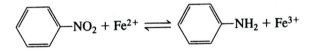

From Table 7.4, the two half-reactions for the overall reaction are

$$\text{—NO}_2 + 6\text{H}^+ + 6e^- \rightleftharpoons \text{—NH}_2 + 2\text{H}_2\text{O} \qquad E^0 = +0.42 \text{ V}$$

$$\text{Fe}^{2+} \rightleftharpoons \text{Fe}^{3+} + e^- \qquad E^0 = -0.77 \text{ V}$$

2. The oxygen, hydrogen, and charge are already balanced in each reaction. Next, multiply the iron (II) reaction by 6 to obtain the same number of electrons in each reaction:

$$6\text{Fe}^{2+} \rightleftharpoons 6\text{Fe}^{3+} + 6e^-$$

3. Add the two half-reactions and determine the overall electromotive force by adding the individual potentials from Table 7.4:

$$\text{—NO}_2 + 6\text{Fe}^{2+} + 6\text{H}^+ \rightleftharpoons \text{—NH}_2 + 6\text{Fe}^{3+} + 2\text{H}_2\text{O} \qquad E^0 = -0.35 \text{ V}$$

4. The equilibrium constant may then be determined using Equation 7.26:

$$E^0 = \frac{RT}{nF} \ln K$$

where

$R = 0.00199$ kcal/mole-°K
$T = 298$°K (25°C)
$n = 6$ electrons
$F = 23.06 \dfrac{\text{kcal}}{\text{volt-mole}}$

Substituting these values into Equation 7.26 and solving yields

$$K = 3.43 \times 10^{-36}$$

Because $K = \dfrac{[\text{products}]}{[\text{reactants}]}$ at equilibrium, the reaction does not proceed based on thermodynamics.

Although thermodynamic calculations may be used to determine whether a reaction will proceed, half-reaction data are unavailable for most of the redox reactions involving hazardous chemicals. However, some qualitative trends toward potential oxidative or reductive pathways can be established based on the oxidation state of the contaminant.

A list of organic compounds, with decreasing reactivity toward oxidative processes, is shown in Figure 7.3. Compounds without an electron-withdrawing substitution group (e.g., methane, ethene, benzene), shown at the top of the table, are easily oxidized because they provide a ready source of electrons. However, they are not effectively reduced because they do not have the potential to draw electrons. Compounds containing halogens, nitro and other electron-withdrawing groups are resistant to oxidative degradation. These substituents on an aliphatic chain or an aromatic ring draw electrons away from the sigma or pi bonds due to their electronegativity and deactivate the compound toward oxidative attack. The more halogens and nitro groups present on the structure, the more electrons are pulled away from the carbon atoms, with a corresponding slower rate of oxidation. The potential for reduction is just the opposite. Electronegative groups (e.g., chlorines, nitro) attract electrons, which may be reduced and lost as chloride, NH_2, and so on. By analogy, the more electron-attracting substituent groups on the molecule, the greater the potential for reductive transformations.

One means of tracking the oxidation or reduction of an organic contaminant as it is transformed is to evaluate its oxidation state. For example, a common pathway of TCE is to either *cis-* or *trans-*DCE:

$$\text{(7.27)}$$

The transformation of TCE to DCE is a reduction because the average carbon oxidation state is reduced from (+I) to (0).

Although a redox reaction may be favored thermodynamically, it may proceed slowly based on the system kinetics. For example, molecular oxygen is the most abundant oxidant in the environment, but it is usually not important as an oxidant because it reacts slowly with organic contaminants, although thermodynamics do favor its reactions. Nonetheless, thermodynamic calculations provide a fundamental approach for assessing the potential for a reaction to proceed.

Second to molecular oxygen, iron (III) and manganese (III)/manganese (IV) oxides are the most common oxidants in the natural environment. Ulrich and Stone [7.26] de-

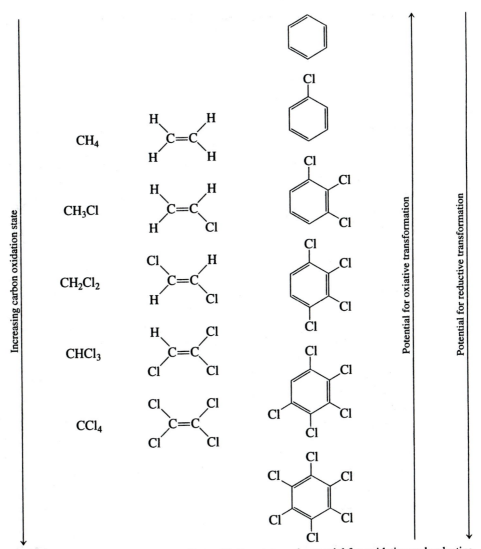

Figure 7.3 Relationship between carbon oxidation state and potential for oxidative and reductive transformations.

scribed mechanisms by which substituted phenols are oxidized on the surfaces of manganese minerals. They found that the rate of chlorophenol transformation was a function of the compound's sorptivity as well as its tendency to be oxidized by the transfer of an electron to the surface of the mineral. Stone [7.27] proposed a mechanism for the oxidation of chlorophenols at manganese oxide surfaces in which (1) the phenol binds to Mn(III) at the mineral surface and (2) an electron is transferred from the phenol to Mn(III), producing a phenoxy radical and Mn(II). The phenoxy radical then undergoes subsequent oxidation to yield a quinone or may form polymeric products.

Oxidation by Mn(III/IV) oxides occurs with only some organic compounds. Stone and Morgan [7.28] reported that saturated alcohols, aldehydes, ketones, and carboxylic acids generally showed no reactivity. For aromatic compounds, ring deactivating substituents (e.g., halogens, nitro groups) lower or halt oxidation rates. However, electron-donating substituents (e.g., —OH, —NH$_2$) significantly increase manganese oxide–promoted oxidation rates.

Iron, manganese, aluminum, and other trace elements associated with soil minerals have also been shown to initiate radical-mediated oxidations in some soils [7.29]. Although clays and iron and manganese oxyhydroxides have documented catalytic activity, some oxidations have been attributed to the nonspecific properties of soils [7.30]. The transformation of nearly 100 compounds by soil-catalyzed reactions has been documented; essentially all of the compounds are highly reactive aromatics substituted with electron-donating hydroxyl, amino, and methyl groups [7.31]. In summary, aromatic contaminants substituted with ring-activating groups are likely to degrade rapidly in soils and the subsurface.

One of the most important oxidants found in air, water, and biological systems is the *hydroxyl radical* (OH·), a species that is characterized by a one-electron deficiency compared to the thermodynamically stable species—OH$^-$. Hydroxyl radicals are ubiquitous in the environment and are found in low concentrations in surface waters and the atmosphere [7.32, 7.33].

In general, one-electron transfer redox reactions yield unstable, radical intermediates, whereas two-electron transfers commonly provide a stable product. Because of the one-electron deficiency of hydroxyl radical, it is a transient, highly oxidizing species that usually reacts with the first chemical with which it comes into contact.

Hydroxyl radicals are formed through a number of mechanisms in the environment. One of the most important of these is the reduction of O$_2$ to superoxide radical (O$_2$·$^-$) followed by its disproportion to hydrogen peroxide:

$$O_2 \xrightarrow{e^-} O_2 \cdot^- \xrightarrow[\text{dismutase}]{\text{superoxide}} H_2O_2 \tag{7.28}$$

The decomposition of hydrogen peroxide may then be catalyzed by transition metals to form hydroxyl radicals via the Fenton reaction:

$$H_2O_2 + Fe^{2+} \longrightarrow OH\cdot + OH^- + Fe^{3+} \tag{7.29}$$

In biological systems, superoxide is commonly formed by the uncoupling of reactions that carry electrons through an energy cascade involving electron transfer reactions between cytochromes (see Section 7.4.2):

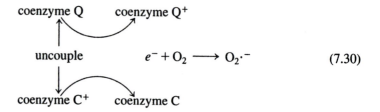

$$ (7.30) $$

Hydroxyl radicals are also formed in the atmosphere by reactions involving ozone and nitrous oxides (NO$_x$). Hydroxyl radicals are important not only in oxidizing sulfur dioxide to sulfuric acid to produce acid rain but also in the oxidation of hazardous organic air pollutants that have been released to the atmosphere [7.34].

Numerous hazardous waste treatment systems have been developed to generate hydroxyl radicals. These *advanced oxidation processes (AOPs)* include ultraviolet light combined with ozone, ozone and hydrogen peroxide, and Fenton's reagent (Equation 7.29). These processes will be discussed in detail in Chapters 12 and 13.

Most reactions of hydroxyl radicals involve electrophilic addition to alkenes and aromatics

$$
\text{(benzene)} + \text{OH·} \longrightarrow \text{(phenol with OH)} + \text{H}^+ \tag{7.31}
$$

$$
\overset{\text{Cl}}{\underset{\text{Cl}}{>}}\text{C}=\text{C}\overset{\text{H}}{\underset{\text{Cl}}{<}} + \text{OH·} \xrightarrow{\text{H}_2\text{O}} \overset{\text{Cl}}{\underset{\text{Cl}}{>}}\text{C}-\overset{\text{O}}{\overset{\|}{\text{C}}}-\text{OH} + \text{Cl}^- \tag{7.32}
$$

or hydrogen abstraction

$$
\text{CH}_3-\text{CH}_2-\text{CH}_2-\text{CH}_2-\text{CH}_2-\text{CH}_3 + \text{OH·}
$$
$$
\longrightarrow \text{CH}_3-\text{CH}_2-\text{CH}_2-\text{CH}_2-\text{CH}_2-\text{CH}_2\text{·} \tag{7.33}
$$
$$
\xrightarrow{\text{H}_2\text{O}} -\text{CH}_3-\text{CH}_2-\text{CH}_2-\text{CH}_2-\text{CH}_2-\text{CH}_2-\text{OH}
$$

The rate of contaminant oxidation by hydroxyl radicals may be described by a second-order reaction:

$$
-\frac{d[C]}{dt} = k_{\text{OH·}}[C][\text{OH·}] \tag{7.34}
$$

where

[C] = contaminant concentration (M)
[OH·] = hydroxyl radical concentration (M)
$k_{\text{OH·}}$ = second-order rate constant for the reaction of OH· with the contaminant
$\quad\quad$ (M^{-1} sec^{-1})

Second-order rate constants, some of which are listed in Table 7.6, provide a good indicator of the reactivity of hydroxyl radicals with organic contaminants. The upper limit of $k_{\text{OH·}}$ is approximately 10^{10} M^{-1} sec^{-1}, which is governed by the rate of diffusion of hydroxyl radicals in water. Reactivities defined by $k_{\text{OH·}}$ above 10^9 M^{-1} sec^{-1} are considered rapid enough to be significant. Rate constants below 10^8 M^{-1} sec^{-1} are deemed insignificant because competing OH· quenching reactions, such as the reaction of OH· with superoxide

$$
\text{OH·} + \text{O}_2\text{·}^- + \text{H}^+ \longrightarrow \text{O}_2 + \text{H}_2\text{O} \tag{7.35}
$$

are in the range of 10^6 to 10^8 M^{-1} sec^{-1}.

Based on the hydroxyl radical rate constants listed in Table 7.6, alkenes and aromatics react rapidly with hydroxyl radicals. Even highly chlorinated alkenes, such as

Table 7.6 Second-Order Rate Constants of Selected Compounds
for Reactivity with Hydroxyl Radicals

Compound	$k_{OH} \cdot (M^{-1} \, sec^{-1})$
Monocyclic Aromatic Hydrocarbons	
Benzene	7.8×10^9
Ethylbenzene	7.5×10^9
Toluene	3.0×10^9
m-Xylene	7.5×10^9
o-Xylene	6.7×10^9
p-Xylene	7.0×10^9
Polycyclic Aromatic Hydrocarbons	$\approx 1 \times 10^{10}$
Chlorinated Solvents	
Carbon tetrachloride	$< 2 \times 10^6$
Chloroform	$\approx 5 \times 10^6$
1,2-Dichloropropane	3.8×10^8 @ pH 2.8
Epichlorohydrin	2.9×10^8
Methylene chloride	$7.3 \times 10^{7*}$
Tetrachloroethylene (PCE)	2.8×10^9
1,2,4-Trichlorobenzene	4×10^9
1,1,2-Trichloroethane	$1.5 \times 10^{8*}$
Trichloroethylene (TCE)	4.0×10^9
Vinyl chloride	1.2×10^{10}
Insecticides	
Aldicarb	8.1×10^9 @ pH 3.5
Carbofuran	7×10^9
Chlordane	6×10^8 @ pH 3.3
1,2-Dibromo-3-chloropropane	$3.4 \times 10^{8*}$
Endrin	$8.9 \times 10^{8*}$
Lindane	$7.8 \times 10^{8*}$
Methoxychlor	2×10^{10}
Toxaphene	$4.7 \times 10^{8*}$ @ pH 1.9
1,2,3-Trichlorobenzene	4×10^9
Herbicides	
Atrazine	2.6×10^9 @ pH 3.6
2,4-D	5×10^9
Dalapon	7.3×10^7 @ pH 3.4
Diquat	8.0×10^8 @ pH 3.1
Endothall	1.5×10^9 @ pH 1
Simazine	2.8×10^9 @ pH 3.5
2,4,5-T	4×10^9

Table 7.6 Continued

Compound	$k_{OH} \cdot (M^{-1} \ sec^{-1})$
Fungicides	
Dinoseb	4×10^9
Pentachlorophenol	4×10^9
Industrial Intermediates	
2-Bromoethanol	3.5×10^8 @ pH 2.7
Bromoform	1.3×10^8 @ pH 8.5
Chlorobenzene	5.5×10^9
1,2-Dichlorobenzene	4.0×10^9
1,3-Dichlorobenzene	5.0×10^9 @ pH 8.7
1,4-Dichlorobenzene	4.0×10^9
trans-1,2-Dichloroethylene	6.2×10^9
Diethyl phthalate	4×10^9
Dimethyl phthalate	4×10^9
Hexachlorocyclopentadiene	$2.4 \times 10^{9*}$
Methyl bromide	$9.5 \times 10^{7*}$
Nitrobenzene	3.9×10^9
Phenol	6.6×10^9
1,1,1-Trichloro-2-methyl-2-propanol	2.7×10^8 @ pH 3
Polychlorinated Biphenyls	
2,3,3',5,6-PCB (1254)	5×10^9
2,2',4,4',5,5'-PCB (1260)	6×10^9
Chlorinated Dioxins	
2,3,7,8-Tetrachlorodibenzo-p-dioxin (TCDD)	4×10^9

*Average rate constant.
Source: Reference 7.35

PCE and hexachlorocyclopentadiene, react rapidly. However, chlorinated alkanes, such as chloroform and carbon tetrachloride, are considered essentially nonreactive. Even though numerous important hazardous chemicals (e.g., DDT, TCDD) react rapidly with hydroxyl radicals, many of these compounds are highly hydrophobic, with log K_{ow} > 5. As noted in Section 7.1.3, these compounds exhibit significantly lower reactivity with hydroxyl radicals if sorbed in solid-water systems. Quenching of hydroxyl radicals by inorganic species, especially bicarbonate ($k_{OH} \cdot = 1.5 \times 10^7 \ M^{-1} \ sec^{-1}$) and carbonate ($k_{OH} \cdot = 4.2 \times 10^8 \ M^{-1} \ sec^{-1}$) may also significantly impact rates of hydroxyl radical concentrations. Although the $k_{OH} \cdot$ for most quenching species is in the range of 10^6 to $10^8 \ M^{-1} \ sec^{-1}$ their concentrations are often up to 1000 times greater than the concentrations of the contaminants. The use of rate data in evaluating hydroxyl radical reactions is demonstrated in Example 7.6.

EXAMPLE 7.6   *Rate of Contaminant Oxidation by Hydroxyl Radicals*

The steady-state concentration of hydroxyl radicals in a UV/ozone reactor is 10^{-11} M. Determine the half life of hexachlorocyclopentadiene in the reactor.

SOLUTION

From Table 7.6, the second order hydroxyl radical rate constant with hexachlorocyclopentadiene is 2.4×10^9 M^{-1} sec^{-1}. Because the concentration of hydroxyl radicals is at steady state in the reactor, a steady-state approximation to the second-order rate expression may be used:

$$k = k_{OH} \cdot [OH \cdot]_{\text{steady state}}$$

$$k = (2.4 \times 10^9 \text{ M}^{-1} \text{ sec}^{-1}) (10^{-11} \text{ M}) = 0.024 \text{ sec}^{-1}$$

A first-order reaction equation can then be used to determine the half-life of hexachlorocyclopentadiene:

$$\frac{C}{C_0} = e^{-kt}$$

where

$C/C_0 = 0.5$
$k = 0.024$ sec^{-1}

Substituting values in the equation yields

$$0.5 = e^{-0.024t}$$

$$\ln(0.5) = -0.024t$$

$$t_{1/2} = 29 \text{ sec.}$$

The presence of bicarbonate or other quenchers increases the steady state hydroxyl radical concentration needed to achieve equally effective treatment. The treatment of most natural waters containing typical concentrations of alkalinity requires a steady state hydroxyl radical concentration of 10^{-9} M.

7.3.4   Reduction

The reduction of hazardous organics has not been studied as extensively as their oxidation; however, the results of recent research have confirmed that highly oxidized compounds, such as carbon tetrachloride and hexachloroethane, may be rapidly re-

Table 7.7 Second-Order Rate Constants
$(M^{-1} sec^{-1})$ of Selected Compounds for Reactivity
with Aqueous Electrons

Compound	$k_{e_{aq}^-}(M^{-1} sec^{-1})$
Benzene	9.0×10^6
Carbon tetrachloride	1.6×10^{10}
Chlorobenzene	5.0×10^8
Chloroform	3.0×10^{10}
1,2-Dichlorobenzene	4.7×10^9
1,3-Dichlorobenzene	5.2×10^9
1,4-Dichlorobenzene	5.0×10^9
trans-1,2-Dichloroethylene	7.5×10^9
Nitrobenzene	3.7×10^{10}
Phenol	2.0×10^7
Tetrachloroethylene (PCE)	1.3×10^{10}
Toluene	1.4×10^7
Trichloroethylene (TCE)	1.9×10^9
Vinyl chloride	2.5×10^8

Source: Reference 7.36.

duced if a source of electrons is available. Although oxidants, such as hydroxyl radicals, react rapidly with electron-rich functional groups (e.g., double bonds), reduction proceeds rapidly only for highly oxidized organic contaminants. Some second-order rate constants for the reaction of electrons with common organic contaminants are listed in Table 7.7. Note that the most oxidized compounds, such as carbon tetrachloride, react rapidly with electrons, whereas the more reduced compounds (e.g., benzene, phenol) react more slowly. An inverse correlation is found between the potentials for chemical oxidation and for reduction. This trend is also confirmed by inspection of Figure 7.3; oxidized organics are most susceptible to reductive processes, and reduced organics are most easily oxidized. Such conceptual knowledge serves as the basis for site and facility assessments as well as for technology screening and the design of hazardous waste treatment systems.

The results of recent investigations of reductive processes have shown that oxidized organic contaminants (e.g., chlorinated alkanes, nitroaromatics) can be reduced in anaerobic soil and subsurface systems. Although the mechanisms and rates of reductive processes require more study, it is now obvious that natural reductive processes do provide a pathway by which oxidized organics are lost through *natural attenuation* mechanisms.

Numerous compounds have been investigated as possible reductants in reducing environments (e.g., subsurface systems free of oxygen) including pyrite (FeS), iron (II) carbonates, and sulfide [7.37]. The reductive degradation pathway of carbon tetrachloride by such a natural reductive process is shown in Figure 7.4. However, the reducing agents found in the highest concentrations in anaerobic natural systems [e.g., iron (II) sulfides, iron (II) carbonates, and hydrogen sulfide] react slowly with oxidized organic contaminants, and their presence alone cannot account for the more rapid rates

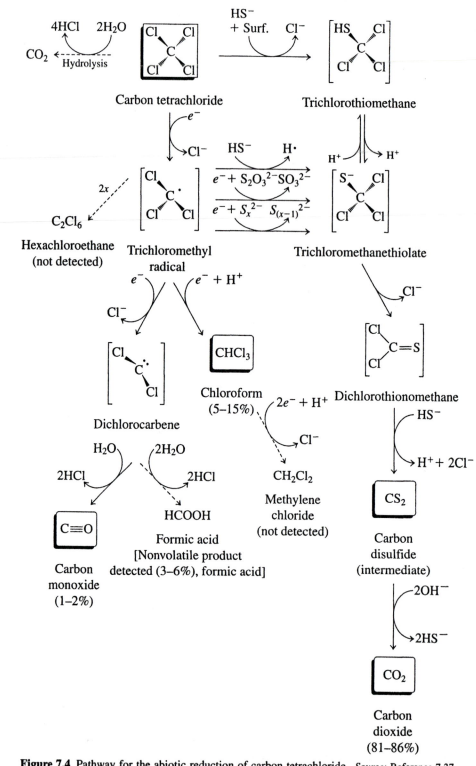

Figure 7.4 Pathway for the abiotic reduction of carbon tetrachloride. *Source:* Reference 7.37.

of reduction observed in the field [7.38]. Therefore, other reactive redox species have recently been studied to provide a basis for the rapid reductions often observed in anaerobic soils and sediments. These intermediates are believed to include iron (II) porphyrins [7.39] and natural organic matter (which is usually modeled using a hydroquinone/quinone couple [7.40]). The most common model of such coupled redox systems is

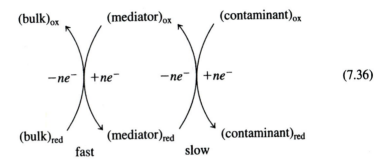

$$\tag{7.36}$$

Tracing the coupled redox reactions from contaminant reduction on the right side of Equation 7.36, a mediator is oxidized as the contaminant is reduced. The most important mediator in soils is believed to be the natural organic matter, in which hydroquinone-like moieties reduce the oxidized organics [7.41]. Cycling of the mediator is promoted by subsequent reduction from a bulk pool of reductants, such as iron (II) or sulfide.

Reductive processes have also been the focus of recent studies applied to remediation and treatment systems. These processes have been studied primarily as the basis for the remediation of dilute aqueous solutions of oxidized organic compounds (e.g., contaminated groundwater) rather than pathways that affect site assessments or facility assessments. *Zero-valent iron* (i.e., iron metal) has been used as a source of electrons in the reduction of chlorinated alkanes such as carbon tetrachloride and chloroform [7.42].

Iron metal is a natural reductant that readily loses electrons through corrosion:

$$Fe^0 \rightleftharpoons Fe^{2+} + 2e^- \tag{7.37}$$

In the presence of water (a proton donor), chlorinated alkanes proceed through reductive dehalogenation:

$$Fe^0 + CCl_4 + H^+ \longrightarrow Fe^{2+} + CCl_3H + Cl^- \tag{7.38}$$

In addition, the iron (II) can also serve as a reductant:

$$2Fe^{2+} + CCl_4 + H^+ \longrightarrow 2Fe^{3+} + CCl_3H + Cl^- \tag{7.39}$$

Although zero-valent metals are not common to soils or subsurface systems, fine-grained iron has been used in packed bed reactors [7.43] and as a permeable barrier, called a tongue-and-gate system, through which contaminated groundwater passes as a basis for in situ remediation [7.44].

Another mechanism for promoting reduction is an electron beam that uses a 5000 Ci mass of ^{60}Co to generate electrons [7.45]. The system not only provides electrons that reduce oxidized organics, it also converts molecular oxygen to superoxide radical ion through one-electron transfers, which may disproportionate to hydrogen peroxide and decompose to OH· via Fenton-like reactions (Equations 7.28 and 7.29).

Yet another innovative process that supplies electrons is the near-UV light illumination of the anatase form of titanium dioxide. In this process, aqueous suspensions of titanium dioxide, when irradiated with light of wavelengths < 400 nm, absorb photons. This absorbed light energy causes a change in the electrical balance on the surface of the anatase in which an electron is shifted from a region called the valence band to a higher-energy location known as the conduction band. This mechanism results in the potential for both reductive and oxidative pathways for contaminant destruction.

7.3.5 Photochemical Reactions

The photochemical transformation of organic hazardous chemicals and some inorganic nonmetals (e.g., cyanides) can be important under certain conditions. Obviously, sunlight-induced photolysis can occur only in systems where light reaches the contaminant. Therefore, the two hazardous waste systems where photolysis occurs are the upper layers of aquatic systems (e.g., surface waters or surface impoundments) and the top few millimeters of soil surfaces.

Rates of photolysis in aquatic systems have been described by Zepp and Cline [7.46], and the photolytic degradation of hazardous compounds on soil surfaces has been investigated by Miller and co-workers [7.47, 7.48]. Photodegradation processes often involve direct light absorption by the contaminant followed by one of many decomposition mechanisms, a process known as *direct photolysis.* However, another common mechanism is also prevalent—*indirect photolysis,* in which intermediate species, such as humic acids, some iron oxides, colored pigments, or soil surfaces, absorb the light. The light energy is then transferred to oxygen or water, resulting in the formation of oxidants such as hydroxyl radicals or singlet molecular oxygen, species that react with organic contaminants as described above.

Photochemical treatment of hazardous aqueous wastes has received a great deal of interest and, as with most of the pathways, serves as a basis for the design of engineered systems. Photochemical hazardous waste treatment processes will be surveyed in Chapter 12.

7.4 BIOTIC TRANSFORMATIONS

Microorganisms are ubiquitous in almost all phases of the natural environment, and many use organic matter as a carbon and energy source. As a result, most organic compounds, including many hazardous chemicals, may be transformed to carbon dioxide and water through microbial metabolism involving biochemical redox reactions. Hazardous organic chemicals are often metabolized by pathways and reactions similar to those for naturally occurring organic compounds, although in many cases their rates of degradation are slower than the metabolism of natural products.

7.4.1 Important Organisms in Hazardous Waste Systems

Biodegradation is mediated by microorganisms that are classified ecologically as *decomposers*. Their natural role is the metabolism of fats, proteins, carbohydrates, nucleic acids, and other energy-rich biological compounds. These natural chemicals are synthesized by primary producing organisms (plants) from carbon dioxide and the energy of sunlight. Consumers (animals) metabolize the natural molecules of plants and also synthesize some new ones. The metabolic pathways of primary producers and consumers are responsible for the synthesis of complex biological molecules, including proteins, nucleic acids, and lipids. Decomposers degrade these biological molecules (which are usually found in the form of dead plant and animal matter or fecal material) to their thermodynamic end points—carbon dioxide and water. This process of degradation, known as *catabolism,* yields energy which the organism uses for growth and motility. As part of the natural growth process, a portion of the biological molecules are incorporated into new cell material of the microorganisms—a process known as *anabolism.*

Most organisms found in soils, groundwater, and hazardous waste treatment systems are unicellular; these organisms are divided into prokaryotes and eukaryotes. *Prokaryotes* do not have a true nucleus or membrane-bound organelles, whereas *eukaryotes* do have membrane-bound internal structures. The organisms responsible for the biodegradation of hazardous chemicals in soils, groundwater, sludges, and treatment facilities are primarily bacteria and fungi. *Bacteria* (which are prokaryotes) are the predominant form of biomass in soils; however, *fungi* (which are eukaryotes) may predominate in acidic soils. Although protozoans, annelids (earthworms), and algae also decompose hazardous chemicals, their role in most systems is usually insignificant. The most important microorganisms responsible for biodegradation in environmental systems are illustrated in Figure 7.5. A micrograph of a colony of nitrophenol-degrading bacteria, *Rhodococcus* spp., is shown in Figure 7.6.

Bacteria represent a morphologically diverse group of microorganisms. Their typical size is 0.5–3 μm, and they are found in spherical, rod, and spiral shapes. Two important structures of bacteria are the membrane and cell wall. The membrane is semipermeable—it regulates the entry of nutrients and water. The cell wall is a structurally complex three-dimensional structure that dictates the shape of the cell. Bacteria are sometimes placed in three groups—*eubacteria, mycobacteria,* and *spirochetes.* These groups are classified based on motility and cell wall structure. Eubacteria, which are probably the most important in hazardous waste biodegradation, are the largest and most diverse group of bacteria. They are found as cocci or spheres (0.5–1.0 μm in diameter), rods (0.5–1.0 μm wide and 1.5–3.0 μm long), and helixes, and are often aggregated in chains. These cells are often motile, using flagella to propel themselves. Although numerous genera of bacteria are responsible for the biodegradation of hazardous contaminants, perhaps the most important include *Pseudomonas, Mycobacterium, Arthrobacter,* and *Bacillus.*

The mycelial bacteria, Actinomycetes, are another significant group in hazardous waste systems. Two of the most important genera of the Actinomycetes are *Nocardia* and *Mycobacterium,* which are important soil organisms that metabolize typical petroleum hydrocarbons.

The enhanced presence of some groups of bacteria in bioremediation systems can be promoted by adjusting environmental conditions to favor their growth. *Methanogens*

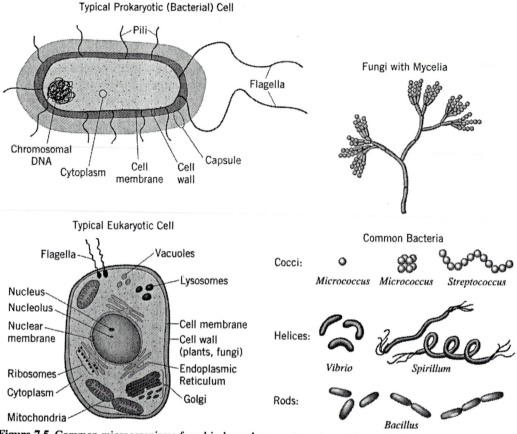

Figure 7.5 Common microorganisms found in hazardous waste systems. *Source:* Reference 7.49.

are strict anaerobes that convert fermentation products (e.g., ethanol, acetate, etc.) to methane while using carbon dioxide as a terminal electron acceptor. Methanogens also use hydrogen as an energy source:

$$CH_3COOH \longrightarrow CH_4 + CO_2 \tag{7.40}$$

$$CO_2 + 4H_2 \longrightarrow CH_4 + 2H_2O \tag{7.41}$$

Methanogens are important in bioremediation because they provide a cometabolic route for the dechlorination of some oxidized hazardous organic compounds.

Metabolically opposite to methanogens are *methanotrophs,* which are ubiquitous, obligate aerobic bacteria that use methane as a carbon and energy source. They occur naturally in transition areas between anaerobic and aerobic zones of soils where they have a supply of both methane and oxygen. Methanotrophs are capable of oxidatively cometabolizing some chlorinated alkenes and alkanes because they possess the enzyme

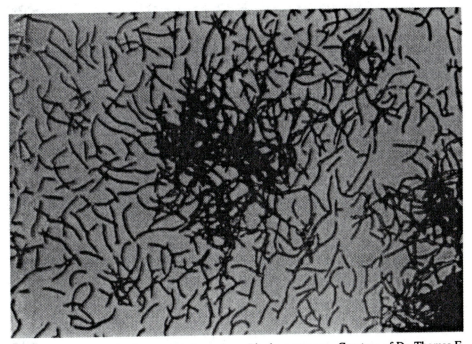

Figure 7.6 Micrograph of a common bacterium, *Rhodococcus* spp. Courtesy of Dr. Thomas F. Hess, University of Idaho.

methane monooxygenase, a relatively nonspecific catalyst for oxidative biodegradation reactions.

Fungi are characterized by long, narrow filaments called *hyphae,* which in turn make up a *mycelium.* One of the most important aspects of fungi is the wide pH range at which they exist. Although most commonly found at low pH regimes of 4–5, fungi can also grow at pH levels as high as 10. Of particular interest is white rot (or wood rot) fungus, such as *Phanerochaete chrysosporium* and *Phanerochaete sordida.* These fungi occur naturally, degrading lignin, a component of wood, which is resistant to most enzymatic reactions because of its β-glycosidic linkages. These fungi have been shown to degrade PAHs, PCBs, pentachlorophenol, DDT, and other biorefractory compounds through the activity of extracellular peroxidase enzymes. *Peroxidase* is a nonspecific enzyme that degrades lignin by producing hydroxyl radicals through a biological Fenton-like reaction. As discussed in Section 7.3.3, hydroxyl radicals readily degrade DDT, PCBs, and many other hazardous contaminants.

Although numerous research investigations have focused on the degradation of a hazardous chemical using a single species of microorganism, a microbial consortium degrading the parent compound as well as its products is the most common phenomenon in natural and engineered hazardous waste systems. Often, a bacterium has the capability of degrading the parent compound but cannot metabolize one or more breakdown products. In contrast, another species may be capable of degrading the products but not the parent compound. This phenomenon becomes even more complex because hazardous chemicals are usually found as complex mixtures, so the need for a consortium is even greater.

7.4.2   Fundamentals of Microbial Growth and Metabolism

One of the most important concepts in biodegradation is energy flow. Not only do microorganisms require carbon and other elements to grow and reproduce, they need energy for reproduction, active transport, and motility. Different classes of microorganisms have evolved wide-ranging metabolic pathways to obtain energy. *Chemotrophs* capture energy from chemical sources, and *phototrophs* use energy from light. Bacteria are also classified based on their source of carbon. *Autotrophic* bacteria obtain carbon from CO_2; bacteria that obtain carbon from organic matter are defined as *heterotrophic*. Furthermore, if the source of electrons (i.e., the electron donor) is an organic compound, the organism is classified as an *organotroph*; *lithotrophs* use inorganic compounds as the electron source. Organisms may be classified based on the carbon source and energy source by linking the names for each metabolic requirement: *Chemoheterotrophs* receive both their energy and carbon from organic matter; *chemoorganotrophs* obtain their energy and electrons from organic compounds. The most common classes of microorganisms found in hazardous waste systems, such as contaminated soils and sludges, are chemoorganotrophs and chemoheterotrophs.

The basis for hazardous waste biotransformations is the metabolism and ultimate cycling of carbon and energy. Heterotrophic microorganisms have evolved over time to degrade such natural products such as fats, proteins, and carbohydrates. As discussed in Section 7.1.1, one of the problems with hazardous organic chemicals is that they are sometimes foreign to the natural enzymology and metabolism of microorganisms, and are therefore called *xenobiotic*. These chemicals are biorefractory because they are metabolized slowly. The fundamental basis for the rate of biodegradation lies in the energy available in the chemical bonds. Therefore, a review of the principles of enzymology, the energy of organic and biological molecules, and fundamental heterotrophic metabolism is appropriate before discussing the biodegradation of hazardous compounds.

Enzymes are proteins that are made up of subunits called *amino acids*. Most proteins consist of hundreds of amino acids linked together by peptide bonds, which are formed by a condensation reaction between the carboxylic group and the amino group of adjacent amino acids. The catalytic properties of enzymes are based on the sequence of amino acids. Furthermore, the order of the amino acids provides a three-dimensional structure, and through twisting and folding, an enzyme is able to catalyze the decomposition of a specific biological substrate. The structure of a common enzyme, ribonuclease, is shown in Figure 7.7. Enzymes contain an active site on their three-dimensional surface where biochemical redox reactions are catalyzed. These sites are usually substrate-specific and have a conceptual analogy to a lock (the enzyme) and key (the substrate). However, not all enzymes are substrate-specific; for example, some enzymes are capable of nonspecific oxidations.

Many biological transformations are similar to reactions that occur abiotically. However, biological reactions have the potential to proceed more rapidly through enzymatic catalysis because enzymes lower the activation energy of biochemical reactions, allowing slow reactions to proceed in seconds. Enzymes are essentially conservative and can be used over and over.

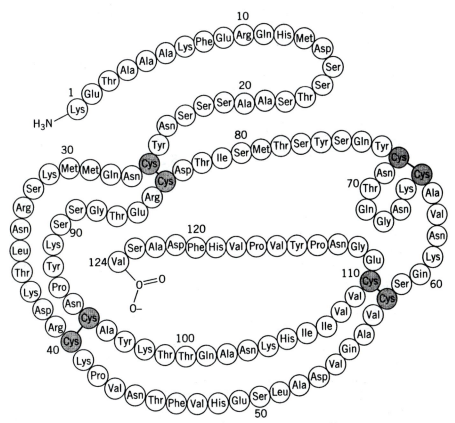

Figure 7.7 Ribonuclease, a common enzyme. *Source:* Reference 7.50.

Names of enzymes end in -*ase;* for example, alcohol dehydrogenase is the enzyme that catalyzes the abstraction of a hydrogen atom from ethyl alcohol:

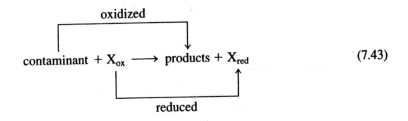

$$CH_3—CH_2OH + NAD^+ \xrightarrow[\text{dehydrogenase}]{\text{alcohol}} CH_3—\overset{\overset{\displaystyle O}{\|}}{C}—H + NADH + H^+ \quad (7.42)$$

A conceptual basis for biodegradation reactions involving hazardous compounds is a coupled biochemical oxidation-reduction reaction:

$$\text{contaminant} + X_{ox} \longrightarrow \text{products} + X_{red} \quad (7.43)$$

oxidized

reduced

where X is a terminal electron acceptor. The energy available for microbial growth and metabolism is directly proportional to the electromotive force ($E°$) of the biochemical redox reaction. Therefore, the reduction reaction (which may involve competing terminal electron acceptors) proceeds according to the rule that the environmental and metabolic conditions that yield the highest energy for the microorganism dictate the biodegradation pathway; that is, the terminal electron acceptor that provides the greatest free energy is favored.

Aerobic reactions occur in the presence of molecular oxygen, which is the terminal electron acceptor. Anaerobic processes occur in the absence of molecular oxygen with three potential reaction pathways—anaerobic respiration, fermentation, and methane fermentation. In *anaerobic respiration,* an oxidized inorganic compound (e.g., nitrate, sulfate, ferric iron) or an oxidized organic compound (e.g., carbon tetrachloride, PCE) serves as the electron acceptor. Organic compounds and intermediates function as both electron donors and acceptors in *fermentation;* most of the substrate energy remains in such intermediates as organic acids, alcohols, and ketones. In *methane fermentation,* CO_2 serves as the terminal electron acceptor, resulting in the formation of methane.

Energy yield is the underlying driving force in the microbial use of an electron acceptor, so free-energy relationships can be used to predict the most effective electron acceptor and the pathway of the microbial metabolism. When more than one electron acceptor is present in the system, the acceptor that is used by the microorganism is based on thermodynamics; that is, the electron acceptor that yields the greatest energy is the one that is favored. The greatest capture of energy by terminal electron acceptors is $O_2 > NO_3^- > SO_4^{2-} > CO_2$. (The most energy captured is through aerobic metabolism with O_2 as the electron acceptor). The environment then dictates the predominant microorganisms based on a hierarchy of available free energy, with the dominant microorganisms being those that are capable of using the most energy-rich electron acceptor. The classes of microorganisms may then be said to characterize the redox conditions of the system. If the O_2 is depleted, nitrate reduction (i.e., denitrification) provides the most free energy as a substitute for O_2 and the system may be characterized as a *denitrifying environment.* If nitrate, in turn, has been depleted, sulfate serves as the electron acceptor through *sulfate reduction* or *sulfate respiration.* Finally, if sulfate is unavailable, carbon dioxide becomes the terminal electron acceptor through *methanogenesis.* The ability of chlorinated aliphatic compounds to serve as electron acceptors is a relatively recent discovery and has opened new possibilities for the bioremediation of chlorinated solvents. A discussion of chlorinated aliphatic compounds as electron acceptors will be presented in Section 7.4.3.

The pathway for the aerobic metabolism of one of the most common natural compounds, glucose, serves as an appropriate example of microbial metabolism. In this pathway, glucose proceeds through stepwise oxidations in which small amounts of energy are released in some of the steps. The first section of metabolism, the *Embden-Meyerhof pathway,* is shown in Figure 7.8. The initial reaction in glucose metabolism is the addition of two high-energy phosphate bonds to the molecule, yielding fructose-1,6-diphosphate, which are provided by the high-energy phosphate bonds of *adenosine triphosphate* (ATP). Adenosine triphosphate is the common energy-rich compound that is generated and used in the biochemical pathways of all organisms. The energy contained in the chemical bonds of carbohydrates, fats, proteins, and any other compound that is metabolized by microorganisms is eventually captured by adenosine diphosphate (ADP) to form ATP:

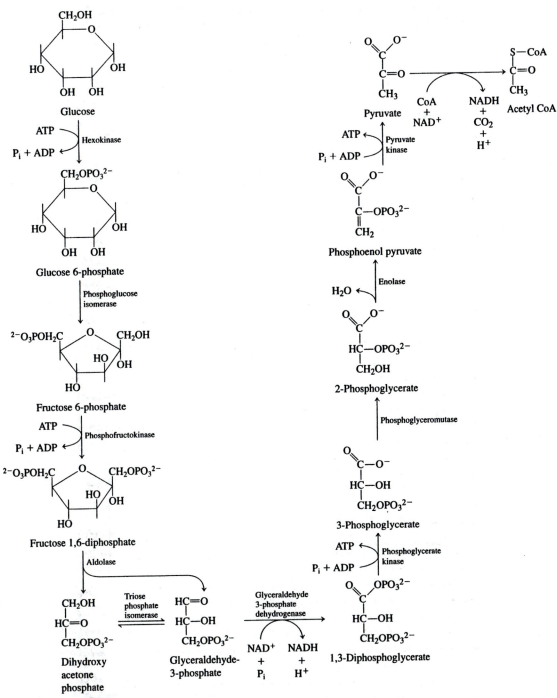

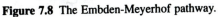

Figure 7.8 The Embden-Meyerhof pathway.

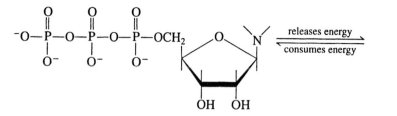

Adenosine triphosphate (ATP)   (7.44)

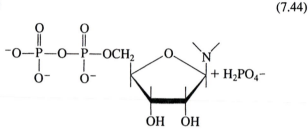

Adenosine diphosphate (ADP)

The energy-rich phosphate bond of ATP contains 7 kcal per mole and is used by organisms for cellular synthesis, motility, and other cellular functions such as the active transport of chemicals through the cell membrane.

The scheme of glucose metabolism involves splitting fructose-1,6-diphosphate into dihydroxy acetone phosphate and glyceraldehyde-3-phosphate. Note that these steps are enzyme-catalyzed and a specific enzyme is involved in each reaction. Aerobic metabolism eventually results in the formation of pyruvate and the net generation of two ATPs per mole of glucose as well as two moles of the energy-rich intermediate nicotinamide adenine dinucleotide (NADH). Biochemical energy is equivalent to reducing potential, and reducing potential is equivalent to an abundance of electrons. The biochemical role of NAD^+ is to capture reducing potential from energy-rich compounds as they are metabolized. The NAD^+ captures both an electron and a proton (i.e., hydrogen atom) as a means of transferring this biochemical energy. The NAD^+ is then reduced to NADH and is termed an energy-rich intermediate. The energy contained in NADH is eventually released to form ATP in the electron transport system.

Pyruvate, the final product of the Embden-Meyerhof pathway under aerobic conditions, is converted to acetyl Coenzyme A (CoA) prior to entering the next major section of metabolism, which is called the *tricarboxylic acid cycle (TCA cycle)* (Figure 7.9). Here, the remaining carbon bonds are metabolized to carbon dioxide and a significant amount of energy is captured in the intermediate NADH. In fact, all of the remaining energy associated with the glucose molecule is captured in the form of the energy-rich intermediates NADH and $FADH_2$. The TCA cycle is semicatalytic; in other words, the cycle must proceed through three passes in order to metabolize the carbon associated with one glucose molecule.

Although considered energy-rich intermediates, NADH and $FADH_2$ cannot be used directly in cellular metabolism. Therefore, the final step in the aerobic metabolism of glucose is the transfer of energy from NADH to form ATP, which occurs in the *electron transport system* (Figure 7.10). Through a series of coupled reactions, the energy contained in NADH is transferred to a series of cytochromes, which are proteins capable of electron transfer reactions. At three points in the process, a sufficient energy

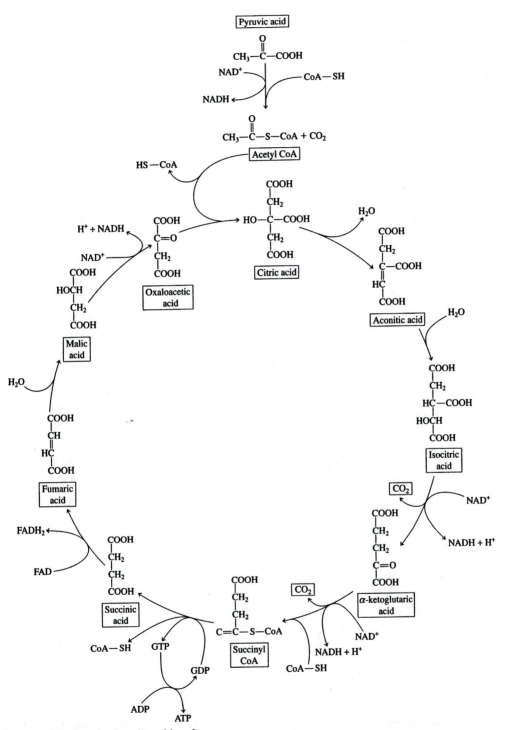

Figure 7.9 The tricarboxylic acid cycle.

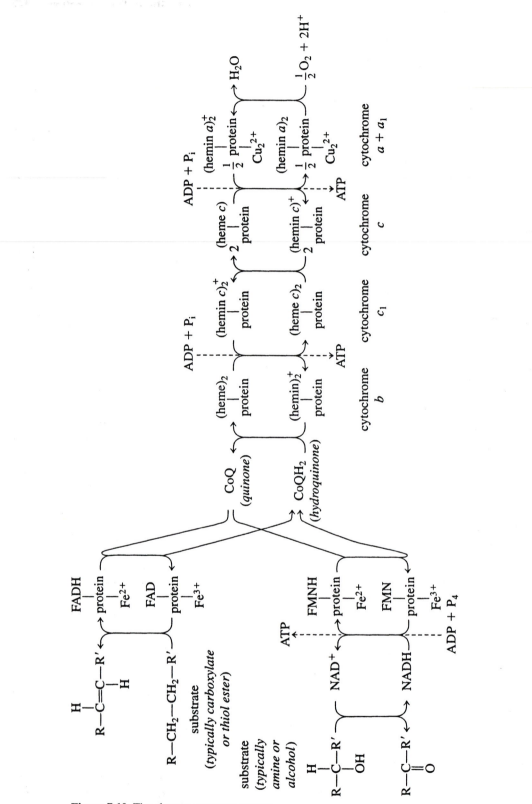

Figure 7.10 The electron transport system.

difference exists between the cytochromes that the energy can be captured through a coupled reaction with ADP to form ATP. In the case of $FADH_2$, only sufficient potential energy exists in $FADH_2$ for the capture of energy into two ATPs through the electron transport chain.

The metabolic routes shown in Figures 7.8 through 7.10 describe the degradation of the common carbohydrate glucose. However, organic molecules are not only degraded to carbon dioxide and water, they also are used as a carbon source from which new cells are synthesized. Intermediates that enter the tricarboxylic acid (TCA) cycle (Figure 7.9) may then proceed to other synthesis pathways because the TCA cycle is linked to routes that synthesize amino acids, nucleic acids, and other building blocks for new bacterial cells.

7.4.3   Biodegradation Reactions and Pathways of Hazardous Contaminants

Although knowledge of the Embden-Meyerhof pathway and the TCA cycle provide an understanding of fundamental microbial metabolism, hazardous organic chemicals do not degrade by the same pathways as glucose. How, then, do these xenobiotic compounds biodegrade? The usual first step in metabolism is passing through the cell membrane. Compounds that are too large to move through the membrane by passive diffusion or active transport may be partially degraded by exoenzymes, which are secreted through the cell wall. A number of reactions may then be used by microorganisms in the cytoplasm to metabolize organic compounds, including *hydroxylation, hydrolysis, dehalogenation, dealkylation,* and *reduction* processes, which are illustrated in Table 7.8.

Note that many of the reactions listed in Table 7.8 are similar to the abiotic reactions described in Section 7.3. The primary difference is that biodegradation reactions are catalyzed by enzymes. For example, enzyme-catalyzed hydrolysis is much the same as abiotic hydrolysis through addition of water. Hydrolysis is also a common reaction catalyzed by exoenzymes. Dehydrohalogenation is the removal of a halogen and a hydrogen atom, similar to elimination reactions described in Section 7.3.3. The enzymes that catalyze these reactions are usually specific for the particular compound and the type of reaction taking place. However, particularly for oxidation reactions, nonspecific enzymes can often carry out a number of reactions. These enzymes are commonly called the *mixed function oxidase* enzymes.

Cometabolism may be considered not as metabolism in a strict sense but as an incidental transformation. Cometabolism is the biodegradation of a contaminant in the presence of another compound that functions as the primary energy source. Contaminants that are cometabolized are usually transformed only to intermediate compounds via the cometabolic mechanism. Cometabolism is highly significant in the biological degradation of many hazardous compounds. Some compounds, such as a group of chlorinated solvents, are biodegradable only through cometabolism.

The mechanisms and pathways of hundreds of hazardous chemicals have been elucidated. The following discussion of microbial degradation focuses on the dominant compounds covered in Chapter 2.

Aliphatic Hydrocarbons. As expected by their reduced nature, aliphatic and aromatic hydrocarbons biodegrade most efficiently under aerobic conditions. A large number of bacterial species uses aliphatic and aromatic hydrocarbons as their sole source of carbon and energy. Biodegradation rates of alkanes are related to their chain length;

Table 7.8 Biochemical Transformation Reactions

I. Substitution

  a. Solvolysis, hydrolysis $\quad RX + H_2O \longrightarrow ROH + HX$

  b. Other nucleophilic reactions $\quad RX + N^- \longrightarrow RN + X^-$

II. Oxidation

  a. α-Hydroxylation

$$-\underset{\underset{H}{|}}{\overset{|}{C}}-X + H_2O \longrightarrow -\underset{\underset{OH}{|}}{\overset{|}{C}}-X + 2H^+ + 2e^-$$

  b. Epoxidation

$$\underset{/}{\overset{\backslash}{C}}=\underset{\backslash}{\overset{X}{C}} + H_2O \longrightarrow \underset{/}{\overset{\backslash}{C}}-\underset{\backslash}{\overset{\overset{O}{\triangle}}{C}}{\overset{X}{}} + 2H^+ + 2e^-$$

III. Reduction

  a. Hydrogenolysis $\quad RX + H^+ + 2e^- \longrightarrow RH + X^-$

  b. Dihaloelimination

$$-\underset{\underset{X}{|}}{\overset{|}{C}}-\underset{\underset{X}{|}}{\overset{|}{C}}- + 2e^- \longrightarrow \underset{/}{\overset{\backslash}{C}}=\underset{\backslash}{\overset{/}{C}} + 2X^-$$

  c. Coupling $\quad 2RX + 2e^- \longrightarrow R{-}R + 2X^-$

IV. Dehydrohalogenation

$$-\underset{\underset{X}{|}}{\overset{|}{C}}-\underset{\underset{H}{|}}{\overset{|}{C}}- \longrightarrow \underset{/}{\overset{\backslash}{C}}=\underset{\backslash}{\overset{/}{C}} + HX$$

Source: Reference 7.51.

alkanes less than 10 carbons are more toxic to microorganisms as a result of their higher water solubility. Alternatively, biodegradation rates of long-chain alkanes ($>C_{10}$) are affected by their low water solubilities and correspondingly high degree of sorption. However, soil and groundwater microorganisms that are actively metabolizing hydrocarbons produce natural surfactants that aid in desorbing the contaminants. These compounds, which are often based on complex hybrid structures of proteins, lipids, and carbohydrates, include rhamnolipids, phospholipids, and lipopolysaccharides [7.16].

As shown in Figure 7.11, alkanes proceed through dehydrogenase reactions with transformations to an alcohol, an aldehyde, and a carboxylic acid. The final step in this pathway is β-oxidation, the same degradation mechanism used in the metabolism of natural fatty acids.

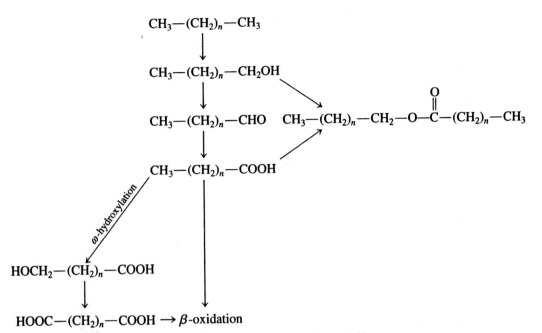

Figure 7.11 Biodegradation pathway for *n*-alkanes. *Source:* Reference 7.52.

Acetic acid, generated through β-oxidation, can then proceed through the TCA cycle after it is converted to acetyl CoA; the alcohol also degrades further via the aldehyde-acid route through another β-oxidation. Branching interferes with β-oxidation (in which carbon atoms are cleaved two at a time) because a one-carbon oxidation (α-oxidation) must occur, a more difficult pathway in microbial metabolism.

Benzene and PAHs. Aromatic compounds degrade by a dual dioxygenase-catalyzed hydroxylation mechanism in which oxygen is added to form a diol. The diol then proceeds through ring cleavage and the products enter standard metabolic pathways, as illustrated in Figure 7.12.

Polycyclic aromatic hydrocarbons follow similar pathways of hydroxylation and ring cleavage (Figure 7.13). However, biodegradation rates of PAH compounds larger than three rings are slowed by their need for specific enzymes to degrade these larger compounds [7.55]. Therefore, the biodegradation of PAHs is a function of the ring size and the enzymatic complement of the microbial consortium. The general consensus is that two- and three-ring PAHs are readily biodegradable, whereas those containing four or more rings are exceedingly difficult, if not biorefractory. Furthermore, large-ring PAHs are highly hydrophobic and sorption often controls their rates of biodegradation [7.56]. Another factor that has been shown to influence the biodegradability of PAHs is the number and positions of substitution groups. Recent studies have documented that the addition of three methyl groups to a number of PAHs resulted in significantly decreased rates of biodegradation [7.54].

Anaerobic biotransformations of PAHs have received less attention relative to aerobic processes. However, results to date have shown that two- and three-ring PAHs are biodegraded under denitrifying, sulfate-reducing, and methanogenic conditions [7.55].

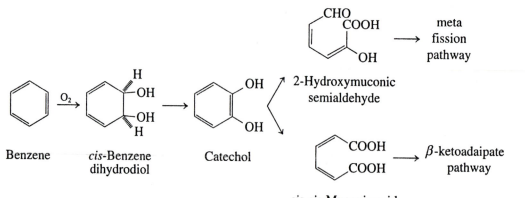

Figure 7.12 Pathway for the aerobic biodegradation of benzene. *Source:* Reference 7.53.

Halogenated Solvents. Biodegradation of halogenated alkanes and alkenes may proceed by a number of metabolic pathways, including (1) aerobic metabolism with the contaminant as an energy source, (2) aerobic cometabolism, and (3) reductive dehalogenation. The pathway is a function of the contaminant oxidation state as well as environmental conditions, such as the dissolved oxygen concentration and the redox potential. Many halogenated aliphatics are biodegraded via both oxidative and reductive pathways. In general, however, more reduced compounds (e.g., chloroethane) proceed through oxidative metabolism and oxidized contaminants (e.g., carbon tetrachloride, PCE) are degraded via reductive processes. Therefore, knowledge of process microbiology is necessary in site and facility assessments and in the design of bioremediation systems.

Some lower halogenated compounds such as 1,2-dichloroethane and methylene chloride have been biodegraded aerobically as the sole carbon and energy source. For example, Van der Wijngaard [7.57] found that *Xanthobacter autotrophicus* and *Ancylobacter aquatias* metabolize 1,2-dichloroethane as a sole carbon source. Similar results were obtained by Semprini et al. [7.58], who documented that *Pseudomonas* and *Hyphomucrobria* degraded methylene chloride and used it as a sole carbon and energy source.

Although some of the halogenated aliphatic compounds with lower halogen substitution can be biodegraded aerobically as the sole carbon and energy source, cometabolism is the most common metabolic process for aliphatic compounds of higher halogen substitution. Cometabolism has been studied extensively as a basis for the aerobic degradation of chlorinated aliphatics. The basis for aerobic cometabolism is the activity of oxygenase enzymes, which catalyze wide-ranging oxidations. Many chlorinated aliphatic compounds, including some that are highly chlorinated, are biodegraded by aerobic cometabolism. Methylene chloride, chloroform, and TCA are cometabolically degraded by the nitrifying bacterium *Nitrosomonas* as it oxidizes ammonia [7.59]. However, in most cometabolic pathways, other reduced organic compounds (e.g., propane, phenol, toluene) serve as the energy source. For example, phenol and toluene have been effective primary substrates in the biodegradation of TCE [7.60]. In situ field studies have also demonstrated that these phenol-degrading bacteria can be stimulated in subsurface systems [7.61].

Two groups of aerobic bacteria that have been the subject of substantial cometabolic research are the methanotrophs and the methylotrophs. *Methanotrophs* metabolize

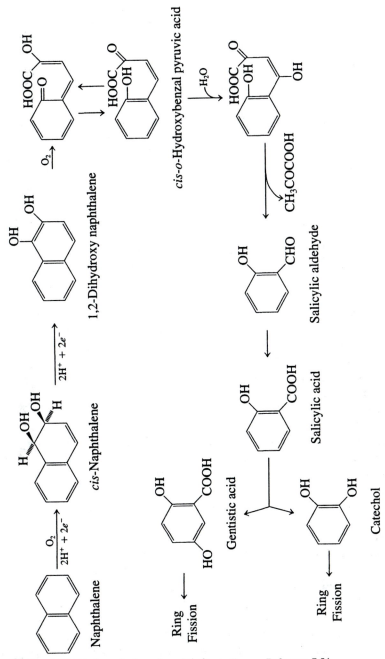

Figure 7.13 Biodegradation of naphthalene. *Source:* Reference 7.54.

methane, although some species also use methanol and formaldehyde. *Methylotrophs* metabolize a range of small organic compounds, but not methane. Methanotrophic bacteria are important in degrading otherwise biorefractory chlorinated aliphatic compounds because their metabolism involves *methane monooxygenase,* an enzyme that catalyzes the oxidation of methane to methanol. Methane monooxygenase is a relatively nonspecific enzyme that also hydroxylates many chlorinated alkenes and alkanes. Many halogenated aliphatic compounds, including *cis*- and *trans*-1,2-DCE and vinyl chloride, have been cometabolized by methanotrophs [7.62]. In addition, the cometabolic aerobic biodegradation of TCE by methanotrophs has been documented in a number of studies [7.63]. Because the reaction catalyzed by methane monooxygenase is oxidative, the effectiveness of methanotrophic cometabolism is decreased as the degree of chlorination (and the contaminant oxidation state) increases. Because of its high degree of chlorination, PCE has not been cometabolically degraded by methanotrophs.

Although the lower chlorinated solvents biodegrade aerobically, reductive processes that occur in the absence of oxygen have been shown to be the most common biodegradation pathway for higher chlorinated aliphatic compounds. The primary mechanism for the anaerobic degradation of halogenated aliphatics is reductive dehalogenation, a process that proceeds almost exclusively through cometabolism. During reductive dehalogenation, the oxidized contaminant (e.g., TCE, PCE, carbon tetrachloride) serves as the terminal electron acceptor. The source of energy (electrons) is a cometabolite (e.g., acetate, toluene, or another hydrocarbon); as it is oxidized, the contaminant receives the electrons (its energy) and is reduced. Therefore, the halogenated aliphatic is the electron acceptor and the reduced organic is the electron donor. It follows that the rate of reduction (and corresponding dehalogenation) of the contaminant is controlled by the rate at which the cometabolite is oxidized. An example of reductive metabolism, perchloroethylene biodegradation, proceeds through dehalogenation in which one chlorine and the one hydrogen atom are displaced resulting in the formation of the products TCE, *cis*-1,2-DCE, and *trans*-1,2,-DCE. The two forms of DCE can proceed by analogous pathways to form vinyl chloride and, consequently, ethylene. Ethylene may then be transformed to carbon dioxide and water (Figure 7.14).

The result of such microbial metabolism is that a large number of degradation products are generated, some of which are more toxic (e.g., vinyl chloride) than the parent compound. During reductive dehalogenation, other electron acceptors (e.g., nitrate, sulfate) can compete with the contaminant for electrons. Therefore, an evaluation of

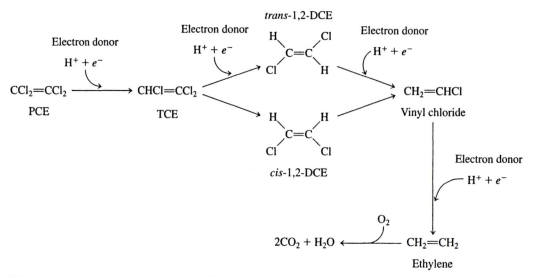

Figure 7.14 Reductive dehalogenation of PCE. *Source:* Reference 7.64.

Table 7.9 Selected Redox Half Reactions for Possible Electron Acceptors and Their Corresponding Reduction Potentials

Reaction	E^0 (V)
$Cl_3C-CCl_3 + 2e^- \longrightarrow Cl_2=CCl_2 + 2Cl^-$	+1.13
$O_2 + 4H^+ + 4e^- \longrightarrow 2H_2O$	+0.82
$2NO_3^- + 12H^+ + 10e^- \longrightarrow N_2 + 6H_2O$	+0.74
$CCl_4 + H^+ + 2e^- \longrightarrow CHCl_3 + Cl^-$	+0.67
$Cl_2C=CCl_2 + H^+ + 2e^- \longrightarrow HClC=CCl_2 + Cl^-$	+0.58
$Cl_3C-CH_3 + H^+ + 2e^- \longrightarrow HCl_2C-CH_3 + Cl^-$	+0.57
$HClC=CCl_2 + H^+ + 2e^- \longrightarrow trans\text{-}HClC=CClH + Cl^-$	+0.54
$trans\text{-}HClC=CClH + H^+ + 2e^- \longrightarrow H_2C=CHCl + Cl^-$	+0.37
$SO_4^{2-} + 9H^+ + 8e^- \longrightarrow HS^- + 4H_2O$	-0.22
$CO_2 + 8H^+ + 8e^- \longrightarrow CH_4 + 2H_2O$	-0.24
$2CO_2 + 8H^+ + 8e^- \longrightarrow CH_3COOH + 2H_2O$	-0.40

Source: References 7.1, 7.21, 7.51, 7.56.

redox potentials of the electron acceptors is necessary to assess the potential effectiveness of reductive dehalogenation. The standard redox potentials for a number of possible electron acceptors, including oxidized aliphatic compounds and natural inorganic electron acceptors (nitrate, sulfate, carbon dioxide) are listed in Table 7.9. Because these half reactions are ordered in decreasing oxidation-reduction potential (E^0), the compounds at the top of the table have more of a tendency to undergo reductive dehalogenation. The reaction of the electron donor can be coupled with the reactions of potential electron acceptors to evaluate which of these are thermodynamically favorable. Such a thermodynamic approach to evaluating electron acceptors is illustrated in Example 7.7.

EXAMPLE 7.7   *Evaluation of Electron Acceptors*

If acetate is the electron donor, evaluate the thermodynamic potential for the reductive dehalogenation of chloroform in the presence of sulfate.

SOLUTION

Compare the thermodynamic potential for the reaction of chloroform and acetate to the reaction of sulfate and acetate. The half-reactions for chloroform (Table 7.4) and acetate (Table 7.9) are

$$CHCl_3 + H^+ + 2e^- \longrightarrow CH_2Cl_2 + Cl^- \qquad E^0 = +0.56$$

$$CH_3COOH + 2H_2O \longrightarrow 2CO_2 + 8H^+ + 8e^- \qquad E^0 = +0.40$$

The atoms and charges are balanced; therefore, multiply the chloroform reaction by 4 to balance the electrons:

$$4CHCl_3 + 4H^+ + 8e^- \longrightarrow 4CH_2Cl_2 + 4Cl^- \qquad E^0 = +0.56$$

$$CH_3COOH + 2H_2O \longrightarrow 2CO_2 + 8H^+ + 8e^- \qquad E^0 = +0.40$$

Adding the reactions yields

$$4CHCl_3 + CH_3COOH + 2H_2O \longrightarrow 4CH_2Cl_2 + 2CO_2 + 4Cl^- + 4H^+ \qquad E^0 = {}^+0.96$$

The redox reactions for sulfate (Table 7.9) and acetate are

$$SO_4^{2-} + 9H^+ + 8e^- \longrightarrow HS^- + 4H_2O \qquad E^0 = -0.22$$

$$CH_3COOH + 2H_2O \longrightarrow 2CO_2 + 8H^+ + 8e^- \qquad E^0 = +0.40$$

The atoms and electrons are already balanced, so adding the reactions yields

$$SO_4^{2-} + CH_3COOH + H^+ \longrightarrow HS^- + 2H_2O + 2CO_2 \qquad E^0 = +0.18$$

The reaction of chloroform with acetate results in a significantly higher electromotive force (0.96 V > 0.18 V); therefore, the reductive dehalogenation of chloroform is more likely to proceed than the reduction of sulfate with acetate as the electron donor.

Pesticides. Persistence was once considered a positive characteristic of pesticides by agricultural chemists because their potency would be maintained on crops [7.51]. The organochlorine insecticides did indeed maintain long-term potency, with detectable concentrations of DDT and dieldrin found in soils after over 20 years [7.67]. However, organophosphate esters and carbamates are much shorter lived, often due to abiotic hydrolysis reactions. Herbicides and fungicides vary widely in their biodegradability, although all are more biodegradable than organochlorine insecticides. Numerous investigations of the biodegradation of pesticides have been reported, and a detailed account of the many studies is beyond the scope of this text. However, the biodegradation pathways of two common pesticides—DDT and 2,4-D—are summarized as examples of the many transformations that have been elucidated.

The biodegradability of DDT has been studied extensively due to the reputation it achieved as an environmentally persistent contaminant. The recognized first step in DDT biodegradation is its reductive dehalogenation under anaerobic conditions to DDD. Through combined aerobic and anaerobic metabolism, DDT can be mineralized (Figure 7.15), although the time required for mineralization may be decades.

The phenoxy herbicide 2,4-D is biodegraded aerobically by *Arthrobacter* spp. and proceeds first through β-oxidation of the aliphatic moiety, yielding 2,4-dichlorophenol (Figure 7.16). Degradation then proceeds through catechol structures to ring cleavage.

Polychlorinated biphenyls. The general consensus in the biodegradability of PCBs is that those substituted with five or more chlorines are biorefractory under aerobic

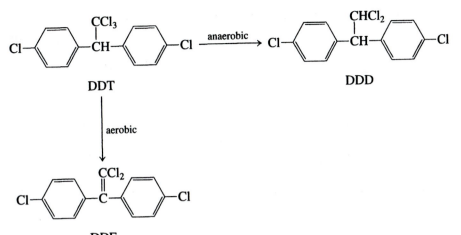

Figure 7.15 Biodegradation pathways of DDT. *Source:* Reference 7.67.

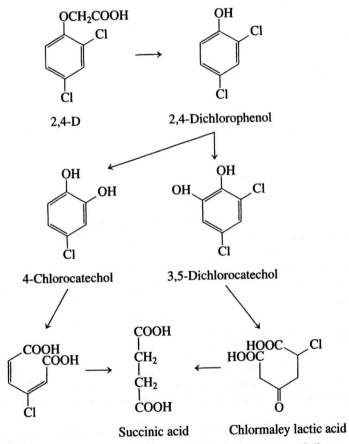

Figure 7.16 Aerobic degradation of 2,4-D. *Source:* Reference 7.68.

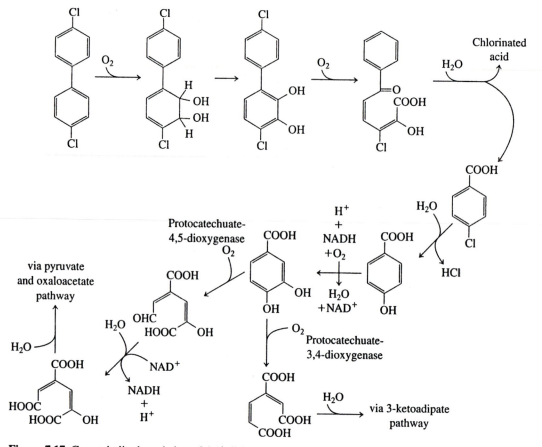

Figure 7.17 Cometabolic degradation of 4,4′-dichlorobiphenyl. *Source:* Reference 7.70.

conditions. For example, Aroclor 1260 congeners are considered essentially non-biodegradable [7.69]. However, the lower PCBs readily proceed through biotransformation reactions. For example, the pathway for the cometabolic degradation of 4,4′-dichlorobiphenyl (Figure 7.17) shows that it proceeds through a series of hydroxylations, the two rings are cleaved, and eventually the cleaved products enter natural pathways. As expected by the oxidation state of compounds with greater halogen substitution, PCBs with more than four chlorines are more readily biodegraded by reductive processes. In fact, reductive dehalogenation under anaerobic conditions is the only documented biodegradation pathway for higher PCBs [7.71].

7.5 COMMON THEMES AND PERSPECTIVES IN ABIOTIC AND BIOTIC TRANSFORMATIONS

Some common trends are found in both abiotic and biotic redox transformations. For both processes, the presence of oxygen dictates whether oxidative or reductive processes occur. Obviously, the presence of oxygen will determine whether conditions favor the growth of anaerobic or aerobic microorganisms. In a parallel manner, oxygen scavenges aqueous electrons in abiotic systems to form superoxide, which can form hydrogen peroxide through catalysis by superoxide dismutase. Many of the common metabolic pathways used by microorganisms (Table 7.8) are similar to the abiotic reactions described in Section 7.3. The obvious difference between abiotic and biotic transformations is the enzymatic catalysis that is inherent in biological processes. However, reaction thermodynamics are relatively consistent for both abiotic and biotic transformations. For example, thermodynamics favor the reduction of oxidized contaminants, such as carbon tetrachloride and PCE, whether the transformation proceeds abiotically or biotically. In contrast, the transformation of reduced compounds (e.g., benzene, BTEX, PAHs) proceeds most effectively via oxidative processes. These trends are illustrated in Table 7.10, which shows that carbon tetrachloride is easily reduced by reductive dehalogenation through cometabolic biodegradation. The same pattern holds for abiotic processes—carbon tetrachloride is rapidly reduced by aqueous electrons, and the trend then proceeds to the most reduced analog, chloromethane, which is not highly reactive via reductive processes. Once again, a similar theme is found in abiotic transformations—chloromethane is not readily reduced by aqueous electrons. Oxidation reactions follow much the same trend (but in reverse) as the reductive transformations listed in Table 7.10. Carbon tetrachloride is not readily oxidized, but chloromethane is easily transformed by both abiotic and biotic oxidative processes.

One reactant that does not follow the general rule of low reactivity with a high degree of halogen substitution is the hydroxyl radical. Although hydroxyl radicals are essentially unreactive with perhalogenated alkanes, highly substituted alkenes and aromatics (e.g., PCE, hexachlorobenzene, hexachlorocyclopentadiene) react rapidly. For example, PCE reacts only slightly less rapidly with hydroxyl radicals (2.6×10^9 M^{-1} sec^{-1}) than TCE (4×10^9 M^{-1} sec^{-1}). The electron-rich pi bonds of alkenes and aromatics are still available to react with hydroxyl radicals, even though electron-withdrawing groups are present. This high, relatively nonspecific reactivity of hydroxyl radicals is the basis for their potential in advanced oxidation processes. The products formed are often biodegradable, which provides the basis for another common theme—coupled abiotic and biotic processes. In such a hazardous waste treatment scheme, oxidation of the biorefractory parent compound by hydroxyl radicals is followed by biological processes to degrade the intermediates, such as organic acids and ketones.

Still another relationship between abiotic and biotic transformations is illustrated in Figure 7.18. The combined abiotic and biotic transformations of TCA via three different pathways shows the potential for two initial abiotic transformations (to either 1,1-

Table 7.10 Relationship Between Abiotic and Biotic Oxidations and Reductions

Compound	Aerobic Biodegradation[a]	Chemical Oxidation[b]	Anaerobic Biodegradation[a]	Chemical Reduction[b]
CCl$_4$	0	0	↑ ↑ ↑	↑ ↑ ↑
CHCl$_3$	↑	0	↑ ↑	↑ ↑
CH$_2$Cl$_2$	↑ ↑ ↑	↑	0	↑
CH$_3$Cl	↑ ↑ ↑	↑ ↑	0	0

0, very small if any potential; ↑, some potential; ↑ ↑, good potential; ↑ ↑ ↑, excellent potential.
[a] From Reference 7.72.
[b] Based on data from Tables 7.6 and 7.7.

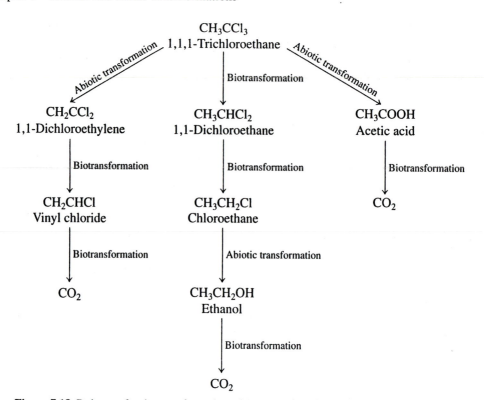

Figure 7.18 Pathways for the transformation of 1,1,1-trichloroethane. *Source:* Reference 7.51.

DCE or acetic acid) or a series of three biotic dehalogenation reactions to ethanol. Each of the three pathways provides a mechanism for TCA mineralization.

Numerous transformation pathways, such as hydrolysis, elimination, redox reactions, and cometabolic biodegradation, are outlined in this chapter. Although many of these processes can *potentially* occur at contaminated sites or as wastes migrate away from a RCRA source, in most cases, hazardous contaminants are not reactive via most of the transformation pathways outlined in this chapter. Many factors, such as the presence of NAPLs, sorption, and the absence of an appropriate reacting species negate the transformation of many hazardous chemicals, which explains why many contaminants (e.g., PAHs, carbon tetrachloride, PCE, DDT, Mirex, PCBs, PCDDs) persist for decades or even centuries in the environment. For example, while significant numbers of soil microorganisms may be present in the upper horizons of soils, lower numbers are usually found in the subsurface and groundwater systems. Furthermore, the microorganisms present may not be capable of degrading the contaminants, especially if they are biorefractory. In addition, the environmental conditions may not support the growth of microorganisms that transform the contaminant (i.e., the *appropriate* transforming species is not present). For example, if PCE is present in an aerobic surface soil, it will not likely be biodegraded.

Although many of the pathways described in this chapter are not prevalent in natural systems, many have been successfully applied to engineered systems, such as advanced oxidation processes, zero-valent iron reactions, and cometabolic bioremediation. A summary of the abiotic and biotic processes covered in Chapter 7, including their importance in natural transformation and hazardous waste treatment processes, is presented in Table 7.11. This information provides perspective on the role of each of the processes covered in this chapter in hazardous waste management. Some abiotic processes, such as hydrolysis, elimination, and oxidation on the surfaces of metal oxides, are important only in the natural environment for contaminants that are reactive. In general, transformations involving hydroxyl radicals and aqueous electrons (e.g., AOPs, zero-valent iron, etc.) are most important as hazardous waste treatment processes. Biodegradation, in which the contaminant is the carbon and energy source, is most common in the natural environment. However, cometabolism is commonly promoted in the bioremediation of recalcitrant contaminants.

7.6   MEASUREMENT OF TRANSFORMATION RATES

Due to the high degree of variability of contaminant and site characteristics, and also to the competition between contaminants in complex wastes, assessment and treatability studies are usually required to evaluate rates of degradation and the potential for physicochemical treatment and bioremediation. Because biodegradation rates are a function of so many variables, the use of tabular data to assess the potential for biotransformation for the purposes of site assessments or hazardous waste treatment processes is unreliable. For example, the time required for ethylbenzene concentrations to biodegrade to undetectable levels in natural soils was reported to be 192 hours in one study [7.73] and nearly 2 years in a similar investigation [7.74]. Such wide-ranging results preclude the use of empirical data in assessing transformation rates. Although physicochemical processes are subject to less variability, they are also highly variable in the complicated matrix of soils. Therefore, Tyre et al. [7.75] recommended that treatability studies should be performed prior to implementing soil remediation using catalyzed hydrogen peroxide. Biodegradation rates should also be determined on a site-by-site basis using laboratory *bioassessment studies,* which are based on natural chemical and physical conditions found at the site. A`modification of such a basic quantitation of biodegradation rates is the *biotreatability study,* in which microbial degradation rates are enhanced using nutrients, cometabolites, and different terminal electron acceptors, in order to accelerate biodegradation as a basis for bioremediation.

The primary goal of bioassessment studies is to determine whether contaminants at a particular site are biodegradable. Other aspects that may be evaluated are (1) the numbers of microorganisms in the system, (2) whether the indigenous microorganisms will biodegrade the contaminants, (3) the rate at which they are degraded, and (4) the identity of the biodegradation products.

Soil, groundwater, and sludges are usually characterized for general microbial activity using plate counts or most probable number (MPN) analysis. In general, microbially healthy soils are characterized by counts $\geq 10,000$ colony forming units (CFU)/g soil. Soils unfavorable to microbial life generally contain microbial numbers < 100 CFU/g [7.76].

Some general requirements are inherent in biodegradation studies, whether the purpose is bioassessment or biotreatability. First, mass balances are necessary on the contaminant carbon, a procedure that is best accomplished using systems spiked with ^{14}C-labeled contaminants; the fate of the carbon can be tracked effectively by monitoring its β emissions through liquid scintillation counting. As the experiment proceeds, aliquots are collected from the aqueous phase, the soil, the biomass, and in a basic (e.g., KOH) water trap in which $^{14}C\text{-}CO_2$ dissolves as $^{14}C\text{-}CO_3^{2-}$.

Table 7.11 Importance of Abiotic and Biotic Processes in the Natural Environment and Hazardous Waste Treatment Systems

Transformation Process	Role and Significance
Abiotic Processes	
Hydrolysis	Important transformation process for alkyl halides and esters (including organophosphorus esters) in soils and water. Limited use as a hazardous waste treatment process.
Elimination	Sometimes significant for a few classes of compounds (e.g., TCA) in natural anaerobic systems (e.g., groundwater). No hazardous waste treatment applications.
Oxidation and reduction	
Oxidation by iron (III) or manganese (III/IV) oxides	An important pathway in the natural environment for relatively hydrophobic compounds with ring activating groups (e.g., phenols, anilines). No hazardous waste treatment applications.
Oxidation by hydroxyl radicals	An important pathway for water-soluble hazardous chemicals in the natural environment where hydroxyl radicals are present (e.g., surface waters and the atmosphere). Most important in hazardous waste treatment processes including UV/O_3, UV/H_2O_2, Fenton's reagent, etc.
Reduction by organic matter coupled with excess reductants	Can be a significant pathway for oxidized contaminants in anaerobic natural systems. No hazardous waste treatment applications.
Zero-valent iron	Not important in the natural environment. A hazardous waste treatment process with high potential.
Titanium dioxide photocatalysis	Not important in the natural environment. Emerging hazardous waste treatment process.
Electron beam	Not important in the natural environment. A potential hazardous waste treatment process.
Photolysis	Sometimes a significant pathway for sunlit waters and soil surfaces. Major role in hazardous waste treatment processes (UV irradiation, titanium dioxide photocatalysis).
Biotic Processes	
Contaminant as carbon and energy source	Primary route of biodegradation in the natural environment. Basis for the bioremediation of hydrocarbons and other reduced contaminants.
Aerobic cometabolism	Some occurrence in the natural environment. An important hazardous waste treatment application in the remediation of recalcitrant chemicals using phenol, toluene, methane, etc. as cometabolites.
Reductive dehalogenation	Minimal occurrence in the natural environment. Important bioremediation application through the addition of electron donors such as acetate.

Another requirement of laboratory biodegradation studies is the use of valid controls. One of the more important controls in biodegradability studies is a parallel abiotic reaction in which mercuric chloride or sodium azide is added to inhibit any biological activity. Other controls may also be necessary depending on the experimental design.

In a bioassessment study, the samples from an unaugmented system are incubated under conditions similar to those at the site from which the sample was taken. In other words, temperature, pH, nutrient levels, and other environmental parameters are assessed at their natural levels. Analogous to bioassessment studies, biotreatability studies are designed to evaluate a number of potential routes that may stimulate biodegradation as a basis for bioremediation process design. Different aliquots of soil or groundwater are often spiked with bacteria or fungi (a procedure known as *bioaugmentation*), cometabolites (e.g., methane, phenol, etc.), electron acceptors (e.g., oxygen, sulfate), or varied nutrient levels. When a range or a matrix of conditions is evaluated, those that provide the highest effective contaminant degradation are used in field implementation.

Bioassessment and biotreatability studies can be simple or complex. Simple studies have been conducted in volatile organic analysis (VOA) vials or biometry flasks (Figure 7.19). More complicated studies may involve bench- or pilot-scale reactors.

Numerous standard procedures have been developed for bioassessment and biotreatability studies under the guidance of the U.S. Environmental Protection Agency. Published in the *Code of Federal Regulations,* some of the procedures that have been developed include Anaerobic Biodegradability of Organic Chemicals, Ready Biodegradability, and Inherent Biodegradability. Detailed summaries of these procedures are provided in Reference 7.78.

The ideal site assessment would use biodegradation rate data for samples collected from the site. In most cases, these data are not available and performing laboratory studies is beyond the budget and the time commitment for most projects. Moreover, an

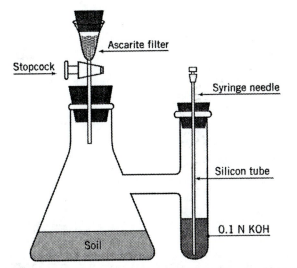

Figure 7.19 Biometry flask system for performing bioassessment and biotreatability studies.
Source: Reference 7.77.

emergency response situation such as a hazardous material spill may require approximate data within a matter of hours or days. In these situations, the estimation of biodegradation rates may be necessary. Although biodegradation rates are quite variable, under some circumstances the estimation of biodegradation rates is necessary. If tabular data are to be used as a basis for biodegradation rates, compilations may be found in References 7.79 and 7.80.

7.7   SUMMARY OF IMPORTANT POINTS AND CONCEPTS

- Abiotic and biotic transformations, which have traditionally been studied separately, have a number of similarities. The most common abiotic and biotic transformations involve oxidation-reduction reactions that are consistent with thermodynamic principles.

- Although the variables affecting abiotic and biotic processes are numerous, the primary variables that control their transformation rates are contaminant structure, presence of reactive species, and availability. The carbon oxidation state of the contaminant provides a thermodynamic basis for the potential occurrence of oxidative or reductive transformations, whether they be abiotic or biotic.

- Transformation reactions may be described by zero-, first-, second-, or mixed-ordered rates. Michaelis-Menton kinetics describe catalyzed reactions in which rates vary from first order to zero order as a function of substrate concentration.

- Abiotic transformations occur by hydrolysis, elimination, substitution, and redox reactions. Of these, hydrolysis and redox reactions are the most important in hazardous waste transformations.

- Hydrolysis reactions are characterized by the nucleophilic attack of water on electron-poor areas of molecules. Alkyl halides, esters of carboxylic acids, organophosphate esters, and carbamates are the most common hazardous compounds that undergo hydrolysis reactions.

- The most common oxidant in nature, O_2, reacts slowly with most organic compounds and is therefore of limited importance. The second most prevalent oxidant, oxidized metal oxides [e.g., iron (III) oxides], can be important in transforming some hazardous organics, such as phenols and anilines substituted with electron-donating groups. Free radical oxidations involving hydroxyl radicals are important under some environmental conditions. However, engineered advanced oxidation processes (AOPs) that generate hydroxyl radicals are the most important treatment application of abiotic oxidation reactions.

- Oxidized organic contaminants, such as carbon tetrachloride and PCE, undergo chemical reductions. The reaction of oxidized contaminants with natural organic matter coupled with reduced iron minerals provides the most common mechanism for chemical reduction in soil and subsurface solids. Engineered reduction processes include zero-valent iron systems and electron beams.

- Biotic transformations, which are usually carried out by bacteria and fungi in the environment, are the most significant routes for the transformation of carbon on a global scale. Although some contaminants are slowly biodegraded, biotic transformations represent the most important pathway by which organic compounds are mineralized in the natural environment.

- Biodegradation can occur through a number of metabolic pathways, including (1) aerobic metabolism with the contaminant as an energy source, (2) aerobic cometabolism, and (3) reductive dehalogenation. The pathway is a function of the contaminant oxidation state as well as environmental conditions. Fundamental knowledge of metabolic pathways serves as the basis for the design of bioremediation systems.

- Due to the extreme variability of abiotic and biotic transformation rates, assessment studies are necessary to evaluate rates of degradation as a basis for site and facility assessments. Treatability studies are an extension of assessment studies in which treatment process parameters are evaluated in the laboratory as a basis for process engineering design.

PROBLEMS

7.1. Determine the average carbon oxidation state of the following compounds.

a. Dibromochloromethane	b. TCA
c. TCE	d. PCE
e. Hexachloroethane	f. Benzene
g. Chlorobenzene	h. 1,2,3-Trichlorobenzene
i. Hexachlorobenzene	j. Isooctane
k. Anthracene	

7.2. Determine the average carbon oxidation state of each of the following pesticides.

a. Methoxychlor	b. DDT
c. Carbaryl	d. Dieldrin
e. Diuron	f. 2,4,5-T

7.3. The following data were obtained for a bioremediation treatability study of petroleum contaminated spills.

Time (days)	TPH (ppm)
0	1780
10	910
20	455
30	233
40	120
50	58
60	33

Using these data, calculate the first-order rate constant for biodegradation and determine the time required for 99% treatment.

7.4. The biodegradation rate of benzo[*a*]pyrene has been described by the expression

$$-dC/dt = k[X][C]$$

where

$k = 3 \times 10^{-15}$ L/cell-h

$[C]$ = benzo[a]pyrene concentration

$[X]$ = biomass concentration

During a bioremediation project of a contaminated groundwater, the biomass concentration reached steady state at 7.1×10^{11} cells/L during treatment and remained at approximately that concentration throughout the project. If the initial concentration of benzo[a]pyrene in the groundwater is 25 μg/L and the hydraulic detention time of the groundwater as it passes through the control volume is 10 days, determine the effluent concentration of benzo[a]pyrene as the water exits the system.

7.5. The biodegradation rate in a slurry reactor for pyridine treatment has been described by the expression

$$-dC/dt = k[X][C]$$

where

$[C]$ = pyridine concentration

$[X]$ = biomass concentration

The rate constant $k = 4 \times 10^{-11}$ L/cell-h, and the biomass concentration in the slurry reactor is constant at 1.5×10^9 cells/L. If the initial pyridine concentration is 48 mg/L and the desired concentration after treatment is 0.1 mg/L, determine the time required for treatment.

7.6. The following data were collected in pilot-scale treatability studies in a 5 m × 5 m (16.4 ft × 16.4 ft) land farm plot.

Loading Rate of Waste Oil (L/day)	Biodegradation Rate (day^{-1})
2	0.08
4	0.09
6	0.07
8	0.08
10	0.001
12	0.0008

At full scale, the site will be 200 m × 200 m (656.2 ft × 656.2 ft) and will openly accept waste oil. Provide a recommendation on the volume of waste oil the facility will accept on a daily basis.

7.7. Epichlorhydrin was transformed by an enzyme and the rate of disappearance measured every 30 sec for 2.5 min. A series of six reactors was arranged with 1.5 μg of enzyme added to the same volume of epichlorhydrin solution, but each with a different concentration of epichlorhydrin. The results are summarized on the following page. Determine the initial velocities for each epichlorhydrin concentration and plot the velocities as a function of the initial concentration.

Time (min)	Epichlorhydrin Reacted (mg/L) Sample No.					
	1	2	3	4	5	6
Initial Epichlorhydrin conc. (mg/L)	2.55	3.60	4.55	6.31	7.43	8.52
0.5	2.29	2.97	3.80	5.35	6.32	7.13
1.0	1.92	2.35	3.09	4.30	5.19	5.75
1.5	1.47	1.75	2.20	3.30	4.02	4.42
2.0	0.97	1.16	1.45	2.19	2.75	3.04
2.5	0.51	0.59	0.73	1.15	1.50	1.72

7.8. The first-order rate constant for the transformation of BTEX in a soil slurry reactor is 0.081 day^{-1} at 20°C. Determine (a) the rate constant and (b) the time required for BTEX to degrade from 1200 mg/kg to 10 mg/kg if the temperature = 12°C and $\theta = 1.088$ h.

7.9. The following data were obtained in an aqueous batch reactor in which hydroxyl radicals are generated to maintain a steady-state concentration of 10^{-11} M. The contaminant, mevinphos, is water soluble. The following data were collected for the degradation of mevinphos (assume volatilization is negligible).

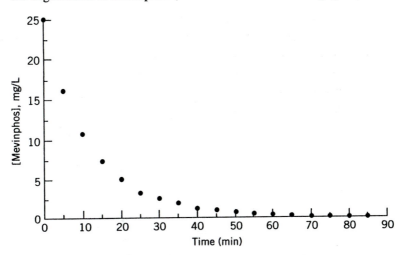

Based on these data, determine the second-order rate constant for the reaction of hydroxyl radicals with mevinphos.

7.10. Describe the most likely abiotic transformation for the following compounds.
   a. Ethyl acetate
   b. Carbon tetrachloride
   c. Parathion
   d. TCA
   e. 1,2,3-Trimethylbenzene
   f. Hexachloroethane
   g. Tetranitromethane

7.11. What is the hydrolysis half-life of chloroform at 25°C at pH 7? What about pH 10?

7.12. How long would it take for 99% of an ethyl acetate solution at 25°C to hydrolyze at (a) pH 3, (b) pH 7, and (c) pH 11?

7.13. Determine the time required for ethyl acetate to hydrolyze from 20 mg/L to 4 mg/L at pH 9. (Assume the temperature = 25°C.)

7.14. You are a private consultant and are interested in obtaining a contract from a pesticide formulator who has a soil pit 20 m × 50 m (65.6 ft × 164 ft) with soil contamination 1 m (3.3 ft) deep. Write a one-page "White Paper" on the feasibility of remediating the surface soil, which is contaminated with a mixture of carbamates, organophosphorus esters, acid amides, and phenoxyacetic acid esters. Describe why you believe hydrolysis may be a cost-effective and practical solution. Feel free to use your engineering judgment and to compare hydrolysis to other pathways. You may also speculate on the conceptual design you would use (e.g., in situ mixing of reactants, excavation followed by treatment in a slurry reactor).

7.15. Determine the time required for a 500 mg/L solution of acetamide (CH_3CONH_2) to hydrolyze to 1 mg/L at
a. pH 4
b. pH 7
c. pH 10

7.16. Graph the first-order rate constants for the hydrolysis of ethyl acetate as a function of pH from pH 2 to 12 at intervals of 1 pH unit.

7.17. Dimethyl phthalate has been disposed of in a surface lagoon at a concentration of 200 mg/L. It is in contact with a sodic saline soil, which results in a lagoon pH of 9. Estimate its half-life in the lagoon. Assume sorption and temperature effects on the hydrolysis rate are negligible.

7.18. Repeat Problem 7.17, but in this case the dimethyl phthalate is in contact with a peat bog at pH 3.

7.19. The following transformation pathway has been documented for the degradation of pentachlorophenol.

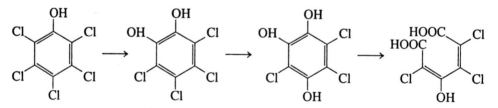

Determine whether this is an oxidative or reductive pathway.

7.20. Which of the following compounds would you expect to be oxidized by manganese (III/IV) oxides?
a. 1,2,3-Trichlorobenzene   b. PCE
c. Methoxychlor   d. 1,3,5-Trimethylbenzene
e. *p*-Chlorophenol

7.21. An engineer at a solvent recovery facility has mixed 1000 L (264 gal) of caustic solution with 100,000 L (26,400 gal) of waste ethyl acetate. The ethyl acetate formed a distinct phase on top of the basic solution. Three days later, only one phase is present. Describe what has occurred in this system.

7.22. Balance the following reactions and determine the equilibrium constant for each.
   a. The oxidation of chloroform by copper (II).
   b. The reduction of carbon tetrachloride by iron (II).
   c. The oxidation of iron (II) by chromium (VI).
   d. The oxidation of methylene chloride by iron (III).

7.23. Which abiotic transformations are most likely to occur in soil and groundwater systems for each of these chemicals?
   a. *n*-Butyl chloride   b. 2,4-Dichlorophenol
   c. Carbon tetrachloride   d. Mirex
   e. Trinitrobenzene   f. Malathion

7.24. Which of the following compounds are transformed most rapidly under oxidative conditions and which under reductive conditions? Justify your answer.
   a. Phenol   b. Carbon tetrachloride
   c. Hexachlorobenzene   d. Mirex
   e. *o*-Xylene   f. Chloroform
   g. 1,3,5-Trinitrobenzene

7.25. A groundwater is contaminated with the following compounds. Which would be effectively oxidized by an oxidation process that generates hydroxyl radicals? Justify your answer using rate constants for each compound's reactivity with hydroxyl radicals.

Benzene	PCE
Toluene	TCE
p-Xylene	TCDD
Carbon tetrachloride	Chloroform

7.26. An advanced oxidation process provides a steady-state concentration of hydroxyl radicals of 4×10^{-12} M to treat 38 mg/L PCE in a pump-and-treat groundwater reactor. What is the time required to reach 5 mg/L? Assume the reactor detention time follows a plug-flow regime in which

$$C/C_0 = e^{-k\theta}$$

where θ = the hydraulic detention time of the reactor. In addition, assume that inorganic hydroxyl radical quenchers, such as bicarbonate, are not present in significant concentrations.

7.27. A completely mixed AOP reactor maintains a steady-state concentration of 3×10^{-9} M hydroxyl radicals. Determine the half-life of (a) phenol, (b) TCE, (c) PCE, and (d) chloroform. Assume that 0.01% of the hyroxyl radicals react with the contaminant, and the remainder are quenched by bicarbonate and carbonate.

7.28. Derive an expression for the rate of change of 0.1 M endothall at t = 0 in a groundwater containing 68 mg/L bicarbonate and 2 mg/L carbonate being treated in an AOP reactor with a steady state hydroxyl radical concentration of 2×10^{-9} M. The detention time of the reactor is 20 minutes.

7.29. Draw or outline the carbon cycle and denote the pathways where primary producers, consumers, and decomposers are involved.

7.30. Define the following metabolic classes of microorganisms. Provide an example of a redox reaction associated with each group.
a. Chemotrophs   b. Autotrophs
c. Heterotrophs   d. Organotrophs
e. Lithotrophs   f. Chemoheterotrophs
g. Chemoautotrophs   h. Chemoorganotrophs

7.31. How many moles of ATP are generated per mole of glucose metabolized? Base your answer on adding the number of moles of NADH and $FADH_2$ reduced by glucose metabolism and the corresponding ADPs that capture the reducing potential in the electron transport system.

7.32. Dodecane labeled with ^{14}C and ^{3}H is introduced into an experimental model soil bioslurry unit, and by scintillation counting, the fate of the labeled carbon and hydrogen is traced in order to follow metabolism. As a basis for setting up a sampling scheme, hypothesize the pathways of the carbon and hydrogen as the dodecane is metabolized during system operation. The possibilities for sampling the products of metabolism include
a. the gas evolved from the bioslurry unit,
b. the aqueous phase of the slurry, and
c. cellular material.
Use your conceptual knowledge of hydrocarbon metabolism, the Embden-Myerhoff pathway, the TCA cycle, the electron transport system, and anabolic pathways to make your estimate and to justify your answer.

7.33. Describe the dynamics of biodegradation in a groundwater system in equilibrium with a readily degradable pool of LNAPLs floating on top of the groundwater. Assume that the rate of dissolution is slower than the rate of biodegradation.

7.34. Describe the metabolic differences of methanotrophs and methanogens.

7.35. Describe the importance of cometabolism in the biodegradation of chlorinated aliphatic solvents.

7.36. Arrange the following compounds in decreasing order of aerobic biodegradability (1 = highest; 5 = lowest).

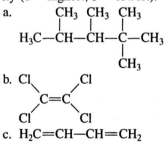

d. $H_3C-CH_2-CH_2-CH_2-CH_2-CH_3$

c. $H_2C=CH-CH=CH_2$

7.37. You are a project manager supervising a team of scientists and engineers on the cleanup of a PCE-contaminated groundwater. A member on the team proposes implementing aerobic in situ bioremediation for cleanup of the contaminated groundwater. Would you approve the proposed research plan?

7.38. A soil formulation and mixing area has been contaminated with Mirex. A consultant proposes using aerobic in situ bioremediation to degrade the Mirex. Do you agree with the proposal? Why or why not?

7.39. A groundwater contaminated with TCA is being studied for in situ bioremediation. Provide a recommendation of the most effective process conditions. What degradation products would you expect?

7.40. Compare the thermodynamics of nitrate and chloroform as terminal electron acceptors when acetate is used as the electron donor.

7.41. Perchloroethylene is being evaluated as an electron acceptor in the presence of sulfate. If acetate is used as the electron donor, compare the competing thermodynamics of the two potential electron acceptors.

7.42. What is the difference between a bioassessment study and a biotreatability study?

7.43. Outline a procedure for assessing the aerobic biodegradability of parathion in soils. Include details on the experimental apparatus, controls, and data collection.

7.44. A groundwater system is contaminated with PCE and TCE. A remediation team has proposed using in situ anaerobic bioremediation. They estimate that cleanup will require eight years, so they propose quarterly sampling for PCE and TCE to monitor process effectiveness and potential public health problems. Do you see any problem with their proposed remediation plan?

7.45. Outline how a bioassessment study would be performed to assess the rate of biodegradation in a surface soil contaminated with PAHs. Try to incorporate typical conditions of a surface soil in your proposed conditions.

7.46. Design a treatability study to assess the biodegradation potential of (a) naphthalene, (b) PCE, and (c) methylene chloride.

7.47. Samples of groundwater and well cuttings containing PCE have been collected from a groundwater system 120 in. deep. Outline a bioassessment study to evaluate the natural rate of biodegradation.

7.48. You are performing a site assessment for a pesticide rinse and formulation area contaminated with atrazine. Using the abiotic and biotic transformation literature, assess the most likely pathways for atrazine degradation and provide an estimate of its potential half-life.

7.49. A deep groundwater system is contaminated with TCE at an average concentration of 1.5 mg/L. As a consultant to the industry that is responsible for the contamination, write a five page miniproposal to the state regulatory agency (including a review of the recent literature) proposing *intrinsic* bioremediation for cleanup of the aquifer.

REFERENCES

7.1. Schwarzenbach, R. P., P. M. Gschwend, and D. M. Imboden, *Environmental Organic Chemistry*, John Wiley & Sons, New York, 1993.

7.2. Lewis, N., K. Topudurti, G. Welshans, and R. Foster, "Control technology: A field demonstration of the UV/oxidation technology to treat groundwater contaminated with VOCs," *J. Air Waste Manage. Assoc.*, **40**, 540–547 (1990).

7.3. Leung, S. W., R. J. Watts, and G. C. Miller, "Degradation of perchloroethylene by Fenton's reagent: Speciation and pathway," *J. Environ. Qual.*, **21**, 377–381 (1992).

7.4. Watts, R. J., M. D. Udell, P. A. Rauch, and S. W. Leung, "Treatment of pentachlorophenol-contaminated soils using Fenton's reagent," *Hazard Waste Hazard Mater.*, **7**, 335–345 (1990).

7.5. Bitton, G. and C. P. Gerba, *Groundwater Pollution Microbiology*, John Wiley & Sons, New York, 1984.

7.6. Scow, K. M., "Rate of biodegradation," Chapter 9 in Lyman, W. J., W. F. Rheel, and D. H. Rosenblatt (Eds.), *Handbook of Chemical Property Estimation Methods*, McGraw-Hill, New York, 1982.

7.7. Van der Wijngaard, A. J., R. G. VanderKleij, R. E. Doornweerd, and D. B. Janssen, "Influence of organic nutrients and cocultures on the competitive behavior of 1,2-dichloroethane-degrading bacteria," *Appl. Environ. Microbiol.*, **59**, 3400–3405 (1993).

7.8. Sedlak, D. L. and A. W. Andren, "Aqueous-phase oxidation of polychlorinated biphenyls by hydroxyl radicals," *Environ. Sci. Technol.*, **25**, 1419–1427 (1991).

7.9. Ogram, A. V., R. E. Jessup, L. T. Ou, and P. S. C. Rao, "Effects of sorption on biological degradation rates of (2,4-dichlorophenoxy) acetic acid in soils," *Appl. Environ. Microbiol.*, **49**, 582–587 (1985).

7.10. Weissenfels, W. D., H.-J. Klewer, and J. Langhoff, "Adsorption of polycyclic aromatic hydrocarbons (PAHs) by soil particles—Influence on biodegradability and biotoxicity," *Appl. Microbiol. Biotechnol.*, **36**, 689–696 (1992).

7.11. Guerin, W. F. and S. A. Boyd, "Differential bioavailability of soil-sorbed naphthalene to two bacterial species," *Appl. Environ. Microbiol.*, **58**, 1142–1152 (1992).

7.12. Steinberg, S. M., J. J. Pignatello, and B. L. Sawhney, "Persistence of 1,2-dibromoethane in soils—Entrapment in intraparticle micropores," *Environ. Sci. Technol.*, **21**, 1201–1208 (1987).

7.13. Manilal, V. B. and M. Alexander, "Factors affecting the microbial degradation of phenanthrene in soil," *Appl. Microbiol. Biotechnol.*, **35**, 401–405 (1991).

7.14. Speitel, G. E., Jr., C.-J. Lu, M. Tarakhia, and X.-J. Zhu, "Biodegradation of trace concentrations of substituted phenols in granular activated soil columns," *Environ. Sci. Technol.*, **23**, 68–74 (1989).

7.15. Watts, R. J., S. Kong, M. Dippre, and W. T. Barnes, "Oxidation of sorbed hexachlorobenzene in soils using catalyzed hydrogen peroxide," *J. Hazard Mater.*, **39**, 33–47 (1994).

7.16. Britton, L. N., "Microbial degradation of aliphatic hydrocarbons," in Gibson, D. T. (Ed.), *Microbial Degradation of Organic Compounds*, Marcel Dekker, New York, 1984.

7.17. Troy, M. A. and D. E. Jerger, "Evaluation of laboratory treatability study data for the full-scale bioremediation of petroleum hydrocarbon contaminated soils," *Proceedings Applied Bioremediation 93*, Fairfield, NJ, Oct. 25–26, 1993.

7.18. Leffler, J. E. and E. Grunwald, *Rates and Equilibria of Organic Reactions,* John Wiley & Sons, New York, 1963.

7.19. Hess, T. F., S. K. Schmidt, J. Silverstein, and B. Howe, "Supplemental substrate enhancement of 2,4-dinitrophenol mineralization by a bacterial consortium," *Appl. Environ. Microbiol.,* **56,** 1551–1556 (1990).

7.20. Dorfman, L. M. and G. E. Adams, "Reactivity of the hydroxyl radical in aqueous solutions," *NSRDS-NBS,* **46,** 1–72 (1973).

7.21. Dragun, J., *The Soil Chemistry of Hazardous Materials,* Hazardous Materials Control Research Institute, Silver Spring, MD, 1988.

7.22. Mabey, W. and T. Mill, "Critical review of hydrolysis of organic compounds in water under environmental conditions," *J. Phys. Chem. Ref. Data,* **7,** 383–415 (1978).

7.23. Hoffman, M., "Catalysis in aquatic environments," in Stumm, W. (Ed.), *Aquatic Chemical Kinetics,* Wiley-Interscience, New York, 1990.

7.24. Stone, A. T., "Enhanced rates of monophenyl terephthalate hydrolysis in aluminium oxide suspensions," *J. Colloid. Interface Sci.,* **127,** 429–441 (1989).

7.25. Harris, J. C., "Rate of hydrolysis," Chapter 7 in Lyman, W. J., W. F. Rheel, and D. H. Rosenblatt (Eds.), *Handbook of Chemical Property Estimation Methods,* McGraw-Hill, New York, 1982.

7.26. Ulrich, H.-J. and A. T. Stone, "Oxidation of chlorophenols absorbed to manganese oxide surfaces," *Environ. Sci. Technol.,* **23,** 421–428 (1989).

7.27. Stone, A. T., "Reductive dissolution of manganese (III/IV) oxides by substituted phenols," *Environ. Sci. Technol.,* **21,** 979–988 (1987).

7.28. Stone, A. T. and J. J. Morgan, "Reduction and dissolution of manganese (III) and manganese (IV) oxides by organics. 2. Survey of reactivity of organics," *Environ. Sci. Technol.,* **18,** 617–624 (1984).

7.29. Solomon, D. H., B. C. Loft, and J. D. Swift, "Reactions catalyzed by minerals. 4. The mechanism of the benzidine blue reaction," *Clay Miner.,* **7,** 389–397 (1968).

7.30. Dragun, J. and C. S. Helling, "Physicochemical and structural relationships of organic chemicals undergoing soil- and clay-catalyzed free-radical oxidation," *Soil Sci.,* **139,** 100–111 (1985).

7.31. Pillai, P., C. S. Helling, and J. Dragun, "Soil-catalyzed oxidation of aniline," *Chemosphere,* **11,** 299–317 (1982).

7.32. Sedlak, D. L. and A. W. Andren, "Oxidation of chlorobenzene with Fenton's reagent," *Environ. Sci. Technol.,* **25,** 777–782 (1991).

7.33. Mill, T., D. G. Hendry, and H. Richardson, "Free-radical oxidants in natural waters," *Science (Wash. DC),* **207,** 886–887 (1980).

7.34. Seinfeld, J. H., "Urban air pollution—State of the science," *Science (Wash. DC),* **243,** 745–752 (1989).

7.35. Haag, W. R. and C. D. D. Yao, "Rate constants for reaction of hydroxyl radicals with several drinking water contaminants," *Environ. Sci. Technol.,* **26,** 1005–1013 (1992).

7.36. Buxton, G. V., C. L. Greenstock, W. P. Helman, and A. B. Ross, "Critical review of rate constants for reactions of hydrated electrons, hydrogen atoms and hydroxyl radicals ($\cdot$OH/$\cdot$O$^-$) in aqueous solution," *J. Phys. Chem. Ref. Data,* **17,** 513–886 (1987).

7.37. Kriegman-King, M. R. and M. Rheinhard, "Transformation of carbon tetrachloride in the presence of sulfide, biotite, and vermiculite," *Environ. Sci. Technol.,* **26,** 2198–2206 (1992).

7.38. Jafvert, C. T. and N. L. Wolfe, "Degradation of selected halogenated ethanes in anoxic sediment-water systems," *Environ. Toxicol. Chem.,* **6,** 827–837 (1987).

7.39. Schwarzenbach, R. P., R. Stierli, K. Lanz, and J. Zeyer, "Quinone and iron porphyrin mediated reduction of nitroaromatic components in homogeneous aqueous solution," *Environ. Sci. Technol.,* **24,** 1566–1574 (1990).

7.40. Curtis, G. P. and M. Rheinhard, "Reductive dehalogenation of hexachloroethane, carbon tetrachloride, and bromoform by anthrahydroquinone disulfonate and humic acid," *Environ. Sci. Technol.,* **28,** 2393–2401 (1994).

7.41. Thurman, E. M., *Organic Geochemistry of Natural Waters,* Martinus Nijhoff, Boston, MA, 1985.

7.42. Matheson, L. J. and P. G. Tratnyek, "Reductive dehalogenation of chlorinated methanes by iron metal," *Environ. Sci. Technol.,* **28,** 2045–2053 (1994).

7.43. Sweeny, K. H., "The reductive treatment of industrial wastewaters. II. Process applications," *AICHE Symp. Ser.,* **77,** 72–78 (1981).

7.44. Gillham, R. W. and S. F. O'Hannesin, "Enhanced degradation of halogenated aliphatics by zero-valent iron," *Ground Water,* **32,** 958–967 (1994).

7.45. Cooper, W. J., D. E. Meacham, M. G. Nickelsen, K. Lin, D. B. Ford, C. N. Kurucz, and T. D. Waite, "The removal of tri- (TCE) and tetrachloroethylene (PCE) from aqueous solution using high energy electrons," *J. Air Waste Manage. Assoc.,* **43,** 1358–1366 (1993).

7.46. Zepp, R. G. and D. M. Cline, "Rates of direct photolysis in aquatic environment," *Environ. Sci. Technol.,* **11,** 359–366 (1977).

7.47. Gohre, K. and G. C. Miller, "Photooxidation of thioether pesticides on soil surfaces," *J. Agric. Food Chem.,* **34,** 709–713 (1986).

7.48. Kieatiwong, S., L. V. Nguyen, U. R. Hebert, M. Hachett, G. C. Miller, M. J. Miille, and R. Mitzel, "Photolysis of chlorinated dioxins in organic solvent and on soils," *Environ. Sci. Technol.,* **24,** 1575–1580 (1990).

7.49. Tchobanoglous, G. and E. D. Schroeder, *Water Quality,* Addison-Wesley, Reading MA, 1985.

7.50. Hirs, C. H. W., S. Moore, and W. H. Stein, "The sequence of the amino acid residues in performic-acid oxidized ribonuclease," *J. Biol. Chem.,* **234,** 633–637 (1960).

7.51. Vogel, T. M., C. S. Criddle, and P. L. McCarty, "Transformations of halogenated aliphatic compounds," *Environ. Sci. Technol.,* **21,** 722–736 (1987).

7.52. Britton, L. N., "Microbial degradation of aliphatic hydrocarbons," in Gibson, D. T. (Ed.), *Microbial Degradation of Organic Compounds,* Marcel Dekker, New York, 1984.

7.53. Gibson, D. T. and V. Subramanian, "Microbial degradation of aromatic hydrocarbons," in Gibson, D. T. (Ed.), *Microbial Degradation of Organic Compounds,* Marcel Dekker, New York, 1984.

7.54. McKenna, E., "Biodegradation of polynuclear aromatic hydrocarbon pollutants by soil and water microorganisms," Final Report, Project No. A-073-ILL, University of Illinois, Water Resources Center, Urbana, IL, 1976.

7.55. Cookson, J. T., Jr., *Bioremediation Engineering: Design and Application,* McGraw-Hill, New York, 1995.

7.56. Mihelcic, J. R. and R. G. Luthy, "The potential effects of sorption processes on the microbial degradation of hydrophobic organic compounds in soil-water suspensions," in Wu, X. C. (Ed.), *International Conference on Physicochemical and Biological Detoxification of Hazardous Wastes,* Vol. II, Technomic Publication, Lancaster, PA, 1988.

7.57. Van der Wijngaard, A. J., R. D. Wind, and D. B. Janssen, "Kinetics of bacterial growth on chlorinated aliphatic compounds," *Appl. Environ. Microbiol.,* **59,** 2041–2048 (1993).

7.58. Semprini, L., D. Grbic-Galic, P. L. McCarty, and P. V. Roberts, "Methodologies for evaluating in-situ bioremediation of chlorinated solvents," EPA/600/R-92/042, U.S. Environmental Protection Agency (March 1992).

7.59. Vanelli, T., M. Logan, D. M. Arciero, and A. B. Hooper, "Degradation of halogenated aliphatic compounds by the ammonia-oxidizing bacterium *Nitrosomonas europaea,*" *Appl. Environ. Microbiol.,* **56,** 1169–1171 (1990).

7.60. Shields, M., "Treatment of TCE and degradation products using *Pseudomonas cepacia,*" in *Symposium on Bioremediation of Hazardous Wastes: EPA's Biosystems Technology Development Program Abstracts,* Falls Church, VA, Apr. 16–18, 1991.

7.61. Hopkins, G. D., L. Semprini, and P. L. McCarty, "Microcosm and in-situ field studies of enhanced biotransformation of trichloroethylene by phenol-utilizing microorganisms," *Appl. Environ. Microbiol.,* **59,** 2277–2285 (1993).

7.62. Strand, S. E., J. V. Woodrich, and D. H. Stensel, "Biodegradation of chlorinated solvents in a sparged, methanotrophic biofilm reactor," *Res. J. Water Pollut. Control Fed.,* **63,** 859–867 (1991).

7.63. Palumbo, A. V., W. Eng, P. A. Boerman, G. W. Strandberg, T. L. Donaldson, and S. E. Herber, "Effects of diverse organic contaminants on trichloroethylene degradation by methanotrophic bacteria and methane-utilizing consortia," In Hinchee, R. E. and R. F. Offenbuttel (Eds.), *On-Site Bioreclamation Processes for Xenobiotic and Hydrocarbon Treatment,* Butterworth-Heinemann, Stoneham, MA, 1991.

7.64. Vogel, T. M. and P. L. McCarty, "Biotransformation of tetrachloroethylene to trichloroethylene, dichloroethylene, vinyl chloride, and carbon dioxide under methanogenic conditions," *Appl. Environ. Microbiol.,* **49,** 1080–1083 (1985).

7.65. Snoeyink, V. L. and D. Jenkins, *Water Chemistry,* John Wiley & Sons, New York (1980).

7.66. Slater, H. J. and D. Lovatt, "Biodegradation and the significance of microbial communities," in Gibson, D. T. (Ed.), *Microbial Degradation of Organic Compounds,* Marcel Dekker, New York, 1984.

7.67. Meikle, R. W., "Decomposition: Qualitative relationships," In Goring, C. A. I. and J. W. Hamaker (Eds.), *Organic Chemicals in the Soil Environment,* Marcel Dekker, New York, 1972.

7.68. Tiedje, J. M. and M. Alexander, "Enzymatic cleavage of the ether bond of 2,4-dichlorophenoxyacetate," *J. Agric. Food Chem.,* **17,** 1080–1084 (1969).

7.69. Quensen, J. F., S. A. Boyd, and J. M. Tiedje, "Dechlorination of four commercial polychlorinated biphenyl mixtures (Aroclors) by anaerobic microorganisms from sediments," *Appl. Environ. Microbiol.,* **56,** 2360–2369 (1990).

7.70. Adriaens, P., H. P. E. Kohler, D. Kohler-Staub, and D. D. Focht, "Bacterial dehalogenation of chlorobenzoates and coculture biodegradation of 4,4'-dichlorobiphenyl," *Appl. Environ. Microbiol.,* **55,** 887–892 (1989).

7.71. Morris, P. J., W. W. Mohn, J. F. Qeunsen III, J. M. Tiedje, and S. A. Boyd, "Establishment of a polychlorinated biphenyl-degrading enrichment culture with predominantly meta dechlorination," *Appl. Environ. Microbiol.,* **53,** 3088–3094 (1992).

7.72. McCarty, P. L. and L. Semprini, "Groundwater treatment for chlorinated solvents," in Norris, R. D., et al., *Handbook of Bioremediation,* Lewis Publishers, Boca Raton, FL, 1994.

7.73. Wilson, B. H., G. B. Smith, and J. F. Rees, "Biotransformations of selected alkylbenzenes and halogenated aliphatic hydrocarbons in methanogenic aquifer material: A microcosm study," *Environ. Sci. Technol.*, **20**, 997–1002 (1986).

7.74. Kappeler, T. and K. Wuhrmann, "Microbial degradation of the water-soluble fraction of gas oil—I," *Water Res.*, **12**, 327–333 (1978).

7.75. Tyre, B. W., R. J. Watts, and G. C. Miller, "Treatment of four biorefractory contaminants in soils using catalyzed hydrogen peroxide," *J. Environ. Qual.*, **20**, 832–838 (1991).

7.76. Brock, T. D., *Biology of Microorganisms*, Prentice-Hall, Englewood Cliffs, NJ, 1974.

7.77. 40 CFR 796.310, "Shake-Flask Test Standard Protocol."

7.78. Baker, K. H. and D. S. Herson, *Bioremediation*, McGraw-Hill, New York, 1994.

7.79. Verschueren, K., *Handbook of Environmental Data on Organic Chemicals*, 2nd Edition, Van Nostrand Reinhold, New York, 1983.

7.80. *The Agrochemical Handbook*, 2nd Edition, The Royal Society of Chemistry, London, 1990.

Chapter **8**

Contaminant Release and Transport from the Source

In hazardous waste management, the assessment of contaminants after they are released from a source is often necessary. The source can be a contaminated site, such as a pit-pond-lagoon, pesticide rinse-formulation area, UST, or a RCRA-type hazardous waste facility, including a 90-day drum storage area, solvent recycling facility, landfill, and so on. As described in Chapter 1, the characterization of contaminants and their concentrations at the source, as well as away from the source, is necessary in order to (1) assess risk, (2) evaluate the need for remediation and treatment, (3) devise health and safety plans, and (4) comply with RCRA, CERCLA, and the Clean Air Act.

Contaminant release and transport dynamics can be evaluated through a number of procedures, each with a range of complexity. Contaminant distributions are commonly quantified using detailed sampling and analysis schemes (Chapter 4); because monitoring data are the most reliable, actual data from the site or facility are often preferred. For small sites, a Phase I study may be conducted using only a conceptual analysis combined with a site visit; that is, no field data or calculations are used. More extensive site assessments and Remedial Investigation/Feasibility Studies may involve relatively simple calculations or complicated groundwater and air dispersion models. Such quantitative methods, which are based on the principles of advection and dispersion, are introduced in this chapter. Some of the more complicated models involve complex analytical solutions or numerical methods. These calculations serve as the basis for quantifying the transport of contaminants as they are released from a source, as well as for developing a conceptual framework for site assessments and facility management.

8.1 THE CONTROLLING PROCESSES IN CONTAMINANT RELEASE AND TRANSPORT: SORPTION, VOLATILIZATION, TRANSFORMATION

The pathways covered in Chapters 5, 6, and 7 may occur directly at the source (e.g., at a surface impoundment, in a solvent recycling vat, in a chromium sludge drying bed) or after the contaminants are released and transported from the source, as illustrated in

Figure 8.1. The two most important pathways of hazardous wastes are through the atmosphere and the subsurface and they are the focus of this chapter. Detailed descriptions of contaminant transport via surface water may be found in References 8.2 and 8.3.

A conceptual understanding of sorption, volatilization, and transformation processes serves as the basis for the qualitative and quantitative evaluation of hazardous waste facilities and contaminated sites. Chemicals characterized by high K_{ow} values (e.g., PAHs, organochlorine pesticides, PCBs, dioxins) have a thermodynamic basis for sorption, and, as described in Chapter 5, sorbed contaminants also have less potential to volatilize, be transformed, and be transported from the source than contaminants that are found primarily in the aqueous phase. For example, OCDD, a highly hydrophobic contaminant, would likely not migrate significantly from a surface soil where it was spilled because it would be strongly sorbed. As detailed in Chapter 6, the variables of vapor pressure and Henry's Law constant govern contaminant volatilization. Vapor pressure is the most important variable in the volatilization of high-strength liquids, such as surface impoundments and solvents being recycled, and Henry's Law governs contaminant volatilization from dilute aqueous solutions. Numerous variables affect the abiotic and biotic transformations of contaminants. Three of the most important are contaminant structure, the presence of reactive species, and accessibility to transformation reactions. As outlined in Chapter 7, oxidized organic contaminants, such as carbon tetrachloride, are transformed primarily through abiotic and biotic reductions. Reduced contaminants, such as benzene and toluene, are transformed via oxidative reactions. A labile contaminant, such as 2,4-dimethylphenol, may degrade even before it is released from the source. These three major pathways—sorption, volatilization, and transformation—now become the processes that control the release and transport of contaminants from the source.

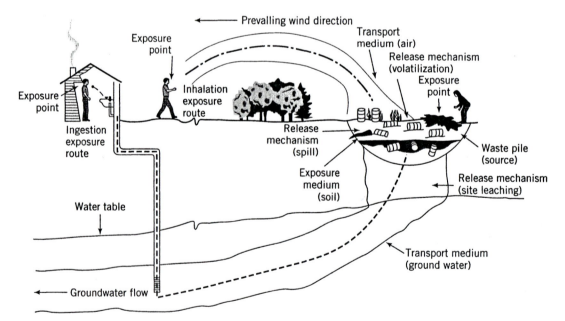

Figure 8.1 Atmospheric and subsurface transport—the most common pathways of contaminants after release from a source. *Source:* Reference 8.1.

A conceptual understanding of contaminant pathways can be invaluable in performing site assessments and facility inspections. For example, a common Phase I site assessment is a property transfer evaluation of an industrial area prior to its sale. An industrial site onto which acetone was disposed will likely have a different outcome than a property transfer evaluation of a site on which 2,4,7,8-TCDD was disposed. In many such short-term Phase I studies, time and budget constraints do not provide sufficient resources for sampling, chemical analysis, and atmospheric or subsurface modeling. Instead, a survey of company records and site characteristics, coupled with a walk-through inspection, is used to assess whether more detailed study is needed [8.4]. The past disposal of biorefractory and toxic chemicals suggests that hazardous wastes may have persisted at the site. In contrast, if only chemicals of relatively low toxicity that had a tendency to volatilize or degrade were used at the facility, the probability of their current presence is likely minimal. The importance of a conceptual knowledge of pathways in site and facility assessments is illustrated in Example 8.1.

EXAMPLE 8.1 *Conceptual Pathway Analysis in a Phase I Site Assessment*

A pesticide rinse and formulation area received approximately 500 kg (1100 lb) of 2,4,5-T and 200 kg (440 lb) of aldrin over a 10-year period ending 15 years ago. The soil in the area is a clay loam with an organic carbon content of 2.4%. Provide a conceptual assessment of the site for a property transfer evaluation.

SOLUTION

Although 2,4,5-T would have likely biodegraded over 15 years, its primary synthesis contaminant, 2,3,7,8-TCDD, is probably still present in the soil. Aldrin is among the most biorefractory contaminants, and is likely still present in the soil. The physical characteristics of the compounds include

$\log K_{ow}$ (Table 5.7):	2,3,7,8-TCDD: 5.77
	Aldrin: 5.17
Vapor pressures (Table 6.1):	2,3,7,8-TCDD: 6.4×10^{-10} mm Hg
	Aldrin: 6×10^{-6} mm Hg
Probable transformation pathways (Chapter 7):	2,3,7,8-TCDD: Reductive dehalogenation
	Aldrin: Reductive dehalogenation

The K_{ow} data indicate that both the TCDD and aldrin have a high tendency for sorption; in addition, the vapor pressure data suggest minimal potential for volatilization. Reductive dehalogenation in an oxygen-rich surface soil, the most likely transformation pathway, would probably provide nominal TCDD and aldrin degradation. Due to the high clay and organic carbon contents of the soil, the contaminants have likely remained in the top meter. Therefore, transformation and loss of the contaminants cannot be assumed, and more detailed study of the site is necessary through a Phase II assessment.

8.2   MASS TRANSFER OF CONTAMINANTS IN THE ATMOSPHERE AND THE SUBSURFACE

More detailed assessment of hazardous waste facilities and contaminated sites than Phase I studies requires quantitative techniques to evaluate contaminant transport away from the source. The analysis of such transport phenomena is based on a series of fundamental principles including mass balances, advection, and dispersion. In any defined system, such as a control volume of the atmosphere or groundwater (Figure 8.2), a mass balance may be described by

$$
\begin{array}{c}
\text{rate of mass} \\
\text{accumulation} \\
\text{within the system} \\
\text{boundary}
\end{array}
=
\begin{array}{c}
\text{rate of mass} \\
\text{flow into the} \\
\text{system}
\end{array}
-
\begin{array}{c}
\text{rate of mass} \\
\text{flow out of the} \\
\text{system}
\end{array}
+/-
\begin{array}{c}
\text{rate of reaction} \\
\text{within the} \\
\text{system}
\end{array}
\qquad (8.1)
$$

Based on the principle of conservation of mass, the movement of contaminants (or other materials) in and out of a control volume is often described by a general mass flux vector:

$$
\frac{\partial m}{\partial t} = \frac{\partial J_x}{\partial x} + \frac{\partial J_y}{\partial y} + \frac{\partial J_z}{\partial z}
\qquad (8.2)
$$

where

J = the mass flux vector in the x, y, or z direction
m = contaminant mass per unit volume
t = time

Two important differences are found between atmospheric and subsurface definitions of mass. First, the volume of air that is used to quantify mass is the total volume of the sample; however, in the subsurface, the volume is normalized by the porosity of

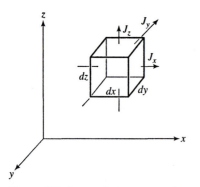

Figure 8.2 A mass balance around a control volume—the basis for mass fluxes in the atmosphere and subsurface.

the subsurface media (n). Second, the mass in the subsurface is distributed between the subsurface media and the pore-water, which is represented by using the retardation factor (R) as a multiplier (see Section 5.7). Therefore, the mass (per unit volume) of contaminants in atmospheric and subsurface systems is

$$m = C \quad \text{(atmosphere)} \tag{8.3}$$

$$m = n \cdot R \cdot C \quad \text{(subsurface)} \tag{8.4}$$

where

m = mass per unit volume
C = contaminant concentration [mg/L (water) or mg/m^3 (air)]
n = porosity of the subsurface medium
R = the retardation factor

Whether hazardous wastes are released from sources into the atmosphere or groundwater, the same fundamental principles and relationships control their transport. Two processes, which occur in both air and water, transport contaminants away from the source—advection and dispersion. In the atmosphere, *advection* is synonymous with wind; in groundwater systems, advection is described by the pore-water velocity or seepage velocity. *Dispersion* is the spreading of a contaminant as it moves downwind or downgradient and results in an associated decrease in concentration.

Advection and dispersion are fluxes that may be considered subdivisions of the total mass flux about the control volume described in Equation 8.2:

$$J = J_{\text{adv}} + J_{\text{disp}} \tag{8.5}$$

A general discussion of advection and dispersion as they apply to both air and groundwater is appropriate before beginning a more detailed analysis of mass flux in and out of a control volume.

Advective Flux. The *advective flux* is the bulk movement (often considered the average movement) of the fluid; in the atmosphere, bulk transport occurs through the movement of air, and in the subsurface, through the average motion of groundwater.

Quantitatively, atmospheric advection may be represented by

$$J_{\text{adv}} = u \cdot C \quad \text{(atmosphere)} \tag{8.6}$$

where

J_{adv} = advective flux $\left(\dfrac{\text{mg}}{\text{m}^2\text{-sec}} \right)$
u = the average wind velocity (m/sec)
C = the contaminant concentration (mg/m^3)

By analogy, the subsurface advection may be quantitatively defined by

$$J_{\text{adv}} = q \cdot C = n \cdot v \cdot C \quad \text{(subsurface)} \tag{8.7}$$

where

J_{adv} = advective flux $\left(\dfrac{mg}{m^2\text{-sec}}\right)$

$q = n \cdot v$ = the Darcy flux of water (m/sec)

C = the contaminant concentration in groundwater (mg/m^3)

The average wind velocity near a hazardous waste facility or contaminated site may be obtained from historical records or direct measurements that are averaged over time. Advective flux in the subsurface must account for lower volume through which flow is transmitted, which is described by the porosity. Therefore, v (similar to u) is the average linear velocity of the groundwater.

Dispersive Flux. Air and water not only move via advection, but also longitudinally and laterally through dispersive fluxes. A common convention is to classify all of the mass transfer exclusive of advection as *dispersive flux*. The fundamental definition of dispersion is based on (1) fluctuations in the velocity field at scales smaller than advection (mechanical dispersion), and (2) molecular diffusion. *Mechanical dispersion* is related to turbulent mixing and varied pathways as the medium moves down-gradient (Figure 8.3); in contrast, *molecular diffusion* is promoted by local concentration gradients. Such mixing of contaminants as they move downgradient results in their movement from areas of higher to lower concentration.

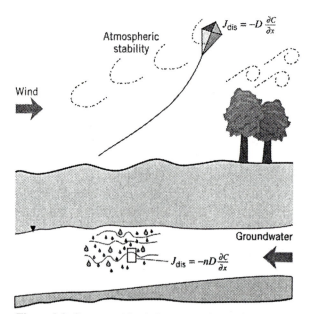

Figure 8.3 Conceptual basis for contaminant dispersion in the atmosphere and the subsurface.

Mass transport by dispersive flux is quantified by Fick's law: In the atmosphere, the dispersive flux is described by the rate at which contaminants move across a plane in the x direction.

$$J_{disp} = -D \frac{\partial C}{\partial x} \quad \text{(atmosphere)} \tag{8.8}$$

where

J_{disp} = the dispersive flux across a plane in the x direction $\left(\dfrac{mg}{m^2\text{-sec}} \right)$

D = the dispersion coefficient (m^2/sec)

C = the contaminant concentration (mg/m^3)

x = the distance through which dispersion of the contaminant is considered (m)

The dispersive flux in the subsurface is analogous to that in the atmosphere, but, like advective flux (Equation 8.7), the flow must be normalized for porosity (n):

$$J_{disp} = -nD \frac{\partial C}{\partial x} \quad \text{(subsurface)} \tag{8.9}$$

Dispersion coefficients of Equations 8.8 and 8.9 may be classified into mechanical dispersion and diffusion:

$$D = D_{mech} + D_0 \tag{8.10}$$

where

D_{mech} = dispersion component associated with mechanical energy

D_0 = dispersion component of molecular diffusion

Mechanical dispersion (D_{mech}) in the atmosphere may be considered turbulent dispersion that results from the eddies that occur from the random motion of air from wind moving around obstructions, and from atmospheric instability. In the subsurface environment, D_{mech} is related to the tortuous path of water through the subsurface solids. In the atmosphere, turbulent (or eddy) diffusion is the predominant form of mechanical energy. In groundwaters, where turbulence and eddy diffusion are not found as they are in the atmosphere, the tortuosity of flow paths that water and dissolved contaminants take result in the longitudinal mixing of contaminants, as well as their movement to areas of decreased concentration [8.5].

Another mechanism by which contaminants move from regions of higher concentration to lower concentration is molecular diffusion, which occurs independently of any advective, dispersive, or turbulent motion. Such *molecular diffusion* is the result of random molecular movements. It also obeys Fick's law, with D in this case called the *molecular diffusivity coefficient* (D_0).

The Advection-Dispersion Equation. A general approach to evaluating the mass transport of contaminants in the atmosphere or the subsurface is based on the control volume of dimensions x, $x + dx$, y, $y + dy$, z, and $z + dz$ shown in Figure 8.2, with a volume of $dx \cdot dy \cdot dz$. The bulk fluid passes through the control volume at corresponding vector velocities of v_x, v_y, v_z. A mass balance on all of the dimensions of the control volume using a Taylor series expansion with subsequent cancellation of like terms results in [8.6]:

$$\frac{\partial C}{\partial t} = -\left[v_x\left(\frac{\partial C}{\partial x}\right) + v_y\left(\frac{\partial C}{\partial y}\right) + v_z\left(\frac{\partial C}{\partial z}\right) \right]$$
$$+ \left[D_x\left(\frac{\partial^2 C}{\partial x^2}\right) + D_y\left(\frac{\partial^2 C}{\partial y^2}\right) + D_z\left(\frac{\partial^2 C}{\partial z^2}\right) \right]$$

(8.11)

or more generally,

$$\frac{\partial C}{\partial t} = -v_i\left(\frac{\partial C}{\partial i}\right) + D_i\left(\frac{\partial^2 C}{\partial i^2}\right)$$

(8.12)

Equation 8.12 is the fundamental expression of mass transfer for atmospheric and subsurface systems; appropriate limits and boundary conditions then provide the basis for integrating Equation 8.12 using analytical or numerical techniques. Specific parameters for the atmosphere involve u (wind velocity) for the advective flux and the incorporation of R (the retardation factor) for subsurface systems:

$$\frac{\partial C}{\partial t} = -u\frac{\partial C}{\partial i} + D_i\frac{\partial^2 C}{\partial i^2} \qquad \text{(atmosphere)}$$

(8.13)

$$R\frac{\partial C}{\partial t} = -v_i\frac{\partial C}{\partial i} + D_i\frac{\partial^2 C}{\partial i^2} \qquad \text{(subsurface)}$$

(8.14)

Equations 8.13 and 8.14 are fundamental forms of advection-dispersion equations and can be applied to a number of hazardous waste release and transport phenomena. Integration of Equations 8.13 and 8.14 for advection and dispersion, which can become complicated for three-dimensional representations, is performed by either (1) analytical methods or (2) numerical methods. Analytical methods involve establishing the governing equations and boundary conditions to provide an equation that can be easily solved. The boundary conditions for the analytical integrations of Equations 8.13 and 8.14 are not detailed in this chapter. However, derivations of these analytical solutions are provided in References 8.7 and 8.8. Analytical solutions are often somewhat limited in their potential to accurately predict contaminant concentrations downgradient as a result of the simple boundary conditions that must be used to provide a closed form solution. However, Bedient et al. [8.9] emphasized that "their simplicity of use and understanding make them valuable and convenient tools."

Equations that use more complex governing equations and boundary conditions must be approximated using numerical methods. The procedure that is commonly used for numerical solutions is dividing the problem into cells, followed by approximating the potential differential equations that govern the system by the difference between parameter values between each cell. Such finite element and finite difference techniques usually require the power of a computer workstation.

8.3 ATMOSPHERIC TRANSPORT FOLLOWING VOLATILIZATION RELEASES

The consequences of the transport of hazardous chemicals from their source depend on the nature of the contaminants and the duration of their release. Acutely toxic contaminants, such as hydrogen cyanide and chlorine released from hazardous materials spills, pose a short-term threat to public health. One of the worst industrial chemical disasters occurred in Bhopal, India, when the acutely toxic chemical intermediate, methyl isocyanate, escaped from a chemical plant. The plume traveled out of the industrial property into the city of Bhopal, resulting in over 2500 deaths [8.10, 8.11]. In contrast, low-concentration, continuous releases of hazardous compounds from RCRA facilities and contaminated sites may pose long-term health threats. A common goal in the assessment of RCRA facilities and hazardous waste sites is to ensure that concentrations of hazardous chemicals meet the hazardous air pollutant (HAP) standards of the Clean Air Act or a specified level of risk under CERCLA.

The first step in evaluating the atmospheric transport of hazardous chemicals is to estimate or quantify the flux of contaminants from the source. These volatilization rates may be estimated using Equation 6.8 for open containers, such as drums or vats in RCRA facilities, and Equations 6.24–6.26 for contaminated soils. Alternatively, volatile source emissions may be quantified using the air sampling methods described in Section 4.6.1. Because of the inherent uncertainty and nominal accuracy of the volatilization calculations, monitoring data are always more reliable. After the rate of release from the source has been quantified, their downwind concentrations may be estimated, a procedure that is based on fundamental dispersion models. Before describing solutions of Equation 8.13 for atmospheric transport, an introduction to the atmospheric environment is appropriate.

Nature of the Atmospheric Environment. A conceptual understanding of many atmospheric characteristics, such as wind, temperature, and cloud cover, is easy to achieve because of our first-hand familiarity with the atmosphere. The mass transport of contaminants in the atmosphere is dictated by the physical nature of this expansive medium. Although the atmosphere contains small particulates and condensed water in the form of clouds, it is still considered a homogeneous medium. Most potential health effects of hazardous chemicals occur in the troposphere, the lowest layer of the atmosphere, which extends to approximately 7500 m to 17,000 m (4.7 miles to 10.6 miles), depending on the latitude. The potential for wind to occur can be quantified by pressure differences in the atmosphere [8.12]; however, the short-range wind velocity data needed to assess localized exposures to hazardous chemicals (e.g., potential exposure to a population in a suburb from an air stripper) are often most effectively obtained by average wind speeds collected using standard meteorological instruments.

Atmospheric stability is a key factor in evaluating fluxes in the atmosphere. The vertical temperature gradient of the atmosphere may or may not have an effect on the movement of air. Maximum atmospheric stability is found when the actual temperature profile is equal to the adiabatic lapse rate (the thermodynamically based temperature profile based on the predicted compressibility of air—9.8°C/1000 m for dry air). Deviations of actual temperature profiles from adiabatic lapse rates result in a troposphere that is relatively well mixed, although local regions (such as in mountain valleys) are found with minimal advective and dispersive transport.

The diurnal cycle of daytime solar radiation and nighttime cooling can have significant effects on atmospheric stability. During the day, the ground and the overlying air are heated causing the air to rise; conversely, warm air is cooled and sinks during the night. Therefore, time of day and the amount of cloud cover can have a significant effect on atmospheric dispersion and serve as the basis for atmospheric stability classes, which are used to estimate dispersive fluxes in the atmosphere.

8.3.1 Development and Use of Fundamental Atmospheric Dispersion Models

The release of hazardous air pollutants (HAPs) to the atmosphere by volatilization is followed by transport and dispersion downwind, a process that may ultimately affect public health. Volatile releases may be classified into two time-related categories: (1) *puff models* that result from a pulse release, such as a hazardous waste spill, and (2) *plume models* as the result of continuous releases from contaminated sites or RCRA waste facilities. The puffs and plumes that form and move downwind from pulse and continuous hazardous waste releases are illustrated in Figures 8.4 and 8.5, respectively.

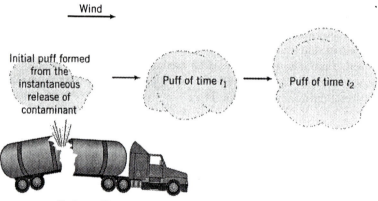

Figure 8.4 Puff release for an instantaneous release of hazardous chemicals.
Source: Reference 8.13.

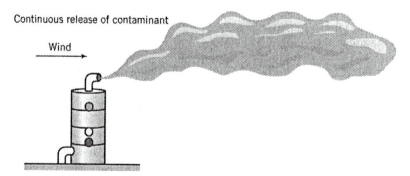

Figure 8.5 Plume formation for a continuous release of hazardous air emissions.

When a contaminant is released through volatilization, its maximum concentration is at the source and then decreases as the plume disperses downwind. Wind is the most important meteorological parameter in the transport of contaminants in the atmosphere. In addition, atmospheric turbulence dilutes the contaminant through dispersion. Plume dispersion is considered to begin after the plume reaches its effective height; that is, minimal dispersion occurs as the plume rises to its height at the source. The rates at which contaminants disperse and are transported in the atmosphere are then affected by (1) the wind velocity, (2) atmospheric turbulence, and (3) plume properties.

Pasquill and Gifford [8.14, 8.15] related the concepts of atmospheric stability to fundamental mass transfer relationships and were able to correlate Equations 8.8 and 8.13 with practical meteorological conditions through the estimation of dispersion coefficients. Use of the Pasquill-Gifford model is based on quantifying the puff or plume along two- and three-dimensional coordinate systems, which are shown in Figure 8.6 for puff models and Figure 8.7 for plume models. The Pasquill-Gifford plume model is based on the dispersion coefficients σ_x, σ_y, and σ_z rather than the more general Fick's Law constants D_x, D_y, and D_z (Equation 8.8). The coefficients σ_i are related to

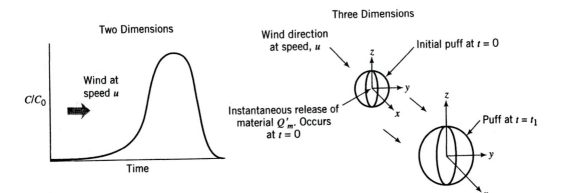

Figure 8.6 Representation of puffs from instantaneous releases in two and three dimensions. *Source:* Reference 8.13.

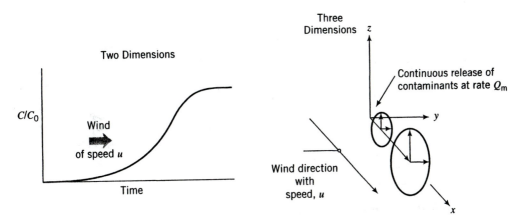

Figure 8.7 Representation of continuous source plumes in two and three dimensions. *Source:* Reference 8.13.

Table 8.1 Atmospheric Stability Classes for Use with the Pasquill-Gifford Dispersion Model

Wind Speed	Day (radiation intensity)			Night (cloud cover)	
(m/s)	Strong	Medium	Slight	Cloudy	Calm and Clear
<2	A	A–B	B		
2–3	A–B	B	C	E	F
3–5	B	B–C	C	D	E
5–6	C	C–D	D	D	D
>6	C	D	D	D	D

Source: Reference 8.15.

the standard deviations of the distribution of the contaminant about the respective axis i around which dispersion occurs.

The numerical values of σ_i are a function of wind velocity and atmospheric stability. Although these dispersion coefficients can be obtained by monitoring tracers released to the atmosphere [8.16], they are commonly estimated using average wind speeds and atmospheric stability classes. The atmospheric stability classes that are used in the Pasquill-Gifford model are listed in Table 8.1. The first step in estimating the dispersion coefficient is to assess the meteorological conditions and compare them to the stability classes in Table 8.1. The stability class—designated by a letter A through F—is selected based on the wind speed and solar radiation. Note that increased wind speed promotes greater atmospheric stability during the day; however, higher wind speeds result in lower atmospheric stability at night.

The two dispersion coefficients σ_y and σ_z for puff models and general stability classes are shown in Figures 8.8 and 8.9, respectively. The corresponding dispersion

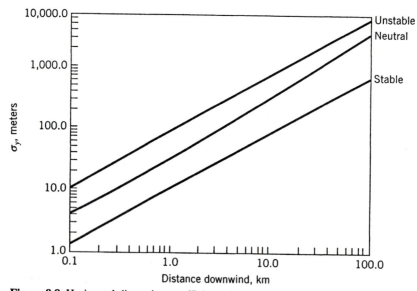

Figure 8.8 Horizontal dispersion coefficients for puff releases. *Source:* Reference 8.15.

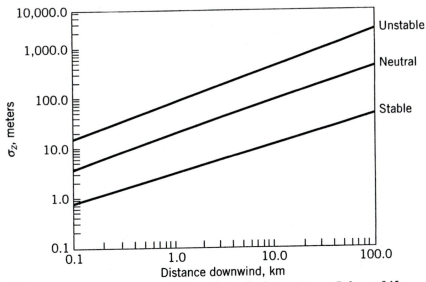

Figure 8.9 Vertical dispersion coefficients for puff releases. *Source:* Reference 8.15.

coefficients (σ_y and σ_z) for continuous hazardous waste releases are illustrated in Figures 8.10 and 8.11. In most cases $\sigma_x = \sigma_y$, so Figures 8.8 and 8.10 may be used to estimate σ_x as well as σ_y [8.8].

Through the use of the dispersion coefficients σ_x, σ_y, and σ_z, Equation 8.13 may be integrated using appropriate limits and boundary conditions into a number of equations that can be applied to specific conditions. The boundary conditions and limits found

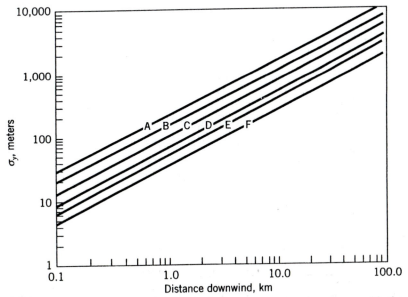

Figure 8.10 Horizontal dispersion coefficients for continuous releases with the Pasquill-Gifford model. *Source:* Reference 8.15.

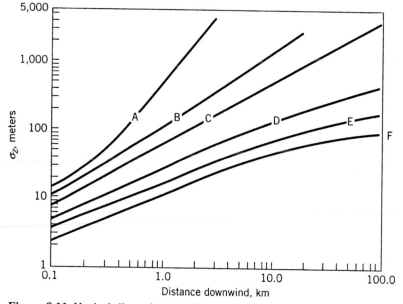

Figure 8.11 Vertical dispersion coefficients for continuous releases with the Pasquill-Gifford model. *Source:* Reference 8.15.

in the derivation of the air dispersion are available in References 8.4 and 8.15. The following description of the Pasquill-Gifford equations is divided into two primary categories—plume and puff models. Nested within the plume category are releases from (1) ground level and (2) a height H above the ground.

Puff Models. Most hazardous wastes and materials are released as a puff at the ground surface as the result of a spill or tanker accident (Figure 8.4). The *puff model for an instantaneous release at ground level* with constant wind velocity u in the x direction at $y = z = 0$ is

$$C(x,t) = \frac{Q'_m}{\sqrt{2}\pi^{3/2}\sigma_x\sigma_y\sigma_z} \exp\left[-\frac{1}{2}\left(\frac{x-ut}{\sigma_x}\right)^2\right] \tag{8.15}$$

where $C(x,t)$ is the contaminant concentration at coordinates x, 0, 0 and time t (mg/m³)

Q'_m = the mass of contaminant released (mg)
u = wind speed (m/s)
$\sigma_x, \sigma_y, \sigma_z$ = the dispersion coefficients in the x, y, and z directions, respectively (m).

The ground-level concentration may also be derived with only $z = 0$:

$$C(x,y,t) = \frac{Q'_m}{\sqrt{2}\pi^{3/2}\sigma_x\sigma_y\sigma_z} \exp\left\{-\frac{1}{2}\left[\left(\frac{x-ut}{\sigma_x}\right)^2 + \frac{y^2}{\sigma_y^2}\right]\right\} \tag{8.16}$$

where $C(x,y,t)$ is the contaminant concentration at coordinates x and y at time t (mg/m³)

The corresponding ground centerline concentrations for coordinates x, y, and z is

$$C(x,y,z,t) = \frac{Q'_m}{\sqrt{2}\pi^{3/2}\sigma_x\sigma_y\sigma_z} \exp\left\{-\frac{1}{2}\left[\left(\frac{x-ut}{\sigma_x}\right)^2 + \frac{y^2}{\sigma_y^2} + \frac{z^2}{\sigma_z^2}\right]\right\} \qquad (8.17)$$

where $C(x,y,z,t)$ is the contaminant concentration at coordinates x, y, z at time t (mg/m³).
    The use of a puff model in assessing a hazardous waste spill is illustrated in Example 8.2.

EXAMPLE 8.2 *Downwind Concentration after an Instantaneous Release*

A hazardous waste spill has occurred, releasing 10 kg (22 lb) of TCE into the air. If the spill was at night under mostly overcast skies and the wind velocity was 7 m/sec in the x direction, estimate the TCE concentration 0.5 km (0.31 miles) downgradient.

SOLUTION

1. Estimate σ_x, σ_y, σ_z.

$$x = 500 \text{ m}$$

From Table 8.1, with wind velocity = 7 m/sec, the atmospheric stability class is D.
From Figure 8.8, $\sigma_y = \sigma_x = 42$ m
From Figure 8.9, $\sigma_z = 20$ m
2. Estimate $C_{0.5 \text{ km}}$.
Equation 8.15 is used

$$C(x,t) = \frac{Q'_m}{\sqrt{2}\pi^{3/2}\sigma_x\sigma_y\sigma_z} \exp\left\{-\frac{1}{2}\left[\left(\frac{x-ut}{\sigma_x}\right)^2\right]\right\}$$

with the following inputs:

$t = x/u = 500 \text{ m}/7 \text{ m/sec} = 71.4 \text{ sec}$
$Q'_m = 10,000,000 \text{ mg}$

$$C(500 \text{ m}, 71.4 \text{ sec}) = \frac{1 \times 10^7 \text{ mg}}{\sqrt{2}\ \pi^{3/2}\ (42 \text{ m})(42 \text{ m})(20 \text{ m})} \exp\left\{-\frac{1}{2}\left[\frac{500 \text{ m} - 7 \text{ m/sec} \cdot 71.4 \text{ sec}}{42 \text{ m}}\right]\right\}$$

$$= 36 \text{ mg/m}^3$$

Plume Models. Using appropriate boundary conditions in a manner similar to puff models, plume models have been derived. The *plume model for a continuous, steady-state release* at ground level with wind moving in the x direction at constant velocity u along the x-axis (at $y = z = 0$) is

$$C(x) = \frac{Q_m}{\pi \sigma_y \sigma_z u} \tag{8.18}$$

where Q_m = the contaminant mass volatilization rate from the source (mg/sec).
  The ground-level concentration at coordinates x and y where $z = 0$ is`

$$C(x,y) = \frac{Q_m}{\pi \sigma_y \sigma_z u} \exp\left[-\frac{1}{2}\left(\frac{y}{\sigma_y}\right)^2\right] \tag{8.19}$$

where $C(x,y)$ = the steady-state concentration at coordinates x, y, and $z = 0$ (mg/m^3).
  The steady-state ground-level concentration at a coordinate x, y, z is

$$C(x,y,z) = \frac{Q_m}{\pi \sigma_y \sigma_z u} \exp\left[-\frac{1}{2}\left(\frac{y^2}{\sigma_y^2} + \frac{z^2}{\sigma_z^2}\right)\right] \tag{8.20}$$

where $C(x,y,z)$ = the steady-state contaminant concentration at coordinates x, y, z (mg/m^3).
  A common problem in hazardous waste management is a release from some height H, such as an air stripper 8 m (26 ft) high or an incinerator stack 24 m (79 ft) high. A modification of Equation 8.18 may be derived for a *continuous, steady-state source release at height H* above the ground surface. As with Equation 8.18, the plume is moving in the wind direction x at constant velocity u at $y = z = 0$:

$$C(x) = \frac{Q_m}{\pi \sigma_y \sigma_z u} \exp\left[-\frac{1}{2}\left(\frac{H}{\sigma_z}\right)^2\right] \tag{8.21}$$

where H = the height at which the contaminant is released (m).
  The corresponding ground-level concentration at $z = 0$ is

$$C(x,y) = \frac{Q_m}{\pi \sigma_y \sigma_z u} \exp\left[-\frac{1}{2}\left(\frac{y}{\sigma_y}\right)^2 - \frac{1}{2}\left(\frac{H}{\sigma_z}\right)^2\right] \tag{8.22}$$

and the concentration along the centerline of the plume directly downwind at x, y, z may be described by

$$C(x,y,z) = \frac{Q_m}{2\pi \sigma_y \sigma_z u} \exp\left[-\frac{1}{2}\left(\frac{y}{\sigma_y}\right)^2\right] \cdot \left\{\exp\left[-\frac{1}{2}\left(\frac{z-H}{\sigma_z}\right)^2\right] + \exp\left[-\frac{1}{2}\left(\frac{z+H}{\sigma_z}\right)^2\right]\right\} \tag{8.23}$$

The use of the plume dispersion equations is demonstrated in Example 8.3.

EXAMPLE 8.3 *Downwind Concentration of Steady-State Volatilization*

The mean flux of parathion from a 20 m² (215 ft²) pesticide formulation area is 4.3 mg/(m²-day). If daytime conditions with moderate solar radiation control the atmospheric stability and the mean wind velocity is 4 m/s, estimate the parathion concentration at a farmhouse 2 km (1.24 miles) downwind.

SOLUTION

1. Determine Q_m.

   Mean flux = 4.3 mg/(m²-day)
   Area = 20 m²
   Q_m = (4.3 mg/m² · day) (20 m²) = 86.0 mg/day = 0.000995 mg/sec

2. Estimate σ_y and σ_z.

   x = 2000 m
   From Table 8.1, the atmospheric stability class is C (a conservative assumption).
   From Figure 8.10, σ_y = 230 m.
   From Figure 8.11, σ_z = 220 m.

3. Determine $C_{2\text{ km}}$.
   Equation 8.20 is used to determine the continuous, steady-state concentration at ground level

   $$C(x) = \frac{Q_m}{\pi \sigma_y \sigma_z u}$$

   by estimating the concentration along the plume centerline at ground level.

   $$C(2000 \text{ m}) = \frac{0.000995 \text{ mg/sec}}{\pi(230 \text{ m})(220 \text{ m})(4 \text{ m/sec})} = 1.57 \times 10^{-9} \text{ mg/m}^3$$

8.4 SUBSURFACE TRANSPORT OF CONTAMINANTS

One of the greatest threats of hazardous waste releases is the contamination of groundwater supplies. Hazardous chemicals released from contaminated sites and RCRA facilities have sometimes migrated through the unsaturated zone to groundwater where they have reached drinking water wells or other receptors. Just as the Bhopal incident served as a significant event in the harmful atmospheric transport of hazardous chemicals, Love Canal was a landmark case in subsurface transport (see Chapter 1). Dozens of chemicals, such as chlorophenols, chlorobenzenes, and hexachlorocyclopentadiene, were disposed of in an unsecure landfill, trenches, and pits. The wastes then migrated through the subsurface and became apparent at a nearby subdivision [8.17]. Because migration through the subsurface is such a common pathway, a quantitative and conceptual knowledge of subsurface contaminant transport is essential in hazardous waste management.

Nature of the Subsurface Environment. The region of the geosphere near the ground surface is classified as *soil* and is defined by influence through plants and other surface inputs. Beneath the soil, the materials that make up the geosphere are sometimes called *subsurface solids* or *subsurface materials.* The subsurface environment is a complex medium that, unlike the atmosphere, is characterized by small- and large-scale heterogeneities. As illustrated in Figure 8.12, the subsurface exists downward from the lowest horizon of surface soils. The subsurface that lies above bedrock (solid rock) consists of porous and unconsolidated gravels, sands, silts, and clays.

Below the soil surface lies the *vadose zone,* which is defined as the subsurface region at which the pore-water pressure is less than atmospheric pressure. The water present in the vadose zone exists as a thin film on the subsurface solids. The *volumetric water content* is the fraction of the pores occupied by water; the remainder of the pore space is filled with air. The pore-water pressures of the vadose zone increase linearly as a function of depth; in contrast, volumetric water content increases in a nonlinear fashion with depth. Under the vadose zone is a layer at which the pore-water is equal to atmospheric pressure—the boundary of the water table. This region below the

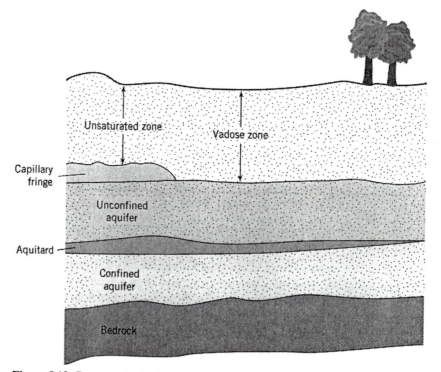

Figure 8.12 Common physical features of subsurface systems.

water table in which the pore spaces are filled with water is the *saturated zone*. Surface tension and capillary forces promote the upward movement of water from the saturated zone, resulting in an area called the *capillary fringe*. A term that is frequently used to describe the vadose zone is the *unsaturated zone*; however, definitions of the vadose zone and unsaturated zone are not entirely synonymous. The definition of the vadose zone is based on pore-water pressures and, as a result, includes the capillary fringe. Based on these definitions, the unsaturated zone does not include the capillary fringe.

Of special concern in subsurface systems is groundwater that can be used for drinking water supplies. An *aquifer* is defined as a saturated permeable geologic unit that can transmit quantities of water sufficient for water supply under ordinary hydraulic gradients. The term *aquitard* is used to describe strata that are less permeable than an aquifer. Beneath the water table, the pore-water pressures are greater than atmospheric and increase linearly with depth. The two main classifications of groundwater systems are unconfined and confined aquifers. *Unconfined aquifers*, also called water table aquifers, have an upper boundary that is unrestricted and free to move as the head in the aquifer changes; *confined aquifers*, as the name implies, are bound both above and below by aquitards.

The traditional conceptual model of the saturated zone is a layered system in which confined and unconfined aquifers are stacked on top of each other. At the top of the system is an unconfined aquifer that reaches from the water table down to the top of an upper aquitard. Below the aquitard may lie a confined aquifer, which has an aquitard above and below it. A perspective on the definitions of aquifers and aquitards is important to achieve a conceptual understanding of these systems. Although characterized by less permeability, aquitards are still permeable and may transmit water and contaminants between aquifers, and are therefore termed *leaky*.

Heterogeneity refers to spatial variations in subsurface characteristics. For example, if the porosity is 0.20 at one location and 0.45 at another location, the material is said to be heterogeneous with respect to porosity. The importance of heterogeneity with the subsurface environment cannot be overemphasized. Heterogeneity is probably one of the most important properties that affect water and contaminant transport in the subsurface environment.

The basis for advection in groundwater is *Darcy's Law*, an empirical mathematical description of the relationship between the flow rate of a fluid through a porous medium and the *head gradient* or driving force. Darcy's velocity, *q*, and the head gradient, *dh/dl*, are both vectors, having magnitude and direction:

$$q = -K\frac{dh}{dl} \tag{8.24}$$

where

q = specific discharge (m^2/sec per m)
K = hydraulic conductivity (m/s)
$-\dfrac{dh}{dl}$ = hydraulic gradient (m/m)

Because gradients, by convention, are measured from low to high, and the direction of flow is from high head to low head, the minus sign in Equation 8.24 is necessary to keep the vector directions consistent.

Hydraulic conductivity is a parameter that describes the ease with which a fluid passes through porous media. The greater the hydraulic conductivity, the more readily the fluid moves through the media. Moreover, hydraulic conductivity is a function not only of the physical media but also of the fluid passing through it. In particular, water will have a higher conductivity than a viscous substance such as motor oil because more energy (in length) is required to drive an equivalent amount of oil through the medium. However, because water is usually the fluid of interest, values for hydraulic conductivity are usually tabulated only for water. Typical hydraulic conductivities are listed in Table 8.2.

Although q has units of velocity, it is not an actual velocity because of the inherent assumption that flow occurs over the entire cross-sectional area. Because flow occurs only through the pore spaces, the actual velocity will be greater than the Darcy velocity. For this reason, it is useful to refer to q as the *specific discharge*, which is the flow rate per unit cross-sectional area, which is computed perpendicular to the direction of flow.

The *pore-water* or *seepage velocity, v*, (i.e., the velocity that accounts for the cross-sectional area) is then defined as

$$v = -\frac{K}{n}\frac{dh}{dl} \tag{8.25}$$

where n = effective porosity of the subsurface material.

An area of recent increased study is the quantitation of subsurface heterogeneities, in which statistical methods are used to assess spatial differences in subsurface systems. The highest variability in groundwater dynamics is related to hydraulic conductivity, which is a function of the heterogeneity of the subsurface strata. Recently, stochastic descriptions of subsurface heterogeneity have been used to evaluate variations in groundwater flow [8.18]. Another approach to evaluating heterogeneity is fractal geometry, which is based on the phenomenon that patterns in nature (such as heterogeneities in the subsurface) are found repeatedly [8.19]. A discussion of subsurface heterogeneities is provided in Reference 8.20.

Table 8.2 Hydraulic Conductivities of Soil Textural Classes

Soil Textural Class	K (m/s)
Clays	$<10^{-8}$
Peat	10^{-8} to 10^{-7}
Silt	10^{-7} to 10^{-6}
Loam	10^{-7} to 10^{-5}
Fine sands	10^{-5} to 10^{-4}
Coarse sands	10^{-4} to 10^{-3}
Sand/gravel	10^{-3} to 10^{-2}
Gravel	$>10^{-2}$

8.4.1   Development and Use of Groundwater Transport Equations

Calculation of subsurface contaminant transport rates is based on the characteristics of the substrata and the groundwater flow rate, as well as the degree of sorption and rates of transformation covered in Chapters 5 and 7. Just as multidimensional transport models may be derived for the transport of atmospheric contaminants after they are released from a source (Equations 8.15 through 8.23), analogous equations may be derived for transport in the subsurface. As these equations are presented, note that many of the groundwater transport equations look similar to the atmospheric equations, which is not surprising considering that the basic mass transport equations for air and groundwater (Equations 8.13 and 8.14) are nearly the same.

The assumptions used in the development and the derivation of groundwater transport equations are that (1) the media are characterized by constant density and viscosity, (2) the flow is incompressible, (3) the media are isotropic, and (4) the system is saturated.

The transport of contaminants in groundwater systems, which is based on the integration of the advection-dispersion equation using specific boundary conditions for the system of interest, provides a number of analytical solutions. Some of these solutions have been derived for (1) one-, two-, or three-dimensional flow, (2) a pulse (puff) contaminant input or a continuous input, and (3) the presence or absence of transformation reactions. Presentation of equations for a matrix of all of these conditions is beyond the scope of this text; however, many groundwater texts and reference books are available that provide extensive coverage of dozens of subsurface transport equations [8.14, 8.17].

Significant advances have been made in the development of analytical and numerical groundwater models [8.21–8.25]. However, the models that have been developed are often limited by estimation of parameters such as dispersion coefficients and rate constants for transformation reactions [8.26].

A number of one-dimensional groundwater transport equations may be derived for pulse inputs and continuous inputs with and without terms for transformation. The following discussion will focus on equations for (1) a pulse input without transformation, (2) a pulse input with a transformation term, (3) a continuous input without transformation, and (4) a continuous input with a transformation term. In addition, a two-dimensional groundwater transport equation will be presented for a pulse input.

One-dimensional advection-dispersion equations are sometimes considered relatively simple models for quantifying contaminant dynamics in groundwater systems. The equations are considered accurate for simple systems, such as soil columns through which water is passed at steady state. Other useful applications of one-dimensional advection-dispersion equations are the evaluation of landfill leachate migrating through porous liner material and assessment of a plume that is migrating toward a point of interest, such as a property boundary [8.27]. Furthermore, the one-dimensional equation provides an effective introductory tool for developing a conceptual basis for the behavior of contaminants in subsurface systems.

Pulse Models. A pulse input of contamination, such as from a hazardous materials spill, has been derived with zero background concentration at $x = 0$:

$$C(x,t) = \frac{M}{\sqrt{4\pi D'_x t}} \exp\left[-\frac{(x - v'_x t)^2}{4D'_x t}\right] \tag{8.26}$$

where

M = the mass spilled per cross-sectional area (g/m^2)
$D'_x = D/R$ (m^2/day)
$v'_x = v/R$ (m/day)
R = the retardation factor

A transformation term can be added to Equation 8.26 to account for the first-order degradation of organic contaminants or exchange of metals on surfaces

$$C(x,t) = \frac{M}{\sqrt{4\pi D'_x t}} \exp\left[-\frac{(x - v'_x t)^2}{4D'_x t}\right] \cdot e^{-k't} \tag{8.27}$$

where

$k' = k/R$ the first-order loss coefficient (day^{-1})

Two- and three-dimensional analytical solutions to the fundamental mass transport equation (Equation 8.14) become computationally complex, with associated intricate terms such as the Bessel function [8.28]. However, Wilson and Miller [8.29] provided relatively straightforward analytical solutions of Equation 8.14 in two dimensions for pulse and continuous source releases. Their two-dimensional analytical solution for a pulse input is

$$C(x,y) = \frac{M'}{4\pi t\sqrt{D'_x D'_y}} \exp\left\{-\left[\frac{(x - v'_x)^2}{4D'_x t} + \frac{y^2}{4D'_y t}\right]\right\} \tag{8.28}$$

where

M' = the mass spilled per unit depth (g/m)

The application of a one-dimensional pulse groundwater transport equation is demonstrated in Example 8.4.

EXAMPLE 8.4 *Estimation of Downgradient Contaminant Concentration Resulting from a Chemical Spill Using One-Dimensional Transport with Transformation*

A spill of a 208-L (55-gal) drum that is 25% full has released TCA into an aquifer over an area of 40 m^2 (430 ft^2). The pore-water velocity in the aquifer is 5 m/day in the x direction, and the porosity of the system is 0.40. The dispersion coefficient in the x direction is 0.8 m^2/day, and the retardation factor has been estimated at 8.0. Abiotic and biotic transformations result in TCA degradation described by a first-order rate constant of 0.0004 day^{-1}. Based on these data, estimate the TCA concentration at $x = 200$ m and $t = 1$ year.

SOLUTION

1. Determine the mass of TCA spilled using its specific gravity (Table 3.6).

$$0.25 \cdot 208 \text{ L} \cdot \frac{1.3390 \text{ g}}{\text{mL}} \cdot \frac{1000 \text{ mL}}{\text{L}} = 69,600 \text{ g}$$

2. Use Equation 8.27 to estimate the TCA concentration at $x = 200$ m and $t = 1$ year.

$$C(x,t) = \frac{M}{\sqrt{4\pi D'_x t}} \exp\left[-\frac{(x - v'_x t)^2}{4D'_x t}\right] \cdot e^{-k't}$$

$$C = \frac{69,600/40}{\sqrt{(4\pi)(0.8/8)(365)}} \exp\left\{-\left[\frac{(200 - (5/8) \cdot 365)^2}{4(0.8/8)(365)}\right]\right\} \cdot e^{-(0.0004/8)(365)}$$

$$C = 0.35 \text{ g/m}^3 = 0.35 \text{ mg/L}$$

Three-dimensional analytical solutions to groundwater transport equations with and without transformation terms have also been developed; many of these are described in References 8.14 and 8.24.

Plume Models. The continuous input of hazardous wastes from a landfill, surface impoundment, or similar sources has been derived with and without terms for first-order contaminant transformation. Equation 8.14 has been integrated using a LaPlace transform in order that contaminant concentration may be calculated at a distance x downgradient at time t:

$$C(x,t) = \frac{C_0}{2}\left[\text{erfc}\left(\frac{x - v'_x t}{2\sqrt{D'_x t}}\right) - \exp\left(\frac{v'_x x}{D'_x}\right)\text{erfc}\left(\frac{x + v'_x t}{2D'_x t}\right)\right] \tag{8.29}$$

where

$C(x,t)$ is the contaminant concentration at point x and time t (mg/L).
erfc = the complementary error function

The *error function* for two variables x and y, which is needed to solve Equation 8.29, is defined as

$$\text{erf}(x) = \frac{2}{\sqrt{\pi}}\int_0^x e^{-x^2}\, d_x \tag{8.30}$$

Also,

$$\text{erf}(-x) = -\text{erf}(x) \tag{8.31}$$

The *complementary error function* is then

$$\text{erfc}(x) = 1 - \text{erf}(x) \tag{8.32}$$

Error functions are listed in Table 8.3.

Under most conditions, the right-hand term becomes negligible and Equation 8.29 may be simplified to

$$C(x,t) = \frac{C_0}{2}\left[\text{erfc}\left(\frac{x - v'_x t}{2\sqrt{D'_x t}}\right)\right] \tag{8.33}$$

Equation 8.33 was developed to quantify diffusion and dispersion in laboratory columns filled with subsurface solids and saturated with water. A steady flow of water is then

Table 8.3 Values for the Error Function

X	0.00	.01	.02	.03	.04	.05	.06	.07	.08	.09
0.0		.01128	.02256	.03384	.04511	.05637	.06762	.07886	.09008	.10128
.1	.11246	.12362	.13476	.14587	.15695	.16800	.17901	.18999	.20094	.21184
.2	.22270	.23352	.24430	.25502	.26570	.27633	.28690	.29742	.30788	.31828
.3	.32863	.33891	.34913	.35928	.36936	.37938	.38933	.39921	.40901	.41874
.4	.42839	.43797	.44747	.45689	.46623	.47548	.48466	.49375	.50275	.51167
.5	.52050	.52924	.53790	.54646	.55494	.56332	.57162	.57982	.58792	.59594
.6	.60386	.61168	.61941	.62705	.63459	.64203	.64938	.65663	.66378	.67084
.7	.67780	.68467	.69143	.69810	.70468	.71116	.71754	.72382	.73001	.73610
.8	.74210	.74800	.75381	.75952	.76514	.77067	.77610	.78144	.78669	.79184
.9	.79691	.80188	.80677	.81156	.81627	.82089	.82542	.82987	.83423	.83851
1.0	.84270	.84681	.85084	.85478	.85865	.86244	.86614	.86977	.87333	.87680
1.1	.88021	.88353	.88679	.88997	.89308	.89612	.89910	.90200	.90484	.90761
1.2	.91031	.91296	.91553	.91805	.92051	.92290	.92524	.92751	.92973	.93190
1.3	.93401	.93606	.93807	.94002	.94191	.94376	.94556	.94731	.94902	.95067
1.4	.95229	.95385	.95538	.95686	.95830	.95970	.96105	.96237	.96365	.96490
1.5	.96611	.96728	.96841	.96952	.97059	.97162	.97263	.97360	.97455	.97546
1.6	.97635	.97721	.97804	.97884	.97962	.98038	.98110	.98181	.98249	.98315
1.7	.98379	.98441	.98500	.98558	.98613	.98667	.98719	.98769	.98817	.98864
1.8	.98909	.98952	.98994	.99035	.99074	.99111	.99147	.99182	.99216	.99248
1.9	.99279	.99309	.99338	.99366	.99392	.99418	.99443	.99466	.99489	.99511
2.0	.99532	.99552	.99572	.99591	.99609	.99626	.99642	.99658	.99673	.99688
2.1	.99702	.99715	.99728	.99741	.99753	.99764	.99775	.99785	.99795	.99805
2.2	.99814	.99822	.99831	.99839	.99846	.99854	.99861	.99867	.99874	.99880
2.3	.99886	.99891	.99897	.99902	.99906	.99911	.99915	.99920	.99924	.99928
2.4	.99931	.99935	.99938	.99941	.99944	.99947	.99950	.99952	.99955	.99957
2.5	.99959	.99961	.99963	.99965	.99967	.99969	.99971	.99972	.99974	.99975
2.6	.99976	.99978	.99979	.99980	.99981	.99982	.99983	.99984	.99985	.99986
2.7	.99987	.99987	.99988	.99989	.99989	.99990	.99991	.99991	.99992	.99992
2.8	.99992	.99993	.99993	.99994	.99994	.99994	.99995	.99995	.99995	.99996
2.9	.99996	.99996	.99996	.99997	.99997	.99997	.99997	.99997	.99997	.99998
3.0	.99998	.99999	.99999	1.0000						

introduced into the column, followed by a nonreactive solute, which is then analyzed as a function of time. Equation 8.33 has also been used to quantify solute transport in simple groundwater systems that behave as one-dimensional aquifers. The use of Equation 8.33 is illustrated in Example 8.5.

EXAMPLE 8.5 *One-Dimensional Contaminant Transport with No Transformation*

A surface impoundment containing cyanide leaches into a groundwater system of high permeability that behaves as a one-dimensional aquifer with a hydraulic conductivity of 4 m/day, effective porosity of 0.3, and a hydraulic gradient of 0.03 m/m. If the initial concentration of cyanide is 42 mg/L with a longitudinal dispersion coefficient (D_x) of 2.1 m^2/day and sorption and transformation processes are negligible, determine the time required for the concentration to reach 1 mg/L at a well 1000 m (3,280 ft) away.

SOLUTION

1. Use Equation 8.25 to determine the pore-water velocity.

$$v = -\frac{K}{n}\frac{dh}{dl}$$

where

$K = 4$ m/day
$n = 0.3$
$dh/dl = -0.03$ m/m
$v_x = -\dfrac{4\text{ m/day}}{0.3}\ (-0.03\text{ m/m})$
$v_x = 0.4$ m/day.

2. Using Equation 8.29, develop an expression for time. Note that the last term can be ignored because the erfc will be zero for this case:

$$C(x,t) = \frac{C_0}{2}\left[\text{erfc}\left(\frac{x - v_x t}{2\sqrt{D_x t}}\right) - \exp\left(\frac{v_x x}{D_x}\right)\text{erfc}\left(\frac{x + v_x t}{2\sqrt{D_x t}}\right)\right]$$

$$C(x,t) = \frac{C_0}{2}\left[\text{erfc}\left(\frac{x - v_x t}{2\sqrt{D_x t}}\right)\right]$$

where

$C = 1$ mg/L
$C_0 = 42$ mg/L
$x = 1000$ m
$v_x = 0.4$ m/day
$D_x = 2.1$ m^2/day

3. An expression in t may be determined by interpolating between values in Table 8.3 converted to erfc.

$$1.401 = \frac{1000 - 0.4t}{2\sqrt{2.1t}}$$

By trial and error,

$$t = 2040 \text{ days} = 5.6 \text{ years}$$

Groundwater transport of a sorbed contaminant that is degrading by abiotic or biotic processes as it moves through the subsurface is based on modifying Equation 8.14, the one-dimensional advection-dispersion equation for a nonreactive contaminant. A significant change is made to Equation 8.14 to account for transformation (through the addition of a new term for first-order contaminant degradation):

$$\frac{\partial C}{\partial t} = \frac{D}{R} \frac{\partial^2 C}{\partial x^2} - \frac{v}{R} \frac{\partial C}{\partial x} - \frac{k}{R} C \tag{8.34}$$

where

C = contaminant concentration in the aqueous phase (mg/L)

R = the retardation factor = $1 + \frac{\rho_B}{n} K_d$

D = groundwater dispersion coefficient (m²/day)

v = pore-water velocity (m/day)

k = first-order rate constant for contaminant degradation (day^{-1})

t = time (day)

x = distance (m)

Inspection of Equation 8.34 shows that the rate of change of a contaminant C at any point x downgradient is a function of not only the advective and the dispersive transport, but also the first-order degradation rate. The effects of retardation and transformation on contaminant profiles at a fixed location downgradient are shown in Figures 8.13 and 8.14, respectively. As with the one-dimensional transport for a conservative species, Equation 8.34 is a simplification of the complex processes that occur in groundwater systems. One-dimensional flow is found only under steady, uniform conditions, which rarely occur in natural systems. Heterogeneities such as boulders and clay lenses result in deviations from steady and uniform flow; in addition, actual flow takes place in three dimensions rather than in one dimension. Furthermore, grouping chemical reaction terms into a first-order rate often oversimplifies the process and the conceptual understanding of the mechanisms taking place. Although first-order reactions are not always the most realistic, they are easy to incorporate into transport models. More realistic, higher-order kinetics are often difficult to measure and model in complex systems.

Although Equation 8.34 does not provide solutions that are sufficiently accurate for most exposure assessments, it has been used for a number of other applications. The use of this relatively simple analytical equation has a number of applications, by providing (1) an efficient model for controlled laboratory experiments and (2) as a simple check to more complicated models that require numerical solutions. More importantly, calculations using Equation 8.34 provide a powerful tool for developing conceptual knowledge of the effects of sorption, transformation, advection, and dispersion on the rate of subsurface transport.

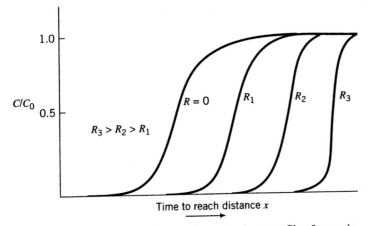

Figure 8.13 Effect of retardation on the contaminant profile of a continuous source input.

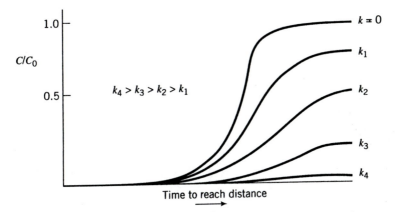

Figure 8.14 Effect of first-order transformation rate constant on contaminant profiles of a continuous input.

The solution to Equation 8.34 is based on its integration using the appropriate boundary conditions for a continuous contaminant input. The analytical solution to Equation 8.34, which was first described by Cho [8.30], is

$$
\frac{C(x,t)}{C_0} = \frac{1}{2}\left\{ \exp\left[\frac{x}{2(D_x/R)} \cdot \left(\frac{v_x}{R} - \sqrt{(v_x/R)^2 + 4(D_x/R) \cdot (k/R)} \right) \right] \right.
$$

$$
\cdot \, \mathrm{erfc}\left[\frac{x - t\sqrt{(v_x/R)^2 + 4(D_x/R) \cdot (k/R)}}{2\sqrt{(D_x/R)t}} \right]
$$

$$
+ \exp\left[\frac{x}{2(D_x/R)} \cdot \left(\frac{v_x}{R} + \sqrt{(v_x/R)^2 + 4(D_x/R) \cdot (k/R)} \right) \right]
$$

$$
\left. \cdot \, \mathrm{erfc}\left[\frac{x + t\sqrt{(v_x/R)^2 + 4(D_x/R) \cdot (k/R)}}{2\sqrt{(D_x/R)t}} \right] \right\}
\tag{8.35}
$$

The solutions provided by Cho [8.30] have been modified to account for multicomponent transport [8.31], a spatially variable retardation factor [8.32], and variable source concentrations [8.33]. Furthermore, Lunn et al. [8.34] extended the LaPlace transform solution to multiple species contaminant transport problems using Fourier sine transformations.

A more manageable form of Equation 8.35 may be obtained by replacing the larger terms with the variables A_1, A_2, B_1, and B_2, and by using the surrogate variables D'_x, V'_x, and k':

$$\frac{C(x,t)}{C_0} = \frac{1}{2}[\exp(A_1) \cdot \mathrm{erfc}(A_2) + \exp(B_1) \cdot \mathrm{erfc}(B_2)] \tag{8.36}$$

where

$$A_1 = \frac{x}{2D'_x}\left(v'_x - \sqrt{v'^2_x + 4D'_x k'}\right)$$

$$A_2 = \frac{x - t\sqrt{v'^2_x + 4D'_x k'}}{2\sqrt{D'_x t}}$$

$$B_1 = \frac{x}{2D'_x}\left(v'_x + \sqrt{v'^2_x + 4D'_x k'}\right)$$

$$B_2 = \frac{x + t\sqrt{v'^2_x + 4D'_x k'}}{2\sqrt{D'_x t}}$$

$$D'_x = D_x/R$$

$$v'_x = v_x/R$$

$$k' = k/R$$

Similar to Equation 8.29, the B terms are often negligible, resulting in the following simplified equation:

$$\frac{C(x,t)}{C_0} = \frac{1}{2}[\exp(A_1) \cdot \mathrm{erfc}(A_2)] \tag{8.37}$$

Equation 8.29 was used to determine the time required for a conservative contaminant of concentration C to reach a point downgradient in a one-dimensional saturated system, which was demonstrated in Example 8.5. Equation 8.37 may be used in a similar manner to determine the time required for a contaminant that is sorbed and degrading as it moves through the groundwater; this analogous calculation is illustrated in Example 8.6.

EXAMPLE 8.6 *One-Dimensional Contaminant Transport with Transformation*

Reevaluation of the subsurface conditions in which a surface impoundment is leaching cyanide at a concentration of 42 mg/L (Example 8.5) has shown that the cyanide is exchanged on soil functional groups with a K_d of 2.2 and that soil-catalyzed oxidation reactions are occurring at a rate described by a first-order k of 0.000014 day^{-1}.

Rework Example 8.5 with consideration of sorption and transformation to determine the time required for the concentration to reach 1 mg/L at a well 1000 m away.

SOLUTION

From Example 8.5, recall that

$$D_x = 2.1 \text{ m}^2/\text{day}$$

$$v_x = 0.4 \text{ m/day}$$

Determine the retardation factor ($n = 0.3$ from Example 8.5 and assume $\rho_B = 1.8$ g/cm^3).

$$R = 1 + \frac{\rho_B}{n}K_d = 1 + (1.8/0.3) \cdot (2.2) = 14.2$$

Obtain the system characteristics accounting for dispersion, sorption, and transformation.

$$D'_x = 2.1/14.2 = 0.148 \text{ m}^2/\text{day}$$

$$v'_x = 0.4/14.2 = 0.0282 \text{ m/day}$$

$$k' = 0.000014/14.2 = 9.86 \times 10^{-7} \text{ day}^{-1}$$

Substitute values and solve with $x = 1000$ m:

$$A_1 = \frac{1000}{2 \cdot 0.148}\left(0.0282 - \sqrt{0.0282^2 + 4 \cdot 0.148 \cdot 9.86 \times 10^{-7}}\right) = -0.0350$$

$$A_2 = \frac{1000 - t\sqrt{0.0282^2 + 4 \cdot 0.148 \cdot 9.86 \times 10^{-7}}}{2\sqrt{0.148t}} = \frac{1000 - 0.0282t}{2\sqrt{0.148t}}$$

Similarly,

$$B_1 = 193.2$$

$$B_2 = \frac{1000 + 0.0282t}{2\sqrt{0.148t}}$$

As in Example 8.5, the last terms (B_1 and B_2) may be ignored because the erfc of B_2 will be zero for this case. Therefore,

$$\frac{C(x,t)}{C_0} = \frac{1}{2}\left[\exp(-0.0350) \cdot \text{erfc}\left(\frac{1000 - 0.0282t}{2\sqrt{0.148t}}\right)\right]$$

where $C/C_0 = 1/42 = 0.0238$.

An expression in t may be determined by interpolating between values in Table 8.3 converted to erfc.

$$1.39 = \frac{1000 - 0.0282t}{2\sqrt{0.148t}}$$

By trial and error, t may be calculated:

$$t = 29{,}000 \text{ days} = 79.5 \text{ years}$$

Equation 8.37 provides an effective tool for estimating the time required for a relative contaminant concentration to reach a distance downgradient as shown in Example 8.6. However, another effective use of the equation is to develop a contaminant concentration profile as a function of time at a set distance or as a function of distance at a set time t. For example, the distance to a well may be 100 m (330 ft), and t can be varied to $t = 0$, $t = 100$ days, $t = 200$ days, $t = 300$ days, and so on. Plotting contaminant profiles provides a meaningful approach to understanding how retardation and transformation affect transport dynamics in the subsurface.

EXAMPLE 8.7 *Development of a Groundwater Profile for a Sorbed, Transforming Contaminant*

An end-of-the-year materials balance revealed that a UST is leaking dimethyl phthalate. Maintenance personnel suspect that a valve stem to the stainless steel tank may have been damaged 9 years ago when the tank was accidentally dropped during its installation. Several samples of water taken from just below the valve stem indicate that a concentration of 900 mg/L of dimethyl phthalate is contaminating the groundwater. Representatives of the chemical company and regulatory personnel need to estimate how long a well located 80 m from the tank has been contaminated (samples taken from the well indicate dimethyl phthalate concentrations of approximately 200 mg/L).

The UST and groundwater system are shown in Figure 8.15; in addition, the following data are available.

Pore-water velocity = 0.06 m/day
D = groundwater dispersion coefficient = 0.085 m²/day

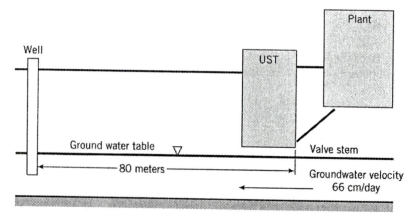

Figure 8.15 Schematic diagram of groundwater contamination from a leaking UST.

K_d = soil distribution coefficient = 0.10 mL/g
ρ_B = soil bulk density = 1.3 g/cm^3
n = 0.11
k = first-order degradation constant = 0.0011 day^{-1}

Estimate how long the well may have been contaminated if the groundwater system behaves as one-dimensional in the x direction.

SOLUTION

Develop a contaminant profile that shows the concentration of dimethyl phthalate as a function of time. Use Equation 8.37

$$\frac{C(x,t)}{C_0} = \frac{1}{2}[\exp(A_1) \cdot \text{erfc}(A_2)]$$

with x = 80 m (held constant) and varying t.

1. Calculate the necessary equation parameters.

$$R = 1 + \frac{\rho_B}{n}K_d = 1 + \frac{1.3}{0.11}0.10 = 2.18$$

$$v' = \frac{v}{R} = 0.06/2.18 = 0.0275 \text{ m/day}$$

2. Use the retardation factor and pore-water velocity to estimate the contaminant travel time to the well.

$$\text{Approximate travel time} = \frac{80 \text{ m}}{0.0275 \dfrac{\text{m}}{\text{day}}} = 2910 \text{ days} = 8 \text{ years}$$

$$D' = \frac{D}{R} = 0.085/2.18 = 0.039 \text{ m}^2/\text{day}$$

$$k' = \frac{k}{R} = 0.0011/2.18 = 0.000504 \text{ day}^{-1}$$

3. Using x = 80 m and t = 3000 days, the terms of Equation 8.37 may be determined.

$$A_1 = \frac{x}{2D'_x}\left(v'_x - \sqrt{v'^2_x + 4D'_x k'}\right) = -1.43$$

$$A_2 = \frac{x - t\sqrt{v'^2_x + 4D'_x k'}}{2\sqrt{D'_x t}} = -0.31$$

$$\text{erfc}(A_2) = 1 - (-\text{erf } 0.31) = 1 + 0.339 = 1.339 \qquad \text{(Table 8.3)}$$

$$\frac{C(80 \text{ m}, 3000 \text{ days})}{C_0} = \frac{1}{2}[\exp(A_1) \cdot \text{erfc}(A_2)] = 0.16$$

Similar calculations may be performed by varying t, as summarized in Table 8.4. The contaminant profile is shown in Figure 8.16; nearly 1800 days were required for the

Table 8.4 Spreadsheet Data Used to Generate Figure 8.16

Distance X (m)	Time (day)	A_1	A_2	erfc (A_2)	C/C_0	C (mg/L)
80	1000	−1.43	4.09	0.00	0.00	0.00
80	1250	−1.43	3.14	0.00	0.00	0.00
80	1500	−1.43	2.40	0.00	0.00	0.00
80	1750	−1.43	1.78	0.01	0.00	1.43
80	2000	−1.43	1.26	0.08	0.01	8.27
80	2250	−1.43	0.80	0.26	0.03	28.0
80	2500	−1.43	0.39	0.58	0.07	62.4
80	2750	−1.43	0.03	0.97	0.12	105
80	3000	−1.43	−0.31	1.34	0.16	144
80	3250	−1.43	−0.62	1.62	0.19	174
80	3500	−1.43	−0.90	1.80	0.22	194
80	3750	−1.43	−1.17	1.90	0.23	205
80	4000	−1.43	−1.42	1.96	0.23	211
80	4250	−1.43	−1.66	1.98	0.24	213
80	4500	−1.43	−1.89	1.99	0.24	215
80	4750	−1.43	−2.10	2.00	0.24	215
80	5000	−1.43	−2.31	2.00	0.24	215
80	5250	−1.43	−2.51	2.00	0.24	215
80	5500	−1.43	−2.69	2.00	0.24	215

dimethyl phthalate to reach the well. Because the spill occurred 9 years ago (3285 days), the well has been contaminated for approximately 4 years (1500 days).

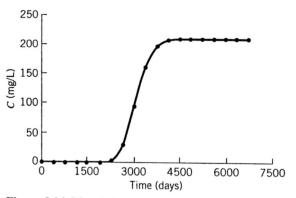

Figure 8.16 Dimethyl phthalate profile for the contaminated well of Example 8.7.

8.4.2 Contaminant Transport in the Vadose Zone

Depending on the geographic location, the vadose zone may be less than a meter to hundreds of meters deep. This region may be characterized by higher organic matter, more metal oxides, and more microbial activity relative to groundwater systems. The transport of hazardous chemicals in the unsaturated zone is inherently more complex than in the saturated zone. The contaminants can move in the vadose zone (1) as a solute in the water phase, (2) as a separate, immiscible (NAPL) phase, and (3) as a gas resulting from volatilization [8.35]. The flow in unsaturated media is complex because fluids are often immiscible (i.e., the fluids displace each other without mixing) and there is a distinct fluid-fluid interface within each pore, even for water-soluble contaminants. This phenomenon is described by multiphase flow phenomena.

Under normal conditions in the unsaturated zone, the voids are partially filled with water. When precipitation events occur, water infiltrates the weathered soil zone and then the vadose zone. Water moves through the unsaturated zone, traveling mainly through large voids in the subsurface matrix. The percolation rate increases as a function of the porosity of the subsurface strata and can range from cm/h to cm/yr.

Numerous models have been used to describe contaminant transport in the unsaturated zone. Many of these have complicated solutions, which often require numerical techniques. One simplified approach has been to use a modification of Equation 8.37 in which the volumetric water content (ϕ) is used in place of porosity.

8.5 SUMMARY OF IMPORTANT POINTS AND CONCEPTS

- Release of contaminants from a hazardous waste source (e.g., a RCRA waste facility or a contaminated site) is a function of their sorptivity, volatilization rate, and rates of abiotic and biotic transformations. Contaminant transport from the source is not only a function of sorption, volatilization, and transformation, but also advective and dispersive processes in air and water.

- The rate of change of contaminants in air may be described in three dimensions by integrating fundamental advection and dispersion equations using appropriate boundary conditions.

- Source releases into the atmosphere can be instantaneous, which is described downgradient by a puff model. Continuous releases to the atmosphere are quantified by plume models.

- The transport of contaminants released from a source to the subsurface may be quantified using the advection-dispersion equation to model the one-dimensional flow of water-soluble, conservative solutes in soil columns and some groundwater systems. Using appropriate initial and boundary conditions, analytical solutions may be obtained for continuous and pulse sources.

- The advection-dispersion equation may be modified for contaminant sorption and transformation using the retardation factor and a term for first-order degradation, respectively.

- Contaminant dynamics in the unsaturated zone are significantly more complex than in groundwater systems. The migration of contaminants in the unsaturated zone is complicated by multiphase flow and the interrelationships between hydraulic conductivity and volumetric water content of the system.

PROBLEMS

8.1. A soil pit received 500 L (130 gal) of waste PCE over 2 years. The porosity of the soil is 0.52 with an organic carbon content of 0.1%. The depth to groundwater (a drinking water supply) is 12 m. Provide a qualitative evaluation for the fate of PCE at the site as a basis for a Phase I site assessment.

8.2. An industrial intermediate, 2,4-dimethylphenol, has been spread on a surface soil as a 500-mg/L solution. The weekly volume placed on the soil was 25% of the pore volume for the top 0.5 m (1.6 ft) of soil. The soil is a clay loam characterized by an iron oxide content of 4.6% and 2.9% organic carbon. Provide a qualitative evaluation of the site as a basis for a Phase I property transfer evaluation.

8.3. Sketch a graph for relative contaminant concentration as a function of time for
   a. an instantaneous release
   b. a continuous, steady state release with no dispersion
   c. a continuous, steady state release with dispersion.

8.4. A spill of hydrogen cyanide has occurred on a concrete pad 50 m × 30 m (164 ft × 98 ft), releasing a 1.42 kg puff of cyanide gas. If the air stability class is D and the wind velocity in the x direction is 3.5 m/sec, plot the cyanide concentration downwind at 1-km (3280-ft) intervals.

8.5. A drum containing 50 kg (110 lb) of methyl isocyanate has spilled with instantaneous release of the chemical. If the wind velocity is 4 m/s during daylight under strong solar radiation, determine the methyl isocyanate concentration 2 km (1.2 miles) downwind.

8.6. Atmospheric sampling has shown that the volatilization rate of benzene from a recycling vat is 142 g/h. If the air is characterized by stability class B and the wind speed is 2.5 m/sec, determine the benzene concentration 1.5 km (0.93 miles) downgradient.

8.7. An air stripper 11 m (36 ft) high is releasing methylene chloride at a rate of 125 g/min. Estimate the downwind concentration at the upward end of a subdivision 1 km (0.62 miles) away and 0.2 km (660 ft) in the y-direction if the wind speed = 12 km/h and the meteorological conditions are clear and daytime.

8.8. A solvent recycling vat 3 m (10 ft) in diameter contains primarily MEK. If $k =$ 0.2, the ventilation rate = 150 m³/h and the air in the recycling facility is exhausted through a vent 6 m (20 ft) high, determine the MEK concentration at a point 2 km (1.24 miles) downgradient if the wind speed is 4 m/sec at night under mostly cloudy skies.

8.9. An open waste vat 2 m (6.6 ft) in diameter contains waste acetone. The temperature is 25°C, the atmospheric pressure is 1 atm, the ventilation rate is 125 L/h, and $k = 0.3$. The vapor is exhausted from the building through a vent at ground level. Assuming no dead-air areas around the building, determine the acetone concentration on the 12th floor of a building ($z = 40$ m) 1 km (0.62 miles) downgradient at a location of $y = 0.3$ km (980 ft) if $u = 15$ km/h. Assume daytime, moderate meteorological conditions.

8.10. Using the principles of mass transfer described in Section 8.2, derive the retardation factor for a contaminant in a soil-water system.

8.11. A 90-kg (200-lb) drum of atrazine has spilled at a pesticide warehouse and distribution center and is leaching into a shallow aquifer. The average bulk density of the aquifer = 1200 kg/m^3, the porosity = 0.40, and the organic carbon content = 0.2%. The average pore velocity = 3.6 m/day, and the longitudinal dispersion coefficient = 0.05 m^2/day. The first-order rate coefficient for atrazine degradation is 0.0012 day^{-1}. Assume that the soil organic matter is the primary sorbent. If the system behaves as a one-dimensional aquifer, estimate the atrazine concentration at a well 2 km (1.24 mi) downgradient in 5 years.

8.12. A tanker trailer containing MIBK has spilled into an aquifer that behaves as a one-dimensional groundwater system. The following data are available to describe the system.

C_o = 180 mg/L
v_x = 0.08 m/day
D_x = 0.04 m^2/day
k = 0.005 day^{-1}
ρ_B = 1500 kg/m^3
Porosity = 0.48
Soil organic carbon (the primary sorbent) = 0.4%

Plot a MIBK concentration profile as a function of time at a well 3 km (1.86 miles) downgradient from the spill.

8.13. A spill of diuron at an initial concentration of 150 mg/L has occurred above a shallow aquifer that behaves as a one-dimensional groundwater system. The average characteristics of the system include

Pore-water velocity = 0.05 m/day
Bulk density = 1600 kg/m^3
Porosity = 0.48
Organic carbon content (the primary sorbent) = 0.3%
Longitudinal dispersion coefficient = 0.003 m^2/day
First-order degradation contaminant = 0.0004 day^{-1}

Using this information, plot a concentration profile at a well 1.5 km (0.93 miles) downgradient.

8.14. A drinking water well is located 1.6 km (1 mile) downgradient from a plume containing dieldrin entering the aquifer. Determine the concentration vs. time response for the dieldrin at the well based on the following data.

Pore-water velocity = 0.58 m/day
D_x = 0.02 m^2/day
n = 0.45
Organic carbon content (the primary sorbent) = 0.1%
Bulk density = 1.8 g/cm^3
k = 0.000001 day^{-1}

8.15. A plume of 20 mg/L 2,4-D has entered a groundwater system characterized by a pore-water velocity of 0.24 m/day, organic carbon content of 0.4%, bulk density of 1800 kg/m^3, porosity = 0.4, and a longitudinal dispersion coefficient (D) equal to 0.015 m^2/day. The 2,4-D degrades at a rate of 0.000045 day^{-1}. Assume that the soil organic matter is the primary sorbent.

a. Develop a concentration vs. time plot of 2,4-D for a well 2 km (1.24 miles) downgradient.

b. Plot concentration as a function of distance at times of 100 days, 1000 days, and 10,000 days.

8.16. A "mom and pop" business is a local distributor for industrial supplies, including solvents used by foundries, dry cleaners, machine shops, automotive shops, etc. The solvents have been stored in underground tanks over the 30-year period mom and pop have operated the business. Separate underground tanks held benzene, MEK, MIBK, TCE, PCE, chloroform, and carbon tetrachloride.

Mom and pop are selling their property, and the purchaser has hired your hazardous waste consulting firm to perform a property transfer site assessment. The job involves viewing company records, site inspections, and collecting soil and groundwater samples, if necessary.

The first part of your effort, a company records search, showed that the rupture of one of the underground tanks 8 years ago resulted in a 1-year pulse input of PCE to the groundwater. (The tank was sitting just above the saturated zone, so the leaking PCE went directly into the groundwater).

The following data have been estimated for the site.

v_x = 0.30 m/day
D_x = 0.06 m^2/day
k = 0.0005 day^{-1}
n = 0.5
Bulk density = 1.2 g/cm^3
Average organic carbon content (the primary sorbent) = 0.1%
C_o = PCE water solubility

The client purchasing the property has a limited budget for the site assessment and can afford the installation of only one monitoring well. However, neither party is in a big hurry to act on the transaction, so the purchaser would prefer to collect monthly samples for PCE analysis over a 1-year time period. Based on this information, determine the distance downgradient from the site to locate the monitoring well in order to verify that the pulse of PCE is present. (*Hint:* Use the one-dimensional transport equation, but set t and vary x.)

8.17. A surface impoundment has been contaminated with alachlor. The contamination was present for 8 years before the sediments and soils were excavated and deposited off site. The groundwater is shallow, with the waste protruding into the saturated zone. The site characteristics include

Pore-water velocity = 0.48 m/day
Bulk density = 1900 kg/m^3
Porosity = 0.35
D_x = 0.6 m^2/day
k = 0.000002 day^{-1}
Mean organic carbon content (the primary sorbent) = 0.3%

Determine the concentration vs. time plot at a monitoring well 500 m (1640 ft) downgradient.

8.18. For the system described in Problem 8.17, a consulting firm has proposed implementing an in situ bioremediation system. Their laboratory treatability studies have provided the following results.

Time (day)	Alachlor (mg/L)
0	32
12	21
25	14
48	6
112	1.2
216	0.2

Assuming these biodegradation results can be realized in the field, rework Problem 8.17 to obtain a concentration vs. time curve 500 m (1640 ft) downgradient.

8.19. A storage basin for aqueous nickel waste with a nickel concentration of 5 mg/L has leaked into a nearby aquifer. The permeable soil on which this steady leak has occurred has a groundwater pore velocity of 0.65 m/day, and a porosity of 0.55. The dispersion coefficient is 0.025 m²/day, and the average bulk density of the soil is 1600 kg/m³. If the distribution coefficient of nickel is 200 mL/g, estimate the time for the nickel to reach a monitoring well 100 m (330 ft) away.

8.20. Industrial wastewater containing cadmium at a concentration of 1.5 mg/L has contaminated local groundwater. As the cadmium ($K_d = 210$ mL/g) moves through the groundwater it partitions irreversibly onto subsurface soils, resulting in first-order disappearance with $k = 0.0042$ day^{-1}. The site is characterized by the following:

Pore water velocity = 0.73 m/day
Bulk density = 1600 kg/m³
Porosity = 0.45
$D_x = 0.0055$ m²/day

Generate the concentration versus time plot at a drinking water well 150 m (490 ft) away.

8.21. Pure vinyl chloride is released from a damaged storage tank into the nearby saturated zone with the following characteristics:

Pore-water velocity = 0.35 m/day
Bulk density = 1.75 g/cm³
Porosity = 0.35
$D_x = 0.9$ m²/day
Mean organic carbon content (the primary sorbent) = 0.09%

Determine the maximum steady-state concentration of vinyl chloride to reach a monitoring well 500 m (1640 ft) away if
a. $k = 1 \times 10^{-6}$ day^{-1} b. $k = 1 \times 10^{-4}$ day^{-1}
c. $k = 1 \times 10^{-3}$ day^{-1} d. $k = 1 \times 10^{-2}$ day^{-1}

8.22. A facility is leaking 1,2,3-trimethylbenzene into an aquifer. A proposal is made to enhance natural biodegradation by injecting nutrients into the subsurface, thereby increasing the first-order degradation rate constant from 7.6×10^{-7} day^{-1} to 5×10^{-4} day^{-1}. Compare the concentration versus time profiles for each case at a monitoring well 650 m (2130 ft) downgradient if the soil is characterized by the following:

Pore water velocity = 0.35 m/day
Bulk density = 1.9 g/cm^3
Porosity = 0.4
D_x = 0.4 m^2/day
Average organic carbon content (the primary sorbent) = 0.15%

8.23. A plume composed of an industrial solvent *sec*-butylbenzene at a concentration of 163 mg/L has entered the saturated zone. To enhance mobility of the contaminant in order to flush it out for pump and treat remediation, a proposal is made to add a surfactant to the subsurface. This will decrease the soil distribution coefficient of the *sec*-butylbenzene by 80%. Compare the time for all of the *sec*-butylbenzene to reach an extraction well 250 m (820 ft) away for each case. The soil has a pore-water velocity of 0.24 m/day, a bulk density of 1800 kg/m^3, a porosity of 0.32, a dispersion coefficient of 0.6 m^2/day, and a mean organic carbon content of 0.35%. *sec*-Butylbenzene is degraded by a first-order rate constant of 0.00035 day^{-1}. Assume that the soil organic matter is the primary sorbent.

8.24. A 1000-L polyethylene mixing tank containing 10% 2,4-D in water has spilled at a pesticide formulation area to a depth of 2 m (6.6 ft). The water table is near the ground surface at the site, and the pore water velocity in the x direction is 8 m/day. The longitudinal dispersion coefficient (in the x direction) is 2 m^2/day, and the lateral (D_y) dispersion coefficients is 0.5 m^2/day. Using this information, estimate the 2,4-D concentration at x = 200 m (656 ft), y = 0.5 m (16.4 ft), 2 years after the spill. The porosity of the soil is 0.4, the bulk density is 1400 kg/m^3, and the foc is 0.01. Assume that the soil organic matter is the primary sorbent and no 2,4-D transformation occurs.

8.25. A spill of 200 kg (440 lb) dimethyl phthalate has occurred at a plastic manufacturing facility to a depth of 1 m (3.3 ft), and the chemical is being transported into shallow groundwater below the site. The longitudinal pore water velocity is 5.5 m/day, the average bulk density is 1400 kg/m^3, the porosity is 0.45, and the organic carbon content is 0.3%. Dispersion coefficients include D_x = 2 m^2/day and D_y = 0.04 m^2/day. Using this information, estimate the dimethyl phthalate concentration after two years 1.5 km (4290 ft) downgradient at y = 5 m (164 ft). Assume that the soil organic matter is the primary sorbent and no dimethyl phthalate transformation occurs.

REFERENCES

8.1. *Risk Assessment Guidance for Superfund: Environmental Evaluation Manual*, EPA/540/1-69/001A, OSWER Directive 9285.7-01, U.S. EPA, Washington, DC, 1989.

8.2. Hemond, H. F. and E. J. Fechner, *Chemical Fate and Transport in the Environment*, Academic Press, San Diego, CA, 1994.

8.3. Chapra, S. C., *Surface Water Quality Monitoring*, McGraw-Hill, New York, 1997.

8.4. Hess, K., *Environmental Site Assessments. Phase I: A Basic Guide*, CRC Press, Boca Raton, FL, 1993.

8.5. Bedient, P. B., H. S. Rifai, and C. J. Newell, *Groundwater Contamination: Transport and Remediation*, Prentice-Hall, Englewood Cliffs, NJ, 1994.

8.6. Haas, C. N. and R. J. Vamos, *Hazardous and Industrial Waste Treatment*, Prentice-Hall, Englewood Cliffs, NJ, 1995.

8.7. Bear, J., *Dynamics of Fluids in Porous Media*, Elsevier, New York, 1972.

8.8. Freeze, R. A. and J. A. Cherry, *Groundwater*, Prentice-Hall, Englewood Cliffs, NJ, 1979.

8.9. Bedient, P. B., N. K. Springer, C. J. Cook, and M. B. Tomson, "Modeling chemical reactions and transport in groundwater systems: A review," *Modeling the Fate of Chemicals in the Aquatic Environment*, Ann Arbor Science Publishers, Ann Arbor, MI, 1982.

8.10. Heylin, M. (Ed.), "News of the week: Bhopal," *Chem. Eng. News*, **63**(3), 4–6 (1975).

8.11. Gunning, A., "News: Bhopal," *Chem. Eng.*, **12**, 578–579, (1994).

8.12. Houghton, D. D., *Handbook of Applied Meteorology*, John Wiley & Sons, New York, 1985.

8.13. Crowl, D. A. and J. F. Louvar, *Chemical Process Safety: Fundamentals with Applications*, Prentice-Hall, Englewood Cliffs, NJ, 1990.

8.14. Pasquill, F., *Atmospheric Diffusion*, Van Nostrand, London, 1962.

8.15. Gifford, F. A., "Turbulent diffusion typing schemes: A review," *Nucl. Saf.*, **17**, 68–86, (1976).

8.16. Wesely, M. L., "Parameterization of surface resistances to gaseous dry deposition in regional-scale numerical models," *Atmos. Environ.*, **23**, 1293–1304 (1989).

8.17. Janerich, D. T., W. S. Burnett, G. Feci, M. Hoff, P. Nasca, A. P. Polednak, P. Greenwald, and N. Vianna, "Cancer incidence in the Love Canal area," *Science (Wash. DC)*, **212**, 1404–1407 (1981).

8.18. Gelhar, L. W., "Stochastic subsurface hydrology from theory to applications," *Water Resour. Res.*, **22**, 135S–145S (1986).

8.19. Mandelbrot, B. B., *The Fractal Geometry of Nature*, W.H. Freeman, New York, 1983.

8.20. Fetter, C. W., *Contaminant Hydrogeology*, Macmillan, New York, 1993.

8.21. Hamed, M. M., P. B. Bedient, and J. P. Conte, "Numerical stochastic analysis of ground water contaminant transport and plume containment," *J. Contam. Hydrol.*, **24**, 1 (1996).

8.22. Rautman, C. A. and J. D. Istok, "Probabilistic assessment of ground-water contamination. 1. Geostatistical framework," *Ground Water*, **34**, 899 (1996).

8.23. Saiers, J. E. and G. M. Hornberger, "Migration of ^{137}Cs through quartz sand: Experimental results and modeling approaches," *J. Contam. Hydrol.*, **22**, 255 (1996).

8.24. Srivastava, R. and M. L. Brusseau, "Nonideal transport of reactive solutes in heterogeneous porous media. 1. Numerical model development and moments analysis," *J. Contam. Hydrol.*, **24**, 117 (1996).

8.25. Therrien, R. and E. A. Sudicky, "Three-dimensional analysis of variably-saturated flow and solute transport in discretely-fractured porous media," *J. Contam. Hydrol.*, **23**, 1 (1996).

8.26. Javandel, I., C. Doughty, and C. F. Tsang, *Groundwater Transport: Handbook of Mathematical Models*, American Geophysical Union, Washington, DC, 1984.

8.27. Sauty, J.-P., "An analysis of hydrodispersive transfer in aquifers," *Water Resour. Res.*, **16,** 145–158 (1980).

8.28. Fetter, C. W., *Applied Hydrogeology*, 2nd Edition, Macmillan, New York, 1988.

8.29. Wilson, J. L. and P. J. Miller, "Two-dimensional plume in uniform groundwater flow," *J. Hydraulic Div.*, **104,** 503–514 (1978).

8.30. Cho, C. M., "Convective transport of ammonium with nitrification in soil," *Can. J. Soil Sci.*, **51,** 339–350 (1971).

8.31. Van Genuchten, M. Th., "Convective-dispersive transport of solutes involved in sequential first-order decay reactions," *Comp. Geosci.*, **11,** 129–147 (1985).

8.32. Chrysikopoulos, P., K. Kitandidis, and P. V. Roberts, "Analysis of one-dimensional solute transport through porous media with spatially variable retardation factors," *Water Resour. Res.*, **26,** 437–466 (1990).

8.33. Fujikawa, Y. and M. Fukui, "Adsorptive solute transport in fractured rock: Analytical solutions for delta-type source conditions," *J. Contam. Hydrol.*, **6,** 85–102 (1989).

8.34. Lunn, M., R. J. Lunn, and R. Mackay, "Determining analytic solutions of multiple species contaminant transport, with sorption and decay," *J. Hydrol.*, **180,** 195–210 (1996).

8.35. Baehr, A. L., "Selective transport of hydrocarbons in the unsaturated zone due to aqueous and vapor phase partitioning," *Water Resour. Res.*, **21,** 19–26 (1987).

Part Three

RECEPTORS

If a contaminant is released from a source via a pathway such as volatilization or leaching to groundwater, it may be transported to receptor organisms. The hazard to the receptors, which are most commonly humans, is evaluated using principles of toxicology—a discipline based on the absorption, distribution, biochemical transformation, and excretion of hazardous chemicals. Fundamental concepts of toxicology are presented in Chapter 9, and these qualitative principles provide a basis for understanding the toxic effects of hazardous wastes. In Chapter 10, the quantitative aspects of hazardous wastes are developed, including acute and chronic toxicity quantification, industrial hygiene values, and reference doses and slope factors, which are used to evaluate toxicological risk. The principles and quantification of toxic responses provide a basis for the protection of public health and the environment as well as performing hazardous waste risk assessments. Chapter 11, risk assessment, is the final topic of the sources-pathways-receptors theme. It is a capstone chapter in which the concepts and computational knowledge of Chapters 1–11 are used to quantify the risks associated with hazardous waste sites and facilities.

Chapter **9**

Concepts of Hazardous Waste Toxicology

Although hazardous wastes and materials may affect public health through explosion, fire, corrosion, or other physical processes, the most common concern is their toxicity. *Toxicology* is the study of poisons and their effects on living organisms. Knowledge of the principles of toxicology is important in hazardous waste management in order to assess the potential biological effects of hazardous chemicals and as a basis for performing risk assessments. The field of toxicology, like many others, is multidisciplinary and includes such topics as exposure routes, distribution of toxic substances within the organism, molecular mechanisms for toxicity, and movement of the chemicals out of the organism (excretion). Toxicology is a broad subject, and many facets of the discipline are beyond the scope of this text. The field of toxicology also includes a number of subdisciplines that cover a range of toxic materials. For example, the discipline encompasses food toxicology, natural products toxicology (e.g., snake venoms), and aquatic toxicology. These subdisciplines, although interesting, are of minimal importance to hazardous waste management and are not covered in this text. More detailed information on these topics may be found in References 9.1–9.4.

The purpose of this chapter is not to provide sufficient information for environmental scientists and engineers to consider themselves toxicologists; rather, it is to provide nontoxicologists with fundamental concepts of toxicology as a basis for collaboration with other professionals in performing site and risk assessments and communicating hazard and risk. Therefore, this chapter presents an introduction to the routes and mechanisms that influence a chemical's toxicity. Such an approach may be termed classical toxicology; it includes material on routes of exposure, distribution within the organism, biotransformations, interactions with receptor molecules, and excretion. The quantitative aspects of toxicology and industrial hygiene are covered in Chapter 10 as a basis for evaluating public health hazards and risk.

9.1 OVERVIEW OF TOXICOLOGICAL MECHANISMS

Toxicity results from a number of dynamic processes, including absorption, distribution, metabolism of the parent compound, storage, and excretion. Toxic agents usually do not exhibit detrimental effects on the surface of the organism. Instead, they proceed through a series of exposure and metabolic pathways in which toxic effects result when a chemical or its transformation products reach a site of action at a *receptor molecule* on a *target organ* at a sufficiently high concentration [9.5].

When an organism is exposed to a hazardous compound, the material is often absorbed, a process that occurs by three common routes: orally, dermally (through the skin), and through inhalation. The *route of exposure* is a function of the media to which the organism is exposed; examples include drinking contaminated water, breathing vapors released from contaminated soils, or wearing clothes contaminated with a pesticide. Once inside the body, the contaminants are distributed through the blood stream. In many cases, the chemical is detoxified through a biochemical mechanism, which may include enzyme-mediated oxidations, dealkylations, and hydrolysis reactions. The outcome of most of these reactions is a more water-soluble compound that can then be more readily excreted by the organism for ultimate removal from the system. However, the metabolic reactions may also transform the compound to a more toxic product. Depending on its characteristics and properties, the contaminant or its metabolites may bind to receptor molecules, which can be proteins or lipoproteins on the surfaces of cell membranes, nucleic acids, or other biological molecules. The interaction with the receptor provides the mechanism of toxicity. For example, when the common air contaminant carbon monoxide is inhaled, it binds to the receptor molecule hemoglobin, which results in toxicity due to the inability of the hemoglobin to carry molecular oxygen to the body.

It is interesting that the basis for approaching hazardous waste problems used in this text—contaminants, pathways, and receptors—had its beginnings in classical toxicology. Hazardous compounds are absorbed into organisms, are distributed, and interact with receptor molecules, in much the same way they behave in the environment or in hazardous waste treatment systems. An organism is analogous to other systems that are separated from the rest of the universe by boundaries, such as an aquifer or an activated sludge system—although comparing the human body to a waste treatment facility is an approach not commonly used in the medical or toxicological literature. Inputs, releases, and reactions occur in organisms just as they do in a natural or engineered system. The "bioreactor" illustrated in Figure 9.1 has potential inputs through oral, inhalation, and dermal routes. The material is then distributed throughout the "reactor" and may proceed through reactions by which it is transformed into other products, is stored, or equilibrates with specific biomolecules, or receptors, within the bioreactor. Finally, the substance can exit the "bioreactor" through excretion by the kidneys. These dynamics, called *toxicokinetics,* can be very complicated; however, the physics, chemistry, and biology of an organism are not too conceptually different from other systems in nature. Furthermore, many of the reaction rates in these physiological systems follow apparent first-order kinetics—similar to the dynamics of chemicals in the environment and in treatment reactors.

9.2 FUNDAMENTALS OF MAMMALIAN PHYSIOLOGY

To understand the mechanisms by which hazardous compounds are toxic to humans or other organisms, an introduction to the functions of organs of mammalian systems is requisite. As with the microorganisms covered in Chapter 7, the most fundamental organizational unit of higher organisms is the cell. A typical human body consists of about 100 trillion cells [9.6]. If an aggregation of cells works in concert in a homogeneous fashion, the group of cells is defined as *tissue.* Common tissues include muscles and skin. An *organ* is an even more complex system in which cells are linked by

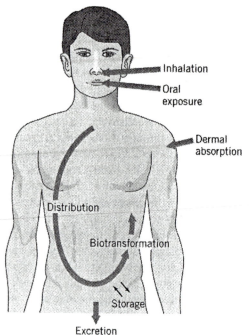

Figure 9.1 Pathways of chemicals in a "bioreactor"—the human body.

intracellular supporting structures, with a number of different types of cells adapted to perform different functions—a feature known as differentiation. Examples of organs include the lungs, liver, and kidneys.

On an organ and tissue basis, most mammals are somewhat similar to humans; that is, all of these higher animals have similar organ systems such as a liver, digestive system, and circulatory system that carries oxygen-containing blood throughout the body. Fluids within the human body can exist inside the cells (intracellular fluid) or outside the cells (extracellular fluid). Extracellular fluid, which contains nutrients and oxygen, is in constant motion through blood and fluid circulation.

The following discussion focuses primarily on the physiology of humans; however, most of the principles are applicable to mammals and other higher organisms, and therefore may be used as a basis for some terrestrial ecological risk assessments. The material is not meant to be a thorough treatise on anatomy and physiology; detailed information on physiology may be found in References 9.6–9.9. The discussion that follows will focus on the skin, the respiratory system, the digestive system (the primary routes of absorption), the circulatory system (the route of distribution), the liver (the organ most important in transformations), and the kidneys (the organ responsible for excretions).

9.2.1 The Skin

The skin not only serves as a covering of the muscles and skeletal formations of the body, it also functions as a protective layer that impedes the entry of harmful agents and chemicals. The skin may be considered an organ because it consists of differentiated tissues, each of which performs specific functions. The largest organ of the body, the skin connects the membranes located inside the eyes, nose, and other openings of the body [9.10]. As illustrated in Figure 9.2, it consists of two main layers—the *epidermis* and *dermis*—and is separated from deeper body tissues by a *subcutaneous layer,* which is sometimes called the hypodermis.

Dermis is dense connective tissue that is joined to the epidermis by a *basement membrane.* During growth and development of the skin, hair follicles, sebaceous glands, and sweat glands grow into the dermis from their origins in the epidermis. Hairs grow from the follicles to the skin surface and secretions from the sebaceous glands lubricate the hairs and skin surfaces. The dermis extends into the subcutaneous layer lying below. The *subcutaneous layer* consists of adipose tissue, skeletal muscle tissue, and stored water. The *vascular system* (the blood stream), which is of concern for the distribution of absorbed hazardous chemicals, extends through the dermis and subcutaneous layers, but not the epidermis. Therefore, the skin functions as a barrier to the entry of many hazardous chemicals.

9.2.2   The Respiratory System

A series of organs and body parts, including the mouth, the nose, the trachea, and the lungs, comprises the respiratory system. From a macroscopic perspective, respiration is the transport of oxygen from the ambient atmosphere to the fluids and cells of the body coupled with the transport of the body's primary metabolic product, carbon dioxide, out of the system.

The lungs are the site of respiration in mammals. The lungs are expanded and contracted by (1) movement of the diaphragm to lengthen and shorten the chest cavity and (2) movement of the ribs to increase and decrease the diameter of the chest cavity. Neural and muscular control is an important factor in the promotion of breathing, so it is not surprising that the toxic response for chemicals that affect nerve transmission (e.g., organophosphate esters, carbamates) is often respiratory failure.

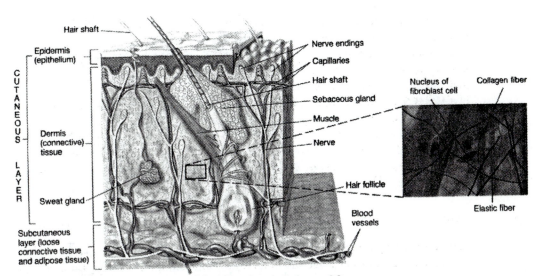

Figure 9.2 Dermal cells and layers of the skin. *Source:* Reference 9.8.

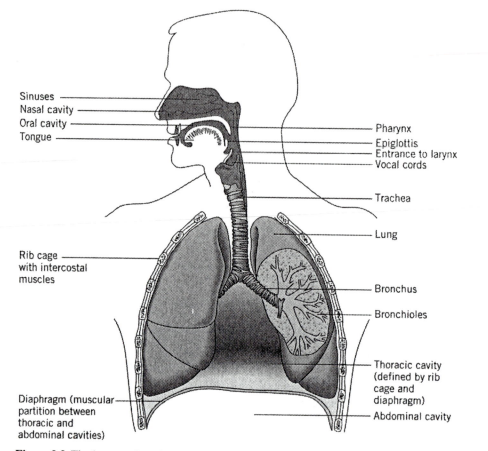

Figure 9.3 The lungs and respiratory system. *Source:* Reference 9.8.

The respiratory system may be divided into two areas—(1) the upper respiratory tract and (2) the gas exchange area in the lungs (Figure 9.3). Only particulate matter less than 10 μm in size, called PM_{10}, can be transported through the upper respiratory system into the lungs. The lungs are intricate organs in which air moves down through small brachials, or air passages, into the *alveoli,* which are tiny air sacs in which oxygen is transferred from the air within the lungs to the bloodstream, which circulates past the alveoli. The core of the respiratory system, including respiratory brachioli, alveolar ducts, atria, and alveoli, is shown in Figure 9.4. The epithelium (lining) of this system is so thin that the gas in the alveoli is in close contact with the blood in the capillaries; this proximity of blood and gases provides the basis for gas exchange in the lungs. The thickness of all of these layers is usually 0.1–1 μm. Two human lungs typically contain approximately 250 million alveoli, each with an average diameter of 0.1 mm [9.6].

9.2.3   The Digestive System

The gastrointestinal tract of humans, which includes the mouth, pharynx, esophagus, stomach, small intestines, large intestine, rectum, and anus, ranges from 6.5 m to 9 m

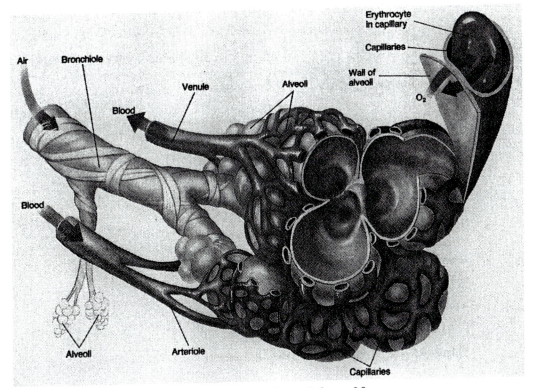

Figure 9.4 Microstructure of the respiratory system. *Source:* Reference 9.8.

(21 ft to 29 ft) in length. Adjacent organs, including the salivary glands, liver, gallbladder, and pancreas, are also involved in digestion. The general features of the gastrointestinal tract are shown in Figure 9.5.

Although the mouth and salivary glands are important in the primary step in food digestion, they do not significantly affect the absorption of hazardous chemicals. During the ingestion of food, muscular contractions control the movement of food to the stomach as it passes through the *pharynx* (a muscular tube continuous with the esophagus). The *stomach* is a muscular sac that stores food, secretes substances that aid in food digestion, and controls the rate at which food moves into the small intestine. Digestion in the stomach is promoted by the release of hydrochloric acid and enzymes known as pepsinogens. Although absorption into the bloodstream can occur in the stomach, it is considered minor relative to that which occurs in the small intestine.

The *small intestine* is the most important organ for absorbing food (as well as hazardous chemicals) along the gastrointestinal tract. Food, which has been digested into carbohydrates, fats, and amino acids from larger biomolecules, is absorbed across the intestinal lining. Carbohydrates and some amino acids are actively transported into blood capillaries lining the intestinal walls. Fatty acids and monoglycerides diffuse passively across epithelial cells and enter lymph vessels, which then drain into the circulatory system. In addition, minerals (e.g., cations and anions) and water are absorbed in the small intestine. The pancreas and liver also play important roles in digestion. The pancreas secretes trypsin and chymotrypsin, which aid in digestion; in addition, it secretes bicarbonate, which neutralizes HCl entering the small intestine from the

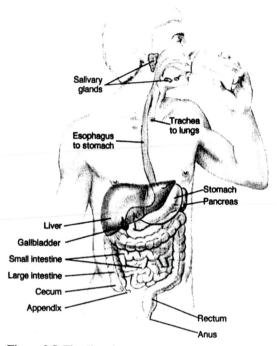

Figure 9.5 The digestive system. *Source:* Reference 9.8.

stomach. The liver secretes bile (a solution containing bile salts, bile pigments, choles-
terol, and lecithin), which enhances the dissolution of fat droplets in the small intestine.

The final major organ of the gastrointestinal tract, the *large intestine,* functions
mainly in storing and concentrating feces (which includes undigested and unabsorbed
material, bacteria, and water).

9.2.4   The Circulatory System

An important aspect of toxicology is the distribution of chemicals after they are absorbed. Distribution of contaminants occurs through the circulatory or vascular system,
which comprises 56% of the mass of the body [9.6]. Blood passes through the entire
circulatory system once per minute for an adult at rest; for an individual engaged in
activity, the blood circulates as much as five times per minute.

Some of the body fluid is found inside cells, and is known as *intracellular fluid.*
Liquids outside cells are *extracellular fluids* and contain nutrients, oxygen, and ions.
The extracellular fluids are in constant motion throughout the body; they are mixed by
blood circulation and by diffusion between the blood and tissue spaces.

As shown in Figure 9.6, the heart pumps blood through two different systems: (1)
the lungs for oxygenation, and (2) the remainder of the body (systemic circulation).
As blood moves through the circulatory system, tissues are supplied with nutrients
and oxygen in two stages. First, the blood moves through the circulatory system; second, nutrients and oxygen diffuse across capillary walls to the cells. Perfusion of cells
with extracellular fluids is fostered by the porous nature of the capillaries, which allows a large flux of fluid to diffuse back and forth between the capillaries and tissues

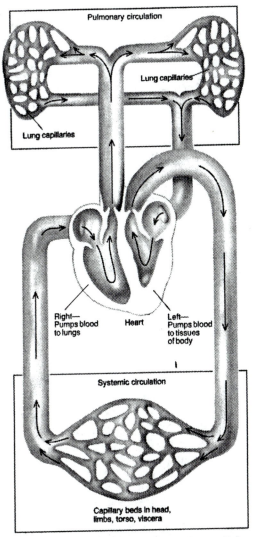

Figure 9.6 The circulatory system. *Source:* Reference 9.8.

(Figure 9.7). Almost all cells are within 50 μm of a capillary, and most substances can diffuse to cells within seconds.

9.2.5 The Liver

The *liver* may be considered a filter for the blood and also a control system for the levels of important nutrients, such as sugars. A large proportion of the blood flow that exits the heart perfuses the intestines, then the spleen, and finally moves into the portal vein system and the liver (Figure 9.8). The flow rate of blood into the liver of an average adult is 1.1 L/min [9.6] and occurs through *hepatic sinuses* (blood channels through the liver), which are in contact with liver *parenchymal cells* (the functional

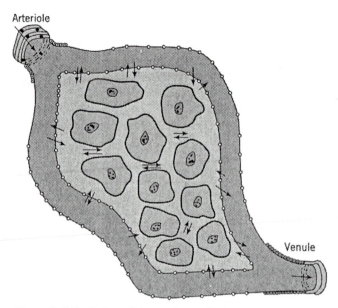

Figure 9.7 Perfusion of tissues by capillaries. *Source:* Reference 9.6.

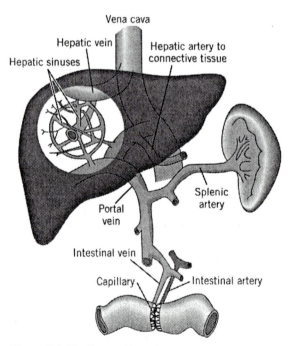

Figure 9.8 The liver and hepatic system. *Source:* Reference 9.9.

cells of the liver where metabolism is controlled and toxic substances are transformed). The blood then flows into the central veins of the liver and on to the vena cava, which leads to the heart. Without considering toxic substances, one of the primary functions of the liver is regulation of carbohydrate metabolism and release of sugars from stored sugar polymers (*glycogen*) when their levels become low in the bloodstream. Glucose and amino acids, absorbed through the intestinal wall, enter capillaries and are then transported to the liver capillary bed. The liver then regulates glucose metabolism— excess glucose is stored as glycogen or converted to fat. (The liver can also release glucose stored as glycogen—gluconeogenesis—when needed by the body). Amino acids may be deaminated in the liver, resulting in the release of ammonia. The ammonia is converted enzymatically to a less toxic product, urea

$$NH_2-\underset{\underset{\displaystyle \|}{O}}{C}-NH_2$$

which is then excreted through the kidneys. Still another liver function is the inactivation of hormones, which are also excreted by the kidneys.

In considering toxic chemicals to which the organism is exposed, the liver is perhaps the most important organ. It is filled with enzymes that catalyze the oxidation, hydrolysis, and other detoxification reactions of toxic compounds in the body, the most important being a group of enzymes called *cytochrome P-450*. Some scientists consider the liver to be the most important organ of the body. Indeed, it is impossible to live without a liver because it has such important functions in metabolism. Its role will be discussed further in Section 9.4.

9.2.6 The Kidneys

Blood also passes through the *kidneys,* where substances not needed by the body (including toxic chemicals and their metabolites) are separated and excreted in the urine. The chemicals that are normally excreted are end products of metabolism and excess electrolytes such as potassium, sodium, and chloride. The two functions of the kidney are (1) the excretion of the end products of metabolism and (2) control of the concentrations of electrolytes and other constituents in the body fluids.

One important morphological feature of the kidney is the *nephron* (a functional kidney unit), which is composed of a *glomerulus* (a filter for the blood) and a *tubule,* which is a long, coiled tube in which the filtrate is converted into urine (Figure 9.9). In the first step of kidney function, blood enters the glomerulus in a network of up to 50 parallel capillaries encased in *Bowman's capsule.* The function of the nephron is to clean the blood plasma of unwanted substances which may include urea, creatine, sulfates, and phenols. Other nonmetabolic substances (e.g., sodium ions, potassium ions) may also be filtered by the nephron.

The mechanism for excreting unneeded substances in the blood is accomplished by first filtering large volumes of blood through the glomeruli into the tubules. The compounds that are required by the body, such as amino acids, sugars, electrolytes, and water, are then reabsorbed by the kidneys in the nephron. Substances that are not needed by the body are not reabsorbed; instead, they pass out of the body in the urine.

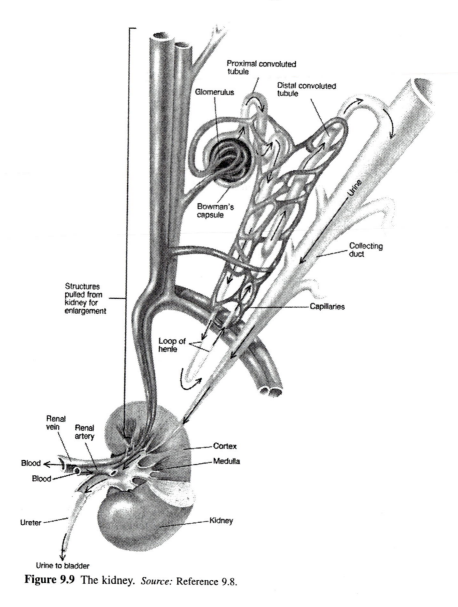

Figure 9.9 The kidney. *Source:* Reference 9.8.

9.3   CONCEPTS AND MECHANISMS OF TOXICITY

The classical approach to studying toxicology focuses on routes of exposure, distribution of the toxic chemical through the body, the biochemical transformation of the compound, toxicant-receptor interactions, storage, and excretion. Each of these subtopics will be covered in this section as a basis for developing a fundamental understanding of hazardous waste toxicology. The physiological and biochemical mechanisms of toxicity are outlined in Figure 9.10; reference to this figure will aid in the following discussion of toxicological mechanisms.

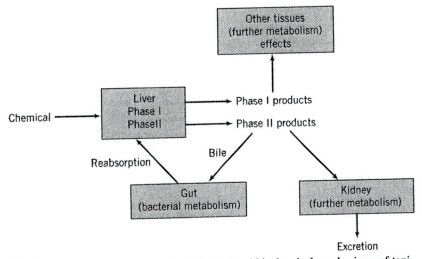

Figure 9.10 An overview of the physiological and biochemical mechanisms of toxicity. *Source:* Reference 9.3.

9.3.1 Routes of Exposure and Absorption

Toxic effects vary substantially depending on the place where the compound is absorbed. For example, asbestos and chromium are highly toxic when inhaled but do not exhibit such high toxicity when ingested, probably due to poor absorption in the gastrointestinal tract [9.11].

The first step in toxicity is exposure to the toxic chemical(s) and absorption into the organism. A few routes of exposure are important only in clinical toxicology, such as injections into the muscle (intramuscular), under the skin (subcutaneous), into the gut cavity (interperitoneal), and into the veins (intravenous). Three different routes of exposure are important in hazardous waste management. The routes by which organisms are exposed to hazardous wastes are through natural processes in which the chemicals flow with the air, water, or food as it enters the body. The corresponding routes of exposure are dermal (i.e., through the skin), inhalation, and oral.

Absorption Processes. As described in Section 9.1, the body is protected by epithelial cells, or membrane tissue. Absorption mechanisms evolved to provide a means of supplying nutrients while protecting the organism from external hazards. For example, oxygen is supplied through the alveoli of the lungs and nutrients are absorbed through the small intestine. Toxic chemicals often enter the organism in the same manner as nutrients, such as during ingestion, when chemicals pass through the membrane that lines the intestinal tract. In a similar sense, hazardous chemicals that are inhaled pass through the thin lining of cells inside the alveoli—the point of air-blood exchange.

The two primary absorption mechanisms into the organism are (1) passive transport (diffusion) or active transport into cells and (2) passage through pores or channels between cells. Chemical factors that influence passive transport are degree of ionization, lipid solubility (K_{ow}), protein binding, and water solubility. Compounds that are characterized by high K_{ow} values have high rates of diffusion through membranes. The correlation of hydrophobicity with high rates of cellular diffusion is due to partitioning of the contaminant into the lipid bilayer of cell membranes. Highly polar or charged species do not have a tendency to partition into cellular membranes. However, small, polar contaminants (those with molecular weights <200) can pass through pores. There are also specialized absorption systems that involve active mechanisms, which include *phagocytosis* and *macrophages*.

Dermal Exposure. The skin serves as a protective barrier for the body; as a result, dermal exposure is considered a relatively minor route of exposure. Dermal absorption has been estimated to be one to two orders of magnitude less important than other routes of exposure. In general, large molecules are not readily absorbed through the skin, although some hazardous contaminants, such as parathion, are. In fact, three drops of parathion applied to the skin will likely kill an adult [9.12]. In addition, absorption through the skin is an important route of exposure for most organic solvents. However, direct contact with concentrated organic solvents usually occurs only in industrial settings, such as RCRA storage areas or waste transfer facilities.

Exposure through the skin is dictated primarily by the chemical properties of the contaminant. The most important factor is its polarity—compounds that have a charge do not dissolve in the epidermal layer and therefore do not readily enter through the skin. In contrast, nonpolar compounds readily penetrate the epidermis into the subcutaneous area, where they enter the bloodstream. For example, metallic mercury is not absorbed by the skin; however, when it is methylated to dimethyl mercury, it is more nonpolar and is readily absorbed through the skin [9.13].

Oral Exposure. The most common route of oral exposure is via contaminated drinking water; other mechanisms include the ingestion of contaminated food or, for small children, contaminated soil. Acute oral toxicity is often lower than inhalation toxicity for most contaminants because the majority of chemicals are not efficiently absorbed through the intestines [9.14]. Furthermore, some foods may interfere with absorption through the stomach and intestines.

Most compounds are absorbed through the lining of the small intestine; however, toxic compounds may also be absorbed into the bloodstream from the stomach and large intestine linings. Absorption mechanisms in the gastrointestinal tract include passive diffusion and facilitated transport. As with epidermal tissue, compounds that are nonpolar are most easily absorbed through the gut lining. In other words, compounds that have a high hydrophobicity also tend to cross the gut boundary and into the bloodstream. Many drugs that are given orally are weak acids; therefore, a classic exercise of toxicology is to determine the degree of absorption that would take place in the pH regime of the stomach or small intestine for the specific chemical's pK_a using the Henderson-Hasselbalch equation (Section 3.2). The effect of charge on the absorption rate of a contaminant is demonstrated in Example 9.1.

EXAMPLE 9.1 *Effect of Protonation on Contaminant Absorption*

If pentachlorophenol is absorbed through the intestinal lining at a flux rate of 0.26 $\dfrac{\mu\text{mole}}{\text{cm}^2\text{-sec}}$, at what rate will a mixture of pentachlorophenol and pentachlorophenate be absorbed if the pH in the microenvironment of the intestinal lining is 6.8? Assume that, because of its negative charge, the rate of pentachlorophenate absorption is negligible.

SOLUTION

From Table 3.2, the pK_a of pentachlorophenol is 4.75. The Henderson-Hasselbalch equation is

$$pH = pK_a + \log \frac{[base]}{[acid]} = pK_a + \log \frac{[pentachlorophenate]}{[pentachlorophenol]}$$

At the pH of the intestinal lining, the molar ratio of pentachlorophenate to pentachlorophenol may be determined.

$$6.8 = 4.75 + \log \frac{[pentachlorophenate]}{[pentachlorophenol]}$$

$$\frac{[pentachlorophenate]}{[penatachlorophenol]} = 112$$

The pentachlorophenate concentration is 112 times that of the pentachlorophenol at the pH of the intestinal lining. Therefore, the rate at which pentachlorophenol is absorbed is

$$0.26 \frac{\mu mole}{cm^2\text{-sec}} \times \frac{1}{112} = 0.0023 \frac{\mu mole}{cm^2\text{-sec}}$$

Inhalation. Inhalation is the primary route of exposure for airborne particulates, common gases, and other volatile and air-suspended hazardous compounds. Not all material that is inhaled is absorbed into the blood. For example, some finely divided particles are readily exhaled.

The primary route of absorption by inhalation is via diffusive transfer of a gaseous compound to the liquid layer of the alveoli. This diffusion is primarily a function of the compound's gas-diffusion coefficient and the concentration of the contaminant in the lungs. After the contaminant dissolves in the liquid layer of the alveoli, it is removed from the liquid layer by blood perfusing the tissues.

9.3.2 Distribution and Storage

After a compound is absorbed, it travels through the vascular system (i.e., the bloodstream). Toxic chemicals rarely exhibit localized effects at the point where they are absorbed; rather, they move through the body where they may attach to receptor biomolecules, and exert in toxic effects. Organs where binding occurs and toxic effects are expressed include the liver, the kidneys, and the blood itself [9.6].

Storage may or may not occur, depending on the partitioning behavior between the contaminant and different biomolecules including fat. Chemicals or their transformation products that are highly water soluble and do not have the propensity to partition into fats are rapidly eliminated in the urine. Organic chemicals characterized by high K_{ow}s are stored in fat, where they may remain for years or decades. Octanol-water partition coefficients were originally developed to model the distribution of pharmaceutical chemicals, and their use then expanded to other applications, such as the correlation to K_{oc} (see Chapter 5). Metals may also be stored by binding to fat and other biological molecules, such as negatively charged sulfur-containing groups of proteins. For example, lead complexes with biological molecules of the central nervous system and cadmium binds to receptor molecules in the kidney, resulting in renal damage.

9.3.3   Bioactivation

Biotransformation is the primary fate of chemicals in organisms, a process that generally yields products with few or no toxicological effects. However, biotransformations sometimes yield toxic metabolites, a process known as *bioactivation.* Research over the past few decades has demonstrated that bioactivation provides the basis for the toxicity and carcinogenicity of most hazardous chemicals.

A wide range of enzymes catalyzes bioactivation reactions, which are grouped into Phase I and Phase II biotransformations. The two groups are defined based on the reactions that are catalyzed. *Phase I* reactions transform hydrophobic chemicals to more polar products via oxidation, hydrolysis, or similar reactions. *Phase II* processes involve conjugation reactions that add polar functional groups, such as glucose or sulfate, to the Phase I products to produce what are often even more polar metabolites that can be readily excreted. Many biotransformation enzymes exhibit broad substrate specificity, providing a mechanism for enhancing the excretion of a wide range of hazardous compounds.

The extent to which a chemical undergoes biotransformations is correlated somewhat with its hydrophobicity [9.15]. Polar organic compounds and ionized, water-soluble inorganic compounds (e.g., ethyl alcohol, sodium chloride) are readily excreted by the natural processes of the kidney and often do not require transformation. More hydrophobic compounds (those with high K_{ow}s) usually require a biotransformation reaction before they can be effectively excreted by the kidneys. Alternatively, a few mechanisms are found for the direct excretion of hydrophobic compounds, such as through the bile system of the liver. It is not surprising, based on the importance of hydrophobicity in biotransformations, that octanol-water partition coefficients were originally developed for these systems.

Essentially all organisms are capable of transforming hazardous chemicals. Bacteria and other lower organisms have been shown to metabolize hundreds of hazardous compounds (Chapter 7). Specific biotransformation reactions of mammals are highly variable and are a function of the enzymatic, tissue, and organ development of the species. These differences are in part responsible for the range of toxicities between species (see Section 10.2).

The liver is the most important organ of bioactivation because of its high concentration of enzymes that catalyze biotransformation reactions. Furthermore, the liver is considered the primary filter of the blood, and splanchnic circulation (the system that transports blood from the gastrointestinal area to the liver) carries all orally absorbed chemicals through the liver before they reach general systemic circulation. Other tissues and organ systems are also involved in bioactivation reactions. Although the liver is the most important, the kidneys and the lungs also contain relatively high concentrations of bioactivation enzymes. Minor activity is also found in the stomach, intestines, gonads, and skin.

The wide range of enzymes that catalyze the Phase I and Phase II metabolic processes can transform most chemicals that are absorbed and distributed in mammalian systems. Almost all toxic metabolites are electrophiles that react with cellular nucleophiles, which are most commonly proteins and DNA. The most common reactions include the formation of bonds to produce covalent adducts or the abstraction of electrons from DNA and proteins to produce oxidation products. In both situations, the structure and function of the receptor molecule is changed, which often leads to toxicity through cellular or tissue dysfunction [9.16].

Phase I Transformations. The first stage of metabolism, Phase I biotransformations, most commonly involves adding a hydroxyl group to a hydrophobic compound to increase its water solubility. Transformation reactions also include reductions, deaminations, dehalogenations, sulfoxidations, and conjugations. Typical Phase I reactions are listed in Table 9.1. Phase I biotransformations are commonly catalyzed by a nonspecific group of enzymes commonly known as the *mixed function oxidase (MFO) system, the cytochrome P-450 system,* or *the cytochrome P-450 dependent monooxygenase system.* The *P-450* is derived from a spectrophotometric band at 450 nm that appears when an enzymatic suspension of the enzyme is treated with carbon monoxide [9.17].

Phase I transformations have been described as *functionalization reactions* because they add functional groups that are required for subsequent Phase II (biosynthetic) transformations [9.5]. In other words, Phase I and Phase II reactions are often complementary, with the product of Phase I becoming the substrate for Phase II reactions.

Some common themes may be found among cytochrome P-450-catalyzed reactions. All of the reactions are oxidations in which the contaminant is hydroxylated and made more water-soluble. The mechanism of oxidation is analogous to the metabolism of natural energy-rich substrates such as glucose, in that flavin nucleotides [i.e., nicotinamide adenine dinucleotide (NAD)] serve as electron transfer agents through coupled reactions. The NADH provides the potential for the reduction of one atom of molecular oxygen to water; the other atom of O_2 is incorporated into the substrate.

The metabolites that result from Phase I biotransformations may or may not cause damage to the organism. Toxic metabolites may be detoxified by subsequent enzymatic biochemical reactions or nonenzymatic chemical reactions to produce less toxic products. Alternatively, molecular damage may be repaired by a variety of mechanisms or the affected cell or tissue may be replaced. However, as expected based on the fundamental dose-response basis for toxicity, high doses, repeated exposures, or increased bioactivation due to enzyme induction increases the potential for cellular injury [9.1].

In addition to the cytochrome P-450 enzymes, numerous other enzymes catalyze Phase I transformations. Some of these include prostaglandin synthetase-hydroperoxidase and other peroxidases, alcohol and aldehyde dehydrogenases, epoxide hydrases, and flavoprotein reductases. A discussion of the reactions of these enzyme systems is beyond the scope of this text. Detailed information on these enzyme systems may be found in References 9.18 and 9.19.

Table 9.1 Common Phase I Biotransformation Reactions

I. Oxidations

Microsomal Oxidations

Aliphatic oxidation

$$RCH_3 \longrightarrow RCH_2OH$$

Aromatic hydroxylation

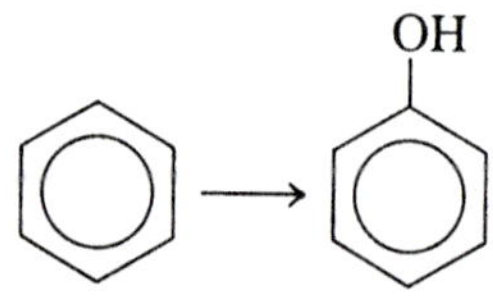

Epoxidation

$$R-CH_2-CH_2-R \longrightarrow R-\overset{\displaystyle O}{\overset{\diagup\!\!\diagdown}{CH-CH}}-R$$

N-hydroxylation

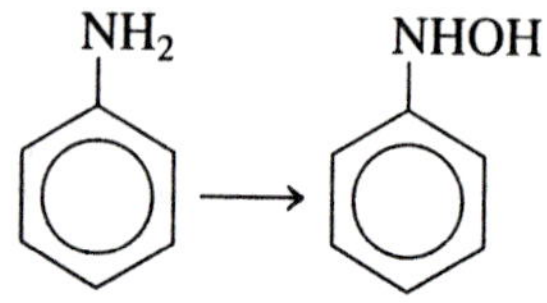

Sulfoxidation

$$R-S-R \longrightarrow R-\overset{\displaystyle O}{\underset{\|}{S}}-R$$

Desulfonation

$$\underset{R}{\overset{R}{\diagdown}}C=S \longrightarrow \underset{R}{\overset{R}{\diagdown}}C=O$$

Dehalogenation

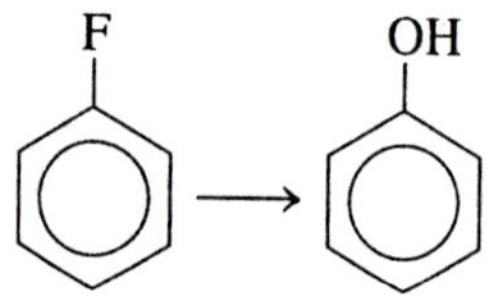

II. Reductions

Microsomal Reductions

Nitro reduction $RNO_2 \longrightarrow RNO \longrightarrow RNHOH \longrightarrow RNH_2$

Azo reduction $RN{=}NR \longrightarrow RNHNHR \longrightarrow RNH_2-RNH_2$

Reductive dehalogenation $R-CCl_3 \longrightarrow R-CHCl_2$

Nonmicrosomal Reductions

Aldehyde reduction

$$\underset{H}{\overset{R}{\diagdown}}C=O \longrightarrow \underset{H}{\overset{R}{\diagdown}}CHOH$$

III. Hydrolysis

Ester hydrolysis $R-CO-O-R_1 \longrightarrow R-COOH + R_1-OH$

Amide hydrolysis $R-CO-NH_2 \longrightarrow R-COOH + NH_3$

462

Phase II Transformations. Phase I oxygenations provide products that are more water soluble; however, many Phase I products still exhibit considerable hydrophobicity. *Phase II biotransformations* are conjugation reactions in which hazardous compounds or metabolites are linked to glucose or sulfate—common species in the body.

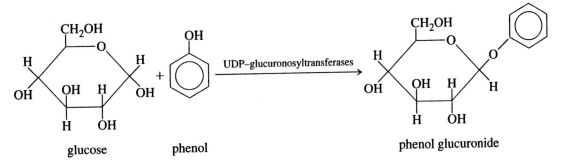

glucose   phenol   phenol glucuronide

Two reactants are required for Phase II transformations: (1) a reactive functional group (e.g., —OH) and (2) a cosubstrate, sulfate or glucose, that can combine with the reactive functional group. Just as cytochrome P-450 is the dominant Phase I enzymatic system, *UDP-glucuronosyltransferase* is the principal Phase II enzyme (UDP stands for uridine diphosphate). These Phase II enzyme systems most commonly convert toxic compounds with free —OH groups into glucose derivatives—*glucuronides*. A common glucuronidation reaction is illustrated in Figure 9.11.

Sulfate conjugation is less important than conjugation to glucose. Some of the most common sulfonation reactions occur with aliphatic and aromatic hydroxyl groups. The mechanism for sulfate conjugation involves an activated sulfate cosubstrate, 3'-phosphoadenosine-5'-phosphosulfate (PAPS). Other classes of biotransformation products that are formed include esters and amides. Some common conjugation reactions are illustrated in Figure 9.12.

Phase II transformations provide conjugation products with higher water solubility that are readily excreted in the urine or bile. Furthermore, the addition of the glucuronic moiety often decreases the toxicity of the Phase I metabolite, by its bulky nature and by blocking the binding of the metabolite to the receptor. Predicting the most probable biotransformation reactions is demonstrated in Example 9.2.

EXAMPLE 9.2   ***Phase I and Phase II Biotransformations***

Outline the most probable Phase I and Phase II biotransformations of 1,3-dichlorobenzene.

SOLUTION

The most likely Phase I biotransformation of 1,3-dichlorobenzene is hydroxylation by cytochrome P-450 to form 2,4-dichlorophenol.

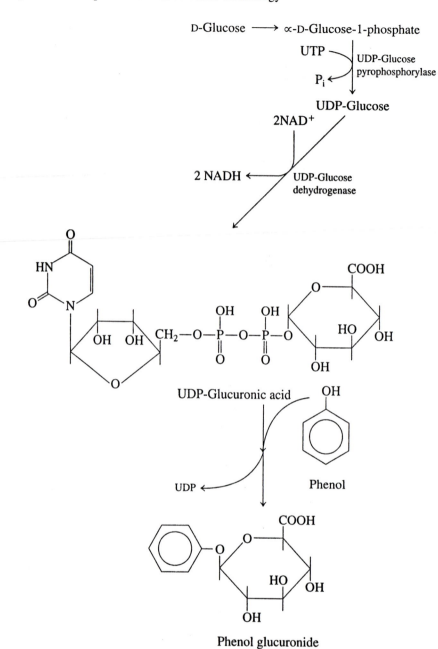

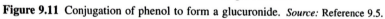

Phenol glucuronide

Figure 9.11 Conjugation of phenol to form a glucuronide. *Source:* Reference 9.5.

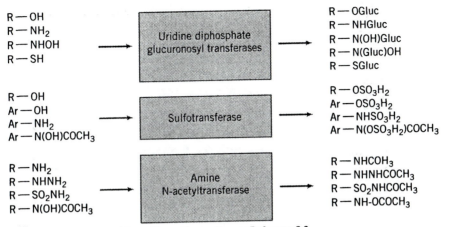

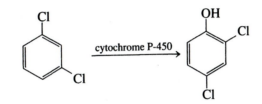

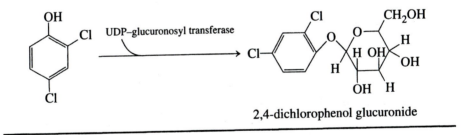

Figure 9.12 Common Phase II reactions. *Source:* Reference 9.3.

The Phase I metabolite is then conjugated with glucose to form the corresponding glucuronide:

9.3.4   Structural Affinity and Toxic Effects

Although its occurrence is not common, localized chemical toxicity does occur as the result of exposure to some chemicals. Perhaps the most common chemicals in this class are strong acids and bases, which destroy cells (and other organic matter) by direct contact. Other chemicals that promote toxicity by direct contact include oxidizers, such as chlorine and hydrogen peroxide.

The mechanisms of toxic action for most hazardous chemicals occur due to binding to receptor molecules on a cell, tissue, or organ. The most common receptors are enzymes; however, receptors also include hemoglobin, membranes, DNA, or other biological molecules. Toxic effects usually involve one organ or tissue system, and numerous terms have been developed to describe toxicity to the target organ, including

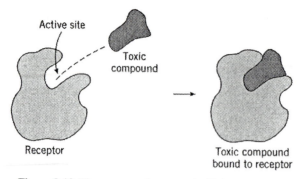

Figure 9.13 The concept of structural affinity demonstrated by the lock and key concept.

hepatoxicity (liver), *nephrotoxicity* (kidney), *genotoxicity* (the chromosomes), and *immunotoxicity* (the immune system).

The driving force in toxic action is the affinity of a hazardous chemical for a specific receptor, a concept that may be viewed as a lock and key mechanism in which toxicity occurs only when the compound locks into a receptor biomolecule, thus disabling it and impeding its metabolic function (Figure 9.13). A good example of the lock and key concept is the competition for hemoglobin by carbon monoxide, which displaces O_2 and causes asphyxiation. The more perfect the fit, the more potent the toxic response. This relationship between chemical structure and binding to receptor molecules is a topic that is being studied extensively through quantitative structure-activity relationships (QSARs).

Some common toxic compounds and corresponding receptors are listed in Table 9.2. One advantage of toxicant-receptor specificity (if one can call it an advantage) is that,

Table 9.2 Common Toxic Compounds and Corresponding Receptors

Compound	Bioactivation Enzyme	Transformation Products	Receptor Tissues	Reference
Hexane	P-450	2,5-Diketone	Peripheral nerves	9.19
Benzene	P-450	Quinones	Bone marrow stem cells	9.20
PAHs	P-450	Arene oxides, quinones	Lung, skin, mammary glands	9.21
Carbon tetrachloride	P-450	Radicals	Liver, lung, kidney	9.22
TCE	P-450	Acyl halides, aldehydes, epoxides	Liver, lung, kidney	9.23
Chlorobenzene	P-450	Arene oxides, quinones	Liver, lung, kidney	9.24
PCBs	P-450	Arene oxides, quinones	Liver, lung	9.25

based on the toxicology studies that have been conducted in the past, a predictable biological response results from chemical exposure. For example, if a spill of parathion occurs as the result of a tanker truck accident, a predictable physiological response (neurotoxicity) resulting from the inhibition of acetyl cholinesterase is the result.

Toxicity to an organism is also expressed at the macroscopic level and includes such broad classes as tissue pathology, aberrant growth processes, and extreme physiologic responses. Specific toxicological responses include

Inflammation: a local response to irritating chemicals such as acids.

Necrosis: tissue or cell death resulting from chemical injury.

Enzyme inhibition: some toxic substances may affect the ability of enzymes to catalyze normal physiological reactions.

Uncoupling of biochemical reactions: enzymes that catalyze the synthesis of ATP and other energy-rich molecules may be uncoupled by such hazardous compounds as dinitrophenol and pentachlorophenol. The organism experiences a loss of energy; furthermore, electrons that normally are accepted by O_2 to produce water will instead form free radicals.

9.4 CARCINOGENICITY

Cancer is a disease in which altered cells divide uncontrollably (called *neoplastic growth*), resulting in tumors (or *neoplasms*). Most chemically induced cancers are characterized by latency periods of up to 40 years, a phenomenon that complicates cause-and-effect relationships of carcinogenicity.

Hundreds of different cancers may be found, each with a range of causes and pathologies. In general, cancers that result from exposure to hazardous wastes are caused by chemically induced mutations. Some of the common types of cancers are

Leukemias: cancers of white blood cells and the tissues from which they are derived.

Lymphomas: cancers of the lymphatic system.

Sarcomas: cancers of connective tissues such as bone and cartilage.

Carcinomas: cancers of the epithelial tissues that form the inside and outside linings of organisms.

The first documentation of chemically induced cancer was in 1775, when the soot to which chimney sweepers were exposed was linked to their scrotum cancer [9.27]. Subsequently, polycyclic aromatic hydrocarbons, such as benzo[*a*]pyrene, were documented as cancer-causing agents. Other direct evidence of chemically induced cancers was documented in the 1930s for azo dyes and aromatic amines [9.28].

Most chemically induced cancers result from an alteration of the genetic material (a mutation); therefore, a discussion of mutagenicity provides a basis for understanding carcinogenicity.

9.4.1 Mutagenicity

Because mutations occur within deoxyribonucleic acid (DNA), an overview of this genetic material is necessary. *Deoxyribonucleic acid,* shown in Figure 9.14, contains four different nitrogenous bases—two purines (adenine and guanine) and two pyrimidines (cytosine and thymine). Chromosomes are made up of two complementary strands of

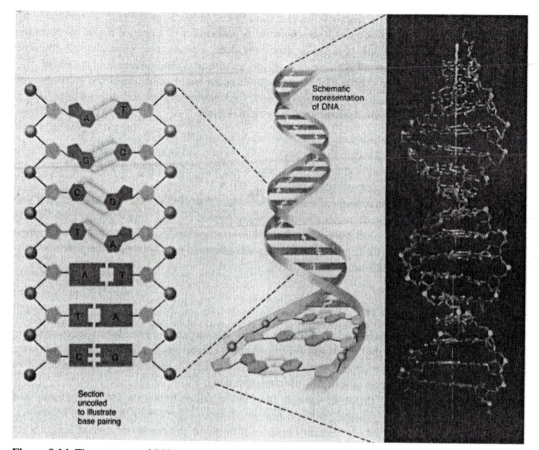

Figure 9.14 The structure of DNA. *Source:* Reference 9.8.

DNA; they are complementary in that a guanine is always paired with a cytosine and an adenine is always paired with a thymine. The order of these bases encodes an organism's genes. What shape an organism is, its color, and its size are a function of the genetic code, which is also called the *triplet code* because three nitrogenous bases code for one amino acid within a protein (see Section 7.4.2) by a process known as *protein synthesis,* which occurs at the ribosomes. The structure of the proteins that are coded and then synthesized eventually dictate what the organism is and what its characteristics are.

A *mutation* is a change in the genetic code, which may or may not have an effect on the organism. Harmful effects from mutations depend on the type of cell that is affected and whether the mutation leads to metabolic malfunctions. Mutations that occur in somatic cells (nonreproductive cells of the body) may or may not prove a threat to the organism. Mutations occur naturally, most commonly from ionizing radiation, and organisms have enzymes that repair DNA (called *repairases*). However, not all mutations are repaired, and some mutations will cause the cell's metabolism to go out of control, which may result in cancer.

Numerous types of mutations have been documented. A *point mutation* is caused by the reaction of a genotoxic substance with DNA that results in a *base substitution—*

the replacement of a nitrogenous base in the DNA sequence. Base substitutions usually do not result in significant mutagenic activity. Because of the redundancy of the genetic code, one base substitution will not likely result in a major change in the translation of the genetic information. Moreover, if a mutation results in an inappropriate amino acid in a protein that functions as an enzyme, it may not even change the enzymatic activity if it is not at or near the active site on the enzyme. Such base substitutions that have no effect on the organism are called *cryptic mutations*.

Frameshift mutations are the result of the addition of a large chemical between base pairs (called an *intercalation*), which shifts the triplet code down the DNA strand. Such shifts essentially change the entire coding of proteins, with consequent high potential for malfunction of the protein. The effect of frameshift mutations are illustrated in Example 9.3.

EXAMPLE 9.3 *Effect of a Frameshift Mutation on the Genetic Code*

A sequence of DNA is characterized by the following base sequence.

a. List the five triplet codes in this sequence that code for amino acids.
b. If a mutagenic species intercalates at $\textcircled{X}$, list the new set of triplet codes.

SOLUTION

a. A T T, G A T, C C A, T G A, G C A
b. A T $\textcircled{X}$, T G A, T C C, A T G, A G C

Another type of mutation is the *covalently bound adduct,* which occurs when a toxic species (or more commonly, its bioactivation product) forms a covalent bond with the nitrogenous bases of DNA. Evidence to date suggests that, of the different categories of mutations, covalently bound adducts provide the most serious threat of cancer [9.29]. For example, azo dyes become covalently bound in the liver [9.30] and benzo[*a*]pyrene becomes covalently bound in the skin of mice [9.31]. Structures of some common DNA adducts are shown in Figure 9.15.

9.4.2 Chemical Carcinogens

The defining element of cancer is a tumor, which originates when a normal cell undergoes a genetic alteration and then grows out of control. Tumors can be *benign* or *malignant*. Benign tumors are inert masses, whereas malignant tumors commonly proliferate and rob the body of life.

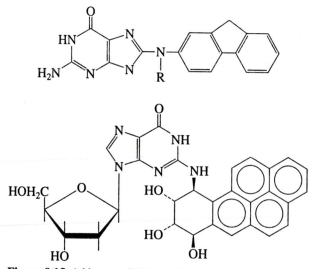

Figure 9.15 Adducts to DNA for 2-naphylamine and benzo-[*a*]pyrene. *Source:* Reference 9.5.

Carcinogens, which are chemicals that cause cancer, are absorbed, bioactivated, metabolized, and excreted like noncarcinogenic chemicals; however, they are different because they result in cumulative and delayed effects. By definition, a *chemical carcinogen* is a compound that, based on experimental studies, has been found to increase the rate of cancer, compared to a corresponding control population. Four general types of responses provide a basis for carcinogenicity: (1) higher incidence of tumors than what would have occurred without exposure, (2) increased incidence of new forms of cancer, (3) an increase in multiple cancers, and (4) a decrease in the time needed for tumors to develop [9.32].

Carcinogens are often classified based on their potential reactions with DNA. Chemicals producing changes in nucleic acids and other genetic material that result in chromosomal dysfunction are termed *genotoxic compounds*. Two important subclasses of genotoxic carcinogens are chemicals that react either directly or indirectly with DNA. *Direct genotoxic chemicals* are usually epoxides, multi-ring PAHs, or heterocycles, which, because of their strong electrophilic character, react rapidly with DNA subunits. Many of these compounds are precarcinogens, because they have carcinogenic potential only after they proceed through Phase I biotransformation reactions.

Indirectly acting genotoxic chemicals are most commonly oxygen radicals that are generated through one-electron transfers as the result of uncoupling of oxidative phosphorylation. The first electron transfer produces superoxide radical:

$$O_2 + e^- \longrightarrow O_2^{\cdot-} \tag{9.1}$$

Superoxide is a genotoxic species, so most cells contain superoxide dismutase (SOD), which converts superoxide to hydrogen peroxide:

$$O_2^{\cdot-} + 2H^+ \xrightarrow{\text{SOD}} H_2O_2 \tag{9.2}$$

Hydrogen peroxide is also toxic, and aerobic organisms possess catalase, which decomposes hydrogen peroxide to molecular oxygen and water:

$$H_2O_2 \xrightarrow{\text{catalase}} \frac{1}{2} O_2 + H_2O \qquad (9.3)$$

However, a competing reaction is the iron-catalyzed decomposition of H_2O_2 to hydroxyl radical (OH·):

$$H_2O_2 + Fe^{2+} \longrightarrow OH· + OH^- + Fe^{3+} \qquad (9.4)$$

The hydroxyl radical is an aggressive electrophile that rapidly reacts with DNA. It has been shown to cause mutations, cancer, lipid peroxidation, and glaucoma, and has been implicated in the aging process. The strong evidence for the pathologic effects of oxygen radicals has promoted the recent trend in the use of antioxidant vitamins (e.g., vitamin C, vitamin E).

Epigenic carcinogens cause cancer by other mechanisms than reaction with DNA. A wide range of mechanisms are involved, but, unfortunately, epigenic carcinogens are not as well understood as genotoxic carcinogens.

9.4.3 Mechanisms of Carcinogenicity

The latency of carcinogenicity is not simply a function of a tumor slowly growing from microscopic to macroscopic size; instead, a series of developmental stages of carcinogenesis appears to be required. In fact, strong evidence exists that cancer is a three-step process: (1) initiation, in which genetic damage occurs through a mutation to DNA, (2) promotion, in which the genetic damage is expressed through the multiplication of cells in which initiation occurred previously, and (3) progression, the spreading of cancer through uncontrolled growth. The three stages of carcinogenesis are illustrated in Figure 9.16.

The first stage of cancer, *initiation,* involves a change in the capacity of DNA to function properly. Initiation is characterized by (1) irreversibility, (2) cumulative effects of repeated exposures, (3) no morphological changes of initiated cells, and (4) no threshold dose [9.3]. The first deleterious alteration that occurs and is part of cancer initiation is a genetic alteration, through a genetic mutation. A special class of genes in the DNA—called *oncogenes*—appears to be involved in cancer initiation [9.33]. Although tens of thousands of genes are found on human chromosomes, mutations in only a small percentage are suspected to initiate cancer. Oncogenes are believed to function in the rapid cell growth necessary during embryological development. Although these genes do not usually function after birth, mutations are believed to initiate their expression, which, in turn, can cause cancer [9.34]. A wealth of evidence has shown that oncogenes are the primary loci of cancer initiation; however, mutations of oncogenes have not been able to explain all cancers. Nonetheless, almost all evidence to date points to mutations of DNA as the mechanism for the initiation of cancer. The oncogene theory provides a good conceptual model for cancer initiation; however, much is still unknown.

The second stage of carcinogenesis, cancer *promotion,* is the increase in growth of initiated cells, a process resulting from the presence of promoters—a relatively small

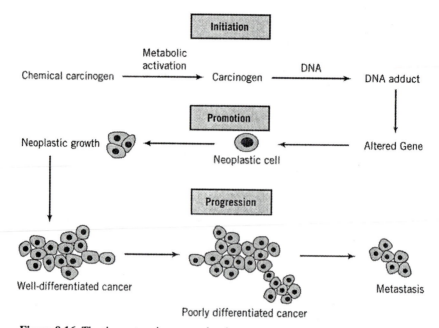

Figure 9.16 The three steps in cancer development. *Source:* Reference 9.3.

class of compounds that are not necessarily genotoxic. A potential mechanism for promotion may be an alteration of gene expression that results in neoplastic (uncontrolled) growth in the initiated cells [9.33]. The mechanism of promotion may also involve new mutations and the triggering of other genes. Promotion is also characterized by (1) reversibility, (2) no additive effects, (3) threshold effects, and (4) interaction with environmental factors [9.1]. Promoters themselves are usually not carcinogens; they include dietary fat, asbestos, saccharin, halogenated solvents, alcohols, and estrogen [9.3].

Many chemical carcinogens are *complete carcinogens;* these chemicals function as both initiators and promoters. Many natural and synthetic chemicals have been classified, and even more may be found in the future.

The final stage in carcinogenicity is the growth of neoplastic cells followed by *progression,* or spreading throughout the body. This final step may be subdivided into two stages—invasion and metastasis. *Invasion* is the localized movement of neoplastic cells into adjoining tissues, and *metastasis* is the more distant movement throughout the body. Invasion and metastasis are consistent with a physiological system that is out of control. Mammals have immune systems (e.g., T cells) that normally protect them from foreign invasion, often by phagocytosis and other means of engulfing the foreign objects. Furthermore, foreign neoplastic cells do not adapt and grow easily in an environment with entirely different tissue characteristics. Substantial evidence has shown that subsequent mutations (beyond initiation) may occur as progression proceeds, further suggesting that the organism is lacking homeostasis (i.e., is "out of sync"). Because of this phenomenon, some epidemiologists have proposed that cancer death rates should not be an indication of overall public health because a high proportion of those who die from cancer are weak and susceptible to the mechanisms of metastasis [9.35].

9.4.4 Classes of Chemical Carcinogens

A common concern in hazardous waste management is whether chemicals to which receptor populations are exposed, indeed cause cancer. Although in vitro and animal studies have provided a wealth of information on potential human carcinogenicity, the only absolute confirmation is through statistical proof by epidemiological research. Numerous government groups and agencies have reported classification schemes for carcinogenicity, most of which are quite similar. The most common systems have been reported by the International Agency for Research on Cancer (IARC) and the U.S. EPA [9.36]. The categories for the carcinogenic groupings are listed in Table 9.3, and some of the Group A chemicals are listed in Table 9.4. Hundreds of hazardous chemicals, including many of those described in Chapter 2, are listed as Group B carcinogens (probable or suspected carcinogens). Although Group A carcinogens must be substantiated by direct human evidence, animal studies support strong evidence for the carcinogenicity of Group B compounds. To provide quantitative extrapolations, slope factors (SFs), which may be used to estimate carcinogenic risk, have been established (see Section 10.7).

Table 9.3 EPA Categories for Carcinogenic Groups

Class	Description
A	Human carcinogen
B	Probable carcinogen
B_1	Linked human data
B_2	No evidence in humans
C	Possible carcinogen
D	No classification
E	No evidence

Table 9.4 Selected Group A Chemicals (Human Carcinogens)

Hazardous Waste Compounds	Other Compounds
Arsenic	Aflatoxins
Asbestos	Analgesic mixtures containing phenacetin
Benzene	Diethylstilbestrol
Benzidine	Erionite
bis-Chloromethylether	Melphalan
Chromium, hexavalent	Mustard gas
2-Naphthylamine	Soots
Nickel	Smokeless tobacco products
Plutonium-239	Tobacco smoke
Radium-226	
Radon-222	
Vinyl chloride	

Source: Reference 9.35.

9.5 TOXIC RESPONSES OF COMMON HAZARDOUS CHEMICALS

Individual hazardous chemicals are characterized by different mechanisms of absorption, distribution, transformation, toxicant-receptor binding, and excretion; therefore, reporting the toxicology of the thousands of hazardous chemicals is an ambitious undertaking, and is beyond the scope of this text. However, a few common toxic compounds demand attention, and a description of the toxicology of five of these—benzene, benzo[a]pyrene, halogenated solvents, parathion, and lead—is presented next. In addition, brief descriptions of the toxic effects of some of the most common hazardous compounds discussed in Chapter 2 are listed in Table 9.5. Detailed toxicological descriptions of other hazardous compounds may be found in Reference 9.3.

Benzene. The primary routes of benzene absorption are ingestion (from contaminated drinking water) and inhalation (releases from contaminated soils, industrial sources, petroleum, and RCRA facilities). Benzene is a confirmed Group A human carcinogen that causes leukemia—the malignant proliferation of white blood cells. Once absorbed, benzene is transformed in the liver by cytochrome P-450, first to phenol and then to polyphenolic metabolites (e.g., hydroquinone), with potential oxidation through peroxidases to biphenols, hydroquinone, and p-benzoquinone.

A dominant enzyme of bone marrow granulocytic cells, myeloperoxidase, which catalyzes the oxidation of hydroquinone to p-benzoquinone, appears to be an important factor in bone marrow toxicity and leukemia [9.37]. The two possible mechanisms in benzene-induced leukemia are (1) adduct formation on DNA, and (2) chromosomal deletions. Although adduct formation has been implicated in most cases of cancer, most evidence to date suggests that the presence of chromosomal aberrations is the most likely mechanism of leukemogenesis. Recent medical chromosomal studies have shown that leukemia cells are characterized by abnormalities in at least one chromosome with DNA adduct formation (a deoxyguanosine adduct) observed only during in vitro studies using purified DNA in the presence of quinone metabolites. However, the exact mechanism of the onset of leukemia remains unproven [9.38].

Polycyclic Aromatic Hydrocarbons. Like benzene, polycyclic aromatic hydrocarbons (PAHs) are a public health concern due to their potency as carcinogens. One of the most powerful PAH carcinogens is benzo[a]pyrene (BaP), a confirmed Group A human carcinogen. Once absorbed, BaP is transformed via cytochrome P-450 to an epoxide at the 7,8 position followed by hydration of the epoxide with subsequent formation of the 7,8-dihydrodiol (Figure 9.17). Another epoxide is then formed (catalyzed by cytochrome P-450) to yield the carcinogenic metabolite of BaP, 7,8-dihydrodiol-9,10-oxide. This final species binds covalently to RNA and DNA (Figure 9.15). As described in Section 9.4.3, such covalently bound DNA adducts result in mutagenic, and subsequently, carcinogenic activity, especially if the mutation initiated is on an oncogene.

Halogenated Solvents. Volatility affects not only the environmental dynamics of halogenated solvents, but also their toxicological dynamics. As with many hazardous compounds, halogenated solvents are commonly ingested in contaminated drinking water and inhaled as hazardous air pollutants. Absorption of these compounds is highly efficient because they rapidly diffuse through epithelial membranes. Once absorbed, they are distributed in the fat, brain, and blood. Of the common halogenated solvents, only

Table 9.5 Toxic Effects of Common Hazardous Compounds

Chemical	Acute Effects	Chronic Effects
Aliphatic Hydrocarbons		
Alkanes	Central nervous system impairment	No known carcinogenicity or other chronic effects
Monocyclic Aromatic Hydrocarbons		
Toluene	Central nervous system depression, including agitation, delirium, coma	Central nervous system impairment; no proven carcinogenic activity
Xylenes	Liver toxicity, including steatosis, hepatic cell necrosis, and partial tract enlargement	
Nonhalogenated Solvents		
Acetone	Low toxicity, some CNS effects	Minimal chronic effects
Pesticides		
Aldrin and Dieldrin	Tremors, seizures, and coma	Carcinogenicity
DDT	Affects sodium-potassium pump of neural membranes to produce headaches, nausea, vomiting, and tremors	Minimal chronic effects in humans; no proven human carcinogenicity
Hydrogen cyanide	Binding to cytochromes of cardiovascular and central nervous systems, resulting in blockage of cellular respiration	Minimal chronic toxicity
Pentachlorophenol	Uncoupling of oxidative phosphorylation	Liver toxicity, including fatty tissue, infiltration and elevated enzymes
Industrial Intermediates		
Phenol	A range of health effects including cardiac dysrhythmia, dermal necrosis, and elevated liver enzymes	Not a human carcinogen; some evidence that phenol is a complete carcinogen in mice
Chlorinated benzenes	Dizziness, headaches	Porphyria cutanea, aplastic anemia, leukemia
Explosives		
TNT	Dermatitis	Liver toxicity, aplastic anemia

Table 9.5 (Continued)

Chemical	Acute Effects	Chronic Effects
Polychlorinated Biphenyls		
PCBs	Minimal acute toxicity (0.5 g/kg to 11.3 g/kg)	Chloracne; increased liver enzymes; possible reproduction effects; act as cancer promoters
Dioxins and Furans		
PCDDs/PCDFs	Chloracne, headaches, peripheral neuropathy	Induction of microsomal enzymes; altered liver metabolism; altered T-cell subsets; immunotoxicity; strongly implicated in carcinogenicity (may be a promoter)
Inorganic Compounds		
Arsenic	Loss of blood, intestinal injuries, acute respiratory failure	Myelogeneous leukemia, cancer of skin, lungs, lymph glands, bladder, kidney, prostate, and liver
Cadmium	Vomiting, cramping, weakness, and diarrhea	Oral ingestion results in renal necrosis and dysfunction; induces lung, prostate, kidney, and stomach cancer in animals; no documented human cancer
Hexavalent chromium	Readily absorbed by the skin, where it acts as an irritant and immune-system sensitizer; oral absorption results in acute renal failure	Lung cancer
Mercury	Central nervous system impairment including injury to motor neurons; renal dysfunction	Central nervous system dysfunction, memory deficits, decrease in psychomotor skills, tremors
Nickel	Not highly toxic; headache, shortness of breath	Immune system effects resulting in allergic contact dermatitis

PCE tends to bioaccumulate [9.39]. Transformations occur primarily via cytochrome P-450 reactions, with some accompanying Phase II reactions, particularly as conjugates of the amino acid glutathione. A representative metabolic scheme, that of TCE, is illustrated in Figure 9.18. In this pathway, TCE is transformed via cytochrome P-450 reactions through an epoxide stage to trichloroacetaldehyde and chloral hydrate, with final products of trichloroethanol (which is excreted as a glucuronide conjugate)

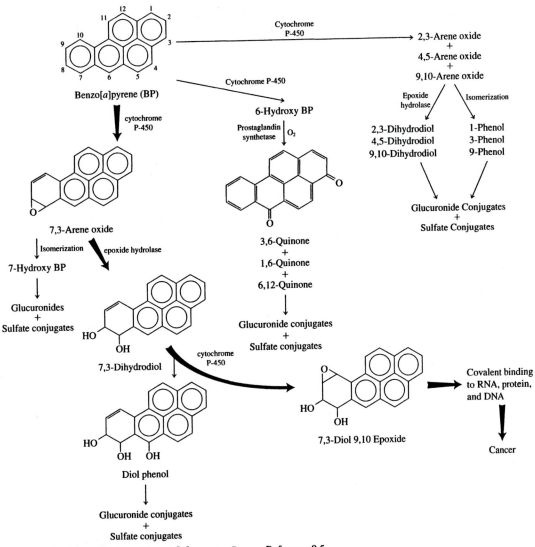

Figure 9.17 Bioactivation of benzo[a]pyrene. *Source:* Reference 9.5.

or trichloroacetic acid, which is metabolized further to carbon dioxide and chloroform. Various reports of halogenated solvent toxicity have been established based on high dose and acute exposure. For example, central nervous system depression develops as a result of inhalation of most halogenated solvents, with a corresponding loss of blood pressure.

The most common concerns of halogenated solvents in hazardous waste management are chronic exposures from drinking contaminated water and the inhalation of toxic air pollutants. Long-term exposure studies have shown that many halogenated solvents (e.g., carbon tetrachloride, chloroform, PCE) promote pathologic changes to the liver and secondarily to the kidneys. The most common pathological effects include the growth of fatty tissues, necrosis (cell death), and cirrhosis (dysfunctional tissue)

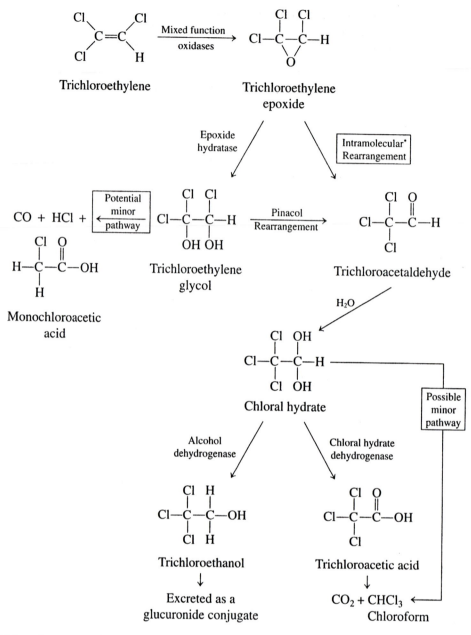

*Major pathway

Figure 9.18 The biotransformation of TCE. *Source:* Reference 9.3.

[9.40]. For example, carbon tetrachloride induces hepatotoxicity as the result of the formation of trichloromethyl free radicals with accompanying lipid peroxidation. While PCE, TCA, and methylene chloride are weak hepatotoxins, liver damage from TCE (which was once believed to be significant) is now considered minimal [9.3].

Perhaps the most common concern related to halogenated solvent exposure is the potential for cancer. Although none of the halogenated solvents is listed as a Group A carcinogen, five (carbon tetrachloride, methylene chloride, chloroform, 1,2-dichloro-ethane, and PCE) are in the Group B category of the Environmental Protection Agency and the International Agency for Research on Cancer [9.36].

Organophosphorus Ester and Carbamate Insecticides. Neurotoxicity is the toxico-logical mechanism of organophosphorus esters and carbamates. Nerve transmissions, which control limb movements, breathing, and motor movements, are promoted by electrical impulses along a neuron (nerve cell)—a process that involves pumping of sodium and potassium in and out of the cell as the impulse moves along the nerve. Nerve cells are discontinuous throughout the body, and it follows that the nerve impulse must be transmitted across the synapse—the gap between two neurons. The impulse is transmitted by acetylcholine, which provides a chemical signal across the synapse.

Acetylcholine is a neurotransmitter at autonomic and neuromuscular synapses; its function and mechanism of inhibition are illustrated in Figure 9.19. In normal physiological functions, acetylcholine is hydrolyzed to acetate and choline via the enzyme acetylcholinesterase. This hydrolysis reaction prevents buildup of acetylcholine to maintain a functioning level at the synapse; that is, hydrolysis of acetylcholine prevents the continuous and prolonged excitation of nerve pulses. Obviously, the activity of acetylcholinesterase is essential to prevent the continuous firing of nerve impulses. Organophosphate ester (e.g., parathion, malathion) and carbamate ester (e.g., carbaryl) insecticides inhibit acetylcholinesterase activity, resulting in a loss of the organism's capacity to hydrolyze acetylcholine. Actually, most organophosphates, such as para-thion, are activated by cytochrome P-450 to the corresponding oxon (i.e., the sulfur is replaced by oxygen). The oxon, such as paraoxon, then binds strongly to acetyl-cholinesterase, resulting in its inhibition and ensuing toxicity. The accumulation of acetylcholine results in neuromuscular dysfunction, including involuntary responses, contractions, and under high doses, death due to respiratory failure (because of associated loss of neurological control of respiratory functions).

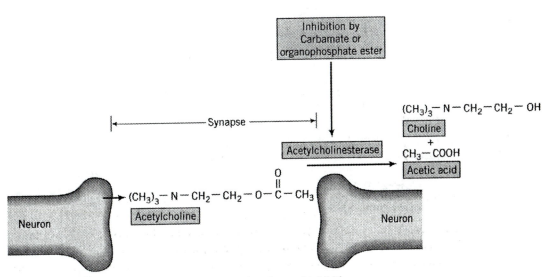

Figure 9.19 Mechanisms of acetylcholinesterase function and inhibition.

Lead. Although its bioavailability is often low because it is sorbed and exchanged in the environment, lead is an insidious poison that is known for its effects on the central nervous system, especially in children. Lead is most effectively absorbed by the lungs if it is in the form of a fume or a particulate <5 μm [9.41]. Fortunately, these physical states of lead are found only in industrial settings, and are therefore not a threat to the general population. Although the most common route of exposure for lead is ingestion, the absorption efficiency is only about 20–30% for adults and 50% for children [9.42].

Once absorbed, lead is carried by red blood cells throughout the body, where it binds to bone, teeth, lungs, kidneys, the brain, and the spleen. Storage in the bones has been implicated as a major sink of lead, which can provide long-term effects by mobilization, even after the exposure has ceased.

Although acute toxicity of lead has been reported from industrial exposure (which includes encephalopathy, renal failure, and gastrointestinal symptoms), the public con-

Table 9.6 Toxic Responses to Increased Lead Concentrations in the Blood

Health Effects Blood Lead Concentration (μg/l)		
	Adults	Children
100	Hypertension may begin to occur	Crosses placenta, developmental toxicity
		Impairment of IQ
		Increased erythrocyte protoporphyrin
200	Increased erythrocyte protoporphyrin	Beginning impairment of nerve conduction velocity
300	Systolic hypertension, decreased hearing	Impaired vitamin D metabolism
400	Infertility in males, renal effects, neuropathy	Hemoglobin synthesis impaired
	Fatigue, headache, abdominal pain	
500	Decreased heme synthesis, decreased hemoglobin, anemia, GI symptoms, headache, tremor	Colicky abdominal pain, neuropathy
		Encephalopathy, anemia, nephropathy, seizures
	Lethargy, seizures	
1000	Encephalopathy	

Source: Reference 9.3.

cerns associated with lead in hazardous waste management are chronic exposure and toxicity.

In the body, lead binds with a number of enzymes, particularly those with sulfhydryl groups. Two of the most common enzymes to which lead absorbs are δ-aminoevulenic acid dehydratase and ferrochelatase. The toxic responses to increased blood lead concentrations in adults and children are listed in Table 9.6. Although some specific physiological responses are listed in this table, the common theme in lead toxicity is its effects on the central nervous system and related systems, such as peripheral nerves. In adults, low levels of lead cause cognitive difficulties and visual motor problems. However, perhaps the greatest threat to the public health is in children who often ingest lead in soil and other solids. Epidemiological studies have shown that low levels of lead impair IQ, school performance, and central nervous system function.

9.6   SUMMARY OF IMPORTANT POINTS AND CONCEPTS

- A knowledge of toxicology is important for environmental scientists and engineers in order to communicate with other hazardous waste professionals and as a basis for performing site and risk assessments.

- Although numerous routes of exposure are used in clinical toxicology (e.g., intramuscular, subcutaneous, intravenous), the most common routes of exposure at hazardous waste sites and RCRA facilities are oral, respiratory, and dermal.

- Although some toxic effects are localized, such as dermal necrosis by acids or hydrogen peroxide, most hazardous chemicals exert toxic action after they have been absorbed, distributed in the body, and bioactivated by enzymes.

- Transformation reactions, which occur primarily in the liver, but also in the kidney, lungs, and intestines, can occur through oxidation and hydrolysis (Phase I) or conjugation reactions (Phase II).

- Chemicals converted to more polar degradation products via Phase I and Phase II transformations are usually more readily excreted. However, Phase I reactions often make the chemical more toxic, a process known as bioactivation.

- Toxicity occurs when the toxic compound or its metabolite binds to a receptor molecule, such as a protein or DNA.

- Chemicals and their metabolites are excreted by the kidneys. Metabolites resulting from Phase I and Phase II transformations are more polar than the parent compound and therefore more easily excreted.

- Carcinogenicity is manifested in the formation of tumors in animals and humans and almost always results from changes in the organism's DNA. It proceeds through three stages—initiation, promotion, and progression.

PROBLEMS

9.1. Describe three primary routes of absorption to which the public is exposed in hazardous waste management. Rank them in order of importance as threats to health and provide a rationale for the ranking.

9.2. List the route(s) of exposure that will be significant for each of the following hazardous waste cases.
   a. Supplying contaminated drinking water to a subdivision
   b. Swimming in a river with PCB-contaminated sediments
   c. Working in a solvent recycling plant
   d. Living in Kellogg, Idaho, a town built on mining tailings containing heavy metals
   e. Living in Times Beach, Missouri (see Chapter 2)
   f. Living at Love Canal in the early 1970s (see Chapter 1)

9.3. Fugitive dust emissions from remedial activities at an abandoned wood-preserving site are characterized by particles 15–38 μm in diameter with sorbed pentachlorophenol. Do these particles represent a significant route for the inhalation of pentachlorophenol?

9.4. For each of the following, provide a brief qualitative assessment of their potential for absorption into the body.
   a. PCE
   b. Trichloroacetic acid at pH 7
   c. Parathion
   d. Chromium (VI)
   e. 1,2,3-Trichlorobenzene

9.5. A worker in a hazardous waste recycling facility inhales and absorbs toluene at a rate of 22 μg/h. He excretes 2 L per day containing 18 μg/L toluene. If the blood volume of the individual is 5.6 L and the toluene is metabolized with a first-order rate constant of 0.08 h^{-1}, what is the steady-state concentration of toluene in the individual's vascular system? Assume that storage is negligible.

9.6. An organism is exposed to a highly water-soluble chemical with a constant ingestion rate of 0.12 mg/h. Its transformation in the liver is described by a first-order rate constant of 0.1 h^{-1}. The compound and its metabolites are cleared through the kidneys at a rate of 0.03 mg/h. If the volume of body fluids is 3 L, determine the steady-state concentration of the chemical in the organism's vascular system.

9.7. If chloroacetic acid diffuses through the stomach lining at a flux rate of 9.2 μmole/cm^2-min and chloroacetate diffuses at 0.17 μmole/cm^2-min, estimate the rate of absorption if the pH regime of the stomach lining is 2.4.

9.8. Pentachlorophenol is absorbed across the lining of the small intestine at a rate of 8.1 μmole/cm^2-min and across the stomach lining at a rate of 0.16 μmole/cm^2-min. The absorption rate of the anion, pentachlorophenate, is negligible. If the stomach pH is 2.3 and the small intestine pH is 6.8, where is absorption of the pentachlorophenol the greatest?

9.9. Provide a synopsis of Phase I and II transformations with emphasis on the physiological end points of these processes.

9.10. Predict the products of cytochrome P-450-catalyzed transformations of each of the following hazardous chemicals.
   a. Benzene
   b. Aniline (Hint: The oxidation is not on the ring).
   c. Toluene
   d. TNT
   e. 1,1-Dichloroethylene

9.11. What products would be expected for each of the compounds by Phase I metabolism?

  a. Bromobenzene   b. Ethanol

  c. 1,2-Dichloroethane   d. Parathion

9.12. Describe the pathway for the sulfonation of toluene by Phase I and Phase II metabolism.

9.13. Propose the steps involved for the Phase I oxidation and Phase II conjugation to glucuronides of the following hazardous chemicals.

  a. Aniline   b. TNT

  c. Ethyl acetate   d. Carbaryl

9.14. List the following contaminants in increasing order of potential for storage. Assume that metabolism is not a significant factor.

  a. Acetone   b. DDT

  c. PCE   d. Hexachlorocyclopentadiene

  e. 1,2,3-Trichlorobenzene   f. TCA

9.15. List the following compounds in increasing order of storage in the body.

  a. Hexachlorobenzene   b. Phenol

  c. PCE   d. Hexachlorocyclopentadiene

9.16. Describe the concept of structural affinity and its role in toxicity.

9.17. A section of DNA has the following sequence of nitrogenous bases.

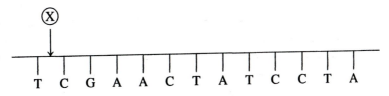

If a frameshift mutation occurs at ⊗, list the triplet codes before and after the mutation.

9.18. Using a genetics book that contains the genetic code, list the four amino acids coded by the DNA (1) before the frameshift mutation and (2) after the frameshift mutation.

9.19. Describe the concepts of cancer initiation, promotion, and progression.

9.20. Describe the biochemical and genetic characteristics of the carcinogenic process, including genotoxic activity, adduct formation, etc.

9.21. The carcinogenicity of 2-naphthylamine is analogous to that of benzo[a]pyrene. Propose a scheme for the activation of 2-naphthylamine and its adduct formation with DNA.

9.22. Describe the criteria for Group A carcinogens and why sufficient data are unavailable for many hazardous compounds to be on this list.

9.23. Using Reference 9.3 and *Dangerous Properties of Industrial Materials* by I. Sax, explain why benzene is a confirmed carcinogen while toluene and xylenes are not.

9.24. A 90-day hazardous waste storage area handles spent sulfuric acid, sodium cyanide, and amyl nitrate. Provide a two-page toxicological assessment of potential toxic hazards if a catastrophic event (e.g., storm or earthquake) caused all of the drum contents to be spilled and mixed.

9.25. In performing a Phase I property evaluation of a pesticide distributor, you experience increased salivation, muscle contractions, and twitching. To what chemical do you suspect you were exposed?

9.26. The day after a facility inspection, a team member feels tired and heated. To what class of compounds was she likely exposed?

9.27. Using *Dangerous Properties of Industrial Materials* by I. Sax, write a one page synopsis of the toxicological characteristics for each of the following chemicals.
a. Hexane   b. Benzene
c. Benzo[*a*]pyrene   d. Carbaryl
e. 2,4,5-T   f. PCB-1260

9.28. Using Reference 9.3, provide a two-page toxicological evaluation of
a. 2,4-D   b. Lindane
c. Paraquat   d. Methyl ethyl ketone

REFERENCES

9.1. Amdur, M. O., J. Doull, and C. D. Klaassen (Eds.), *Casarett and Doull's Toxicology—The Basic Science of Poison*, 4th Edition, Pergamon Press, Elmsford, New York, 1991.

9.2. Loomis, T. A., *Essentials of Toxicology*, Henry Kimpton Publishers, London, 1974.

9.3. Sullivan, J. B. and G. R. Krieger, *Hazardous Materials Toxicology: Clinical Principles of Environmental Health*, Williams & Wilkins, Baltimore, MD, 1992.

9.4. Matsumura, F., *Toxicology of Insecticides*, 2nd Edition, Plenum Press, New York, 1985.

9.5. Hayes, A. W. (Ed.), *Principles and Methods of Toxicology*, Raven Press, New York, 1989.

9.6. Guyton, A. C., *Basic Human Physiology: Normal Function an Mechanisms of Disease*, W. B. Saunders Company, Philadelphia, PA, 1971.

9.7. Berne, R. M. and M. N. Levy, *Physiology*, 3rd Edition, Mosby Year Book, St. Louis, 1993.

9.8. Brum, G., L. McKane, and G. Karp, *Biology: Exploring Life*, 2nd Edition, John Wiley & Sons, New York, 1994.

9.9. Dienhart, C. M., *Basic Human Anatomy and Physiology*, W. B. Saunders and Company, Philadelphia, 1973.

9.10. Goldsmith, L. A., *Biochemistry and Physiology of the Skin*, Oxford University Press, New York, 1983.

9.11. Morgan, W. K., and A. Seaton (Eds.), *Occupational Lung Disease*, 2nd Edition, W. B. Saunders, Philadelphia, PA, 1984.

9.12. Minton, N. and V. Murray, "A review of organophosphate poisoning," *Med. Toxicol.*, **3**, 350–375 (1988).

9.13. Wright, N., W. B. Yeoman, and G. F. Carter, "Massive oral ingestion of elemental mercury without poisoning," *Lancet*, **1**, 206 (1980).

9.14. Oehme, F. W., "Absorption, biotransformation, and excretion," *Clin. Toxicol.,* **17,** 147–158 (1980).

9.15. Anders, M. W. (Ed.), *Bioactivation of Foreign Compounds,* Academic Press, Orlando, FL, 1985.

9.16. Guengerich, F. P., and D. C. Liebler, "Enzymatic activation of chemicals to toxic metabolites," *Crit. Rev. Toxicol.,* **14,** 259–307 (1985).

9.17. Wislocki, P. G., G. T. Miwa, and A. Y. H. Lu, "Reactions catalyzed by the cytochrome P-450 system," in Jakoby, W. B. (Ed.), *Enzymatic Basis of Detoxification,* Vol. 1, Academic Press, New York, 1980.

9.18. Boyd, M. R., "Biochemical mechanisms of chemical-induced lung injury: Roles of metabolic activation," *Crit. Rev. Toxicol.,* **7,** 103–176 (1980).

9.19. Eisen, H. J., "Induction of hepatic P-450 isoenzymes: Evidence for specific receptors," in Ortiz de Montellano, P. R. (Ed.), *Cytochrome P-450: Structure, Mechanism and Biochemistry,* Plenum Press, New York, 1986.

9.20. Kravasage, W. J., J. L. O'Donoghue, G. D. DiVincenzo, and C. J. Terhaar, "The relative neurotoxicity of methyl *n*-butyl ketone, *n*-hexane and their metabolites," *Toxicol. Appl. Pharmacol.,* **56,** 433–441 (1980).

9.21. Eastmond, D. A., M. T. Smith, and R. D. Irons, "An interaction of benzene metabolites reproduces the myelotoxicity observed in benzene exposure," *Toxicol. Appl. Pharmacol.,* **91,** 81–91 (1987).

9.22. Thakker, D. R., H. Yagi, W. Levin, A. W. Wood, A. H. Conney, and D. M. Jerina, "Polycyclic aromatic hydrocarbons: Metabolic activation to ultimate carcinogens," in Anders, M. W. (Ed.), *Bioactivation of Foreign Compounds,* Academic Press, Orlando, FL, 1985.

9.23. Macdonald, T. L., "Chemical mechanisms of halocarbon metabolism," *Crit. Rev. Toxicol.,* **11,** 85–120 (1982).

9.24. Guengerich, F. P. and D. C. Liebler, "Enzymatic activation of chemicals to toxic metabolites," *Crit. Rev. Toxicol.,* **14,** 259–307 (1985).

9.25. Jollow, D. J., J. R. Mitchell, N. Zampaglione, and J. R. Gillette, "Bromobenzene induced liver necrosis. Protective role glutathione and evidence for 3,4-bromobenzene oxide as the hepatotoxic intermediate," *Pharmacology (Basel),* **11,** 151–169 (1974).

9.26. Boyd, M. R., "Biochemical mechanisms of chemical-induced lung injury: Roles of metabolic activation," *Crit. Rev. Toxicol.,* **7,** 103–176 (1980).

9.27. Pott, P., *Chirurgical Observations Relative to the Cataract, the Polypus of the Nose, the Cancer of the Scrotum, the Different Kinds of Ruptures, and the Mortification of the Toes and Feet,* Hawkes, Clarke, and Collins, London, 1775.

9.28. Miller, E. C. and J. A. Miller, "The presence and significance of bound amino azo dyes in the livers of rats fed *p*-dimethylaminoazobenzene," *Cancer Res.,* **7,** 468 (1947).

9.29. Shelby, M. D., "The genetic toxicity of human carcinogens and its implications," *Mutat. Res.,* **240,** 3–15 (1988).

9.30. Miller, J. A., J. W. Cramer, and E. C. Miller, "The *n*- and ring-hydroxylation of 2-acetylaminofluorene during carcinogenesis in the rate," *Cancer Res.,* **20,** 950–962 (1960).

9.31. Miller, E. C., "Studies on the formation of protein-bound derivatives of 3,4-benzopyrene in the epidermal fraction of mouse skin," *Cancer Res.,* **11,** 100 (1951).

9.32. Krisch-Volders, M., *Mutagenicity, Carcinogenicity and Teratogenicity of Industrial Pollutants,* Plenum Press, New York, 1984.

9.33. Brusick, D., *Principles of Genetic Toxicology,* Plenum Press, New York, 1985.

9.34. Goodenough, U. and R. P. Levine, *Genetics,* Holt, Rinehart and Winston, New York, 1974.

9.35. Rothman, K. J., *Modern Epidemiology,* Little, Brown, Boston, 1986.

9.36. *Overall Evaluations of Carcinogenicity,* IARC Monographs on the Evaluation of Carcinogenic Risks, An Updating of International Agency on Research of Cancer Monographs, Vols. 1 to 42, Suppl. 7, International Agency on Research of Cancer, 1987.

9.37. Karchmer, R. K., M. Amare, W. E. Larson, et al., "Alkylating agents as luekemogens in multiple myeloma," *Cancer,* **33,** 1103–1107 (1974).

9.38. Alderson, M., "The epidemiology of luekemia," *Adv. Cancer Res.,* **31,** 1–76 (1980).

9.39. Pegg, D. G., J. A. Zempel, W. H. Braun, et al., "Distribution of tetra(^{14}C)ethylene following oral and inhalation exposure in rats," *Toxicol. Appl. Pharmacol.,* **51,** 465–474 (1979).

9.40. Monster, A. C., "Biological monitoring of chlorinated hydrocarbon solvents," *J. Occup. Med.,* **28,** 583–588 (1986).

9.41. Kehoe, R. A., "Toxicological appraisal of lead in relation to the tolerable concentration in the ambient air," *J. Air Pollut. Control Assoc.,* **19,** 690–700 (1969).

9.42. Watson, W. S., R. Hume, and M. R. Moore, "Oral absorption of lead and iron," *Lancet,* **2,** 236–237 (1989).

Chapter 10

Quantitative Toxicology

The ability to quantify the toxicity of hazardous chemicals is essential in hazardous waste management. Industrial personnel and the public can be exposed to hazardous wastes as a result of chemical spills, breathing fugitive vapors, or the consumption of contaminated drinking water. The exposure may occur by oral, dermal, and inhalation routes, depending on the nature of the release and the contaminant characteristics; furthermore, the duration of the exposure can vary from seconds for chemical spills to a lifetime for chronic exposures to dilute concentrations of contaminants in drinking water.

This chapter will focus on the quantification of toxicity, using values that describe acute (short-term) exposures and data that have been developed for long periods of exposure. For example, the LD_{50}—the lethal dose to 50% of the population—is used to quantify acute toxicity; in contrast, Threshold Limit Values (TLVs) have been developed as guidelines for lifetime exposure limits to chemicals. Some of the systems used in quantifying toxic responses have been developed for specific environmental programs. Maximum Contaminant Levels (MCLs) are used in regulating drinking water under the Safe Drinking Water Act, while Reference Doses and Slope Factors have been developed for assessing risk at contaminated sites and RCRA facilities.

10.1 CLASSIFICATION OF TOXIC RESPONSES

The most important factor that influences toxicity is the dose, and it has been said that *the dose makes the poison*. A second important factor is the time period of exposure. Toxicity is often classified by the number or duration of exposures. For example, a worker breathing air above a tank of sulfuric acid and sodium cyanide would be exposed to a one-time, high dose of a toxic substance that has the potential to cause biochemical or cellular damage. Repeated exposures may be classified as *acute* (<5% of the organism's life span), *subchronic* (5–20% of the life span), and *chronic* (>20% of the life span). Based on the average life expectancy of humans, the traditional corresponding exposures are acute (1 day), subchronic (2 weeks to 7 years), and chronic (>7 years). A common chronic exposure to hazardous chemicals is that which occurs over a lifetime from drinking water contaminated with trace levels of hazardous chemicals.

For chemicals that are both acutely and chronically toxic, the mechanisms of toxicity for these two categories are often different. For example, acute toxicity from a large dose of chloroform is caused by effects on the central nervous system, which results in dizziness and narcosis. However, consumption of drinking water containing trace concentrations of chloroform over a lifetime causes liver damage and cancer [10.1].

Many chemicals that are acutely toxic are not chronically toxic and vice versa. For example, vitamin D, if taken in pure form, exhibits high acute toxicity. However, low doses (such as in the normal intake of milk) are not only nontoxic, but essential for good health.

Variability is inherent in biological communities, in which height, weight, and other characteristics follow a Gaussian distribution. Furthermore, variability is found within an individual's anatomy, physiology, and biochemistry (e.g., cell size, enzymatic activity, etc.). The differences in frequency of a given characteristic are usually based on genetics but may also be the result of environmental factors. Because toxicological responses result from detrimental effects on cellular or tissue systems, it follows that variability will also be found in toxic responses. The first step in quantifying such a variable parameter as toxicity is designating the response that will be measured. In classification by *end point*, the toxic effect is the basis for classification; for example, the end point can be cancer or a noncancerous effect. Toxicity may be *temporary* or *permanent*, or it may be *latent*, which means that its onset is delayed. Some effects may be classified by *irreversible* versus *reversible* effects. For example, cholinesterase inhibition induced by parathion is reversible, whereas the onset of cancer is irreversible.

In classifying toxic actions based on *developmental effects*, the life stage at which toxicity occurs, from fertilization until maturation to an adult, is the basis for classification. *Reproductive effects* are associated with the process of reproduction through *teratogenicity*, which is toxicity to the fetus during pregnancy.

In quantifying toxic responses, numerical values may be absolute or normalized for body weight. *Dosage* is often defined as the total mass of chemical to which an organism is exposed. *Dose*, on the other hand, is the chemical dosage normalized for body weight. The difference between these terms is illustrated in Example 10.1.

EXAMPLE 10.1   *Dose vs. Dosage*

A dose of 8 mg/kg of lindane is being administered to a 3-kg (6.6-lb) animal. What is the dosage?

SOLUTION

The dosage is weight-specific. Based on a dose of 8 mg/kg and 3 kg body mass, the dosage = 8 mg/kg · 3 kg = 24 mg.

Some definitions used for quantifying dose-response relationships are related to absorption. *Administered dose* is the concentration to which the organism is exposed. *Intake* or *uptake dose* is the actual amount of actual chemical absorbed by the organism. *Target* or *effective dose* is the amount of chemical reaching the target organ. Unless otherwise specified, the dose in hazardous waste management is assumed to be administered dose; that is, the total mass of chemical to which the organism is exposed. Although other dose conventions are used, administered dose is the most practical for environmental applications because it is a value that is easily attainable through sampling or modeling releases from a source.

10.2   ACUTE TOXICITY: THE LD$_{50}$

Because of the inherent diversity of biological populations, an absolute value of acute toxicity is impossible. For a given dose, some organisms are more susceptible to the exposure, while others are less affected. Regardless of the stimulus, populations of organisms normally respond as a Gaussian, or bell-shaped, distribution (Figure 10.1), and quantitation of acute toxicity is based on this distribution. Toxicity data are obtained through animal studies in which the concentration of a chemical given to groups of organisms is varied. A plot of percent mortality as a function of dose then provides a means for determining acute toxicity, a procedure that is described in Section 10.3. The effect of dose on a population may be viewed by plotting the cumulative response as a function of dose or the natural logarithm of dose (Figure 10.2). Such a plot is characterized by a sigmoidal shape with a log normal distribution symmetrical about the midpoint of the curve. The midpoint is defined as the *LD$_{50}$*—the dose that results in death to 50% of the population. In a more general sense, the midpoint of the dose-response curve may be considered the median *effective dose (ED$_{50}$)* for the characteris-

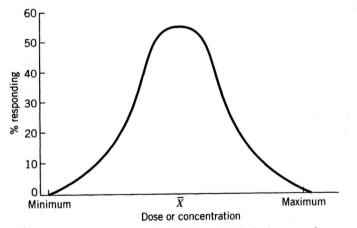

Figure 10.1 Variation of characteristics in a biological community or biochemical system described by a Gaussian distribution.

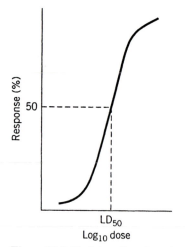

Figure 10.2 Toxicity effects based on a natural log cumulative Gaussian distribution.

tic being tested (e.g., onset of dizziness, promotion of chloracne). Names are given to groups found at the extreme ends of the dose-response curve. The portion of the population that responds to low dosages is called *hyperactive* or *hypersensitive*, and the group that does not respond until high doses are reached is *hypoactive* or *hyposensitive*. The standard measure of acute toxicity, the LD_{50}, is expressed in mg/kg body weight, and is, therefore, independent of the organism's weight. A wide range of LD_{50} values are commonly found from 1 mg/kg to grams/kg. Representative ranges of LD_{50} are listed in Table 10.1, and typical oral LD_{50} data for rats are listed in Table 10.2.

Although the LD_{50} is the most common measure of acute toxicity, other regions of the dose-response curve are sometimes quantified (e.g., LD_5, LD_{10}, LD_{90}, LD_{99}). Ratios of some of these values are used to evaluate the margin of safety and potential effects on hyposensitive and hypersensitive populations. More detailed information on the evaluation of acute toxicity data may be found in References 10.3–10.5.

Another measure of toxicity is the LC_{50}, which is used to quantify the lethality of contaminants in the air and water to which organisms are exposed. The units of LC_{50} are mg/m^3 for air or mg/L for water.

Table 10.1 Hodge-Sterner Table for Degree of Toxicity

Experimental LD_{50} Dose per kg Body Weight	Degree of Toxicity	Probable Lethal Dose for a 70-kg Person
<1.0 mg	Dangerously toxic	A taste
1.0–50 mg	Seriously toxic	A teaspoonful
50–500 mg	Highly toxic	An ounce
0.5–5 g	Moderately toxic	A pint
5–15 g	Slightly toxic	A quart
>15 g	Extremely low toxicity	More than a quart

Source: Reference 10.2.

Table 10.2 LD_{50} Values (Oral Rat) of Some Common Hazardous Compounds

Compound	LD_{50} (oral rat) (mg/kg)
Aliphatic Hydrocarbons	
Cyclohexane	29,820
Cyclohexene	460
Monocyclic Aromatic Hydrocarbons	
Benzene	3,306
Toluene	5,000
Ethylbenzene	3,500
o-Xylene	5,000
m-Xylene	5,000
p-Xylene	5,000
Polycyclic Aromatic Hydrocarbons	
Naphthalene	490
Phenanthrene	700 (mouse)
Pyrene	2,700
Nonhalogenated Solvents	
Acetone	5,800
Methyl butyl ketone	2,590
Methyl ethyl ketone	2,737
Methyl isobutyl ketone	2,080
Chlorinated Solvents	
Carbon tetrachloride	2,350
Chloroform	908
Methylene chloride	1,600
Perchloroethylene (PCE)	2,629
1,1,2-TCA	836
TCE	22,402 (mouse)
Insecticides	
Aldrin	39
Carbaryl	230
Dieldrin	38.3
DDT	87
Hexachlorobenzene	10,000
Lindane	100
Malathion	290
Parathion	2

Table 10.2 (Continued)

Compound	LD$_{50}$ (oral rat) (mg/kg)
Herbicides	
Atrazine	672
2,4-D	831
Trifluralin	5,000 (mouse)
Fungicides	
Pentachlorophenol	27
Industrial Intermediates	
Chlorobenzene	2,290
2,4-Dichlorophenol	580
Dimethyl phthalate	6,800
Di-*n*-octyl phthalate	6,510 (mouse)
Hexachlorocyclopentadiene	1,300
Phenol	317
2,4,6-Trichlorophenol	820
Explosives	
2,4,6-Trinitrotoluene	795
Polychlorinated Biphenyls	
PCB 1016	2,300
PCB 1232	4,470
PCB 1248	11,000
PCB 1254	1,010
PCB 1260	1,320
Chlorinated Dioxins	
2,3,7,8-Tetrachlorodibenzo-*p*-dioxin (TCDD)	0.020

In addition to the LD$_{50}$, other classifications for acute toxicity have been described, which are listed in Table 10.3. Some of these other classifications include a dose that promotes a toxic response (TD) or causes death (LD). Similar data have been obtained for toxic concentrations in air (TC and LC). The subscript *Lo* is the lowest dose to produce the toxic effect.

Although the use of animals is the only practical means of evaluating human toxicity, many shortcomings are inherent in the use of animals as surrogates, including (1) the low degree of precision that results from the logistics of using small sample sizes of test animals, (2) uncertainty whether the same biochemical and physiological responses occur in animal as humans, and (3) extrapolation of animal dose-response

Table 10.3 Classifications for Acute Toxicity

Parameter	Description
TD$_{Lo}$ (toxic dose low)	The lowest dose of a material introduced by any route, other than inhalation, over any given period of time and reported to produce any toxic effect in humans or to produce carcinogenic, neoplastigenic, or teratogenic effects in animals or humans.
TC$_{Lo}$ (toxic concentration low)	The lowest concentration of a material in air to which humans or animals have been exposed for any given period of time that has produced any toxic effect in humans or produced a carcinogenic, neoplastigenic, or teratogenic effect in animals or humans.
LD$_{50}$ (lethal dose fifty)	A calculated dose of a material that is expected to cause the death of 50% of an entire defined experimental animal population. It is determined from the exposure to the material by any route other than inhalation of a significant number from that population. Other lethal dose percentages, such as LD$_1$, LD$_{30}$, and LD$_{99}$, may be published in the scientific literature for the specific purposes of the author. Such data would be published if these figures, in the absence of a calculated lethal dose (LD$_{50}$), were the lowest found in the literature.
LC$_{Lo}$ (lethal concentration low)	The lowest concentration of a material in air, other than LC$_{50}$, that has been reported to have caused death in humans or animals. The reported concentrations may be entered for periods of exposure that are less then 24 hours (acute) or greater than 24 hours (subacute and chronic).

Source: Reference 10.2.

relationships to humans. For most chemicals, the LD$_{50}$ has been measured on a variety of animals by a number of routes of exposure. Very few LD$_{50}$ data are available for humans because of the ethics of such toxicity testing, although some direct estimates are available from homicides and poisonings. How, then, are we able to relate animal toxicity data to humans? Animal data for LD$_{50}$ by the same route of exposure are often used as an *estimate* of human toxicity, although in many cases different enzyme systems and metabolic pathways will render such estimates invalid. As a general rule, if a chemical has the same degree of toxicity for all animal species tested, it will likely exhibit similar toxicity in humans [10.6]. If the LD$_{50}$ values from different species vary widely, an estimate of human toxicity is difficult without detailed metabolic studies. The consensus among toxicologists is that humans are more susceptible to toxicity than the most sensitive species tested. Many toxicologists divide the animal LD$_{50}$ by 10 to account for safety in assessing potential hazards to humans [10.7].

Epidemiological studies, which statistically relate hazardous chemical exposure to a toxicological response, can provide evidence confirming the data from animal studies. However, epidemiological studies, in and of themselves, cannot provide a toxicological quantification. Numerous uncontrollable factors are inherent in human populations (e.g., smoking, alcohol consumption, etc.) that can confound epidemiological results [10.8].

The use of LD_{50} and LC_{50} information is somewhat limited in most hazardous waste investigations (e.g., site assessments, risk assessments, RCRA facility siting, etc.) because exposures to hazardous chemicals are usually chronic. However, the assessment of hazardous materials spills and other emergency response incidents relies strongly on acute toxicity data. For example, if a highway tanker carrying parathion is involved in an accident, knowledge of its high acute toxicity ($LD_{50} = 2$ mg/kg) is essential to emergency response plans at the site.

10.3 QUANTITATIVE EVALUATION OF ACUTE TOXICITY

The general approach used in toxicity testing is to divide a population of animals into a number of subgroups and administer succeedingly higher dosages of the chemical to each group. After the animals are dosed, the number of deaths within each group is recorded over a specified time period (e.g., 14 days). At low doses, some groups may show no toxic effect; however, as the dose increases, a few toxic effects. will be seen. Finally, at high doses, all of the organisms in the test group will experience the toxic response.

Experimental data from toxicity testing may be displayed in a number of ways. A common method of displaying toxicity data is a *frequency distribution*:

$$\text{frequency}_n = \%\ \text{deaths}_n - \%\ \text{deaths}_{n-1} \tag{10.1}$$

where n = the test group number.

The mean and standard deviation (σ) are normally determined for each dose. The mean and standard deviation of the distribution are quantified by

$$\bar{x} = \frac{\sum_{i=1}^{n} x_i f(x_i)}{\sum_{i=1}^{n} f(x_i)} \tag{10.2}$$

$$s^2 = \frac{\sum_{i=1}^{n} (x_i - \bar{x})^2 f(x_i)}{\sum_{i=1}^{n} f(x_i)} \tag{10.3}$$

where

$s = \sqrt{s^2}$ = standard deviation

$\bar{x}$ = the mean

A dose-response curve may then constructed by plotting the cumulative mean at each dose with error bars (±s) at each point. As an alternative, the data may be plotted as the natural logarithm of the dose, which results in a near-straight line. The point on the curve that has most toxicological significance is the dose that results in lethality to 50% of the population—the LD_{50}.

The frequency data are sometimes plotted as a function of the logarithm of dose, and this dose-response function is quantitatively described as a Gaussian distribution. The normal distribution of such a population is described by

$$f(x) = \frac{1}{s\sqrt{2\pi}} e^{-\frac{1}{2}\left(\frac{x-\bar{x}}{s}\right)^2}$$ (10.4)

The apex of the curve is the dose that represents 50% lethality, and this dose is defined as the LD_{50}. The statistical analysis of toxicological data is demonstrated in Example 10.2.

EXAMPLE 10.2   *Statistical Analysis of Toxicity Data*

The following acute toxicity data were obtained in evaluating a new insecticide.

Dose (mg/kg)	No. of Lethal Responses
0	0
5	4
10	9
15	14
20	16
25	11
30	10
35	7
40	6
45	2
50	1

1. Determine the mean and standard deviation for the data.
2. Plot a histogram of the data and the normal distribution for the data.

SOLUTION

1. The mean is calculated using Equation 10.2.

$$\bar{x} = \frac{\displaystyle\sum_{i=1}^{n} x_i f(x_i)}{\displaystyle\sum_{i=1}^{n} f(x_i)}$$

$$\bar{x} = \frac{(0 \cdot 0) + (5 \cdot 4) + (10 \cdot 9) + (15 \cdot 14) + (20 \cdot 16) + (25 \cdot 11) + (30 \cdot 10) + (35 \cdot 7) + (40 \cdot 6) + (45 \cdot 2) + (50 \cdot 1)}{0 + 4 + 9 + 14 + 16 + 11 + 10 + 7 + 6 + 2 + 1}$$

$$\bar{x} = 23$$

The standard deviation can be determined by using Equation 10.3.

$$s^2 = \frac{\sum_{i=1}^{n} (x_i - \bar{x})^2 f(x_i)}{\sum_{i=1}^{n} f(x_i)}$$

$$s^2 = [(5 - 23)^2(4) + (10 - 23)^2(9) + (15 - 23)^2(14) + (20 - 23)^2(16) + (25 - 23)^2(11) + (30 - 23)^2(10) + (35 - 23)^2(7) + (40 - 23)^2(6) + (45 - 23)^2(2) + (50 - 23)^2(1)] / 80 = 110.4$$

$$s = \sqrt{s^2} = 10.5$$

2. The histogram is generated by plotting the number of lethal responses as a function of dose; the normal distribution may be shown on the same graph.

$$f(x) = \frac{1}{s\sqrt{2\pi}} e^{-\frac{1}{2}\left(\frac{x - \bar{x}}{s}\right)^2}$$

$$f(x) = \frac{1}{(10.5)\sqrt{2\pi}} e^{-\frac{1}{2}\left(\frac{x - 23}{10.5}\right)^2} = 0.0380e^{-0.00454(x-23)^2}$$

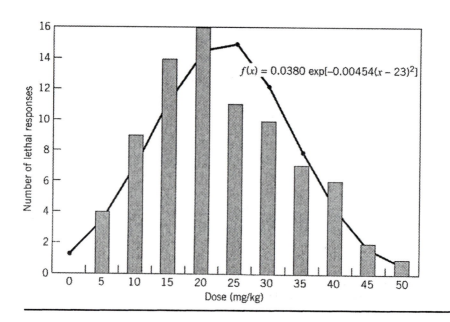

The analysis of experimental toxicity data using the Gaussian distribution to obtain an LD_{50} is somewhat limited. A more effective approach is the use of cumulative frequency data, which are plotted as a function of dose (see Figure 10.2). Cumulative frequency data are usually linear in the middle region of the plot and are asymptotic at the two extremes of the curve. As a result, a no-effect threshold, if one exists, cannot be quantified from such experimental data.

Probit Analysis. In computing the LD_{50} and other toxicological parameters, use of a cumulative dose-response curve provides only an approximate solution; therefore, other analytical solutions have been developed. One of the most common algorithms used for evaluating dose-response relationships is the probit analysis. The *probit*, or probability unit, provides a basis for transforming the sigmoidal curves of dose-response relationships to straight lines (Figure 10.3). The probability unit Y may be related to the probability P by

$$P = \frac{1}{\sqrt{2\pi}} \int_{-\infty}^{Y-5} e^{-x^2/2} dx \qquad (10.5)$$

The relationships between probability and the probit variable are illustrated in Figure 10.4. For most analyses, a probit table (Table 10.4) is used to transform raw dose-response data into probits. The mean, or LD_{50}, is assigned a value of 5; each $+s$ adds a value of 1 to the probit and each $-s$ results in -1 to the initial value of 5. The advantage of the probit analysis is the complete linear transformation of the dose-response relationship, making linear regression and statistical analyses straightforward to determine the LD_{50}, slope, and confidence limits. The use of probit analysis in evaluating dose-response relationships is demonstrated in Example 10.3.

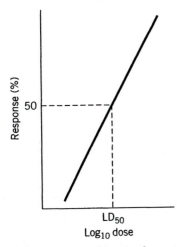

Figure 10.3 Probit transformation of the cumulative dose-response curve.

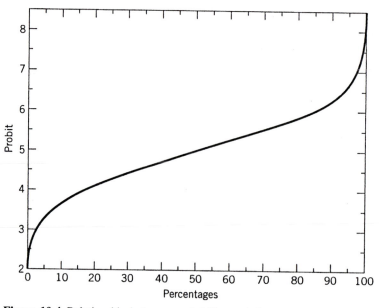

Figure 10.4 Relationship between probability and the probit value. *Source:* Reference 10.9.

Table 10.4 The Transformation from Percentages to Probits

%	0	1	2	3	4	5	6	7	8	9
0	—	2.67	2.95	3.12	3.25	3.36	3.45	3.52	3.59	3.66
10	3.72	3.77	3.82	3.87	3.92	3.96	4.01	4.05	4.08	4.12
20	4.16	4.19	4.23	4.26	4.29	4.33	4.36	4.39	4.42	4.45
30	4.48	4.50	4.53	4.56	4.59	4.61	4.64	4.67	4.69	4.72
40	4.75	4.77	4.80	4.82	4.85	4.87	4.90	4.92	4.95	4.97
50	5.00	5.03	5.05	5.08	5.10	5.13	5.15	5.18	5.20	5.23
60	5.25	5.28	5.31	5.33	5.36	5.39	5.41	5.44	5.47	5.50
70	5.52	5.55	5.58	5.61	5.64	5.67	5.71	5.74	5.77	5.81
80	5.84	5.88	5.92	5.95	5.99	6.04	6.08	6.13	6.18	6.23
90	6.28	6.34	6.41	6.48	6.55	6.64	6.75	6.88	7.05	7.33

%	0.0	0.1	0.2	0.3	0.4	0.5	0.6	0.7	0.8	0.9
99	7.33	7.37	7.41	7.46	7.51	7.58	7.65	7.75	7.88	8.09

Source: Reference 10.9.

EXAMPLE 10.3 *Evaluating Dose-Response Relationships Using Probit Analysis*

The following data were obtained in a toxicity study on grasshoppers using malathion. Each test group contained 50 grasshoppers.

Dose of Malathion Applied to Grasshoppers (mg/kg)	Number of Deaths
0	0
4	7
7	17
12	26
18	43
23	47

Determine the LD$_{50}$ for these data using probit analysis.

SOLUTION

Convert these data to a probit value (Table 10.4) and plot the probit as a function of log dose. Fit the resulting data to a straight line and determine the equation.

Dose of Malathion Applied to Grasshoppers (mg/kg)	Percentage of Deaths	Probit Value
0	0	—
4	14	3.92
7	34	4.59
12	52	5.05
18	86	6.08
23	94	6.55

The LD$_{50}$ is determined at a probit value of 5 (or 50%). Substituting this value into a linear equation with slope = 3.42 and y-intercept = 1.73 yields

$$5 = 3.42 \cdot (\log \text{LD}_{50}) + 1.73$$

$$\text{LD}_{50} = 9.1 \text{ mg/kg}$$

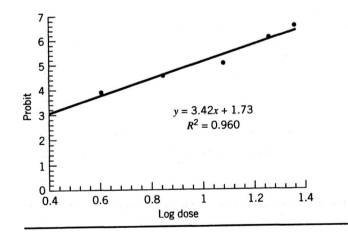

$y = 3.42x + 1.73$
$R^2 = 0.960$

The slope of the dose-response curve is important, because it provides a general indication of a compound's margin of safety. The higher the slope, the greater the change in toxicity with dose, and less the margin of safety. Dose-response data may be a lethal response or some physiological end point such as appearance of lesions, increase in heart rate or onset of chloracne. The dose that provides such a nonlethal response is called the *effective dose*. A related parameter is the *therapeutic index*, which is defined as

$$TI_{50} = \frac{ED_{50}}{LD_{50}} \tag{10.6}$$

where ED_{50} = the effective dose for 50% of the population.

The therapeutic index was developed for drug responses as a margin of safety; that is a larger margin for a dose to provide an effective response without causing lethality results in a higher therapeutic index, or margin of safety.

Potency is the relative toxicity between two or more compounds based on their dose-response curves. The relationship between the two curves in Figure 10.5 shows that the two compounds have different potencies. Although the minimal toxicity is found for compound A at lower doses, the relative toxicity is reversed at higher concentrations and the potency of compound B is higher. Ratios from different regions of the dose-response curve (e.g., LD_{90}/LD_{10}) are often used to describe the potency of different compounds [10.10].

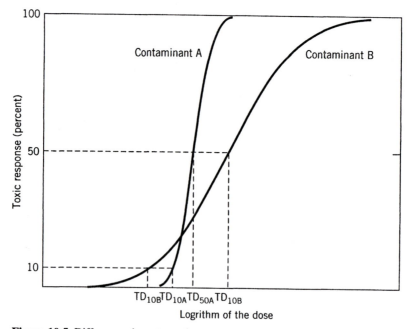

Figure 10.5 Differences in potency between two compounds based on slopes of dose-response functions.

10.4   CHRONIC INDUSTRIAL EXPOSURE: THE THRESHOLD LIMIT VALUE

Chronic toxicity in hazardous waste management is most often caused by long-term, low-level exposure to hazardous chemicals. Chronic toxicity is difficult to quantify, because less is known about the long-term effects of chemicals compared to acute toxicity evaluations. A typical dose-response curve for chronic toxicity is shown in Figure 10.6. At low contaminant concentrations, essentially no evidence of toxicity is exhibited, and this is called the *no-effect region.* The basis for the lack of response from zero dose to the threshold dose is a biochemical or physiological defense, such as detoxification or excretion, that prevents the effect from occurring. The middle segment rises from a point where some toxicity is shown (the threshold), and then increases in a sigmoidal fashion until a maximum effect is found. As the exposure to the chemical is increased, the detoxification mechanisms are overwhelmed and the effects increase as a function of dose. Eventually, all of the toxicological effects are exhibited.

A parameter used by industrial hygienists, the *Threshold Limit Value (TLV)*, is the concentration of a chemical in air to which workers may be exposed safely over their occupational lifetimes (that is, for 8 hours per day, 5 days per week, over a working lifetime). Threshold Limit Values have been defined for hundreds of airborne contaminants. Some contaminants (e.g., carcinogens) have been given a zero TLV because no threshold exists. The TLV is a copyrighted term of the American Conference of Governmental Industrial Hygienists (ACGIH), a group that develops policy and continually updates TLV data on an annual basis.

Epidemiological data have been used extensively to develop TLVs. Individual cases involving human exposures are also used; however, perhaps the most common data used for TLV development have been from animal toxicity studies. The ACGIH has preferred the use of data from chronic inhalation studies using several animal species, while the least desirable data are from acute oral toxicity studies. Documentation of TLV values (i.e., how these values have been determined) is available in the ACGIH publication *Threshold Limit Values and Biological Exposure Indices* [10.11], which is

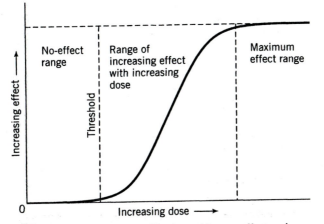

Figure 10.6 Dose-response function with a no-effect region.

revised periodically. Because TLVs are updated so often, the year is usually reported along with the TLV value.

The intention of the ACGIH has been that TLVs should be used as guidelines for worker safety but not as strict regulatory standards. Nonetheless, TLVs have been incorporated into state and federal regulations. For example, the 1968 TLVs became part of the Occupational Safety and Health Act as *Permissible Exposure Limits (PELs)*. The data for PELs are not as widespread or up to date as TLVs.

Three categories of TLVs have been established: (1) the Time-Weighted Average (TLV-TWA), (2) the Short-Term Exposure Limit (TLV-STEL), and (3) the ceiling (TLV-C). The *Time-Weighted Average* is the TLV that is averaged for the 8-hour work day or 40-hour work week to which individuals may be exposed over long periods of time without adverse effects. The *Short-Term Exposure Limit* is the maximum concentration to which individuals may be exposed for repeated periods of up to 15 minutes without suffering from (1) irritation, (2) chronic tissue damage, or (3) narcosis or dizziness. The STEL guideline states that an individual's exposure should not be greater than four exposures of 15 minutes within 1 day. The TLV *ceiling value* is the concentration that should not be exceeded at any time.

The use of the Time Weighted Average (TWA), which has been adopted by OSHA, aids in assessing exposure at different periods of time:

$$\text{TWA} = \frac{C_a T_a + C_b T_b + \cdots + C_n T_n}{8} \qquad (10.7)$$

where

T_a = time of the first exposure period (hours)
C_a = contaminant concentration during the first period
T_b = succeeding time periods (hours)
C_b = contaminant concentrations during the succeeding time periods
T_n = final time period (hours)
C_n = contaminant concentration during final time period

Not only are TLV data available for gaseous and particulate airborne chemicals, they are also available for physical factors such as heat, ultraviolet light, and ionizing radiation. Threshold limit values are usually listed in units of both ppm and mg/m^3, with particulates listed as mg/m^3 only. More TLV-TWA data are generally available than for TLV-STELs and TLV-Cs. The ACGIH has also developed *excursion limits* as a general rule for assessing compounds for which no STELs are available. One excursion limit guideline is at no time should the exposure exceed five times the TLV-TWA. Threshold Limit Values for common hazardous chemicals are listed in Table 10.5.

Although more applicable to industrial monitoring, there may be cases in hazardous waste management when individuals are exposed to hazardous chemicals during site assessments, monitoring, solvent recycling, and the management of RCRA facilities. The most common use of TLVs involves collecting air monitoring data, which are then compared to the TLV-TWA. The actual TWA may be obtained from air monitoring data collected over discrete periods of time.

Most areas containing hazardous wastes have more than one chemical present. The basis for evaluating TLVs for mixtures is to assume that their toxic effects are additive.

Table 10.5 Threshold Limit Values of Some Common Hazardous Compounds

Compound	Threshold Limit Value (TWA)
Aliphatic Hydrocarbons	
Cyclohexane	300 ppm
Cyclohexene	300 ppm
n-Heptane	400 ppm
n-Octane	300 ppm
Monocyclic Aromatic Hydrocarbons	
Benzene	10 ppm
Toluene	100 ppm
Ethylbenzene	100 ppm
o-Xylene	100 ppm
m-Xylene	100 ppm
p-Xylene	100 ppm
Polycyclic Aromatic Hydrocarbons	
Chrysene	Suspected human carcinogen
Naphthalene	10 ppm
Nonhalogenated Solvents	
Acetone	750 ppm
Methyl butyl ketone	5 ppm
Methyl ethyl ketone	200 ppm
Methyl isobutyl ketone	50 ppm
Chlorinated Solvents	
Carbon tetrachloride	5 ppm
Chloroform	10 ppm
Methylene chloride	50 ppm
Perchloroethylene (PCE)	50 ppm
1,1,2-TCA	10 ppm
TCE	50 ppm
Insecticides	
Aldrin	0.25 mg/m^3
Carbaryl	5 mg/m^3
Dieldrin	0.25 mg/m^3
DDT	1 mg/m^3
Hexachlorobenzene	0.025 mg/m^3
Malathion	10 mg/m^3
Parathion	0.1 mg/m^3

Table 10.5 (Continued)

Compound	Threshold Limit Value (TWA)
Herbicides	
Atrazine	5 mg/m^3
2,4-D	0.1 mg/m^3
Fungicides	
Pentachlorophenol	0.5 mg/m^3
Industrial Intermediates	
Chlorobenzene	10 ppm
Dimethyl phthalate	5 mg/m^3
Hexachlorocyclopentadiene	0.01 ppm
Phenol	5 ppm
Explosives	
Picric acid	0.1 mg/m^3
2,4,6-Trinitrotoluene	0.5 mg/m^3
Polychlorinated Biphenyls	
PCB 1254	0.5 mg/m^3

The TLV-TWA for mixtures of hazardous chemicals may be calculated by

$$\text{TLV-TWA}_{\text{mix}} = \frac{\sum\limits_{i=1}^{n} C_i}{\sum\limits_{i=1}^{n} \dfrac{C_i}{\text{TLV-TWA}}} \qquad (10.8)$$

where

n = the number of hazardous chemicals
C_i = the concentration of chemical i

The value for TLV-TWA$_{\text{mix}}$ obtained in Equation 10.8 is compared to the sum of the concentrations of the same compounds found in the area. If the sum is greater than the TLV-TWA$_{\text{mix}}$ calculated in Equation 10.8, then the exposure is unhealthful based on Threshold Limit Values. The evaluation of TLVs is demonstrated in Example 10.4.

EXAMPLE 10.4   *Determination of TLV-TWA for a Mixture of Hazardous Chemicals*

Determine the mixture TLV-TWA for a worker exposed to 42 ppm PCE, 38 ppm TCE, 86 ppm MEK, and 35 ppm MIBK in the air of a RCRA solvent recycling operation. Are the concentrations of chemicals exceeding the TLV-TWA$_{mixture}$?

SOLUTION

From Table 10.5, the TLVs for the chemicals are

Compound	TLV (ppm)
PCE	50
TCE	50
MEK	200
MIBK	50

Using Equation 10.8, these values are used to determine the TLV-TWA$_{mixture}$ over an 8-h work day.

$$\text{TLV-TWA}_{mix} = \frac{42 + 38 + 86 + 35}{\dfrac{42}{50} + \dfrac{38}{50} + \dfrac{86}{200} + \dfrac{35}{50}}$$

$$\text{TLV-TWA}_{mix} = 73.6 \text{ ppm}$$

The total exposure $= 42 + 38 + 86 + 35 = 201$ ppm

Because the actual exposure of 201 ppm is greater than the TLV-TWA$_{mix}$ of 73.6 ppm, the air in the solvent recycling facility is unhealthful based on Threshold Limit Values.

10.5   MAXIMUM CONTAMINANT LEVELS

As part of the Safe Drinking Water Act, *Maximum Contaminant Levels (MCLs)* were established to protect the health of the public who drink water over a lifetime. The regulated contaminants and their corresponding MCLs are listed in Table 10.6. Note that the MCLs of many of these common hazardous chemicals are in the low μg/L range, which makes analytical sensitivity and quality control a necessity in their chemical analyses.

The MCLs are based on Maximum Contaminant Level Goals (MCLGs), which are nonenforcable goals based on extremely low risk. The EPA has been given the directive to set MCLs as close to MCLGs as possible. In many cases, MCLs do not correspond to the 1×10^{-6} level of cancer risk [based on an intake of 2 L (0.53 gal) of water per day], but instead include a level of pragmatism dictated by water treatment

Table 10.6 Maximum Contaminant Levels for Selected
SDWA Compounds

Compound	MCL (μg/L)
Inorganic Chemicals	
Antimony	6.0
Arsenic	50
Asbestos	7 MFL*
Barium	2,000
Beryllium	1.0
Cadmium	5.0
Chromium	100
Cyanide	200
Fluoride	4,000
Mercury	2.0
Nickel	100
Selenium	50
Thallium	2.0
Chlorinated Solvents and Related Chemicals	
Carbon tetrachloride	5.0
1,2-Dichloroethane	5.0
1,1-Dichloroethylene	7.0
cis-1,2-Dichloroethylene	70
trans-1,2-Dichloroethylene	100
Dichloromethane	5.0
Perchloroethylene	5.0
1,1,1-Trichloroethane	200
1,1,2-Trichloroethane	5.0
Trichloroethylene	5.0
Vinyl chloride	2.0
Aromatic Hydrocarbons	
Benzene	5.0
Benzo[a]pyrene	0.2
Ethylbenzene	700
Styrene	100
Toluene	1,000
Xylenes (total)	10,000
Chlorinated Organic Compounds	
Chlorobenzene	100
o-Dichlorobenzene	600
p-Dichlorobenzene	75
Hexachlorobenzene	1.0
Hexachlorocyclopentadiene	50

Table 10.6 (Continued)

Compound	MCL (μg/L)
Chlorinated Organic Compounds (cont.)	
PCBs	0.5
2,3,7,8-TCDD	5.0×10^{-5}
1,2,4-Trichlorobenzene	70
Pesticides and Herbicides	
Alachlor	2.0
Aldicarb	3.0
Aldicarb sulfone	3.0
Aldicarb sulfoxide	4.0
Atrazine	3.0
Carbofuran	40
Chlordane	2.0
2,4-D	70
Dibromochloropropane (DBCP)	0.2
1,2-Dichloropropane	5.0
Endrin	0.2
Ethylene dibromide (EDB)	0.05
Heptachlor	0.4
Heptachlor epoxide	0.2
Lindane	0.2
Methoxychlor	40
Pentachlorophenol	1.0
Picloram	500
Simazine	4.0
Toxaphene	3.0
2,4,5-TP (Silvex)	50
Other Organic Compounds	
Dalapon	200
Di(2-ethylhexyl)adipate	500
Di(2-ethylhexyl)phthalate	6.0
Dinoseb	7.0
Endothall	100
Glyphosate	700
Oxamyl (Vydate)	200

*Million fibers per liter

technologies and analytical detection limits [10.12]. The primary use of MCLs in hazardous waste management is their application as Applicable or Relevant and Appropriate Requirements (ARARs) as groundwater cleanup criteria at Superfund sites. In addition, many state and local regulatory authorities use MCLs as de facto cleanup criteria for contaminated groundwater.

10.6 QUANTIFYING THE CHRONIC TOXICITY OF NONCARCINOGENS

Detrimental effects caused by all chemicals, with the exception of those that induce cancer, are classified as noncarcinogenic toxicities. The toxic end points resulting from such chronic toxic actions vary widely as described in Section 9.5. The most common effects are due to interactions between the chemical and biological receptor molecules, especially enzymes; that is, the toxic chemical binds to an important enzyme, resulting in a negative effect on its function. Important noncarcinogenic toxicant-enzyme interactions include the inhibition of acetylcholinesterase by organophosphate ester and carbamate insecticides, the binding of carbon monoxide to hemoglobin, and the binding of cyanide to cytochromes.

The toxicity of noncarcinogens is based on the concept of threshold in a manner similar to the development of TLVs (i.e., there is a dose below which there is no observable effect and therefore no short- or long-term effects on the organism). The mechanism may be visualized on a molecular basis as millions of receptor biomolecules being available for a given function (e.g., nerve transmission, transport of oxygen) and the binding of a toxic chemical to a small number of these receptors not producing a measurable effect. Once again, a good example of no-effect toxicity is the inactivation of a very small number of molecules of acetylcholinesterase.

The No Observed Adverse Effect Level. An important parameter in the management of hazardous wastes is a no-effect level to which a population is exposed. Such a quantitative value is difficult to measure and, as a result is also difficult to define accurately. Therefore, no-effect levels are estimated based on epidemiological data and controlled animal experiments. These studies are designed to determine the highest dose that will not produce an effect, which is defined as the *No Observed Effect Level (NOEL)*. The *No Observed Adverse Effect Level (NOAEL)* is a variation of the NOEL in that it classifies only toxicological effects. Other measures related to the NOEL and the NOAEL are the LOEL *(Lowest Observed Effect Level)* and the LOAEL *(Lowest Observed Adverse Effect Level)*, a stricter version of the LOEL.

Acceptable Daily Intakes and Reference Doses. The *Acceptable Daily Intake (ADI)* is a level of daily ingestion or inhalation of a toxic compound that does not produce an adverse health effect. The ADI is based on NOAELs but is not considered an absolute physiological threshold. The development of ADIs has been based on safety factors that reflect variations in the population. Therefore, values for ADIs are significantly lower than corresponding NOAELs [10.13].

Reference doses (RfDs) are a regulatory parameter based on NOAELs. The RfD is a surrogate used by the EPA in place of the ADI (which sometimes results in lower values). The use of safety factors to account for hypersensitivity is inherent in the derivation of RfDs. Reference doses usually include two safety factors: (1) a factor of 10 for variation among individuals and (2) another factor of 10 for extrapolation from animals to humans during experimental determination [10.14]. Another factor sometimes used with the RfD is a "modifying factor" ranging from 1 to 10 based on professional judgment. The procedure for establishing reference doses is somewhat more detailed than for ADI development including use of the most sensitive species, the appropriate route of exposure, and the most sensitive end point [10.15]. The procedure for assessing noncarcinogenic risk using RfDs is covered in Chapter 11.

Reference doses, which have become a widely used indicator of chronic toxicity, have been established for oral and inhalation routes. Reference doses for over 400 chemicals have been reported by the EPA [10.16] and are listed in Appendix L. Reference doses for the most common hazardous compounds are listed in Table 10.7.

Table 10.7 Oral Reference Doses and Slope Factors of Some Common Hazardous Compounds

Compound	Reference Dose (RfD) $(mg/kg \cdot d)$	Slope Factor (SF) $(mg/kg \cdot d)$
Monocyclic Aromatic Hydrocarbons		
Benzene	Pending	0.029
Toluene	0.2	—
Ethylbenzene	0.1	—
Xylenes	2.0	—
Polycyclic Aromatic Hydrocarbons		
Anthracene	0.3	—
Fluorene	0.04	None
Naphthalene	Pending	None
Pyrene	0.03	None
Nonhalogenated Solvents		
Acetone	0.1	—
Methyl ethyl ketone	0.6	None
Methyl isobutyl ketone	Withdrawn	—
Chlorinated Solvents		
Carbon tetrachloride	7.0×10^{-4}	0.13
Chloroform	0.01	0.0061
Perchloroethylene (PCE)	0.01	Pending
1,1,2-TCA	0.004	0.057
Trichloroethylene (TCE)	Pending	Withdrawn
Insecticides		
Aldrin	3.0×10^{-5}	17
Carbaryl	0.1	—
Dieldrin	5.0×10^{-5}	16
DDT	5.0×10^{-4}	0.34
Hexachlorobenzene	8.0×10^{-4}	1.6
Lindane	3.0×10^{-4}	—
Malathion	0.02	—
Parathion	Pending	—

Table 10.7 (Continued)

Compound	Reference Dose (RfD)	Slope Factor (SF)
Herbicides		
Atrazine	0.035	Pending
2,4-D	0.01	—
2,4,5-T	0.01	—
Trifluralin	0.0075	0.0077
Fungicides		
Pentachlorophenol	0.03	0.12
Industrial Intermediates		
Chlorobenzene	0.02	—
2,4-Dichlorophenol	0.003	—
Dimethyl phthalate	Pending	—
Hexachlorocyclopentadiene	0.007	None
Phenol	0.6	None
2,4,6-Trichlorophenol	—	0.011
Explosives		
2,4,6-Trinitrotoluene	5.0×10^{-4}	0.03
Polychlorinated Biphenyls		
PCB 1016	7.0×10^{-5}	—
PCB 1248	Pending	—
PCB 1254	2.0×10^{-5}	—

Source: Reference 10.16.

10.7 DOSE-RESPONSE RELATIONSHIPS FOR CARCINOGENS

Two methods have been used to evaluate human carcinogenicity—epidemiological studies and animal studies. Of these two types of evaluations, the use of animals is considered more reliable among toxicologists. Epidemiological studies, which are usually based on random data, provide less evidence of cause-effect relationships. Because of the difficulty in confirming human carcinogenicity, the International Agency for Research on Cancer (IARC), an organization that is part of the World Health Organization (WHO) of the United Nations, has developed three categories for evidence of carcinogenicity. These include (1) sufficient evidence of human cancer definitely caused by exposure, (2) limited evidence due to inadequate data, and (3) inadequate evidence (i.e., no data available) [10.17].

The use of quantitative risk assessment has become widespread as a regulatory and management tool. A significant concern in quantifying the risk of carcinogens is the actual hazard imposed when the public is exposed to very low concentrations of these chemicals. Evaluating chemical carcinogenicity using the contaminant concentrations to which the public is typically exposed would require chronic toxicity studies with enormously high numbers of test animals. Alternatively, dose-response relationships may be determined experimentally at higher doses with the development of mathematical relationships to relate these high-dose results to the lower doses to which human populations are typically exposed. Therefore, the most common guidelines for animal studies are based on using higher doses than commonly found in the environment. Other criteria include the use of lifetime studies on at least 50 animals × two sexes × two species × at least three doses (a control and two doses) [10.18]. One factor that must be included is the background cancer level, which may be a function of the environment and the genetic makeup of the population being evaluated.

The extrapolation of empirical laboratory animal data to lower doses in humans is an important exercise that requires careful assessment of assumptions and mechanisms. Two classes of models have been developed to quantify the probability of cancer using extrapolations from higher doses to lower doses. *Tolerance models* (e.g., log-probit, log-logistic) are statistically based. *Mechanistic models* (e.g., one-hit, gamma, multihit, and Weibull) are based on the biochemical and physiological processes. Mechanistic models are usually favored over tolerance models among those who quantify low-level carcinogenicity. More information on carcinogenic dose-response models is available in References 10.19 and 10.20.

Two criteria must be considered when evaluating the carcinogenicity of hazardous chemicals. First, an assessment of the carcinogenic potential should be addressed; second, the availability of data on the carcinogens of concern must be determined. As with all toxicological evaluations, some type of dose-response relationship is the basis for quantifying a toxicological response. Dose-response evaluations are usually based on animal studies and provide a quantitative relationship of tumor incidence as a function of dose (Figure 10.7). The procedures for such assays involve providing doses to test

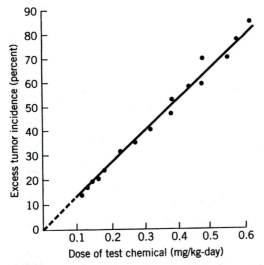

Figure 10.7 Dose-response relationship for carcinogens.

animals as an administered dose per day [mg/(kg-day)]. The response (y-axis) is the incidence of cancer corrected for background cancer rates; however, the cancer for animals must then be extrapolated to humans. The basis for linking animal data to humans is the assumption that one molecule can initiate cancer (i.e., no threshold exists for cancer-causing chemicals); therefore, any concentration of a carcinogen is a risk. What level of risk is acceptable? The most common value of risk used in hazardous waste assessments is 1×10^{-6}, or one in one million; however, cancer bioassays cannot evaluate such a low dose by any method currently known. Therefore, the data collected from animal studies are extrapolated to 10^{-6} risk. Such an extrapolation of carcinogen risk uses a *carcinogen potency factor (CPF)*, which is the slope of dose-response curves at low exposures. The EPA has adopted the use of such slopes of carcinogenic dose-response relationships and, not surprisingly, such a curve has become commonly known as a *slope factor (SF)*, with units of $[mg/(kg-day)]^{-1}$. As with reference doses, oral and inhalation slope factors have been reported by the EPA [10.16], and these are listed in Appendix L. Slope factors for the most common hazardous chemicals are also listed in Table 10.7.

10.8   SOURCES OF TOXICITY INFORMATION

Although slope factors and reference doses provide the best overall quantitative values of toxicity, particularly as related to risk, more detailed evaluations of contaminant toxicity are sometimes needed. The *Merck Index* [2.54] contains cursory information on toxicity; however, the best single source of detailed toxicity information is *Sax's Dangerous Properties of Industrial Materials* [10.2]. Information in this reference text includes physical properties, CAS registry numbers, molecular formula, molecular weights, water solubility, flammability data, and DOT hazard codes. A range of synonyms are provided, including trade names, international nomenclature, and RCRA manifest numbers. Some standards and recommended exposure concentrations are also provided, such as the OSHA PEL and the TLV.

A toxicity rating code developed by Sax [10.2] is provided at the top of each entry. These toxicity and hazard characteristics are listed in Table 10.8. Toxicity data for a number of different routes of exposure, test species, and criteria (e.g., LD_{Lo}, TC_{Lo}) are then provided with accompanying references. Finally, an overall safety profile is pro-

Table 10.8 Toxicity and Hazardous Characteristics

Toxicity Rating	Characteristics
0—No toxicity	Chemicals that produce no toxic effects under conditions of normal usage.
1—Slight toxicity	Slight acute and chronic local and systemic effects; changes are readily reversible once the exposure ceases.
2—Moderate toxicity	Moderate acute local and systemic effects; reversible and irreversible changes in the body not severe enough to cause serious physical impairment or threaten life.
3—Severe toxicity	Acute or chronic exposures result in permanent physical impairment, disfigurement, or threat to life.

Source: Reference 10.2.

vided, including carcinogenic status, reproductive effects, and potential hazards from flammability, corrosion, and reactivity. The safety profile may be brief for some compounds and may be extensive for common hazardous chemicals such as benzene and TCE.

10.9 SUMMARY OF IMPORTANT POINTS AND CONCEPTS

- Toxic effects are classified based on duration of exposure; the most common exposure periods are acute, subchronic, and chronic.

- Numerous end points have been used to evaluate toxicological responses. End points have been based on developmental effects, organs or tissues where the toxic action is found, reproductive effects, and local versus systemic effects.

- The lethal dose to 50% of the population (LD_{50}) is the most common indicator of acute toxicity. The LD_{50} is used to evaluate potential toxicity during one-time exposures to high concentrations of contaminants, such as hazardous material spills.

- Determination of the LD_{50} is accomplished by dosing population sets of organisms with succeedingly higher concentrations of a toxic chemical. The sigmoidal cumulative response to the dosages given is usually transformed to probits as a basis for determining the LD_{50}.

- Threshold Limit Values (TLVs) were developed for exposures to individuals in the workplace. They are a policy-related parameter based on a no-effect threshold during an 8-hour daily exposure over a working lifetime.

- Maximum Contaminant Levels (MCLs) are regulatory values under the Safe Drinking Water Act for over 50 chemicals to which the public may be exposed without significant health effects. The MCLs are important in hazardous waste management because they are often used as Applicable or Relevant and Appropriate Requirements (ARARs) at Superfund sites.

- The No Observed Adverse Effect Level (NOAEL), based on epidemiological and animal toxicity data, is the concentration of a chemical to which a population may be exposed with no toxicological effects. The NOAEL serves as the basis for reference doses (RfDs), which are used to quantify noncarcinogenic risk.

- Carcinogenic risk is based on the concept of no safe level of carcinogens. The slope factor (SF), which is the slope of the carcinogen potency factor (CPF), is used to quantify carcinogenic risk.

PROBLEMS

10.1. The LD_{50} of a hazardous chemical is 32 mg/kg. What is the corresponding dosage for a 62-kg (137-lb) adult?

10.2. The following data were obtained in an acute toxicological evaluation of OCDD exposure to white rabbits. Using groups of 15 rabbits, the toxicological end point was the visual presence of chloracne anywhere on the rabbit's skin.

Dose (mg/kg)	No. of Rabbits Showing Chloracne
0	0
3	0
6	1
9	3
12	5
15	8
18	12
21	14
24	15
27	15
30	15

a. Using these data, plot a cumulative function.

b. Using probit analysis, determine the ED_{50} for the data.

10.3. The following data have been obtained in a toxicological evaluation of a landfill leachate to *Daphnia magna*. Ten organisms were in each test group.

Total Contaminant Concentration (mg/L)	Lethality
0	0
6	0
12	1
18	2
24	3
30	5
36	7
42	9
48	10

Determine the LC_{50} for this landfill leachate.

10.4. An emergency response coordinator in a rural area receives two calls within minutes of each other reporting separate tanker truck spills. One is a tanker carrying methyl isocyanate; the other involves a spill of dodecane. Only one emergency response crew is available. Which accident should receive the higher priority?

10.5. What is the difference between a PEL and TLV?

10.6. Describe the criteria that are used to develop TLVs.

10.7. Determine the TWA for a RCRA waste management facility in which personnel are exposed to benzene for 6 h/day, TCE for 3 h/day, and MEK for 4 h/day.

10.8. A RCRA landfill is characterized by diurnal variations in the volatilization of contaminants as a result of changes in temperatures, wind speed, and activity at the site. The most prevalent contaminant, benzene, follows the pattern:

	Benzene Concentration, ppm
8 A.M.–9 A.M.	5
9 A.M.–10 A.M.	8
10 A.M.–11 A.M.	8
11 A.M.–12 P.M.	9
12 P.M.–1 P.M.	11
1 P.M.–2 P.M.	14
2 P.M.–3 P.M.	11
3 P.M.–4 P.M.	10

Determine the TWA for benzene concentrations at the landfill.

10.9. Using Reference 10.2, find the TLV and STEL of
a. Benzene
b. Vinyl chloride
c. p-Dichlorobenzene
d. Toluene
e. Nonane

10.10. An air sample collected at a RCRA 90-day storage facility contains 8 ppm toluene, 12 ppm PCE, and 25 ppm MEK.
a. What is the TLV-TWA for the mixture?
b. Has the TLV been exceeded?

10.11. The ambient concentration of benzene in a RCRA drum storage area is 0.56 ppm. Based on the TLV for benzene, is this level a concern?

10.12. Workers at a soil remediation site are exposed to varying concentrations of benzene as a function of time of day (and related diurnal changes in volatilization rates based on ambient temperature). Estimated concentrations are 18 ppm for 3 h, 41 ppm for 4 h, and 26 ppm for 1 h. Determine the 8-h TWA for exposure to the benzene.

10.13. Describe the conceptual basis for RfDs and how they are derived.

10.14. Determine the maximum daily oral dose of dimethyl phthalate to which a person could be exposed to have no noncarcinogenic toxicity.

10.15. Provide a synopsis of the assumptions and concepts used in the development of slope factors.

10.16. With your knowledge of the classes of hazardous wastes from Chapter 2, scan Appendix L. What are the most carcinogenic classes of compounds?

10.17. Based on the relative values and availabilities of the slope factors for benzene, toluene, ethylbenzene, xylenes, hexane, and cyclohexane, what qualitative assessment may be made about the carcinogenic potential of these gasoline compounds?

10.18. A hazardous waste spill of parathion has occurred. Is there a concern?

10.19. A spill of hexadecane has occurred. Is the acute toxicity of this chemical a concern?

10.20. Using Reference 10.2, rank each of the following chemicals into the four toxicity classes listed in Table 10.8.

a. Acetone   b. 2,4-D
c. DDT   d. Carbaryl
e. TCA   f. Toluene
g. Mirex   h. 2,4,6-TNT

10.21. Using Reference 10.2, provide a two-page toxicological assessment of the following hazardous chemicals:

a. Benzene   b. Toluene
c. Hexane   d. Naphthalene
e. Benzo[a]pyrene   f. MEK
g. PCE   h. Pentachlorophenol
i. TNT   j. 2,3,7,8-TCDD

For each compound, provide the value for Sax's hazard code, an assessment of acute toxicity values (including variation among species), a summary of its mutagenic, carcinogenic, and teratogenic potential, and an overall synopsis of its hazard to public health.

REFERENCES

10.1. Stewart, R. D., "Methyl chloroform intoxication: Diagnosis and treatment," *JAMA*, **215**, 1789–1792 (1971).

10.2. Sax, N. I., *Dangerous Properties of Industrial Materials*, 7th Edition, Van Nostrand Reinhold, New York, 1989.

10.3. Finney, D. J., "The median lethal dose and its estimation," *Arch. Toxicol.*, **56**, 215–218 (1985).

10.4. Lorke, D., "A new approach to practical acute toxicity testing," *Arch. Toxicol.*, **54**, 275–287 (1983).

10.5. Sperling, F., "Quantitation of toxicology—the dose-response relationship," in Sperling, F. (Ed.), *Toxicology: Principles and Practice*, Vol. 2, John Wiley & Sons, New York, 1984.

10.6. Zbinden, G. and M. Flury-Roversi, "Significance of the LD_{50} test for the toxicological evaluation of chemical substances," *Arch. Toxicol.*, **47**, 77–99 (1981).

10.7. Calabrese, E. J., *Principles of Animal Extrapolation*, John Wiley & Sons, New York, 1983.

10.8. Rothman, K. J., *Modern Epidemiology*, Little, Brown, Boston, 1986.

10.9. Finney, D. J., *Probit Analysis*, Cambridge University Press, Cambridge, UK, 1971.

10.10. Sullivan, J. B., Jr. and G. R. Krieger, *Hazardous Materials Toxicology: Clinical Principles of Environmental Health*, Williams & Wilkins, Baltimore, MD, 1992.

10.11. *Documentation of the Threshold Limit Values and Biological Exposure Indices*, 5th Edition, American Conference of Governmental Industrial Hygienists, Cincinnati, OH, 1986.

10.12. Travis, C. C., E. A. C. Crouch, R. Wilson, and E. D. Klema, "Cancer risk management: A review of 132 federal regulatory cases," *Environ. Sci. Technol.*, **21**, 415–420 (1987).

10.13. U.S. Environmental Protection Agency, "Guidelines for carcinogen risk assessment," *Fed. Reg.*, **51**, 33992–34003.

10.14. Barnes, D. G. and M. Dourson, "Reference dose (RfD): Description and use in health risk assessments," *Regul. Toxicol. Pharmacol.*, **8**, 471–486 (1988).

10.15. "Guidelines for development toxicity risk assessment," *Federal Register*, **51**(185), 34028–34040 (1986).

10.16. *Integrated Risk Information System (IRIS)*, U.S. Environmental Protection Agency, June, Washington, DC, 1995.

10.17. "Overall evaluations of carcinogenicity," *IARC Monographs on the Evaluation of Carcinogenic Risks to Humans*, An Updating of International Agency on Research of Cancer Monographs, Vol. 1 to 42, Suppl. 7, International Agency on Research of Cancer, 1987.

10.18. Interagency Regulator Liaison Group, "Scientific bases for identification of potential carcinogens and estimation of risks," *J. Natl. Cancer Inst.*, **63**, 241–268 (1979).

10.19. Rai, K. and J. Van Ryzin, "Risk assessment of toxic environmental substances using a generalized multi-hit dose response model," in Breslow, N. E. and A. S. Whittemore, *Energy and Health*, Society for Industrial and Applied Mathematics, Philadelphia, PA, 1979.

10.20. Gaylor, D. W. and R. L. Kodell, "Linear interpolation algorithm for low dose risk assessment of toxic substances," *J. Environ. Pathol. Toxicol.*, **4**, 305–312 (1980).

Chapter 11

Hazardous Waste Risk Assessment

The assessment of health effects on workers, the general public, and the environment is often required in hazardous waste management. Potential hazards to public health and the environment are often evaluated using a multistep risk assessment process. The ultimate goal in the use of risk assessment is to provide a quantitative basis for making decisions involving hazardous waste treatment, remediation, and disposal options, waste minimization, and siting new facilities, such as incinerators and RCRA landfills. Hazardous waste risk assessments are a key part of CERCLA and SARA; not only are risk assessments performed to assess health and ecological risks at Superfund sites, they are also used to evaluate the effectiveness of remedial alternatives in attaining a Record of Decision (ROD). For most contaminants that are considered hazardous substances or pollutants and contaminants under CERCLA, specific cleanup requirements have not been established. For example, not all soils at sites containing pentachlorophenol (PCP) are cleaned up to a certain level, such as 1 mg/kg. Rather, each site is assessed on an individual basis and cleaned up to a predetermined level of risk, such as one cancer case per one million people. The rationale for using risk assessment is illustrated by comparing the two very different sites shown in Figure 11.1. The high-risk site, although containing only 10 mg/kg PCP, is near a school and drinking water supply. Cleanup to low levels would be necessary to protect public health. At the low-risk site, a relatively high concentration (10,000 mg/kg) of PCP-bearing sludge has been buried, but there are no receptors nearby and the slow rate of PCP migration in the low permeability soils would further reduce the risk.

Based on commonly used risk assessment models, the best remedial option for the low-risk site may be to leave the contaminated sludge in place, where *natural attenuation* processes will eventually result in its degradation. Excavating the contaminated material with disposal in an RCRA landfill or ex situ treatment may result in high risk due to the release of windblown dusts and exposure of workers at the site. Furthermore, if the material is disposed of in an RCRA landfill, it adds to the landfill loading and may very well pose more of a threat as landfill leachate than if it were left in place at the site.

518

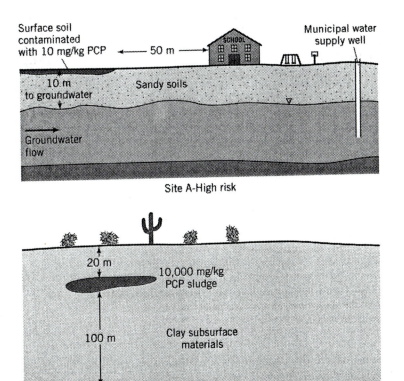

Figure 11.1 Two extremes of potential risk from contaminated sites. Site A is a high-risk site with potential for migration from the source to nearby receptors. Site B, although characterized by a higher source concentration, has minimal potential for contaminant migration and risk.

The level of contamination in soils and the subsurface often decreases as a function of depth, as illustrated in Figure 11.2. In evaluating such a site, the depth to which cleanup is conducted must often be determined; that is, to what degree is cleanup necessary, or *how clean is clean?* The degree to which the source material will be cleaned up is often evaluated by first determining the allowable concentration to which receptors may be exposed. Then, pathway analysis is conducted to determine the acceptable source concentration. Using a predetermined or regulatory level of risk, pathway calculations can be used (beginning at the receptor, rather than the source, and then working backward), to determine the release from the source that is acceptable.

Other hazardous waste risk assessment applications include the evaluation of RCRA hazardous waste management plans and the assessment of risk from the transportation of hazardous wastes. In RCRA facility management, risk assessments are often used to determine the health effects resulting from the release of air pollutants from solvent recycling facilities and the release of groundwater pollutants from RCRA landfills. For example, if fugitive emissions from a hazardous waste transfer facility re-

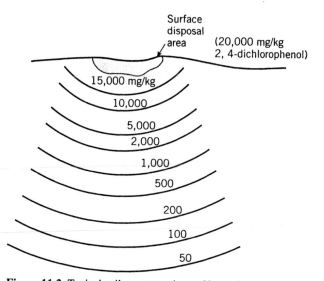

Figure 11.2 Typical soil concentrations of hazardous chemicals away from the source provides a conceptual basis for addressing "how clean is clean?"

quire some degree of control by the installation of scrubbers, the rate of acceptable emissions may be evaluated by allowing a release that will provide a predetermined level of risk. Heath [11.1] emphasized that risk assessment strategies can be used to focus RCRA corrective action activities, resulting in compliance with regulations and cost-effective protection of human health and the environment. Risk assessment methodologies have also been applied to RCRA hazardous waste incinerators, boilers, and industrial furnaces [11.2]. Furthermore, risk assessment may be used to evaluate the potential spills occurring during transportation accidents by analyzing the frequency of accidents, exposure resulting from chemical spills, and ambient concentrations in the air resulting from the spills [11.3]. Routes taken by transporters of hazardous wastes are of concern to generators because of the cradle-to-grave liability. Krukowski [11.4] reviewed factors important in selecting routes, such as avoidance of population centers, type of vehicle, and type of highways. He emphasized that risk analysis will likely become more important in selecting routes for hazardous waste transportation.

An introduction to hazardous waste risk assessments is provided in this chapter; the EPA has also published a series of documents [11.5–11.10], many of which offer detailed procedures and outlines for performing hazardous waste risk assessments. In this chapter, the concepts of sources, pathways, and receptors are brought together to assess the risk of exposure to hazardous wastes and facilities. Looking back to Section 1.5 and Table 1.5, the analysis of hazardous waste risks follows the logical progression of defining and characterizing the contaminants at the source, evaluating the pathways the wastes may take to reach receptors, and then assessing any toxicological effect on receptor organisms. Although humans have been the receptor organisms most commonly evaluated, plants and animals are also common receptors for hazardous chemicals. Therefore, both human health and ecological risk assessments are covered in this chapter.

11.1 PRINCIPLES, DEFINITIONS, AND PERSPECTIVES OF HAZARDOUS WASTE RISK ASSESSMENTS

Risk, as it applies to hazardous waste, may be defined as the chance of encountering the potential adverse effects of human or ecological exposures to environmental hazards. In general terms, risk is the probability of harm or loss, which may also be considered as a product of probability and the severity of consequences.

Risk is inherent in all forms of life—humans, animals, and plants. Natural disasters, such as tornadoes, landslides, hurricanes, and earthquakes, present a risk of injury or death to any living thing in their path. Human-caused risks also occur at varying levels of severity, including automobile accidents, plane crashes, and nuclear disasters.

Some specific definitions apply to risk assessment in hazardous waste management. *Background risk* in hazardous waste management is the risk to which a population is exposed without considering hazardous chemicals. *Incremental risk* is the risk caused by the presence of hazardous chemicals. *Total risk,* then, is the background risk plus the incremental risk. For example, the background risk of cancer to the average U.S. citizen is one in four, or 0.25 [11.11]. Superfund risk assessments are based on multiple worst-case-scenarios to the "most exposed individual" proposed by the EPA, with an incremental target risk of 1×10^{-6}. The total lifetime risk for contaminant exposure to carcinogens at Superfund sites is then 0.25 plus 1×10^{-6}. The total risk analysis often involves evaluation of all aspects of quantitative risk assessment, including an analysis of the uncertainties of the assessment and acceptable risks of exposure to hazardous wastes. Nonetheless, Raider [11.12] suggested that multiple conservative assumptions should apply to risk assessment, but the cleanup of Superfund sites could be more efficient and less costly if future risk-based assumptions were more realistic.

Hazard, which is different than risk, is a descriptive term that characterizes the intrinsic capability of a waste to cause harm. Hazard may be thought of as a source of risk and is a function of the persistence, mobility, and toxicity of the contaminants.

Four steps have been defined by both the National Academy of Sciences and the EPA for the assessment of risk from hazardous wastes [11.5, 11.13]:

1. *Hazard identification* (the chemicals present at the site or facility and their characteristics; i.e., source analysis).
2. *Exposure assessment* (potential transport of the chemicals to receptors and levels of intake; i.e., pathway analysis).
3. *Toxicity assessment* including the determination of numerical indices of toxicity (i.e., receptor analysis).
4. *Risk characterization* involving the determination of a number that expresses risk, such as one in one hundred (0.01) or one in one million (1×10^{-6}).

Different criteria may be used for risk assessments depending on the nature of the source, the toxicity of the contaminants, and what the receptors are. For many sites, the lifetime risk for contracting cancer is the criterion for which risk is calculated, a procedure that is used when humans are the receptors and many of the chemicals described in Chapter 2 are the primary contaminants. Other calculations using reference doses for noncarcinogens are used when source contaminants include lead, cyanide, and other chemicals that do not cause cancer. In ecological risk assessments, the effect on plants and animals is assessed and quantified. Ecological risk assessments are often difficult because toxicity data for animal and plant species are usually lacking.

The science and engineering that serve as the basis for most risk assessments are often not exact and are subject to a great deal of variability and uncertainty, which often results in only order-of-magnitude quantitation of risk. Defining acceptable risk is an intricate exercise that requires understanding the procedures, uncertainties, and values of the risk assessment process. Use of risk assessment in hazardous waste evaluations and remedial decisions must be tempered by a knowledge of its basic nature, uncertainties, and limitations; that is, risk assessments are based on a number of assumptions and the numerical value obtained is not absolute. Furthermore, those making decisions based on risk assessments are often required to account for the uncertainties and incorporate professional judgments related to economic, political, and social factors.

Related to the uncertainties and value judgments that are inherent in risk assessments is the need for experienced personnel. Experience and judgment are required for hazardous waste risk assessments perhaps more than for any other topic covered in this text. The material of this chapter provides only an introduction to hazardous waste risk assessments; true experience on a risk assessment team is the key to learning the intangibles and appreciating the uncertainties of the risk assessment process.

11.2   THE RISK ASSESSMENT PROCESS

Based on the four-step procedure for hazardous waste risk assessments and the dynamics occurring at a hazardous waste site or facility, a standard approach using hazard identification exposure assessment, toxicity assessment, and risk characterization has been established for hazardous waste risk assessments. A description of each of the four processes follows.

11.2.1   Hazard Identification

The first step in hazardous waste risk assessments is a detailed evaluation of the source. Procedures for *hazard identification* involve the use of the concepts described in Part I, *Sources*; that is, the identity, nomenclature, concentration, and properties are obtained for all contaminants at a site or facility. Some of the specific tasks that are part of hazard identification include sampling, installation of monitoring wells, chemical analysis, quality assurance/quality control plans, and data analysis (see Chapter 4). Sampling and analysis of the source provide site characteristics and input data for exposure assessment models. The hazard identification usually focuses more on chemicals of concern (those with the highest hazard). For example, more detailed source information would be required for 2,3,7,8-TCDD than for dodecane.

Most hazardous waste sites and facilities contain tens to hundreds of chemicals; therefore, data collection can become overwhelming and unrealistic. Many risk assessment efforts screen the hundreds of chemicals for those that pose the greatest risk, which then serve as surrogates. The use of surrogates reduces the data input for pathway studies, which significantly streamlines computing efforts. The two most important source characteristics that are used in screening a large number of chemicals at a hazardous waste site or facility are their concentrations and toxicities. Therefore, screening for surrogates has been called *Concentration-Toxicity Screening* by the EPA [11.5]. The basis for the screening procedure is to obtain a risk factor, called a *Chemical Score (R)*, for each chemical, and to separate the scores by medium. The units for the chemical score are a function of the medium that is evaluated. The units do not usually matter, as long as they are consistent. If toxicity data are available for both oral and inhalation routes, the most conservative number (i.e., the most toxic) should be used in concentration-toxicity screening. Surrogate chemicals should account for 99% of the risk at the site. The procedure for the initial screening of contaminants to identify surrogates is outlined in Table 11.1 and illustrated in Example 11.1.

Table 11.1 Procedure for Identifying Surrogates in Risk Assessment

1. Sort the contaminant data by medium.

   - Soil

   - Miscellaneous solids (e.g., sludges)

   - Surface water

   - Groundwater

   - Air

2. Tabulate the mean and range of concentrations at the site or facility.
3. List the reference doses for noncarcinogens and slope factors for carcinogens.
4. Determine the chemical score for each contaminant.
   In general, the chemical score is

$$R_{ij} = C_{ij} \cdot T_{ij} \tag{11.1}$$

where

R_{ij} = chemical score for chemical i in medium j
C_{ij} = concentration of chemical i in medium j
T_{ij} = toxicity value for chemical i in medium j (SF or 1/RfD)

A. For noncarcinogens,

$$R_{ij} = \frac{C_{max}}{RfD} \tag{11.2}$$

where
C_{max} = maximum concentration
RfD = reference dose $\left(\dfrac{mg}{kg\text{-}day}\right)$

B. For carcinogens,

$$R_{ij} = SF \times C_{max} \tag{11.3}$$

where
SF = slope factor $\left(\dfrac{kg\text{-}day}{mg}\right)$

The total chemical score is

$$R_j = R_{1j} + R_{2j} + R_{3j} + R_{ij} \tag{11.4}$$

Table 11.1 (Continued)

where

R_j = total risk factor for medium j

$R_{1j} + \cdots R_{ij}$ = risk factors for chemicals 1 through i in medium j

5. Rank the compounds by chemical scores for each exposure route.

6. Select chemicals comprising 99% of the total score.

EXAMPLE 11.1 *Evaluation of Contaminants Using Concentration-Toxicity Screening*

The following contaminants and corresponding concentrations are found at an abandoned pesticide formulation area. Select the surrogate contaminants that account for 99% of the risk.

Contaminant	Concentration (mg/kg)
Aldicarb	140
Captan	86
Fonofos	280
Malathion	65
Naled	90

SOLUTION

1. Tabulate the reference doses for noncarcinogens and slope factors for carcinogens from Appendix L.

Contaminant	RfD	SF
Aldicarb	0.001	N/A
Captan	0.13	N/A
Fonofos	0.002	N/A
Malathion	0.02	N/A
Naled	0.002	N/A

2. Determine the chemical score for each contaminant in the soil (all are noncarcinogens, so use Equation 11.2) and rank each contaminant.

$$R = \frac{C_{max}}{RfD}$$

Contaminant	R	Rank
Aldicarb	140,000	1
Captan	662	4
Fonofos	140,000	1
Malathion	3250	3
Naled	45,000	2

The two contaminants that should be selected as surrogates are aldicarb and fonofos, both with the highest chemical scores (140,000); however, the surrogate chemicals must comprise 99% of the total chemical score. The total R is 328,912; the two selected surrogates make up only 85% of the total R. Therefore, naled must also be considered a surrogate contaminant for the site. Adding the chemical score of naled to the other two selected surrogates results in 99% of the total chemical score. Therefore, the surrogates selected for the site should be aldicarb, fonofos, and naled.

11.2.2 Exposure Assessment

Exposure is defined as contact of an organism, such as humans or an endangered species, with a contaminant. *Exposure assessment* is the estimation of the magnitude, frequency, duration, and route of exposure [11.14]. The purpose of exposure assessment is the estimation of the contaminant concentrations and dosages to the populations at risk. More specifically, the primary tasks in exposure assessments include (1) identifying potentially exposed populations, (2) identifying potential exposure pathways, (3) estimating exposure concentrations, and (4) estimating chemical intakes. Identification of exposure pathways and the estimation of exposure concentrations is based on the analysis of *Pathways* (covered in Part II), including chemical release mechanisms (leaching and volatilization), transformations (e.g., hydrolysis, biotransformation), transport (subsurface or atmospheric transport), and determination of contaminant concentrations at the exposure point such as a drinking water well. Exposure concentrations are estimated by using (1) data from sampling and analysis and (2) the results of fate and transport models, which are often supplemented with at least some sampling data. As emphasized in Chapter 4, actual chemical analysis data are the most reliable in hazardous waste management; however, data from soil, groundwater, and air sampling and analysis are not always available. Use of sampling data is most applicable when exposure involves direct contact with the medium, such as contaminated soils [11.5]. Sampling data are also used when exposure occurs directly at the source. Nonetheless, monitoring data provide estimates of exposure concentrations only from current exposures. Future exposure concentrations, particularly in drinking water supplies, may only be attainable through the use of pathway calculations (Chapters 5–8) and exposure models. Furthermore, the risk from potential hazardous waste spills and accidents requires modeling techniques in order to predict spatial and temporal concentrations of the hazardous wastes released, sometimes before they occur based on hypothetical situations.

An important aspect of risk assessment is identifying the population that may be exposed to hazardous chemicals. Characterization of potentially exposed populations can be an intricate and difficult process [11.15]. As expected, such characterizations involve site visits, population screenings near the site, evaluation of land use and housing maps, and a survey of recreational data. In addition, future land use patterns must be evaluated because exposures to hazardous chemicals may occur into the future. Another aspect, which is often difficult to quantify, is the activity pattern of the population, such as time spent in the area relative to time spent at work, time spent outside versus inside, seasonal activity patterns, and so on. The receptor populations may include the general population of a residential area, sensitive populations (e.g., children, elderly, infirmed), or site workers and personnel. Identifying receptor populations may be a qualitative exercise in which the number, location, and sensitive populations are evaluated. In many cases professional judgment is required, and in complex systems the aid of a geographer and geographic information systems (GISs) may be necessary.

Related to the need for identifying the receptors is necessary information of population behavior such as exposure duration, mean body weight of the individuals, frequency of exposures, and future demographics, including population increases or decreases. The EPA recommends the use of exposure data normalized for time and body weight, which is formulated into an equation for intake (*I*). Three categories of variables are used to estimate intake: (1) a contaminant-related variable (exposure concentration), (2) exposed population variables (contact rate, exposure frequency, duration, and body weight), and (3) an assessment-determined variable (averaging time). The *exposure concentration* (*C*) term of the intake equations is the arithmetic mean of the contaminant concentration over the exposure period. The *contact rate* (CR) is the amount of contaminant contacted per time. As a conservative estimate, the EPA [11.5] recommends the use of the 95th percentile of data obtained for the contact rate. The total time of exposure is estimated using *exposure frequency* (EF) and *exposure duration* (ED), terms that are site- and population-specific. Conservative assumptions are usually used, such as the 95th percentile value for exposure time [11.5]. The body weight (BW) value used is the average weight over the exposure period. If exposure occurs primarily for children, the average child's body weight is used. The value of 70 kg is commonly used for adult exposures. The *averaging time* (AT) is a value that is based on the mechanism of toxicity. The most commonly used averaging times are 70 years × 365 days/year for carcinogens and a pathway-specific period of exposure for noncarcinogenic effects (i.e., the ED × 365 days/year).

The mean exposure concentration of contaminants is used with exposed population variables and the assessment-determined variables to estimate contaminant intake. The general equation for chemical intake is

$$\text{CDI} = \frac{C \times \text{CR} \times \text{EFD}}{\text{BW}} \times \frac{1}{\text{AT}} \qquad (11.5)$$

where

CDI = chronic daily intake (the amount of chemical at the exchange boundary) $\left(\frac{\text{mg}}{\text{kg-day}}\right)$

C = the average exposure concentration over the period (e.g., mg/L for water or mg/m^3 for air)

CR = contact rate, the amount of contaminated medium contacted per unit time (L/day or m^3/day)

EFD = exposure frequency and duration, a variable that describes how long and how often exposure occurs. The EFD is usually divided into two terms:
  EF = exposure frequency (days/year)
  ED = exposure duration (years)

BW = the average body mass over the exposure period (kg)

AT = averaging time; the period over which the exposure is averaged (days)

Determination of accurate intake data is sometimes difficult; for example, exposure frequency and duration vary among individuals and must often be estimated; site-specific information may be available, or professional judgement may be necessary. Equations for estimating daily contamination intake rates from drinking water, the air, and contaminated food, and by dermal exposure while swimming have been reported by the EPA [11.10]. Two of the most common routes of exposure are through drinking contaminated water and breathing contaminated air. The intake for ingestion of water-borne chemicals is

$$CDI = \frac{CW \times IR \times EF \times ED}{BW \times AT} \tag{11.6}$$

where

CDI = chronic daily intake by ingestion (mg/kg-day)
CW = chemical concentration in water (mg/L)
IR = ingestion rate (L/day)
EF = exposure frequency (days/year)
ED = exposure duration (years)
BW = body weight (kg)
AT = averaging time (period over which the exposure is averaged—days)

Some of the values used in Equation 11.6 are

CW: site-specific measured or modeled value
IR: 2 L/day (adult, 90th percentile) [11.10]
     1.4 L/day (adult, average) [11.10]
EF: Pathway-specific value (dependent on the frequency of exposure-related activities)
ED: 70 years (lifetime; by convention)
     30 years [national upper-bound time (90th percentile) at one residence] [11.10]
     9 years [national median time (50th percentile) at one residence] [11.10]
BW: 70 kg (adult, average) [11.10]
     Age-specific values [11.9, 11.10]
AT: Pathway-specific period of exposure for noncarcinogenic effects (i.e., ED × 365 days/year), and 70-year lifetime for carcinogenic effects (i.e., 70 years × 365 days/year)

The intake for inhalation of airborne contaminants is

$$CDI = \frac{CA \times IR \times ET \times EF \times ED}{BW \times AT} \tag{11.7}$$

where

CDI = chronic daily intake by inhalation (mg/kg-day)
CA = contaminant concentration in air (mg/m^3)
IR = inhalation rate (m^3/hour)
ET = exposure time (hours/day)
EF = exposure frequency (days/year)
ED = exposure duration (years)
BW = body weight (kg)
AT = averaging time (period over which the exposure is averaged—days)

Some of the values used in Equation 11.7 are

CA: site-specific measured or modeled value

IR: 30 m^3/day (adult, suggested upper bound value) [11.10]
    20 m^3/day (adult, average) [11.10]

ET: Pathway-specific values (dependent on the duration of exposure-related activities)

EF: Pathway-specific value (dependent on the frequency of exposure-related activities)

ED: 70 years (lifetime; by convention)
    30 years [national upper-bound time (90th percentile) at one residence] [11.9]
    9 years [national median time (50th percentile) at one residence] [11.10]

BW: 70 kg (adult, average) [11.10]
    Age-specific values [11.9, 11.10]

AT: Pathway-specific period of exposure for noncarcinogenic effects (i.e., ED × 365 days/year), and 70-year lifetime for carcinogenic effects (i.e., 70 years × 365 days/year)

EXAMPLE 11.2   *Estimation of an Oral Chronic Daily Intake*

The mean concentration of 1,2-dichlorobenzene in a water supply is 1.7 μg/L. Determine the chronic daily intake for a 70-kg adult. Assume that 2 L of water are consumed per day.

SOLUTION

The chronic daily intake (*CDI*) may be calculated using Equation 11.6:

$$CDI = \frac{C \times CR \times EF \times ED}{BW \times AT}$$

where

CDI = chronic daily intake $\left(\dfrac{mg}{kg\text{-}day}\right)$

$C = 1.7$ μg/L $= 0.0017$ mg/L

CR = 2 L/day

EF = 365 days/year

ED = 30 years (standard exposure duration for an adult exposed to a noncarcinogen)

BW = 70 kg

AT = 365 days/year · 30 years = 10,950 days

Substituting values into the equation yields the chronic daily intake.

$$CDI = \frac{0.0017 \times 2 \times 365 \times 30}{70 \times 10,950} = 4.86 \times 10^{-5} \frac{mg}{kg\text{-}day}$$

11.2.3 Toxicity Assessment and Risk Characterization

Toxicity assessment is the acquisition and evaluation of toxicity data for each contaminant, a procedure that is performed for both noncarcinogens and carcinogens. As described in Chapter 10, reference doses (RfDs) are used to quantify noncarcinogenic toxicity; similarly, slope factors (SFs) are used for carcinogens. Almost all toxicity assessments make use of existing data such as the list of reference doses and slope factors listed in Appendix L; therefore, toxicity assessment often involves obtaining tabular RfD and SF data for the chemicals to which receptors are exposed. Risk assessments are sometimes conducted on contaminants for which reference doses and slope factors have not been determined. In such cases, toxicity data may be derived—a procedure based on a series of assumptions and safety factors. Furst [11.17] discussed issues in the interpretation of toxicological data used in risk assessment. He suggested that animal toxicity data should only be used for risk calculation when experiments employ routes that mimic human exposure (i.e., oral, inhalation, or dermal). Such toxicity assessments are normally performed by toxicologists and are beyond the scope of this text. Detailed procedures for deriving toxicity data are available in Reference 11.16.

Risk characterization is the calculation of risk for both noncarcinogens and carcinogens for all receptors that may be exposed to hazardous wastes. Some of the general requirements include calculating risk for all of the exposure routes to hazardous chemicals (oral and inhalation) for both noncarcinogens and carcinogens. Noncarcinogenic risk is calculated as a *hazard index* (HI), which is the ratio of the chronic daily intake to the reference dose (RfD):

$$HI = \frac{CDI}{RfD} \tag{11.8}$$

where

HI = the hazard index (dimensionless)

CDI = chronic daily intake $\left(\dfrac{mg}{kg\text{-}day}\right)$

RfD = reference dose $\left(\dfrac{mg}{kg\text{-}day}\right)$

By inspection of Equation 11.8, a hazard index <1.0 provides acceptable risk; however, the cumulative acceptable risk for all contaminants and routes of exposure must be <1.0. If the hazard index is <1.0, the receptors are exposed to concentrations that do not present a hazard. In such cases, detoxification and other mechanisms allow the receptor exposure to the contaminant with no toxic effects. Note that the quantitative value obtained for the HI is not a value of risk; that is, it does not provide a value for the probability of harm as the result of exposure. Instead, the hazard index quantifies the absence of effects from exposure to noncarcinogens. The use of Equation 11.8 in evaluating noncarcinogenic risk is illustrated in Example 11.3.

EXAMPLE 11.3 *Estimation of Noncarcinogenic Risk*

A population is exposed to MEK with an intake of 0.0046 $\dfrac{mg}{kg\text{-}day}$. Estimate the hazard index.

SOLUTION

The Hazard Index (HI) is calculated using Equation 11.8.

$$HI = \frac{CDI}{RfD}$$

The RfD for MEK is 0.6 $\frac{mg}{kg\text{-}day}$ (Table 10.6 and Appendix L); substituting into Equation 11.8 yields

$$HI = \frac{0.0046}{0.6} = 0.0077$$

Because MEK is the only contaminant to which the population is exposed, the final measure of hazard (total cumulative HI) is 0.0077, which is less than 1.0. This result indicates that no adverse health effects will result from exposure to this noncarcinogen.

Carcinogenic risk is a function of the chronic daily intake (e.g., Equation 11.5) and the slope factor:

$$Risk = CDI \times SF \tag{11.9}$$

where

Risk = the probability of carcinogenic risk (dimensionless)

CDI = chronic daily intake $\left(\frac{mg}{kg\text{-}day}\right)$

SF = carcinogenic slope factor $\left(\frac{kg\text{-}day}{mg}\right)$

The risk term is the quantitative end point sought in risk assessment calculations and is commonly used in regulatory and management decisions in assessing and controlling hazardous wastes. As with noncarcinogens, the risk term is calculated for each contaminant, for each route of exposure, and for all sets of receptor populations. The estimation of carcinogenic risk is demonstrated in Example 11.4.

EXAMPLE 11.4 *Estimation of Carcinogenic Risk*

If the CDI for a population to TCA is 6.8×10^{-4} $\frac{mg}{kg\text{-}day}$, what is its carcinogenic risk?

SOLUTION

The carcinogenic risk is the product of the CDI and the slope factor. The slope factor for TCA from oral exposure is $0.057 \frac{\text{kg-day}}{\text{mg}}$ (Appendix L). Substituting these values into Equation 11.9 yields

$$\text{Risk} = 6.8 \times 10^{-4} \frac{\text{mg}}{\text{kg-day}} \times 0.057 \frac{\text{kg-day}}{\text{mg}} = 3.9 \times 10^{-5}$$

The carcinogenic risk is greater than the minimum 1×10^{-6} carcinogenic risk that is often allowed. The calculated risk indicates that excess cancer incidence will be about 39 times greater than the common goal of 1×10^{-6} and is therefore unacceptable.

11.3 ECOLOGICAL RISK ASSESSMENTS

Although much of the emphasis in hazardous waste management has been in protecting public health, the key provisions in RCRA and CERCLA refer to "public health and *the environment*." Therefore, assessing the risk of hazardous wastes to the environment is also important.

Ecological risk assessments are conducted in a manner similar to human health risk assessments through a four-step process: (1) hazard identification, (2) exposure assessment, (3) toxicity assessment, and (4) risk characterization. Ecological risk assessments can be complicated and intricate because of the variability between species, the difficulty of describing ecosystems, and the paucity of data. The following discussion provides an introduction to the topic; more detailed coverage of ecological risk assessments may be found in References 11.18 and 11.19.

11.3.1 Hazard Assessment

As with human health risk assessments, the first step in assessing ecological risks is exposure assessment, a procedure that focuses on the determination of source concentrations, physical and chemical properties of the contaminants, and concentrations in all potential phases to which the receptor organisms are exposed. Surrogates are often used in ecological risk assessments, just as they are in human health risk assessments.

In addition to evaluating the concentrations and properties of hazardous chemicals, the ecosystem itself requires definition, which often includes identifying endangered species, but may (and should) also include such fundamental ecological data as species composition, diversity indices, and predator-prey interactions [11.20]. As an example of the importance of evaluating the entire ecosystem rather than one sensitive species, the prey of an endangered species may be more sensitive to hazardous waste exposure than the endangered predator. If its food source is eliminated from the ecosystem, the predator may also be affected through such a predator-prey interaction.

11.3.2 Exposure Assessment

Just as subsurface and air dispersion calculations are used to estimate downgradient concentrations for evaluating human health exposures, environmental fate models for soils, sediments, and surface waters are often used to predict concentrations in these media as a basis for ecological risk assessments. Such fugacity models, some of the most common of which are EXAMS II [11.21], SARAH2 [11.22], and TOXI4 [11.23], are based on the concentrations to which contaminants partition between phases in the environment. A significant difference between human health and ecological exposure assessments is the emphasis on transport through surface water rather than groundwater—because most higher organisms are in contact with surface media.

Routes of exposure and uptake rates vary depending on the organism, its habitat, and whether it is a terrestrial or aquatic organism. For example, dermal contact is an important route of exposure for reptiles (who are in contact with the soil). In contrast, oral exposure is the most common route for birds. In general, ingestion is the most common route of exposure for animals; contaminant uptake rates can be estimated based on food sources, ingestion rates, and contaminant concentrations of the food (from sampling data). Feeding dynamics have been reported for many species in the ecological literature.

Exposure for aquatic organisms is not only a function of ingestion, but also the organisms' constant contact with the water containing the contaminant. Two models, FGET (Food and Gill Exchange Model of Toxic Substances [11.24]) and WASP (Water Quality Analysis Simulation Program [11.25]), have been used to predict the intake of contaminants by aquatic organisms. Other, more empirical approaches have been developed in which the octanol-water partition coefficient (K_{ow}) has been correlated with the bioconcentration factor (BCF), which is the partitioning of contaminants between the water and the organism. The BCF does not estimate ingestion—another predictor the *bioaccumulation factor* accounts for ingestion. Some common correlations between K_{ow} and the BCF are listed in Table 11.2. Like the correlations between K_{ow} and K_{oc} (Section 5.6), the equations listed in Table 11.2 are compound-specific.

In aquatic toxicology, other toxicological end points have been established, including the *Maximum Acceptable Toxicant Concentration* (MATC); over the last few years, the nomenclature of this term has been changed to the *final chronic value* (fCv) [11.30], although the latter term is still not widely used. Another term recently proposed is the maximum acceptable tissue concentration, which Calabrese and Baldwin [11.31] provided with the acronym MATissueC. Still another value used in ecological risk assessments is the *Terrestrial Reference Value* (TRV), which is a no-effect contaminant exposure rate for terrestrial animals, a term analogous to the RfD [11.32].

11.3.3 Toxicity Assessment and Risk Characterization

When endangered species are considered in ecological risk assessments, Congress has dictated that individual organisms, in addition to populations, must be protected. In such cases, ecological risk assessments are similar to human risk assessments. Ecological risk assessment end points may include effects on growth, metabolism, health, or reproduction [11.33]. Although the effects of hazardous wastes on endangered species are important in some cases, the most common objective of toxicity assessments is to evaluate the effects of hazardous contaminants (as a function of dose) on

Table 11.2 Common Correlation Equations for K_{ow} and BCF

Equation	Equation	Compounds	Species	Reference
$\log \text{BCF} = 0.761 \log K_{ow} - 0.23$	11.10	Wide range (PCBs, pesticides, chlorinated aliphatics, chlorinated aromatics, nitroaromatics)	Flathead minnow, bluegill, rainbow trout, mosquitofish	11.26
$\log \text{BCF} = 0.542 \log K_{ow} + 0.124$	11.11	Chlorinated aliphatics and aromatics	Rainbow trout	11.27
$\log \text{BCF} = 0.935 \log K_{ow} - 1.495$	11.12	Pesticide, chlorinated aromatics, PCBs	Bluegill, flathead minnow, rainbow trout, mosquitofish	11.28
$\log \text{BCF} = 0.819 \log K_{ow} - 1.146$	11.13	Nitrogen heterocycles	*Daphnia pulex*	11.29

ecological effects. Common dose-response effects range from no effect, to a quantifiable effect on biological functions of ecosystems, to mortality. As in human health risk assessments, the appropriate route of exposure must be quantified, a procedure that is based on exposure assessment data (i.e., the compartments where the contaminants exist) and uptake patterns of the organisms.

Criteria and databases associated with aquatic toxicology and risk assessment are well developed; however, only a paucity of risk assessment information is available for terrestrial systems. Two approaches are used for evaluating toxicity to organisms and ecosystems—measurement at the site, and prediction using bioassay data in the literature [11.34]. Published data sources include the EPA's *Quality Criteria for Water* [11.35] and the U.S. Fish and Wildlife Service publications series *Hazards to Fish, Wildlife, and Invertebrates* [11.36].

If sufficient published toxicity data are not available, site-specific toxicity testing and bioassays are sometimes conducted in situ or in the laboratory. In situ evaluations often involve placing caged animals or vegetation at the site with subsequent toxicity monitoring. Laboratory studies focus on moving contaminated water and sediment from the site and evaluating their effects on laboratory animals.

Although each of the four components of risk assessment is conceptually similar for human and ecological assessments, there are differences. A general distinction may be made in ecological versus human health risk assessment—ecological risk assessments usually evaluate the effects of hazardous wastes on populations and communities, and human risk assessments focus on the effects of hazardous wastes on individuals. An exception to this general scheme is related to the presence of endangered species exposed to hazardous wastes; regulations mandate that ecological risk assessments involving endangered species must focus on both populations and individuals.

11.4   SOURCES OF UNCERTAINTIES IN RISK ASSESSMENT

One of the limitations in performing both human health and ecological risk assessments is the uncertainty that occurs at almost all levels of the analysis. Although the risk assessment process provides a numerical value for the probability of hazard that is relatively straightforward, a significant degree of uncertainty is inherent in risk calculations.

Suter [11.33] emphasized that at the current level of knowledge, detrimental effects on ecosystems cannot be adequately predicted; that is, the current methods can only assess risks in a simplified manner. He concluded that current methods allow only relative ranking of risk—from chemical to chemical or site to site. Nonetheless, such relative risk ranking provides a basis for prioritizing environmental hazards, particularly if the data are analyzed by qualified risk assessors.

Because most risk assessments use data based on assumptions and extrapolations, uncertainties result from the lack of knowledge or data. A high degree of caution is often used by risk assessors in assigning absolute numbers to risk values and hazard indices because significant uncertainty is inherent in each of the four steps of risk assessment which is then compounded into the final risk value. Descriptions of some of the sources of uncertainty follow.

Source Characterization. Examples of uncertainties in source characterization include inaccurate source sampling, limitations in analytical results, and improper selection of the contaminants used to calculate risk. For example, priority pollutant analyses are often used to screen chemicals at hazardous waste sites. However, contaminants other than the 129 priority pollutants, as well as their metabolites, may be found at these sites, which are not detected by EPA Methods 624, 625, or other standard analytical schemes. The problems of including nondetected chemicals in a health risk assessment were discussed by Seigneur et al. [11.37]. They emphasized that such a practice may lead to estimated risks that exceed regulatory thresholds because the detection limit or half of the detection limit must be used in risk calculations.

Lack of Available Data. Unavailability of data is often a significant source of uncertainty in risk assessments. Missing data may include source concentrations or source contaminants such as those not quantified by standard analyses such as EPA Methods 624 and 625. Reference doses and slope factors are currently available for fewer than two hundred chemicals; with thousands of chemicals potentially present at contaminated sites and hazardous waste facilities, a significant amount of uncertainty may result from the paucity of available data.

Exposure Assessment Models and Methods. Whether contaminant concentrations to which receptor organisms are exposed are determined by predictive models or sampling and analysis of the media, uncertainties are inevitable. Although the uncertainty of models is obvious, sampling data are also variable. Furthermore, some standard analytical procedures are subject to false positive and false negative errors; for example, natural soil organic matter appears in total petroleum hydrocarbon (TPH) analyses in the same manner as hydrocarbons, resulting in a false positive error. Some contaminants are not efficiently extracted into organic solvents for analysis from soils and sludge, resulting in a false negative error.

Many risk assessments focus on possible future contaminant concentrations at a point of exposure using exposure assessment models. The uncertainty of such modeling of contaminant concentrations projected into the future is obvious. The possible occurrence of hazardous waste spills and traffic accidents involving hazardous materials is also predicted using stochastic modeling. These predictive models also contain a significant degree of uncertainty.

Toxicological Data. Allowable daily intakes (ADIs), No Observable Adverse Effect Levels (NOAELs), reference doses (RfDs), and slope factors (SFs) are based on animal studies with extrapolation of the animal toxicity data to humans with the use of safety factors. These factors are characterized by a high variability of up to several orders of magnitude (1000–10,000), which significantly affects the outcome of risk assessments.

Toxicity data for ecological risk assessments are often subject to even more uncertainty compared to human health toxicity data. Furthermore, toxicological effects on communities, through disruption of predator-prey relationships and changes to species diversity and community structure, are unavailable and estimates are likely to have a high degree of uncertainty.

One approach to evaluating uncertainty of risk assessments is the use of stochastic modeling. Ünlü [11.38] used Monte Carlo, first-order, and point-estimate methods to evaluate uncertainties in contaminant concentrations downgradient from a waste pit. A comparison analysis showed that for conservative contaminants, the first-order method is comparable with the accuracy of the Monte Carlo method and can be used as an alternative to the Monte Carlo analyses. In general, the accuracy of the first-order and point-estimate methods were sensitive to the fate and transport of the contaminants. Risks from groundwater contamination have also been evaluated with a model that addresses the effects of source-, chemical-, and aquifer-related uncertainties [11.39]. The application of the method was demonstrated with a nonreactive solute and with *o*-xylene as a reactive contaminant in groundwater. The model results were checked against those from a Monte Carlo simulation method, and were found to be in good agreement except for low-probability events.

Shevenell and Hoffman [11.40] used uncertainty analyses to improve the reliability of risk assessment modeling, a procedure that helps to moderate the sometimes unknown model assumptions and the uncertainty associated with some model parameters. Finally, Doyle and Young [11.41] recommended that the uncertainty in a risk analysis should be included in the assessment report. In most cases, the effect of uncertainties on the potential negative public health effects are addressed by the use of conservative assumptions. For example, conservative estimates of attenuation and lower natural degradation rates than normal may be used to counter the uncertainty of exposure assessment models.

11.5 RISK MANAGEMENT AND RISK COMMUNICATION

Risk assessment alone provides a quantitative value of potential hazards from hazardous waste sites and facilities; however, the risk evaluation is usually carried further through risk management. *Risk management* is the decision-making process that is based on the quantitative value obtained from risk assessment models coupled with insight, experience, and judgment. Risk assessment and risk management are multifaceted methodologies that require the consideration of quantitative values, qualitative assessments, and professional judgment. The lines between science, engineering, economics, and policy become indistinct, and, therefore, the ultimate decision in a risk analysis needs to be balanced with professional judgment.

Scientists and engineers involved in hazardous waste risk assessment often have an appreciation of the risks associated with hazardous wastes. However, the public, especially those who may live near a hazardous waste site or facility, require information on the risk assessment process, its uncertainties, and the value judgments, all of which are an integral part of *risk communication.*

A requirement of the hazardous waste risk assessment team is effective communication of the procedures used and the results obtained that are of concern to the local population. Some specific aspects of risk communication of which the public must be aware are (1) the steps of the risk assessment process, (2) acceptable levels of risk, and (3) the uncertainties and value judgments that are inherent in the risk assessment process. The four fundamental steps in hazardous waste risk assessments (source characterization, exposure assessment, toxicity assessment, and risk estimation) are described to the public—a process that is easily presented in outline form at citizens' meetings and hearings. Typical acceptable levels of risk, such as the 1×10^{-6} typically used in Superfund risk assessments, are often discussed in relation to other natural and anthropogenic risks that are part of the world in which we live. In addition to communicating the risk assessment procedure and commonly used levels of risk, the public must be aware of the uncertainties at the different stages of the risk assessment process and that the quantitative value of risk that is obtained is bounded above and below that number by a range of uncertainty. Furthermore, the public must be aware that these uncertainties are a part of quantifying such complex processes and absolute certainty in such assessments will probably never be realized. The public should also be aware that the experience and value judgments of the risk assessment team are an important and necessary part of the risk assessment process and that the use of such qualitative tools is a common procedure.

11.6 SUMMARY OF IMPORTANT POINTS AND CONCEPTS

- Risk assessment is most commonly used to evaluate the level of cleanup necessary at contaminated sites and to aid in the selection of remedial technologies. It is also used to evaluate potential releases from RCRA facilities, hazardous waste spills, and other potentially hazardous events.

- Risk assessments are divided into four phases: hazard identification, exposure assessment, toxicity assessment, and risk characterization.

- Hazard identification is based on source analysis (Chapters 1–4), in which the identity, concentrations, characteristics, and properties of the contaminants are evaluated.

- Exposure assessment is an extension of pathways analysis (Chapters 5–8), which examines the routes through which the contaminants proceed when released from the site; such an analysis provides a measure of the contaminant concentration at the point of exposure.

- Toxicity evaluation most commonly involves obtaining reference doses (RfDs) and slope factors (SFs), which are used in subsequent risk characterization. In some cases, RfDs and SFs must be derived for contaminants for which data are not available.

- Risk characterization is the final step in the risk assessment process; it involves obtaining a quantitative risk value based on the daily intake rate reference doses, for noncarcinogens, and slope factors for carcinogens.

- Ecological risk assessments are performed in a manner analogous to human health risk assessments and have been used to evaluate the effects of hazardous wastes on sensitive ecosystems, and endangered species.

PROBLEMS

11.1. Describe, using a paragraph for each, the four steps in human health risk assessments.

11.2. The following compounds have been found at a hazardous waste site.

Compound	Soil Concentration (mg/kg)
Aldrin	42
Acetophenone	240
Cadmium	8
Di-*n*-butylphthalate	56
Diquat	110
PCE	450
Pyrene	106

Determine the least number of surrogates that will account for 99% of the risk at the site.

11.3. The following hazardous chemicals have been found migrating from a hazardous waste landfill.

Chemical	Leachate Concentration (mg/L)	Air Concentration ($\mu g/m^3$)
Acetone	240	8
Bis(2-chloroethyl)ether	40	3
trans-1,2-Dichloroethylene	12	12
1,4-Dichlorobenzene	16	1
MEK	320	22
Hexachlorocyclopentadiene	0.080	ND
Dieldrin	0.002	ND
PCE	40	10
TCE	70	14
1,2,4-Trichlorobenzene	2	2
ND = nondetectable		

Using both water and air data, determine the least number of surrogate compounds that will account for 99% of the health risk at the site.

11.4. Describe some of the processes that affect the intake rate of organisms and how intake is quantified during ecological risk assessments.

11.5. A drinking water supply is contaminated with 85 $\mu g/L$ of MEK. Determine the hazard index for exposure to a 70-kg (154 lb) adult over 30 years, if the intake rate = 2 L/day, and the exposure frequency = 365 days/year.

11.6. A drinking water well is contaminated with 8 $\mu g/L$ of PCE. Determine the CDI for a 70-kg (154 lb) adult using typical intake parameters.

11.7. Determine the hazard index for each of the following chemicals if the daily oral intake is 0.98 mg/(kg-day).
a. Acephate   b. Baygon
c. Bidrin   d. Captan
e. Chromium (III)   f. Chromium (VI)
g. Paraquat   h. Terbacil

11.8. Determine the individual cancer risk for a daily oral intake of 0.024 mg/(kg-day).

    a. Aniline   b. Azobenzene

    c. 1,1,2-TCA   d. Prochloraz

    e. Benzene   f. Toxaphene

11.9. Benzene is being released from an air stripper 8 m (26.2 ft) high used in a pump-and-treat groundwater cleanup operation at a rate of 0.5 m³/sec and at a rate of 40 mg/sec at the top of the air stripper. A housing development is located 2.5 km downwind from the site. The site is located at the mouth of a canyon, where afternoon downcanyon winds provide the only significant transport of the benzene to the housing development. Meteorological data show that winds occur from 1 P.M. to 4 P.M. daily at a velocity of 12 km/h under strong solar radiation. A 1×10^{-6} cancer risk factor has been established for emissions from the air stripper. Risk calculations are based on a 5-kg child who breathes the air at a rate of 0.25 m³/h and is exposed for 3 h per day for 20 years. Determine the carcinogenic risk from the air stripper.

11.10. Through sampling and chemical analysis, the following contaminants have been found in an urban housing area.

Compound	Air Concentration (μg/m³)	Drinking Water Concentration (μg/L)
2,4-Dichlorophenol	8	15
Hexachlorocyclopentadiene	4	8
Nitrobenzene	17	40

Determine the total risk from these chemicals. Use an exposure frequency of 365 days per year for noncarcinogens. Assume a 70-kg body weight and an exposure period of 30 years for all chemicals.

11.11. Routine air monitoring at a housing development downwind from a RCRA solvent recycling facility documented mean concentrations of 4 μg/m³ MEK, 12 μg/m³ acetone, 0.5 μg/m³ 1, 1, 2-TCA, and 1 μg/m³ PCE. Determine the total health risk to 70-kg adults in the development if their inhalation rate is 0.83 m³/h.

11.12. Chloroform is volatilizing from a contaminated soil at a rate of 0.45 $\dfrac{\text{g}}{\text{cm}^2\text{-h}}$

The site is 42 m². An elementary school is located 4 km downgradient, with students at the school from 8 A.M. through 3 P.M., 200 days per year. If the average wind speed is 8 km/h and the air stability class is C, what is the risk to 23-kg (51 lb) children if their mean inhalation rate is 0.46 m³/h?

11.13. A soil pit is leaching benzene into a shallow aquifer, and the primary concern is a drinking water well 800 m (0.5 miles) away. The soil pit is saturated with benzene, so assume a continuous input controlled by the water solubility of benzene. The following data have been obtained for the site.

Pore-water velocity = 0.4 m/day
$D = 0.020$ m²/day
Porosity = 0.4
Bulk density = 1.4 g/cm³
Average soil organic carbon content (the primary sorbent) = 0.2%
First order biodegradation rate constant = 0.006 day⁻¹

The following receptor information has been obtained for the population drinking from the well.

Body weight = 10 kg (small child)
Exposure frequency = 350 days/year
Contact rate = 2 L/day
Exposure duration = 20 years
Averaging time = 25550 days

Using the information given, determine the risk to the population drinking the water.

11.14. Waste 2,4,6-trichlorophenol (TCP) from a surface impoundment is released into a shallow groundwater system with the following characteristics.

$V = 0.14$ m/day
Organic carbon content (the primary sorbent) = 0.6%
Porosity = 0.45
$D = 0.010$ m²/day
$\rho_B = 1900$ kg/m³

Biodegradability studies have shown that TCP degrades by first-order kinetics with $k = 0.00065$ day⁻¹. The primary receptor for the system is a drinking water well 250 m (820 ft) downgradient. What is the maximum concentration of TCP that can be released from the site into groundwater to provide a 1×10^{-6} cancer risk at the well?

11.15. A waste tip from an abandoned landfill, from which primarily 1,1,2-TCA is released, extends into groundwater. The 1,1,2-TCA is released from the site at a constant rate of 1% of the concentration sorbed on the soil, and a concentration gradient exists ranging from undetectable to 1200 mg/kg in the subsurface below the site. Other site characteristics include pore water velocity = 30 cm/day, subsurface organic carbon content (the primary sorbent) = 0.2%, $D = 0.005$ m²/day, porosity = 0.4, $\rho_B = 1700$ kg/m³, and $k = 0.002$ day⁻¹. The distance to a downgradient drinking water supply well is 1.2 km (0.75 miles). If the maximum carcinogenic risk possible from drinking water at the site is 1×10^{-6}, determine the cleanup concentration of 1, 1, 2-TCA in the soil that corresponds to the 1×10^{-6} risk at the well. For the risk calculation, assume an ingestion rate of 2 L/day and body mass of a 10-kg child.

11.16. Describe the two sources of data for toxicity assessment in ecological risk assessments. Which source of data should be used first?

11.17. What are the similarities and differences between human health risk assessments and ecological risk assessments?

REFERENCES

11.1. Heath, J., "Applying risk assessment to RCRA," *Water Environ. Technol.,* **5,** 36 (1993).

11.2. Smith, D., and others, "Incinerator risk assessments: Change is in the air," *Environ. Eng.* (Supplement to June 1994 *Chem. Eng.*), **101,** 26 (1994).

11.3. Vergison, E., "A quality-assurance guide for the evaluation of mathematical models used to calculate the consequences of major hazards," *J. Hazard. Mater.,* **49,** 281 (1996).

11.4 Krukowski, J., "Reducing risks on Route 66," *Pollut. Eng.,* **26,** 60–65 (1994).

11.5. *Risk Assessment Guidance for Superfund: Environmental Evaluation Manual,* EPA/540/1-69/001A, OSWER Directive 9285.7-01, U.S. Environmental Protection Agency, Washington, DC, 1989.

11.6. *Risk Assessment Guidance for Superfund. Volume I—Human Health Evaluation Manual (Part B, Development of Risk-Based Preliminary Remediation Goals) Interim,* Publication 9285.7-01B, U.S. Environmental Protection Agency, Office of Emergency and Remedial Response, Washington, DC, 1991.

11.7. U.S. Environmental Protection Agency, "Proposed guidelines for exposure-related measurements," *Federal Register,* 48830, December 1988.

11.8. *Guidance for Conducting Remedial Investigations and Feasibility Studies under CERCLA,* OSWER Directive 9355.3-01, OSWER Directive 9355.3-01, U.S. Environmental Protection Agency, Interim Final, Office of Emergency and Remedial Response, Washington, DC, 1988.

11.9. *Development of Statistical Distributions or Ranges of Standard Factors Used in Exposure Assessments,* U.S. Environmental Protection Agency, Office of Health and Environmental Assesment, Washington, DC, 1985.

11.10. *Exposure Factors Handbook,* Publication EPA/600/8-89/043, U.S. Environmental Protection Agency, Washington, DC, 1989.

11.11. Guidotti, T. L., "Exposure to hazard and individual risk: When occupational medicine gets personal," *J. Occup. Med.,* **30,** 57–577 (1988).

11.12. Raider, R., "A solid approach to risk assessment: Other programs could learn from part 503's risk assessment," *Water Environ. Technol.,* **6,** 40–44 (1994).

11.13. National Academy of Sciences, *Risk Assessment in the Federal Government: Managing the Process,* National Academy Press, Washington, DC, 1983.

11.14. Patton, D. E., "The ABCs of risk assessment," *EPA Journal,* **19,** 10–15 (1993).

11.15. van Leeuwen, C. J. and J. L. M. Hermens, *Risk Assessment of Chemicals: An Introduction,* Kluwer Academic Publishers, Dordrecht, The Netherlands, 1995.

11.16. Kroes, R. and V. J. Feron, "General toxicity testing: Sense and non-sense, science and policy," *Fundam. Appl. Toxicol.,* **4,** 298–308 (1984).

11.17. Furst, A., "Issues in interpretation of toxicological data for use in risk assessment," *J. Hazard. Mater.,* **39,** 143–148 (1994).

11.18. Maughan, J. T., *Ecological Assessment of Hazardous Waste Sites,* Van Nostrand Reinhold, New York, 1993.

11.19. Calow, P., *Handbook of Ecotoxicology,* Blackwell Science, London, 1993.

11.20. Cairns, J., Jr. and J. R. Pratt, "Ecotoxicological effect indices: A rapidly evolving system," *Water Sci. Technol.,* **19,** 1–12 (1987).

11.21. Burns, L. A. and D. M. Cline, *Exposure Analysis Modeling System: Reference Manual for EXAMS II*, EPA/600/3-85/038, U.S. Environmental Protection Agency, Washington, DC, 1985.

11.22. Vandergrift, S. B. and R. B. Ambrose, *SARAH2, A Near Field Exposure Assessment Model for Surface Waters*, EPA/600/3-88/020, U.S. Environmental Protection Agency, Washington, DC, 1988.

11.23. Ambrose, R. B. and T. O. Barnwell, "Environmental software at the U.S. Environmental Protection Agency's center for exposure assessment modeling," *Environ. Software*, **4**, 76–93 (1989).

11.24. Barber, M. C., L. A. Suarez, and R. R. Lassiter, *FGETS (Food and Gill Exchange of Toxic Substances): A Simulation Model for Predicting Bioaccumulation of Nonpolar Organic Pollutants by Fish*, EPA/600/3-87/038, U.S. Environmental Protection Agency, Washington, DC, 1988.

11.25. Connolly, J. P. and R. V. Thomann, *WASTOX, A Framework for Modeling the Fate of Toxic Chemicals in Aquatic Environments. Part 2: Food Chain*, EPA/600/4-85/040, U.S. Environmental Protection Agency, Washington, DC, 1985.

11.26. Veith, G. D., D. DeFoe, and B. Bergstedt, "Measuring and estimating the bioconcentration factor of chemicals in fish," *J. Fish. Res. Board Can.*, **36**, 1040–1048 (1980).

11.27. Neely, W. B., D. R. Branson, and G. E. Blau, "Partition coefficient to measure bioconcentration potential of organic chemicals in fish," *Environ. Sci. Technol.*, **8**, 1113–1115 (1974).

11.28. Kenaga, E. E. and C. A. I. Goring, "Relationship between water solubility, soil-sorption, octanol-water partitioning, and bioconcentration of chemicals in biota," prepublication copy of paper dated Oct. 13, 1978, given at the Third Aquatic Toxicology Symposium, American Society for Testing and Materials, New Orleans, LA, October 17–18, 1978.

11.29. Southworth, G. R., J. J. Beauchamp, and P. K. Schmieder, "Bioaccumulation potential and acute toxicity of synthetic fuels effluents in fresh water biota: Azaarenes," *Environ. Sci. Technol.*, **12**, 1062–1066 (1978).

11.30. Stephen, C. E., D. I. Mount, D. J. Hansen, J. H. Gentile, G. A. Chapman, and W. H. Brungs, *Guidelines for Deriving Numeric National Water Quality Criteria for the Protection of Aquatic Organisms and Their Uses*, NTIS PB85-227049, U.S. Environmental Protection Agency, Washington DC, 1985.

11.31. Calabrese, E. J. and L. A. Baldwin, *Performing Ecological Risk Assessments*, Lewis Publishers, Chelsea, MI, 1993.

11.32. Fordham, C. L. and D. P. Regan, "Pathway analysis method for estimating water and sediment criteria at hazardous waste sites," *Environ. Toxicol. Chem.*, **10**, 949–960, 1991.

11.33. Suter, F. W., II, "Endpoints for regional ecological risk assessments," *Environ. Manage.*, **14**, 9–23 (1990).

11.34. Duncan, P. B., "Using toxicity data to evaluate ecological effects at Superfund sites," *J. Hazard. Mater.*, **35**, 255–272 (1993).

11.35. *Quality Criteria for Water*, Office of Water Regulation and Standards, EPA/440/5-86/001, U.S. Environmental Protection Agency, Washington, DC, 1986.

11.36. Eisler, R., *Polychlorinated Biphenyl Hazards to Fish, Wildlife, and Invertebrates: A Synoptic Review*, Biological Report 85(1.7), U.S. Fish and Wildlife Service, Washington DC, 1986.

11.37. Seigneur, C., E. Constantinou, M. Fencl, L. Levin, L. Gratt, and C. Whipple, "The use of health risk assessment to estimate desirable sampling detection limits," *J. Air Waste Manage. Assoc.*, **45**, 823–830 (1995).

11.38. Ünlü, K., "Assessing risk of ground-water pollution from land-disposed wastes," *J. Environ. Eng.,* **120,** 1578–1597 (1994).

11.39. Hamed, M. M., J. P. Conte, and P. B. Bedient, "Probabilistic screening tool for ground-water contamination assessment," *J. Environ. Eng.,* **121,** 767–775 (1995).

11.40. Shevenell, L. and F. O. Hoffman, "Necessity of uncertainty analyses in risk assessment," *J. Hazard. Mater.,* **35,** 369–386 (1993).

11.41. Doyle, M. E. and J. C. Young, "Human health risk assessments," Chapter 5 in Maughan, J. T. (Ed.), *Ecological Assessment of Hazardous Waste Sites,* Van Nostrand Reinhold, New York, 1993.

Part Four

MANAGEMENT AND DESIGN APPLICATIONS

The theme developed throughout Parts I through III is that conceptual and quantitative knowledge of sources, pathways, and receptors serves as the fundamental basis for approaching almost all hazardous waste problems, such as Phase I site assessments, RCRA facility management, development of sampling plans, and performing human health risk assessments. The information of Chapters 1–11 also serves as a conceptual basis for the management and control of hazardous wastes through minimization/pollution prevention, remediation, treatment, and disposal. Part IV serves only as a primer to these important topics (which will receive more extensive coverage in a forthcoming design text). In Chapter 12, an overview of waste minimization/pollution prevention is followed by a conceptual analysis of remediation and treatment processes. In addition, brief qualitative descriptions are provided of the most common hazardous waste treatment processes. The detailed design of four pathway applications—granular activated carbon, air stripping, soil treatment using Fenton's reagent, and bioslurry treatment of soils—is presented in Chapter 13, a propitious introduction to a second-semester hazardous waste design course.

Chapter 12

Approaches to Hazardous Waste Minimization, Remediation, Treatment, and Disposal

Multifaceted problems, in which concentrated organic liquids, dilute aqueous liquids, sludges, soils, and groundwaters containing organic, inorganic, and mixed radioactive wastes are concerns, are inherent in hazardous waste management. Over the past twenty years, particularly under RCRA, a hierarchy of priorities has been developed to manage hazardous wastes. Based on the costs, threats to public health and the environment, and potential liability of some disposal technologies, the highest priority is waste minimization and pollution prevention to decrease the volume, mass, and toxicity of hazardous wastes that are generated. If an industry can effectively reduce the quantity of its waste from 20,000 kg to 8000 kg (44,000 lb to 17,600 lb) per year while maintaining equal production levels, then a substantial decrease in waste management efforts and costs may be realized.

Not all wastes from industrial production can be minimized or recycled; therefore, the second tier of waste management is treatment, with emphasis on destruction of the hazardous chemicals. The selection of treatment processes becomes somewhat complex and is a function of the waste properties and dynamics (see Chapters 3, 5, 6, 7), their concentrations, and the complexity of the matrix (i.e., whether it includes organic compounds of various properties, organics and metals, or radioisotopes mixed with organics and metals). Some of the more common treatment processes and operations include air stripping, granular activated carbon, traditional biological processes, advanced oxidation processes, metals precipitation, ion exchange, and thermal processes (such as incineration).

The final option in the management of hazardous waste is disposal—long-term containment with no treatment, in which landfill disposal is the most common method. The disposal of hazardous residues has a number of problems, the most prevalent of which is the potential for long-term environmental releases. Although current RCRA regulations require only 30-year monitoring of landfills after their closure, wastes disposed of in these facilities will likely persist for centuries because landfills are not designed to promote waste degradation. Therefore, landfill disposal represents a long-term threat of potential environmental releases into the distant future, which results in its low priority as a hazardous waste management alternative. Nonetheless, hazardous residues will continue to be generated for which no other alternatives exist. Wastes such as incinerator fly ash, dioxin-laden sludges, and metal sludges will continue to be disposed of in landfills because it is the only option for the management of these wastes.

This chapter serves only as an overview and introduction to the approaches that are used in evaluating and designing hazardous waste control systems. The concepts in this chapter are applications of the principles outlined in Chapters 1 through 10—particularly Part II, "Pathways"—and provide the scientific basis for the design of hazardous waste control systems. This chapter provides perspectives on a wide range of hazardous waste management and design topics, and Chapter 13 focuses on the in-depth design of four pathway applications: granular activated carbon, air stripping, modified Fenton's reagent, and a soil slurry bioreactor.

12.1   CONCEPTS OF WASTE MINIMIZATION AND POLLUTION PREVENTION

Pollution is sometimes defined as *resources gone to waste.* The focus of the recent past has been end-of-pipe treatment technologies, but the current trends are toward waste minimization and pollution prevention, which are implemented in a variety of ways, including process changes to prevent waste generation and separation and recovery of economically valuable components of waste streams. For example, Randall [12.1] documented reduced raw mercury requirements in the electrical and electronics industries as a result of minimization efforts. The largest decrease was in the use of mercury in batteries, with a 99% reduction over 10 years. Numerous terms have been applied to waste minimization, including pollution prevention, waste reduction, and a host of others. Terms such as *waste minimization* and *waste reduction* are more specific identifiers and focus almost solely on reduction processes compared to *pollution prevention* practices, which tend to be more comprehensive. Because of its widespread use, the acronym *P2* has been established for pollution prevention.

The widespread implementation of hazardous waste minimization and prevention confirms its importance in solving future hazardous waste problems. As minimization and pollution prevention become more a part of an industry's process configuration rather than an expensive, "add-on" requirement, more opportunities for meeting target environmental standards will be achieved with corresponding increases in production efficiency, less waste, and more cost-efficient operations. Quinn [12.2] provided a few examples of the positive results of P2 activities, such as a 95% reuse of wastewater from a painting/coating manufacturer with an annual savings of $375,000. Archer et al. [12.3] documented an economic cost-benefit analysis of P2 measures for four processes at a foundry. The four pollution prevention systems—sand reclamation, energy recovery, naphtha recovery, and coolant recycling—showed a positive net economic benefit of $2.4 million for fiscal 1992. In another successful P2 case history, process changes and alternative coatings were identified and implemented in the manufacture of over-the-road platform truck trailers [12.4]. They recycled 100% of paint-related wastes and eliminated all of their hazardous waste. The company went from a large quantity generator to a conditionally exempt small quantity generator and saved more than $135,000 in one year.

Increases in P2 practices have been emphasized through the 1984 Hazardous and Solid Waste Amendments (HSWA) to RCRA. Specific clauses within RCRA require pollution prevention plans as part of a generator's responsibility. Furthermore, a total of seven Executive Orders and nine major environmental statutes affect federal facilities and several states and local entities now require P2 implementation [12.5].

Numerous classification schemes have been developed for waste minimization and P2 efforts. Allen and Rosselot [12.6] provided an analysis of pollution prevention opportunities from three perspectives: (1) the raw materials required for industrial processes, (2) the products produced, and (3) the wastes generated. Another scheme of P2 practices has been developed based on (1) volume reduction, (2) toxicity reduction, and (3) recycling [12.7]. This classification focuses on two primary emphases of waste minimization—volume reduction and toxicity reduction. Volume reduction is usually implemented by minimizing the amount of nonhazardous materials with the waste—the most common of which include water and solids such as sweepings and soils. In contrast, toxicity reduction reduces the amount of hazardous waste generated, which usually results in volume reduction as well. Another view that may be taken for P2 efforts is shown in Figure 12.1, which is based on the hierarchy of waste management alternatives introduced in Section 4.1 combined with recent advances in P2 efforts. The highest priority is in source reduction, followed by volume reduction and recycling. The least attractive options are treatment and disposal.

Reducing the volume of waste does not necessarily reduce its mass, so early waste minimization efforts placed minimal emphasis on volume reduction. However, larger volumes of waste [e.g., 10,000 L (2,640 gal) of water containing 1% acetone vs. 200 L (53 gal) of water with 50% acetone] present more difficult waste challenges. The 200 L (53 gal) of waste could be transported more economically, recycled more efficiently, or landfilled more inexpensively. Therefore, volume reduction has received a significantly greater emphasis in P2 implementation. Volume reduction can be promoted through a number of routes, many of which parallel source reduction. For example, process modifications may be implemented that require less solvent, whether it is water or an organic component.

Hazardous waste P2 efforts are usually based on an audit, sometimes a cursory one, such as a plant tour with a visual inspection of waste management and housekeeping practices. Hazardous waste audits, however, are usually detailed balances of inputs and outputs of wastes to assess all potential processes by which P2 efforts can be established (Figure 12.2).

The first two P2 levels shown in Figure 12.1 (source reduction and volume reduction) are often implemented together and are the easiest to initiate because they do not usually require changes in piping, process layout, and so on. The most common categories of source and volume reductions are materials substitution, improved housekeeping practices, source segregation, reuse, process modifications, and recycling.

Materials Substitution. One of the most effective practices in waste minimization is the substitution of less toxic source materials, especially chemicals that are not subject to RCRA authority. Some of the easiest and most common substitutions include the

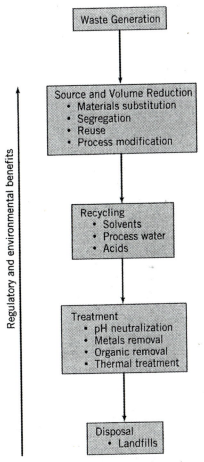

Figure 12.1 Priorities in hazardous waste management, minimization, and prevention.

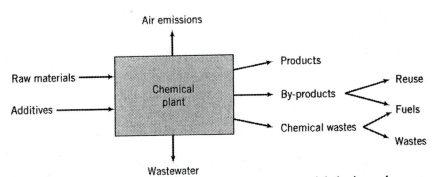

Figure 12.2 Potential options for wastes in hazardous waste minimization and prevention. *Source:* Reference 12.8.

use of natural products for solvents. For example, limonene, a natural product found in citrus, has been used as a substitute for TCE, MEK, and other solvents listed as hazardous wastes under RCRA.

Housekeeping Practices. A simple management decision can provide an almost instant implementation of P2 through more effective housekeeping procedures. A number of relatively simple practices can be established, including proper labeling (to ensure that wastes are disposed of in proper containers—e.g., F001 wastes), minimizing leakage from drums and containers, and keeping solid and liquid wastes separate. Other improved housekeeping practices are outlined in Reference 12.9.

Segregation. Perhaps one of the most logical P2 practices, as well as one that is easy to implement, is waste separation. Wastes that contain many different chemicals are not only difficult to manage and dispose of under normal RCRA hazardous waste management schemes, they may also make their reduction difficult, if not impossible. For example, if a 10-L (2.6-gal) container of spent, concentrated acetone is placed into a 208-L (55-gal) drum of dilute 200 mg/L pentachlorophenol (a waste difficult to dispose of due to the presence of dioxins), then the acetone waste becomes part of the larger PCP disposal problem. If the acetone waste had been segregated, it could have been reused, recycled, or burned as an energy source.

Reuse. A corollary to keeping wastes segregated is their potential for reuse on site. Segregated wastes that have a low degree of contamination (e.g., solvents that are still relatively clean or acids that do not contain high concentrations of metals or other contaminants) may be used in an application where a clean stock of chemical is not necessary. For example, moderately clean waste solvents could be used to degrease, on a first-cut basis, a heavy tar coating a metal part. In a similar sense, waste acid from a pickling bath could be used for pH adjustments in an application where a pure form of the acid is not required.

Process Modifications. Perhaps the most common practice that correlates with pollution prevention is changing of some of the chemical process characteristics and dynamics in order that (1) waste generation is reduced or (2) reactions associated with chemical processing are more efficient or proceed further to completion so that potentially hazardous by-products are reduced. For example, reactor type, reaction temperature, residence time, and entering feed temperature have been studied as parameters of process design modifications used to reduce and minimize waste products [12.10]. The implementation of process modifications requires knowledge of process dynamics and control; therefore, environmental engineers and scientists often work in teams with chemical engineers to implement these goals.

Recycling. Recently published reports from the EPA Toxics Release Inventory (TRI) document that the U.S. chemical industry has made significant progress in reducing hazardous chemical wastes. According to data compiled by the Chemical Manufactures Association (CMA) from 1553 member facilities preparing TRI reports, the U.S. chemical industry recycled or treated 93% of the 5.9×10^9 kg (13×10^9 lb) of chemical by-products produced [12.11]. Although not as effective a practice of pollution prevention as source reduction and volume reduction, recycling is an effective practice for recovering spent hazardous chemicals. Furthermore, it is more environmentally sound and cost-effective than waste treatment and disposal in RCRA landfills. Water is often the largest-volume material that can be recycled; water containing low concentrations of hazardous chemicals (which may be considered a characteristic hazardous waste) is often treated using standard environmental engineering operations and processes, and then recycled. Solvents are also effectively recycled, usually via distillation, which can be accomplished by designed or packaged units.

Hazardous waste minimization and P2 activities represent some of the most rapidly changing areas of hazardous waste management. The preceding discussion serves as only a brief overview of the topic. More detailed annual reviews of hazardous waste minimization may be found in the June issue of *Water Environment Research* under the topic "Hazardous Wastes: Assessment, Management, Minimization" [12.12].

12.2   CONCEPTS IN HAZARDOUS WASTE REMEDIATION AND TREATMENT

The remediation and treatment of hazardous wastes presents a number of challenges not common to municipal and industrial wastewaters. The concentration of hazardous compounds varies from the $\mu g/L$ range in contaminated groundwater to drums and tanks containing spent nonaqueous waste solvents. Nyer and Morello [12.13] proposed different definitions for hazardous waste treatment and hazardous waste remediation. They defined *treatment* as the application of a technology to a specific medium (e.g., water, soil, or air) to remove contaminants by such processes as partitioning or destruction. Alternatively, they defined *remediation* as "cleaning the environment," or an entire environmental system such as a Superfund site, a wetland, or an area surrounding an abandoned chemical plant. They also noted that the treatment of biorefractory contaminants in a single medium (e.g., a stream of water pumped from an aquifer as part of pump-and-treat scheme) is now relatively straightforward. In contrast, cleaning an entire environmental system, such as a vadose-groundwater system (i.e., remediation) is still a difficult undertaking due to the presence of NAPL phases, sorption, and the asymptotic progression to low levels of contamination.

Another difference between hazardous waste treatment processes compared to wastewater treatment is the range of media that are involved. In wastewater treatment, most of the contamination is in water, and most of the compounds that are treated are water soluble. Hazardous wastes, however, require treatment in a variety of media, including the aqueous phase, nonaqueous liquid waste materials (e.g., a drum of spent, concentrated TCE), contaminated soils and sludges, vapors and off-gases from contaminated sites, hazardous waste treatment facilities, and groundwater remediation systems. In addition, hazardous wastes are often present in slurries and other mixed media systems, which further complicates their treatment.

As a result of the presence of such complicated hazardous waste matrices (i.e., such wide variations in waste strength, waste properties, and multimedia composition), the conceptualization, selection, and design of hazardous waste treatment systems are complicated by more variables and difficulties than most other environmental engineering process designs. Some conceptual design selections are logical—e.g., the use of incineration for the destruction of mixed, nonhalogenated solvent waste and the selection of air stripping to remove $\mu g/L$ concentrations of trichloroethylene from a contaminated groundwater. However, consider the design of a treatment system for a lagoon sludge that is 5% solids and 95% water containing heavy metals, chlorinated solvents, chlorinated phenols (with the possible presence of chlorinated dioxins), and PCBs. Obviously, the treatment of such a waste would be much more complex, potentially involving a number of unit operations and processes in series.

Because of the widely varying properties of the waste chemicals and the complications inherent in the matrices, process selection becomes more manageable by having a fundamental knowledge of contaminant properties and dynamics. In Chapters 5–7, the octanol-water partition coefficient was presented as a predictor of sorption and vapor pressure and Henry's Law as predictors of volatilization. Biodegradation is affected by the oxidation state of a compound. Furthermore, these properties, as well as the rate of subsurface transport of these chemicals (quantified in Chapter 8), significantly influence the effectiveness and rates of hazardous waste treatment processes. Hazardous waste remediation and treatment may therefore be conceptualized as the same mechanisms that occur naturally in the environment (i.e., sorption and distribution, volatilization, abiotic and biotic transformations), but optimized in order that the waste can be treated within a small volume and as economically as possible. Although numerous data bases are available that list applicable treatment processes for specific wastes, there is no substitute for fundamental knowledge of contaminant properties and dynamics as a basis for remediation and treatment design.

Ex Situ and In Situ Processes. The treatment of contaminated soils and groundwater can be designed on an in situ or ex situ basis. In situ groundwater remediation processes are conducted "in place" without pumping groundwater from the subsurface to the surface for treatment. In situ soil and vadose zone treatment does not involve excavating the contaminated materials, but instead keeping them in their natural place for treatment.

Ex situ treatment of groundwater, vadose zone subsurface solids, and surface soils involves taking the contaminated media out of its natural place, treating it, and then placing it back from where it came or putting it somewhere else. Contaminated groundwater is often cleaned up by "pump-and-treat" schemes (Figure 12.3) in which, through a series of wells, contaminated water is pumped to the surface and passed through a treatment process, such as an air stripping tower or an activated carbon column. The treated water is then used on the surface, discharged to a surface water or publicly owned treatment works (POTW), or injected back to the subsurface. Contaminated soils and subsurface materials are excavated using back hoes, caterpillars, and other construction equipment, and then treated in reactors, which may be as simple as a diked, lined area or a highly engineered system such as a rotary kiln incinerator. After treatment, the excavated soils or subsurface materials are placed back into the excavated area or used for some other purpose, such as construction fill.

Ex situ treatment technologies are more expensive and require more labor and costs to accomplish. However, ex situ treatment is easier to control and has more potential for success. Ex situ processes are also more conservative in that there is less chance that contaminants or their degradation products will be released into a subsurface area that is not contaminated. In situ processes, although they have the risk of spreading the contamination beyond the original source, have been embraced by scientists and engineers because of the potential for energy and labor savings and minimal site disturbance.

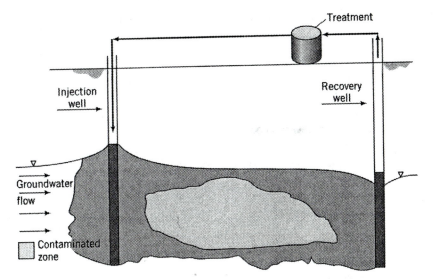

Figure 12.3 General scheme of pump-and-treat groundwater remediation. *Source:* Reference 12.13.

Removal or Treatment of Source Material. Treatment of wastes managed under RCRA is usually initiated only after the source material is minimized. By analogy, the source contamination must be removed before other media are treated in the remediation of contaminated sites (Table 12.1). For example, if drums containing waste solvents are discovered during excavation at a construction site, the first step in site remediation would be to remove the drums, followed by cleanup of the contaminated soil and, finally, treatment of the contaminated groundwater. At a pesticide formulation and rinse area, if 20-kg (44-lb) bags of DDT are found buried a few centimeters beneath the soil surface, the first step in site remediation would be removal of the bags and any of the solid DDT, followed by cleanup of the contaminated soils and then treatment of the contaminated groundwater.

Environmental releases of LNAPLs and DNAPLs often result in subsurface pools and lenses of the free product. These subsurface pools must also be removed before pump-and-treat or in situ groundwater treatment is implemented. Otherwise, the LNAPLs and DNAPLs will serve as a continuous source of groundwater contamination; that is, as the groundwater concentration decreases as the result of treatment, it is replenished by the subsurface pool of NAPL contamination. Because lenses of LNAPLs float on groundwater, they can be removed by pumping the floating layer from a well field, as shown in Figure 12.4. The removal of DNAPLs from groundwater systems has proved to be a difficult task. These liquids are found in pockets and depressions at the bottom of groundwater systems and, as a result, are extremely difficult to locate using geophysical methods (Figure 12.5). To date, no DNAPL-contaminated groundwater systems have been remediated [12.15].

Effect of Sorption. In the remediation of contaminated soils and groundwater, a number of transport and partitioning processes regulates the rate of treatment, the primary example being sorption. Hydrophobic contaminants partition onto soil, sludges, and

Table 12.1 Hierarchy of Source Removal and
Remediation Methods

First Priority

- Drums
- Tanks
- Sludges
- Other containers of source materials (e.g., bags, bins, etc.)

Second Priority

- Contaminated surface soils
- Contaminated subsurface solids
- Light nonaqueous phase liquids (LNAPLs)
- Dense nonaqueous phase liquids (DNAPLs)

Third Priority

- Contaminated groundwater
- Contaminated surface waters

subsurface solids; however, most chemical and biochemical reactions occur in the aqueous phase (i.e., in the soil-water), and in many cases, sorbed contaminants are physically unavailable for such chemical and biochemical reactions [12.16, 12.17]. Therefore, desorption almost always controls the rate of treatment of organic and inorganic contaminants. The dynamics of treatment processes, in which the rate of reaction is controlled by desorption, are illustrated in Figure 12.6. As the contaminant is degraded in the water, a deficit is created in its concentration in the aqueous phase, resulting in an increased driving force for desorption. The rate of treatment of the contaminant is governed by the rate desorption, which is a function of a mass transfer coefficient and a concentration gradient that is the difference between the saturation concentration in the aqueous phase and the concentration that results from the reaction occurring in the aqueous phase [12.18]:

$$-\frac{dC}{dt} = K(C_s - C) \tag{12.1}$$

where

$-\dfrac{dC}{dt}$ = contaminant degradation rate [mg/(L-sec)]

K = a mass transfer coefficient for contaminant desorption (sec^{-1})
C_s = the aqueous phase contaminant saturation concentration (mg/L)
C = the contaminant concentration in the aqueous phase (mg/L)

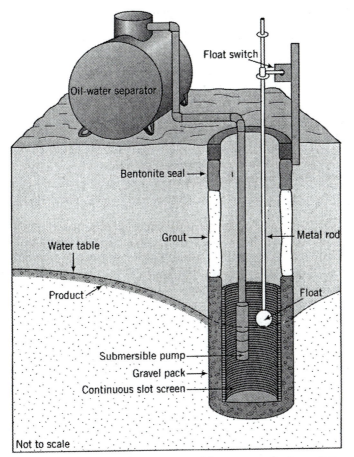

Figure 12.4 LNAPL and source removal from a groundwater system.
Source: Reference 12.13.

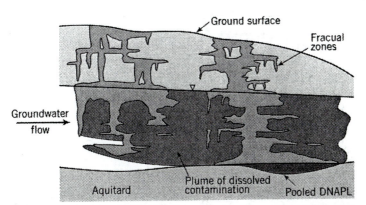

Figure 12.5 Pockets of DNAPLs in a groundwater system. *Source:* Reference 12.14.

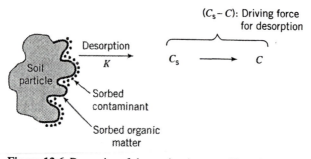

Figure 12.6 Dynamics of desorption in controlling the remediation and treatment of hazardous wastes.

Compounds that are degraded slowly in the aqueous phase desorb slowly due to the low deficit of the $(C_s - C)$ term. Conversely, compounds that desorb slowly because they are strongly sorbed (with a corresponding low value for K) will be also characterized by low corresponding rates of degradation in the aqueous phase.

Because sorption controls the rate of treatment in water-solid systems, pump-and-treat groundwater remediation has been shown to be ineffective in most cases. Even if the groundwater system is continually flushed with clean water, the MCLs commonly used as part of ARARs (e.g., 5 µg/L for PCE) are often reached only after hundreds of pore volumes have passed through the system (Figure 12.7). Furthermore, after the pump-and-treat operation is shut down, equilibrium conditions are often reestablished at aqueous phase concentrations that are sometimes close to their origi-

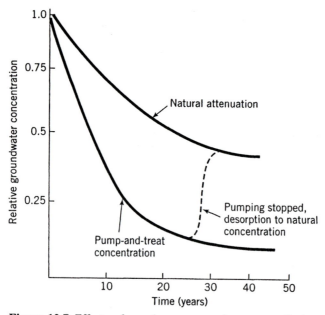

Figure 12.7 Effects of sorption on groundwater remediation through (1) asymptotic approach to reaching cleanup levels and (2) the release of contaminants to the aqueous phase after the pump-and-treat process has stopped.

nal levels, a phenomenon known as a *rebound effect*. Because of the overlying influence of sorption on pump-and-treat groundwater remediation, its use has been in decline.

12.3   REACTOR ANALYSIS APPLIED TO HAZARDOUS WASTE SYSTEMS

A fundamental approach to the analysis and design of traditional environmental engineering processes (e.g., activated sludge reactors, chlorine contact chambers, etc.) is based on mass balances and reactor analysis. The most common models for reactors are batch systems, continuous-flow stirred tank reactors (CFSTRs), and plug-flow reactors (PFRs) (Figure 12.8). Detailed analyses and derivations of different reactor configurations have been described in a number of sources [12.19, 12.20]. *Batch reactors* are characterized by no influent or effluent. The wastes in these systems are treated by adding reagents, but without any movement of contaminants (or the media of which they are part) in or out of the system boundaries. If the reaction in the system is first order, it may be described by

$$\frac{C}{C_0} = e^{-kt} \tag{12.2}$$

where,

C = contaminant concentration at time t (mg/L)
C_0 = contaminant concentration at time 0 (mg/L)
k = first-order rate constant (min^{-1})
t = time (min)

The effluent concentration of a *continuous flow stirred tank reactor (CFSTR)* is equal to the concentration in the reactor. An equation for these systems may be easily derived using a system mass balance:

$$\frac{C}{C_0} = \frac{1}{1 + k\theta} \tag{12.3}$$

where θ = the hydraulic detention time.

Plug-flow reactors are characterized by no mixing or dispersion; that is, the water moves in a "plug" through the reactor. The equation that describes plug flow reactors may be derived by performing a mass balance on an infinitesimally small cross section, *dz*, of the reactor:

$$\frac{C}{C_0} = e^{-k\theta} \tag{12.4}$$

Pure plug-flow and continuous-flow stirred tank reactors are essentially nonexistent because they represent ideal systems. Rather, most reactors may be considered as plug-flow reactors with some degree of dispersion. As the degree of dispersion increases, these reactors approach complete mix reactors.

Essentially all hazardous waste management and design cases may be approached as reactor systems. In Chapter 4, hazardous waste management facilities were evaluated as reactors for the purpose of conducting hazardous waste audits. To extend the analysis to pollution prevention using a mass balance, the goal is to reduce the *QC*

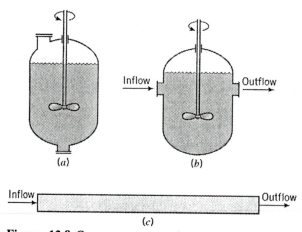

Figure 12.8 Common reactors in waste management: (a) batch reactor, (b) continuous-flow stirred tank reactor (CFSTR), (c) plug-flow reactor (PFR). *Source:* Reference 12.19.

term out of the facility (Figure 12.9). This reduction may be based on decreasing QC_0 (i.e., source material reduction, volume reduction) or recycling QC (e.g., solvent recycling, reuse of process water).

Almost all hazardous waste treatment systems (e.g., granular activated carbon contactors, ozone reactors, ex situ slurry reactors) are designed using reactor fundamentals. A common example is the analysis of air strippers, in which a mass balance on the air and water of the system serves as the basis for an equation used to size these systems (Figure 12.10). Although ex situ treatment systems are approached conceptually as reactors, remediation systems (i.e., treating the environment per Reference 12.13) have seen little application of reactor analysis. In reality, almost all environmental systems, such as the vadose zone, groundwater systems, and wetlands, may be

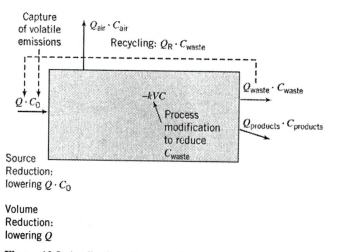

Figure 12.9 Application of reactor theory to pollution prevention implementation.

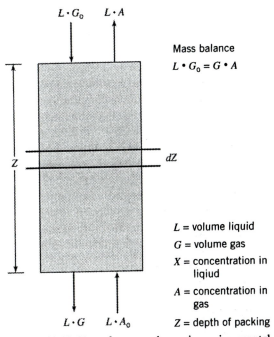

Mass balance

$$L \cdot G_0 = G \cdot A$$

L = volume liquid

G = volume gas

X = concentration in liqiud

A = concentration in gas

Z = depth of packing

Figure 12.10 Use of reactor theory in environmental engineering design: analysis of an air stripper. *Source:* Reference 12.21.

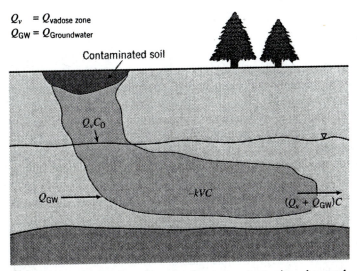

Figure 12.11 Application of reactor theory to a contaminated groundwater system.

analyzed as large reactor systems. For example, an in situ groundwater bioremediation system, shown in Figure 12.11, may be thought of as a large fixed-film biological reactor, similar to a huge trickling filter. The critical aspect in such a reactor analysis is drawing the system boundaries while accounting for all of the inputs and outputs to the system.

12.4 CLASSIFICATION OF REMEDIATION AND TREATMENT PROCESSES

Environmental engineering treatment systems have traditionally been classified as physicochemical and biological operations and processes. For example, coagulation-flocculation, sedimentation, and filtration are common physicochemical processes, and activated sludge and trickling filters are the prevalent biological processes. Hazardous waste treatment process selection is significantly more complex than municipal or even industrial waste treatment process selection due to (1) the thousands of contaminants that can potentially be present at a hazardous waste site or facility, (2) the widely varying waste concentrations and characteristics, and (3) the different media that may require treatment (concentrated waste liquids, surface soils, vadose zone, groundwater, surface water, sludges, air). Therefore, more specific classifications of hazardous waste treatment processes may be appropriate. One approach to a more specific classification of remediation and treatment is based on pathway analysis, that is, dividing processes into sorption, volatilization, abiotic, and biotic processes with the addition of a few other operations and processes (e.g., neutralization, stabilization, thermal processes).

Numerous hazardous waste treatment processes based on the four major pathways of Part II have been developed since the promulgation of RCRA and CERCLA, and some have seen more widespread use than others. Conceptual and qualitative descriptions of some of the more commonly employed processes are described here. The design of four common processes (granular activated carbon sorption, air stripping, soil treatment with Fenton's reagent, and slurry bioreactors) is developed in Chapter 13.

12.4.1 Sorption Processes

Granular Activated Carbon. Because of its high surface area and hydrophobic surface characteristics, activated carbon is an excellent sorbent for removing contaminants from water. Granular activated carbon (GAC) is prepared from a number of carbon sources—wood, bituminous coal materials, coconut shells, and lignite. The carbon is first dehydrated by heating it to 170°C; the temperature is then increased further, resulting in carbonization—which turns the material into a charcoal-like substance. The final step is activation, or the addition of superheated steam, which enlarges pores, removes ash, and increases the surface area. The resulting activated carbon has an extremely high surface area of 1000–1400 m^2/g and is composed of both macropores and micropores; the microstructure influences sorption through diffusion and mass transfer processes [12.21]. The microstructure of GAC is illustrated in Figure 12.12.

Although many process configurations have been developed for the treatment of aqueous waste streams, such as upflow and fluidized bed systems, the most common process configuration is gravity flow through a packed bed column. The dynamics of a gravity-flow GAC system are illustrated in Figure 12.13. As water is distributed at the top of the column, the contaminants are sorbed at the top of the bed. These sorption sites become saturated, and the contaminants are then sorbed further down the column. The area that is saturated (i.e., the zone of saturation) eventually moves all the way through the column until the entire column is saturated. At this point, *breakthrough* occurs and the column (which is classified as *exhausted*) has lost its effectiveness. Note that, as the zone of saturation passes down the column, an area of nonuniform saturation exists, which is due to dispersion similar to that encountered in subsurface systems (Chapter 8).

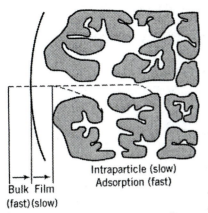

Intraparticle (slow)
Adsorption (fast)

Bulk   Film
(fast) (slow)

Figure 12.12 The microstructure of granular activated carbon. *Source:* Reference 12.21.

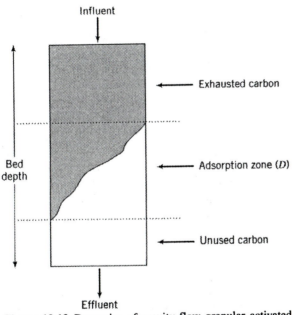

Figure 12.13 Dynamics of gravity-flow granular activated carbon treatment. *Source:* Reference 12.22.

After the carbon has been spent (i.e., exhausted), it can be regenerated by heating it in an oxygen-rich environment. Although regeneration is an effective procedure, the carbon loses some of its sorptive qualities. Furthermore, about 10% of the carbon is lost with each regeneration [12.23]. Although some industries have constructed their own regeneration facilities, the procedure is often performed at centralized locations. The design of gravity-flow GAC contactors is presented in Chapter 13.

Ion Exchange. Dilute aqueous metal-bearing waste streams may be effectively treated using packed bed reactors containing ion exchange resins. In addition, some anions can be removed from water by ion exchange. Although clays and other minerals are effective exchange materials, most ion exchange treatment systems are composed of synthetic exchange resins. These materials are most commonly polystyrene and divinyl benzene polymers, which are characterized by an abundance of negatively charged groups, such as SO_3^- [12.7]. The negatively charged groups on the surface of the polystyrene matrix must exhibit electroneutrality; that is, some positively charged species must be present to maintain a neutral charge. As described in Section 5.7, multivalent cations are selectively exchanged over monovalent cations and this selectivity can be used to optimize and design ion exchange systems. For example, the exchange of cadmium on a sodium-saturated resin may be described by

$$
\begin{array}{l}
X\!-\!Na^+ \\
X\!-\!Na^+
\end{array}
+ Cd^{2+} \longrightarrow
\begin{array}{l}
X \\
\quad Cd^{2+} + 2Na^+ \\
X
\end{array}
\tag{12.5}
$$

where X = the surface of the exchange resin.

Ion exchange resins are classified by their structure and properties. The first level of a resin's classification is based on its capability of acting as a cation or anion exchange resin. More specifically, exchange resins are classified as strong acid, weak acid, strong base, or weak base resins. Strong acid resins contain sulfonic exchange sites, and weak acid resins are characterized by carboxylic groups. Basic resins are made up of amino groups that can exchange anions.

Because of the complexity of the waste streams that are treated by ion exchange (e.g., competition for exchange sites), pilot studies are usually conducted. Breakthrough curves then serve as the basis for the design of full-scale systems.

Stabilization. *Stabilization* is the addition of materials to hazardous wastes that, through sorption, exchange, and encapsulating mechanisms, render the waste less mobile and less toxic. Stabilization, which is often also referred to as *solidification* or *fixation,* has been used in treating (1) liquid and slurry organic and inorganic hazardous wastes generated under RCRA, (2) hazardous wastes at contaminated sites, and (3) residuals from other treatment processes, such as the fly ash remaining after hazardous wastes have been incinerated. Stabilization has been applied extensively as part of the land disposal restrictions (LDRs) of hazardous wastes. Some wastes can be landfilled only after they have been treated, and many wastes that otherwise could not be disposed of in landfills may be placed in there after stabilization.

Some specific definitions have been applied to stabilization processes. *Stabilization* is the addition of reagents to a waste that reduce the potential for contaminant migration and reduce contaminant toxicity. *Solidification* is the modification of a liquid or slurry waste to a solid material by adding solids or other reagents.

The mechanisms by which stabilization occurs include macroencapsulation, microencapsulation, absorption, adsorption, and precipitation. In *macroencapsulation,* hazardous compounds are caught with the structural framework of the stabilizing material. At a smaller level, contaminants may also be trapped within the crystalline microstructure of the stabilizing material, a mechanism known as *microencapsulation.* Hazardous chemicals are also stabilized by the mechanisms of absorption, adsorption, exchange, and precipitation that are described in Chapter 5.

Stabilization often employs cement technology, in which a material such as *portland cement* (a mixture of di- and tricalcine silicates) is used. Mixing the portland cement with water provides a solid matrix that is an effective stabilization agent for metal-laden hazardous wastes. *Organically modified clays* have also been used successfully in stabilizing organic wastes. These clays are prepared by replacing inorganic cations of the inner structure of clays with organic cations, such as quarternary ammonium salts. Solidification of hazardous wastes is accomplished by adding the organoclays to the waste, followed by other additives to improve the solid properties.

Other stabilization materials include organically modified lime, organic polymers, and thermoplastic materials, such as polyethylene and polypropylene, which are mixed with wastes as they are produced at high temperatures. *Vitrification,* a process developed at Department of Energy sites, involves passing an electrical current through a contaminated soil, sludge, or other material. The current melts the solid, which cools into a glasslike material in which the contaminants are entrained.

Soil Washing and Thermal Desorption. Soils and sludges contaminated with hydrophobic organic compounds and metals are often difficult to treat because the waste compounds are strongly sorbed or exchanged on the solids. For organic compounds, transformation processes, such as biological treatment or advanced oxidation processes, are often controlled by the rate of desorption. Therefore, an important area in hazardous waste treatment is enhanced desorption through such processes as soil washing and thermal desorption. These desorption processes have not only been used as a treatment method themselves, they have also been combined with treatment processes such as bioremediation. One of the most common processes, soil washing, is based on using water or a water-surfactant solution to wash the soil or solids. Surfactants, which are commonly used in detergents, are long-chain amphibolic compounds; that is, they possess both a polar and a nonpolar end. One of the most common surfactants is sodium lauryl sulfate:

$$C_{12}H_{25}-O-SO_3Na$$

Sodium lauryl sulfate

The sulfonic end is attracted to water, and the hydrocarbon end partitions weakly with hydrophobic contaminants, bringing them into solution. Almost without exception, an excess of surfactant molecule is required to desorb contaminants [12.24]. The concentration required (called the *critical micelle concentration*—CMC) is necessary in order to form micelles, or small globule-like units that solubilize and desorb contaminants (Figure 12.14). Common CMCs are in the range of 500–1500 mg/L, which can result in significant surfactant costs for the remediation of contaminated soils. Soil washing has most commonly been performed through ex situ processes [12.24], as in the mobile unit shown in Figure 12.15; however, soil flushing (the equivalent of in situ soil washing) has been conducted with some success.

Desorption can also be enhanced by heating a soil or sludge—a process known as *thermal desorption.* Increasing the temperature enhances the release and volatilization of soil contaminants, providing an effective method for their desorption. Thermal desorption systems are operated at temperatures less than 550°C in the absence of oxygen. The contaminants, which desorb and volatilize, are collected and treated by a

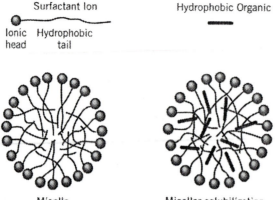

Figure 12.14 The role of micelles in desorbing and solubilizing soil contaminants. *Source:* Reference 12.24.

Figure 12.15 An ex situ soil washing unit. *Source:* U.S. Department of Energy, Richland, WA office.

transformation process. Some of the full-scale processes that have been developed include rotary kilns, heated screw conveyors, and fluidized bed systems.

12.4.2 Volatilization Processes

Air Stripping. The design of air stripping towers is an application of volatilization just as the design of GAC columns is based on sorption. Air stripping has been used for the treatment of contaminated groundwater, landfill leachates, and dilute RCRA wastes. Perhaps the most common use of air strippers has been in pump-and-treat groundwater remediation.

The selection of air stripping as an effective treatment technology is based on the tendency of contaminants to volatilize, which was covered in Chapter 6. Compounds that do not have the potential to volatilize from water are not effectively treated by air stripping. In fact, some very nonvolatile chemicals, such as dieldrin, volatilize more slowly than water and will actually become more concentrated if the attempt is made to strip them from water [12.25]. Compounds not strippable include those of high polarity, water solubility, and molecular weight. For example, acetone, a highly polar compound, is not strippable, and benzo[*a*]pyrene is also not strippable. What physicochemical variable, then, is the best predictor of the potential for air stripping? Recall from Chapter 6 that the Henry's Law constant is a thermodynamic property that describes the partitioning of contaminants between air and water. Because air stripping is a partitioning phenomenon between water and air, it follows that compounds characterized by high Henry's Law constants also have a high potential for strippability. The guideline that has been used most commonly is if a contaminant's dimensionless Henry's law constant $(H') \geq 0.01$, air stripping is a viable process option. Based on field experience, air stripping has been effective only for compounds in concentrations less than about 200 mg/L [12.26].

Air stripping has been used for decades for the removal of ammonia, sulfur dioxide, and hydrogen sulfide from water. For example, air strippers were installed at the South Lake Tahoe, Nevada tertiary wastewater treatment facility. However, the ammonia that was stripped from the water never left the Lake Tahoe watershed and still posed a eutrophication hazard, so air stripping was discontinued and breakpoint chlorination was used in its place. A similar problem has occurred in the air stripping of hazardous chemicals. When stripped from water into the air, contaminants may become hazardous air pollutants. Limitations are often placed on the mass of contaminants that can be emitted from air strippers, and process modifications to control volatiles, such as GAC scrubbers, are often installed. A detailed air stripping design that serves as a representative volatilization treatment process is described in Chapter 13.

Soil Vapor Extraction. Soil vapor extraction (SVE) systems are used to remove VOCs from the vadose zone (as an in situ process) or from piles of excavated soils (an ex situ process). The basis for SVE operation is placing a vacuum on the soil which promotes the transfer of contaminants from soil or soil-water to air. A diagram of a typical in situ SVE system is shown in Figure 12.16, and an operating ex situ SVE system is illustrated in Figure 12.17. The physical facilities of SVE systems include vapor extraction wells that are lined with perforated piping. Air is drawn through the piping emanating from the wells by a vacuum pump, which is sized based on the draw needed for the site. Vacuum gauges and flow control valves are also installed throughout the system. The contaminated vapor is then removed from the evacuated air, first passing through a knockout drum to remove moisture, and then vapor phase treatment, such as gas phase GAC sorption or biofilters. The effective radius of influence of vapor extraction wells is 6 m to 45 m (20 ft to 148 ft), which depends on the permeability characteristics of the vadose zone. The maximum depth is considered to be 7 m (23 ft) [12.27].

Soil vapor extraction is used for volatile contaminants in unsaturated soils, so the process may be considered an application of the volatilization processes covered in

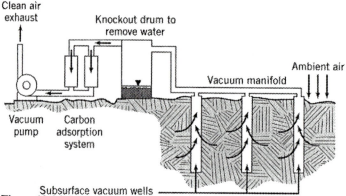

Figure 12.16 Schematic of an in situ soil vapor extraction (SVE) system.
Source: Reference 12.7.

Chapter 6. The induced volatilization in the vadose zone promotes the evacuation of soil gases, which then enhances movement of contaminants from solid and liquid phases. The selection of SVE as a remedial process is based on a number of contaminant and site characteristics; the most important variables in its process selection include soil permeability/porosity and contaminant volatility.

The design and operational criteria include (1) extraction well spacing, (2) air flow rate induced in the remediation zone, and (3) subsurface vacuum. System design is based on pilot studies, in which the loss of vacuum is measured away from the vapor extraction well, providing data for spacing extraction wells (Figure 12.18).

Figure 12.17 An operating ex situ SVE system. *Source:* Kleinfelder, Inc.

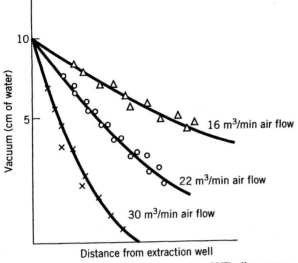

Figure 12.18 Data from the testing of an SVE pilot system.

12.4.3 Abiotic Transformation Processes

Although many abiotic processes (e.g., hydrolysis, photolysis, oxidation-reduction) can potentially be applied to hazardous waste remediation and treatment, the most common design applications have been *advanced oxidation processes* (AOPs), which include ozone, ultraviolet light and ozone, hydrogen peroxide and ozone, ultraviolet light and hydrogen peroxide, Fenton's reagent (hydrogen peroxide and catalysts), and titanium dioxide–mediated photocatalysis.

Engineered processes based on sorption and volatilization are applications of phase transfer mechanisms and do not destroy contaminants [12.28]. (Although carbon regeneration may be considered a process of contaminant destruction, spent carbon contaminated with some contaminants, such as PCBs, cannot be regenerated because of the potential for the formation of dioxins. Furthermore, some regeneration processes do not effectively destroy contaminants and the regeneration processes themselves produce waste streams.) Therefore, processes that transform and eventually mineralize hazardous compounds (i.e., convert them to carbon dioxide and water) have received increased emphasis.

The common reactant among AOPs is the hydroxyl radical (OH·), a transient species that lacks one electron relative to related thermodynamically stable species, which include water and hydroxide (see Section 7.3.3). Hydroxyl radicals can reach stability by adding to electron-rich areas of molecules

$$
\text{⬡} + OH\cdot \longrightarrow \text{⬡OH} + H^+ \tag{12.6}
$$

or by abstracting a hydrogen atom

$$CH_3—CH_2—CH_2—CH_3 + OH· \longrightarrow CH_3—CH_2—CH_2—CH_2· + H_2O \qquad (12.7)$$

As described in Chapter 7, hydroxyl radicals react with most, but not all, organic contaminants. Electron-rich compounds, such as aromatics and alkenes, react rapidly with hydroxyl radicals, but many oxidized organic compounds are considered unreactive with OH· ($K_{OH·} < 10^8$) (see Table 7.6).

Although biological processes generally provide the most cost-effective mechanism for treating organic contaminants, AOPs are, in many cases, more effective for destroying biorefractory compounds. For example, although PCE can be degraded through anaerobic cometabolism (see Section 7.4), it reacts rapidly with hydroxyl radicals and may be transformed in a matter of minutes. The primary disadvantage of AOPs is their cost; however, proponents of AOPs believe that the increased cost may be offset by the shorter time period required for treatment.

Numerous processes have been developed that generate hydroxyl radicals, most of which involve reactions of ozone, hydrogen peroxide, and ultraviolet light. These processes have been developed for the treatment of aqueous solutions; however, new applications include the in-situ and ex-situ remediation of soils and groundwater.

Reactants in AOPs. The most common reactants in AOPs are hydrogen peroxide, ozone, and UV light. Hydrogen peroxide, although considered a reactive material, is relatively stable and can be purchased from a number of manufacturers in quantities ranging from small polyethylene drums to railroad tank cars. Ozone is unstable and must be generated on site. The most common ozone-generating process is corona discharge, in which dry molecular oxygen is passed through a high-potential electrode, splitting O_2 into two oxygen (O·) radicals. An oxygen radical then combines with O_2 to form O_3. Ultraviolet lights are commercially available and are usually fit into quartz tubular reactors, which do not absorb the light that is generated.

Ozone is a gas that is injected into aqueous solutions either by itself or with other species, such as hydrogen peroxide or ultraviolet light, to increase the production of hydroxyl radicals. Ozone reacts more slowly than hydroxyl radicals primarily by adding to double bonds. It has been used extensively for disinfecting drinking water in Europe; in the last decade, ozone has received increased interest for the treatment of toxic and refractory organics in contaminated groundwater.

The decomposition of ozone has been studied extensively [12.29], and the mechanisms of ozone oxidation have been elucidated by Hoigné and Bader [12.30–12.32]. They proposed two possible mechanisms: (1) direct reaction of O_3 with organics (D reactions), and (2) free radical reactions involving hydroxyl radicals (R reactions):

$$\longrightarrow O_3 \longrightarrow \text{direct oxidation}$$
$$\downarrow \text{(alkaline conditions)} \qquad (12.8)$$
$$OH· \longrightarrow \text{radical reaction}$$

Data have shown that a portion of the ozone reacts directly with organic contaminants. The faster this direct rate, the faster the D mechanisms. Another portion of the O_3 decomposes to form OH·, a mechanism that is predominant at higher pH because it is catalyzed by OH⁻.

$$O_3 + OH^- \longrightarrow O_3^- + OH\cdot \tag{12.9}$$

$$OH\cdot + RH \longrightarrow R\cdot + H_2O \tag{12.10}$$

$$R\cdot + O_2 \longrightarrow RO_2\cdot \tag{12.11}$$

$$RO_2\cdot + RH \longrightarrow ROOH + R\cdot \tag{12.12}$$

Initiators are species that promote the decomposition of ozone, often into hydroxyl radicals and other transient species. Hydroxide ion is a common initiator in water. Other initiators include hydrogen peroxide, ultraviolet light, and organic compounds such as formic acid and humic acids [12.33].

Hundreds of chemicals that are not oxidized by ozone can be degraded when ozone decomposes to hydroxyl radicals. For example, TCA is oxidized slowly with ozone only (direct mechanism) but is rapidly oxidized by the radical mechanism [12.34]. Two processes are commonly used to promote ozone decomposition to hydroxyl radicals— ozone and hydrogen peroxide and activation of ozone by UV light.

Hydrogen Peroxide/Ozone. Hydrogen peroxide and ozone may each be fed into a reactor, a process that also promotes the formation of hydroxyl radicals:

$$H_2O_2 + 2O_3 \longrightarrow 2OH\cdot + 3O_2 \tag{12.13}$$

The primary initiator of ozone decomposition in the ozone-hydrogen peroxide process is actually hydroperoxide ion (HO_2^-), the weak base that is the conjugate of hydrogen peroxide—a weak acid:

$$H_2O_2 \rightleftharpoons HO_2^- + H^+ \qquad pK_a = 11.75 \tag{12.14}$$

Hydrogen peroxide reacts slowly with ozone, whereas hydroperoxide ion reacts rapidly to produce hydroxyl radicals:

$$O_3 + HO_2^- \longrightarrow OH\cdot + O_2^- + O_2 \tag{12.15}$$

$$k = 2.2 \times 10^6 \ M^{-1}sec^{-1}$$

The rate constant for Equation 12.15 is significantly greater than the rate constant for decomposition by hydroxide ion (Equation 12.9). Although only a small fraction of the total hydrogen peroxide is in the form of hydroperoxide, a void in the presence of hydroperoxide is caused by its rapid reaction with ozone and it is rapidly replaced by a shift in equilibrium. Therefore, the rate of change of hydrogen peroxide is

$$\frac{-d[H_2O_2]}{dt} = k[H_2O_2][O_3] \tag{12.16}$$

These process equations serve as the basis for ozone/hydrogen peroxide treatment systems, including a number of commercial package processes.

UV/Ozone. Solutions containing ozone may be irradiated with ultraviolet light (λ = 220 nm), resulting in the following reactions:

$$O_3 + h\nu + H_2O \longrightarrow H_2O_2 + O_2 \tag{12.17}$$

$$H_2O_2 + h\nu \longrightarrow 2OH\cdot \tag{12.18}$$

Another mechanism of promoting the decomposition of ozone to molecular oxygen and an oxygen atom by UV light has been developed for water-saturated air:

$$O_3 + h\nu \longrightarrow O_2 + O\cdot \tag{12.19}$$

$$O\cdot + H_2O_2 \longrightarrow 2OH\cdot \tag{12.20}$$

The hydroxyl radicals may recombine to form hydrogen peroxide, which in turn promotes a hydrogen peroxide–ozone mechanism based on Equation 12.19. Because the photochemical decomposition of hydrogen peroxide is slow, the predominant mechanism in UV/ozone systems is probably the reaction of hydrogen peroxide with atomic oxygen (Equation 12.20).

Process selection of AOPs is based on (1) contaminant reaction rates with ozone and hydroxyl radicals and (2) the concentration of scavengers. Glaze [12.35] concluded that ozone–hydrogen peroxide systems provide efficient stoichiometry for the generation of hydroxyl radicals with relatively low cost and can be easily engineered. Some of the most important process parameters in advanced oxidation processes are reactivity of the contaminant, pH, concentration of the oxidant, phase of the contaminant (e.g., NAPL, sorbed), and scavenger concentrations.

EXAMPLE 12.1 *Oxidation of a Contaminant Using Ozone and Hydroxyl Radicals*

If (a) O_3 is present at 10^{-5} mM or (b) OH· at 10^{-5} mM, what is the time required to oxidize 10 mg/L TCE to 1 μg/L TCE? The rate constant for the reaction of ozone with TCE is 17 m^{-1} sec^{-1}.

SOLUTION

Assume that the oxidant concentration is constant.

$$\frac{-d[\text{TCE}]}{dt} = k[\text{oxidant}][\text{TCE}]$$

Separate variables in the rate expression.

$$-\int_{[TCE]_0}^{[TCE]} \frac{d[TCE]}{[TCE]} = k' \int_0^t dt$$

where $k' = k[\text{oxidant}]$. Integrating yields

$$\ln \frac{[TCE]}{[TCE]_0} = -k't$$

(a) In the case of O_3,

$$k' = (17 \ M^{-1} \ sec^{-1})(1 \times 10^{-8} \ M) = 1.7 \times 10^{-7} \ sec^{-1}$$

Substituting k' into the first-order equation yields

$$\ln(0.001/10) = (1.7 \times 10^{-7})t$$

$$t = 15,100 \ h$$

(b) For $OH\cdot$, $K_{OH\cdot} = 4 \times 10^9 \ M^{-1} \ sec$ (Table 7.6)

$$k' = (4 \times 10^9 \ M^{-1} \ sec^{-1})(1 \times 10^{-8} \ M) = 40 \ sec^{-1}$$

Substituting and solving yields

$$\ln(0.001/10) = (40)t$$

$$t = 0.23 \ sec$$

Therefore, hydroxyl radicals react significantly faster with TCE than ozone.

Supercritical Water Oxidation. Two high-temperature–high-pressure processes have been used to oxidize organic hazardous wastes. Wet air oxidation is a technology that has been available since the 1950s. A more recent development, supercritical water oxidation, is characterized by higher temperature and pressure to change the physical characteristics of water to a state known as supercritical with accompanying conditions that greatly enhance oxidation. These technologies are finding increased applications for treating concentrated organic wastes, contaminated soils, and sludges.

Wet air oxidation (WAO) was originally developed for high-strength wastewaters and sludges. Typical temperatures range from 150°C to 300°C with corresponding pressures of 10 atm to 70 atm. Although these conditions have been effective in oxidizing refractory organics, mineralization is often incomplete, resulting in the need for activated carbon for effluent polishing. Therefore, more effective oxidation has been accomplished by extending WAO to supercritical water oxidation. *Supercritical water oxidation* (sometimes called *supercritical wet oxidation*) is another abiotic oxidation application that generates strong oxidants. Supercritical processes are based on the phenomenon of a *critical point* for water where no phase boundary exists between liquid water and water vapor. In other words, at temperatures and pressures above the critical point (which is 374.2°C and 218.4 atm), water exists as a supercritical fluid (a single phase with properties of both liquid water and water vapor). Organic contaminants are characterized by high solubilities in supercritical water and partition into it from sorbed and nonaqueous phases; therefore, if transformation reactions take place in a system with supercritical water, sorption effects may be negligible.

The same conditions that promote the formation of supercritical water (i.e., high temperature and high pressure) also enhance the potential for molecular oxygen to oxidize organic contaminants. Based on thermodynamics, molecular oxygen should react with organic compounds at an ambient temperature and pressure, resulting in their mineralization to carbon dioxide, water, and inorganic salts; however, the rates are so slow in the natural environment that molecular oxygen may be considered almost unreactive with most organic compounds. However, at the elevated temperatures and pressures of supercritical water, the kinetics of oxidation become rapid and most contaminants are oxidized within minutes. Therefore, contaminants (1) partition into and (2) are oxidized in a supercritical water phase. As a result of these favorable conditions, >99% destruction has been reported for TCE, TCA, organochlorine insecticides, PCBs, chlorinated dioxins, and other biorefractory contaminants.

Supercritical water oxidation has been used in ex situ reactors as well as for in situ applications. High-pressure reactors that are used for in situ systems and in situ applications are promoted deep in the subsurface by taking advantage of the hydraulic head to provide high enough pressures to achieve supercritical conditions.

The application of another AOP, Fenton's reagent, in the treatment of hazardous compounds in soils and solids will be discussed in Chapter 13.

12.4.4 Biotic Processes

The application of biological processes to the treatment of hazardous wastes has received more emphasis than most other pathway applications. Hazardous waste treatment using biological processes has been given a number of names including *bioremediation, biorestoration,* and *bioreclamation. Intrinsic biodegradation* (allowing natural biodegradation to take place in contaminated soils and groundwater rather than actively promoting the process) has sardonically been termed "bioprocrastination."

As described in Chapter 7, electron-withdrawing groups, sorption/bioavailability, and toxicity are significant process variables in the design and operation of bioremediation systems. Most contaminants slowly biodegrade in the vadose zone and groundwater because of these factors as well as the paucity of microorganisms, electron acceptors, and nutrients. The conceptual design of in situ and ex situ biological treatment systems is often based on overcoming these controlling variables.

In Situ Bioremediation. The in situ biological treatment of contaminated soils and groundwater, which is based on the fundamental microbial processes described in Chapter 7, was developed extensively in the 1980s. The processes can be conceptualized by considering the entire contaminated subsurface system as a reactor. Enhanced growth of attached biofilms is promoted in the reactor using the contaminants for carbon and energy, just as a community of microorganisms grow as a fixed film in trickling filters. The primary advantages of in situ bioremediation include lower pumping costs and minimal disposal of water or a need for reinjection. Disadvantages of the process include its difficulty to control and mass transfer limitations.

The goal in designing and operating an in situ bioremediation system is to promote microbial growth and metabolism in the subsurface system. Process considerations in such an in situ reactor include (1) hydraulic (or plume) control, (2) delivery and recovery systems, and (3) sources of oxygen or other terminal electron acceptors.

A typical bioremediation system, illustrated in Figure 12.19, consists of an injection well, recovery well, and nutrient addition systems. Hydraulic, or plume control, is usually conducted to isolate and control the subsurface contamination. Plume control can be used to increase or decrease the groundwater flow and, hence, the movement of contaminants. Furthermore, it also serves as a basis for process control in which nutrients and electron acceptors are delivered through the system. In summary, a system of injection wells and recovery wells (and potentially physical barriers, such as slurry walls) promote plume control and, in turn, in situ bioremediation.

The recovered groundwater can be treated and reinjected or the recovered water can be discharged at the surface. The water that is injected is usually amended with nutrients, an electron acceptor, and possibly microorganisms, a process known as *bioaugmentation*.

Delivery systems are designed in conjunction with hydraulic subsurface controls. A number of injection systems has been developed, including gravity and forced injection systems. Water containing nutrients and terminal electron acceptors can also be introduced using flooding, trenches, and infiltration galleries.

Gravity-feed systems are more effective for the vadose zone than for groundwater due to limited mixing in the saturated zone. Forced injection is promoted using grout seals (which minimize water flowing up the casing when high pressures are used). Forced injection systems provide more effective delivery to groundwater because the rate and pressure of injection can be controlled.

One of the most important aspects of bioremediation process designs is the terminal electron acceptor, which is based on the principles outlined in Chapter 7. Stoi-

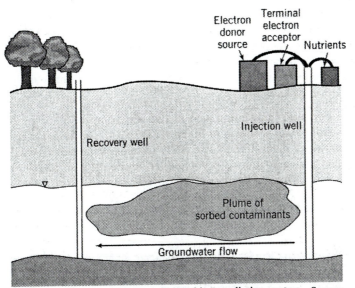

Figure 12.19 An in situ groundwater bioremediation system. *Source: Reference 12.36.*

chiometric nutrient and electron acceptor requirements are required for in situ, as well as ex situ, bioremediation. Procedures for stoichiometric nutrient calculations are outlined in Section 13.4.

The most common terminal electron acceptor used in bioremediation has been oxygen, which has limited solubility in water (<10 mg/L); therefore, oxygen delivery systems have been the subject of extensive study. The most common methods of adding oxygen to groundwater have included sparging boreholes with compressed air or compressed pure oxygen, injection of oxygenated water, and the addition of stabilized hydrogen peroxide. One of the problems in sparging or injecting oxygenated water is that, due to the limited solubility of oxygen in water, most of the dissolved oxygen is consumed within a few meters of the point of injection. Therefore, adding hydrogen peroxide achieved popularity during the 1980s because of its potential for delivering oxygen a significant distance downgradient. Hydrogen peroxide decomposition is catalyzed by the bacterial enzyme catalase or by iron (III) at neutral pH:

$$H_2O_2 \xrightarrow{\text{catalase}} H_2O + \frac{1}{2} O_2 \qquad (12.21)$$

$$H_2O_2 \xrightarrow{\text{Fe(III)}} H_2O + \frac{1}{2} O_2 \qquad (12.22)$$

However, Spain et al. [12.36] documented that even highly stabilized hydrogen peroxide decomposes rapidly to O_2 in biofilms around injection wells, negating its potential for transporting oxygen downgradient. Since then, the use of stabilized hydrogen peroxide as an oxygen source has declined, although it is still used in some situations.

12.4.5 Bioventing

In the operation of some SVE systems, such as those constructed at Hill Air Force Base in Utah, biodegradation of some of the compounds was found. As a result, SVE systems evolved into processes with almost the same physical characteristics, but in which biodegradation is promoted—hence, a bioventing system. Unlike groundwater systems in which oxygen transfer is limited, air drawn through the vadose zone can be rich in oxygen, which provides significant potential for enhanced aerobic biodegradation.

A typical operating bioventing system that has been designed to treat aviation fuel in the vadose zone is shown in Figure 12.20. At the surface, the hardware used in bioventing looks much like that of SVE systems, but the process dynamics, in which aerobic biodegradation is promoted, are quite different from soil vapor extraction. Bioventing has a number of advantages over SVE because it effectively treats compounds of low volatility. Capture of volatile emissions is a costly process in SVE or any other remediation process; if bioventing systems are operated properly, minimal emissions to the atmosphere are released, negating the need to capture VOCs.

Obviously, bioventing is used only for compounds that are biodegraded aerobically. For example, it is highly effective in the treatment of hydrocarbons (e.g., BTEX) but ineffective for highly oxidized organic contaminants such as PCE and PCBs.

Figure 12.20 An operating bioventing system.

12.4.6   Landfarming

Bioremediation of surface soils, either in situ or in a lined cell, is commonly known as *landfarming*. It has been used to treat contaminated soils and sludges, and as an engineered system in which RCRA wastes (e.g., drums of spent solvents or motor oil) are added to a soil as a land-based treatment process.

A landfarming operation is shown in Figure 12.21. The process is based on developing a high microbial biomass in the soil by adding nutrients and supplying sufficient oxygen. In most cases, the contaminant serves as the carbon and energy source. The systems are tilled and mixed to promote oxygen transfer and mass transfer using typical agricultural equipment—hence, the label landfarming.

12.4.7   Thermal Processes—Incineration

Although it can have high operation and capital costs, one of the most effective means of destroying biorefractory organic contaminants is incineration. The basis for incineration is burning the contaminants to carbon dioxide, water, and inorganic end products. Some of the obvious requirements for incineration are fuel and oxygen. Incinerators must always use excess air to achieve complete combustion; too little air results in pyrolysis. Conceptually, the principles of incineration are straightforward; the contaminants are burned like conventional fuels. In reality, incineration is more complex. For example, every chemical that is incinerated has a heating value (kJ/kg). Organic wastes usually have sufficient fuel value to support combustion, and fuel is only needed for ignition. However, some compounds, such as highly chlorinated contaminants, have low fuel values and need an amendment, such as natural gas, propane, or

Figure 12.21 Construction of a small-scale landfarming operation. *Source:* Blasland, Bouck and Lee.

fuel oil, to supplement combustion. In addition, the effectiveness of incineration is governed by time, temperature, and turbulence because most incineration systems do not behave ideally. Therefore, wastes are normally characterized for chemical composition, heat of combustion, viscosity, and corrosivity before they are incinerated.

Numerous incinerators have been developed for different applications. *Liquid injection incinerators* are used for pumpable liquid wastes. *Hearth incinerators* (e.g., multiple hearth and rotary kiln) and fluidized bed incinerators have been developed for soil and liquid applications. In the United States, the operation of incinerators is regulated under RCRA, including a requirement for a *trial burn* (including a report on the material to be burned). Part of the permitting process for incinerators is a Trial Burn Plan, which, among other things, includes identifying the Principal Organic Hazardous Constituents (POHCs).

Other Treatment Operations and Processes. Numerous other pathway and miscellaneous operations and processes are used in hazardous waste management. The most common unit operation in hazardous waste management is acid-base neutralization. The vast majority of wastes manifested under RCRA are the D002 category—corrosive wastes in which treatment can be easily accomplished by neutralization of acidic or basic wastes.

Many traditional environmental engineering processes (described in Reference 12.1) have also been applied to the treatment of hazardous wastes regulated under RCRA.

These include screening, coagulation/flocculation, precipitation, reverse osmosis, steam distillation, sedimentation, and filtration.

Many treatment processes have been used for both RCRA waste treatment and site remediation. For example, the removal of VOCs from aqueous wastes through air stripping is applied to (1) recycling process waters in industrial facilities, (2) removing VOCs from landfill leachate, and (3) treating contaminated groundwater. Similar multiple applications have been documented with other systems, such as GAC, AOPs, and various biological processes. Hundreds of different applications of hazardous waste treatment processes have been documented in both RCRA and CERCLA applications. These applications are often site-specific and range from the cleanup of soluble contaminants to the biological treatment of sludges containing biorefractory compounds.

Other hazardous waste treatment processes have been developed, especially for site remediation. *Air sparging* is an in situ volatilization process in which wells are aerated to strip volatile organic compounds from groundwater. Zero-valent iron (see Section 7.3.4) has been used ex situ and in situ to dechlorinate oxidized halogenated organic contaminants. *Phytoremediation* is the use of vascular plants to remediate contaminated surface soils.

Remediation and treatment technologies continue to be the subject of engineering research, with new processes continually being evaluated. Recent advances in hazardous waste treatment process development may be found in the June annual literature review issue of *Water Environment Research,* "Hazardous Wastes: Remediation and Treatment" [12.37].

Treatment Processes in Series. A consideration in technology selection and process design is the use of multiple unit processes to treat wastes containing a large number of contaminants of varying physical and chemical properties. For example, an aqueous waste stream containing PCE, TCE, and lead could be effectively treated by air stripping in conjugation with ion exchange. The critical issues in the process design of complex waste streams are (1) the most effective unit processes and (2) the order of the processes. Because hazardous waste remediation and treatment design may be considered an application of hazardous waste pathways, the knowledge acquired in Part II serves as the basis for a rational approach to the design of hazardous waste treatment systems. Such an approach is illustrated in Example 12.2.

EXAMPLE 12.2 *Conceptual Process Design of a Contaminated Groundwater*

Provide a conceptual process design of an ex situ pump-and-treat groundwater treatment system to remove carbon tetrachloride, TCE, pentachlorophenol, and pyrene.

SOLUTION

Carbon tetrachloride and TCE are characterized by high Henry's Law constants and low K_{ow}s. In contrast, PCP and pyrene have opposite characteristics:

Compound	Henry's Law Constant (atm-m^3/mole)	Log K_{ow}
Carbon tetrachloride	0.0302	2.71
TCE	0.0091	2.31
PCP	3.4×10^{-6}	5.02
Pyrene	1.87×10^{-5}	5.12

Based on these physical characteristics, carbon tetrachloride and TCE would be most effectively treated by a volatilization process (e.g., air stripping) and, because of the high K_{ow}, pentachlorophenol and pyrene would be effectively removed by a partitioning process (e.g., granular activated carbon).

The second consideration in the conceptual design is whether the air stripper or GAC should be the first process in the treatment train. Such a design analysis may be made by considering the two design options. If the water passes through the GAC first, followed by air stripping, the GAC would become saturated by all four compounds and use of the air stripper would not be needed. Furthermore, the more volatile compounds (each with lower K_{ow}) would migrate more rapidly through the carbon (just as in a subsurface system). Conversely, placing the air stripper first will remove the volatile carbon tetrachloride and TCE; the PCP and pyrene would then be effectively removed by the GAC.

Another design option would be the use of advanced oxidation processes. However, the low reaction rate of hydroxyl radicals with carbon tetrachloride would make this process ineffective.

Compound	$k_{OH\cdot}$ (M^{-1} sec^{-1})
Carbon tetrachloride	$<2 \times 10^6$
TCE	3×10^9
PCP	4×10^9
Pyrene	1×10^{10}

Biological treatment in an ex situ slurry reactor would not be practical because carbon tetrachloride would only be effectively biodegraded under anoxic or anaerobic conditions; pyrene would be most effectively degraded under aerobic conditions. Although anaerobic-aerobic bioreactors in series could be used, the kinetics would likely prove ineffective.

12.5 ULTIMATE DISPOSAL—HAZARDOUS WASTE LANDFILLS

Although the primary goals of hazardous waste management are (1) minimization and pollution prevention followed by (2) treatment (especially contaminant destruction), wastes will continue to be generated that cannot be minimized or treated. For example, some PCBs are not effectively destroyed and must be landfilled, and some metal-bearing soils and sludges can only be managed by landfilling. In addition, some residues from other treatment processes, such as fly ash from incinerators, must be landfilled. Based on these examples (and hundreds of others), landfill disposal will remain an option for at least a small segment of wastes that are generated.

When RCRA was first promulgated in 1980, hazardous waste landfills were of low quality—from both technical and management perspectives. The consensus among scientists and engineers in the 1980s was that all of hazardous waste landfills managed under RCRA would eventually leak and become another generation of Superfund sites. Landfill technology has progressed over the past 15 years and, although not a perfect option, hazardous waste landfills represent a satisfactory means for the disposal of non-minimizable and untreatable hazardous wastes. Landfills are designed to contain waste, while minimizing releases to environment. The basis for their design is simply containing the waste rather than treating it to innocuous products.

Hazardous Waste Landfill Design Considerations. The goal in the design of hazardous waste landfills is to contain the wastes in a safe manner and minimize their migration off site. The design is performed with a significant amount of redundancy to account for the ubiquitous potential for release to the subsurface and the atmosphere. Inherent in the design are a liner and leachate collection system to minimize downward movement of the waste. In addition, a cover system is used to minimize air emissions and infiltration resulting from rainfall.

Site selection is an important predesign decision for hazardous waste landfills. Numerous factors are involved in site selection. First, a geological evaluation of the site with emphasis on features that provide waste containment is initiated; the most favorable geological aspects in selecting a site include a clay region or an aquitard. Another alternative is to select a site above unusable groundwater, such as a saline aquifer.

Hazardous waste landfills are usually divided into cells, each of which is managed as a separate unit (Figure 12.22). Different wastes are usually placed in each cell; for

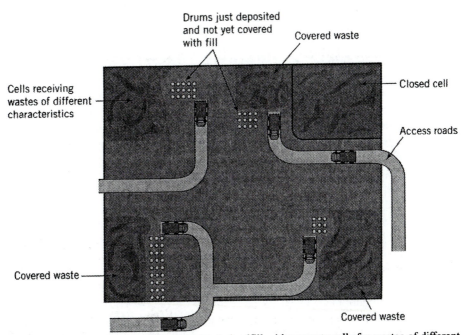

Figure 12.22 Plan view of a hazardous waste landfill with separate cells for wastes of different characteristics.

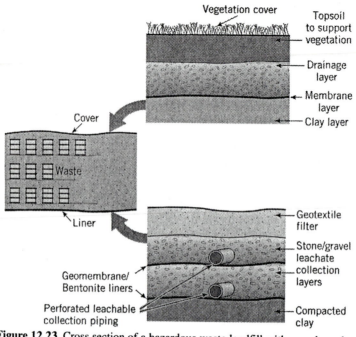

Figure 12.23 Cross section of a hazardous waste landfill with an enlarged view of the cover and liner. *Source:* Reference 12.38.

example, the leachate from cells in which chlorinated solvents are disposed would be managed as a unit (e.g., leachate from such a cell would be treated using air stripping).

Design aspects of hazardous waste landfills include the liner, the leachate collection system, waste segregation cells, and the cover/runoff mitigation system (Figure 12.23). Landfill leachate results primarily from precipitation on the landfill; therefore, cover systems are designed to minimize infiltration into the system. After a cell is completely filled, it is graded to provide a slope to enhance the runoff of water, then a series of layers is constructed to cover the waste. Directly above the waste is a subgrade layer followed by a gas collection layer of porous sands with gas collection piping. One or more barriers are then installed, including compacted clay and a geomembrane barrier. Above the barrier layers are drainage controls, and finally grass to minimize erosion.

Liners are one of the most important design features of hazardous waste landfills due to the need to contain leachate—the liquid that is composed of infiltrating rainfall and liquid waste components. Hazardous waste landfills are constructed with multiple liners and leachate collection systems. Nested within each barrier layer are smaller layers of geomembranes, bentonite, and geotextiles. Hazardous waste landfill liners, commonly known as *geomembranes,* have improved substantially over the past two decades. Between the barriers are layers of porous gravels through which leachate may drain into perforated piping. This *leachate collection system* is usually treated after it passes from under the landfill.

After wastes are placed in a landfill cell, soil cover is applied daily, usually at a depth of 30 cm (1 ft), to minimize contaminant volatilization and odor emissions. Research to date has shown that no landfill liner is 100% effective in preventing contaminant migration. As a result, a high degree of redundancy is usually incorporated into landfill designs. Some of the design features of hazardous waste landfills include alternating layers of different liner materials and underlying piping systems to collect the leachate, which consists of liquids that were disposed of in the landfill combined with water from infiltration.

12.6 SUMMARY OF IMPORTANT POINTS AND CONCEPTS

- The priorities of managing hazardous wastes, in decreasing order of importance, are minimization/prevention, treatment/remediation, and disposal.

- Hazardous waste minimization efforts hold the potential of decreasing the mass, volume, and toxicity of wastes at the source (i.e., before the chemicals enter into the RCRA waste manifest process).

- Hazardous waste remediation and treatment processes may be considered applications of hazardous waste pathways. Therefore, treatment processes may be grouped into sorption, volatilization, abiotic transformation, and biotic transformation processes. Another important class of mechanism, thermal processes (e.g., incineration), is important in hazardous waste management.

- Hazardous waste remediation and treatment processes may be classified by a number of schemes, such as in situ and ex situ processes or as RCRA wastes or CERCLA-type hazardous waste sites. Nonetheless, treatment process selection and design requires consideration of the contaminant characteristics and the matrix of the waste (i.e., liquid, soil, sludge, etc.).

- Almost every hazardous waste management system, such as a contaminated aquifer, soil pits, or ex situ treatment systems, may be conceptualized as a reactor as a basis for analysis and design.

- A number of remediation and treatment process modifications have been developed based on sorption, volatilization, abiotic, and biotic processes. The dynamics of these processes are a function of the contaminants being treated as well as the contaminant matrix.

PROBLEMS

12.1. What are the priorities in the management of RCRA wastes? What is the rationale for this priority?

12.2. A large company uses 50,000 L (13,200 gal) per year of MEK for cleaning metal parts. Discuss options for reducing the quantity of MEK used.

12.3. Describe the advantages of using in situ remediation technologies over ex situ processes. What are some risks associated with in situ remediation?

12.4. Explain the shortcomings of pump-and-treat groundwater remediation.

12.5. Briefly describe bioventing, including where it is used (groundwater? soils?), and the contaminant classes on which it can be used.

12.6. Describe the layout of a landfarming system and for what type of wastes it is used.

12.7. 2,4,6-Trinitrotoluene is present in a groundwater system with $f_{oc} = 0.02$ (the primary sorbent). How many pore volumes are required to desorb 99% of the compound? (Assume the porosity is 0.50 and equilibrium is reached with each change in pore volume.)

12.8. Consider a landfarming system, which consists of a microbial population of approximately 10^7 CFU/g soil metabolizing diesel contamination in the soil. Construct a mass balance on the system.

12.9. Perform a qualitative mass balance on the volatile solvent in the recovery/recycling facility shown below, with the open drums and the open recovery vat. Assume that the exhaust stack is the primary point of output from the system.

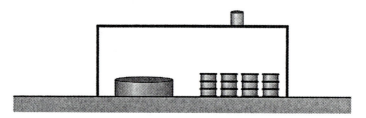

12.10. A contaminated groundwater contains mean concentrations of 120 μg/L TCE, 88 μg/L benzene, 52 μg/L pentachlorophenol, and 41 μg/L hexachlorocyclopentadiene. The Record of Decision requires pump-and-treat remediation. Propose a process scheme for ex situ pump-and-treat remediation.

12.11. A surface soil is contaminated to a depth of 1 m (3.3 ft) with Aroclor 1260. Propose a method for its treatment.

12.12. A surface impoundment contains the following contaminants in aqueous solution.

Carbon tetrachloride	5 mg/L
Cyanide	215 mg/L
Lead	18 mg/L
MEK	4800 mg/L

Propose a treatment scheme for the aqueous solution.

12.13. A sandy/gravely surface soil is contaminated with acetone, MEK, xylenes, and *n*-butylchloride. Which pathway should be engineered for removal? List three possible remedial processes for remediation of the site.

12.14. A hazardous waste site contaminated with wood preservation wastes to a depth 10 m (33 ft), with groundwater 20 m (66 ft) deep, is being evaluated for ex situ and in situ soil remediation.
a. Propose one ex situ process and one in situ process.
b. Discuss the advantages and disadvantages of the ex situ process vs. the in situ process.

12.15. A plating sludge contains cadmium, nickel, and lead. Local landfills will not accept the sludge; propose an alternative to landfill disposal for the metals.

12.16. An aqueous waste stream contains PCE, TCE, chloroform, and BTEX. A UV/ozone system has been proposed to treat the water. Do you agree with this process selection?

12.17. A hazardous waste management team is evaluating processes to treat process water in a chemical manufacturing facility in order to recycle the water for a variety of gray water uses (i.e., uses that do not require water of drinking water quality such as in cooling jackets or wash down). The water contains hexachlorocyclopentadiene, chlorobenzene, toluene, and aniline. The team recommends AOPs for treatment. Do you agree? Justify your answer.

12.18. A groundwater containing 560 μg/L of toluene is to be treated to 5 μg/L in a plug-flow ultraviolet light/hydrogen peroxide reactor. If the steady-state hydroxyl radical concentration is 2 × 10^{-10} M, determine the required detention time in the reactor.

12.19. Injection of ozone and hydrogen peroxide into an aqueous waste stream provides a steady-state hydroxyl radical concentration of 10^{-9} M. Determine the detention time of a reactor to treat 1100 μg/L PCE to the SDWA MCL in a CFSTR O_3/H_2O_2 reactor. Assume that 92% of the hydroxyl radicals are quenched by bicarbonate.

12.20. A quartz tubular reactor with a diameter of 2.0 m (6.6 ft) is to be designed to treat groundwater contaminated with 140 μg/L dimethyl phthalate to 10 μg/L. The flow rate is 2000 m^3/day (0.5 mgd). The manufacturer's data document that a concentration of 10^{-9} M hydroxyl radicals can be produced under normal conditions. Using this information, determine the necessary hydraulic detention time and volume of the reactor. Assume plug-flow hydraulics in the tubular reactor. Assume that 99.5% of the hydroxyl radicals are quenched by organic matter and other scavengers.

12.21. A contaminated wash water from an industrial facility contains the following compounds in dilute concentrations of less than 5 mg/L.

PCE
TCE
Chloroform
Carbon tetrachloride
Methylene chloride
TCA

Propose a treatment process for the water. Justify your answer.

12.22. Spent recycle water in a chemical plant is regulated under RCRA and contains the following chemicals.

2 mg/L TCE
4.1 mg/L Cd^{2+}
0.023 mg/L pyrene
1.1 mg/L hexachlorocyclopentadiene
3.7 mg/L toluene
0.3 mg/L Pb^{2+}
1.4 mg/L 2,4-dichlorophenol

Propose a treatment scheme based on the properties of the waste chemicals.

12.23. A hazardous waste site received wastes for 10 years. After extensive site assessment, the following characteristics of the site have been established:

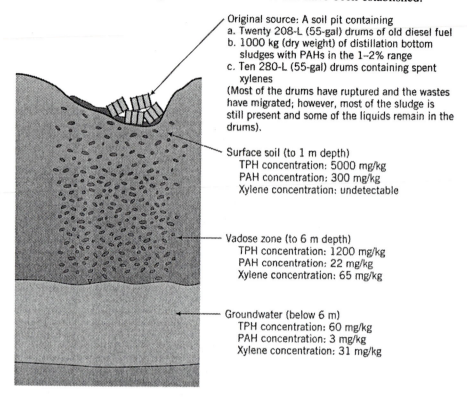

Original source: A soil pit containing
a. Twenty 208-L (55-gal) drums of old diesel fuel
b. 1000 kg (dry weight) of distillation bottom sludges with PAHs in the 1–2% range
c. Ten 280-L (55-gal) drums containing spent xylenes
(Most of the drums have ruptured and the wastes have migrated; however, most of the sludge is still present and some of the liquids remain in the drums).

Surface soil (to 1 m depth)
TPH concentration: 5000 mg/kg
PAH concentration: 300 mg/kg
Xylene concentration: undetectable

Vadose zone (to 6 m depth)
TPH concentration: 1200 mg/kg
PAH concentration: 22 mg/kg
Xylene concentration: 65 mg/kg

Groundwater (below 6 m)
TPH concentration: 60 mg/kg
PAH concentration: 3 mg/kg
Xylene concentration: 31 mg/kg

The soil is a sandy loam, and subsurface materials consist of sand and gravel, providing high permeability. Based on these data, list (a) the steps that should be taken to clean up the site and (b) the process you would propose for the treatment of (1) the soil in the pit, (2) the vadose zone, and (3) the groundwater. Provide a rationale for your decision based on a conceptual pathway analysis of the waste.

12.24. A large, multinational chemical manufacturing corporation generates the following wastes every 3 months.

Chemical	Number of 208-L (55-gal) Drums
Acetone	9
Benzene	4
Cadmium sulfide	1
Carbon tetrachloride	8
Chloroform	7
p-Dichlorobenzene	6
Hexachlorocyclopentadiene	4
Hexane	4
Hydrochloric acid	2

Chemical	Number of 208-L (55-gal) Drums
MEK	9
PCE	8
Sodium cyanide	2
2,4,6-Trichlorophenol	3

Hazardous waste landfills are designed with cells, each of which holds chemicals that are compatible and that can be treated by the same engineering processes. The design life for the landfill is 20 years; assume that 50% of the landfill material will be fill material, which is used to cover layers of waste as they are placed in the landfill. Assume that the drums will be placed in the landfill with the waste chemicals contained within them. (The size of one of the drums is 60 cm in diameter × 90 cm high).

Provide an initial design of a landfill to contain these wastes. The design is open ended and should include (1) the number of cells that will contain compatible and chemically similar wastes, (2) which waste materials will go into each cell, (3) the volume of each cell and the total volume of the landfill, and (4) a leachate treatment system for each cell (which is based on the waste characteristics).

REFERENCES

12.1. Randall, P. M., "Mercury reductions in products and processes: A review of the electrical and electronic industries," *Environ. Prog.*, **14,** 232–239 (1995).

12.2. Quinn, B., "The surface coating industries try on new coats," *Pollut. Eng.*, **27**(2), 67 (1995).

12.3. Archer, H. V., M. L. Wolff, F. Smyser, and A. Leppo, "Foundry calculates the value of pollution prevention," *Water Environ. Technol.*, **6**(6), 59–61 (1994).

12.4. Walpole, D., "Recycling paint and solvents and reducing use of 1,1,1-trichloroethane," *Waste Manage.*, **13,** 195 (1993).

12.5. Bridges, J. S., and N. T. Hoagland, "Using pollution prevention tools for compliance in the federal community," *Environ. Prog.*, **14,** 273–279 (1995).

12.6. Allen, D. T. and K. S. Rosselot, "Pollution prevention at the macro scale: Flows of wastes, industrial ecology and life cycle analysis," *Waste Manage.*, **14,** 317 (1994).

12.7. LaGrega, M. D., P. L. Buckingham, and J. C. Evans, *Hazardous Waste Management*, McGraw-Hill, New York, 1994.

12.8. Freeman, H. M. and J. Lounsbury, "Waste minimization as a waste management strategy in the United States," in Freeman, H. M. (Ed.), *Hazardous Waste Minimization*, McGraw-Hill, New York, 1990.

12.9. *Waste Minimization Opportunity Assessment Manual*, EPA/625/7-88/003, U.S. Environmental Protection Agency, Washington, DC, 1988.

12.10. Hopper, J. R., C. L. Yaws, M. Vichailik, and T. C. Ho, "Pollution prevention by process modification: Reactions and separations," *Waste Manage.*, **14,** 187 (1994).

12.11. Mullins, M. L., "Pollution prevention progress," *Water Environ. Technol.*, **5**(9), 90 (1993).

12.12. Watts, R. J., M. E. Nubbe, and T. F. Hess, "Hazardous wastes: Management, assessment, minimization," *Water Environ. Res.,* **68**, 509–520 (1996).

12.13. Nyer, E. K., and B. Morello, "Trichloroethylene remediation and treatment," *Ground Water Monit. & Remed.,* **13**(2), 98–103 (1993).

12.14. Clarke, J. H., D. D. Reible, and R. D. Mutch, Jr., "Contaminant transport and behavior in the subsurface," In Wilson, D. J. and A. N. Clarke (Eds.), *Hazardous Waste Site Soil Remediation: Theory and Application of Innovative Technologies,* Marcel Dekker, New York, 1994.

12.15. Hunt, J. R., N. Sitar, and K. D. Udell, "Nonaqueous phase liquid transport and cleanup, analysis and mechanisms," *Water Resour. Res.,* **24**, 1247–1258 (1991).

12.16. Sedlak, D. L. and A. W. Andren, "Aqueous phase oxidation of polychlorinated biphenyls by hydroxyl radicals," *Environ. Sci. Technol.,* **25**, 1419–1427 (1991).

12.17. Ogram, A. V., R. E. Jessup, L. T. Ou, and P. S. C. Rao, "Effects of sorption on biological degradation rates of (2,4-dichlorophenoxy) acetic acid in soils," *Appl. Environ. Microbiol.,* **49**, 582–587 (1985).

12.18. Watts, R. J., S. Kong, M. Dippre, and W. T. Barnes, "Oxidation of sorbed hexachlorobenzene in soils using catalyzed hydrogen peroxide," *J. Hazard. Mater.,* **39**, 33–47 (1994).

12.19. Tchobanoglous, G. and E. D. Schroeder, *Water Quality,* Addison-Wesley, Reading, MA, 1985.

12.20. Levenspiel, O., *Chemical Reaction Engineering,* 2nd Edition, John Wiley & Sons, New York, 1972.

12.21. Weber, W. J., Jr., "Evolution of a technology" *J. Environ. Engr.,* **110**, 899–917 (1984).

12.22. Noonan, D. C. and J. T. Curtis, *Groundwater Remediation and Petroleum: A Guide for Underground Storage Tanks,* Lewis Publishers, Chelsea, MI, 1990.

12.23. Crittenden, J. C., D. W. Hand, H. Arora, and B. W. Lykins, Jr., "Design considerations for GAC treatment of organic chemicals," *J. Am. Water Works Assoc.,* **79**, 74–82 (1987).

12.24. Wilson, D. J. and A. N. Clarke, "Soil surfactant flushing/washing," in Wilson, D. J. and A. N. Clarke (Eds.), *Hazardous Waste Site Soil Remediation: Theory and Application of Innovative Technologies,* Marcel Dekker, New York, 1994.

12.25. Lyman, W. J., W. F. Rheel, and D. H. Rosenblatt, *Handbook of Chemical Property Estimation Methods,* McGraw-Hill, New York, 1982.

12.26. Ball, W. P., M. D. Jones, and M. C. Kavanaugh, "Mass transfer of volatile organic compounds in packed aeration," *J. Water Pollut. Control Fed.,* **56**, 127–136 (1984).

12.27. Hoeppel, R. E., R. E. Hinchee, and M. F. Arthur, "Bioventing soils contaminated with petroleum hydrocarbons," *J. Ind. Microbiol.,* **8**, 141–146 (1991).

12.28. Tchobanoglous, G. and F. L. Burton, *Wastewater Engineering: Treatment, Disposal, and Reuse,* 3rd Edition, Metcalf & Eddy, McGraw-Hill, New York, 1991.

12.29. Glaze, W. H. and J. W. Kang, "Advanced oxidation processes for treating groundwater contaminated with TCE and PCE: Laboratory studies," *J. Am. Water Works Assoc.,* **80**, 57–63 (1988).

12.30. Hoigné, J. and H. Bader, "The role of hydroxyl radical reactions in ozonation processes in aqueous solutions," *Water Res.,* **10**, 377–386 (1976).

12.31. Hoigné, J. and H. Bader, "Rate constants of reactions of ozone with organic and inorganic compounds in water. I. Non-dissociating organic compounds," *Water Res.,* **17**, 173–183 (1983).

12.32. Hoigné, J. and H. Bader, "Rate constants of reactions of ozone with organic and inorganic compounds in water. II. Dissociating organic compounds," *Water Res.,* **17,** 185–194 (1983).

12.33. Staehelin, J and J. Hoigné, "Decomposition of ozone in the presence of organic solutes acting as promoters and inhibiters of radical chain reactions," *Environ. Sci. Technol.,* **19,** 1206–1213 (1985).

12.34. Topudurti, K. V., N. M. Lewis, and S. R. Hirsch, "The applicability of UV/oxidation technologies to treat contaminated groundwater," *Environ. Prog.,* **12,** 54–60 (1993).

12.35. Glaze, W. H., "Drinking water treatment with ozone," *Environ. Sci. Technol.,* **21,** 224–230 (1987).

12.36. Spain, J. C., J. D. Milligan, D. C. Downey, and J. K. Slaughter, "Excessive bacterial decomposition of H_2O_2 during enhanced biodegradation," *Ground Water,* **27,** 163–167 (1990).

12.37. Cha, D. K., J. S. Song, D. Sarr, and B. J. Kim, "Hazardous waste treatment technologies," *Water Environ. Res.,* **68,** 575–586 (1996).

12.38. Landfill and Surface Impoundment Performance Evaluation, SW-869, U.S. Environmental Protection Agency Office of Water and Waste Management, Washington, DC, 1980.

Chapter 13

Design of Selected
Pathway Applications

From the concepts presented in Chapter 12, a number of conclusions may be drawn about the selection and design of hazardous waste treatment systems. First, no one technology or design effectively treats all hazardous wastes; second, the selection of the most effective and economical technology is often based on the contaminant properties as well as site characteristics, which can be evaluated by applied pathway analysis. Finally, a single hazardous chemical is rarely found in soils, groundwater, landfill leachate, and RCRA wastes; rather, they are encountered as complex mixtures of varying hydrophobicity, density, and chemical structures, which sometimes dictate the need for treatment processes designed in series.

Regardless of the complexity of the waste to be treated, a conceptual understanding of how chemicals behave is fundamental to designing systems for their treatment. Therefore, engineered systems for hazardous waste remediation and treatment may be considered adaptations of natural processes in which a pathway is optimized to remove the contaminants from soil, water, air, or sludge. In this chapter, two physical phase-change processes (sorption and volatilization) will be applied through the design of granular activated carbon (GAC) contactors and air stripping towers, respectively. In addition, an abiotic transformation process and a biotic process will be applied to the treatment of contaminated soils using Fenton's reagent and ex situ biological soil treatment, respectively. The design of all of the hazardous waste treatment systems surveyed in Chapter 12 is beyond the scope of this text. More information on the design of hazardous waste treatment processes may be found in References 13.1–13.3 and the forthcoming text *Hazardous Wastes: Minimization, Remediation, Treatment, Disposal.*

13.1 SORPTION DESIGN APPLICATION: GRANULAR ACTIVATED CARBON

Hazardous chemicals that have a tendency to partition onto soils may also be effectively removed from aqueous solutions, such as groundwater or dilute RCRA wastes, by sorption and ion exchange processes. One of the most common engineered sorption systems for the treatment of organic contaminants is granular activated carbon

(GAC), a high-surface area sorbent commonly prepared from natural carbonaceous materials (see Section 12.4.1).

Process Description. Granular activated carbon treatment is usually conducted in cylindrical contactors, such as the one shown in Figure 13.1. The physical features of GAC adsorption systems include the contactor, distributor, and plenum plate (Figure 13.2). The contactor may be constructed of steel, polyethylene, or fiberglass, and is usually cylindrical with depth to diameter ratios of 3:1 to 10:1 [13.1]. The contaminated water is delivered to the top of the column, where it is distributed, usually by a perforated ring system over the top of the carbon bed. The carbon is held in place by a plenum plate with an outlet port at the bottom of the contactor. A low concentration of contaminant often passes through the column due to partitioning and diffusion limitations. In order to meet typically low cleanup requirements (e.g., ARARs based on MCLs under the Safe Drinking Water Act), GAC contactors are often configured in series. The use of three columns in series often provides a lower effluent contaminant concentration throughout the service life of the carbon while also maintaining long contactor runs before regeneration is required.

Numerous transport phenomena affect sorption of contaminants in GAC columns (Figure 13.3). *Bulk fluid transport* is a rapid process that sweeps the contaminants past the GAC and provides contact. *Film transport* occurs at the boundary layer of the carbon. Although both bulk fluid and film processes provide a mechanism of transport, the film process is generally much slower. *Intraparticle (pore) diffusion* is also a slow

Figure 13.1 A typical granular activated carbon unit in operation. *Source:* Kleinfelder, Inc.

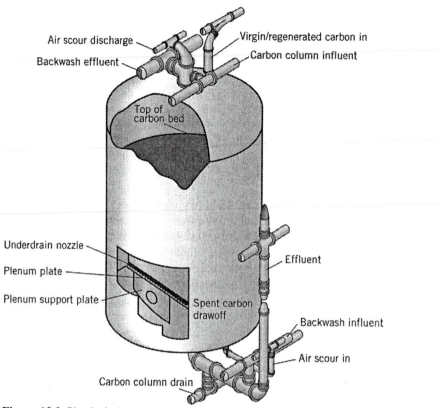

Figure 13.2 Physical characteristics of granular activated carbon contactors. *Source:* Reference 13.4.

process, limited by the pore size of the carbon and increasing molecular weight of the contaminant. As described in Chapter 5, the ultimate step is sorption, which may occur by a number of mechanisms including hydrogen bonding, Van der Waals forces, and electrostatic attraction.

The dynamics of GAC treatment are conceptually similar to one-dimensional subsurface contaminant transport (Figure 13.4). The contaminant concentration in the effluent is minimal over the first stages of treatment. As the volume of water treated increases, the *adsorption zone* or *mass transfer zone* (a region of rapidly varied concentration of contaminant on the carbon) moves down the column; that is, as the carbon at the top of the column becomes saturated, the adsorption zone in Figure 13.4 migrates down the column. After some volume of water has passed through the column, the adsorption zone is positioned at the bottom of the column; subsequently, the relative contaminant concentration in the column effluent begins to increase as the adsorption zone migrates out of the column. For the purpose of quantitation and design, *breakthrough* is defined as a relative term such as $C/C_0 = 0.05$ or 0.10. At the point where the relative concentration increases to 0.90 or 0.95, the carbon is *exhausted* (or saturated) and the contaminants are essentially moving through the column unadsorbed (actually, there is adsorption-desorption of contaminants with the same result—$C/C_0 = 1.0$).

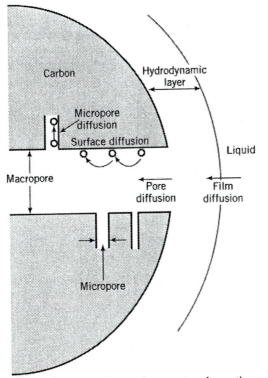

Figure 13.3 The microenvironment and sorption mechanisms of granular activated carbon. *Source:* Reference 13.2.

13.1.1 Granular Activated Carbon Design Procedures

The end points in the process design for GAC systems are (1) the mass and volume of carbon necessary to treat a given volume of water before regeneration is required, (2) the system configuration (e.g., the number of columns in series), and (3) the proposed dimensions of the system. Design procedures have been developed ranging from simple loadings to eloquent analytical models [13.3]. One of the simpler methods involves the use of published loading values or isotherm data for specific contaminants (Table 13.1). These values are usually specific to the type of carbon, so they are often provided by manufacturers. The design of GAC systems using loading or isotherm data has a number of shortcomings. These design procedures do not account for nonequilibrium conditions (i.e., the contaminant moves past the carbon before equilibrium takes place), the presence of an adsorption zone, and the effects of biological activity. Mass loadings also do not take into account competitive sorption or mass transfer limitations and are therefore best suited for small applications, such as the treatment of a low volume (e.g., 10,000 L) of contaminated water from a surface impoundment or a storage tank.

Sophisticated processes have also been developed for GAC design. The *operating line method* accounts for diffusion and mass transfer into pores within the carbon to provide a thorough basis for design. Models using finite element analysis have been

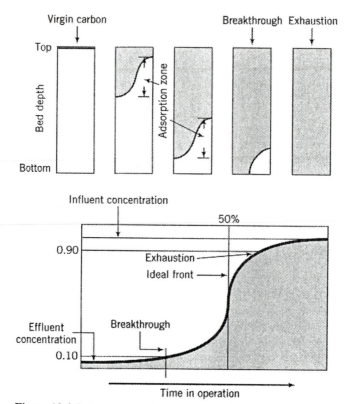

Figure 13.4 Relative contaminant concentration, C/C_0, as a function of service time during GAC contactor operation. *Source:* Reference 13.2.

developed by Crittenden et al. [13.6]. Their models account for mass transfer, dispersion, and competitive sorption by different contaminants.

Pilot Studies. An effective procedure for designing full-scale GAC contactors is the use of pilot columns in which the relative contaminant concentration is measured as a function of time (Figure 13.5). In a typical pilot study, the water to be treated

Table 13.1 Granular Activated Carbon Mass Loading
Values for Selected Hazardous Compounds

Compound	Adsorption Capacity (mg/g)
Methylene chloride	1.3
1,2-Dichloroethane	3.6
Benzene	1.0
Toluene	26.0
Ethylbenzene	53
p-Xylene	85
Naphthalene	132
Phenol	161

Source: Reference 13.5.

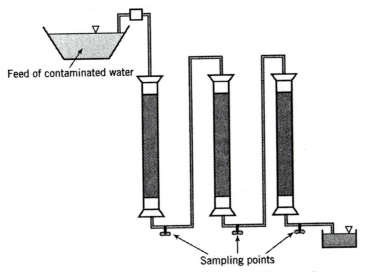

Figure 13.5 System configuration for pilot-scale GAC evaluation.

is pumped through a series of columns that are 5–10 cm (2–4 in.) in diameter and 2–3 m (6.6–10 ft) deep. The contaminant concentration is measured as a function of time, and the data are converted to relative contaminant concentrations (C/C_0). Typical data for a pilot column study are illustrated in Figure 13.6.

Analysis of Pilot System Data. The most common method for evaluating pilot column data is the graphical analysis of the breakthrough curve. Bohart and Adams [13.7] developed a procedure for analyzing data obtained from GAC pilot studies. Their procedure, known as the *bed depth–service time (BDST)* analysis, has been streamlined by Hutchins [13.8] and has since been used in drinking water treatment [13.9, 13.10] and industrial and hazardous waste treatment [13.1, 13.11, 13.12]. The first step in the procedure involves drawing horizontal lines through each of the curves at defined points on Figure 13.6 at $C/C_0 = 0.90$ and 0.10. The $C/C_0 = 0.10$ value represents *breakthrough*, and $C/C_0 = 0.90$ is often defined as *exhaustion*. The horizontal difference be-

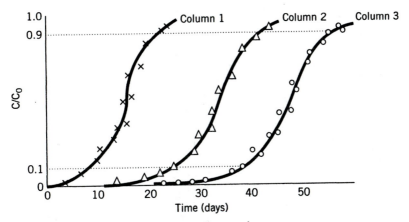

Figure 13.6 Data from a GAC pilot-scale operation.

tween the exhaustion line (C/C_0 = 90%) and the breakthrough line (C/C_0 = 10%) is defined as the *height of the adsorption zone* (*D*).

The C/C_0 = 0.90 and C/C_0 = 0.10 (i.e., service) times are each plotted as a function of bed depth (Figure 13.7) and fit to a straight line by linear regression:

$$t = a \cdot x + b \tag{13.1}$$

From the fit of the data, the constants *a* and *b* may be analyzed further to develop parameters that can be used for full-scale design.

The *a* intercept is

$$a = \text{slope (h/m)} = \frac{10^3 \cdot N}{C_0 \cdot v} \tag{13.2}$$

where

N = sorptive capacity of the carbon = $\dfrac{\text{mass of contaminant removed (kg)}}{\text{volume of carbon (m}^3)}$

C_0 = influent contaminant concentration (mg/L)

v = superficial velocity through column [m³/(m²-day) = (m/day)]

The *b*-intercept is

$$b = \text{intercept (h)} = -\left(\frac{10^3}{K \cdot C_0}\right) \times \ln\left[\left(\frac{C_0}{C}\right) - 1\right] \tag{13.3}$$

where

K = the adsorption rate constant [m³/(kg-day)]

C = contaminant concentration at breakthrough (mg/L)

After the results of the regression analysis have been completed and *a* and *b* have been determined, the procedure may be carried further. The column at $t = 0$ (the

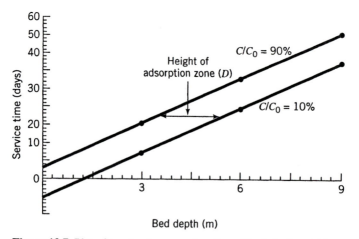

Figure 13.7 Plot of service time as a function of bed depth—the basis for GAC design.

x-intercept for the $C/C_0 = 0.10$ effluent concentration in Figure 13.7) is the *critical bed depth*, which is the minimum column depth to obtain an acceptable effluent concentration at time zero. The velocity of the adsorption zone as it moves down the column and the time for a column to become exhausted is

$$\text{adsorption velocity (m/h)} = \frac{1}{a(\text{h/day})} \tag{13.4}$$

$$\text{exhaustion time (h)} = \frac{d(\text{m})}{\text{adsorption velocity (m/day)}} \tag{13.5}$$

where d = height of the pilot columns (m).

Equation 13.4 may be expanded to determine the rate at which carbon becomes spent on a mass basis:

carbon exhaustion rate (kg/day)

$$= \frac{1}{a\ (\text{h/m})} \cdot (\text{cross-sectional area})(\text{m}^2) \cdot (\text{bulk density of carbon})(\text{kg/m}^3) \tag{13.6}$$

System Scale-Up. If GAC columns are being designed in series, the total number of columns (based on the height of the adsorption zone) is

$$n = \frac{D}{d} + 1 \tag{13.7}$$

where

n = number of columns in series required
D = height of the adsorption zone (m)

The dimensions of the full-scale GAC contactors may be determined using straightforward loading rates. The loading rate on the pilot columns may be calculated:

$$V = \frac{Q_P}{A_P} \tag{13.8}$$

where

V = column surface loading rate $\left(\dfrac{\text{m}^3}{\text{m}^2\text{-day}} \right)$

Q_P = pilot column flow rate $\left(\dfrac{\text{m}^3}{\text{day}} \right)$

A_P = cross-sectional area of the pilot columns (m^2)

The loading rate may then be used to determine the area of the full-scale columns:

$$A_F = \frac{Q_F}{V} \tag{13.9}$$

where

Q_F = full-scale flow rate (m^3/day)
A_F = area of the full-scale contactor (m^2)

Changes in Operating Criteria. After the system is designed and in operation, a change of flow rate may be accommodated by

$$t = a' \cdot X + b \tag{13.10}$$

$$a' = a\left(\frac{Q}{Q'}\right) \tag{13.11}$$

where Q' = new flow rate (m^3/day).

Equation 13.7 may also be modified to account for a change in the feed concentration:

$$a' = a \times \frac{C_0}{C_0'} \tag{13.12}$$

$$b' = b \times \frac{C_0}{C_0'} \times \frac{\ln[(C_0'/C') - 1]}{\ln[(C_0/C) - 1]} \tag{13.13}$$

where C' and C_0' are the effluent and influent concentrations for the new conditions. The use of BDST analysis is illustrated in Example 13.1.

EXAMPLE 13.1   ***Design of an Activated Carbon Contactor Using Bed Depth–Service Time Analysis***

A landfill leachate with a flow rate of 1500 m^3/day contains a mixed organic waste with a total organic carbon (TOC) concentration of 228 mg/L. A GAC system will be designed to remove 90% of the TOC. A pilot study is conducted using three columns 5 cm (2 in.) in diameter and 2 m (6.6 ft) high with a flow rate of 5.5 m^3/day. The following data were collected from the pilot study.

Time (days)	Column #1	Concentration of TOC (mg/L) Column #2	Column #3
2	0		
10	10.3		
16	34.2		
20	68.4		
22	91.2		
25	160		
28	189	0	

| | Concentration of TOC (mg/L) | | |
Time (days)	Column #1	Column #2	Column #3
33	212	6.8	
38	221	20.5	
41	223	34.2	
43	226	45.6	
45	228	79.8	
49	228	148	
53		178	0
58		205	6.8
63		217	18.2
66		221	31.9
69		226	52.4
72		227	91.2
75		228	125
78			182
83			210
87			221
94			228
100			228

Based on the pilot data, determine (1) the height of the adsorption zone, (2) the configuration and dimensions of the full-scale system, and (3) the carbon exhaustion rate. Assume the bulk density of the GAC is 481 kg/m^3.

SOLUTION

1. Plot the pilot data as relative concentration as a function of time (breakthrough curves) for each column.

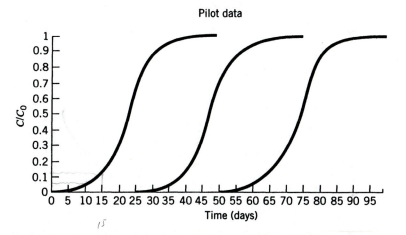

Pilot data

Next, determine the time at which each breakthrough curve crosses the points corresponding to 10% breakthrough ($C/C_0 = 0.10$) and 90% breakthrough ($C/C_0 = 0.90$):

		Service Time (days)	
Column	Depth (m)	10%	90%
1	2	13.9	31.0
2	4	38.8	57.8
3	6	64.3	81.0

Plot the service times as a function of cumulative bed depth.

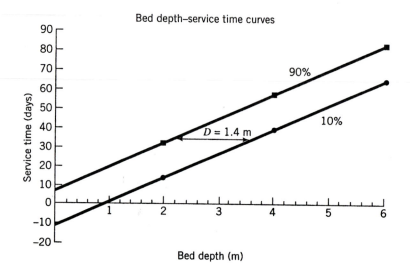

1. From the bed depth–service time plot, the horizontal distance between the break-through and exhaustion lines is 1.4 m, which is the height of the adsorption zone (D).
2. The height of the adsorption zone is 1.4 m. From the preceding bed depth–service time curves, the number of full-scale columns needed may be determined using Equation 13.7:

$$n = (D/d) + 1$$

where

$D = 1.4$ m
$d = 2$ m
$n = (1.4/2) + 1 = 1.7$

Rounding up, $n = 2$ columns. Next, the size of the full-scale columns may be calculated.

Area of pilot columns $= \pi (0.025)^2 = 0.00196$ m^2

Loading rate $V = Q_P/A_P = (5.5 \text{ m}^3/\text{day}) / (0.00196 \text{ m}^2) = 2810 \text{ m}^3/(\text{m}^2\text{-day})$

Use the same volumetric loading rate for the full-scale columns.

$A_F = Q_F/V = (1500 \text{ m}^3/\text{day}) / [2{,}810 \text{ m}^3/(\text{m}^2/\text{day})] = 0.534 \text{ m}^2$

Therefore, the diameter=0.825 m; round up to 0.9 m.

The full scale system will have 2 columns, each with a diameter=0.9 m (3.0 ft). The adsorption zone=1.4 m (4.6 ft).

3. The carbon exhaustion rate may be found from Equation 13.12:

$$\text{carbon exhaustion rate} = \frac{1}{a} \times \text{cross-sectional area} \times \text{bulk density of carbon}$$

From the bed depth–service time curve ($C/C_0 = 0.10$), the slope and intercept are

$$\text{slope } a = 12.6 \text{ days/m}$$

$$\text{intercept } b = -11.4 \text{ days}$$

$$\text{area of full scale contactor} = \pi(0.45)^2 = 0.636 \text{ m}^2$$

The carbon exhaustion rate = (0.0794 m/day) (0.636 m^2) (481 kg/m^3) = 24.3 kg/day.

13.2 VOLATILIZATION DESIGN APPLICATION: AIR STRIPPING

Physical Characteristics. Based on the information presented in Chapter 6 and Section 12.4.2, volatile contaminants ($H' \geq 0.01$) can be effectively removed from groundwater, landfill leachate, and dilute aqueous RCRA wastes using air stripping. Although the process engineering is the most critical aspect of air stripper design, the physical materials and appurtenances of the stripper, such as the housing materials, distribution system, and so on, also require consideration. The physical characteristics of an air stripper are illustrated in Figure 13.8. Air stripping is usually carried out in towers in

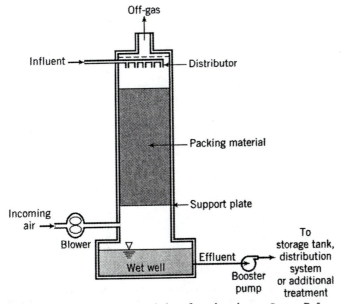

Figure 13.8 Physical characteristics of an air stripper. *Source:* Reference 13.2.

which the water is pumped to the top of a system, where it is distributed over slats, rings, or corrugated surfaces. Air is blown counter-current or cross-current to the water. Stripping towers are commonly constructed of aluminum, which is characterized by good structural properties and light weight. Fiberglass has also been shown to be a good tower material but is difficult to construct. An operating air stripper is shown in Figure 13.9.

The internal components of an air stripper are selected to ensure that mass transfer takes place under effective and economical conditions. The packing material is one of the most important factors in stripper design; it provides surface area for the air and water to interact and creates turbulence in the water stream to expose water surfaces to the air. The most effective packings are molded plastic (e.g., polypropylene), which do not degrade, are nontoxic, and are often the least expensive. A distributor at the top of the tower ensures that water is evenly distributed across the surface of the packing while providing smooth, unimpeded air flow to the top of the tower. Air exhaust ports are located around the circumference of the tower and are sized to permit the air to escape.

Air stripping design equations and procedures have been the subject of thorough development. The following discussion is based on derivations and procedures described in References 13.2, 13.13, and 13.14. The derivation of the primary air stripper design equation involves a mass balance on a cross section of a stripping tower (Figure 13.10). Some assumptions are made in the derivation, including (1) that there is a plug-flow regime in the stripping tower, (2) that the conditions follow Henry's law, (3) that the concentration of the contaminant in the effluent is negligible, and (4) that the liquid and air volumes remain constant [13.13].

Figure 13.9 An operating air stripper.

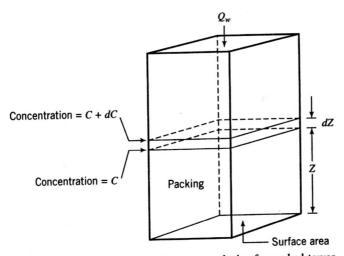

Figure 13.10 Mass balance and reactor analysis of a packed tower air stripper. *Source:* Reference 13.1.

Based on a mass balance around a horizontal section, dz, the depth of air stripping towers may be determined by

$$Z = \left(\frac{L}{K_L a}\right)\left(\frac{R}{R-1}\right)\ln\left[\frac{(C_0/C)(R-1)+1}{R}\right] = (HTU)(NTU) \qquad (13.14)$$

where

$$HTU = \left(\frac{L}{K_L a}\right)$$

$$NTU = \left(\frac{R}{R-1}\right)\ln\left[\frac{(C_0/C)(R-1)+1}{R}\right]$$

Z = depth of packing (m)
L = volumetric loading rate [m³/(m²-sec)]
$K_L a$ = mass transfer coefficient (sec⁻¹)
R = stripping factor (dimensionless)
C_0 = influent concentration (mg/L)
C = design effluent concentration (mg/L)

The number of transfer units (NTU) is, conceptually, the difficulty of removing a compound from solution. The height of a theoretical transfer unit (HTU) is a function of the rate of transfer from the liquid phase to the gas phase.

The tower diameter may be calculated based on the relationship between flow rate and the liquid loading rate:

$$D = \left(\frac{4Q}{\pi L}\right)^{1/2} \qquad (13.15)$$

where

Q = the design water flow rate (m³/sec)
L = the volumetric loading rate [m³/(m²-sec)]

The tower dimensions are interrelated by a number of variables that are used in the design. However, in general, the tower diameter is a function of the design flow rate and the tower height is based on the required contaminant removal from the water.

Use of Equations 13.14 and 13.15 involves the evaluation and determination of a number of design parameters, including the mass transfer coefficient ($K_L a$), the air-water ratio (G/L), and the stripping factor (R). Some of these values vary as a function of the gas pressure drop; therefore, the design procedure requires selection of some of the design parameters while others are determined by specific procedures, such as the use of a gas pressure drop curve.

The Air-Water Ratio and Stripping Factor. The basis for the design of air stripping towers is a mass balance on the contaminant in the air versus the water phases:

$$L(C_0 - C) = G(A - A_0) \tag{13.16}$$

where

> L = water flow rate, normalized for the cross-sectional area of the air stripper (m^3/m^2-sec)
> C_0 = influent contaminant concentration in the water (g/m^3)
> C = effluent contaminant concentration in the water (g/m^3)
> G = air flow rate, normalized for the cross-sectional area of the air stripper (m^3/m^2-sec)
> A_0 = influent contaminant concentration in the air (g/m^3)
> A = effluent contaminant concentration in the air (g/m^3)

Some assumptions are used in applying the mass balance to the derivation of the design equations. First, the contaminant concentration in the influent air (A_o) is assumed to be zero and the effluent concentration in the water (C) is also assumed to be zero (or below the level of detection). The contaminant concentration in the effluent air (A) is calculated based on the dimensionless Henry's Law constant (i.e., equilibrium conditions are assumed); therefore

$$A = H' \cdot C_0 \tag{13.17}$$

where H' = the dimensionless Henry's Law constant. Equation 13.17 may then be substituted into Equation 13.16:

$$L(C_0) = G(H' \cdot C_0) \tag{13.18}$$

Canceling C_0 yields

$$L = G(H') \tag{13.19}$$

or

$$H' \left(\frac{G}{L} \right) = 1 \tag{13.20}$$

The term $H'(G/L)$ is defined as the *stripping factor* (R), a measure of the potential stripping of the contaminant from water to air. Values of $R > 1$ are necessary for stripping to occur effectively. The value G/L is the *air-to-water ratio*, an operating parameter for air stripping towers. The stripping factor may also be defined in practical terms as

$$\text{Stripping factor} = \frac{\text{actual operating air-water ratio}}{\text{theoretical minimum air-water ratio}} \qquad (13.21)$$

or

$$R = \frac{(G/L)_{\text{operating}}}{(G/L)_{\text{min}}} \qquad (13.22)$$

where

G = air volumetric loading rate [m^3/(m^2-s)]
L = water volumetric loading rate [m^3/(m^2-s)]

The stripping factor is often a selected value in the range of 2–10. The minimum gas-liquid ratio, $(G/L)_{\text{min}}$, may be derived based on a mass balance for 100% removal of the contaminant:

$$(G/L)_{\text{min}} = \frac{C_0 - C}{C_0} \cdot \frac{1}{H'} \qquad (13.23)$$

One of the first steps in the design procedure is to determine a value for the stripping factor based on an actual air-water ratio used in pilot studies and the minimum air-water ratio from Equation 13.23. As an alternative, the stripping factor may be selected and the corresponding operating air-water ratio may be calculated [13.2]:

$$\left(\frac{G}{L}\right)_{\text{operating}} = R\left(\frac{G}{L}\right)_{\text{min}} \qquad (13.24)$$

The air-to-water ratio is interrelated with a parameter known as the *gas pressure drop*, a physical property that describes the resistance the blower air must overcome in the tower. Its value is usually found from a gas pressure drop curve, which is often supplied by vendors for a specific packing material; otherwise, a generic curve may be used, as shown in Figure 13.11. The gas pressure drop is a function of the gas and water flow rates and the size and type of packing. Furthermore, because the gas pressure drop curves slope downward as a function of the air-water ratio (part of the *x*-axis term), the actual air flow rate is specific for the preselected gas pressure drop and air-water ratio.

An air stripper designed to operate at a high pressure drop will require a smaller tower volume, which reduces the capital cost of the tower but increases blower costs. The opposite holds for strippers designed for low pressure drops. Often a matrix of pressure drops and air-water ratios (i.e., R values) are evaluated to determine the most cost-effective design. Systems designed with lower pressure drops maintain the flexibility to increase the gas flow rate (and air-water ratio). Then, if contaminant concentrations increase, the air-water ratio can be increased to strip the contaminant. The primary use of the pressure drop curve is in finding absolute values for G and L, rather

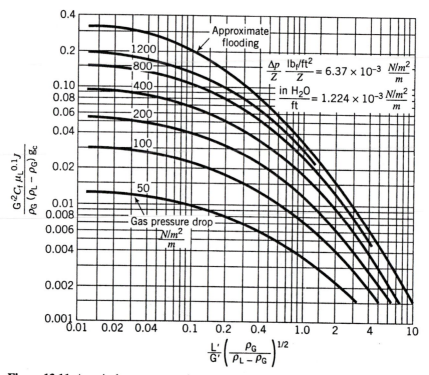

Figure 13.11 A typical gas pressure drop curve. *Source:* Reference 13.15.

than a ratio of the two. The first step in this procedure is conversion of the volume/volume G/L to a mass/mass G/L:

$$\left(\frac{G}{L}\right)_{\text{operating}} \cdot \frac{\rho_G}{\rho_W} = \frac{G'}{L'} \qquad (13.25)$$

where

ρ_G = density of air = 1.21 kg/m³ (20°C)
ρ_W = density of water (kg/m³)
G' = air mass loading rate (kg/m²-sec)
L' = water mass loading rate (kg/m²-sec)

Values for the density of water at different temperatures, as well as its viscosity and surface tension, are listed in Table 13.2.

The ratio G'/L' is dimensionless, and can be used in the pressure drop curve (Figure 13.11). The x-axis value of the gas pressure drop curve may be calculated using the values of Equation 13.26:

$$x\text{-axis} = \frac{L'}{G'}\left(\frac{\rho_G}{\rho_W - \rho_G}\right)^{1/2} \qquad (13.26)$$

The x-axis value and pressure drop curve are then used to obtain the value for the y-axis from the pressure drop curve:

Table 13.2 Physical Characteristics of Water

Temperature (°C)	Density, ρ (kg/m³)	Dynamic viscosity, μ_L (N-sec/m²) $\times 10^{-3}$	Surface tension, σ (N/m)
0	999.8	1.781	0.0765
5	1000.0	1.518	0.0749
10	999.7	1.307	0.0742
15	999.1	1.139	0.0735
20	998.2	1.002	0.0728
25	997.0	0.890	0.0720
30	995.7	0.798	0.0712
40	992.2	0.653	0.0696
50	988.0	0.547	0.0679

$$y\text{-axis} = \frac{G'^2 C_f \mu_L^{0.1} J}{\rho_G (\rho_w - \rho_G) g_c} \qquad (13.27)$$

where

G' = air mass loading rate [kg/(m²-sec)]
C_f = packing factor (m⁻¹) (from Table 13.3)
μ_L = dynamic viscosity of water [kg/(m-sec)]
$J = 1$ for air-water systems
$g_c = 1$ for metric units

The mass air loading rate may be converted to an actual volumetric loading rate by once again using the density of air:

$$G\left(\frac{m^3}{m^2\text{-sec}}\right) = \frac{G'\left(\frac{kg}{m^2\text{-sec}}\right)}{1.21\left(\frac{kg}{m^3}\right)}$$

After obtaining G, L may be determined by

$$L = \frac{G}{(G/L)_{\text{operatiry}}} \qquad (13.28)$$

The importance of Equation 13.28 is the capability of determining L, which is the design area liquid loading rate, which can be used in Equation 13.15 to determine the tower diameter.

The Mass Transfer Coefficient. One of the more important parameters in air stripping is the mass transfer coefficient ($K_L a$), which is a measure of the flux of a contaminant across an air-water interface. The rate of mass transfer of a contaminant is

$$J = -K_L a(C_s - C) \qquad (13.29)$$

Table 13.3 Characteristics of Common Packing Materials

Type*	Size (mm)	Size (in.)	Packing Factor (m^{-1})	Area/Volume Ratio (m^2/m^3)
Intalox saddles	25	1.0	108	207
	51	2.0	69	108
	76	3.0	52	89
Pall rings	16	0.6	318	341
	25	1.0	171	207
	38	1.5	131	128
	64	2.5	82	102
	89	3.5	52	85
Raschig rings	13	0.5	1903	364
	19	0.75	837	262
	25	1.0	509	190
	38	1.5	312	125
	51	2.0	213	92
	76	3.0	121	62
Jaeger Tri-Packs	25	1.0	92	279
	51	2.0	52	157
	89	3.5	39	125

*All data are for plastic shapes except Raschig rings (ceramic).
Source: Reference 13.17.

where

J = rate of contaminant mass transfer [mg/(L-sec)]
$K_L a$ = mass transfer coefficient (sec^{-1})
C_s = equilibrium liquid-phase concentration (i.e., the saturation concentration) (mg/L)
C = operating liquid-phase concentration (mg/L)

The mass transfer coefficient is a function of the tower packing, the temperature, the total dissolved solids concentration of the water, and the air-water ratio. The $K_L a$ may be determined from thermodynamic relationships using equations described by Onda [13.16]. However, an effective procedure for $K_L a$ determination is the use of a pilot system (Figure 13.12). A small diameter column (e.g., 30 cm or 12 inches) is loaded with packing material, connected to a blower, and supplied with contaminated water. Pilot systems do not need to be extravagant; they can be constructed of simple tubing and a small pump and blower. Pilot-scale testing is usually conducted over a range of liquid loading rates and air-water ratios.

The $K_L a$ is determined at the air and water loading rates calculated from the pressure drop curve by plotting NTU (which varies based on C/C_0) as a function of tower depth. The slope is then 1/HTU, from which $K_L a$ can be obtained. Using a range of conditions (air-water ratios, liquid loadings, packing materials, etc.), the contaminant

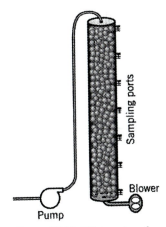

Figure 13.12 Pilot system for the determination of $K_L a$.

concentrations are obtained at each sampling port once steady state is established. These data then serve as the basis for the determination of $K_L a$. The mass transfer coefficient is then

$$K_L a = \frac{L}{\text{HTU}} \tag{13.30}$$

The pilot-scale determination of $K_L a$ is demonstrated in Example 13.2.

EXAMPLE 13.2   *Pilot-Scale Determination of $K_L a$*

A pilot study was conducted to strip TCE from a contaminated groundwater to 5 μg/L using 50-mm-diameter Intalox saddles. A column 30 cm (1 ft) in diameter was used with sampling ports at 0, 2, 4, 6, and 8 m (0, 6.6, 13, 20, and 26 ft) depths. The volumetric air loading rate was 0.78 m^3/(m^2-sec) and the loading of water on the column was 0.039 m^3/(m^2-sec). The following pilot data were collected.

Depth (m)	[TCE] (μg/L)
0	230
2	143
4	82
6	48
8	28

Based on these pilot data, determine $K_L a$.

SOLUTION

First, the dimensionless Henry's Law constant for TCE is calculated.

Assume $T = 25°C$,

$$H'_{TCE} = \frac{0.0088 \frac{\text{atm-m}^3}{\text{mole}}}{8.21 \times 10^{-5} \frac{\text{m}^3\text{-atm}}{\text{mole-°K}} \cdot 298 \text{ °K}} = 0.360$$

The minimum air-water ratio is determined by Equation 13.23.

$$\left(\frac{G}{L}\right)_{min} = \frac{C_0 - C}{C_0} \cdot \frac{1}{H'} = \frac{230 - 5}{230} \cdot \frac{1}{0.360} = 2.72$$

The operating air-water ratio is

$$\left(\frac{G}{L}\right)_{operating} = \frac{0.78}{0.039} = 20$$

The stripping factor may then be calculated using Equation 13.22:

$$R = \frac{(G/L)_{operating}}{(G/L)_{min}} = \frac{20}{2.72} = 7.4$$

Using the stripping factor and the influent and effluent concentrations, the NTU may be determined for each sampling port depth. For example, using the contaminant concentrations for a depth of 2 m

$$NTU = \left(\frac{R}{R-1}\right)\ln\left[\frac{(C_0/C)(R-1)+1}{R}\right] = \left(\frac{7.4}{6.4}\right)\ln\left[\frac{(230/143)(6.4)+1}{7.4}\right] = 0.489$$

The calculation of NTU is then repeated for depths of 0, 4, 6, and 8 m, with a corresponding plot of NTU as a function of tower depth, the slope of which is 1/HTU. The

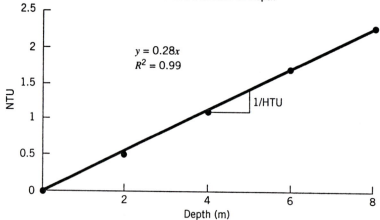

slope=0.282; because slope=1/HTU, the HTU=3.55 m. The K_La may then be calculated using Equation 13.30:

$$K_La = \frac{L}{\text{HTU}} = \frac{0.039\dfrac{m^3}{m^2\text{-sec}}}{3.55\ m} = 0.0110\ \text{sec}^{-1}$$

Onda Equations. As an alternative to pilot evaluations of K_La, a series of thermodynamic equations were developed by Onda et al. [13.16]:

$$\frac{1}{K_La} = \frac{1}{H'k_Ga} + \frac{1}{k_La} \tag{13.31}$$

$$\frac{a_w}{a_t} = 1 - \exp\left[-1.45\left(\frac{\sigma_c}{\sigma}\right)^{0.75}\left(\frac{L'}{a_t\mu_L}\right)^{0.1}\left(\frac{L'^2 a_t}{\rho_L^2 g}\right)^{-0.05}\left(\frac{L'^2}{\rho_L \sigma a_t}\right)^{0.2}\right] \tag{13.32}$$

$$k_L\left(\frac{\rho_L}{\mu_L g}\right)^{1/3} = 0.0051\left(\frac{L'}{a_w\mu_L}\right)^{2/3}\left(\frac{\mu_L}{\rho_L D_L}\right)^{-0.5}(a_t d_p)^{0.4} \tag{13.33}$$

$$\frac{k_G}{a_t D_G} = 5.23\left(\frac{G'}{a_t\mu_G}\right)^{0.7}\left(\frac{\mu_G}{\rho_G D_G}\right)^{1/3}(a_t d_p)^{-2} \tag{13.34}$$

where

K_La = overall mass transfer coefficient (sec^{-1})
H' = dimensionless Henry's law constant
k_G = gas phase mass transfer (m/sec)
k_L = liquid phase mass transfer (m/sec)
a = area-to-volume ratio (equal to a_w for K_La determination) (m^2/m^3)
a_w = wetted packing area (Table 13.3) (m^2/m^3)
a_t = total packing area (Table 13.3) (m^2/m^3)
σ_c = critical surface tension of packing material (Table 13.4) (N/m)
σ = surface tension of water (Table 13.2) (N/m)
L' = liquid mass loading rate (kg/m^2-sec)
μ_L = viscosity of water (Table 13.2) (kg/m-sec)
ρ_L = liquid density (Table 13.2) (kg/m^3)

Table 13.4 Critical Surface Tension of
Packing Materials

Material	σ_c (N/m)
Carbon	0.056
Ceramic	0.061
Glass	0.073
Paraffin	0.020
Polyethylene	0.033
Polyvinyl chloride (PVC)	0.040
Steel	0.075

Source: Reference 13.2.

g = acceleration due to gravity (m/sec²) = 9.81
D_L = liquid diffusivity (m²/sec)
d_p = nominal packing diameter (m)
D_G = gas diffusivity (m²/sec)
G' = gas mass loading rate (kg/m²-sec)
μ_G = viscosity of air $\left(\dfrac{\text{N} \cdot \text{sec}}{\text{m}^2}\right)$ = 1.81 × 10⁻⁵ at 20°C
ρ_G = density of air (kg/m³) = 1.21 at 20°C (sea level)

Necessary data for use of the Onda equations are listed in Tables 13.3 and 13.4. Use of the Onda equations involves a series of detailed calculations, which are illustrated in Example 13.3.

EXAMPLE 13.3 *Estimation of K_La Using the Onda Equations*

An air stripping tower is being designed to treat groundwater contaminated with benzene, which is characterized by a diffusion coefficient in air of 0.0923 cm²/sec. The column loading rate is 20 kg/(m²-sec), the liquid diffusion coefficient is 6 × 10⁻⁶ cm²/sec, the air-water ratio is 80:1, and the temperature = 20°C. If 25-mm Raschig rings are to be used as packing, determine the K_La using the Onda equations.

SOLUTION

The overall mass transfer coefficient can be calculated using Equation 13.31:

$$\frac{1}{K_La} = \frac{1}{H'k_Ga} + \frac{1}{k_La}$$

First, determine H' for benzene using tabulated constants.

$$A = 5.53 \quad \text{and} \quad B = 3190$$

$$H = \exp(A - B/T) = 0.00475 \text{ (atm-m}^3\text{)/mole}$$

$$H' = \frac{0.00475 \dfrac{\text{atm-m}^3}{\text{mole}}}{\left(8.21 \times 10^{-5}\dfrac{\text{m}^3\text{-atm}}{\text{mole-K}}\right)(293°\text{K})} = 0.197$$

The Onda equations are used to determine the unknown quantities of k_G, k_L, and a in equation 13.31. Calculate a ($a_w = a$ in the determination of K_La) using Equation 13.32:

$$\frac{a_w}{a_t} = 1 - \exp\left[-1.45\left(\frac{\sigma_c}{\sigma}\right)^{0.75}\left(\frac{L'}{a_t\mu_L}\right)^{0.1}\left(\frac{L'^2 a_t}{\rho_L^2 g}\right)^{-0.05}\left(\frac{L'^2}{\rho_L \sigma a_t}\right)^{0.2}\right]$$

where

$\dfrac{L'}{a_t \mu_L}$ is the Reynolds number

$\dfrac{L'^2 a_t}{\rho_L^2 g}$ is the Froude number

$\dfrac{L^2}{\rho_L \sigma a_t}$ is the Weber number

$a_t = 190 \text{ m}^{-1}$ (Table 13.3)

$\sigma_c = 0.061 \text{ N/m}$ for ceramic Raschig rings (Table 13.4)

$\sigma = 0.073 \text{ N/m}$ for water at 20°C (Table 13.2)

$L' = 20 \dfrac{\text{kg}}{\text{m}^2\text{-sec}}$

$\mu_L = 1.002 \times 10^{-3} \text{ kg/m-sec}$ @ 20°C (Table 13.2)

$\rho_L = 998.2 \text{ kg/m}^3$ @ 20°C (Table 13.2)

$g = 9.81 \text{ m/sec}^2$

$$\frac{a_w}{190} = 1 - \exp\left[-1.45\left(\frac{0.061}{0.073}\right)^{0.75}\left(\frac{20}{190 \cdot 1.002 \times 10^{-3}}\right)^{0.1}\left(\frac{20^2 \cdot 190}{998.2^2 \cdot 9.81}\right)^{-0.05}\left(\frac{20^2}{998.2 \cdot 0.073 \cdot 190}\right)^{0.2}\right]$$

Therefore, $a_w = a = 136.5 \text{ m}^{-1}$.

Next, determine k_L using Equation 13.33.

$$k_L\left(\frac{\rho_L}{\mu_L g}\right)^{1/3} = 0.0051\left(\frac{L'}{a_w \mu_L}\right)^{2/3}\left(\frac{\mu_L}{\rho_L D_L}\right)^{-0.5}(a_t d_p)^{0.4}$$

$$D_L = 6 \times 10^{-6} \text{ cm}^2/\text{sec} = 6 \times 10^{-10} \text{ m}^2/\text{sec} @ 20°C$$

$$d_p = 0.025 \text{ m}$$

$$k_L\left(\frac{998.2}{1.002 \times 10^{-3} \cdot 9.81}\right)^{1/3} = 0.0051\left(\frac{20}{136.5 \cdot 1.002 \times 10^{-3}}\right)^{2/3}\left(\frac{1.002 \times 10^{-3}}{998.2 \cdot 6 \times 10^{-10}}\right)^{-0.5}(190 \cdot 0.025)^{0.4}$$

Therefore, $k_L = 1.383 \times 10^{-4} \text{ m/sec}$.

Determine k_G using Equation 13.34.

$$\frac{k_G}{a_t D_G} = 5.23\left(\frac{G'}{a_t \mu_G}\right)^{0.7}\left(\frac{\mu_G}{\rho_G D_G}\right)^{1/3}(a_t d_p)^{-2}$$

$$D_G = 9.23 \times 10^{-6} \text{ m}^2/\text{sec} \text{ for benzene}$$

$$\mu_G = 1.81 \times 10^{-5} \text{ kg/(m-sec)} @ 20°C$$

Determine the air loading rate.

$$\text{air-water ratio} = 80$$

Convert the air and water loadings to volumetric flow rates.

$$L = 20 \text{ kg/(m}^2\text{-sec)}/998.2 \text{ kg/m}^3 = 0.0200 \text{ m}^3/\text{(m}^2\text{-sec)}$$

$$G = 0.0200 \times 80 = 1.60 \text{ m}^3/\text{(m}^2\text{-sec)}$$

and since $\rho_A = 1.21$ kg/m^3

$$G' = 1.60 \text{ m}^3/\text{(m}^2\text{-sec)} \cdot 1.21 \text{ kg/m}^3 = 1.94 \text{ kg/(m}^2\text{-sec)}$$

$$\frac{k_G}{190 \cdot 9.234 \times 10^{-6}} =$$

$$5.23\left(\frac{1.94}{190 \cdot 1.81 \times 10^{-5}}\right)^{0.7}\left(\frac{1.81 \times 10^{-5}}{1.205 \cdot 9.234 \times 10^{-6}}\right)^{1/3} (190 \cdot 0.025)^{-2}$$

Therefore, $k_G = 0.0402$ m/sec.
Finally, $K_L a$ may be determined from Equation 13.31.

$$\frac{1}{K_L a} = \frac{1}{0.197 \cdot 0.0402 \cdot 136.5} + \frac{1}{1.383 \times 10^{-4} \cdot 136.5}$$

$$K_L a = 0.0186 \text{ sec}^{-1}$$

From the preceding discussion, it is evident that the design of a packed tower air strippers is an intricate process; therefore, the design procedure is summarized in Table 13.5 and demonstrated in Example 13.4.

Table 13.5 Design Procedure for Packed Tower Air Strippers

1. Select the packing material. Each packing material will be characterized by a different $K_L a$ and pressure drop. A higher $K_L a$ and lower pressure drop provide the most efficient design.
2. Select an air-water ratio and calculate the stripping factor using Equation 13.22. Alternatively, select a stripping factor (2–10 are most common) and calculate the operating air-water ratio.
3. Based on the selection of a reasonable gas pressure drop (20–100 N/m^2/m), determine the value for the x-axis dimensionless group. From this value, read the corresponding y-axis value from the pressure drop curve (Figure 13.11) and calculate the air flow rate for the selected air-water ratio.
4. Determine the liquid loading rate from the air-water ratio using Equation 13.28.
5. Using the calculated values of G and L, conduct pilot studies, sampling contaminant concentrations as a function of pilot tower depth. Develop NTU data from C/C_0.
6. Plot NTU as a function of Z; the slope of the line is 1/HTU. The mass transfer coefficient, $K_L a$, may be determined from Equation 13.30.

Table 13.5 (Continued)

7. Based on the design flow rate and C/C_0, determine the tower height and diameter.
8. Repeat example using a matrix of stripping factors to obtain the lowest-cost design.

EXAMPLE 13.4   *Packed Column Air Stripper Design*

An air stripper is being designed to remove toluene from a contaminated groundwater. (1) Design a pilot study to determine the $K_L a$ for a stripping factor of 5 using 50-mm Raschig rings and a pressure drop of 100 N/m²/m. (2) Based on the pilot data, determine the dimensions of the full-scale air stripping tower if the flow rate is 3000 m³/day, the initial toluene concentration is 114 μg/L, and the design effluent concentration is 1 μg/L. Assume that the temperature of the system is 20°C.

SOLUTION

1. Determine the minimum air-water ratio for the pilot columns in order to calculate the operating air-water ratio. First, find the dimensionless Henry's Law constant for toluene (at a temperature of 20°C):

$$H'_{\text{toluene}} = \frac{0.0066 \dfrac{\text{atm-m}^3}{\text{mole}}}{8.21 \times 10^{-5} \dfrac{\text{m}^3\text{-atm}}{\text{mole-°K}} \cdot 293°\text{K}} = 0.274$$

Therefore, the minimum air-water ratio for design toluene removal is

$$\left(\frac{G}{L}\right)_{\text{min}} = \frac{C_0 - C}{C_0} \cdot \frac{1}{H'} = \frac{114 - 1}{114} \cdot \frac{1}{0.274} = 3.62$$

The operating (actual) air-water ratio may then be calculated:

$$\left(\frac{G}{L}\right)_{\text{operating}} = R\left(\frac{G}{L}\right)_{\text{min}} = (5)(3.62) = 18.1$$

The x-axis (Figure 13.11) dimensionless group is then calculated using Equations 13.25 and 13.26 ($\rho_G = 1.21$ kg/m³ and $\rho_w = 998.2$ kg/m³):

$$\frac{G'}{L'} = \left(\frac{G}{L}\right)_{\text{operating}} \cdot \frac{\rho_G}{\rho_w} = (18.1) \cdot \left(\frac{1.21}{998.2}\right) = 0.0219$$

$$x\text{-axis} = \frac{L'}{G'}\left(\frac{\rho_G}{\rho_w - \rho_G}\right)^{1/2} = \left(\frac{1}{0.0219}\right) \cdot \left(\frac{1.21}{998.2 - 1.21}\right)^{1/2} = 1.59$$

Using a gas pressure drop of 100 N/m^2/m on Figure 13.11, the y-axis value corresponding to 1.59 is 0.0041. The mass air loading rate may then be calculated from Equation 13.25 [$C_f = 213.3$ m^{-1} (Table 13.3), $\mu_L = 1.002 \times 10^{-3}$ kg/(m-sec), $J = 1$ (air-water system), $g_c = 1$ (metric units)]:

$$y\text{-axis} = 0.0041 = \frac{G'^2\, C_f\mu_L^{0.1}\, J}{\rho_G(\rho_w - \rho_G)g_c} = \frac{G'^2(213.3)(1.002 \times 10^{-3})^{0.1}\ (1)}{(1.21)(998.2 - 1.21)(1)}$$

From this calculation, $G' = 0.215$ kg/(m^2-sec). The value G' may be converted to a volumetric loading rate by dividing by 1.21 kg/m^3; therefore, $G = 0.178$ m^3/(m^2-sec). The liquid loading rate may then be calculated using Equation 13.28:

$$\text{air-water ratio} = G/L$$

$$18.1 = 0.178/L$$

$$L = 0.00983\ \text{m}^3/(\text{m}^2\text{-sec})$$

Using these values for the air and liquid loading rates, a pilot study can be established. Concentration measurements are taken as a function of column depth, and NTU calculated for each depth down the column, using the water to be treated. The NTU data are plotted as a function of depth (Example 13.2) and the slope 1/HTU is determined. The K_La is determined from this value using Equation 13.30.

For example, a column 30 cm (1 ft) in diameter can be constructed with sampling ports at 0, 2, 4, 6, and 8 m. Assume the following pilot data were collected.

Depth (m)	[Toluene] (μg/L)
0	114
2	24
4	6
6	1.8
8	0.5

Following the procedure used in Example 13.2, the NTU is determined for each sampling port depth. For example, at a sampling depth of 2 m:

$$NTU = \left(\frac{R}{R-1}\right)\ln\left[\frac{(C_0/C)(R-1)+1}{R}\right] = \left(\frac{5}{4}\right)\ln\left[\frac{(114/24)(4)+1}{5}\right] = 1.73$$

The calculation is repeated for depths of 0, 4, 6, and 8 m. A line may then be generated from a plot of NTU as a function of tower depth, the slope of which is 1/HTU. (See the illustration at the top of the next page.)

The slope of the line for NTU as a function of depth = 0.83; because slope = 1/HTU, the value of HTU = 1.20. The K_La may then be calculated using Equation 13.30:

$$K_La = \frac{L}{HTU} = \frac{0.00983\dfrac{\text{m}^3}{\text{m}^2\text{-sec}}}{1.20\ \text{m}} = 0.00819\ \text{sec}^{-1}$$

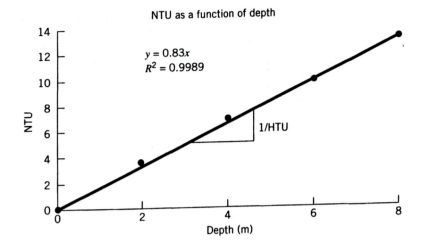

NTU as a function of depth

2. The diameter of the full-scale column may be calculated from Equation 13.15, where $Q = 3000$ m³/day $= 0.0347$ m³/sec:

$$D = \left(\frac{4Q}{\pi L}\right)^{0.5} = \left(\frac{4 \cdot 0.0347}{\pi \cdot 0.00963}\right)^{0.5} = 2.14 \text{ m} = 7 \text{ ft}$$

The tower height may then be determined from the product of the HTU and NTU.

$$\text{NTU} = \left(\frac{R}{R-1}\right)\ln\left[\frac{(C_0/C)(R-1)+1}{R}\right] = \left(\frac{5}{4}\right)\ln\left[\frac{(114/1)(4)+1}{5}\right] = 5.64$$

$$Z = \text{(HTU) (NTU)} = (1.20)\,(5.64) = 6.77 \text{ m} = 22.2 \text{ ft; round up to 23 ft.}$$

Therefore, the air stripper dimensions are 7 ft in diameter and 23 ft high. Some engineers will multiply the column height by a safety factor, such as 1.25, to ensure contaminant removal.

13.3   ABIOTIC TRANSFORMATION DESIGN APPLICATION: SOIL REMEDIATION USING FENTON'S REAGENT

Advanced oxidation processes (AOPs) have been used in hazardous waste management primarily for the treatment of contaminated groundwater by pump and treat technologies. Recently, the use of some AOPs has been extended to treating contaminated soils and groundwater through the in situ injection of oxidizing reagents. A process that has been investigated because of its potential for transport of the oxidizing reagents through soils and groundwater is *Fenton's reagent*—the catalyzed decomposition of hydrogen peroxide by transition elements such as iron (II) or copper (II), which, like other AOPs, generates hydroxyl radicals:

$$H_2O_2 + Fe^{2+} \longrightarrow OH\cdot + OH^- + Fe^{3+} \tag{13.35}$$

Hydroxyl radicals react rapidly with many organic contaminants, such as TCE, PCE, PAHs, PCBs, and chlorinated dioxins, as described in Chapters 7 and 12. Dilute hydrogen peroxide is used in laboratory Fenton's reactions to generate hydroxyl radicals, and many organic compounds can be degraded very efficiently under these conditions (e.g., with 2–10 moles of hydrogen peroxide required to oxidize one mole of the substrate). However, soils and groundwater contaminated with hazardous compounds require significantly more hydrogen peroxide than pure laboratory systems. Many competing reactions are involved in such modified Fenton's systems including the production of perhydroxyl radical through the reaction of hydrogen peroxide and iron (III), cycling of iron (III) to iron (II) by short-lived reactants, and quenching of hydroxyl radicals by iron (II) and hydrogen peroxide:

$$OH\cdot + Fe^{2+} \longrightarrow OH^- + Fe^{3+} \tag{13.36}$$

$$HO_2\cdot + Fe^{3+} \longrightarrow O_2 + H^+ + Fe^{2+} \tag{13.37}$$

$$H_2O_2 + OH\cdot \longrightarrow H_2O + HO_2\cdot \tag{13.38}$$

where $HO_2\cdot$ = perhydroxyl radical.

Fenton's reagent was discovered over a century ago (not surprisingly, by H. J. H. Fenton). Numerous theoretical Fenton's reagent studies were published through the twentieth century, and modified Fenton's reagent has been used at full scale for the treatment of dilute industrial waste streams [13.18, 13.19]. For example, Fenton's reagent has been used to treat spent solvent-water mixtures from the washing of aircraft wings [13.20] and for the treatment of chlorophenols in industrial effluents to reduce toxicity prior to biological treatment [13.21].

Fenton's reactions used to treat water-soluble organic contaminants in industrial waste streams are less efficient than pure laboratory systems but are still characterized by relatively mild conditions [10–1000 mg/L H_2O_2 and excess iron (II)]. Because iron (II) is a weak acid, the solution pH is approximately 3 to 4; raising the pH results in precipitation of the iron as an amorphous ferric hydroxide floc with loss of catalytic activity.

The use of modified Fenton's reactions to treat contaminated soils is affected by significantly different conditions relative to the treatment of water due to (1) contaminant sorption, (2) quenching and scavenging of hydroxyl radicals, and (3) the presence of naturally occurring minerals, which act as Fenton's catalysts.

Effect of Sorption on Fenton's Soil Treatment. Because most contaminants in soils are sorbed, higher stoichiometric dosages of hydrogen peroxide are required; in fact, recent data suggest that the required hydrogen peroxide dosages are directly proportional to the contaminant octanol-water partition coefficient [13.22]. The high hydrogen peroxide requirement is related to the need for aggressive reaction conditions to desorb the contaminant, which may not involve hydroxyl radicals. The ability of aggressive Fenton's reactions to treat a soil significantly faster than natural contaminant desorption takes place has been described in a number of studies [13.23, 13.24]. For example, Watts and Stanton [13.25] reported that 80% mineralization of hexadecane was achieved in 24 h whereas desorption over 72 h was undetectable in parallel experiments. Therefore, the application of Fenton's reagent for soil treatment provides the equivalent of a combined soil washing–oxidation process.

Catalyst–Hydrogen Peroxide–Sorption Interactions: The Basis for Process Design.
Recent studies [13.26, 13.27] have documented that the addition of iron (II) may not
be necessary in the Fenton's treatment of contaminated soils because naturally occurring iron minerals effectively catalyze Fenton-like reactions. However, mineral-
catalyzed reactions are often slower than reactions catalyzed by soluble iron and do
not appear to effectively desorb contaminants. In contrast, if soluble iron is used as the
catalyst, the combined desorption-oxidation mechanism is enhanced, resulting in rapid
reaction rates in which treatment of the soil may be completed within hours. Two
process modifications have been proposed by Spencer et al. [13.28] based on the tendency of contaminants to desorb in the soils that are treated. The process design for
contaminants that desorb over relatively short time periods (e.g., TCE, BTEX) uses
mineral-catalyzed reactions, which are more efficient (~100 moles of hydrogen peroxide per mole of contaminant oxidized) but require higher volumes in a slurry reactor and reaction times that are controlled by the rate of desorption, which may require
days to weeks. Contaminants that desorb slowly (e.g., PAHs, PCBs, dioxins) are most
effectively treated using soluble iron and high hydrogen peroxide dosages to promote
the combined desorption-oxidation of the contaminant.

Although iron (II) is the most effective Fenton's catalyst, iron (III) is preferred when
hydrogen peroxide concentrations in the percent range are required to promote contaminant desorption. Such a Fenton-like reaction is more efficient because, under the
strong oxidizing conditions of the system, iron (II) is immediately oxidized to iron (III)
as soon as hydrogen peroxide is added to the system. A number of forms of iron (III)
may be used. Although ferric perchlorate and ferric nitrate are the most effective catalysts in the laboratory because perchlorate and nitrate ions do not quench hydroxyl
radicals, ferric perchlorate is unavailable in large quantities and the addition of nitrate
to soils may result in groundwater contamination (the Safe Drinking Water Act MCL
for NO_3–N is 10 mg/L). Therefore, ferric sulfate, which is readily available and moderately inexpensive, is the preferred catalyst for field applications.

Typical concentrations of hydrogen peroxide in the mineral catalyzed systems are
100 mg/L to 5000 mg/L, with liquid-to-solid ratios of 5:1 to 20:1 (mass/mass) [13.28].
In the systems in which soluble iron is added to promote contaminant desorption-
oxidation, hydrogen peroxide concentrations range from 1% to 30% and iron (III) is
added in concentrations ranging from 100 mg/L to 1000 mg/L [13.29].

One of the advantages of using Fenton's reactions for the treatment of contaminated
soils is the potential for the in situ remediation of contamination through the injection
or perfusion of hydrogen peroxide [13.30, 13.31]. Alternatively, the process can be
conducted in batch reactors, which is a more conservative approach (Figure 13.13). As
with most soil remediation designs, bench studies, pilot studies, or both may be required prior to full-scale design. The process design of a Fenton's soil treatment system using bench-scale treatability data is illustrated in Example 13.5.

EXAMPLE 13.5 *Conceptual Design of Fenton's Soil Treatment System*

Twenty cubic meters (26 yd^3) of soil contaminated with 5000 μg/kg of 2,3,7,8-tetra-
chlorodibenzo-p-dioxin (TCDD) must be treated to 5 μg/kg. The soil has been exca-

Figure 13.13 A batch Fenton's slurry reactor.

vated and placed in a lined pit. Samples have been collected, and an initial bench-scale analysis, in which TCDD desorption was measured, provided the following data.

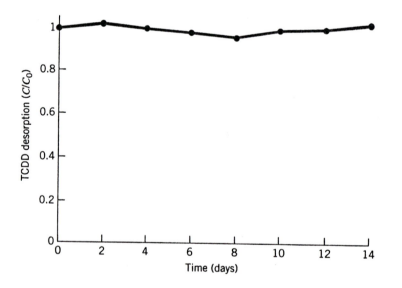

From the desorption data, select the most effective process design and propose a bench-scale treatability study to oxidize the TCDD in the soil.

SOLUTION

Because desorption was negligible over 14 days, a mineral-catalyzed process design using dilute hydrogen peroxide would be ineffective. The process design should focus on the addition of soluble iron with a small volume of concentrated hydrogen peroxide.

Set up a treatability study using a matrix of ferric sulfate and hydrogen peroxide concentrations and sufficient liquid volume to cover the soil (a mass:volume, solid:liquid ratio of 5:1). Vary the hydrogen peroxide concentrations from 1% to 10% and the ferric sulfate concentrations from 50 mg/L to 500 mg/L. Plot the results in three dimensions to provide a graphical representation of the treatability study results.

The following three-dimensional plot of the data (a response surface) was developed from the treatability study. Isoresponse lines on the three-dimensional plot represent the TCDD residual after treatment.

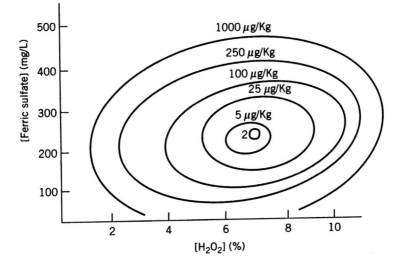

The optimum conditions are found by locating an isoresponse line that provides TCDD treatment to below 5 μg/kg. Reading the corresponding concentrations results in a required hydrogen peroxide concentration of 6% and a ferric sulfate concentration of 200 mg/L in the soil solution. Use 6% hydrogen peroxide and 200 mg/L iron (III) with a mass loading of $\dfrac{0.2 \text{ L H}_2\text{O}_2}{\text{kg soil}}$. Based on the soil bulk density of 1200 kg/m³, the mass of hydrogen peroxide required is:

$$(26 \text{ m}^3)\left(\frac{1200 \text{ kg}}{\text{m}^3}\right)\left(\frac{0.2 \text{ L H}_2\text{O}_2}{\text{kg soil}}\right) = 6240 \text{ L of 6\% H}_2\text{O}_2$$

Hydrogen peroxide is available commercially at a concentration of 30%; therefore, the volume of hydrogen peroxide required is

$$(6240 \text{ L})\left(\frac{6\%}{30\%}\right) = 1248 \text{ L (329 gal) of 30\% H}_2\text{O}_2$$

The quantity of ferric sulfate required would be 4800 L · 200 mg/L · 10^6 kg/mg = 0.96 kg (2.1 lb). The site could be treated using five 208-L (55-gal) drums of 30% hydrogen peroxide diluted 20% and 1 kg of ferric sulfate.

13.4   BIOTIC TRANSFORMATION DESIGN APPLICATION: EX SITU SLURRY BIOREACTOR

An effective approach to the treatment of surface soils contaminated with chemicals susceptible to biodegradation is excavation followed by biological treatment in slurry reactors. The process provides an inexpensive means of treating soils contaminated with compounds that are rapidly metabolized by microbial communities. Slurry treatments are conducted in existing lagoons or in bioreactors designed and constructed for a specific solid hazardous waste that requires treatment. Slurry phase biological treatment has been used to treat soils and solids containing PAHs, pentachlorophenol, 2,4-D, diesel, and gasoline [13.32, 13.33]. The following discussion is based on concepts and procedures described in References 13.33, 13.35, and 13.38.

Process Description. Slurry phase treatment processes are usually designed as batch reactors. The sophistication of design and operation of slurry reactors varies substantially. Simple systems may consist of a lined pit that is mixed occasionally using a tiller. Sophisticated systems may include eloquently designed reactors with automated process control. Specific reactor designs include open constructed reactors, closed constructed reactors, and bermed/lined areas and lagoons. A typical slurry reactor is shown in Figure 13.14. After excavation and screening of rocks greater than 0.2 cm (0.08 in.) in diameter, water is added (60–95% m/m). A sedimentation basin is often used to separate the slurry water after treatment.

The treatment of soils and slurries presents some unique problems. A practical consideration is that thick solids cannot be easily pumped or mixed. Therefore, materials handling is important in the design and operation of slurry reactor treatment systems and may include front-end loaders, screw conveyors, back hoes, and trucks. In the design of these systems, runoff control must be implemented including channels, berm construction, and so on.

In considering the process dynamics, the primary limitations in slurry phase treatment are desorption and biodegradation kinetics. Contaminant sorption is a primary consideration in the process design of bioslurry reactors. Because most biodegradation occurs in the aqueous phase, the desorption rate often controls the bioavailability of the contaminants. For example, Milhelcic and Luthy [13.34] showed that biodegradation of PAHs in soil-water systems is controlled by their desorption rates. Because of the complexity of the linked desorption-biodegradation process in slurry reactors, reaction rates often follow mixed-order kinetics. Zero, first, and fractional orders have been reported in slurry reactors and other soil treatment systems [13.35].

The use of nutrient amendments and bioaugmentation has been shown to increase rates of slurry phase bioremediation [13.36]. In addition, the injection of methane is sometimes used as a cometabolite in the degradation of chlorinated aliphatic contaminants. Treatment times for slurry reactors are less than those for solid phase bioreme-

Figure 13.14 Aerial view of an ex situ soil slurry bioreactor. *Source:* Blasland, Bouck and Lee

diation systems due to increased homogeneity and potential for desorption. For example, the half-life of phenanthrene was 8 days in a slurry reactor and 32 days in a solid phase bioremediation system [13.37].

Process Design. The process design requirements for slurry phase bioreactors include (1) the terminal electron acceptor, (2) the electron donor, and (3) nutrient requirements. Selection of the appropriate electron acceptor is based on the mechanisms described in Section 7.4. The first step in determining the nutrient requirements for the system is to write a balanced redox reaction, which is a function of the mass of the contaminant in the soil, sludge, or solids. Balancing the redox reactions for the system is accomplished by (1) writing the balanced half reactions, (2) normalizing the equations so that electron losses and gains are equal, and (3) summing the two half-reactions. In evaluating the biodegradation oxidation-reductions, the total reaction must include the organic contaminant that is transformed, the electron donors and acceptors, and the nutrients incorporated into cell growth. Some generic half-reactions for three electron acceptors and two synthesis reactions are listed in Table 13.6.

The total reaction for contaminant transformation and nutrient assimilation in biological treatment reactions is

$$H_D + f_c H_A + f_s C_s \tag{13.46}$$

Table 13.6 Generic Half-Reactions for Organic Redox Reactions

Half-Reaction of Electron Donor (H_D)

$$(1/Z)\ C_aH_bO_cN_d + [(2a - c)/Z]\ H_2O \longrightarrow (a/Z)\ CO_2 + (d/Z)\ NH_3 + H^+ + e^- \quad (13.39)$$

where

$Z = 4a + b - 2c - 3d$

a, b, c, d = the average number of atoms for C, H, O, and N, respectively, in the organic contaminants

Half-Reactions of Electron Acceptor (H_A)

Aerobic: When oxygen is the electron acceptor,

$$\frac{1}{4}\ O_2 + H^+ + e^- \longrightarrow \frac{1}{2}\ H_2O \qquad (13.40)$$

Anoxic: When nitrate is the electron acceptor,

$$\frac{1}{6}\ NO_3^- + H^+ + \frac{5}{6}\ e^- \longrightarrow \frac{1}{12}\ N_2 + \frac{1}{2}\ H_2O \qquad (13.41)$$

When sulfate is the electron acceptor,

$$\frac{1}{8}\ SO_4^{2-} + H^+ + e^- \longrightarrow S^{2-} + \frac{1}{2}\ H_2O \qquad (13.42)$$

When carbon dioxide is the electron acceptor,

$$\frac{1}{8}\ CO_2 + H^+ + e^- \longrightarrow \frac{1}{8}\ CH_4 + \frac{1}{4}\ H_2O \qquad (13.43)$$

Cell Synthesis Equation (C_s)

When ammonia is the nitrogen source,

$$\frac{1}{4}\ CO_2 + \frac{1}{20}\ NH_3 + H^+ + e^- \longrightarrow \frac{1}{20}\ C_5H_7O_2N + \frac{2}{5}\ H_2O \qquad (13.44)$$

When nitrate is the nitrogen source,

$$\frac{5}{28}\ CO_2 + \frac{1}{28}\ NO_3^- + \frac{29}{28}\ H^+ + e^- \longrightarrow \frac{1}{28}\ C_5H_7O_2N + \frac{11}{28}\ H_2O \quad (13.45)$$

Source: Reference 13.38.

where

H_D = half-reaction for contaminant oxidation
f_c = fraction of contaminant converted to energy
H_A = half-reaction for the electron acceptor
f_s = cell yield (Table 13.7)
C_s = reaction for stoichiometric cell yield

Table 13.7 Energy Distribution Factors for
Biomass Synthesis

Electron Acceptor	Ranges of f_s
O_2	0.12–0.60 (mean 0.5)
NO_3	0.1–0.5
SO_4	0.04–0.2
CO_2	0.04–0.2

The factor for cell yield and the fraction of contaminant converted to energy must sum to unity:

$$f_s + f_c = 1 \tag{13.47}$$

Common cell yield values for systems as a function of different electron acceptors are listed in Table 13.7. The procedure for determining the stoichiometric electron acceptor and nutrient requirements is listed in Table 13.8.

Treatment Times for Slurry Bioreactors. Because slurry treatment systems are operated as batch reactors, the time required for treatment of the waste components is a significant design criterion. Because the degradation of many wastes follows first-order kinetics in slurry bioreactors, the time required for treatment is

$$\frac{C}{C_0} = e^{-kt} \tag{13.48}$$

or

$$t = -\frac{\ln\left(\dfrac{C}{C_0}\right)}{k} \tag{13.49}$$

where

C_0 = initial contaminant concentration (mg/kg)
C = design concentration for bioreactor (mg/kg)
k = first order rate constant (day^{-1})
t = time requirements for treatment (days)

The delivery of electron acceptors, nutrients, and cometabolites is proportional to the rate of microbial degradation, which occurs at an exponential rate. Unfortunately, the exponential delivery of nutrients and electron acceptors with pumps and metering devices to correspond to microbial rates is not practical. Therefore, Cookson [13.35]

Table 13.8 Procedure for Determining the Stoichiometric Electron Acceptor and Nutrient Requirements for Biological Treatment

1. Determine the most effective electron acceptor and metabolic pathway.
2. Select the source of nitrogen.
3. List the half-reactions and the equation for cell growth based on f_c and f_s.
4. List the total reaction by summing all of the half-reactions.
5. Evaluate the mass ratios.
6. Determine the total mass of electron acceptor and nutrients required for treatment.

proposed using a stepped decrease for the delivery of nutrients. In addition, he proposed using a delivery rate significantly lower than what is initially required because the demand declines so rapidly at the beginning of the exponential decrease of the contaminants. The design of a batch soil bioslurry system is demonstrated in Example 13.6.

EXAMPLE 13.6 **DESIGN OF A SLURRY BIOREACTOR**

A stockpile of 200 m³ (262 yd³) of soil contaminated with 3800 mg/kg xylenes requires treatment to 100 mg/kg. Pilot studies document that the reaction rate follows first-order kinetics with $k = 0.032$ day^{-1}. Design an ex situ slurry bioreactor to treat the soil including (1) the most effective electron acceptor, (2) the stoichiometric nutrient requirements, (3) the reactor volume, and (4) the delivery rates of the terminal electron acceptor and nutrients.

SOLUTION

1. The most effective electron acceptor in the microbial metabolism of xylenes is O_2. Xylenes are reduced hydrocarbons and are readily degraded under oxic conditions.
2. The chemical structure of the contaminants is C_8H_{10}. Ammonia is selected as the nitrogen source. The appropriate range for the fraction associated with conversion to microbial cells (f_s) is 0.12 to 0.60 using oxygen as an electron acceptor. Therefore, a documented mean value of 0.5 (and a corresponding value of 0.5 for f_c) will be used. The H_D may then be determined from the values of a, b, c, and d:

$$a = 8$$

$$b = 10$$

$$c = 0$$

$$d = 0$$

$$Z = 4 \cdot 8 + 10 = 42$$

Using equations from Table 13.6:

$$H_D: \quad \frac{1}{42} C_8H_{10} + \frac{16}{42} H_2O \longrightarrow \frac{8}{42} CO_2 + H^+ + e^-$$

$$f_e H_A: \quad \frac{1}{8} O_2 + \frac{1}{2} H^+ + \frac{1}{2} e^- \longrightarrow \frac{1}{4} H_2O$$

$$f_s C_s: \quad \frac{1}{8} CO_2 + \frac{1}{40} NH_3 + \frac{1}{2} H^+ + \frac{1}{2} e^- \longrightarrow \frac{1}{40} C_5H_7O_2N + \frac{2}{10} H_2O$$

The sum of the half-reactions yields

$$\frac{1}{42} C_8H_{10} + \frac{16}{42} H_2O + \frac{1}{8} O_2 + \frac{1}{2} H^+ + \frac{1}{2} e^- + \frac{1}{8} CO_2 + \frac{1}{40} NH_3 + \frac{1}{2} H^+ + \frac{1}{2} e^- \longrightarrow$$

$$\frac{8}{42} CO_2 + H^+ + e^- + \frac{1}{4} H_2O + \frac{1}{40} C_5H_7O_2N + \frac{2}{10} H_2O$$

Combining terms yields

$$\frac{1}{42} C_8H_{10} + \frac{1}{8} O_2 + \frac{1}{40} NH_3 \longrightarrow \frac{29}{420} H_2O + \frac{11}{168} CO_2 + \frac{1}{40} C_5H_7O_2N$$

or

$$C_8H_{10} + 5.25\, O_2 + 1.05\, NH_3 \longrightarrow 2.9\, H_2O + 2.75\, CO_2 + 1.05\, C_5H_7O_2N$$

Next, establish mass ratios to calculate nutrient requirements. The molecular weights for each compound are

$$C_8H_{10} = 106$$

$$O_2 = 32$$

$$NH_3 \text{ as } N = 14$$

Mass ratios may then be determined (assuming that the phosphorus requirement is approximately one-sixth that of nitrogen based on the mass of nitrogen).

$$C_8H_{10}:O_2:N:P$$

$$1:5.25 \cdot (32/106):1.05 \cdot (14/106):1/6 \cdot 1.05 \cdot (14/106)$$

$$1:1.58:0.139:0.023$$

Assuming a soil bulk density of 1850 kg/m^3, the nutrient requirements to treat 200 m^3 (262 yd^3) of soil contaminated with 3800 mg/kg of xylenes are

$$\frac{3800 \text{ mg}}{\text{kg}} \cdot \frac{1850 \text{ kg}}{\text{m}^3} \cdot \frac{\text{kg}}{10^6 \text{ mg}} \cdot 200 \text{ m}^3 = 1410 \text{ kg of xylenes}$$

Oxygen: $1410 \cdot 1.58 = 2230$ kg O_2

Nitrogen: $1410 \cdot 0.139 = 196$ kg N

Phosphorus: $1410 \cdot 0.023 = 32.4$ kg P

3. The time required for the reaction may be quantified by using the first-order degradation rate expression:

$$t = -\frac{\ln\left(\dfrac{C}{C_0}\right)}{k}$$

$$t = -\frac{\ln\dfrac{100}{3800}}{0.032 \text{ day}^{-1}}$$

$$t = 114 \text{ days}$$

4. The delivery rates of O_2, N, and P may be determined by assuming that the electron acceptor and nutrients are delivered at a rate proportional to microbial degradation. In order to make delivery rates practical in the field, the metering will be adjusted biweekly. Because there are an odd number of weeks, make the first period (with the highest rate of change of reactants) a single week. The soil must be treated to 100 mg/kg xylenes, or $C/C_0 = 0.026$. The delivery rate of O_2 is

Biweekly Period	t (days)	$-kt$	Fraction of Xylenes Remaining	O_2 Requirement Remaining (kg)	Maximum O_2 Update Rate (kg/day)
0	0	—	—	2230	—
1	7	−0.22	0.799	1780	57.80
2	21	−0.67	0.511	1140	36.44
3	35	−1.12	0.326	727	23.20
4	49	−1.57	0.208	464	15.00
5	63	−2.02	0.133	270	9.55
6	77	−2.46	0.085	190	5.91
7	91	−2.91	0.054	120	3.82
8	105	−3.36	0.035	78.1	2.00
9	119	−3.81	0.022	49.1	1.59

The delivery rate of nitrogen is

Biweekly Period	t (days)	$-kt$	Fraction of Xylenes Remaining	N Requirement Remaining (kg)	Maximum N Update Rate (kg/day)
0	0	—	—	196	—
1	7	−0.22	0.799	156	5.00
2	21	−0.67	0.511	100	3.18
3	35	−1.12	0.326	64.1	2.05
4	49	−1.57	0.208	40.9	1.32
5	63	−2.02	0.133	25.9	0.82
6	77	−2.46	0.085	16.8	0.55
7	91	−2.91	0.054	10.5	0.74
8	105	−3.36	0.035	6.82	0.22
9	119	−3.81	0.022	4.32	0.14

The delivery rate of phosphorus is

Biweekly Period	t (days)	$-kt$	Fraction of Xylenes Remaining	P Requirement Remaining (kg)	Maximum P Update Rate (kg/day)
0	0	—	—	32.4	—
1	7	−0.22	0.799	25.9	0.82
2	21	−0.67	0.511	16.3	0.55
3	35	−1.12	0.326	10.5	0.34
4	49	−1.57	0.208	6.81	0.22
5	63	−2.02	0.133	4.32	0.14
6	77	−2.46	0.085	2.77	0.091
7	91	−2.91	0.054	1.77	0.055
8	105	−3.36	0.035	1.14	0.036
9	119	−3.81	0.022	0.73	0.023

13.5 SUMMARY OF IMPORTANT POINTS AND CONCEPTS

- Four process designs were presented in this chapter, each based on a different pathway covered in Part II. Granular activated carbon contactor design is an application of sorption; air stripping is an extension of volatilization processes; modified Fenton's reagent and slurry bioreactors are applications of abiotic and biotic processes, respectively.

- Granular activated carbon contactors may be designed using mass loadings, bed depth–service time (BDST) analysis, operating line methodology, and finite element models. The BDST analysis was used in this chapter to analyze pilot-scale data as a basis for full-scale design.

- Bed depth–service time analysis is based on the graphical analysis of S-shaped curves of relative contaminant concentration as a function of operating time in pilot columns. Using straight lines for both contaminant breakthrough and carbon exhaustion developed from plots of service time as a function of bed depth, full-scale systems may be designed.

- Air stripper design equations are derived from mass balances around a stripping unit. Two of the more important terms, the number of treatment units (NTU) and height of treatment units (HTU) serve as the basis for determining the air stripper height. The diameter of air stripping towers is related primarily to the design flow rate. Many of the variables used in air stripper design (e.g., the air-water ratio and the stripping factor) are interrelated; therefore, cost analyses are often used to determine the optimum design.

- Fenton's reagent (hydrogen peroxide and catalysts) has the capability of oxidizing most hazardous organic compounds, often more rapidly than they are desorbed from soils and solids. The conceptual design of a Fenton's soil treatment system is governed by desorption rates from soils. If the contaminant desorbs over a short time period (e.g., hours to days), the most effective process design is desorption of the contaminants into a dilute hydrogen peroxide slurry to promote its aqueous phase oxidation. If the rate of desorption is slow (e.g., weeks to months), aggressive Fenton's conditions using higher hydrogen peroxide concentrations are used to oxidize the sorbed contaminants.

- Biological slurry reactors are effective ex situ treatment systems for soils and solids contaminated with compounds that are susceptible to microbial biodegradation. These systems are often designed as batch reactors using first-order biodegradation kinetics. The terminal electron acceptor, the proton donor, and nutrient requirements are the most important process variables.

PROBLEMS

13.1. A dilute aqueous RCRA waste containing 32 mg/L hexachlorocyclopentadiene has been evaluated at a flow rate of 20 L/min using four pilot columns in series that are 5 cm (2 in.) in diameter and 3 m (10 ft) long. Plots of exhaustion ($C/C_0 = 0.90$) and breakthrough ($C/C_0 = 0.10$) lines resulted in slopes of 5.37 days/m and 5.42 days/m, and intercepts of 5.1 days and -19.8 days, respectively. Based on these pilot-scale data, determine (1) the number of columns required, and (2) the full-scale pilot system dimensions if the flow rate is 7800 m^3/day (2 mgd).

13.2. A groundwater contaminated with 18 mg/L atrazine has been evaluated for GAC treatment using three pilot columns in series that are 5 cm (2 in.) in diameter and 3 m (10 ft) long at a flow rate of 15 L/min. The pilot data were plotted with lines for exhaustion ($C/C_0 = 0.90$) and breakthrough ($C/C_0 = 0.10$) having slopes of 4.87 days/m and 4.91 days/m and intercepts of 6.2 days and -17.5 days, respectively. The bulk density of the carbon is 450 kg/m^3. Based on these pilot-scale data, determine (1) the carbon exhaustion rate, (2) the time until the carbon is exhausted, (3) the adsorptive capacity of the carbon, (4) the number of columns required, and (5) the full-scale pilot system dimensions if the flow rate is 7800 m^3/day (2 mgd).

13.3. For the system design of Problem 13.2, determine the time for the carbon to reach exhaustion if the full-scale system flow rate is increased to 9750 m³/day (2.5 mgd)

13.4. A pilot study was conducted on a groundwater contaminated with 2,4,6-trichlorophenol using three columns 5 cm (2 in.) in diameter and 3 m (9.9 ft) long at a flow rate of 8.2 L/min. The following data were obtained from the pilot study. From the data, (1) develop bed depth–service time curves and determine the height of the adsorption zone, (2) determine the number and size of columns required for a full-scale system, and (3) determine the carbon utilization rate. The groundwater is flowing at a rate of 4560 m³/day (1.2 mgd) and assume the unit weight of the carbon is 481 kg/m³.

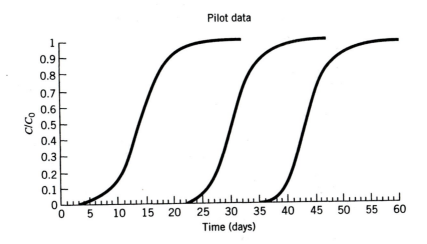

Pilot data

13.5. A groundwater contaminated with 480 μg/L TNT will require treatment to 50 μg/L. A pilot study has been conducted using four GAC columns 10 cm (4 in.) in diameter and 2.5 m (8.2 ft) long with a flow rate of 80 L/day (21 gal/day) and the following data have been collected.

	Concentration of TNT (μg/L)			
Time (days)	Col. #1	Col. #2	Col. #3	Col. #4
1	0			
3	12			
5	24			
7	48			
9	120			
11	297.6			
13	398.4			
15	446.4	0		
17	465.6	9.6		
19	475.2	33.6		
21	480	96		
23	480	288		
25		422.4		

Time (days)	Concentration of TNT (μg/L)			
	Col. #1	Col. #2	Col. #3	Col. #4
27		456	0	
29		470.4	14.4	
31		480	38.4	
33		480	105.6	
35			312	
37			403.2	
39			441.6	0
41			460.8	12
43			475.2	38.4
45			480	124.8
47			480	336
49				422.4
51				451.2
53				468
55				472.8
57				480

Based on the pilot data, design a full-scale system to treat the groundwater at a flow rate of 7.5 m³/day (2700 gel/day).

13.6. An air stripping tower is to be designed to treat carbon tetrachloride (the diffusion coefficient in air is 0.08451 cm²/sec) in a contaminated groundwater using 64 mm (2.5 in.) Pall rings $D_G = 8.541 \times 10^{-6} \dfrac{m^2}{sec}$. The design flow rate will be 3790 m³/day (1 mgd), the air:water ratio is 40:1, the column diameter is 1 m, and the temperature is 20°C. Determine the mass transfer coefficient, $K_L a$, using the Onda equations.

13.7. A landfill leachate must be treated to remove PCE (characterized by a diffusion coefficient in air of 0.07729 cm²/sec). The flow rate for the leachate to be treated is 15,000 m³/day (4 mgd), the column diameter is 2 m (6.6 ft), and 51-mm Tri-Pack packing will be used. If $D_L = 6 \times 10^{-6}$ cm²/sec, and the air-water ratio is 50:1, determine $K_L a$ using the Onda equations.

13.8. Using a mass balance on an air stripper, derive the equation for the height of an air stripper:

$$Z = \frac{L}{K_L a} \cdot \left(\frac{R}{R-1}\right) \ln\left[\frac{(C_0/C)(R-1)+1}{R}\right]$$

13.9. Using the 25-mm (1 in.) Raschig rings, design an air stripper to treat 300 μg/L methylene chloride to the SDWA MCL. The design flow rate is 2800 m³/day (0.74 mgd). Use a stripping factor of 5, a gas pressure drop of 80 N/m²/m, and assume a temperature of 20°C. The following pilot data are also available for the design.

Pilot Column Height (m)	[Methylene chloride μg/L]
0	290
2	115
4	32
6	4

13.10. Design an air stripper to treat

100 μg/L TCE
80 μg/L benzene
90 μg/L toluene
98 μg/L chloroform

to 5 μg/L. The design flow rate is 4000 m^3/day (1.05 mgd). Use 51 mm (2 in.) Tri-Pack packing with an air to water ratio of 35:1, pressure drop of 50 N/m^2/m, 9.404 cm^2/sec, and D_L = 6 × 10^{-6} cm^2/sec at a temperature of 20°C. D_G = 9.404 × 10^{-2} $\dfrac{\text{cm}^2}{\text{sec.}}$

13.11. A groundwater contaminated with 142 μg/L ethylbenzene is to be treated to 5 μg/L by air stripping. The flow rate = 8000 m^3/day (2.1 mgd). Using 89 mm (3.5 in) Jaeger Tri-Pack packing, evaluate a 2 × 2 matrix of designs with two stripping factors (R = 3 and 6) and two pressure drops, 50 N/m^2/m and 100 N/m^2/m. Assume a temperature of 20°C. The diffusion coefficient of ethylbenzene in air is 0.0707 cm^2/sec.

13.12. Soil containing 450 mg/kg PCE has been excavated, and treatment with a modified Fenton's reagent is proposed. If the PCE desorbs into soil water at a first-order rate with k = 0.35 day^{-1}, select the most appropriate process condition and design a treatability study to determine the most effective Fenton's process conditions.

13.13. The use of Fenton's reagent is being considered for the treatment of 1,2,3-trichlorobenzene in a surface soil to 1 mg/kg, which is currently present at an average concentration of 32 mg/kg. The trichlorobenzene desorbs relatively quickly from soil, and the soil is rich in iron oxides, so a mineral-catalyzed Fenton's reaction at pH 4 has been selected as the process configuration. The following bench-scale data have been collected in treating the soil using a liquid:soil ratio of 3 L:1 kg. The porosity of the soil = 0.5.

Time (days)	1,2,3-Trichlorobenzene Residual (mg/kg)				
	Desorption	0.05% H$_2$O$_2$	0.1% H$_2$O$_2$	0.2% H$_2$O$_2$	0.3% H$_2$O$_2$
0	32	31	32	33	33
2	24	30	25	24	25
4	18	28	17	18	19
6	13	26	13	14	13
8	9	22	9	8	10
10	4	20	5	4	4
12	2	18	3	2	3
14	1	17	1	1	1

Determine the optimum hydrogen peroxide concentration based on the treatment stoichiometry (the moles of hydrogen peroxide consumed per moles of trichlorobenzene degraded). Assume that all of the hydrogen peroxide is consumed during the batch treatability studies. Using the most efficient hydrogen peroxide concentration, determine the dimensions of a lined earthen reactor to treat 20 m^3 (26 yd^3) of the contaminated soil. Assume a soil bulk density of 1450 kg/m^3 for the soil. In addition, report (1) the treatment stoichiometry, (2) the quantity of 30% hydrogen peroxide to treat the soil, and (3) the total cost of the hydrogen peroxide if its unit cost is \$0.54/L for 30% sold in drums.

13.14. Modified Fenton's reagent is being evaluated for treating 50 m^3 of a soil at a pesticide formulation area contaminated with 78 mg/kg of dieldrin. The soil, which has a bulk density of 1400 kg/m^3, must be treated to 0.5 mg/kg of dieldrin. Initial measurements have shown that desorption is negligible over 3 weeks. Therefore, enhanced desorption-oxidation of the dieldrin is being evaluated. The following data have been obtained from a bench-scale evaluation of a matrix of ferric sulfate and hydrogen peroxide concentrations using a liquid:solid ratio of 0.5 L:kg soil.

$Fe_2(SO_4)_3$ (mg/L)	Dieldrin Residual (mg/kg)				
	2% H_2O_2	5% H_2O_2	8% H_2O	12% H_2O_2	15% H_2O_2
0	23	24	22	21	13
50	22	11	3	1	3
100	18	5	1	2	4
150	18	2	0.1	1	2
200	19	1	0.1	0.2	1
250	13	2	0.3	1	4
300	14	3	2	4	6

Using these data, determine the concentrations of hydrogen peroxide and ferric sulfate required to treat the soil to 0.5 mg/kg dieldrin. Determine the volume of hydrogen peroxide required, the mass of ferric sulfate, and the chemical costs per cubic meter of soil if the price of hydrogen peroxide is \$0.64 per liter for 50% stock and the unit price of ferric sulfate is \$0.18 per kg.

13.15. A groundwater containing 18 mg/L MIBK is being studied for bioremediation. Determine (a) the most appropriate electron acceptor and (b) the stoichiometric nutrient requirements.

13.16. A soil is contaminated with petroleum hydrocarbons of average molecular formula $C_{10}H_{22}$ at a concentration of 540 mg/kg. Determine (a) the most suitable electron acceptor, (b) the stoichiometric nitrogen and phosphorus requirements, and (c) the electron acceptor requirements.

13.17. Design an ex situ batch bioslurry reactor to treat 5000 mg/kg BTEX to 5 mg/kg. (Hint: Use an average molecular formula for the six compounds and assume they are present in equal concentrations.) Biodegradation is expected to follow the rate of desorption, which proceeds at first-order kinetics with $k = 0.052$ day^{-1}. The process design should include the time required to treat the soil and nutrient requirements for the 100 m³ (130 yd³) batches that will be treated. The bulk density of the soil is 1200 kg/m³. In addition, provide a feed schedule for the biweekly addition of nutrients and the terminal electron acceptor until treatment is completed.

13.18. Design an ex situ soil bioslurry reactor to treat batches of 300 m³ (390 yd³) of soil contaminated with 1000 mg/kg of *o*-cresol. The bulk density of the soil is 1350 kg/m³. Preliminary studies have shown that the biodegradation rate will follow first-order kinetics with a rate constant of 0.08 day^{-1} and that the desorption rate will not be a factor in the design. In the process design, (a) determine the time required to treat the soil to 5 mg/kg, (b) the optimum terminal electron acceptor, (c) nutrient requirements, and (d) nutrient and electron acceptor feed rates for weekly additions.

REFERENCES

13.1. LaGrega, M. D., P. L. Buchingham, and J. C. Evans, *Hazardous Waste Management*, McGraw-Hill, New York, 1994.

13.2. Noonan, D. C. and J. T. Curtis, *Groundwater Remediation and Petroleum: A Guide for Underground Storage Tanks*, Lewis Publishers, Chelsea, MI, 1990.

13.3. Haas, C. N. and R. J. Vamos, *Hazardous and Industrial Waste Treatment*, Prentice-Hall, Englewood Cliffs, NJ, 1995.

13.4. Tchobanoglous, G. and F. L. Burton, *Wastewater Engineering: Treatment, Disposal, and Reuse*, 3rd Edition, Metcalf & Eddy, McGraw-Hill, New York, 1991.

13.5. Dobbs, R. A. and J. M. Cohen, "Carbon adsorption isotherms for toxic organics," U.S. Environmental Protection Agency, Office of Research and Development, Cincinnati, OH, 1980.

13.6. Crittenden, J. C., P. Luft, D. W. Hand, J. L. Oravitz, S. W. Loper, and M. Arl, "Prediction of mutlicomponent adsorption equilibrium using ideal adsorbed solution theory," *Environ. Sci. Technol.*, **19**, 1037–1043 (1985).

13.7. Bohart, G. S., and E. Q. Adams, "Some aspects of the behavior of charcoal with respect to chlorine," *J. Am. Chem. Soc.*, **42**, 523–544 (1920).

13.8. Hutchins, R. A., "New method simplifies design of activated carbon systems," *Chem. Eng.*, **80**, 133–138 (1973).

13.9. Sanks, R. L., *Water Treatment Plant Design*, Ann Arbor Science, Ann Arbor, MI, 1978.

13.10. American Water Works Association, F. W. Pontius (Ed.), *Water Quality and Treatment*, McGraw-Hill, New York, 1990.

13.11. Eckenfelder, W. W., *Industrial Water Pollution Control*, McGraw-Hill, New York, 1989.

13.12. Low, K. S., C. K. Lee, and L. L. Heng, "Sorption of basic dyes by *Hydrilla ventricillata*," *Environ. Technol.*, **15**, 115–224 (1994).

13.13. Kavanaugh, M. C. and R. R. Trussell, "Design of aeration towers to strip volatile contaminants from drinking water," *J. Am. Water Works Assoc.*, **72**, 684–692 (1980).

13.14. Ball, W. P., M. D. Jones, and M. C. Kavanaugh, "Mass transfer of volatile organic compounds in packed aeration," *J. Water Pollut. Control Fed.*, **56**, 127–136 (1984).

13.15. Treybal, R. E., *Mass Transfer Operations*, 3rd Edition, McGraw-Hill, New York, 1980.

13.16. Onda, K., H. Takeuchi, and Y. Okumoto, "Mass transfer coefficients between gas and liquid phases in packed columns," *J. Chem. Eng. Jpn*, **1**, 56–62 (1968).

13.17. Perry, R. H. and D. W. Green (Eds.), *Perry's Chemical Engineers' Handbook*, 6th Edition, McGraw-Hill, New York, 1984.

13.18. Davidson, C. A. and A. W. Busch, "Catalyzed chemical oxidation of phenol in aqueous solution," *Proc. Div. Refin. A.P.I.*, **46**, 299–302 (1966).

13.19. Murphy, A. P., W. J. Boegli, M. K. Price, and C. D. Moody, "A Fenton-like reaction to neutralize formaldehyde waste solutions," *Environ. Sci. Technol.*, **23**, 166–169 (1989).

13.20. Trussler, S., D. Yonge, C. Claiborn, and R. J. Watts, "Mulitmedia trade off: Material substitution results in reduced VOC emissions and increased organic loading to liquid phase treatment processes," Water Environment Federation 68th Annual Conference, Miami, FL, Oct. 21–25, 1995.

13.21. Bowers, A. R., P. Gaddipati, W. W. Eckenfelder, and R. M. Monsen, "Treatment of toxic or refractory wastewater with hydrogen peroxide," *Water Sci. Technol.*, **21**, 477–486 (1989).

13.22. Watts, R. J., P. C. Stanton, C. J. Spencer, S. Kong, and T. M. Haeri-McCarroll, "Process conditions for the treatment of contaminated soils using catalyzed hydrogen peroxide," Contaminated Sites Specialty Conference of the Water Environment Federation, Miami, FL, March, 1994.

13.23. Watts, R. J., S. Kong, M. Dippre, and W. T. Barnes, "Oxidation of sorbed hexachlorobenzene in soils using catalyzed hydrogen peroxide," *J. Hazard. Mater.*, **39**, 33–47 (1994).

13.24. Tyre, B. W., R. J. Watts, and G. C. Miller, "Treatment of four biorefractory contaminants in soils using catalyzed hydrogen peroxide," *J. Environ. Qual.*, **20**, 832–838 (1991).

13.25. Watts, R. J. and P. C. Stanton, "Mineralization of Sorbed and NAPL-phase hexadecane by catalyzed hydrogen peroxide," *Water Res.*, In Press. (1998).

13.26. Watts, R. J., M. D. Udell, and R. M. Monsen, "Use of iron minerals in optimizing the peroxide treatment of contaminated soils," *Water Environ. Res.*, **69**, 839–845 (1993).

13.27. Ravikumar, J. X. and M. Gurol, "Chemical oxidation of chlorinated organics by hydrogen peroxide in the presence of sand," *Environ. Sci. Technol.*, **28**, 394–400 (1994).

13.28. Spencer, C. J., P. C. Stanton, and R. J. Watts, "A central composite rotatable design for the catalyzed hydrogen peroxide remediation of contaminated soils," *J. Air Waste Manage. Assoc.*, **46**, 1067–1074 (1996).

13.29. Watts, R. J. and S. E. Dilly, "Effect of iron catalysts on a Fenton-like process for remediation of diesel-contaminated soil," *J. Hazard. Mater*, **51**, 209–224 (1997).

13.30. Gates, D. D. and R. L. Siegrist, "In situ chemical oxidation of trichloroethylene using hydrogen peroxide," *J. Environ. Eng.*, **121**, 639–644 (1995).

13.31. Ho, C. L., M. A. Shebl, and R. J. Watts, "Development of an injection system for in situ catalyzed peroxide remediation of contaminated soils," *Hazard. Waste Hazard. Mater.*, **12**, 15–25 (1994).

13.32. Ross, D., T. P. Maziarz, and A. W. Bourguin, "Bioremediation of hazardous waste sites in the USA—Case histories," HMCRI 9th National Conference, The Hazardous Materials Control Research Institute, Washington, DC, 395–397 (1988).

13.33. Strou, H. F., W. Mahaffey, and A. W. Bourguin, "Development of an in situ bioremediation system for a creosote-contaminated site," in Wu, X. C. (Ed.), *International Conference on Physicochemical and Biological Detoxification of Hazardous Wastes*, Vol. II, Technomic Publications, Lancaster, PA, 1988.

13.34. Milhelcic, J. R. and R. G. Luthy, "The potential effects of sorption processes on the microbial degradation of hydrophobic organic compounds in soil-water suspensions," in

Wu, X. C. (Ed.), *International Conference on Physicochemical and Biological Detoxification of Hazardous Wastes*, Vol. II, Technomic Publications, Lancaster, PA, 1988.

13.35. Cookson, J. T., Jr., *Bioremediation Engineering: Design and Application*, McGraw-Hill, New York, 1995.

13.36. Janssen, D. B., A. J. Van den Wijngaard, J. J. Van der Waarde, and R. Oldenhuis, "Biochemistry and kinetics of aerobic degradation of chlorinated aliphatic hydrocarbons," in Hinchee, R. E. and R. F. Offenbuttel (Eds.), *On-Site Bioremediation: Processes for Xenobiotic and Hydrocarbon Treatment*, Butterworth-Heinemann, Boston, MA, 1991.

13.37. Yare, B. S., "A comparison of soil-phase and slurry-phase bioremediation of PNA-containing soils," in Hinchee, R. E. and R. F. Offenbuttel (Eds.), *On-Site Bioremediation: Processes for Xenobiotic and Hydrocarbon Treatment*, Butterworth-Heinemann, Boston, MA, 1991.

13.38. McCarty, P. L., "Bioengineering issues related to in situ remediation of contaminated soils and groundwater," in Omenn, G. S. (Ed.), *Environmental Biotechnology*, Plenum Press, New York, 1988.

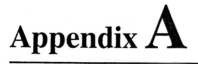

Appendix A

Hazardous Waste Lists

Table A.1 Hazardous Wastes from Nonspecific Sources (The F List)

Industry and EPA Hazardous Waste No.	Hazardous Wastes from Nonspecific Sources	Hazard Code*
F001	The following spent halogenated solvents used in degreasing: tetrachloroethylene, trichloroethylene, methylene chloride, 1,1,1-trichloroethane, carbon tetrachloride, and chlorinated fluorocarbons; all spent solvent mixtures/blends used in degreasing containing, before use, a total of 10 percent or more (by volume) of one or more of the above halogenated solvents or those solvents listed in F002, F004, and F005; and still bottoms from the recovery of these spent solvents and spent solvent mixtures.	(T)
F002	The following spent halogenated solvents: tetrachloroethylene, methylene chloride, trichloroethylene, 1,1,1-trichloroethane, chlorobenzene, 1,1,2-trichloro-1,2,2-trifluoroethane, ortho-dichlorobenzene, trichlorofluoromethane, and 1,1,2-trichloroethane; all spent solvent mixtures/blends containing, before use, a total of 10 percent or more (by volume) of one or more of the above halogenated solvents or those listed in F001, F004, or F005; and still bottoms from the recovery of these spent solvents and spent solvent mixtures.	(T)
F003	The following spent nonhalogenated solvents: xylene, acetone, ethyl acetate, ethyl benzene, ethyl ether, methyl isobutyl ketone, n-butyl alcohol, cyclohexanone, and methanol; all spent solvent mixtures/blends containing, before use, a total of 10 percent or more (by volume) of one or more of the above nonhalogenated solvents, listed in F001, F002, F004, and F005; and still bottoms from the recovery of these spent solvents and spent solvent mixtures.	(I)

636

Table A.1 (Continued)

Industry and EPA Hazardous Waste No.	Hazardous Wastes from Nonspecific Sources	Hazard Code*
F004	The following spent nonhalogenated solvents: cresols, cresylic acid, and nitrobenzene; all spent solvent mixtures/blends containing, before use, a total of 10 percent or more (by volume) of one or more of the above nonhalogenated solvents or those solvents listed in F001, F002, and F005; and still bottoms from the recovery of these spent solvents and spent solvent mixtures.	(T)
F005	The following spent nonhalogenated solvents: toluene, methyl ethyl ketone, carbon disulfide, isobutanol, pyridine, benzene, 2-exthoxyethanol, and 2-nitro-propane; all spent solvent mixtures/blends containing, before use, a total of 10 percent or more (by volume) or one or more of the above nonhalogenated solvents or those solvents listed in F001, F002, or F004; and still bottoms from the recovery of these spent solvents and spent solvent mixtures.	(I,T)
F006	Wastewater treatment sludges from electroplating operations except from the following processes: (1) sulfuric acid anodizing or aluminum; (2) tin plating on carbon steel; (3) zinc plating (segregated basis) on carbon steel; (4) aluminum or zinc-and-aluminum plating on carbon steel; (5) cleaning/stripping associated with tin, zinc, and aluminum plating on carbon steel; and (6) chemical etching and milling of aluminum.	(T)
F019	Waste treatment sludges from the chemical conversion coating of aluminum.	(T)
F007	Spent cyanide plating bath solutions from electroplating operations.	(R,T)
F008	Plating bath residues from the bottom of plating baths from electroplating operations where cyanides are used in the process.	(R,T)
F009	Spent stripping and cleaning bath solutions from electroplating operations where cyanides are used in the process.	(R,T)
F010	Quenching bath residues from oil baths from metal heat-treating operations where cyanides are used in the process.	(R,T)
F011	Spent cyanide solutions from salt bath pot cleaning from metal heat-treating operations.	(R,T)
F012	Quenching wastewater treatment sludges from metal heat-treating operations where cyanides are used in the process.	(T)

Table A.1 (Continued)

Industry and EPA Hazardous Waste No.	Hazardous Wastes from Nonspecific Sources	Hazard Code*
F024	Wastes, including but not limited to, distillation residues, heavy ends, tars, and reactor cleanout wastes from the production of chlorinated aliphatic hydrocarbons, having carbon content from one to five, utilizing free radical catalyzed processes. (This listing does not include light ends, spent filters and filter aids, spent dessicants, wastewater, wastewater treatment sludges, spent catalysts, and wastes.	(T)
F020	Wastes (except wastewater and spent carbon from hydrogen chloride purification) from the production or manufacturing use (as a reactant, chemical intermediate, or component in a formulating process) or tri- or tetrachlorophenol, or of intermediates used to produce their pesticide derivatives. (This listing does not include wastes from the production of hexachlorophene from highly purified 2,4,5-trichlorophenol.)	(H)
F021	Wastes (except wastewater and spent carbon from hydrogen chloride purification) from the production or manufacturing use (as a reactant, chemical intermediate, or component in a formulating process) of pentachlorophenol, or of intermediates used to produce its derivatives.	(H)
F022	Wastes (except wastewater and spent carbon from hydrogen chloride purification) from the manufacturing use (as a reactant, chemical intermediate, or component in a formulating process) of tetra-, penta-, or hexachlorobenzenes under alkaline conditions.	(H)
F023	Wastes (except wastewater and spent carbon from hydrogen chloride purification) from the production of materials on equipment previously used for the production or manufacturing use (as a reactant, chemical intermediate, or component in a formulating process) of tri- and tetrachlorophenols. (This listing does not include wastes from equipment used only for the production or use of hexachlorophene from highly purified 2,4,5-trichlorophenol.)	(H)
F026	Wastes (except wastewater and spent carbon from hydrogen chloride purification) from the production of materials on equipment previously used for the manufacturing use (as a reactant, chemical intermediate, or component in a formulating process) of tetra-, penta-, or hexachlorobenzene under alkaline conditions.	(H)

Table A.1 (Continued)

Industry and EPA Hazardous Waste No.	Hazardous Wastes from Nonspecific Sources	Hazard Code*
F027	Discarded unused formulations containing tri-, tetra-, or pentachlorophenol or discarded unused formulations containing compounds derived from these chlorophenols. (This listing does not include formulations containing hexachlorophene synthesized from prepurified 2,4,5 trichlorophenol as the sole component.)	(H)
F028	Residues resulting from the incineration or thermal treatment of soil contaminated with EPA hazardous waste numbers F020, F021, F022, F023, F026, and F027.	(T)

*Hazard code definitions: C=Corrosivity; E=EP toxicity; H=Acute hazardous waste; I=Ignitability; R=Reactivity; T=Toxic waste.

Table A.2 Hazardous Wastes from Specific Sources (The K List)

Industry and EPA Hazardous Waste No.	Hazardous Wastes	Hazard Code*
Wood Preservation		
K001	Bottom sediment sludge from the treatment of wastewater from wood preserving processes that use creosote and/or pentachlorophenol.	(T)
Inorganic Pigments		
K002	Wastewater treatment sludge from the production of chrome yellow and orange pigments.	(T)
K003	Wastewater treatment sludge from the production of molybdate orange pigments.	(T)
K004	Wastewater treatment sludge from the production of zinc yellow pigments.	(T)
K005	Wastewater treatment sludge from the production of chrome green pigments.	(T)
K006	Wastewater treatment sludge from the production of chrome oxide green pigments (anhydrous and hydrated).	(T)
K007	Wastewater treatment sludge from the production of iron blue pigments.	(T)
K008	Oven residue from the production of chrome oxide green pigments.	(T)

Table A.2 (Continued)

Industry and EPA Hazardous Waste No.	Hazardous Wastes	Hazard Code*
Organic Chemicals		
K009	Distillation bottoms from the production of acetaldehyde from ethylene.	(T)
K010	Distillation side cuts from the production of acetaldehyde from ethylene.	(T)
K011	Bottom stream from the wastewater stripper in the production of acrylonitrile.	(R,T)
K013	Bottom stream from the acetonitrile column in the production of acrylonitrile.	(R,T)
K014	Bottoms from the acetonitrile purification column in the production of acrylonitrile.	(T)
K015	Still bottoms from the distillation of benzyl chloride.	(T)
K016	Heavy ends or distillation residues from the production of carbon tetrachloride.	(T)
K017	Heavy ends (still bottoms) from the purification column in the production of epichlorohydrin.	(T)
K018	Heavy ends from the fractionation column in ethyl chloride production.	(T)
K019	Heavy ends from the distillation of ethylene dichloride in ethylene dichloride production.	(T)
K020	Heavy ends from the distillation of vinyl chloride in vinyl chloride monomer production.	(T)
K021	Aqueous spent antimony catalyst waste from fluoromethanes production.	(T)
K022	Distillation bottom tars from the production of phenol/acetone from cumene.	(T)
K023	Distillation light ends from the production of phthalic anhydride from naphthalene.	(T)
K024	Distillation bottoms from the production of phthalic anhydride from naphthalene.	(T)
K093	Distillation light ends from the production of phthalic anhydride from o-xylene.	(T)
K094	Distillation bottoms from the production of phthalic anhydride from o-xylene.	(T)
K025	Distillation bottoms from the production of nitrobenzene by the nitration of benzene.	(T)
K026	Stripping still tails from the production of methyl ethyl pyridines.	(T)
K027	Centrifuge and distillation residues from toluene diisocyanate production.	(R,T)
K028	Spent catalyst from the hydrochlorinator reactor in the production of 1,1,1-trichloroethane.	(T)
K029	Waste from the product steam stripper in the production of 1,1,1-trichloroethane.	(T)

Table A.2 (Continued)

Industry and EPA Hazardous Waste No.	Hazardous Wastes	Hazard Code*
K095	Distillation bottoms from the production of 1,1,1-trichloroethane.	(T)
K096	Heavy ends from the heavy ends column from the production of 1,1,1-trichloroethane.	(T)
K030	Column bottoms or heavy ends from the combined production of trichloroethylene and perchloroethylene.	(T)
K083	Distillation bottoms from aniline productions.	(T)
K103	Process residues from aniline extraction from the production of aniline.	(T)
K104	Combined wastewater streams generated from nitrobenzene/aniline production.	(T)
K035	Wastewater treatment sludges generated in the production of creosote.	(T)
K036	Still bottoms from toluene reclamation distillation in the production of disulfoton.	(T)
K037	Wastewater treatment sludges from the production of disulfoton.	(T)
K038	Wastewater from the washing and stripping of phorate production.	(T)
K039	Filter cake from the filtration of diethylphosphorodithioic acid in the production of phorate.	(T)
K040	Wastewater treatment sludge from the production of phorate.	(T)
K041	Wastewater treatment sludge from the production of toxaphene.	(T)
K098	Untreated process wastewater from the production of toxaphene.	(T)
K042	Heavy ends or distillation residues from the distillation of tetrachlorobenzene in the production of 2,4,5-T.	(T)
K043	2,6-Dichlorophenol waste from the production of 2,4-D.	(T)
K099	Untreated wastewater from the production of 2,4-D.	(T)
K123	Process wastewater (including supernates, filtrates, and washwaters) from the production of ethylenebis(dithiocarbamic acid) and its salt.	(T)
K124	Reactor vent scrubber water from the production of ethylenebis(dithiocarbamic acid) and its salts.	(C,T)
K125	Filtration, evaporation, and centrifugation solids from the production of ethylenebis(dithiocarbamic acid) and its solids.	(T)
K126	Baghouse dust and floor sweepings in milling and packaging operations from the production or formulation of ethylenebis(dithiocarbamic acid) and its salts.	(T)

Table A.2 (Continued)

Industry and EPA Hazardous Waste No.	Hazardous Wastes	Hazard Code*
Explosives		
K044	Wastewater treatment sludges from the manufacturing and processing of explosives.	(R)
K045	Spent carbon from the treatment of wastewater containing explosives.	(R)
K046	Wastewater treatment sludges from the manufacturing, formulation, and loading of lead-based initiating compounds.	(T)
K047	Pink/red water from TNT operations.	(R)
Petroleum Refining		
K048	Dissolved air flotation (DAF) float from the petroleum refining industry.	(T)
K049	Slop oil emulsion solids from the petroleum refining industry.	(T)
K050	Heat exchanger bundle cleaning sludge from the petroleum refining industry.	(T)
K051	API separator sludge from the petroleum refining industry.	(T)
K052	Tank bottoms (leaded) from the petroleum refining industry.	(T)
K085	Distillation or fractionation column bottoms from the production of chlorobenzenes.	(T)
K105	Separated aqueous stream from the reactor product washing step in the production of chlorobenzenes.	(T)
K111	Product washwaters from the production of dinitrotoluene via nitration of toluene.	(C,T)
K112	Reaction by-product water from the drying column in the production of toluenediamine via hydrogenation of dinitrotoluene.	(T)
K113	Condensed liquid light ends from the purification of toluenediamine in the production of toluenediamine via hydrogenation of dinitrotoluene.	(T)
K114	Vicinals from the purification of toluenediamine in the production of toluenediamine via hydrogenation of dinitrotoluene.	(T)
K115	Heavy ends from the purification of toluenediamine in the production of toluenediamine via hydrogenation of dinitrotoluene.	(T)
K116	Organic condensate from the solvent recovery column in the production of toluene diisocyanate via phosgenation of toluenediamine.	(T)
K117	Wastewater from the reactor vent gas scrubber in the production of ethylene dibromide via bromination of ethene.	(T)

Table A.2 (Continued)

Industry and EPA Hazardous Waste No.	Hazardous Wastes	Hazard Code*
K118	Spent adsorbent solids from purification of ethylene dibromide in the production of ethylene dibromide via bromination of ethene.	(T)
K136	Still bottoms from the purification of ethylene dibromide in the production of ethylene dibromide via bromination of ethene.	(T)
Inorganic Chemicals		
K071	Brine purification muds from the mercury cell process in chlorine production, where separately prepurified brine is not used.	(T)
K073	Chlorinated hydrocarbon waste from the purification step of the diaphragm cell process using graphite anodes in chlorine production.	(T)
K106	Wastewater treatment sludge from the mercury cell process in chlorine production.	(T)
Pesticides		
K031	By-product salts generated in the production of MSMA and cacodylic acid.	(T)
K032	Wastewater treatment sludge from the production of chlordane.	(T)
K033	Wastewater and scrub water from the chlorination of cyclopentadiene in the production of chlordane.	(T)
K034	Filter solids from the filtration of hexachlorocyclo-pentadiene in the production of chlordane.	(T)
K097	Vacuum stripper discharge from the chlordane chlorinator in the production of chlordane.	(T)
Iron and Steel		
K061	Emission control dust/sludge from the primary production of steel in electric furnaces.	(T)
K062	Spent pickle liquor generated by steel finishing operations of facilities within the iron and steel industry (SIC Codes 331 and 332).	(C,T)
Secondary Lead		
K069	Emission control dust/sludge from secondary lead smelting.	(T)
K100	Waste leaching solution from acid leaching of emission control dust/sludge from secondary lead smelting.	(T)

Table A.2 (Continued)

Industry and EPA Hazardous Waste No.	Hazardous Wastes	Hazard Code*
Veterinary Pharmaceuticals		
K084	Wastewater treatment sludges generated during the production of veterinary pharmaceuticals from arsenic or organoarsenic compounds.	(T)
K101	Distillation tar residues from the distillation of aniline-based compounds in the production of veterinary pharmaceuticals from arsenic or organoarsenic compounds.	(T)
K102	Residue from the use of activated carbon for decolorization in the production of veterinary pharmaceuticals from arsenic or organoarsenic compounds.	(T)
Ink Formulation		
K086	Solvent washes and sludges, caustic washes and sludges, or water washes and sludges from cleaning tubs and equipment used in the formulation of ink from pigments, driers, soaps, and stabilizers containing chromium and lead.	(T)
Coking		
K060	Ammonia still lime sludge from coking operations.	(T)
K087	Decanter tank tar sludge from coking operations.	(T)

*Hazard code definitions: C=Corrosivity; E=EP toxicity; H=Acute hazardous waste; I=Ignitability; R=Reactivity; T=Toxic waste.

Table A.3 Hazardous Wastes from Commercial Products, Intermediates, and Residues
(The P and U Lists)

Hazardous Waste No.	Chemical Abstracts No.	Hazardous Substance
P List		
P023	107-20-0	Acetaldehyde, chloro-
P002	591-08-2	Acetamide, n-(aminothioxomethyl)-
P057	640-19-7	Acetamide, 2-fluoro-
P058	62-74-8	Acetic acid, fluoro-, sodium salt
P066	16752-77-5	Acetimidic acid, n-[(methylcarbamoyl)oxy]thio-, methyl ester

Table A.3 (Continued)

Hazardous Waste No.	Chemical Abstracts No.	Hazardous Substance
P002	591-08-2	1-Acetyl-2-thiourea
P003	107-02-8	Acrolein
P070	116-06-3	Aldicarb
P004	309-00-2	Aldrin
P005	107-18-6	Allyl alcohol
P006	20859-73-8	Aluminum phosphide (R,T)
P007	2763-96-4	5-(Aminomethyl)-3-isoxazolol
P008	504-24-5	4-α-Aminopyridine
P009	131-74-8	Ammonium picrate (R)
P119	7803-55-6	Ammonium vanadate
P010	7778-39-4	Arsenic acid
P012	1327-53-3	Arsenic oxide As$_2$O$_3$
P011	1303-28-2	Arsenic oxide As$_2$
P011	1303-28-2	Arsenic pentoxide
P012	1327-53-3	Arsenic trioxide
P038	692-42-2	Arsine, diethyl
P036	696-28-6	Arsonous dichloride, phenyl-
P054	151-56-4	Aziridine
P013	542-62-1	Barium cyanide
P024	106-47-8	Benzenamine, 4-chloro-
P077	100-01-6	Benzenamine, 4-nitro-
P028	100-44-7	Benzene, (chloromethyl)
P042	51-43-4	1,2-Benzenediol, 4-[1-hydroxy-2-(methylamino) ethyl]-, (R)
P046	122-09-8	Benzeneethanamine, α,α-dimethyl-
P014	108-98-5	Benzenethiol
P001	81-81-2	2H-1-Benzopyran-2-one, 4-hydroxy-3- (3-oxo-1-phenylbutyl)-, and salts
P028	100-44-7	Benzyl chloride
P015	7440-41-7	Beryllium dust
P016	542-88-1	bis(chloromethyl) ether
P017	598-31-2	Bromoacetone
P018	357-57-3	Brucine
P021	592-01-8	Calcium cyanide
P022	75-15-0	Carbon bisulfide
P095	75-44-5	Carbonic dichloride
P023	107-20-0	Chloroacetaldehyde
P024	106-47-8	p-Chloroaniline
P029	544-92-3	Copper cyanide
P030		Cyanides (soluble cyanide salts), not otherwise specified
P031	460-19-5	Cyanogen
P033	506-77-4	Cyanogen chloride
P034	131-89-5	2-Cyclohexyl-4,6-dinitrophenol

Table A.3 (Continued)

Hazardous Waste No.	Chemical Abstracts No.	Hazardous Substance
P036	696-28-6	Dichlorophenylarsine
P037	60-57-1	Dieldrin
P038	692-42-2	Diethylarsine
P041	311-45-5	Diethyl-*p*-nitrophenyl phosphate
P040	297-97-2	*O,O*-Diethyl *O*-pyrazinyl phosphorothioate
P043	55-91-4	Diisopropyl fluorophosphate (DEP)
P004	309-00-2	1,4:5,8-Dimethanonaphthalene, 1,2,3,4,10,10-hexachloro-1,4,4*a*,5,8,8*a*-hexahydro-, (1α,4α,4β,5α,8α,8*a*β)
P060	465-73-6	1,4:5,8-Dimethanonaphthalene, 1,2,3,4,10,10-hexachloro-1,4,4*a*,5,8,8*a*-hexahydro-, (1α,4α,4β,5β,8β,8β)-
P037	60-57-1	2,7:3,6-Dimethanonaphth[2,3*b*]oxirane, 3,4,5,6,9,9-hexachloro-1*a*,2,2*a*,3,6,6*a*,7,7*a*-octahydro-, (1*a*α,2β,2*a*α,3β,6β,6*a*α,7β,7*a*α)-
P051	72-20-8	2,7:3,6-Dimethanonaphth[2,3*b*]oxirane, octahydro-, (1*a*α,2β,2*a*α,3α,6α,6*a*β,7β,7*a*α)-
P044	60-51-5	Dimethoate
P045	39196-18-4	3,3-Dimethyl-1-(methylthio)-2-butanone, *O*-[(methylamino)carbonyl]oxime
P046	122-09-8	α,α-Dimethylphenethylamine
P047	534-52-1	4,6-Dinitro-*o*-cresol and salts
P048	51-28-5	2,4-Dinitrophenol
P020	88-85-7	Dinoseb
P085	152-16-9	Diphosphoramide, octamethyl-
P039	298-04-4	Disulfoton
P049	541-53-7	2,4-Dithiobiuret
P050	115-29-7	Endosulfan
P088	145-73-3	Endothal
P051	72-20-8	Endrin
P042	51-43-4	Epinephrine
P101	107-12-0	Ethyl cyanide
P054	151-56-4	Ethyleneimine
P097	52-85-7	Famphur
P056	7782-41-4	Fluorine
P057	640-19-7	Fluoroacetamide
P058	62-74-8	Fluoroacetic acid, sodium salt
P065	628-86-4	Fulminic acid, mercury (2+) salt (R,T)
P059	76-44-8	Heptachlor
P062	757-58-4	Hexaethyltetraphosphate
P116	79-19-6	Hydrazinecarbothioamide
P068	60-34-4	Hydrazine, methyl-

Table A.3 (Continued)

Hazardous Waste No.	Chemical Abstracts No.	Hazardous Substance
P063	74-90-8	Hydrocyanic acid
P063	74-90-8	Hydrogen cyanide
P096	7803-51-2	Hydrogen phosphide
P064	624-83-9	Isocyanic acid, methyl ester
P060	465-73-6	Isodrin
P007	2763-96-4	3(2H)-Isoxazolone, 5-(aminomethyl)-
P092	62-38-4	Mercury, (acetato)phenyl-
P065	628-86-4	Mercury fulminate (R,T)
P082	62-75-9	Methamine, *n*-methyl-*n*-nitroso-
P016	542-88-1	Methane, oxybis[chloro-]
P112	509-14-8	Methane, tetranitro- (R)
P118	75-70-7	Methanethiol, trichloro-
P050	115-29-7	6,9-Methano-2,4,3-benzodioxathiepen, 6,7,8,9,10, 10-hexachloro-1,5,5*a*,6,9,9*a*-hexahydro-, 3-oxide
P059	76-44-8	4,7-Methano-1H-indene, 1,4,5,6,7,8,8-heptachloro-3*a*,4,7, 7*a*-tetrahydro-
P066	16752-77-5	Methomyl
P067	75-55-8	2-Methylaziridine
P068	60-34-4	Methyl hydrazine
P064	624-83-9	Methyl isocyanate
P069	75-86-5	2-Methyllactonitrile
P071	298-00-0	Methyl parathion
P072	86-88-4	α-Naphthylthiourea
P073	13463-39-3	Nickel carbonyl
P073	13463-39-3	Nickel carbonyl, (T)
P075	54-11-5	Nicotine and salts
P076	10102-43-9	Nitric oxide
P077	100-01-6	*p*-Nitroaniline
P078	10102-44-0	Nitrogen dioxide
P076	10102-43-9	Nitrogen oxide NO
P081	55-63-0	Nitroglycerine (R)
P082	62-75-9	*N*-Nitrosodimethylamine
P084	4549-40-0	*N*-Nitrosomethylvinylamine
P074	557-19-7	Nickel cyanide
P085	152-16-9	Octamethylpyrophosphoramide
P087	20816-12-0	Osmium oxide
P087	20816-12-0	Osmium tetroxide
P088	145-73-3	7-Oxabicyclo[2.2.1]heptane-2,3-dicarboxylic acid
P089	56-38-2	Parathion
P034	131-89-5	Phenol, 2-cyclohexyl-4,6-dinitro-
P048	51-28-5	Phenol, 2-4-dinitro-
P047	534-52-1	Phenol, 2-methyl-4,6-dinitro- and salts

Table A.3 (Continued)

Hazardous Waste No.	Chemical Abstracts No.	Hazardous Substance
P020	88-85-7	Phenol, 2-(1-methylpropyl)-4,6-dinitro
P009	131-74-8	Phenol, 2,4,6-trinitro-, ammonium salt (R)
P092	62-38-4	Phenylmercury acetate
P093	103-85-5	Phenylthiourea
P094	298-02-2	Phorate
P095	75-44-5	Phosgene
P096	7803-51-2	Phosphine
P041	311-45-5	Phosphoric acid, diethyl 4-nitrophenyl ester
P039	298-04-4	Phosphorodithioic acid, O,O-diethyl S-[2-(ethylthio)ethyl] ester
P094	298-02-2	Phosphorodithioic acid, O,O-diethyl S-[(ethylthio)methyl] ester
P044	60-51-5	Phosphorodithioic acid, O,O-dimethyl S-[2-(methylamino)-2-oxoethyl] ester
P043	55-91-4	Phosphorofluoric acid, bis(1-methylethyl)-ester
P089	56-38-2	Phosphorothioic acid, O,O-diethyl O-(4-nitrophenyl) ester
P040	297-97-2	Phosphorothioic acid, O,O-diethyl O-pyrazinyl ester
P097	52-85-7	Phosphorothioic acid, O-[4-[(dimethylamino)sulfonyl]phenyl] O,O-dimethyl ester
P071	298-00-0	Phosphorothioic acid, O,O-dimethyl O-(4-nitrophenyl) ester
P110	78-00-2	Plumbane, tetraethyl-
P098	151-50-8	Potassium cyanide
P099	506-61-6	Potassium silver cyanide
P070	116-06-3	Propanal, 2-methyl-2-(methylthio)-, O-[(methylamino) carbonyl]oxime
P101	107-12-0	Propanenitrile
P027	542-76-7	Propanenitrile, 3-chloro-
P069	75-86-5	Propanenitrile, 2-hydroxy-2-methyl-
P081	55-63-0	1,2,3-Propanetriol, trinitrate (R)
P017	598-31-2	2-Propanone, 1-bromo-
P102	107-19-7	Propargyl alcohol
P003	107-02-8	2-Propenal
P005	107-18-6	2-Propen-1-ol
P067	75-55-8	1,2-Propylenimine
P102	591-08-2	2-Propyn-1-ol
P008	504-24-5	Pyridinamine
P075	54-11-5	Pyridine, (S)-3-(1-methyl-2-pyrrolidinyl)-, and salts
P111	107-49-3	Pyrophosphoric acid, tetraethyl ester
P103	630-10-4	Selenourea
P104	506-64-9	Silver cyanide
P105	26628-22-8	Sodium azide
P106	143-33-9	Sodium cyanide
P107	1314-96-1	Strontium sulfide

Table A.3 (Continued)

Hazardous Waste No.	Chemical Abstracts No.	Hazardous Substance
P108	57-24-9	Strychnidin-10-one, and salts
P018	357-57-3	Strychnidin-10-one, 2,3-dimethoxy-
P108	57-24-9	Strychnine and salts
P115	10031-59-1	Sulfuric acid, thallium (I) salt
P109	3689-24-5	Tetraethyldithiopyrophosphate
P110	78-00-2	Tetraethyl lead
P111	107-49-3	Tetraethylpyrophosphate
P112	509-14-8	Tetranitromethane (R)
P062	757-58-4	Tetraphosphoric acid, hexaethyl ester
P113	1314-32-5	Thallic oxide
P113	1314-32-5	Thallium (III) oxide
P114	12039-52-0	Thallium (I) selenite
P115	10031-59-1	Thalium (I) sulfate
P109	3689-24-5	Thiodiphosphoric acid, tetraethyl ester
P045	39196-18-4	Thiofanox
P049	541-53-7	Thioimidodicarbonic diamide
P014	108-98-5	Thiophenol
P116	79-19-6	Thiosemicarbazide
P026	5344-82-1	Thiourea, (2-chlorophenyl)-
P072	86-88-4	Thiourea, 1-naphthalenyl-
P093	103-85-5	Thiourea, phenyl-
P123	8001-35-2	Toxaphene
P118	75-70-7	Trichloromethanethiol
P119	7803-55-6	Vanadic acid, ammonium salt
P120	1314-62-1	Vanadium (V) oxide
P084	4549-40-0	Vinylamine, N-methyl-N-nitroso-
P001	81-81-2	Warfarin
P121	557-21-1	Zinc cyanide
P122	1314-84-7	Zinc phosphide (R,T)

U List

U001	75-07-0	Acetaldehyde (I)
U034	75-87-6	Acetaldehyde, trichloro-
U187	62-44-2	Acetamide, N-(4-ethoxyphenyl)-
U005	53-96-3	Acetamide, N-9H-fluoren-2-yl
U112	141-78-6	Acetic acid, ethyl ester (I)
U144	301-04-2	Acetic acid, lead salt
U214	563-68-8	Acetic acid, thallium (I+) salt
U232	93-76-5	Acetic acid, (2,4,5-trichlorophenoxy)-
U002	67-64-1	Acetone (I)
U003	75-05-8	Acetonitrile, (I,T)
U004	98-86-2	Acetophenone
U005	53-96-3	2-Acetylaminofluorene

Table A.3 (Continued)

Hazardous Waste No.	Chemical Abstracts No.	Hazardous Substance
U006	75-36-5	Acetyl chloride (C,R,T)
U007	79-06-1	Acrylamide
U008	79-10-7	Acrylic acid (I)
U009	107-13-1	Acrylonitrile
U011	61-82-5	Amitrole
U012	62-53-3	Aniline (I,T)
U014	492-80-8	Auramine
U015	115-02-6	Azaserine
U010	50-07-7	Azirino (2′,3′:3,4)pyrrolo[1,2-a]indole-4,7-dione, 6-amino-8-[((aminocarbonyl)oxy)methyl]-1,1a,2,8,8a,8b-hexahydro-8a-methoxy-5-methyl-
U157	50-49-5	Benzo[j]aceanthrylene, 1,2-dihydro-3-methyl-
U016	225-51-4	3,4-Benzacridine
U017	98-87-3	Benzal chloride
U047	91-58-7	β-Chloronaphthalene
U048	95-57-8	o-Chlorophenol
U049	3165-93-3	4-Chloro-o-toluidine, hydrochloride
U032	13765-19-0	Chromic acid, calcium salt
U050	218-01-9	Chrysene
U051	8021-39-4	Creosote
U052	1319-77-3	Cresols (cresylic acid)
U053	4170-30-3	Crotonaldehyde
U055	98-82-8	Cumene (I)
U246	506-68-3	Cyanogen bromide
U197	106-51-4	2,5-Cyclohexadiene-1,4-dione
U056	110-82-7	Cyclohexane (I)
U057	108-94-1	Cyclohexanone (I)
U130	77-47-4	1,3-Cyclopentadiene, 1,2,3,4,5,5-hexachloro-
U058	50-18-0	Cyclophosphamide
U240	94-75-7	2,4-D, salts and esters
U059	20830-81-3	Daunomycin
U060	72-54-8	DDD
U061	50-29-3	DDT
U062	2303-16-4	Diallate
U063	53-70-3	Dibenz[a,h]anthracene
U064	189-55-9	Dibenzo[a,i]pyrene
U066	96-12-8	1,2-Dibromo-3-chloropropane
U069	84-74-2	Dibutyl phthalate
U070	95-50-1	o-Dichlorobenzene
U071	541-73-1	m-Dichlorobenzene
U072	106-46-7	p-Dichlorobenzene
U073	91-94-1	3,3′-Dichlorobenzidine
U074	764-41-0	1,4-Dichloro-2-butene (I,T)

Table A.3 (Continued)

Hazardous Waste No.	Chemical Abstracts No.	Hazardous Substance
U075	75-71-8	Dichlorodifluoromethane
U078	75-35-4	1,1-Dichloroethylene
U079	156-60-5	1,2-Dichloroethylene
U025	111-44-1	Dichloroethyl ether
U081	120-83-2	2,4-Dichlorophenol
U082	87-65-0	2,6-Dichlorophenol
U240	94-75-7	2,4-Dichlorophenoxyacetic acid, salts and esters
U083	78-87-5	1,2-Dichloropropane
U084	542-75-6	1,3-Dichloropropene
U085	1464-53-5	1,2:3,4-Diepoxybutane (I,T)
U108	123-91-1	1,4-Diethyleneoxide
U086	1615-80-1	*N,N*-Diethylhydrazine
U087	3288-58-2	*O-O*-Diethyl-*S*-methyl-dithiophosphate
U088	84-66-2	Diethyl phthalate
U089	56-53-1	Diethylstilbestrol
U090	94-58-6	Dihydrosafrole
U091	119-90-4	3,3'-Dimethoxybenzidine
U092	124-40-3	Dimethylamine (I)
U093	60-11-7	Dimethylaminoazobenzene
U094	57-97-6	6,12-Dimethylbenz[*a*]anthracene
U095	119-93-7	3,3'-Dimethylbenzidine
U096	80-15-9	α,α-Dimethylbenzylhydroperoxide (R)
U097	79-44-7	Dimethylcarbamoyl chloride
U098	57-14-7	1,1-Dimethylhydrazine
U099	540-73-8	1,2-Dimethylhydrazine
U101	105-67-9	2,4-Dimethylphenol
U102	131-11-3	Dimethyl phthalate
U103	77-78-1	Dimethyl sulfate
U105	121-14-2	2,4-Dinitrotoluene
U106	606-20-2	2,6-Dinitrotoluene
U107	117-84-0	Di-*n*-octyl phthalate
U108	123-91-1	1,4-Dioxane
U109	122-66-7	1,2-Diphenylhydrazine
U110	142-84-7	Dipropylamine (I)
U111	621-64-7	Di-*n*-propylnitrosamine
U001	75-07-0	Ethanal (I)
U174	55-18-5	Ethanamine, *N*-ethyl-*N*-nitroso-
U155	91-80-5	1,2-Ethanediamine, *N,N*-dimethyl-*N'*-2-pyridinyl-*N'*-(2-thienylmethyl)-
U067	106-93-4	Ethane, 1,2-dibromo-
U076	75-34-3	Ethane, 1,1-dichloro-
U077	107-06-2	Ethane, 1,2-dichloro-
U131	67-72-1	Ethane, hexachloro-

Table A.3 (Continued)

Hazardous Waste No.	Chemical Abstracts No.	Hazardous Substance
U024	111-91-1	Ethane, 1,1'-[methylenebis(oxy)]bis[2-chloro]-
U117	60-29-7	Ethane, 1,1'-oxybis- (I)
U025	111-44-4	Ethane, 1,1'-oxybis[2-chloro-
U184	76-01-7	Ethane, pentachloro-
U208	630-20-6	Ethane, 1,1,1,2-tetrachloro-
U209	79-34-5	Ethane, 1,1,2,2-tetrachloro-
U218	62-55-5	Ethanethioamide
U227	110-80-5	Ethanol, 2-ethoxy-
U359	79-00-5	Ethane, 1,1,2-trichloro-
U173	1116-54-7	Ethanol, 2,2'-(nitrosoimino)bis-
U004	98-86-2	Ethanone, 1-phenyl-
U043	75-01-4	Ethene, chloro-
U042	110-75-8	Ethene, (2-chloroethyoxy)
U078	75-35-4	Ethene, 1,1-dichloro-
U079	156-60-5	Ethene, 1,2-dichloro- (E)
U210	127-18-4	Ethene, tetrachloro-
U228	79-01-6	Ethene, trichloro-
U112	141-78-6	Ethyl acetate (I)
U113	140-88-5	Ethyl acrylate (I)
U238	51-79-6	Ethyl carbamate
U038	510-15-6	Ethyl 4,4'-dichlorobenzilate
U114	111-54-6	Ethylenebis[dithiocarbamic acid], salts and esters
U067	106-93-4	Ethylene dibromide
U077	107-06-2	Ethylene dichloride
U359	110-80-5	Ethylene glycol monoethyl ether
U115	75-21-8	Ethylene oxide (I,T)
U116	96-45-7	Ethylene thiourea
U117	60-29-7	Ethyl ether (I)
U076	75-34-3	Ethylidene dichloride
U118	97-63-2	Ethyl methacrylate
U119	62-50-0	Ethylmethanesulfonate
U120	206-44-0	Fluoranthene
U122	50-00-0	Formaldehyde
U123	64-18-6	Formic acid (C,T)
U124	110-00-9	Furan (I)
U125	98-01-1	2-Furancarboxyaldehyde (I)
U147	108-31-6	2,4-Furandione
U213	109-99-9	Furan, tetrahydro- (I)
U125	98-01-1	Furfural (I)
U124	110-00-9	Furfuran (I)
U206	18883-66-4	D-Glucopyranose, 2-deoxy-2(3-methyl-3-nitrosoureido)-
U126	765-34-4	Glycidylaldehyde
U163	70-25-7	Guanidine, N-methyl-N'-nitro-N-nitroso-

Table A.3 (Continued)

Hazardous Waste No.	Chemical Abstracts No.	Hazardous Substance
U127	118-74-1	Hexachlorobenzene
U128	87-68-3	Hexachlorobutadiene
U129	58-88-9	Hexachlorocyclohexane (gamma isomer)
U130	77-47-4	Hexachlorocyclopentadiene
U131	67-72-1	Hexachloroethane
U132	70-30-4	Hexachlorphene
U243	1888-71-7	Hexachloropropene
U133	302-01-2	Hydrazine (R,T)
U086	1615-80-1	Hydrazine, 1,2-diethyl-
U098	57-14-7	Hydrazine, 1,1-dimethyl-
U099	540-73-8	Hydrazine, 1,2-dimethyl-
U109	122-66-7	Hydrazine, 1,2-diphenyl-
U134	7664-39-3	Hydrofluoric acid (C,T)
U134	7664-39-3	Hydrogen fluoride (C,T)
U135	7783-06-4	Hydrogen sulfide
U096	80-15-9	Hydroperoxide, 1-methyl-1-phenylethyl- (R)
U136	75-60-5	Hydroxydimethylarsine oxide
U116	96-45-7	2-Imidazolidinethione
U137	193-39-5	Indeno[1,2,3,-c,d]pyrene
U139	9004-66-4	Iron dextran
U190	85-44-9	1,3-Isobenzofurandione
U140	78-83-1	Isobutyl alcohol (I,T)
U141	120-58-1	Isosafrole
U142	143-50-0	Kepone
U143	303-34-4	Lasiocarpine
U144	301-04-2	Lead acetate
U146	1335-32-6	Lead, bis(acetato-O)tetrahydroxytri-
U145	7446-27-7	Lead phosphate
U146	1335-32-6	Lead subacetate
U129	58-89-9	Lindane
U147	108-31-6	Maleic anhydride
U148	123-33-1	Maleic hydrazide
U149	109-77-3	Malononitrile
U150	148-82-3	Melphalan
U151	7439-97-6	Mercury
U152	126-98-7	Methacrylonitrile (I,T)
U092	124-40-3	Methanamine, N-methyl- (I)
U029	74-83-9	Methane, bromo-
U045	74-87-3	Methane, chloro- (I,T)
U046	107-30-2	Methane, chloromethoxy-
U068	74-95-3	Methane, dibromo-
U080	75-09-2	Methane, dichloro-
U075	75-71-8	Methane, dichlorodifluoro-

Table A.3 (Continued)

Hazardous Waste No.	Chemical Abstracts No.	Hazardous Substance
U138	74-88-4	Methane, iodo-
U119	62-50-0	Methanesulfonic acid, ethyl ester
U211	56-23-5	Methane, tetrachloro-
U153	74-93-1	Methanethiol (I,T)
U225	75-25-2	Methane, tribromo-
U044	67-66-3	Methane, trichloro-
U121	75-69-4	Methane, trichlorofluoro-
U123	64-18-6	Methanoic acid (C,T)
U154	67-56-1	Methanol (I)
U155	91-80-5	Methapyrilene
U142	143-50-0	1,3,4-Metheno-2H-cyclobuta[c,d]pentalen-2-one, 1,1a,3,3a,4,5,5,5a,5b,6-decachlorooctahydro-
U247	72-43-5	Methoxychlor
U154	67-56-1	Methyl alcohol (I)
U029	74-83-9	Methyl bromide
U186	504-60-9	1-Methylbutadiene (I)
U045	74-87-3	Methyl chloride (I,T)
U156	79-22-1	Methylchlorocarbonate (I,T)
U226	71-55-6	Methylchloroform
U157	56-49-5	3-Methylcholanthrene
U158	101-14-4	4,4'-Methylenebis(2-chloroaniline)
U068	74-95-3	Methylene bromide
U080	75-09-2	Methylene chloride
U159	78-93-3	Methyl ethyl ketone (MEK) (I,T)
U160	1338-23-4	Methyl ethyl ketone peroxide (R,T)
U138	74-88-4	Methyl iodide
U161	108-10-1	Methyl isobutyl ketone (I)
U162	80-62-6	Methyl methacrylate (I,T)
U163	70-25-7	N-Methyl-N'-nitro-N-nitrosoguanidine
U161	108-10-1	4-Methyl-2-pentanone (I)
U164	56-04-2	Methylthiouracil
U010	50-07-7	Mitomycin C
U059	20830-81-3	5,12-Naphthacenedione, (8-S-cis)-8-acetyl-10-[(3amino-2,3,6-tridexoy)-α-L-lyxo-hexo(pyranosyl)oxy]-7,8,9,10-tetrahydro-6,8,11-trihydroxy-1-methoxy-
U165	91-20-3	Naphthalene
U047	91-58-7	Naphthalene, 2-chloro-
U166	130-15-4	1,4-Naphthalenedione
U236	72-57-1	2,7-Naphthalenedisulfonic acid, 3,3'-[(3,3'dimethyl-(1,1'biphenyl)-4,4'-diyl)]-bis(azo)bis(5-amino-4-4-hydroxy)-, tetrasodium salt
U166	130-15-4	1,4-Naphthoquinone

Table A.3 (Continued)

Hazardous Waste No.	Chemical Abstracts No.	Hazardous Substance
U167	134-32-7	α-Naphthylamine
U168	91-59-8	β-Naphthylamine
U026	494-03-1	2-Naphthylamine, N,N'-bis(2-chloromethyl)-
U167	134-32-7	1-Naphthylenamine
U168	91-59-8	2-Naphthylenamine
U217	10102-45-1	Nitric acid, thallium (1+) salt
U169	98-95-3	Nitrobenzene (I,T)
U170	100-02-7	p-Nitrophenol
U171	79-46-9	2-Nitropropane (I,T)
U172	924-16-3	N-Nitrosodi-n-butylamine
U173	1116-54-7	N-Nitrosodiethanolamine
U174	55-18-5	N-Nitrosodiethylamine
U176	759-73-9	N-Nitroso-N-ethylurea
U177	684-93-5	N-Nitroso-N-methylurea
U178	615-53-2	N-Nitroso-N-methylurethane
U179	100-75-4	N-Nitrosopiperidine
U180	930-55-2	N-Nitrosopyrrolidine
U181	99-55-8	5-Nitro-o-toluidine
U193	1120-71-4	1,2-Oxathiolane, 2,2-dioxide
U058	50-18-0	2H-1,3,2-Oxazaphosphorin-2-amine, N,N-bis(2-chloroethyl)tetrahydro-2-oxide
U115	75-21-8	Oxirane (I,T)
U126	765-34-4	Oxiranecarboxyaldehyde
U041	106-89-8	Oxirane, (chloromethyl)-
U182	123-63-7	Paraldehyde
U183	608-93-5	Pentachlorobenzene
U184	76-01-7	Pentachloroethane
U185	82-68-8	Pentachloronitrobenzene (PCNB)
U242	87-86-5	Pentachlorophenol
U186	504-60-9	1,3-Pentadiene (I)
U187	62-44-2	Phenacetin
U188	108-95-2	Phenol
U048	95-57-8	Phenol, 2-chloro-
U039	59-50-7	Phenol, 4-chloro-3-methyl-
U081	120-83-2	Phenol, 2,4-dichloro-
U082	87-65-0	Phenol, 2,6-dichloro-
U089	56-53-1	Phenol, 4,4'-(1,2-diethyl-1,2-ethenediyl)bis-,
U101	105-67-9	Phenol, 2,4-dimethyl-
U052	1319-77-3	Phenol, methyl-
U132	70-30-4	Phenol, 2,2'-methylenebis[3,4,6-trichloro-]
U170	100-02-7	Phenol, 4-nitro-
U242	87-86-5	Phenol, pentachloro-
U212	58-90-2	Phenol, 2,3,4,6-tetrachloro-

Table A.3 (Continued)

Hazardous Waste No.	Chemical Abstracts No.	Hazardous Substance
U230	95-94-4	Phenol, 2,4,5-trichloro-
U231	88-06-2	Phenol, 2,4,6-trichloro-
U150	148-82-3	L-Phenylalanine, 4-[bis(2-chloroethyl)amino]-
U145	7446-27-7	Phosphoric acid, lead salt
U087	3288-58-2	Phosphorodithioic acid, O,O-diethyl-S-methyl-, ester
U189	108-95-2	Phosphorous sulfide (R)
U190	85-44-9	Phthalic anhydride
U191	109-06-8	2-Picoline
U179	100-75-4	Piperidine, 1-nitroso-
U192	23950-58-5	Pronamide
U194	107-10-8	1-Propanamine (I,T)
U111	621-64-7	1-Propanamine, N-nitroso-N-propyl-
U110	142-84-7	1-Propanamine, N-propyl- (I)
U066	96-12-8	Propane, 1,2-dibromo-3-chloro-
U149	109-77-3	Propanedinitrile
U171	79-46-9	Propane, 2-nitro- (I,T)
U027	39638-32-9	Propane, 2,2'-oxybis[2-chloro-
U193	1120-71-4	1,3-Propane sulfone
U235	126-72-7	1-Propanol, 2,3-dibromo-, phosphate (3:1)
U140	78-83-1	1-Propanol, 2-methyl- (I,T)
U002	67-64-1	2-Propanone (I)
U084	542-75-6	1-Propane, 1,3-dichloro-
U152	126-98-7	2-Propanenitrile, 2-methyl- (I,T)
U007	79-06-1	2-Propenamide
U243	1888-71-7	1-Propene, hexachloro-
U009	107-13-1	2-Propenenitrile
U008	79-10-7	2-Propenoic acid (I)
U113	140-88-5	2-Propenoic acid, ethyl ester (I)
U118	97-63-2	2-Propenoic acid, methyl-, ethyl ester
U162	80-66-2	2-Propenoic acid, 2-methyl-, methyl ester (I,T)
U233	93-72-1	Propionic acid, 2-(2,4,5-trichlorophenoxy)-
U194	107-10-8	n-Propylamine (I,T)
U083	78-87-5	Propylene dichloride
U148	123-33-1	3,6-Pyridazinedione, 1,2-dihydro-
U196	110-86-1	Pyridine
U191	109-06-8	Pyridine, 2-methyl-
U237	66-75-1	2,4(1H,3H)-Pyrimidinedione, 5-[bis(2-chloroethyl)amino]-
U164	56-04-2	4-(1H)-Pyrimidinone, 2,3-dihydro-6-methyl-2-thioxo-
U180	930-55-22	Pyrrolidine, 1-nitroso-
U200	50-55-5	Reserpine
U201	108-46-3	Resorcinol
U202	81-07-2	Saccharin and salts
U203	94-59-7	Safrole

Table A.3 (Continued)

Hazardous Waste No.	Chemical Abstracts No.	Hazardous Substance
U204	7783-00-8	Selenious acid
U204	7783-00-8	Selenium dioxide
U205	7446-34-6	Selenium sulfide (R,T)
U015	115-02-6	L-Serine, diazoacetate (ester)
U233	93-72-1	Silvex
U206	18883-66-4	Streptozotocin
U103	77-78-1	Sulfuric acid, dimethyl ester
U189	1314-80-3	Sulfur phosphide (R)
U232	93-76-5	2,4,5-T
U207	95-94-3	1,2,4,5-Tetrachlorobenzene
U208	630-20-6	1,1,1,2-Tetrachloroethane
U209	79-34-5	1,1,2,2-Tetrachloroethane
U210	127-18-4	Tetrachloroethylene
U212	58-90-2	2,3,4,6-Tetrachlorophenol
U213	109-99-9	Tetrahydrofuran (I)
U214	15843-14-8	Thallium(I) acetate
U215	6533-73-9	Thallium(I) carbonate
U216	7791-12-0	Thallium chloride
U217	10102-45-1	Thallium(I) nitrate
U218	62-55-5	Thioacetamide
U153	74-93-1	Thiomethanol (I,T)
U244	137-26-8	Thioperoxydicarbonic diamide, tetramethyl-
U219	62-56-6	Thiourea
U244	137-26-8	Thiuram
U220	108-88-3	Toluene
U221	25376-45-8	Toluenediamine
U223	26471-62-5	Toluene diisocyanate (R,T)
U328	95-53-4	o-Toluidine
U353	106-49-0	p-Toluidine
U222	636-21-5	o-Toluidine hydrochloride
U011	61-82-5	1H-1,2,4-Triazol-3-amine
U226	71-55-6	1,1,1-Trichloroethane
U227	79-00-5	1,1,2-Trichloroethane
U228	79-01-6	Trichloroethylene
U121	75-69-4	Trichloromonofluoromethane
U230	95-95-4	2,4,5-Trichlorophenol
U231	88-06-2	2,4,6-Trichlorophenol
U234	99-35-4	1,3,5-Trinitrobenzene (R,T)
U182	123-63-7	1,3,5-Trioxane, 2,4,6-trimethyl-
U235	126-72-7	Tris (2,3-dibromopropyl) phosphate
U236	72-57-1	Trypan blue
U237	66-75-1	Uracil mustard
U176	759-73-9	Urea, N-ethyl-N-nitroso-

Table A.3 (Continued)

Hazardous Waste No.	Chemical Abstracts No.	Hazardous Substance
U177	684-93-5	Urea, *N*-methyl-*N*-nitroso-
U043	75-01-4	Vinyl chloride
U248	81-81-2	Warfarin, when present at concentrations of 0.3% or less
U239	1330-20-7	Xylene (I)
U200	50-55-5	Yohimban-16-carboxylic acid, 11,17-dimethoxy-18-[(3,4,5-trimethoxybenzoyl)oxy]-, methyl ester
U249	1314-84-7	Zinc phosphide, when present at concentrations of 10% or less

Appendix **B**

Water Solubilities of Common Hazardous Compounds

Compound	Mean Water Solubility (mg/L)	Range	Reference
Acenaphthene	3.93		111
Acenaphthylene	3.93		177
Acetaldehyde	Miscible		139
Acetic Acid	Miscible		126
Acetone	Miscible		127
Acetonitrile	Miscible		126
Acrolein	200,000		4
Acrylamide	2,050 g/L		177
Acrylonitrile	76,800	73,500–79,000	25,108
Alachlor	242		91
Aldicarb	7,800		91
Aldrin	0.011		178
Allyl alcohol	Miscible		63
Allyl chloride	3,800	3,600–4,000	97,133
Allyl glycidyl ether	141,000		177
4-Aminobiphenyl	842		121
n-Amyl acetate	1,800 @ 20°C		177
sec-Amyl acetate	2,000 @ 20°C		133
Aniline	35.35	34–36.65	73,128,177
p-Anisidine	3.3		91
Anthracene	0.059	0.045–0.073	111,115
ANTU(α-naphthyl-thiourea)	600 @ 20°C		133,186
Arochlor 1016 (PCB-1016)	0.42		25,108

Compound	Mean Water Solubility (mg/L)	Range	Reference
Aroclor 1221 (PCB-1221)	0.20		154
Aroclor 1232 (PCB-1232)	1.45		172
Aroclor 1242 (PCB-1242)	0.24	0.23–0.24	108,170
Aroclor 1248 (PCB-1248)	0.036	0.017–0.054	91,125
Aroclor 1254 (PCB-1254)	0.011	0.01–0.012	25,91
Aroclor 1260 (PCB-1260)	0.0027		125
Atrazine	33		91,107
Benzene	1770	1740–1800	5,11,15,77,116
Benzidine	450	400–500	130,178
Benzo[a]anthracene	0.012	0.0094–0.014	111,115
Benzo[b]fluoranthene	0.0012		172
Benzo[k]fluoranthene	0.00055		178
Benzo[g,h,i]perylene	0.00026		111
Benzo[a]pyrene	0.0038		111
Benzo[e]pyrene	0.0039	0.0038–0.0040	111,157
Benzyl alcohol	42,900		10
Benzyl butyl phthalate	2.80	2.69–2.90	78,177
Benzyl chloride	493 @ 20°C		178
Biphenyl	6.71	5.94–7.48	5,15
Bis(2-chloroethoxy) methane	81,000		129
Bis(2-chloroethyl)ether	10,200		25
Bis(2-chloroisopropyl) ether	1,700		25
Bis(2-ethylhexyl) phthalate	0.4		25
Bis(chloromethyl)ether	22,000		25
Bromobenzene	409		174
Bromochloromethane	16,700		169
Bromodichloromethane	4,500 @ 0°C		159
Bromoform	3,120	3,100–3,130	4,174
4-Bromophenyl phenyl ether	4.8		108
1,3-Butadiene	735		116
n-Butane	61.4		116
1-Butene	222		116
2-Butoxyethanol	Miscible		144
n-Butyl acetate	6,170 @ 20°C	5,000–6,800	133,177
n-Butyl alcohol	71,400	63,300–77,000	4,144,169

Compound	Mean Water Solubility (mg/L)	Range	Reference
sec-Butyl alcohol	169,000 @ 20°C	125,000–201,000	133,139,177
tert-Butyl alcohol	Miscible		139
n-Butylamine	Miscible		87
n-Butylbenzene	24	1.26–58	5,111,138,167,169, 180
sec-Butylbenzene	163.3	17.6–309	5,167
tert-Butylbenzene	32	29.5–34	5,167
Butylbenzylphthalate	2.9		25,60,108
n-Butyl mercaptan	590 @ 22°C		177
Camphor	1,250		186
Carbaryl	76.5	70–83	101,168
Carbofuran	700		186
Carbon disulfide	1,440	1,190–1,700	4,165
Carbon tetrachloride	911	757–1,160	11,87,174,177
Chlordane	0.009		36
α-Chloroacetophenone	Miscible		126
4-Chloroaniline	3,900 @ 20–25°C		92
Chlorobenzene	460	295–503	5,10,77,169,172,188
4-Chlorobiphenyl	1.65		91
p-Chloro-*m*-cresol	3,850		177
Chloroethane	4,700		4
2-Chloroethyl vinyl ether	15,000		159
Chloroform	7,870	7,100–9,300	4,11,34,177
1-Chloronaphthalene	39		190
2-Chloronaphthalene	6.74		172
p-Chloronitrobenzene	241 @ 20°C	30–453	38,133
1-Chloro-1-nitropropane	6		177
Chloropicrin	1,970	1,620–2,300	90,177,186
2-Chlorophenol	18,500	11,400–25,700	11,28
4-Chlorophenyl phenyl ether	3.3		17
Chlorpyrifos	2.00	0.3–2.0	186
o-Cresol	26,000		106
p-Cresol	18,000		106
Chrysene	0.00327	0.0018–0.006	111,115,177
Crotonaldehyde	155,000		177
Cycloheptane	30		116
Cyclohexane	59.3	58.4–66.5	116,145
Cyclohexanol	36,700 @ 20°C	36,000–38,200	133,139,177
Cyclohexanone	19,000 @ 20°C	15,000–23,000	133,177
Cyclohexene	213		116
Cyclopentadiene	681 @ 20°C		166
Cyclopentane	160	156–164	62,116,145
Cyclopentene	535		116

Compound	Mean Water Solubility (mg/L)	Range	Reference
2,4-D	697	522–890	7,29,63,101
DDD	0.013	0.005–0.02	25,91,108
DDE	0.038	0.006–0.04	25,83,91,108,168
DDT	0.0024	0.0012–0.02	25,91,107,108,117, 170
Decahydronapthalene	0.889		145
n-Decane	0.021	0.020–0.022	8,50
Diazinon	40		91
Dibenzo[*a,h*] anthracene	0.0005		178
Dibenzofuran	10		11
1,4-Dibromobenzene	18.3	16.5–20.0	5,111
Dibromochloromethane	4,000		159
1,2-Dibromo-3- chloropropane	1,270		91
1,2-Dibromomethane	4,000		112
Di-*n*-butyl phthalate	13		25,108
1,2-Dichlorobenzene	125	93–145	7,10,122
1,3-Dichlorobenzene	130	123–143	10,122,177
1,4-Dichlorobenzene	77.5	65.3–91	5,10,11,134,177,188
3,3′-Dichlorobenzidine	3.11		11
Dichlorodifluoro- methane	280		25,108
1,1-Dichloroethane	5,060		61
1,2-Dichloroethane	8,310	8,000–8,650	11,61,179
1,1-Dichloroethylene	3,920	273–6,500	4,107,178
trans-1,2-Dichloroethene	6,280	6,260–6,300	34,87
Dichlorofluoromethane	10,000 @ 20°C		133
2,4-Dichlorophenol	9,750	4,500–15,000	28,177
1,2-Dichloropropane	2,800		61,165
cis-1,3-Dichloropropene	2,700		177
Dichlorvos	13,000	10,000–16,000	90,91
Dieldrin	0.198	0.195–0.20	14,178
Diethylamine	815,000 @ 14°C		177
2-Diethylaminoethanol	Miscible		126
Diethyl phthalate	992	896–1,080	25,78,178
1,2-Difluorotetrachloro- ethane	100 @ 20°C		133
Diisobutyl ketone	500 @ 20°C		133
Diisopropylamine	Miscible		126
Dimethylamine	Miscible		126
N,N-Dimethylacetamide	Miscible		126
p-Dimethylaminoazo- benzene	13.6		121
Dimethylaniline	1,110		29

Compound	Mean Water Solubility (mg/L)	Range	Reference
2,2-Dimethylbutane	21.1	18.4–23.8	116,143,145
2,3-Dimethylbutane	20.8	19.1–22.5	143,145
cis-1,2-Dimethylcyclo-hexane	6		116
trans-1,4-Dimethylcyclo-hexane	3.84		145
Dimethylformamide	Miscible		87
1,1-Dimethylhydrazine	Miscible		80
2,3-Dimethylpentane	5.25		145
2,4-Dimethylpentane	4.66	4.41–5.5	116,143,145
3,3-Dimethylpentane	5.94		145
2,4-Dimethylphenol	7,870		11
Dimethyl phthalate	4,160	4,000–4,320	25,78
2,2-Dimethylpropane	33.2		116
2,7-Dimethylquinoline	1,800		145
Dimethyl sulfate	28,000 @ 20°C		133
1,2-Dinitrobenzene	150 @ 20°C		133
1,3-Dinitrobenzene	500 @ 20°C		133
1,4-Dinitrobenzene	100 @ 20°C		133
4,6-Dinitro-o-cresol	250		178
2,4-Dinitrophenol	5,600 @ 18°C		177
1,3-Dinitrotoluene	498		107
2,4-Dinitrotoluene	270 @ 22°C		177
2,6-Dinitrotoluene	180		123
Di-n-octyl phthalate	3		25
Dioxane	Miscible		139
1,2-Diphenylhydrazine	221		172
Diphenylnitrosamine	40		108
Dipropylnitrosamine	9,900		25,108
Diuron	42		7,186
n-Dodecane	0.008		50
α-Endosulfan	0.530		182
β-Endosulfan	0.280		182
Endosulfan sulfate	0.17		2
Endrin	0.243	0.22–0.26	14,123,172
Endrin aldehyde	0.26		182
Epichlorohydrin	62,000 @ 20°C	60,000–64,000	133,177
Ethanolamine	Miscible		126
2-Ethoxyethanol	Miscible		145
2-Ethoxyethyl acetate	230,000 @ 20°C		133,177
Ethyl acetate	74,800	64,000–80,400	3,10,169
Ethyl acrylate	17,500	15,000–20,000	133,177,186
Ethylamine	Miscible		87
Ethylbenzene	181	152–208	5,15,116,138,167, 177

Compound	Mean Water Solubility (mg/L)	Range	Reference
Ethyl bromide	9,140 @ 20°C		186
Ethyl chloride	5,740		25
Ethylcyclopentane	245		64
Ethylene chlorohydrin	Miscible		70
Ethylenediamine	Miscible		144
Ethylene dibromide	3,370		91
Ethylenimine	Miscible		126
Ethyl ether	60,400	60,300–60,500	12,177
Ethyl formate	118,000		177
Ethyl mercaptan	15,000		177
4-Ethylmorpholine	Miscible		70
2-Ethylthiophene	292		145
Fluoranthene	0.233	0.206–0.26	111,115
Flourene	1.84	1.69–1.98	111,115
Formaldehyde	Miscible		153
Formic acid	Miscible		144
Furfural	77,800		169
Furfural alcohol	Miscible		126
Glycidol	Miscible		127
Heptachlor	0.118	0.056–0.180	14,80
Heptachlor epoxide	0.275	0.20–0.35	14,179,182
n-Heptane	2.91	2.24–3.57	116,145,169
2-Heptanone	4,150	4,080–4,300	59,169,180
3-Heptanone	14,300 @ 20°C		177
cis-2-Heptene	15		116
trans-2-Heptene	15		116
Hexachlorobenzene	0.005		188
Hexachlorobutadiene	3.23		11
α-Hexachloro-cyclohexane	1.81	1.63–2.0	25,108
β-Hexachloro-cyclohexane	0.22	0.20–0.24	25,108
δ-Hexachloro-cyclohexane	26.2	21–31.4	25,108
Hexachlorocyclo-pentadiene	1.8		193
Hexachloroethane	38.6	27.2–50	4,107
n-Hexane	10.5	9.47–12.4	116,143,145
1-Hexene	50		116
sec-Hexyl acetate	130 @ 20°C		133
Hydroquinone	70,000		177
Indan	99.0	88.9–109.1	111,145
Indeno[1,2,3-c,d]pyrene	0.062		160
Indole	3,560		145

Compound	Mean Water Solubility (mg/L)	Range	Reference
Indoline	10,800		145
1-Iodopropane	1,070 @ 23.5°C		156
Isoamyl acetate	2,000 @ 20°C		133
Isoamyl alcohol	20,000 @ 20°C		133
Isobutyl acetate	6,300		177
Isobutyl alcohol	86,000 @ 20°C	85,000–87,000	133,144
Isobutylbenzene	21.9	10.1–33.7	29,145
Isophorone	12,000		4
Isopropyl acetate	29,660 @ 20°C	29,000–30,900	133,144,177
Isopropylamine	Miscible		126
Isopropylbenzene	59.3	48.3–73	5,116,145,167
Isopropyl ether	6,500		70
Kepone	2.7		92
Lindane	7.3	6.8–7.8	14,182
Malathion	145		51,91
Mesityl oxide	29,000 @ 20°C	28,000–30,000	133,177
Methoxychlor	0.275	0.10–0.45	14,80
Methyl acetate	268,000 @ 20°C	240,000–319,000	133,177
Methyl acrylate	57,500 @ 20°C	55,000–60,000	133,186
Methylal	330,000 @ 20°C		133,177
Methyl alcohol	Miscible		139
Methylaniline	5,620		29
2-Methylanthracene	0.0302	0.0213–0.039	111,115
Methyl bromide	13,200	13,000–13,400	63,107
2-Methyl-1,3-butadiene	642		116
2-Methylbutane	48.5	47.8–49.6	116,143,145
3-Methyl-1-butene	130		116
Methyl butyl ketone	35,000		177
Methyl cellosolve	Miscible		144
Methyl cellosolve acetate	Miscible		126
Methyl chloride	6,550	4,800–7,400	107,108,164,170
Methylcyclohexane	15.6	14–16.7	62,116,145
1-Methylcyclohexene	52		116
Methylcyclopentane	41.9	41.8–42.0	116,145
Methylene chloride	18,400	16,700–19,400	4,34,177
Methyl ethyl ketone (MEK)	256,000		59
Methyl formate	302,000 @ 20°C	300,000–304,000	133,177
3-Methylheptane	0.792		145
5-Methyl-3-heptanone	2,600 @ 20°C		133
2-Methylhexane	2.54		145
3-Methylhexane	3.8	2.64–4.95	143,145
Methylhydrazine	Miscible		126

Compound	Mean Water Solubility (mg/L)	Range	Reference
Methyl iodide	17,000	14,000–20,000	177,186
Methyl isobutyl ketone (MIBK)	19,100		59
Methyl isocyanate	67,000 @ 20°C		133
Methyl mercaptan	23,700	23,300–24,000	133,186
Methyl methacrylate	15,000 @ 20°C		154
2-Methylnaphthalene	25	24.6–25.4	39,145
4-Methyloctane	0.115		145
Methyl parathion	57		91
2-Methylpentane	14.2	13.8–15.7	116,143,145
3-Methylpentane	14.6	12.8–17.9	116,143,145
2-Methyl-1-pentene	78		116
4-Methyl-1-pentene	48		116
1-Methylphenanthrene	0.269		115
2-Methylpropane	48.9		116
2-Methylpropene	263		116
Mevinphos	Miscible		63
Morpholine	Miscible		126
Naphthalene	31.9	30.0–34.4	5,15,32,145
1-Naphthylamine	1,700		177
2-Naphthylamine	586		121
Nitrapyrin	40		91,177
2-Nitroaniline	1,260		177
3-Nitroaniline	890		177
4-Nitroaniline	800 @ 128.5°C		32
Nitrobenzene	2,390	1,930–3,829	4,5,11,169,179
Nitroethane	45,000 @ 20°C		133
Nitromethane	95,000 @ 20°C		133
2-Nitrophenol	2,500		77
4-Nitrophenol	16,000		177
1-Nitropropane	14,000 @ 20°C		133
2-Nitropropane	17,000 @ 20°C		133
N-Nitrosodimethylamine	Miscible		124
N-Nitrosodiphenylamine	35.1		11
N-Nitrosodi-*n*-propylamine	9,900		124
2-Nitrotoluene	600 @ 20°C		133
3-Nitrotoluene	500 @ 20°C		133
4-Nitrotoluene	50 @ 20°C		133
n-Nonane	0.122		145
n-Octane	1.39	0.431–3.37	116,143,145,169
1-Octene	2.7		116
Oxalic acid	98,100		186
Parathion	24		43
Pentachlorobenzene	0.91	0.561–1.33	11,122,188

Compound	Mean Water Solubility (mg/L)	Range	Reference
Pentachloroethane	769		176
Pentachlorophenol	14 @ 20°C		63
1,4-Pentadiene	558		116
n-Pentane	41.9	38.5–47.6	116,143,145
2-Pentanone	55,100		59
1-Pentene	148		116
cis-2-Pentene	203		116
trans-2-Pentene	203		116
Pentylcyclopentane	0.115		145
Perchloroethylene (PCE)	275	150–400	11,177
Phenanthrene	1.09	0.994–1.29	5,39,111,115
Phenol	77,900	67,000–84,699	28,106,179
p-Phenyldiamine	38,000 @ 24°C		177
Phenyl ether	18		11
Phthalic anhydride	6,200 @ 20°C		133
Picric acid	14,000 @ 20°C		133,177
Pindone	18		63
Propane	62.4		116
n-Propyl acetate	20,400		169
n-Propyl alcohol	Miscible		139
n-Propylbenzene	57.6	55–60.24	5,29
Propylcyclopentane	2.04		145
Propylene oxide	405,000 @ 20°C	400,000–410,000	63,133,177,186
Propyne	3,640 @ 20°C		177
Pyrene	0.134	0.132–0.135	111,115
Pyridine	Miscible		46
p-Quinone	15,000 @ 20°C		133
Ronnel	42	40–44	63,186
Simazine	3.5		43
Strychnine	200 @ 20°C		133
Styrene	230	160–300	5,11
Sulfotepp	25 @ 20°C		133,186
2,4,5-T	273	268–278	63,177
Terbufos	12		91
1,2,4,5-Tetrabromobenzene	0.0406	0.040–0.04122	93,188
1,1,2,2-Tetrabromoethane	700 @ 20°C		133
1,2,3,4-Tetrachlorobenzene	7.48	4.32–12.2	10,122,188
1,2,3,5-Tetrachlorobenzene	3.86	2.9–5.19	10,122,188
1,2,4,5-Tetrachlorobenzene	1.14	0.465–2.35	10,122,188
2,3,7,8-Tetrachloro-dibenzo p-dioxin (TCDD)	0.00032		147
1,1,2,2-Tetrachloroethane	2,910	2,870–2,970	11,164,165
Tetraethyl pyrophosphate	Miscible		126
Tetrahydrofuran	Miscible		46,133,139

Compound	Mean Water Solubility (mg/L)	Range	Reference
1,2,4,5-Tetramethyl-benzene	3.48		145
Tetryl	200 @ 20°C		133
Thiophene	3,020		145
Thiram	30		43
Toluene	546	515–627	10,116,119,178
o-Toluidine	15,700	15,000–16,330	29,177
Toxaphene	1.25	0.74–1.75	179,182
1,3,5-Tribromobenzene	0.79		188
Tribromomethane	3,010 @ 20°C		108
Tributyl phosphate	1,000 @ 20°C		133
1,2,3-Trichlorobenzene	24.8	18.0–31.6	10,188
1,2,4-Trichlorobenzene	35.5	30.0–46.1	10,122,178,188
1,3,5-Trichlorobenzene	5.57	4.12–6.58	10,122,188
1,1,1-Trichloroethane	657		80,107,159
1,1,2-Trichloroethane	4,500 @ 20°C		177
Trichloroethylene (TCE)	1,312	1,100–1,470	107,169,177
Trichlorofluoromethane	1,100		177
2,4,5-Trichlorophenol	1,200	1,190–1,200	106,177
2,4,6-Trichlorophenol	800		177
1,1,2-Trichlorotrifluoro-ethane	200 @ 20°C		133
Tri-o-cresyl phosphate	1.26	0.34–3.1	134,151
Triethylamine	70,900 @ 20°C	15,000–142,000	46,133,177
Trifluralin	1.00		101
1,2,3-Trimethylbenzene	67.7	62.5–75.2	26,152,169
1,2,4-Trimethylbenzene	56	51.9–59.0	116,145,167
1,3,5-Trimethylbenzene	73.2	48–97	5,128,152,167
1,1,3-Trimethylcyclo-hexane	1.77		145
1,1,3-Trimethylcyclo-pentane	3.73		145
2,2,5-Trimethylhexane	0.845	0.54–1.15	116,143
2,2,4-Trimethylpentane (Isooctane)	1.25	0.56–2.46	143,145,177
2,3,4-Trimethylpentane	1.83	1.36–2.3	143,145
2,4,6-Trinitrotoluene	130 @ 20°C		133
Triphenyl phosphate	0.73 @ 24°C		177
Vinyl acetate	25,000		4
Vinyl chloride	1,900	1,100–2,700	34,177
Warfarin	17 @ 20°C		63
o-Xylene	221	170.5–297.3	5,143,167,169
m-Xylene	162.7	146–173	4,5,143,167,169
p-Xylene	189	156–214	5,10,15,143,167,169

All water solubilities at 25°C unless noted.

Appendix C

Specific Gravities of Common Hazardous Compounds

Compound	Specific Gravity	Temperature*	Reference
Acenaphthene	1.024	(90/4°C)	183
Acenaphthylene	0.899	(16/2°C)	183
Acetaldehyde	0.783	(18/4°C)	183
Acetic acid	1.049		183
Acetic anhydride	1.082		183
Acetone	0.789		183
Acetonitrile	0.786		183
Acrolein	0.841		183
Acrylamide	1.122	(30/4°C)	186
Acrylonitrile	0.806		183
Aldrin	1.70		177
Allyl alcohol	0.854		183
Allyl chloride	0.938		183
Allyl glycidyl ether	0.970		154
4-Aminobiphenyl	1.160	(20/20°C)	154
2-Aminopyridine	1.073		107
n-Amyl acetate	0.876		183
sec-Amyl acetate	0.864	(20/20°C)	70
Aniline	1.022		183
o-Anisidine	1.092		183
p-Anisidine	1.096		177
Anthracene	1.283	(25/4°C)	183
ANTU (α-naphthylthiourea)	1.895		107
Aroclor 1016 (PCB-1016)	1.33	(25/4°C)	125
Aroclor 1221 (PCB-1221)	1.15	(25/4°C)	79
Aroclor 1232 (PCB-1232)	1.24	(25/4°C)	125

Compound	Specific Gravity	Temperature*	Reference
Aroclor 1242 (PCB-1242)	1.392	(15/4°C)	164
Aroclor 1248 (PCB-1248)	1.41	(25/4°C)	125
Aroclor 1254 (PCB-1254)	1.505	(15.5/4°C)	164
Aroclor 1260 (PCB-1260)	1.566	(15.5/4°C)	164
Benzene	0.877		183
Benzidine	1.250		177
Benzo[a]anthracene	1.274		183
Benzoic acid	1.316	(28/4°C)	164
Benzo[a]pyrene	1.351		98
Benzo[e]pyrene	0.877		186
Benzyl alcohol	1.045		154
Benzyl butyl phthalate	1.12		184
Benzyl chloride	1.100		183
Biphenyl	0.866		183
Bis(2-chloroethoxy)methane	1.234	(20/20°C)	70
Bis(2-chloroethyl)ether	1.220		183
Bis(2-chloroisopropyl)ethyl	1.103		183
Bis(2-ethylhexyl)phthalate	0.986	(20/20°C)	70
Bromobenzene	1.495		183
Bromochloromethane	1.934		183
Bromodichloromethane	1.980		183
Bromoform	2.890		183
4-Bromophenyl phenyl ether	1.421		183
Bromotrifluoromethane	1.455		107
1,3-Butadiene	0.621		183
n-Butane	0.579		183
1-Butene	0.595		183
2-Butoxyethanol	0.902		183
n-Butyl acetate	0.883		183
sec-Butyl acetate	0.876		183
tert-Butyl acetate	0.867		183
n-Butyl alcohol	0.810		183
sec-Butyl alcohol	0.806		183
tert-Butyl alcohol	0.789		183
n-Butylamine	0.741		183
n-Butylbenzene	0.860		183
sec-Butylbenzene	0.862		183
tert-Butylbenzene	0.867		183
n-Butyl mercaptan	0.834		183
Camphor	0.990	(25/4°C)	183
Carbaryl	1.232	(20/20°C)	186
Carbon disulfide	1.263		183
Carbofuran	1.18	(20/20°C)	177
Carbon tetrachloride	1.594		183
Chlordane	1.59–1.63		120
Chloroacetaldehyde	1.236		27

Compound	Specific Gravity	Temperature*	Reference
α-Chloroacetophenone	1.324	(15/4°C)	183
4-Chloroaniline	1.429	(19/4°C)	183
Chlorobenzene	1.106		183
o-Chlorobenzylidenemalononitrile	1.472		107
Chloroethane	0.898		183
2-Chloroethyl vinyl ether	1.048		183
Chloroform	1.483		183
2-Chloronaphthalene	1.266	(16/4°C)	153
p-Chloronitrobenzene	1.520	(18/4°C)	177
1-Chloro-1-nitropropane	1.209	(20/20°C)	183
2-Chlorophenol	1.263		183
4-Chlorophenyl phenyl ether	1.203	(15/4°C)	183
Chloropicrin	1.656		186
Chloroprene	0.958		183
Chlorpyrifos	1.398	(43.5/4°C)	177
Chrysene	1.274		183
o-Cresol	1.027		183
p-Cresol	1.018		183
Crotonaldehyde	0.853	(20/20°C)	183
Cycloheptane	0.810		183
Cyclohexane	0.779		183
Cyclohexanol	0.962		183
Cyclohexanone	0.948		183
Cyclohexene	0.810		183
Cyclopentadiene	0.802		183
Cyclopentane	0.746		183
Cyclopentene	0.772		183
2,4-D	1.416	(25/4°C)	177
p,p'-DDD	1.476		184
DDT	1.56	(15/4°C)	184
cis-Decahydronapthalene	0.897		183
trans-Decahydronapthalene	0.870		183
n-Decane	0.730		177
Diacetone alcohol	0.939		183
Dibenzo[a,h]anthracene	1.282		80
Dibenzofuran	1.089	(99/4°C)	183
p-Dibromobenzene	1.574		183
Dibromochloromethane	2.451		183
1-2-Dibromo-3-chloropropane	2.05		70
Dibromodiflouromethane	2.288	(15/4°C)	70
Di-n-butylphthalate	1.046	(20/20°C)	183
1,2-Dichlorobenzene	1.305		183
1,3-Dichlorobenzene	1.288		183
1,4-Dichlorobenzene	1.248		183
Dichlorodifluoromethane	1.35	(15/4°C)	184
1,3-Dichloro-5,5-dimethylhydantoin	1.5	(20/20°C)	186

Compound	Specific Gravity	Temperature*	Reference
1,1-Dichloroethane	1.176		183
1,2-Dichloroethane	1.235		183
1,1-Dichloroethylene	1.218		183
trans-1,2-Dichloroethylene	1.257		177
Dichlorofluoromethane	1.75	(−115/4°C)	184
sym-Dichloromethyl ether	1.328	(15/4°C)	183
2,4-Dichlorophenol	1.40	(15/4°C)	184
1,2-Dichloropropane	1.560		183
cis-1,3-Dichloropropylene	1.224		120
trans-1,3-Dichloropropylene	1.182		183
Dichlorvos	1.415	(25/4°C)	118
Dieldrin	1.75		184
Diethylamine	0.706		183
2-Diethylaminoethanol	0.884		27
Diethyl phthalate	1.118		183
1,1-Difluorotetrachloroethane	2.191		107
1,2-Difluorotetrachloroethane	1.645	(25/4°C)	183
Diisobutyl ketone	0.805		183
Diisopropylamine	0.717		183
N,N-Dimethylacetamide	0.937	(25/4°C)	183
Dimethylamine	0.680	(0/4°C)	183
p-Dimethylaminoazobenzene	1.212		107
Dimethylaniline	0.956		183
2,2-Dimethylbutane	0.649		183
2,3-Dimethylbutane	0.662		183
cis-1,2-Dimethylcyclohexane	0.796		183
trans-1,4-Dimethylcyclohexane	0.763		183
Dimethylformamide	0.949		183
1,1-Dimethylhydrazine	0.791	(22/4°C)	183
2,3-Dimethylpentane	0.695		183
2,4-Dimethylpentane	0.673		183
3,3-Dimethylpentane	0.694		183
2,4-Dimethylphenol	0.965		183
Dimethyl phthalate	1.191		183
2,2-Dimethylpropane	0.614		183
2,7-Dimethylquinoline	1.054		107
Dimethyl sulfate	1.328		183
1,2-Dinitrobenzene	1.565	(17/4°C)	183
1,3-Dinitrobenzene	1.575	(18/4°C)	183
1,4-Dinitrobenzene	1.625	(18/4°C)	183
2,4-Dinitrophenol	1.68		184
2,4-Dinitrotoluene	1.379		184
2,6-Dinitrotoluene	1.283	(111/4°C)	183
Dioxane	1.0334		183
Di-n-octylphthalate	0.99	(20/20°C)	183
1,2-Diphenylhydrazine	1.158	(16/4°C)	183

Compound	Specific Gravity	Temperature*	Reference
Diuron	1.385		107
n-Dodecane	0.7489		183
α-Endosulfan	1.745		153
β-Endosulfan	1.745	(20/20°C)	153
Endrin	1.65	(25/4°C)	184
Epichlorohydrin	1.180		183
EPN	1.27	(25/4°C)	183
Ethanolamine	1.018		183
2-Ethoxyethanol	0.930		183
2-Ethoxyethyl acetate	0.975	(20/20°C)	186
Ethyl acetate	0.9003		183
Ethyl acrylate	0.923		183
Ethylamine	0.683		183
Ethylbenzene	0.867		183
Ethyl bromide	1.460		183
Ethylcyclopentane	0.767		183
Ethylene chlorohydrin	1.200		183
Ethylenediamine	0.900	(20/20°C)	183
Ethylene dibromide	2.179		183
Ethylenimine	0.832		183
Ethyl ether	0.714		183
Ethyl formate	0.917		183
Ethyl mercaptan	0.839		183
4-Ethylmorpholine	0.989		183
2-Ethylthiophene	0.993		183
Fluoranthene	1.252	(0/4°C)	183
Fluorene	1.203	(0/4°C)	183
Formaldehyde	0.815	(−20/4°C)	186
Formic acid	1.220		183
Furfural	1.159		183
Furfuryl alcohol	1.130		183
Glycidol	1.114	(25/4°C)	183
Heptachlor	1.66		184
n-Heptane	0.684		183
2-Heptanone	0.811		183
3-Heptanone	0.818		183
cis-2-Heptene	0.708		183
trans-2-Heptene	0.701		183
Hexachlorobenzene	1.569	(23.6/4°C)	183
Hexachlorocyclopentadiene	1.702	(25/4°C)	183
Hexachloroethane	2.091		183
n-Hexane	0.660		183
1-Hexene	0.673		183
sec-Hexyl acetate	0.866		183
Hydroquinone	1.358		154
Indan	0.964		183

Compound	Specific Gravity	Temperature*	Reference
Indole	1.22		183
Indoline	1.069		183
1-Iodopropane	1.749		183
Isoamyl acetate	0.867		183
Isoamyl alcohol	0.809		183
Isobutyl acetate	0.871		183
Isobutyl alcohol	0.802		183
Isobutylbenzene	0.853		183
Isophorone	0.923		183
Isopropyl acetate	0.879		183
Isopropylamine	0.689		183
Isopropylbenzene	0.862		183
Isopropyl ether	0.724		183
Lindane	1.891	(19/4°C)	184
Malathion	1.23	(25/4°C)	186
Maleic anhydride	1.48		154
Mesityl oxide	0.865		183
Methoxychlor	1.41	(25/4°C)	177
Methyl acetate	0.933		183
Methyl acrylate	0.956		186
Methylal	0.859		183
Methyl alcohol	0.791		183
Methylamine	0.663		183
Methylaniline	0.990		183
2-Methylanthracene	1.164		107
Methyl bromide	1.676		183
2-Methyl-1,3-butadiene	0.681		183
2-Methylbutane	0.620		183
3-Methyl-1-butene	0.627		183
Methyl butyl ketone	0.811		183
Methyl cellosolve	0.965		183
Methyl cellosolve acetate	1.009	(19/19°C)	183
Methyl chloride	0.916		183
1-Methylcyclohexane	0.810		183
o-Methylcyclohexanone	0.925		183
1-Methylcyclohexene	0.810		183
Methylcyclopentane	0.749		183
Methylene chloride	1.327		183
Methyl ethyl ketone (MEK)	0.805		183
Methyl formate	0.974		183
3-Methylheptane	0.706		183
5-Methyl-3-heptanone	0.822	(20/20°C)	186
2-Methylhexane	0.679		183
3-Methylhexane	0.686		183
Methylhydrazine	0.874	(25/4°C)	183
Methyl iodide	2.279		183

Compound	Specific Gravity	Temperature*	Reference
Methyl isobutyl ketone (MEK)	0.798		183
Methyl isocyanate	0.923	(27/4°C)	183
Methyl mercaptan	0.867		183
Methyl methacrylate	0.944		183
2-Methylnaphthalene	1.006		183
4-Methyloctane	0.720		183
Methyl parathion	1.358		183
2-Methylpentane	0.653		183
3-Methylpentane	0.665		183
2-Methyl-1-pentene	0.680		183
4-Methyl-1-pentene	0.664		183
1-Methylphenanthrene	1.161		107
2-Methylpropane	0.549		183
2-Methylpropene	0.594		186
α-Methylstyrene	0.908		183
Mevinphos	1.25		186
Morpholine	1.001		183
Naled	1.96	(25/4°C)	186
Naphthalene	1.145		184
1-Naphthylamine	1.123	(25/25°C)	183
2-Naphthylamine	1.061	(98/4°C)	183
Nitrapyrin	1.744		107
2-Nitroaniline	1.44		184
3-Nitroaniline	0.901	(25/4°C)	153
4-Nitroaniline	1.424		183
Nitrobenzene	1.204		183
Nitroethane	1.045	(25/4°C)	183
Nitromethane	1.137		183
2-Nitrophenol	1.495		153
4-Nitrophenol	1.479		164
1-Nitropropane	1.008	(24/4°C)	183
2-Nitropropane	0.988		183
N-Nitrosodimethylamine	1.006		183
N-Nitrosodi-n-propylamine	0.916		80
2-Nitrotoluene	1.163		183
3-Nitrotoluene	1.157		183
4-Nitrotoluene	1.104	(75/4°C)	183
n-Nonane	0.718		183
n-Octane	0.703		183
1-Octene	0.715		183
Oxalic Acid	1.653	(18/4°C)	186
Parathion	1.26	(25/4°C)	186
Pentachlorobenzene	1.834	(16.5/4°C)	183
Pentachloroethane	1.678		183
Pentachlorophenol	1.978	(22/4°C)	183
1,4-Pentadiene	0.661		183

Compound	Specific Gravity	Temperature*	Reference
n-Pentane	0.626		183
2-Pentanone	0.809		183
1-Pentene	0.641		183
cis-2-Pentene	0.656		183
trans-2-Pentene	0.648		183
Perchloroethylene (PCE)	1.623		183
Phenanthrene	0.980	(4/4°C)	183
Phenol	1.058		183
Phenyl ether	1.075		183
Phenylhydrazine	1.099		183
Phthalic anhydride	1.53		186
Picric acid	1.763		177
Propane	0.584	(−45/4°C)	183
β-Propiolactone	1.146	(20/5°C)	183
n-Propyl acetate	0.888		183
n-Propyl alcohol	0.804		183
n-Propylbenzene	0.862		183
Propylcyclopentane	0.776		183
Propylene oxide	0.859	(0/4°C)	183
n-Propyl nitrate	1.054		183
Propyne	0.678	(−27/4°C)	177
Pyrene	1.271	(23/4°C)	183
Pyridine	0.982		183
p-Quinone	1.318		183
Ronnel	1.48	(25/4°C)	177
Simazine	1.302		43
Strychnine	1.36		183
Styrene	0.906		183
Sulfotepp	1.196	(25/4°C)	118
2,4,5-T	1.80	(20/20°C)	186
1,1,2,2-Tetrabromoethane	2.875		183
1,2,3,5-Tetrachlorobenzene	1.858	(21/4°C)	177
2,3,7,8-Tetrachlorodibenzo-*p*-dioxin (TCDD)	1.827		155
1,1,2,2-Tetrachloroethane	1.595		183
Tetraethyl pyrophosphate	1.184		186
Tetrahydrofuran	0.889		183
1,2,4,5-Tetramethylbenzene	0.838	(81/4°C)	183
Tetranitromethane	1.638		183
Tetryl	1.57	(10/4°C)	183
Thiophene	1.065		183
Thiram	1.29		70
Toluene	0.867		183
2,4-Toluene diisocyanate	1.224		186
o-Toluidine	0.998		183
Toxaphene	1.6		120

Compound	Specific Gravity	Temperature*	Reference
Tributyl phosphate	0.973	(25/4°C)	183
1,2,3-Trichlorobenzene	1.69		186
1,2,4-Trichlorobenzene	1.454		183
1,1,1-Trichloroethane (TCA)	1.339		183
1,1,2-Trichloroethane	1.440		183
Trichloroethylene (TCE)	1.464		183
Trichlorofluoromethane	1.484	(17.2/4°C)	153
2,4,5-Trichlorophenol	1.678	(25/4°C)	153
2,4,6-Trichlorophenol	1.490	(75/4°C)	183
1,2,3-Trichloropropane	1.389		183
1,1,2-Trichlorotrifluoroethane	1.564		183
Tri-o-cresyl phosphate	1.955		183
Triethylamine	0.728		183
1,2,3-Trimethylbenzene	0.894		183
1,2,4-Trimethylbenzene	0.876		183
1,3,5-Trimethylbenzene	0.865		183
1,1,3-Trimethylcyclohexane	0.766		183
1,1,3-Trimethylcyclopentane	0.770		183
2,2,5-Trimethylhexane	0.707		183
2,2,4-Trimethylpentane (Isooctane)	0.692		183
2,3,4-Trimethylpentane	0.719		183
2,4,6-Trinitrotoluene	1.654		183
Triphenyl phosphate	1.206	(50/4°C)	183
Vinyl acetate	0.932		183
Vinyl chloride	0.911		183
o-Xylene	0.880		183
m-Xylene	0.864		183
p-Xylene	0.881		183

*Measured at 20/4°C (i.e., the compound at 20°C and water at 4°C) unless noted.

Supplementary Chemical Incompatibility Data

Table D.1 Explosive Combinations of Some Common Chemicals

Acetone with chloroform in the presence of base
Acetylene with copper, silver, mercury, or their salts
Ammonia (including aqueous solutions) with Cl_2, Br_2, or I_2
Carbon disulfide with sodium azide
Chlorine with an alcohol
Chloroform or carbon tetrachloride with powdered Al or Mg
Decolorizing carbon with an oxidizing agent
Diethyl ether with chlorine (including a chlorine atmosphere)
Dimethyl sulfoxide with an acyl halide, $SOCl_2$, $POCl_3$, or CrO_3
Ethanol with calcium chlorate (I) or silver nitrate (V)
Nitric (V) acid with acetic anhydride or acetic acid
Picric acid with a heavy-metal salt, such as that of Pb, Hg, or Ag
Silver oxide with ammonia with ethanol
Sodium with a chlorinated hydrocarbon
Sodium chlorate (I) with an amine

Table D.2 Water Reactive Chemicals

Alkali metals
Alkali metal hydrides
Alkali metal amides
Metal alkyls, such as lithium alkyls and aluminum alkyls
Grignard reagents
Halides of nonmetals, such as BCl_3, BF_3, PCl_3, PCl_5, $SiCl_4$, S_2Cl_2
Inorganic acid halides, such as $POCl_3$, $SOCl_2$, SO_2Cl_2
Anhydrous metal halides, such as $AlCl_3$, $TiCl_4$, $ZnCl_4$, $SnCl_4$
Phosphorus (V) oxide
Calcium carbide
Organic acid halides and anhydrides of low molecular weight

Table D.3 Generation of Toxic Products from Incompatible Chemicals

Column A	Column B	Product
Arsenic compounds	Reducing agents	Arsine
Azides	Acids	Hydrogen azide
Cyanides	Acids	Hydrogen cyanide
Hypochlorites	Acids	Hypochlorous acid or chlorine
Metal halides	Sulfuric acid	Hydrogen halides (especially HF)
Nitrates	Sulfuric acid	Oxides of nitrogen
Nitric acid	Copper, brass, heavy metals	Oxides of nitrogen
Nitrites	Acids	Oxides of nitrogen
Phosphorus	Reducing agents or caustic alkalis	Phosphine
Selenides	Reducing agents	Hydrogen selenide
Sulfides	Acids	Hydrogen sulfide
Tellurides	Reducing agents	Hydrogen telluride

Appendix E

Values for Student's t Distribution

Percentage points of the _t_ distribution

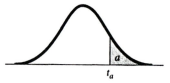

t_a

df	$\alpha = .10$	$\alpha = .05$	$\alpha = .025$	$\alpha = .010$	$\alpha = .005$
1	3.078	6.314	12.706	31.821	63.657
2	1.886	2.920	4.303	6.965	9.925
3	1.638	2.353	3.182	4.541	5.841
4	1.533	2.132	2.776	3.747	4.604
5	1.476	2.015	2.571	3.365	4.032
6	1.440	1.943	2.447	3.143	3.707
7	1.415	1.895	2.365	2.998	3.499
8	1.397	1.860	2.306	2.896	3.355
9	1.383	1.833	2.262	2.821	3.250
10	1.372	1.812	2.228	2.764	3.169
11	1.363	1.796	2.201	2.718	3.106
12	1.356	1.782	2.179	2.681	3.055
13	1.350	1.771	2.160	2.650	3.012
14	1.345	1.761	2.145	2.624	2.977
15	1.341	1.753	2.131	2.602	2.947
16	1.337	1.746	2.120	2.583	2.921
17	1.333	1.740	2.110	2.567	2.898

df	$\alpha = .10$	$\alpha = .05$	$\alpha = .025$	$\alpha = .010$	$\alpha = .005$
18	1.330	1.734	2.101	2.552	2.878
19	1.328	1.729	2.093	2.539	2.861
20	1.325	1.725	2.086	2.528	2.845
21	1.323	1.721	2.080	2.518	2.831
22	1.321	1.717	2.074	2.508	2.819
23	1.319	1.714	2.069	2.500	2.807
24	1.318	1.711	2.064	2.492	2.797
25	1.316	1.708	2.060	2.485	2.787
26	1.315	1.706	2.056	2.479	2.779
27	1.314	1.703	2.052	2.473	2.771
28	1.313	1.701	2.048	2.467	2.763
29	1.311	1.699	2.045	2.462	2.756
inf.	1.282	1.645	1.960	2.326	2.576

Source: "Table of Percentage Points of the *t*-distribution." *Biometrika,* Vol. 32 (1941).

Appendix F

Random Numbers

Line/Col.	(1)	(2)	(3)	(4)	(5)	(6)	(7)	(8)	(9)	(10)	(11)	(12)	(13)	(14)
1	10480	15011	01536	02011	81647	91646	69179	14194	62590	36207	20969	99570	91291	90700
2	22368	46573	25595	85393	30995	89198	27982	53402	93965	34095	52666	19174	39615	99505
3	24130	48360	22527	97265	76393	64809	15179	24830	49340	32081	30680	19655	63348	58629
4	42167	93093	06243	61680	07856	16376	39440	53537	71341	57004	00849	74917	97758	16379
5	37570	39975	81837	16656	06121	91782	60468	81305	49684	60672	14110	06927	01263	54613
6	77921	06907	11008	42751	27756	53498	18602	70659	90655	15053	21916	81825	44394	42880
7	99562	72905	56420	69994	98872	31016	71194	18738	44013	48840	63213	21069	10634	12952
8	96301	91977	05463	07972	18876	20922	94595	56869	69014	60045	18425	84903	42508	32307
9	89579	14342	63661	10281	17453	18103	57740	84378	25331	12566	58678	44947	05585	56941
10	85475	36857	53342	53988	53060	59533	38867	62300	08158	17983	16439	11458	18593	64952
11	28918	69578	88231	33276	70997	79936	56865	05859	90106	31595	01547	85590	91610	78188
12	63553	40961	48235	03427	49626	69445	18663	72695	52180	20847	12234	90511	33703	90322
13	09429	93969	52636	92737	88974	33488	36320	17617	30015	08272	84115	27156	30613	74952
14	10365	61129	87529	85689	48237	52267	67689	93394	01511	26358	85104	20285	29975	89868
15	07119	97336	71048	08178	77233	13916	47564	81056	97735	85977	29372	74461	28551	90707
16	51085	12765	51821	51259	77452	16308	60756	92144	49442	53900	70960	63990	75601	40719
17	02368	21382	52404	60268	89368	19885	55322	44819	01188	65255	64835	44919	05944	55157
18	01011	54092	33362	94904	31273	04146	18594	29852	71585	85030	51132	01915	92747	64951
19	52162	53916	46369	58586	23216	14513	83149	98736	23495	64350	94738	17752	35156	35749
20	07056	97628	33787	09998	42698	06691	76988	13602	51851	46104	88916	19509	25625	58104
21	48663	91245	85828	14346	09172	30168	90229	04734	59193	22178	30421	61666	99904	32812
22	54164	58492	22421	74103	47070	25306	76468	26384	58151	06646	21524	15227	96909	44592
23	32639	32363	05597	24200	13363	38005	94342	28728	35806	06912	17012	64161	18296	22851
24	29334	27001	87637	87308	58731	00256	45834	15398	46557	41135	10367	07684	36188	18510
25	02488	33834	28834	07351	19731	92420	60952	61280	50001	67658	32586	86679	50720	94953

Source: Handbook of Tables for Probability and Statistics, Second Edition, edited by William H. Beyer, 1968. CRC Press, Inc.

Appendix **G**

Log K_{ow} of Common Hazardous Compounds

Compound	Mean Log K_{ow}	Range	References
Acenaphthene	3.92		11
Acenaphthylene	4.07		191
Acetaldehyde	0.43		177
Acetic acid	−0.34	−0.53 to −0.17	103,126,136
Acetone	−0.24		103
Acetonitrile	−0.34		69
2-Acetylamino- fluorene	3.28		121
Acrolein	−0.09		178
Acrylonitrile	0.25		108
Alachlor	2.92		108
Aldrin	5.17		177
Allyl alcohol	−0.03	−0.22–0.17	103,121
Allyl chloride	1.79		68
Allyl glycidal ether	0.63		68
4-Aminobiphenyl	2.78		121
2-Aminopyridine	−0.22		177
n-Amyl acetate	2.35		68
sec-Amyl acetate	5.26		68
Aniline	0.93	0.78–1.09	19,55,91,95,177
o-Anisidine	1.065	0.95–1.18	26,103
p-Anisidine	0.95		103
Anthracene	4.34		91
Aroclor 1016 (PCB-1016)	5.58		91
Aroclor 1221 (PCB-1221)	2.8		172
Aroclor 1232 (PCB-1232)	3.87	3.20–4.54	25,117

Compound	Mean Log K_{ow}	Range	References
Aroclor 1242 (PCB-1242)	5.58		91
Aroclor 1248 (PCB-1248)	6.11		25,91
Aroclor 1254 (PCB-1254)	6.31		91
Aroclor 1260 (PCB-1260)	6.91		25
Atrazine	2.68		108
Benzene	2.05	1.95–2.15	37,103
Benzidine	1.58	1.34–1.81	108,178
Benz[*a*]anthracene	5.91	5.90–5.91	26,191
Benzo[*a*]pyrene	6.06		117
Benzo[*e*]pyrene	6.75		189
Benzo[*b*]fluoranthene	6.57		178
Benzo[*g,h,i*]perylene	7.1		109
Benzo[*k*]fluoranthene	6.45	6.06–6.84	25,117
Benzoic acid	1.85	1.81–1.88	52,103
Benzyl alcohol	1.10		52
Benzyl butyl phthalate	4.78	4.77–4.78	60,177
Benzyl chloride	2.30		103
Bis(2-chloroethoxy) methane	1.26		103
Bis(2-chloroethyl)ether	1.39	1.12–1.58	172,176
Bis(2-chloroisopropyl) ether	2.58		172
Bis(2-ethylhexyl) phthalate	4.66	4.20–5.11	56,108
Bis(chloromethyl)ether	0.00	−0.38–0.38	25,108
Biphenyl	3.85	3.16–4.09	19,26,37,56,84,91,103, 122,150
Bromobenzene	2.99	2.96–3.01	68,91,180,181
Bromochloromethane	1.41		169
Bromodichloromethane	1.88		123
Bromoform	2.34	2.30–2.38	123,174
Bromoethane	1.10	1.09–1.10	25,108
4-Bromophenyl phenyl ether	5.15		178
Bromotrifluoromethane	1.54		66
1,3-Butadiene	1.99		66
Butane	2.89		66
1-Butene	2.40		66
Butylbenzylphthalate	5.68	5.56–5.80	25,108
2-Butoxyethanol	0.45		68
n-Butyl acetate	1.82		169
sec-Butyl acetate	1.66		68

Compound	Mean Log K_{ow}	Range	References
n-Butyl alcohol	0.85	0.79–0.88	103,169
sec-Butyl alcohol	0.61		69
tert-Butyl alcohol	0.36	0.35–0.37	66,69
n-Butylamine	0.79	0.68–0.88	103
n-Butylbenzene	4.37	4.26–4.64	19,21,94,169,180,186
sec-Butylbenzene	4.24		68
tert-Butylbenzene	4.11		103
n-Butyl mercaptan	2.28		177
Camphor	2.42		68
Carbaryl	2.38	2.31–2.56	16,18,56,103,168
Carbofuran	1.62	1.60–1.63	16,91
Carbon disulfide	2.00	1.84–2.16	103
Carbon tetrachloride	2.73		11
Chlordane	6.00		171
cis-Chlordane	5.93		91
trans-Chlordane	9.17	8.69–9.65	91
4-Chloroaniline	2.31	1.83–2.78	56,103
Chlorobenzene	2.84	2.71–2.98	52,169,191
4-Chlorobiphenyl	4.90		91
p-Chloro-*m*-cresol	3.03	2.95–3.10	103,178
Chloroethane	1.43		175
Chloroethene	1.23		117
2-Chloroethyl vinyl ether	1.28		103
Chloroform	1.94	1.90–1.97	69,176
1-Chloronaphthalane	4.08		190
2-Chloronaphthalene	4.07		82
p-Chloronitrobenzene	2.40	2.39–2.41	52,177
1-Chloro-1-nitropropane	4.25		91
4-Chlorophenyl phenyl ether	4.08		17
Chloropicrin	1.03		90
2-Chlorophenol	2.17	2.15–2.19	52,103,176
Chlorpyrifos	5.13	4.99–5.27	21,91
Chrysene	5.71	5.60–5.91	123,178,191
o-Cresol	1.94	1.93–1.95	77,103
p-Cresol	1.93	1.92–1.94	52,103
Cycloheptane	2.64		68
Cyclohexane	3.44		192
Cyclohexanol	1.23		69
Cyclohexanone	0.81		103
Cyclohexene	2.86		66
Cyclopentadiene	2.34		68
Cyclopentane	3.00		86
Cyclopentene	2.45		68
2,4-D	2.94	1.57–4.88	56,103,148

Compound	Mean Log K_{ow}	Range	References
DDD	6.01	5.99–6.02	25,91
DDE	5.76	5.69–5.83	51,91,171
DDT	6.11	5.76–6.36	29,51,56,91,171
Decahydronaphthalene	4.00		68
n-Decane	6.69		22
Dibenzo[a,h]anthracene	6.17	5.97–6.36	29,160
Dibenzofuran	4.15	4.12–4.17	11,103
1,4-Dibromobenzene	3.79		181
Di-n-butyl phthalate	4.79		78
1,2-Dichlorobenzene	3.44	3.38–3.55	11,95,103
1,3-Dichlorobenzene	3.49	3.38–3.60	10,95,103,122,181
1,2-Dibromo-3-chloropropane	2.63		68
Dibromochloromethane	2.08		123
1,4-Dichlorobenzene	3.48	3.37–3.62	11,95,103
3,3'-Dichlorobenzidine	3.51		11
Dichlorodifluoromethane	2.16		67
1,2-trans-Dichloroethane	1.79	1.48–2.09	25,108
1,1-Dichloroethane	1.79	1.78–1.79	66,123
1,2-Dichloroethane	1.47	1.45–1.48	11,96
1,1-Dichloroethylene	2.13		108
trans-1,2-Dichloro-ethylene	2.09		108
Dichlorofluoromethane	1.55		66
2,4-Dichlorophenol	3.08		97
1,2-Dichloropropane	2.28		123
1,3-Dichloropropylene	1.78	1.36–2.00	25,91,108
cis-Dichloropropylene	1.41		97
trans-Dichloropropylene	1.41		97
Dichlorvos	1.4		91,103
Dieldrin	5.16		171
Diethylamine	0.60	0.43–0.81	37,103,177
Diethyl phthalate	2.28	1.40–3.00	78,82,108,176
Diisobutyl ketone	2.58		68
N,N-Dimethylacetamide	−0.77		177
2,4-Dimethylphenol	2.41	2.30–2.50	108,172,176
p-Dimethylaminoazo-benzene	4.58		177
Dimethylaniline	2.47	2.31–2.62	103
2,2-Dimethylbutane	3.82		66
2,3-Dimethylbutane	3.85		66
cis-1,2-Dimethyl-cyclohexane	3.26		68
trans-1,4-Dimethyl-cyclohexane	3.41		68
Dimethylformamide	−1.01		87

Compound	Mean Log K_{ow}	Range	References
1,1-Dimethylhydrazine	−2.42		121
Dimethylnitrosamine	−0.37	−0.68 to −0.06	25,108
2,3-Dimethylpentane	3.26		68
2,4-Dimethylpentane	3.24		68
3,3-Dimethylpentane	3.22		68
Dimethyl phthalate	1.70	1.47–2.00	37,78,82,108,176
2,2-Dimethylpropane	3.11		66
Dimethyl sulfate	−1.24		121
1,2-Dinitrobenzene	1.58		103
1,3-Dinitrobenzene	1.49		103
1,4-Dinitrobenzene	1.48	1.46–1.49	103
4,6-Dinitro-o-cresol	2.12		123
2,4-Dinitrophenol	1.53	1.51–1.54	103,172
2,4-Dinitrotoluene	1.98		108
2,6-Dinitrotoluene	2.00		123
Di-n-octyl phthalate	9.54	9.20–9.87	25,108
Dioxane	−0.35	−0.27 to −0.42	66,103
1,2-Diphenylhydrazine	2.94		108
Diphenylnitrosamine	2.85	2.57–3.13	25,108
Diproplynitrosamine	1.40	1.31–1.49	25,108
Diuron	2.42	1.97–2.60	18,41,91
n-Dodecane	6.44	5.64–7.24	22,56
α-Endosulfan	3.55		2
β-Endosulfan	3.62		2
Endosulfan sulfate	3.66		2
Endrin	5.02	4.56–5.34	82,91,171
Endrin aldehyde	5.60		132
Epichlorohydrin	0.45		33
EPN	3.85		88
Ethanolamine	−1.31		103
2-Ethoxyethanol	−0.54		103
2-Ethoxyethyl acetate	0.50		68
Ethyl acetate	0.69	0.66–0.73	69,103,169
Ethyl acrylate	1.20		68
Ethylamine	−2.74		91
Ethylbenzene	3.11	3.05–3.15	68,178,190
Ethyl bromide	1.57		68
Ethyl chloride	1.54		25
Ethylcyclopentane	1.90		68
Ethylene dibromide	1.76		149
Ethylenimine	−1.01		121
Ethyl ether	0.83	0.77–0.89	66,103
Ethyl formate	0.36		68
Ethyl mercaptan	1.49		68
2-Ethylthiophene	2.83		68
Fluoroanthene	5.22		1

Compound	Mean Log K_{ow}	Range	References
Fluororene	4.38		82
Formaldehyde	0.00		121
Formic acid	−0.54		103
Furfural	0.52		169
Glycidol	−0.95		33
Heptachlor	4.92	4.40–5.44	171,172
Heptachlor epoxide	4.53	3.65–5.40	171,172
n-Heptane	4.66		169
2-Heptanone	1.98		169,180
3-Heptanone	1.32		68
cis-2-Heptene	2.88		68
trans-2-Heptene	2.88		68
Hexachlorobenzene	5.65	5.45–6.18	29,82,108,171,172
Hexachlorobiphenyl	6.34		89
Hexachlorobutadiene	4.78		11
α-Hexachlorocyclohexane	3.85	3.81–3.89	25,108
β-Hexachlorocyclohexane	3.85	3.80–3.89	25,108
δ-Hexachlorocyclohexane	4.15	4.14–4.15	25,108
Hexachlorocyclo-pentadiene	4.52	4.00–5.04	56,123
Hexachloroethane	4.28	3.93–4.62	11,82
n-Hexane	4.0	3.90–4.11	87,169
1-Hexene	2.25		30
sec-Hexyl acetate	3.37		68
Hydroquinone	0.55	0.50–0.59	56,103
Indan	2.38		68
Indeno[1,2,3-c,d]pyrene	6.84	5.97–7.70	123,160
Indole	2.05	1.81–2.25	37,69,103
Indoline	0.16		91
1-Iodopropane	2.49		68
Isoamyl acetate	2.30		68
Isoamyl alcohol	1.16		103
Isobutyl acetate	1.76		68
Isobutyl alcohol	0.74	0.65–0.83	177
Isobutyl benzene	4.11		29
Isophorone	1.69	1.67–1.70	123,176
Isopropyl acetate	1.03		68
Isopropylamine	−0.03		103
Isopropylbenzene	3.59	3.51–3.66	54,103
Isopropyl ether	1.68		68
Kepone	4.07		91
Lindane	3.76	3.66–3.85	82,171
Malathion	2.84		16
Mesityl oxide	1.25		68
Methoxychlor	3.40		187
Methyl acetate	0.18		103

Compound	Mean Log K_{ow}	Range	References
Methyl acrylate	0.67		68
Methylal	0.00		103
Methyl alcohol	−0.73	−0.82 to −0.66	56,69,103
Methylamine	−0.57		103
Methylaniline	1.74	1.66–1.82	103
2-Methylanthracene	5.52		189
Methyl bromide	1.10	1.00–1.19	66,123,178
2-Methyl-1,3-butadiene	1.76		68
2-Methylbutane	2.23		30
3-Methyl-1-butene	2.30		68
Methyl butyl ketone	1.38		103
Methyl chloride	0.91	0.90–0.91	66,123
Methylcyclohexane	2.86		68
1-Methylcyclohexene	2.44		68
Methylcyclopentane	2.47		68
Methylene chloride	1.28	1.25–1.30	67,123
Methyl ethyl ketone (MEK)	0.28	0.26–0.29	103
Methyl formate	−0.18		68
3-Methylheptane	3.97		68
5-Methyl-3-heptanone	1.96		68
2-Methylhexane	3.30		30
3-Methylhexane	3.41		30
Methyl iodide	1.6	1.51–1.69	66,103
Methyl isobutyl ketone (MIBK)	1.09		68
Methyl methacrylate	1.33		68
2-Methylnaphthalene	3.99	3.86–4.11	1,191
4-Methyloctane	4.69		68
Methyl parathion	1.91		91
2-Methylpentane	2.77		30
3-Methylpentane	2.88		30
2-Methyl-1-pentene	2.54		68
4-Methyl-1-pentene	2.70		68
1-Methylphenanthrene	5.27		189
2-Methylpropane	2.29		68
2-Methylpropene	1.99		68
Morpholine	−1.08		103
Naphthalene	3.51	3.01–4.70	56,82,91,103
1-Naphthylamine	2.07		121
2-Naphthylamine	2.07		121
2-Nitroaniline	1.69	1.44–1.83	103
3-Nitroaniline	1.37		52
4-Nitroaniline	1.39		52
Nitrapyrin	3.22	3.02–3.41	18,91
Nitrobenzene	1.85	1.83–1.88	11,52,56,103

Compound	Mean Log K_{ow}	Range	References
Nitroethane	0.18		69
Nitromethane	−0.11	−0.35–0.17	66,69,103,177
1,4-Nitrophenol	1.91		83,107
2-Nitrophenol	1.77	1.73–1.79	52,103,123,178
4-Nitrophenol	1.74		77
1-Nitropropane	0.76	0.65–0.87	66,69
N-Nitrosodimethylamine	0.06		146
N-Nitrosodiphenylamine	3.13		176
N-Nitrosodi-*n*-propylamine	1.31		172
2-Nitrotoluene	2.30		103
3-Nitrotoluene	2.42	2.40–2.45	52,103
4-Nitrotoluene	2.40	2.37–2.42	103
n-Nonane	4.67		68
n-Octane	5.18		169
1-Octene	2.79		30
Oxalic acid	−0.62	−0.43 to −0.81	177,183
Parathion	3.43	2.15–3.93	16,18,44,103,177
Pentachlorobenzene	5.03	4.88–5.182	11,21,56,91,95,122, 176,181
Pentachloroethane	2.89		176
Pentachlorophenol	4.41	3.81–5.01	103,105
1,4-Pentadiene	1.48		66
n-Pentane	3.41	3.23–3.62	66,169
2-Pentanone	0.91		86
1-Pentene	2.26		68
cis-2-Pentene	2.15		68
trans-2-Pentene	2.15		68
Pentacyclopentane	4.90		68
Perchloroethylene (PCE)	2.79	2.53–2.88	25,91,108,176
Phenanthrene	4.52		91,191
Phenol	1.47	1.46–1.48	52,103
Phenyl ether	4.11	3.79–4.36	11,22,103
Phenylhydrazine	1.25		133
Phthalic anhydride	−0.62		91
Picric acid	1.69	1.34–2.03	69,177
Pindone	3.18		91
Propane	2.36		86
n-Propyl acetate	1.24		169
n-Propyl alcohol	0.30	0.25–0.34	66,69
n-Propylbenzene	3.65	3.57–3.72	26,54,68,103,169,180
Propylcyclopentane	3.63		68
Propylene oxide	0.08		33
Propyne	1.61		48
Pyrene	5.32		178
Pyridine	0.90	0.64–1.28	37,103,177

Compound	Mean Log K_{ow}	Range	References
p-Quinone	0.20		103
Ronnel	4.85	4.67–5.07	16,21,91
Strychnine	1.93		121
Styrene	3.16		11
Sulfotepp	3.02		91
2,4,5-T	3.40		148
1,2,4,5-Tetrabromo-benzene	5.13		181
1,1,2,2-Tetrabromoethane	2.91		68
1,2,3,4-Tetrachloro-benzene	4.50	4.37–4.64	21,95,122,181
1,2,3,5-Tetrachloro-benzene	4.57	4.46–4.66	21,95,122,176
1,2,4,5-Tetrachloro-benzene	4.57	4.51–4.67	21,91,95,122,181
2,3,7,8-Tetrachlorodibenzo-*p*-dioxin (TCDD)	5.77	5.38–6.15	147,171
1,1,2,2-Tetrachloroethane	2.48	2.39–2.56	123,191
Tetrahydrofuran	0.46		66
1,2,4,5-Tetramethyl-benzene	4.00		26
Tetryl	2.04		91
Thiophene	1.81		103
Toluene	2.58	2.21–2.79	52,68,103,169,176,190
o-Toluidine	1.31	1.29–1.32	103
Toxaphene	3.30		25,108,141
1,3,5-Tribromobenzene	4.51		181
Tribromoethane	2.38		108
1,2,3-Trichlorobenzene	4.07	4.04–4.11	95,122,181
1,2,4-Trichlorobenzene	4.09	3.93–4.23	1,91,95
1,3,5-Trichlorobenzene	4.12	4.02–4.19	95,122,181
1,1,1-Trichloroethane (TCA)	2.33	2.18–2.47	123,176
1,1,2-Trichloroethane	2.18		123
Trichloroethylene (TCE)	2.33	2.29–2.42	25,83,91
Trichlorofluoromethane	2.53		67
2,4,5-Trichlorophenol	3.72		103
2,4,6-Trichlorophenol	3.24		82,103
1,1,2-Trichlorotrifluoro-ethane	2.57		104
Tri-*o*-cresyl phosphate	5.11		151
Triethylamine	1.45	1.44–1.45	86,177
Trifluralin	5.31	5.28–5.34	20,91
1,2,3-Trimethylbenzene	3.61	3.55–3.66	26,169
1,2,4-Trimethylbenzene	3.78		26
1,3,5-Trimethylbenzene	3.42		29

Compound	Mean Log K_{ow}	Range	References
2,2,5-Trimethylhexane	3.88		68
2,2,4-Trimethylpentane (Isooctane)	5.83		22
2,3,4-Trimethylpentane	3.78		68
2,4,6-Trinitrotolulene	2.25		91
Triphenyl phosphate	5.27		151
Vinyl acetate	0.73		77
Vinyl chloride	0.60		146
Warfarin	3.20		91
o-Xylene	3.11	2.77–3.13	54,82,103,169
m-Xylene	3.20		103
p-Xylene	3.17	3.15–3.18	103,169

Appendix **H**

Estimation of Octanol-Water Partition Coefficients

The primary sources of K_{ow} data are tabulated values that have been determined experimentally. However, some hazardous waste projects inevitably involve chemicals for which K_{ow} data have not been determined. When octanol-water partition coefficients are not available, the best source of these data is the estimation technique of Hansch and Leo. Their method is based on experimentally determined atomic/group *fragment constants* (f) and *structural factors* (F)—Hansch constants—which are analogous to the Hammett constants used to predict chemical reactivity. Factors are used to normalize K_{ow} based on molecular features such as double bonds, multiple halogenation, and branching.

There are two common methods for estimating K_{ow} with fragment constants. First, a K_{ow} can be synthesized completely from fragment constants and factors:

$$\log K_{ow} = \Sigma f + \Sigma F \tag{H.1}$$

Alternatively, one or more groups can be substituted on a parent compound for which the K_{ow} has previously been determined experimentally:

$$\log K_{ow} \text{ (chemical A)} = \log K_{ow} \text{ (Chemical B)} \pm \Sigma f \pm \Sigma F \tag{H.2}$$

This approach aims to decrease the difference between experimental and mathematically derived values.

FRAGMENTS

A *fragment* (f) is an atom or group of atoms that have empirically determined Hansch constant values. A fragment cannot be bonded just anywhere, it must be attached to what is called an *isolating carbon atom*. An isolating carbon is characterized by (1) having four single bonds, at least two of which must be to nonhetero atoms (i.e. C or H), or (2) being bonded to other carbon atoms. A *hetero* atom is an atom other than a carbon or a hydrogen. Procedures for defining an isolating carbon are shown in Example H.1.

Hansch constants for a number of different types of fragments have been developed. A *single-atom fragment* can be an isolating carbon, a hydrogen, or any other (i.e., hetero) atom. Single-atom fragments must be bonded to isolating carbons. A *multiple-atom fundamental fragment* consists of a group of atoms, which may be made of a non-isolating carbon, hydrogen, or hetero atoms. A multi-atom fragment, which often contains two bonding points, must also be bonded to isolating carbons. An *H-polar fragment* is a group involved in hydrogen bonding and includes amino, hydroxyl, and carboxyl groups. The most important aspect of H-polar fragments is that a factor must be added to account for hydrogen bonding. An *S-polar fragment* is electrophilic and has a tendency to attract electrons; the most common S-polar groups are halogens. As with H-polar fragments, a factor must be added to account for intermolecular forces. Underlining a fragment denotes that the group is bonded to an aromatic ring. Common fragment constants are listed in Table H.1.

EXAMPLE H.1

Determine which of the following are bonded to isolating carbons.
1. $-Cl$ bonded in the CH_3-Cl (methyl chloride) molecule
2. $-NO_2$ bonded in the $CH_3-CH_2-O-NO_2$ (ethyl nitrate) molecule
3. $-O-$ in the $CH_3-CH_2-O-CH_2-CH_3$ (diethyl ether) molecule

$$CH_3-CH_2-\underset{\underset{O}{\|}}{C}-O-CH_3$$

4. $-O-$ in the   (methyl ethyl ketone) molecule

SOLUTION

1. The $-Cl$ is bonded to an isolating carbon because it has four single bonds and two of the bonds are to nonhetero atoms (H).
2. The $-NO_2$ is not bonded to an isolating carbon; it is bonded to an oxygen atom.
3. The $-O-$ atom is bonded to two isolating carbons (each isolating carbon has three bonds to nonhetero atoms).
4. The $-O-$ atom is bonded to one isolating carbon (the methyl group). The carbonyl carbon is not an isolating carbon because it has only one bond to a nonhetero atom.

STRUCTURAL FACTORS

A structural factor (F) is an additional value given to a molecule to account for bond rotations, branching, multiple halogenation, and other intramolecular characteristics. Some commonly used factors are described here.

Table H.1 Common Fragment Constants

Fragment	f	$f\phi$	$f\phi\phi$	Miscellaneous
—F	−0.38	0.37		$f^{\phi/2} = 0.00$
—Cl	0.06	0.94		$f^{\phi/2} = 0.50$
—Br	0.20	1.09		$f^{1R} = 0.48$, $f^{\phi/2} = 0.64$
—I	0.59	1.35		$f^{\phi/2} = 0.97$
—N	−2.18	−0.93	−0.50	$f^{1R} = 1.76$
—O—	1.82	−0.61	0.53	$f^{x1} = -0.22$, $f^{x2} = 0.17$
				$f^{\phi/2} = -1.21$
—S—	−0.79	−0.03	0.77	
—NO$_2$	−1.16	−0.03		$f^{x2} = 0.09$
—N=N—			0.14	
—H	0.23	0.23		
—NH—	−2.15	−1.03	−0.09	$f^{x1} = -0.37$
—NH$_2$	−1.54	−1.00		$f^{x1} = -0.23$, $f^{1R} = -1.35$
—OH	−1.64	−0.44		$f^{x1} = -0.23$, $f^{1R} = 1.34$
—SH	−0.23	0.62		
—C—	0.20	0.20		
—CN	−1.27	−0.34		$f^{1R} = -0.88$
—SCN	−0.48	0.64		$f^{1R} = -0.45$
—CH$_3$	0.89	0.89		
—C$_6$H$_5$ (phenyl)	1.90			
—C—H	−1.10	−0.42		
—C—OH	−1.11	−0.03		$f^{1R} = -1.03$

Key to Subscripts:

Superscript Attachment

None	Aliphatic attachment
ϕ	Aromatic ring attachment. If fragment has two bonding points, the fragment is attached to the right side of the ring.
$1/\phi$	Same as ϕ, but fragment is attached to the left side of the ring.
$\phi\phi$	Fragment with two bonding points and an aromatic attached to each point.
1R	Benzyl (C$_6$H$_5$CH$_2$—) attachment
$\phi/2$	Attached to a vinyl carbon (C=C)

F_b: Bond Factor

A bond factor of −0.12 is added for chains or −0.09 for nonaromatic rings. It is applied $(n - 1)$ times, where n = number of bonds in molecule, within the directives that (a) bonds are not counted between H and any other atom, (b) bonds are not counted with any multi-atom fragment, (c) double and triple bonds are considered as single bonds, and (d) as a chain is counted, a ring stops the count.

F_{cBr}, F_{gBr}, F_{rCl}: **Branch Factors**

These are one-time factors used if there is branching of the chain. Specifics related to this rule include these: (a) Each branch must be only one or two carbons long, or more than one of the branches must contain polar groups, (b) the rule is used for each branch of the molecule, and (c) it is not used if the F_{mhG} or F_{mbv} (see below) is used.

Specific one-time branch factors include

1. $F_{cBr} = -0.13$ if the branching is on an alkane or single α-fragment.
2. $F_{gBr} = -0.22$ if the branching is on an H-polar fragment.
3. $F_{rCl} = -0.45$ if the branching is on fused rings (e.g., PAHs, terpenes, steroids).

$F_=$: **Double-Bond Factor**

The double-bond factor is used for every double bond (exclusive of those in multiple-atom fragments). It is not used for aromatic double bonds if f values are counted for fragments in an aromatic ring.

$F_=$ values include

$F_= = -0.55$ for isolated double bonds

$F^{\phi}_= = -0.42$ for double bonds conjugated to an aromatic ring (e.g., styrenes)

$F_= = -0.38$ for a double bond conjugated to another double bond (e.g., butadienes).

$F^{\phi\phi}_= = 0.0$ for double bonds conjugated to two aromatic rings (e.g., stilbenes)

$F_{\equiv}$: **Triple-Bond Factor**

$F_{\equiv} = -1.42$ is applied for normal triple bonds (acetylenes).

F_{mhG}: **Multiple Geminal Halogenation Factor**

This factor is applied when two or more halogens are bonded to the same carbon atom.

$F_{mhG} = 0.30$ per halogen when two halogens are bonded to a single carbon

$F_{mhG} = 0.53$ per halogen when three halogens are bonded

$F_{mhG} = 0.72$ per halogen when four halogens are bonded.

Note: No branching factor (F_{cBr} or F_{gBr}) is required when the F_{mhG} factor is used.

F_{mhV}: **Multiple Vicinal Halogen Factor**

This factor is applied when two or more halogens are bonded to adjacent carbon atoms. $F_{mhV} = 0.28$ is taken $(n - 1)$ times, where n = total number of vicinal halogens.

Note: F_{mhV} applies only to single-bonded carbon atoms. No branching factor is required when F_{mhV} is used.

F_{p1}, F_{p2}, F_{p3}: Proximity Factors for Two H-Polar Fragments

These factors are used when $-NH_2$, $-OH$, $-O-$, or $-COOH$ groups are separated by one, two, or three carbons.

For polar fragments on a chain

$$F_{p1} = -0.42 \, (f_1 + f_2)$$
$$F_{p2} = -0.26 \, (f_1 + f_2)$$
$$F_{p3} = -0.10 \, (f_1 + f_2)$$

For polar fragments in an alicyclic ring

$$F_{p1} = -0.32 \, (f_1 + f_2)$$
$$F_{p2} = -0.20 \, (f_1 + f_2)$$

For polar fragments on an aromatic ring

$$F_{p1}^{\phi} = -0.16 \, (f_1 + f_2)$$
$$F_{p1}^{\phi} = -0.08 \, (f_1 + f_2)$$

Note: The factor must be applied to each hydrophobic chain or fragment on a chain.

Rules for estimating K_{ow} by the fragment constant method are listed in Table H.2. Although the K_{ow} for almost all organic compounds can be estimated by adding all of the fragments and factors from scratch (Equation H.1), by far the most accurate, as well as the easiest, approach is to estimate a K_{ow} from experimental values available for the most structurally similar analog. Such an approach is effective, but if a factor calculation is involved, this method may become cumbersome.

For example, to estimate the K_{ow} of bromopentachlorobenzene, the easiest approach

---

Table H.2 Rules for Estimating K_{ow} Using Hansch and Leo's Fragment Method

---

1. Draw the structural formula for the compound.
2. Find the most structurally similar analog for which a tabulated K_{ow} is available. If possible, choose an analog that differs only by the group being substituted.
3. Select the fragments and/or factors to be added or subtracted to calculate the K_{ow} for the new compound.
4. Calculate log K_{ow} using Equation 5.7.
5. If no structural analog with a tabulated K_{ow} is available, identify all of the fragments and factors characteristic of the chemical. Add all of the necessary fragments and factors (Equation 5.6) to obtain log K_{ow}.

---

would be to use the log K_{ow} for hexachlorobenzene, which is readily available, and substitute the required bromine atom:

log K_{ow} (bromopentachlorobenzene) =

$$\log K_{ow} \text{ (hexachlorobenzene)} - f_{Cl} + f_{Br} \qquad \text{(H.3)}$$

EXAMPLE H.2

Estimate the log K_{ow} for chlorobenzene and compare your answer to the tabulated value.

SOLUTION

Use the phenyl group, $-C_6H_5$, as the starting compound. Then simply add on the factor for chlorine.

$$\log K_{ow} \text{ (chlorobenzene)} = f_{C_6H_5} + f_{Cl}$$

From Table 5.2, $f_{C_6H_5} = 1.90$ and $f_{Cl} = 0.94$.

$$\log K_{ow} = 1.90 + 0.94 = 2.84$$

The tabulated value = 2.84, which, in this instance, is exactly equal to the estimated value.

EXAMPLE H.3

Estimate the log K_{ow} for 1,4-chloronitrobenzene using fragment constants.

SOLUTION

The structure of 1,4-chloronitrobenzene is:

The estimation is based solely on fragments.

$$f_{C_6H_5} = 1.90$$
$$-f_H = -0.23$$
$$+f_{NO_2}^\phi = -0.03$$
$$+f_{Cl}^\phi = 0.94$$
$$\log K_{ow} = 2.58$$

EXAMPLE H.4

Estimate the log K_{ow} for phenoxyacetic acid using fragment constants.

SOLUTION

The structure is

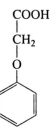

The estimation procedure uses both fragments and factors:

$$f_{C_6H_5} = 1.90$$
$$+f_{-O-}^\phi = -0.61$$
$$+f_c = 0.20$$
$$+2f_H = 2 \cdot 0.23 = 0.46$$
$$+f_{COOH} = -1.11$$
$$-(3-1)F_b = 2 \cdot (-0.12) = -0.24$$
$$+F_{pl} = (-0.42)(-0.61 - 1.11) = 0.72$$
$$\log K_{ow} = 1.32$$

Appendix I

Measured K_{oc} Values of Some Common Hazardous Compounds

Compound	K_{oc}
Acetophenone	35
Alachlor	190
Aldrin	410
Ametryn	392
2-Aminoanthracene	33,500
6-Aminochrysene	162,900
Anthracene	26,000
Anthracene-9-carboxylic acid	517
Asulam	300
Atrazine	148
	162
	216
Benefin	10,700
Benzene	83
2,2'-Biquinoline	10,500
Bromacil	72
Butralin	8,200
Carbaryl	229
	310
Carbofuran	105
	29
Carbophenothion	45,400
Chloramben	21
Chloramben, methyl ester	507
Chlorobromuron	460
Chloroneb	1,159

Compound	K_{oc}
6-Chloropicolinic acid	9
Chlorothiamid	107
Chloroxuron	3,200
Chlorpropham	589
Chlorpyrifos	13,490
	6,070
Chlorpyrifos-methyl	3,300
Crotoxyphos	170
Cyanazine	200
	183
Cycloate	345
2,4-D	57
	20
DDT	240,000
	150,000
Diallate	1,900
Diamidaphos	32
1,2,5,6-Dibenzanthracene	2,029,000
Dibenzothiophene	11,220
Dicamba	0.4
Dichlobenil	235
1,2-Dichlorobenzene	347
1,2-Dichloroethane	32
3,6-Dichloropicolinic acid	2
cis-1,3-Dichloropropene	23
trans-1,3-Dichloropropene	26
Diflubenzuron	6,790
7,12-Dimethylbenzanthracene	235,700
Dinitramine	4,000
Dinoseb	124
Dipropetryn	1,170
Disulfoton	1,780
Diuron	398
	380
	387
Ethion	15,400
Ethylenedibromide	44
Fenuron	27
	43
Fluchloralin	3,600
Fluometuron	174
Glyphosphate	2,640
13H-dibenzo[a,i]carbazole	1,500,000
2,2′,4,4′,5,5′-Hexachlorobiphenyl	417,000
2,2′,4,4′,6,6′-Hexachlorobiphenyl	1,200,000

Compound	K_{oc}
Ipazine	1,660
	813
Isocil	130
Isopropaline	75,250
Leptophos	9,300
Lindane	1,995
Linuron	813
	871
Malathion	1,780
Methazole	2,620
Methomyl	160
Methoxychlor	80,000
2-Methoxy-3,5,6-trichloropyridine	920
9-Methylanthracene	65,000
3-Methylcholanthrene	1,789,000
Methyl isothiocyanate	6
2-Methylnaphthalene	8,500
Methylparathion	5,129
	9,772
Metobromuron	60
Metribuzin	95
Monolinuron	200
	282
Monuron	100
	182
Naphthalene	1,300
Napropamide	680
Neburon	2,300
Nitralin	960
Nitrapyrin	458
	420
Norfluorazon	1,914
Oxadiazon	3,241
Parathion	4,786
	10,715
Pebulate	630
2,2',4,5,5'-Pentachlorobiophenyl	42,500
Pentachlorophenol	900
Perchloroethylene	363
Phenanthrene	23,000
Phenol	27
Phorate	3,200
Picloram	17
Profluralin	8,600
Prometon	350
Prometryn	48

Compound	K_{oc}
Pronamide	200
Propachlor	265
Propazine	158
	155
	363
Propham	51
Pyrazon	120
Pyrene	62,700
	84,000
Pyroxychlor	3,000
Silvex	2,600
Simazine	135
	138
	215
2,4,5-T	53
Tebuthiuron	620
Terbacil	51
	41
Terbutryn	700
Tetracene	650,000
2,3,7,8-Tetrachlorodibenzo-p-dioxin	481,340
1,1,2,2-Tetrachloroethane	79
Thiabendazole	1,720
Triallate	2,220
1,1,1-Trichloroethane (TCA)	178
3,5,6-Trichloro-2-pyridinol	130
Triclopyr	27
Trietazine	547
Trifluralin	4,340
	30,550
	13,700

Appendix J

Vapor Pressures and Henry's Law Constants for Common Hazardous Compounds

Compound	Vapor Pressure (mm Hg @ 20°C)	Henry's Law Constant (atm-m^3/mole) @ 25°C	References
Acenaphthene	1×10^{-3}	7.92×10^{-5}	75
Acenaphthylene	2.9×10^{-2}	2.80×10^{-4}	107,161
Acetaldehyde	750	6.61×10^{-5}	24,133
Acetic acid	11	1.23×10^{-3}	75,133
Acetone	266 @ 25°C	3.97×10^{-5}	75,133
Acetonitrile	73	3.46×10^{-6}	75,133
Acrolein	220	4.4×10^{-6}	77,177
Acrylamide	7×10^{-3}	3.03×10^{-9}	126,133
Acrylonitrile	83	1.1×10^{-4}	77,133
Aldrin	6×10^{-6}	1.4×10^{-6}	133,179
Allyl alcohol	20	5.0×10^{-6}	75,177
Allyl chloride	340	1.08×10^{-2}	34,177
Allyl glycidal ether	3.6	3.83×10^{-6}	126,177
4-Aminobiphenyl	6×10^{-5}	3.89×10^{-10}	121
Ammonia	8.7	2.91×10^{-4}	126,177
n-Amyl acetate	4	3.88×10^{-4}	75,133
sec-Amyl acetate	10 @ 35.2 °C	4.87×10^{-4}	126,183
Aniline	0.6	0.136	65,133
o-Anisidine	1 @ 61°C	1.25×10^{-6}	126,177
Anthracene	1.7×10^{-5}	1.77×10^{-5}	75,108
Aroclor 1016 (PCB-1016)	4×10^{-4}	3.3×10^{-4}	108,172

Compound	Vapor Pressure (mm Hg @ 20°C)	Henry's Law Constant (atm-m³/mole) @ 25°C	References
Aroclor 1221 (PCB-1221)	6.7×10^{-3}	3.24×10^{-4}	172
Aroclor 1232 (PCB-1232)	4.60×10^{-3}	8.64×10^{-4}	172
Aroclor 1242 (PCB-1242)	0.001	5.6×10^{-4}	40,133
Aroclor 1248 (PCB-1248)	4.9×10^{-4} @ 25°C	0.0035	107,110
Aroclor 1254 (PCB-1254)	6×10^{-5}	0.0027	40,133
Aroclor 1260 (PCB-1260)	4.1×10^{-5}	0.0071	107,110
Benzene	76	0.00548	75,133
Benzidine	0.83	3.88×10^{-11}	77,160
Benzo[a]anthracene	2.2×10^{-8}	6.6×10^{-7}	107,108
Benzo[b]fluoranthene	5×10^{-7}	1.2×10^{-5}	6,127
Benzo[k]fluoranthene	9.59×10^{-11} @ 25°C	0.00104	146,172
Benzoic acid	0.0045 @ 25°C	7.02×10^{-8}	77,172
Benzo[g,h,i]perylene	1.01×10^{-10} @ 25°C	1.4×10^{-7}	127,146
Benzo[a]pyrene	5.0×10^{-7}	$<2.4 \times 10^{-6}$	160,162
Benzo[e]pyrene	2.4×10^{-9}	4.84×10^{-7}	119,126
Benzyl alcohol	1 @ 58°C		183
Benzyl butyl phthalate	8.6×10^{-6}	1.3×10^{-6}	77,78
Benzyl chloride	0.9	3.04×10^{-4} @ 20°C	126,133
Biphenyl	0.0497 @ 25°C	4.15×10^{-4}	48,75
Bis(2-chloroethoxy)methane	1 @ 53°C	3.78×10^{-7}	172,183
Bis(2-chloroethyl)ether	0.71	1.3×10^{-5}	159,177
Bis(2-chloroisopropyl)ether	0.85	1.1×10^{-4}	140,177
Bis(2-ethylhexyl)phthalate	2×10^{-7}	1.1×10^{-5}	76,77
Bromobenzene	3.3	2.08×10^{-3}	75,177
Bromochloromethane	141 @ 24.05°C	1.44×10^{-3}	99,126
Bromodichloromethane	50	0.0024	123,159
Bromoform	4	5.32×10^{-4}	131,179
4-Bromophenyl phenyl ether	0.0015	1.0×10^{-4}	35,140
Bromotrifluoromethane	149	0.500	75,133
1,3-Butadiene	1840 @ 21°C	6.3×10^{-2}	31,75
n-Butane	1820 @ 25°C	0.93	23,75
1-Butene	3480 @ 21°C	0.25	75,154
2-Butoxyethanol	0.6	2.36×10^{-6}	126,133
n-Butyl acetate	10	3.3×10^{-4}	75,133
n-Butyl alcohol	4.4	8.48×10^{-6}	75,177
sec-Butyl alcohol	13	1.02×10^{-5}	75,133
tert-Butyl alcohol	31	1.20×10^{-5}	75,133
n-Butylamine	82	1.51×10^{-5}	75,133
n-Butylbenzene	1 @ 22.7°C	0.0125	75,183
sec-Butylbenzene	1.1	0.0114	75,177
tert-Butylbenzene	1.5	0.0117	75,177
n-Butyl mercaptan	35	0.00704 @ 20–22°C	126,133
Camphor	0.18	3.00×10^{-5} @ 20°C	126,133
Carbaryl	0.005	1.27×10^{-5} @ 20°C	126,133

Compound	Vapor Pressure (mm Hg @ 20°C)	Henry's Law Constant (atm-m³/mole) @ 25°C	References
Carbofuran	2×10^{-5} @ 33°C	3.88×10^{-8} @ 30–33°C	126,177
Carbon disulfide	298	0.0133	164,172
Carbon tetrachloride	90	0.0302	177,179
Chlordane	1×10^{-5}	4.8×10^{-5}	133,179
Chloroacetaldehyde	100		133
α-Chloroacetophenone	0.004		177
4-Chloroaniline	0.015	1.07×10^{-5}	77,177
Chlorobenzene	9.0	0.0037	164,174
p-Chloro-m-cresol	0.05	1.78×10^{-6}	108,127
Chloroethane	1011	0.0085	75,164
2-Chloroethyl vinyl ether	26.8	2.5×10^{-4}	140,172
Chloroform	160	0.0032	34,177
2-Chloronaphthalene	0.017 @ 25°	6.12×10^{-4}	172
p-Chloronitrobenzene	<1	$< 6.91 \times 10^{-3}$ @ 20°C	126,133
1-Chloro-1-nitropropane	5.8 @ 25°C	0.157	126,133
o-Chlorophenol	1.42 @ 25°C	8.28×10^{-6}	77,172
4-Chlorophenyl phenyl ether	0.0027 @ 25°C	2.2×10^{-4}	17,140
Chloropicrin	20	0.084	90,133
Chloroprene	200	0.032	75,177
Chloropyrifos	1.87×10^{-5}	4.16×10^{-6}	45,186
Chrysene	6.3×10^{-7}	7.26×10^{-20}	127,161
o-Cresol	0.24 @ 25°C	1.23×10^{-6}	75,177
p-Cresol	0.04	7.92×10^{-7}	75,177
Crotonaldehyde	19	1.96×10^{-5}	53,177
Cyclohexane	95	0.194	75,133
Cyclohexanol	1	5.74×10^{-6}	75,133
Cyclohexanone	4	1.2×10^{-5}	71,177
Cyclohexene	67	0.046	75,154
Cyclopentane	400 @ 31°C	0.186	75,183
Cyclopentene		0.063	75
2,4,-D	0.0047	1.95×10^{-2} @ 20°C	126,148
DDD	1.02×10^{-6} @ 30°C	2.16×10^{-5}	163,172
DDE	6.49×10^{-6} @ 30°C	2.34×10^{-5}	163,172
DDT	7.26×10^{-7} @ 30°C	5.20×10^{-5}	85,163
cis-Decahydronaphthalene	1 @ 22.5°C	39.2	126,183
$trans$-Decahydronaphthalene	10 @ 47.2°C	39.2	126,183
n-Decane	2.7	0.187	126,177
Diacetone alcohol	0.8		133
1,4-Dibromobenzene	0.134 @ 35°C	5.0×10^{-4}	74,75
Dibenz[a,h]anthracene	$\approx 10^{-10}$	7.33×10^{-9}	25,127
Dibromochloromethane	76	9.9×10^{-4}	159
1,2-Dibromo-3-chloropropane	0.8	2.49×10^{-4} @ 20°C	126,133
Dibromodifluoromethane	688		99
Di-n-butyl phthalate	1×10^{-5} @ 25°C	6.3×10^{-5}	108,142

Compound	Vapor Pressure (mm Hg @ 20°C)	Henry's Law Constant (atm-m^3/mole) @ 25°C	References
1,2-Dichlorobenzene	1	0.0019	140,183
1,3-Dichlorobenzene	2.30 @ 25°C	3.60×10^{-3}	140,159
1,4-Dichlorobenzene	0.6	0.00445	75,183
3,3'-Dichlorobenzidine	4.2×10^{-7} @ 25°C	4.5×10^{-8}	77
Dichlorodifluoromethane	4,310	0.425	75,172
1,1-Dichloroethane	182	0.00587	75,164
1,2-Dichloroethane	64	9.1×10^{-4}	159,164
1,1-Dichloroethylene	495	0.021	159,164
trans-1,2-Dichloroethylene	265	0.00674	75,164
Dichlorofluoromethane	1,520 @ 28.4°C	0.0242	127,184
2,4-Dichlorophenol	0.12	6.66×10^{-6}	25,172
1,2-Dichloropropane	42	0.00294	75,177
cis-1,3-Dichloropropylene	25	0.0013	140,159
trans-1,3-Dichloropropylene	25	0.0013	140,159
Dichlorvos	0.012	0.0050	90,186
Dieldrin	1.78×10^{-7}	5.8×10^{-5}	172,179
Diethylamine	195	2.56×10^{-5}	75,133
2-Diethylaminoethanol	1		133
Diethyl phthalate	0.05 @ 70°C	8.46×10^{-7}	47,172
1,2-Difluorotetrachloroethane	40 @ 19.8°C	0.107 @ 20°C	126,183
Diisobutyl ketone	1.7	6.36×10^{-4} @ 20°C	126,177
Diisopropylamine	60		133
N,N-Dimethylacetamide	1.5		133
Dimethylamine	1.7	1.77×10^{-5}	75,133
Dimethylaniline	<1	4.98×10^{-6} @ 20°C	127,133
2,2-Dimethylbutane	319 @ 25°C	1.943	75,185
2,3-Dimethylbutane	200	1.18	126,177
cis-1,2-Dimethylcyclohexane	14.5 @ 25°C	0.354	75,185
trans-1,4-Dimethylcyclohexane	10 @ 13°C	0.870	126,183
Dimethylformamide	2.7		133
1,1-Dimethylhydrazine	103	2.45×10^{-9}	121,133
2,3-Dimethylpentane	68.9 @ 25°C	1.73	126,185
2,4-Dimethylpentane	98.4 @ 25°C	3.152	75,185
3,3-Dimethylpentane	82.8 @ 25°C	1.84	126,185
2,4-Dimethylphenol	0.098 @ 25°C	6.55×10^{-6}	106
Dimethyl phthalate	<0.01	4.2×10^{-7}	47,142
2,2-Dimethylpropane	1,070 @ 19.5°C	2.18	75,137
Dimethyl sulfate	0.5	2.96×10^{-6} @ 20°C	126,183
1,2-Dinitrobenzene	<1	$< 1.47 \times 10^{-3}$ @ 20°C	126,133
1,3-Dinitrobenzene	8.15×10^{-4} @ 35°C	2.75×10^{-7} @ 35°C	75,126
1,4-Dinitrobenzene	2.25×10^{-4} @ 35°C	4.79×10^{-7} @ 35°C	75,126
4,6-Dinitro-o-cresol	3.2×10^{-4}	4.3×10^{-4}	158
2,4-Dinitrophenol	0.00039	1.57×10^{-8} @ 18–20°C	127,158
2,4-Dinitrotoluene	1.1×10^{-4}	8.67×10^{-7}	77

Compound	Vapor Pressure (mm Hg @ 20°C)	Henry's Law Constant (atm-m^3/mole) @ 25°C	References
2,6-Dinitrotoluene	3.5×10^{-4}	2.17×10^{-7}	77
Di-n-octyl phthalate	1.4×10^{-4} @ 25°C	1.41×10^{-12}	108,127
Dioxane	29	4.88×10^{-6}	75,133
1,2-Diphenylhydrazine	2.6×10^{-5} @ 25°C	4.11×10^{-11}	108,127
Diuron	2×10^{-7} @ 30°C	1.46×10^{-9}	70,126
n-Dodecane	0.3	24.2	126,177
α-Endosulfan	10^{-5} @ 25°C	1.01×10^{-4}	113,127
β-Endosulfan	10^{-5} @ 25°C	1.91×10^{-5}	113,127
Endrin	7×10^{-7}	5.0×10^{-7}	172,177
Endrin aldehyde	2×10^{-7} @ 25°C	3.86×10^{-7}	114,127
Epichlorohydrin	12.5	2.46×10^{-5} @ 20°C	70,126
EPN	0.0003 @ 100°C		177
Ethanolamine	0.48		70
2-Ethoxyethanol	4		133
2-Ethoxyethyl acetate	2	9.07×10^{-7} @ 20°C	126,133
Ethyl acetate	72.8	1.34×10^{-4}	75,177
Ethyl acrylate	29.5	2.26×10^{-3} @ 20°C	126,133
Ethylamine	897	1.07×10^{-5}	75,177
Ethylbenzene	7.08	8.68×10^{-3}	23,75
Ethyl bromide	375	7.56×10^{-3}	75,133
Ethylcyclopentane	39.9 @ 25°C	0.0210	126,185
Ethylene chlorohydrin	5		133
Ethylenediamine	10.7	1.73×10^{-9}	75,154
Ethylene dibromide	11	7.06×10^{-4}	75,133
Ethylenimine	160	1.33×10^{-7}	121,177
Ethyl ether	439.8	1.28×10^{-3}	75,186
Ethyl formate	194	2.23×10^{-4}	75,133
Ethyl mercaptan	442	2.74×10^{-3}	75,133
4-Ethylmorpholine	5		133
2-Ethylthiophene	60.9 @ 60.3°C		42
Formaldehyde	760 @ −19.5°C	1.67×10^{-7}	53,183
Formic acid	35	1.67×10^{-7} @ pH 4	53,177
Fluoranthene	0.01	0.0169	75,153
Fluorene	0.005	2.1×10^{-4}	108,142
Furfural	1	2.25×10^{-6} @ 20°C	126,183
Furfuryl alcohol	0.4		177
Glycidol	0.9 @ 25°C		133
Heptachlor	3×10^{-4}	0.0023	142,160
Heptachlor epoxide	2.6×10^{-6}	3.2×10^{-5}	80,179
n-Heptane	40	2.04	75,133
2-Heptanone	2	1.44×10^{-4}	24,133
3-Heptanone	4	4.20×10^{-5} @ 20°C	126,133
cis-2-Heptene	40 @ 21.5°C	0.413 @ 20°C	126,183
trans-2-Heptene	40 @ 21.5°C	0.422	41,126

Compound	Vapor Pressure (mm Hg @ 20°C)	Henry's Law Constant (atm-m³/mole) @ 25°C	References
Hexachlorobenzene	1.089×10^{-5}	0.0017	81,179
Hexachlorobutadiene	0.15	0.026	118,140
α-Hexachlorocyclohexane	2.15×10^{-5}	5.3×10^{-6} @ 20°C	127,161
β-Hexachlorocyclohexane	2.8×10^{-7}	2.3×10^{-7} @ 20°C	9,127
Hexachlorocyclopentadiene	0.081 @ 25°C	0.016	140,173
Hexachloroethane	0.18	0.0025	131,140
n-Hexane	124	1.184	75,133
1-Hexene	186.0 @ 25°C	0.435	75,185
sec-Hexyl acetate	4	0.00515 @ 20°C	126,133
Hydroquinone	<0.001	$<2.07 \times 10^{-9}$	126,133
Indeno[1,2,3-c,d]pyrene	10^{-10} @ 25°C	2.96×10^{-20}	127,172
1-Iodopropane	40 @ 23.6°C	0.00909	75,183
Isoamyl acetate	4	0.0587	75,133
Isoamyl alcohol	2.3	8.89×10^{-6} @ 20°C	126,177
Isobutyl acetate	13	4.85×10^{-4}	126,133
Isobutyl alcohol	9.0	9.25×10^{-6} @ 20°C	126,133
Isobutylbenzene	1 @ 14.1°C	0.0109	75,183
Isophorone	0.38	5.8×10^{-6}	172,177
Isopropyl acetate	47.5	2.81×10^{-4}	75,177
Isopropylamine	478		49
Isopropylbenzene	3.2	0.0147	75,177
Isopropyl ether	130	9.97×10^{-3}	75,177
Kepone	2.25×10^{-7} @ 25°C	0.0311	92,126
Lindane	9.4×10^{-6}	3.25×10^{-6}	85,177
Malathion	4×10^{-5}	4.89×10^{-9}	45,91
Mesityl oxide	8.7	4.01×10^{-6} @ 20°C	126,177
Methyl acetate	173	9.09×10^{-5}	75,133
Methyl acrylate	65	1.31×10^{-4} @ 20°C	126,154
Methylal	76	1.73×10^{-4}	75,133
Methyl alcohol (methanol)	97.6	4.66×10^{-6}	58,75
Methylamine	2,360	0.0181	126,177
Methylaniline	0.3	1.19×10^{-5}	126,177
Methyl bromide	1,630 @ 25°C	0.20	77,140
2-Methyl-1,3-butadiene	550 @ 25°C	0.077	75,185
2-Methylbutane	687 @ 25°C	1.35	126,186
3-Methyl-1-butene	902 @ 25°C	0.535	75,185
Methyl butyl ketone	2	0.00175	127,177
Methyl cellosolve	6.2		177
Methyl cellosolve acetate	2		133
Methyl chloride	3,790	0.010	75,120
Methylcyclohexane	37	0.435	75,133
o-Methylcyclohexanone	≈ 1		133
Methylcyclopentane	138 @ 25°C	0.362	75,185
Methylene chloride	455 @ 25°C	0.00269	34,174

Compound	Vapor Pressure (mm Hg @ 20°C)	Henry's Law Constant (atm-m³/mole) @ 25°C	References
Methyl ethyl ketone (MEK)	77.5	4.66×10^{-5}	75,177
Methyl formate	476	2.23×10^{-4}	75,133
3-Methylheptane	19.5 @ 25°C	3.7	126,185
5-Methyl-3-heptanone	2	1.30×10^{-4} @ 20°C	127,133
2-Methylhexane	65.9 @ 25°C	3.42	126,145
3-Methylhexane	61.6 @ 25°C	1.60	126,185
Methylhydrazine	36		133
Methyl iodide	375	5.48×10^{-3}	75,133
Methyl isobutyl ketone (MIBK)	15	1.49×10^{-5}	127,133
Methyl isocyanate	348	3.89×10^{-4} @ 20°C	126,133
Methyl mercaptan	1,520 @ 25°C	3.01×10^{-3}	75,183
Methyl methacrylate	35	2.46×10^{-4} @ 20°C	126,133
4-Methyloctane	7 @ 25°C	10.27	126,185
Methyl parathion	9.77×10^{-6}		43
2-Methylpentane	212 @ 25°C	1.732	75,185
3-Methylpentane	190 @ 25°C	1.693	75,185
2-Methyl-1-pentene	195 @ 25°C	0.277	126,185
4-Methyl-1-pentene	271 @ 25°C	0.615	75,185
2-Methylpropane	1,520 @ 7.5°C	1.171	75,183
2-Methylpropene	2,270 @ 25°C	0.21	75,185
α-Methylstyrene	1.9		133
Mevinphos	0.003		133
Morpholine	7		133
Naphthalene	0.054	4.6×10^{-4}	140,164
1-Naphthylamine	6.5×10^{-5}	1.27×10^{-10}	121
2-Naphthylamine	2.56×10^{-4}	2.01×10^{-9}	121
Nitrapyrin	0.0028	2.13×10^{-3}	126,177
2-Nitroaniline	8.1 @ 25°C	9.72×10^{-5}	108,127
3-Nitroaniline	1 @ 119.3°C		183
4-Nitroaniline	0.0015	1.14×10^{-8}	127,177
Nitrobenzene	0.15	2.45×10^{-5}	85,177
Nitroethane	15.6	4.66×10^{-5}	75,133
Nitromethane	27.8	2.86×10^{-5}	53,177
2-Nitrophenol	0.20 @ 25°C	3.5×10^{-6}	77
4-Nitrophenol	1×10^{-4}	3×10^{-5} @ 20°C	156
1-Nitropropane	7.5	8.68×10^{-5}	75,133
2-Nitropropane	12.9	1.23×10^{-4}	75,133
N-Nitrosodimethylamine	8.1 @ 25°C	0.143	108,127
N-Nitrosodiphenylamine	0.1 @ 25°C	2.33×10^{-8}	108,127
2-Nitrotoluene	0.15	4.51×10^{-5} @ 20°C	126,133
3-Nitrotoluene	0.15	5.41×10^{-5} @ 20°C	126,133
4-Nitrotoluene	0.12	5.0×10^{-5}	75,133

Compound	Vapor Pressure (mm Hg @ 20°C)	Henry's Law Constant (atm-m^3/mole) @ 25°C	References
n-Nonane	3.22	5.95	126,177
Octachloronaphthalene	<1		133
n-Octane	10.37	3.225	23,75
1-Octene	17.4 @ 25°C	0.952	75,185
Oxalic acid	<0.001	1.43×10^{-10} @ pH 4	53,133
Parathion	4×10^{-4}	8.56×10^{-8}	45,133
Pentachlorobenzene	0.0060	0.0071 @ 20°C	121,135
Pentachloroethane	3.4	2.45×10^{-3}	75,177
Pentachlorophenol	1.4×10^{-4}	3.4×10^{-6}	107
1,4-Pentadiene	735 @ 25°C	0.120	75,185
n-Pentane	426	1.255	75,133
2-Pentanone	12	6.44×10^{-5}	75,177
1-Pentene	638 @ 25°C	0.406	75,185
cis-2-Pentene	495 @ 25°C	0.225	126,185
trans-2-Pentene	506 @ 25°C	0.234	75,185
Perchloroethylene (PCE)	14	0.0153	118,140
Phenanthrene	2.1×10^{-4}	2.56×10^{-4}	75,107
Phenol	0.2	3.97×10^{-7}	75,177
Phenyl ether	0.02	2.13×10^{-4} @ 20°C	126,177
Phthalic anhydride	2×10^{-4}	6.29×10^{-9} @ 20°C	126,177
Picric acid	<1	$<2.15 \times 10^{-5}$ @ 20°C	126,133
Propane	6,540	0.706	75,133
β-Propiolactone	3.4 @ 25°C	7.63×10^{-7}	126,133
n-Propyl acetate	25	1.99×10^{-4}	75,177
n-Propyl alcohol	14.5	6.74×10^{-6}	75,177
n-Propylbenzene	2.5	0.010	75,177
Propylcyclopentane	12.3 @ 25°C	0.890	126,185
Propylene oxide	445	8.34×10^{-5} @ 20°C	70,126
n-Propyl nitrate	18		133
Pyrene	2.5×10^{-6} @ 25°C	1.87×10^{-5}	108,162
Pyridine	14	1.2×10^{-5}	71,177
p-Quinone	0.1	9.48×10^{-7} @ 20°C	126,133
Ronnel	5.29×10^{-5}	8.46×10^{-6}	51,126
Simazine	0.0061		43
Styrene	5	0.00261	172,177
Sulfotepp	0.00017	2.88×10^{-6} @ 20°C	126,186
2,4,5-T	3.75×10^{-5}	4.87×10^{-8}	126,148
1,1,2,2-Tetrabromoethane	0.1	6.4×10^{-5}	126,177
1,2,3,4-Tetrachlorobenzene	0.026 @ 25°C	6.9×10^{-3} @ 20°C	13,135
1,2,3,5-Tetrachlorobenzene	1 @ 58.2°C	0.00158	75,183
1,2,4,5-Tetrachlorobenzene	<0.1 @ 25°C	0.010 @ 20°C	135,154
2,3,7,8-Tetrachlorodibenzo-*p*-dioxin	6.4×10^{-10}	5.40×10^{-23} @ 18–22°C	127,147
1,1,2,2-Tetrachloroethane	5	4.56×10^{-4}	75,177

Compound	Vapor Pressure (mm Hg @ 20°C)	Henry's Law Constant (atm-m^3/mole) @ 25°C	References
Tetrahydrofuran	145	7.06×10^{-5}	75,133
1,2,4,5-Tetramethylbenzene	1 @ 45°C	0.0249	126,183
Tetranitromethane	8.4		133
Tetryl	<1	$<1.89 \times 10^{-3}$ @ 20°C	126,133
Thiophene	60	0.00293	126,177
Toluene	22	0.00674	75,177
2,4-Toluene diisocyanate	0.01		177
o-Toluidine	0.1	1.88×10^{-6}	126,177
Toxaphene	1×10^{-6}	0.063	142,160
1,2,3-Trichlorobenzene	1 @ 40.0°C	0.0089 @ 20°C	135,183
1,2,4-Trichlorobenzene	0.29 @ 25°C	0.00232	174,179
1,3,5-Trichlorobenzene	10 @ 78°C	0.0019 @ 20°C	135,183
1,1,1-Trichloroethane (TCA)	96	0.018	107,172
1,1,2-Trichloroethane	19	9.09×10^{-4}	75,177
Trichloroethylene (TCE)	58	0.0091	140,159
Trichlorofluoromethane	687	1.73	75,177
2,4,5-Trichlorophenol	0.022 @ 25°C	1.76×10^{-7}	106
2,4,6-Trichlorophenol	0.017 @ 25°C	9.07×10^{-8}	106
1,2,3-Trichloropropane	2	3.18×10^{-4}	34,177
1,1,2-Trichlorotrifluoroethane	270	0.333 @ 20°C	126,177
Triethylamine	54	4.79×10^{-4} @ 20°C	126,133
Trifluralin	1.1×10^{-4} @ 25°C	4.84×10^{-5} @ 23°C	36,45
1,2,3-Trimethylbenzene	1 @ 16.8°C	0.00318	126,183
1,2,4-Trimethylbenzene	1 @ 13.6°C	0.0057	75,183
1,3,5-Trimethylbenzene	1 @ 9.6°C	0.00393	126,183
1,1,3-Trimethylcyclopentane	39.7 @ 25°C	1.57	126,185
2,2,5-Trimethylhexane	16.5 @ 25°C	2.42	126,185
2,2,4-Trimethylpentane (Isooctane)	40 @ 20.7°C	3.01	75,183
2,3,4-Trimethylpentane	27.0 @ 25°C	2.98	126,185
2,4,6-Trinitrotoluene	0.00426 @ 54.8°C		102
Triphenyl phosphate	<0.1	0.0588	126,177
Vinyl acetate	83	4.81×10^{-4}	77,177
Vinyl chloride	2580	0.056	75,107
o-Xylene	10 @ 25.9°C	0.00535	75,183
m-Xylene	8.29 @ 25°C	0.0063	15,75
p-Xylene	8.76 @ 25°C	0.0063	15,75

Appendix K

Saturation Concentrations in Air and Heats of Vaporization for Some Common Hazardous Compounds

Compound	Saturation Concentration in Air (g/m^3)		Heat of vaporization (kJ/mol)
	@ 20°C	@ 30°C	
Acetaldehyde	1,800		25.8
Acetic acid	38	63	23.7
Acetic anhydride	19	38	41.2
Acetone	553	825	29.1
Acetonitrile	163	249	29.8
Acetophenone	1.96	3.80	38.8
Acrolein	671	974	28.3
Acrylic acid	12.6	22.8	44.1
Acrylonitrile	257	383	32.6
Allyl alcohol	57	98	40.0
Allyl chloride	1,230	1,770	29.0
Allyl glycidyl ether	22	35	
Aniline	1.5	3.4	42.4
Benzene	319	485	30.7
Benzoyl chloride	3.1	5.2	
Benzyl chloride	6.2	11	
Benzyl cyanide		0.06	
n-Butanol	20	39	43.3

Compound	Saturation Concentration in Air (g/m^3)		Heat of vaporization (kJ/mol)
	@ 20°C	@ 30°C	
sec-Butanol	52	94	40.8
t-Butanol	121	219	
α-Caprolactam	0.006	0.021	
Carbon tetrachloride	754	1,109	29.8
α-Chloroacetophenone	0.034	0.11	
p-Chloroaniline	0.01	0.34	44.4
Chlorobenzene	54	89	35.2
2-Chloroethanol	24	42	41.4
Chloroform	1,030	1,540	29.2
p-Chlorophenol	0.70	1.39	
Chloropicrin	170	286	
Chloroprene	964		
o-Chlorotoluene	18.6	33.3	
o-Cresol	1.2	2.8	42.7
m-Cresol	0.24	0.68	61.7
p-Cresol	0.24	0.74	43.2
Cyclohexane	357	532	30.0
Cyclohexanol	4.9	10.0	45.5
Cyclohexanone	19	32	40.3
Diacetone alcohol	5.7	10	28.5
1,2-Dichlorobenzene	8.0	15	40.6
1,4-Dichlorobenzene	4.8	14	38.8
1,1-Dichloroethane	986	1,410	28.9
1,1-Dichloroethylene	2,640	3,780	26.1
1,2-Dichloropropane	258	393	31.8
Diethylamine	757	1,120	29.1
Diethylene triamine	2.1		
Diisobutyl ketone	9.3	17	
Dimethoxymethane	1,370		
N,N-Dimethylaniline	3.3	7.0	
N,N-Dimethylformamide	12		38.3
1,1-Dimethylhydrazine	505 @ 25°C		32.6
Dimethylsulfide	1420	2,030	27.0
4,6-Dinitro-o-cresol	0.001 @ 25°C		
Diphenyl ether	0.56	1.1	47.1
Epichlorohydrin	60	107	
Ethanol	105	182	38.6
2-Ethoxyethanol	18	33	39.2
2-Ethoxyethyl acetate	14	26	
Ethyl acetate	336	533	31.9
Ethyl acrylate	158	258	34.6
Ethylbenzene	40	67	35.6
Ethylene bromide	113	168	34.8
Ethylenediamine	29	51	38.0

Compound	Saturation Concentration in Air (g/m³)		Heat of vaporization (kJ/mol)
	@ 20°C	@ 30°C	
Ethylene dichloride	350	537	32.0
Ethylene glycol	<0.34	0.65	49.6
Ethyleneimine	375	567	30.3
Ethyl formate	774	1,170	29.9
Ethyl mercaptan	1,490	2,093	26.8
Fluorotrichloromethane	5,110		25.0
Formamide		<0.24	75.4
Formic acid	80	131	22.7
Furfuryl alcohol	2.1	4.1	53.6
Glycerol		<0.5	61.0
n-Heptane	196	306	31.8
2-Heptanone	6.8	13	39.8
Hexachloroethane	5.2	10	45.9
n-Hexane	564	862	28.9
Hydrazine	28	42	45.3
Isoamyl alcohol	11	22	44.1
Isoprene	178	2,510	25.9
Methacrylic acid	3.0	6.3	44.4
Methacrylonitrile	208	318	31.8
Methanol	166	270	35.2
Methyl acetate	665	994	30.3
Methyl acrylate	319	499	34.5
Methyl cellosolve	33	56	45.2
Methyl cellosolve acetate	47		41.1
Methylcyclohexane	192	295	31.3
Methylene chloride	1,550	2,240	28.1
Methyl formate	1,570	2,210	27.9
Methyl isobutyl ketone	27	53	34.5
Methyl isothiocyanate	75.6	115	
Methyl methacrylate	164	258	36.0
4-Nitroaniline	0.011	0.051	
Nitrobenzene	1.0	2.3	40.8
Nitromethane	90	148	34.0
2-Nitrotoluene	0.75	1.8	16.5
3-Nitrotoluene	0.75	1.8	15.0
4-Nitrotoluene	0.75	1.6	15.5
n-Octane	62	108	34.4
2-Octanone	5.2	10	
Pentachloroethane	37	64	37.2
n-Pentane	1,690	2,355	25.8
2-Pentanone	52	95	33.4
Perchloroethylene (PCE)	126	210	34.7
Phenol	0.77	2.0	45.9
o-Phthalic anhydride	0.0016	0.0078	

Compound	Saturation Concentration in Air (g/m³)		Heat of vaporization (kJ/mol)
	@ 20°C	@ 30°C	
n-Propanol	46	85	41.4
n-Propyl acetate	139	226	33.9
n-Propylamine	788	1,180	29.6
Pyridine	65	108	35.1
Styrene	31	52	37.0
1,1,2,2-Tetrachloroethane	46	75	37.6
Tetrahydrofuran	557		29.8
1,2,3,4-Tetrahydro- naphthalene	2.2	4.2	43.9
Tetranitromethane	90	154	40.7
Thiophene	275		31.5
Toluene	110	184	33.2
o-Toluidine	0.58	1.7	44.6
1,1,1-Trichloroethane (TCA)	726	1,090	29.9
1,1,2-Trichloroethane	136	225	34.8
Trichloroethylene (TCE)	415	643	31.4
1,2,3-Trichloropropane	16	31	8.9
1,1,2-Trichloro-1,2,2- trifluoroethane	2,754	3,950	27.0
n-Valeric acid	0.83	2.2	44.1
Vinyl acetate	398	634	34.4
o-Xylene	29	50	36.2
m-Xylene	35	61	36.7
p-Xylene	38	67	35.7

Appendix L

Oral and Inhalation Slope Factors and RfDs

Compound	Oral RfD	Oral SF	Inhalation RfD	Inhalation SF	Carcinogen Class
Acenaphthene	0.06	—	—	—	—
Acephate	0.004	0.0087	—	—	C
Acetochlor	0.02	—	—	—	—
Acetone	0.1	—	—	—	D
Acetonitrile	0.006	—	Pending	—	—
Acetophenone	0.1	—	Pending	—	D
Acifluorfen, sodium	0.013	—	—	—	—
Acrylamide	2.0×10^{-4}	4.5	—	4.5	B2
Acrylic acid	0.5	—	8.58×10^{-5}	—	—
Acrylonitrile	Pending	0.54	—	—	—
Alachlor	0.01	—	Pending	Pending	—
Alar	0.15	—	Pending	Pending	—
Aldicarb	0.001	—	—	—	D
Aldicarb sulfone	0.001	—	—	—	—
Aldrin	3.0×10^{-5}	17	—	17	B2
Allyl alcohol	0.005	—	—	—	—
Aluminum phosphide	4.0×10^{-4}	—	—	—	—
Amdro	3.0×10^{-4}	—	—	—	—
Ametryn	0.009	—	—	—	—
Amitraz	0.0025	—	—	—	—
Ammonium sulfamate	0.2	—	—	—	—
Aniline	—	0.0057	2.86×10^{-4}	—	B2
Anthracene	0.3	—	—	—	D
Antimony	4.0×10^{-4}	—	—	—	—
Apollo	0.013	—	—	—	C
Aramite	Pending	0.025	—	0.025	B2
Aroclor 1016 (PCB-1016)	7.0×10^{-5}	—	—	—	—

Compound	Oral RfD	Oral SF	Inhalation RfD	Inhalation SF	Carcinogen Class
Aroclor 1248 (PCB-1248)	Pending	—	—	—	—
Aroclor 1254 (PCB-1254)	2.0×10^{-5}	—	—	—	—
Arsenic, inorganic	3.0×10^{-4}	1.5	—	50	A
Arsine	—	—	—	—	—
Asbestos	—	—	—	0.23	A
Assure	0.009	—	—	—	D
Asulam	0.05	—	—	—	—
Atrazine	0.035	Pending	—	Pending	—
Avermectin	4.0×10^{-4}	—	—	—	—
Azobenzene	—	0.11	—	0.11	B2
Barium	0.07	—	—	—	—
Barium cyanide	W/D	—	—	—	—
Baygon	0.004	—	—	Pending	—
Bayleton	0.03	—	—	—	—
Baythroid	0.025	—	—	—	—
Benefin	0.3	—	—	—	—
Benomyl	0.05	Pending	—	Pending	—
Bentazon	0.0025	—	—	—	—
Benz[a]anthracene	—	—	—	—	B2
Benzaldehyde	0.1	0.029	—	—	—
Benzene	Pending	0.029	—	0.029	A
Benzidine	0.003	230	—	230	A
Benzo[a]pyrene	—	7.3	—	—	B2
Benzo[e]pyrene	—	Pending	—	Pending	—
Benzo[b]fluoranthene	—	—	—	—	B2
Benzo[g,h,i]perylene	—	—	—	—	D
Benzo[j]fluoranthene	—	Pending	—	Pending	—
Benzo[k]fluoranthene	—	—	—	—	B2
Benzoic acid	4.0	—	—	—	D
Benzotrichloride	—	13	—	—	B2
Benzyl chloride	—	0.17	—	—	B2
Beryllium	0.005	4.3	—	8.4	B2
Bidrin	1.0×10^{-4}	—	—	—	—
Biphenthrin	0.015	Pending	—	—	—
1,1-Biphenyl	0.05	—	—	—	D
Bis(2-chloroisopropyl) ether	0.04	—	—	—	—
Bis(chloroethyl)ether (BCEE)	—	1.1	—	1.1	B2
Bis(chloromethyl)ether (BCME)	—	220	—	220	A
Bisphenol	0.05	—	—	—	—
Boron (boron and borates only)	0.09	Pending	—	Pending	—

Compound	Oral RfD	Oral SF	Inhalation RfD	Inhalation SF	Carcinogen Class
Bromodichloromethane	0.02	0.062	—	—	B2
Bromoform	0.02	0.0079	Pending	7.7×10^{-7}	B2
Bromomethane	0.0014	—	0.00143	—	D
Bromoxynil	0.02	—	—	—	—
Bromoxynil octanoate	0.02	—	—	—	—
1,3-Butadiene	—	No risk	—	1.8	B2
n-Butanol	0.1	—	—	—	D
Butyl benzyl phthalate	0.2	—	—	—	C
Butylate	0.05	—	—	—	—
Butylphthalyl butyl glycolate	1.0	—	—	—	D
Cacodylic acid	Pending	—	—	—	D
Cadmium	5.0×10^{-4}	—	Pending	6.1	B1
Calcium cyanide	0.04	—	—	—	—
Caprolactam	0.5	—	—	—	—
Captafol	0.002	Pending	—	Pending	—
Captan	0.13	Pending	—	Pending	—
Carbaryl	0.1	—	—	—	—
Carbofuran	0.005	—	—	—	—
Carbon disulfide	0.1	—	Pending	—	—
Carbon tetrachloride	7.0×10^{-4}	0.13	—	0.13	B2
Carbosulfan	0.01	—	—	—	—
Carboxin	0.1	—	—	—	—
Chloral	0.002	—	—	—	—
Chloramben	0.015	Pending	—	Pending	—
Chlordane	6.0×10^{-5}	1.3	Pending	1.3	B2
Chlorimuron-ethyl	0.02	—	—	—	—
Chlorine	0.1	Pending	Pending	Pending	—
Chlorine cyanide	0.05	—	—	—	—
Chlorine dioxide	Pending	Pending	5.72×10^{-5}	—	—
p-Chloroaniline	0.004	—	—	—	—
Chlorobenzene	0.02	—	Pending	—	D
Chlorobenzilate	0.02	—	—	—	—
Chloroform	0.01	0.0061	Pending	0.081	B2
β-Chloronaphthalene	0.08	—	—	—	—
2-Chlorophenol	0.005	—	—	—	—
Chlorothalonil	0.015	—	—	Pending	—
o-Chlorotoluene	0.02	—	—	—	—
Chlorpropham	0.2	—	—	—	—
Chlorpyrifos	0.003	—	—	—	—
Chlorsulfuron	0.05	—	—	—	—
Chromium (III), insoluble salts	1.0	Pending	Pending	Pending	—
Chromium (VI)	0.005	—	Pending	41	A
Copper cyanide	0.005	—	—	—	—

Compound	Oral RfD	Oral SF	Inhalation RfD	Inhalation SF	Carcinogen Class
o-Cresol	0.05	None	—	—	C
m-Cresol	0.05	None	—	—	C
p-Cresol	W/D	None	—	—	C
Crotonaldehyde	Pending	—	—	—	C
Cumene	0.04	—	Pending	—	—
Cyanazine	W/D	Pending	—	—	—
Cyanide, free	0.02	—	—	—	D
Cyanogen	0.04	—	—	—	—
Cyanogen bromide	0.09	—	—	—	—
Cyclohexanone	5.0	—	—	—	—
Cyclohexylamine	0.2	—	—	—	—
Cyhalothrin/karate	0.005	—	—	—	—
Cypermethrin	0.01	—	—	—	—
Cyromazine	0.0075	—	—	—	—
Dacthal	0.01	—	—	—	—
Dalapon, sodium salt	0.03	—	—	—	—
Danitol	0.025	—	—	—	—
Decabromodiphenyl ether (DBDPE)	0.01	—	—	—	C
Demeton	4.0×10^{-5}	—	—	—	—
Di(2-ethylhexyl)adipate	0.6	0.0012	—	—	C
Di(2-ethylhexyl) phthalate (DEHP)	0.02	0.014	—	—	B2
1,2-Dibromo-3-chloropropane	—	Pending	5.72×10^{-5}	Pending	—
1,4-Dibromobenzene	0.01	—	—	—	—
Dibromochloromethane	0.02	0.084	—	—	C
1,2-Dibromoethane	—	85	Pending	0.77	B2
Dibutyl phthalate	0.1	—	—	—	D
Dicamba	0.03	Pending	—	—	—
1,2-Dichlorobenzene	0.09	—	—	—	D
1,3-Dichlorobenzene	Pending	—	—	—	D
3,3'-Dichlorobenzidine	—	0.45	—	—	B2
Dichlorodifluoromethane	0.2	—	—	—	—
DDD	—	0.24	—	—	B2
DDE	—	0.34	—	—	B2
DDT	5.0×10^{-4}	0.34	—	0.34	B2
1,1-Dichloroethane	Pending	—	Pending	—	C
1,2-Dichloroethane	—	0.091	—	0.091	B2
1,1-Dichloroethylene	0.009	0.6	Pending	0.180	C
cis-1,2-Dichloroethylene	Pending	—	—	—	D
trans-1,2-Dichloroethylene	0.02	—	—	—	—

Compound	Oral RfD	Oral SF	Inhalation RfD	Inhalation SF	Carcinogen Class
Dichloromethane	0.06	0.0075	Pending	0.00165	B2
2,4-Dichlorophenol	0.003	—	—	—	—
4-(2,4-Dichloro-phenoxy)butyric acid (2,4-DB)	0.008	—	—	—	—
2,4-D	0.01	—	—	—	—
1,2-Dichloropropane	—	—	0.00114	—	—
2,3-Dichloropropanol	0.003	—	—	—	—
1,3-Dichloropropene	3.0×10^{-4}	—	0.00572	—	B2
Dichlorvos	5.0×10^{-4}	0.29	Pending	—	B2
Dicofol	—	W/D	—	—	C
Dieldrin	5.0×10^{-5}	16	—	16	B2
Diethyl phthalate	0.8	—	—	—	D
Difenzoquat	0.08	—	—	—	—
Diflubenzuron	0.02	Pending	—	—	—
1,1-Difluoroethane	—	—	Pending	—	—
Diisopropyl methyl-phosphonate (DIMP)	0.08	—	—	—	D
Dimethipin	0.02	—	—	—	C
Dimethoate	2.0×10^{-4}	Pending	—	Pending	—
Dimethyl phthalate	Pending	—	—	—	D
Dimethyl terephthalate (DMT)	0.1	—	—	—	—
Dimethylamine	Pending	—	—	—	—
N,N-Dimethylaniline	0.002	—	—	—	—
2,4-Dimethylphenol	0.02	—	—	—	—
2,6-Dimethylphenol	6.0×10^{-4}	—	—	—	—
3,4-Dimethylphenol	0.001	—	—	—	—
4,6-Dinitro-o-cyclohexyl phenol	0.002	—	—	—	—
m-Dinitrobenzene	1.0×10^{-4}	None	—	—	D
o-Dinitrobenzene	—	None	—	Pending	D
2,4-Dinitrophenol	0.002	—	—	—	—
2,4-/2,6-Dinitrotoluene mixture	—	0.68	—	—	B2
2,4-Dinitrotoluene	0.002	—	—	—	—
Dinoseb	0.001	—	—	—	D
1,4-Dioxane	—	0.011	—	—	B2
Diphenamid	0.03	—	—	—	—
Diphenylamine	0.025	Pending	—	—	—
1,2-Diphenylhy-drazine	—	0.8	—	0.8	B2
Diquat	0.0022	Pending	—	—	—
Disulfoton	4.0×10^{-5}	—	—	—	—
1,4-Dithiane	0.01	None	—	—	D

Compound	Oral RfD	Oral SF	Inhalation RfD	Inhalation SF	Carcinogen Class
Diuron	0.002	—	—	—	—
Dodine	0.004	—	—	—	—
Endosulfan	0.006	Pending	—	—	—
Endothall	0.02	Pending	—	Pending	—
Endrin	3.0×10^{-4}	—	—	—	D
Epichlorohydrin	W/D	0.0099	2.86×10^{-4}	0.0042	B2
Ethephon	0.005	—	—	—	—
Ethion	5.0×10^{-4}	—	—	—	—
Ethyl acetate	0.9	—	—	—	—
Ethyl chloride	—	Pending	2.86	—	—
S-Ethyl dipropylthio- carbamate (EPTC)	0.025	—	—	—	—
Ethyl ether	0.2	—	—	—	—
Ethyl p-nitrophenyl phenylphosphono- thioate (EPN)	1.0×10^{-5}	—	—	—	—
Ethylbenzene	0.1	—	0.286	—	D
Ethylene diamine	—	None	—	Pending	—
Ethylene glycol	2.0	—	—	—	—
Ethylene thiourea (ETU)	8.0×10^{-5}	Pending	—	Pending	—
Ethyleneimine	—	—	—	—	—
Ethylphthalyl ethyl- glycolate (EPEG)	3.0	—	—	—	—
Express	0.008	Pending	—	Pending	—
Fenamiphos	2.5×10^{-4}	—	—	—	—
Fluometuron	0.013	—	—	—	—
Fluoranthene	0.04	None	—	—	D
Fluorene	0.04	None	—	—	D
Fluorine (soluble fluoride)	0.06	—	—	—	—
Fluridone	0.08	Pending	—	Pending	—
Flurprimidol	0.02	—	—	—	—
Flutolanil	0.06	—	—	—	—
Fluvalinate	0.01	—	—	—	—
Folpet	0.1	0.0035	—	—	B2
Fomesafen	—	0.19	—	—	C
Fonofos	0.002	—	—	—	—
Formaldehyde	0.2	None	—	0.045	B1
Formic acid	W/D	Pending	—	Pending	—
Fosetyl-al	3.0	—	—	—	C
Furan	0.001	—	—	—	—
Furfural	0.003	—	Pending	—	—
Furmecyclox	—	0.03	—	—	B2
Glufosinate- ammonium	4.0×10^{-4}	—	—	—	—

Compound	Oral RfD	Oral SF	Inhalation RfD	Inhalation SF	Carcinogen Class
Glycidaldehyde	4.0×10^{-4}	None	—	—	B2
Glyphosate	0.1	—	—	—	D
Haloxyfop-methyl	5.0×10^{-5}	Pending	—	Pending	—
Harmony	0.013	—	—	—	—
Heptachlor	5.0×10^{-4}	4.5	—	4.5	B2
Heptachlor epoxide	1.3×10^{-5}	9.1	—	9.1	B2
n-Heptane	—	None	—	—	D
Hexabromobenzene	0.002	—	—	—	—
Hexabromodiphenyl ether	—	None	—	—	D
Hexachlorobenzene	8.0×10^{-4}	1.6	—	1.6	B2
Hexachlorobutadiene	W/D	0.078	—	0.078	C
α-Hexachloro-cyclohexane (α-HCH)	—	6.3	—	6.3	B2
β-Hexachloro-cyclohexane (β-HCH)	—	1.8	—	1.8	C
γ-Hexachloro-cyclohexane (γ-HCH)	3.0×10^{-4}	—	Pending	Pending	—
technical Hexachloro-cyclohexane (t-HCH)	—	1.8	—	1.8	B2
Hexachlorocyclo-pentadiene (HCCPD)	0.007	None	—	—	D
Hexachlorodibenzo-p-dioxin, mixture (HxCDD)	—	0.0062	—	0.0062	B2
Hexachloroethane	0.001	0.014	Pending	0.014	C
Hexachlorophene	3.0×10^{-4}	—	—	—	—
Hexahydro-1,3,5-trinitro-1,3,5-triazine (RDX)	0.003	0.11	—	—	C
n-Hexane	—	Pending	0.0572	Pending	—
Hexazinone	0.033	—	—	—	—
Hydrazine/hydrazine sulfate	—	3.0	—	17.1	B2
Hydrogen cyanide	0.02	—	—	—	—
Hydrogen sulfide	0.003	—	2.57×10^{-4}	—	—
Imazalil	0.013	—	—	—	—
Imazaquin	0.025	—	—	—	—
Iprodione	0.04	—	—	—	—
Isobutyl alcohol	0.3	—	—	—	—

Compound	Oral RfD	Oral SF	Inhalation RfD	Inhalation SF	Carcinogen Class
Isophorone	0.2	9.5×10^{-4}	—	—	C
Isopropalin	0.015	—	—	—	—
Isopropyl methyl phosphonic acid (IMPA)	0.1	None	—	—	D
Isoxaben	0.05	—	—	—	C
Lactofen	0.002	—	—	—	—
Linuron	0.0023	—	—	—	C
Londax	0.2	—	—	—	—
Malathion	0.02	—	Pending	—	—
Maleic anhydride	0.1	—	—	—	—
Maleic hydrazide	0.5	Pending	—	Pending	—
Maneb	0.005	—	—	—	—
Manganese	0.005	—	1.14×10^{-4}	—	D
Mepiquat chloride	0.03	—	—	—	—
Mercuric chloride (HgCl$_2$)	3.0×10^{-4}	None			
Mercury, elemental	—	None	—	—	D
Merphos	3.0×10^{-5}	—	—	—	—
Merphos oxide	3.0×10^{-5}	—	—	—	—
Metalaxyl	0.06	—	—	Pending	—
Methacrylonitrile	1.0×10^{-4}	—	—	—	—
Methamidophos	5.0×10^{-5}	—	—	—	—
Methanol	0.5	—	—	—	—
Methidathion	0.001	—	—	—	C
Methomyl	0.025	Pending	—	Pending	—
Methoxychlor	0.005	—	—	—	D
2-Methoxyethanol	Pending	—	0.00572	—	—
Methyl acrylate	—	None	—	—	D
Methyl chlorocarbonate	W/D	Pending	—	Pending	—
Methyl ethyl ketone (MEK)	0.6	None	0.286	—	D
Methyl iodide	—	Pending	Pending	—	—
Methyl isobutyl ketone (MIBK)	W/D	—	Pending	—	—
Methylmercury (MeHg)	1.0×10^{-4}	None	—	—	—
Methyl parathion	2.5×10^{-4}	—	—	—	—
Methyl *tert*-butyl ether (MTBE)	Pending	—	0.143	—	—
4-(2-Methyl-4-chlorophenoxy) butyric acid (MCPB)	0.01	—	—	—	—
2-(2-Methyl-4-chlorophenoxy) propionic acid (MCPP)	0.001	—	—	—	—

Compound	Oral RfD	Oral SF	Inhalation RfD	Inhalation SF	Carcinogen Class
2-Methyl-4-chloro- phenoxyacetic acid (MCPA)	5.0×10^{-4}	—	—	—	—
4,4'-Methylene bis(N,N'-dimethyl) aniline	—	0.046	—	—	B2
Metolachlor	0.15	—	—	—	C
Metribuzin	0.025	—	—	Pending	—
Mirex	2.0×10^{-4}	Pending	—	Pending	—
Molinate	0.002	—	—	—	—
Molybdenum	0.005	—	—	—	—
Monochloramine	0.1	—	—	Pending	—
Naled	0.002	—	—	—	—
Naphthalene	Pending	None	—	—	D
Napropamide	0.1	—	—	—	—
Nickel, soluble salts	0.02	—	Pending	—	—
Nitrapyrin	W/D	—	—	—	—
Nitrate	1.6	Pending	—	—	—
Nitric oxide	W/D	—	—	—	—
Nitrite	0.1	Pending	—	Pending	—
Nitrobenzene	5.0×10^{-4}	None	Pending	—	D
Nitrogen dioxide	W/D	Pending	—	Pending	—
Nitroguanidine	0.1	None	—	—	D
p-Nitrophenol	Pending	—	—	—	—
N-Nitroso-di-n- butylamine	—	5.4	—	5.4	B2
N-Nitroso-n-methyl- ethylamine	—	22	—	—	B2
N-Nitrosodi-n- propylamine	—	7.0	—	—	B2
N-Nitrosodi- ethanolamine	—	2.8	—	—	B2
N-Nitrosodiethyl- amine	—	150	—	150	B2
N-Nitrosodimethyl- amine	—	51	—	51	B2
N-Nitrosodiphenyl- amine	—	0.0049	—	—	B2
N-Nitrosopyrrolidine	—	2.1	—	2.1	B2
Nonabromodiphenyl ether	—	None	—	—	D
Norflurazon	0.04	Pending	—	Pending	—
NuStar	7.0×10^{-4}	—	—	—	—
Octabromodiphenyl ether	0.003	None	—	—	D

Compound	Oral RfD	Oral SF	Inhalation RfD	Inhalation SF	Carcinogen Class
Octahydro-1,3,5,7-tetranitro-1,3,5,7-tetrazocine (HMX)	0.05	—	—	—	D
Oryzalin	0.05	—	—	—	C
Oxadiazon	0.005	Pending	—	Pending	—
Oxamyl	0.025	—	—	—	—
Oxyfluorfen	0.003	Pending	—	Pending	—
Paclobutrazol	0.013	—	—	—	—
Paraquat	0.0045	—	—	—	C
Parathion	Pending	—	—	—	C
Pendimethalin	0.04	—	—	—	—
Pentabromodiphenyl ether	0.002	None	—	—	D
Pentachlorobenzene	8.0×10^{-4}	None	—	—	D
Pentachlorocyclo-pentadiene	—	None	—	—	D
Pentachloronitro-benzene (PCNB)	0.003	Pending	—	Pending	—
Pentachlorophenol	0.03	0.12	Pending	—	B2
Perchloroethylene	0.01	Pending	—	Pending	—
Permethrin	0.05	—			
Phenanthrene	—	None	—	—	D
Phenmedipham	0.25	—	—	—	—
Phenol	0.6	None	—	—	D
m-Phenylenediamine	0.006	—	—	—	—
Phenylmercuric acetate	8.0×10^{-5}	—	—	—	—
Phosalone	W/D	—	—	—	—
Phosmet	0.02	—	—	—	—
Phosphine	3.0×10^{-4}	None	Pending	—	—
Phthalic anhydride	2.0	Pending	—	Pending	—
Picloram	0.07	Pending	—	Pending	—
Pirimiphos-methyl	0.01	—	—	—	—
Potassium cyanide	0.05	—	—	—	—
Potassium silver cyanide	0.2	—	—	—	—
Prochloraz	0.009	0.15	—	—	C
Prometon	0.015	—	—	—	—
Prometryn	0.004	—	—	—	—
Pronamide	0.075	—	—	Pending	—
Propachlor	0.013	—	—	—	—
Propanil	0.005	—	—	—	—
Propargite	0.02	—	—	—	—
Propargyl alcohol	0.002	—	—	—	—
Propazine	0.02	Pending	—	Pending	—

Compound	Oral RfD	Oral SF	Inhalation RfD	Inhalation SF	Carcinogen Class
Propham	0.02	—	—	—	—
Propiconazole	0.013	Pending	—	Pending	—
β-Propiolactone	—	Pending	—	—	—
Propylene oxide	—	0.24	0.00853	0.026	B2
Pursuit	0.25	—	—	—	—
Pydrin	0.025	—	—	—	—
Pyrene	0.03	None	—	—	D
Pyridine	0.001	—	—	—	—
Quinalphos	5.0×10^{-4}	—	—	—	—
Radium 226,228	—	W/D	—	W/D	—
Radon 222	—	W/D	—	W/D	—
Resmethrin	0.03	—	—	—	—
Rotenone	0.004	Pending	—	Pending	—
Savey	0.025	Pending	—	Pending	—
Selenious acid	0.005	None	—	—	D
Selenium and compounds	0.005	None	—	—	D
Selenium sulfide	W/D	—	—	—	B2
Selenourea	0.09	—	—	—	—
Sethoxydim	0.09	—	—	—	—
Silver	0.005	—	—	—	D
Silver cyanide	0.1	—	—	—	—
Simazine	0.005	Pending	—	Pending	—
Sodium azide	0.004	—	—	—	—
Sodium cyanide	0.04	—	—	—	—
Sodium diethyldithio-carbamate	0.03	—	—	—	—
Sodium fluoroacetate	2.0×10^{-5}	—	—	—	—
Strontium	0.6	—	—	—	—
Strychnine	3.0×10^{-4}	—	—	—	—
Styrene	0.2	Pending	0.286	Pending	—
Systhane	0.025	—	—	—	—
Tebuthiuron	0.07	—	—	—	—
Terbacil	0.013	—	—	—	—
Terbutryn	0.001	—	—	—	—
Tetrabromodiphenyl ether	—	None	—	—	D
1,2,4,5-Tetrachloro-benzene	3.0×10^{-4}	—	—	—	—
Tetrachlorocyclo-pentadiene	—	None	—	—	D
1,1,1,2-Tetrachloro-ethane	0.03	0.026	—	0.026	C
1,1,2,2-Tetrachloro-ethane	Pending	0.2	—	0.2	C

Compound	Oral RfD	Oral SF	Inhalation RfD	Inhalation SF	Carcinogen Class
2,3,4,6-Tetrachloro-phenol	0.03	—	—	—	—
Tetrachlorovinphos	0.03	Pending	—	Pending	—
Tetraethyl lead	1.0×10^{-7}	—	—	—	—
Tetraethyldithiopyro-phosphate	5.0×10^{-4}	—	—	—	—
Thallic oxide	W/D	None	—	—	D
Thallium acetate	9.0×10^{-5}	None	—	—	D
Thallium carbonate	8.0×10^{-5}	None	—	—	D
Thallium chloride	8.0×10^{-5}	None	—	—	D
Thallium nitrate	9.0×10^{-5}	None	—	—	D
Thallium selenite	W/D	None	—	—	D
Thallium (I) sulfate	8.0×10^{-5}	None	—	—	D
Thiobencarb	0.01	—	—	—	—
Thiophanate-methyl	0.08	—	—	—	—
Thiram	0.005	Pending	1.4	Pending	—
Toluene	0.2	—	—	1.1	D
Toxaphene	—	1.1	—	—	B2
Tralomethrin	0.0075	—	—	—	—
Triallate	0.013	—	—	—	—
Triasulfuron	0.01	—	—	—	—
1,2,4-Tribromobenzene	0.005	—	—	—	—
Tribromochloromethane	—	None	—	—	D
Tribromodiphenyl ether	—	None	—	—	D
Tributyltin oxide (TBTO)	300,000	—	—	—	—
1,1,2-Trichloro-1,2,2-trifluoroethane (CFC-113)	30	—	Pending	—	—
1,2,4-Trichlorobenzene	0.01	—	Pending	—	D
Trichlorocyclo-pentadiene	—	None	—	—	D
1,1,1-Trichloroethane	W/D	—	Pending	—	D
1,1,2-Trichloroethane	0.004	0.057	Pending	0.057	C
Trichloroethylene	Pending	W/D	—	W/D	—
Trichlorofluoromethane	0.3	—	—	—	—
2,4,5-Trichlorophenol	0.1	Pending	—	Pending	—
2,4,6-Trichlorophenol	—	0.011	—	0.011	B2
2-(2,4,5-Trichloro-phenoxy) propionic acid (2,4,5-TP)	0.008	—	—	—	D
2,4,5-T	0.01	—	—	—	—
1,1,2-Trichloropropane	0.005	—	—	—	—
1,2,3-Trichloropropane	0.006	Pending	—	—	—
Tridiphane	0.003	Pending	—	Pending	—

Compound	Oral RfD	Oral SF	Inhalation RfD	Inhalation SF	Carcinogen Class
Trifluralin	0.0075	0.0077	—	—	C
1,3,5-Trinitrobenzene	5.0×10^{-5}	Pending	—	Pending	—
2,4,6-Trinitrotoluene (TNT)	5.0×10^{-4}	0.03	—	—	C
Uranium, natural	—	W/D	—	W/D	—
Uranium, soluble salts	0.003	—	—	—	—
Vanadium pentoxide	0.009	—	—	—	—
Vernam	0.001	—	—	—	—
Vinclozolin	0.025	—	—	—	—
Vinyl acetate	—	Pending	0.0572	Pending	—
Warfarin	3.0×10^{-4}	—	—	—	—
White phosphorus	2.0×10^{-5}	None	—	—	D
Xylenes	2.0	—	Pending	—	D
Zinc and compounds	0.3	None	—	—	D
Zinc cyanide	0.05	—	—	—	—
Zinc phosphide	3.0×10^{-4}	—	—	—	—
Zineb	0.05	Pending	—	Pending	—

Appendix

References

1. Abdul, S. A., T. L. Gibson, and D. N. Rai, "Statistical correlations for predicting the partition coefficient for nonpolar organic contaminants between aquifer organic carbon and water," *Hazard. Waste Hazard. Mater.*, **4**, 211–222 (1987).
2. Ali, S., "Degradation and environmental fate of endosulfan isomers and endosulfan sulfate in mouse, insect, and laboratory ecosystem," PhD Thesis, University of Illinois, Ann Arbor, MI, 1978.
3. Altshuller, A. P. and H. E. Everson, "The solubility of ethyl acetate in water," *J. Am. Chem. Soc.*, **75**, 1727 (1953).
4. Amoore, J. E. and E. Hautala, "Odor as an aide to chemical safety: Odor thresholds compared with threshold limit values and volatilities for 214 industrial chemicals in air and water dilution," *J. Appl. Toxicol.*, **3**, 272–290 (1983).
5. Andrews, L. J. and R. M. Keefer, "Cation complexes of compounds containing carbon-carbon double bonds. IV. The argentation of aromatic hydrocarbons," *J. Am. Chem. Soc.*, **71**, 3644–3647 (1949).
6. *Aquatic Fate Process Data for Organic Priority Pollutants*, EPA-440/4-81-014, U.S. Environmental Protection Agency, Washington, DC, 1982.
7. Bailey, G. W. and J. L. White, "Herbicides: A compilation of their physical, chemical, and biological properties," *Res. Rev.*, **10**, 97–122 (1965).
8. Baker, E. G., "Origin and migration of oil," *Science (Wash., DC)*, **129**, 871–874 (1959).
9. Balson, E. W., "Studies in vapour pressure measurement. Part III. An effusion manometer sensitive to 5×10^{-6} millimetres of Mercury: Vapour pressure of DDT and other slightly volatile substances," *Trans. Faraday Soc.*, **43**, 54–60 (1947).
10. Banerjee, S., "Solubility of organic mixtures in water," *Environ. Sci. Technol.*, **18**, 587–591 (1984).
11. Banerjee, S., S. H. Yalkowsky, and S. C. Valvani, "Water solubility and octanol/water partition coefficients of organics: Limitations of the solubility-partition coefficient-correlation," *Environ. Sci. Technol.*, **14**, 1227–1229 (1980).
12. Bennett, G. M. and W. G. Philip, "The influence of structure on the solubilities of ethers. Part I. Aliphatic ethers," *J. Chem. Soc.*, **131**, 1930–1937 (1928).
13. Bidleman, T. F., "Estimation of vapor pressures for nonpolar organic compounds by capillary gas chromatography," *Anal. Chem.*, **56**, 2490–2496 (1984).
14. Biggar, J. W. and I. R. Riggs, "Apparent solubility of organochlorine insecticides in water at various temperatures," Hilgardia, 42, 383-391 (1974).
15. Bohon. R. L. and W. F. Claussen, "The solubility of aromatic hydrocarbons in water," *J. Am. Chem Soc.*, **73**, 1571–1578 (1951).
16. Bowman, B. T. and W. W. Sans, "Determination of octanol/water partitioning coefficients (K_{ow}) of 61 organophosphorous and carbamate insecticides and their relationship to respective water solubility (S) values," *J. Environ. Sci. Health*, **B16**, 667–683 (1983).

17. Branson, D. R., "Predicting the fate of chemicals in the aquatic environment from laboratory data," in *Estimating the Hazard of Chemical Substances to Aquatic Life*, American Society for Testing and Materials, 55–70, 1978.

18. Briggs, G. G., "Theoretical and experimental relationships between soil-adsorption, octanol-water partition coefficients, water solubilities, bioconcentration factors and the parachor," *J. Agric. Food Chem.*, **29**, 1050–1059 (1981).

19. Brooke, D., I. Nielsen, J. de Bruijn, and J. Hermens, "An interlaboratory evaluation of the stir-flask method for the determination of octanol-water partition coefficients (Log K_{ow})," *Chemosphere*, **21**, 119–133 (1990).

20. Brown, D. S. and E. W. Flagg, "Empirical prediction of organic pollutant sorption in natural sediments," *J. Environ. Qual.*, **10**, 382–386 (1981).

21. de Bruijn, J., F. Busser, W. Seinen, and J. Hermens, "Determination of octanol/water partition coefficients for hydrophobic organic chemicals with the slow stirring method," *Environ. Toxicol. Chem.*, **8**, 499–512 (1989).

22. Burkhard, L. P., D. W. Kuehl, and G. D. Veith, "Evaluation of reverse phase liquid chromatography/mass spectrometry for estimation of *n*-octanol/water partition coefficients for organic chemicals," *Chemosphere*, **14**, 1551–1560 (1985).

23. Burris, D. R. and W. G. MacIntyre, "A thermodynamic study of solutions of liquid hydrocarbon mixtures in water," *Geochim. Cosmochim. Acta.*, **50**, 1545–1549 (1986).

24. Buttery, R. G., L. C. Ling, and D. G. Guadagni, "Volatilities of aldehydes, ketones, and esters in dilute water solution," *J. Agric. Food Chem.*, **17**, 385–389 (1969).

25. Callahan, M. A., *Water-Related Fate of 129 Priority Pollutants*, EPA-440/4-79-029b, U.S. Environmental Protection Agency, Washington, DC, 1979.

26. Camilleri, P., S. A. Watts, and J. A. Boraston, "A surface area approach to determination of partition coefficients," *J. Am. Chem Soc.*, **9**, 1699–1707 (1988).

27. *Catalog Handbook of Fine Chemicals*, Aldrich Chemical Co., Milwaukee, WI, 1990.

28. Caturla, F., J. M. Martin-Martinez, M. Molina-Sabio, R. Rodriquez-Reinoso, and R. Torregrosa, "Adsorption of substituted phenols on activated carbon," *J. Colloid Interface Sci.*, **124**, 528–534 (1988).

29. Chiou, C. T., D. W. Schedmedding, and M. Manes, "Partitioning of organic compounds in octanol-water systems," *Environ. Sci. Technol.*, **16**, 4–10 (1982).

30. Coates, M., D. W. Connell, and D. M. Barron, "Aqueous solubility and octanol to water partition coefficients of aliphatic hydrocarbons," *Environ. Sci. Technol.*, **19**, 628–632 (1985).

31. Cox, R. A., K. F. Patrick, and S. A. Chant, "Mechanism of atmospheric photooxidation of organic compounds. Reactions of alkoxy radicals in oxidation of *n*-butane and simple ketones," *Environ. Sci. Technol.*, **15**, 587–592 (1981).

32. Dean, J. A. (Ed.), *Lange's Handbook of Chemistry*, 11th Edition, McGraw-Hill, New York, 1973.

33. Deneer, J. W., T. L. Sinnige, W. Seinen, and J. L. M. Hermens, "A quantitative structure activity relationship for the acute toxicity of some epoxy compounds to the guppy," *Aquat. Toxicol.*, **13**, 195–204 (1988).

34. Dilling, W. L., "Interphase transfer processes. Evaporation rates of chloromethanes, ethanes, ethylenes, propanes, and propylenes from dilute aqueous solutions: Comparisons with theoretical predictions," *Environ. Sci. Technol.*, **11**, 405–409 (1977).

35. Dreisbach, R. R., *Pressure-Volume-Temperature Relationships of Organic Compounds*, Handbook Publishers, Sandusky, OH, 1952.

36. *Drinking Water Health Advisory. Pesticides*, Lewis Publishers, Chelsea, MI, 1989.

37. Eadsforth, C. V., "Application of reverse-phase HPLC for the determination of partition coefficients," *Pestic. Sci.*, **17**, 311–325 (1986).

38. Eckert, J. W., "Fungistatic and phytotoxic properties of some derivatives of nitrobenzene," *Phytopathology*, **52**, 642–649 (1962).

39. Eganhouse, R. P. and J. A. Calder, "The solubility of medium weight aromatic hydrocarbons and the effect of hydrocarbon co-solutes and salinity," *Geochim. Cosmochim. Acta.*, **40**, 555–561 (1976).

40. Eisenreich, S. J., B. B. Looney, and J. D. Thorton, "Airborne organic contaminants in the Great Lakes ecosystem," *Environ. Sci. Technol.*, **15**, 30–38 (1981).

41. Ellgehausen, H., C. D'Hondt, and R. Fuerer, "Reversed phase chromatography as a general method for determining 1-octanol/water partition coefficients," *Pestic. Sci.*, **12**, 219–227 (1981).

42. Eon, C. and G. Guichon, "Vapor pressures and second viral coefficients of some five-membered heterocyclic derivitives," *J. Chem Eng. Data*, **16**, 408–410 (1971).

43. *Farm Chemicals Handbook*, Meister Publishing, Willoughby, OH, 1988.

44. Felsot, A. and P. A. Dahm, "Sorption of organophosphorous and carbamate insecticides by soil," *J. Agric. Food Chem.*, **27**, 557–563 (1979).

45. Fendinger, N. J. and D. E. Glotfelty, "Henry's law constants for selected pesticides, PAHs, and PCBs," *Environ. Toxicol. Chem.*, **9**, 731–735 (1990).

46. Fischer, I. and L. Ehrenberg, "Studies of the hydrogen bond. II. Influence of the polarizability of the heteroatom," *Acta. Chem. Scand.*, **2**, 669–677 (1948).

47. Fishbein, L. and P. W. Albro, "Chromatographic and biological aspects of the phthalate esters," *J. Chromatogr.*, **70**, 365–412 (1972).

48. Foreman, W. T. and T. F. Bidleman, "Vapor pressure estimates of individual polychlorinated biphenyls and commercial fluids using gas chromatographic retention data," *J. Chromatogr.*, **330**, 203–216 (1985).

49. Franklin, J. L., J. G. Dillard, H. M. Rosenstock, J. T. Herron, K. Draxl, and F. H. Field, *Ionization Potentials, Appearance Potentials and Heats of Formation of Gaseous Positive Ions*, National Bureau of Standards Report NSRDS-NBS26, U.S. Government Printing Office, Washington, DC, 1969.

50. Franks, F., "Solute-water interactions and solubility behavior of long chain paraffin hydrocarbons," *Nature (London)*, **210**, 87–88 (1966).

51. Freed V. H., C. T. Chiou, and R. Haque, "Chemodynamics: Transport and behavior of chemicals in the environment—A problem in environmental health," *Environ. Health Perspective*, **20**, 55–70 (1977).

52. Fujita, T., J. Iwasa, and C. Hansch, "A new substituent constant, π, derived from partition coefficients," *J. Am. Chem. Soc.*, **86**, 5175–5180 (1964).

53. Gaffney, J. S., G. E. Streit, W. D. Spall, and J. H. Hall, "Beyond acid rain," *Environ. Sci. Technol.*, **21**, 519–524 (1987).

54. Galassi, S., M. Mingazzini, L. Vigano, D. Cesareo, and M. L. Tosato, "Approaches to modeling toxic responses of aquatic organisms to aromatic hydrocarbons," *Ecotoxicol. Environ. Saf.*, **16**, 158–169 (1988).

55. Geyer, H., A. S. Kraus, W. Klein, E. Richter, and F. Korte, "Relationship between water solubility and bioaccumulation potential of organic chemicals in rats," *Chemosphere*, **9**, 277–294 (1980).

56. Geyer, H., G. Politzki, and D. Freitag, "Prediction of ecotoxicological behaviour of chemicals: Relationship between *n*-octanol/water partition coefficient and bioaccumulation of organic chemicals by *Alga chlorella*," *Chemosphere*, **13**, 269–284 (1984).

57. Geyer, H., I. Scheunert, and F. Korte, "Correlation between the bioconcentration potential of organic environmental chemicals in humans and their *n*-octanol/water partition coefficients," *Chemosphere*, **16**, 239–252 (1987).

58. Gibbard, H. F. and J. L. Creek, "Vapor pressure of methanol from 288.15 to 337.65°K" *J. Chem. Eng. Data*, **19**, 308–310 (1974).

59. Ginnings, P. M., D. Plonk, and E. Carter, "Aqueous solubilities of some aliphatic ketones," *J. Am. Chem Soc.*, **62**, 1923–1924 (1940).

60. Gledhill, W. E., "An environmental safety assessment of butyl benzyl phthalate," *Environ. Sci. Technol.*, **14**, 301–305 (1980).

61. Gross, P., "The determination of the solubility of slightly soluble liquids in water and the solubilities of the dichloro- ethanes and propanes," *J. Am. Chem. Soc.*, **51**, 2362–2366 (1929).

62. Groves, F., Jr., "Solubility of cycloparaffins in distilled water and salt water," *J. Chem. Eng. Data*, **33**, 136–138 (1988).

63. Gunther, F. A., W. E. Westlake, and P. S. Jaglan, "Reported solubilities on 738 pesticide chemicals in water," *Resid. Rev.*, **20**, 1–148 (1968).

64. Guseva, A. N. and E. I. Parnov, "Isothermal cross-sections of the systems cyclohexanes-water," *Vestn. Mosk. Univ., Ser. II Khim*, **19**, 77 (1964).

65. Hakuta, T., A. Negishi, T. Goto, J. Kato, and S. Ishizaki, "Vapor-liquid equilibria of some pollutants in aqueous and saline solutions," *Desaliniation*, **21**, 11–21 (1977).

66. Hansch, C. and A. Leo, *Substituent Constants for Correlation Analysis in Chemistry and Biology*, John Wiley & Sons, New York, 1979.

67. Hansch, C., A. Vittoria, C. Silipo, and P. Y. C. Jow, "Partition coefficients and the structure-activity relationship of the anesthetic gases," *J. Med. Chem.*, **18**, 546–548 (1975).

68. Hansch, C., J. E. Quinlan, and G. L. Lawrence, "The linear free-energy relationship between partition coefficients and aqueous solubility of organic liquids," *J. Org. Chem.*, **33**, 347–350 (1968).

69. Hansch, C., and S. M. Anderson, "The affect of intermolecular hydrophobic bonding on partition coefficients," *J. Org. Chem.*, **32**, 2583–2586 (1967).

70. Hawley, G. G., *The Condensed Chemical Dictionary*, Van Nostrand Reinhold, New York, 1981.

71. Hawthorne, S. B., R. E. Sievers, and R. M. Barkely, "Organic emissions from shale oil wastewaters and their implications for air quality," *Environ. Sci. Technol.*, **19**, 992–997 (1985).

72. *Hazardous Waste Management Law, Regulations, and Guidelines for the Handling of Hazardous Waste*, California Department of Health, Sacramento, CA, 1975.

73. Hill, A. E. and R. Macy, "Ternary Systems. II. Silver perchlorate, aniline and water," *J. Am. Chem. Soc.*, **46**, 1132–1143 (1924).

74. Hine, J., H. W. Haworth, and O. B. Ramsey, "Polar effects on rates and equilibria. VI. The effect of solvent on the transmission of polar effects," *J. Am. Chem. Soc.*, **85**, 1473–1475 (1963).

75. Hine, J. and P. K. Mookerjee, "The intrinsic hydrophilic character of organic compounds: Correlation in terms of structural contributions," *J. Org. Chem.*, **40**, 292–298 (1975).

76. Hirzy, J. W., W. J. Adams, W. E. Gledhill, and J. P. Mieure, "Phthalate esters: The environmental issues," Seminar Document, Monsanto Industrial Chemicals Co., St. Louis, MO, (1978).

77. Howard, P. H., *Handbook of Environmental Fate and Exposure Data for Organic Chemicals. Volume I. Large Production and Priority Pollutants*, Lewis Publishers, Chelsea, MI, 1989.

78. Howard, P. H., S. Banerjee, and K. H. Robillard, "Measurement of water solubilities, octanol/water partition coefficients and vapor pressures of commercial phthalate esters," *Environ. Toxicol. Chem.*, **4**, 653–661 (1985).

79. Hutzinger, O., S. Safe, and V. Zitko, *The Chemistry of PCB's*, CRC Press, Boca Raton, Fl. (1974).

80. *IARC Monographs on the Evaluation of Carcinogenic Risk of Chemicals to Man. Some Organochlorine Pesticides*, International Agency for Research on Cancer, Vol. 4, Lyon, France, 1974.

81. Isensee, A. R., E. R. Holden, E. A. Woolson, and G. E. Jones, "Soil persistence and aquatic bioaccumulation potential of hexachlorobenzene," *J. Agric. Food Chem.*, **24**, 1210–1214 (1976).

82. Isnard, S. and S. Lambert, "Estimating bioconcentration factors from octanol-water partition coefficient and aqueous solubility," *Chemosphere*, **17**, 21–34 (1988).

83. Jaffe, P. R. and R. A. Ferrara., "Desorption kinetics in modeling of toxic chemicals," *J. Environ. Eng.*, **109**, 859–867 (1983).

84. Johnsen, S., I. S. Gribbstead, and S. Johansen, "Formation of chlorinated PAH's— A possible health hazard from water chlorination," *Sci. Total Environ.*, **81/82**, 231–238 (1989).

85. Jury, W. A., W. F. Spencer, and W. J. Farmer, "Behavior assessment model for trace organics in soil. III. Application of screening model," *J. Environ. Qual.*, **13**, 573–579 (1984).

86. Kamlet, M. J., M. H. Abraham, R. M. Doherty, and R. W. Taft, "Solubility properties in polymers and biological media. 4. Correlation of octanol/water partition coefficients with solvatochromic parameters," *J. Am. Chem. Soc.*, **106**, 464–466 (1984).

87. Kamlet, M. J., R. M. Doherty, M. H. Abraham, P. W. Carr, R. F. Doherty, and R. W. Taft, "Linear solvation energy relationships. 41 important differences between aqueous solubility relationships for aliphatic and aromatic solutes," *J. Phys. Chem.*, **91**, 1996–2004 (1987).

88. Kanazawa, J., "Relationship between the soil sorption constants for pesticides and their physiochemical properties," *Environ. Toxicol. Chem.*, **8**, 477–484 (1989).

89. Karickhoff, S. W., D. S. Brown, and T. A. Scott, "Sorption of hydrophobic pollutants on natural sediments," *Water Res.*, **13**, 241–248 (1979).

90. Kawamoto, K. and K. Urano, "Parameters for predicting the fate of organochlorine pesticides in the environment. I. Octanol-water and air-water partition coefficients," *Chemosphere*, **18**, 1987–1996 (1989).

91. Kenaga, E. E. and C. A. Goring, "Relationship between water solubility, soil sorption, octanol-water partitioning, and concentration of chemicals in biota," *Aquat. Toxicol.*, ASTM STP 707, (1980).

92. Kilzer, L., I. Scheunert, H. Geyer, W. Klein, and F. Korte, "Laboratory screening of the volatilization rates of organic chemicals from soil and water," *Chemosphere*, **8**, 751–761 (1979).

93. Kim, I.-Y. and F. Y. Saleh, "Aqueous solubilities and transformations of tetrahalogenated benzenes and effects of aquatic fulvic acids," *Bull. Environ. Contam. Toxicol.*, **44**, 813–818 (1990).

94. Klein, W., W. Kordel, M. Weib, and H. J. Poremski, "Updating of the OECD test guideline 107 partition coefficient *n*-octanol/water: OECD laboratory intercomparison test of the HPLC method," *Chemosphere*, **21**, 119–133 (1988).

95. Konemann, H., R. Zelle, and F. Busser, "Determination of Log P_{oct} values of chlorosubstituted benzenes, toluenes, and anilines by high-performance liquid chromatography on ODS-silica," *J. Chromatogr.*, **178**, 559–565 (1979).

96. Konietzko, H., "Chlorinated ethanes: Sources, distribution, environmental impact, and health effects," in *Hazard Assessment of Chemicals*, Vol. 3, 404–448 (1984).

97. Krijgsheld, K. R. and A. van der Gen, "Assessment of the impact of the emission of certain organochlorine compounds on the aquatic environment," *Chemosphere*, **15**, 861–880 (1986).

98. Kronberger, H. and J. Weiss, "Formation and structure of some organic molecular compounds. III. The dielectric polarization of some solid crystalline molecular compounds," *J. Chem. Soc. (Lond.)*, 464–469 (1944).

99. Kudchaker, A. P., S. A. Kudchaker, R. P. Shukla, and P. R. Patnaik, "Vapor pressures and boiling points of selected halomethanes," *J. Phys. Chem. Ref. Data*, **8**, 499–517 (1979).

100. Kurihara, N., M. Uchida, T. Fujita, and M. Nakajima, "Studies on BHC isomers and related compounds. V. Some physicichemical properties of BHC isomers," *Pestic. Biochem. Physiol.*, **2**, 383–390 (1973).

101. LaFleur, K. S., "Sorption of pesticides by model soils and agronomic soils: Rates and equilibria," *Soil Sci.*, **127**, 94–101 (1979).

102. Lenchitz, C. and R. W. Velicky, "Vapor pressure and heat of sublimation of three nitrotoluenes," *J. Chem Eng. Data*, **15**, 401–403 (1970).

103. Leo, A., C. Hansch, D. Elkins, "Partition coefficients and their uses," *Chem. Rev.*, **71**, 525–616 (1971).

104. Lesage, S., R. E. Jackson, M. W. Priddle, and P. G. Riemann, "Occurrence and fate of organic solvent residues in anoxic groundwater at the Gloucester landfill, Canada," *Environ. Sci. Technol.*, **24**, 559–566 (1990).

105. Lu, P.-Y and R. L. Metcalf, "Environmental fate and biodegradability of benzene derivatives as studied in a model ecosystem," *Environ. Health Perspect.*, **10**, 269–284 (1975).

106. Luenberger, C., M. P. Ligocki, and J. F. Pankow, "Trace organic compounds in rain. 4. Identities, concentrations, and scavenging mechanisms for phenols in urban air and rain," *Environ. Sci. Technol.*, **19**, 1053–1058 (1985).

107. Lyman, W. J., W. F. Rheel, and D. H. Rosenblatt (Eds.), *Handbook of Chemical Property Estimation Methods*, McGraw-Hill, New York, 1982.

108. Mabey, W. R., J. H. Smith, R. T. Podoll, H. L. Johnson, T. Mill, T. W Chou, J. Gates, I. W. Partridge, H. Jaber, and D. Vandenberg, *Aquatic Fate Process Data for Organic Priority Pollutants*, EPA-440/4-81-014, U.S. Environmental Protection Agency, Washington, DC, 1982.

109. Mackay, D., A. Bobra, W. Y. Shiu, and S. H. Yalkowsky, "Relationships between aqueous solubility and octanol-water partition coefficients," *Chemosphere*, **9**, 701–711 (1980).

110. Mackay, D. and A. W. Wolkoff, "Rate of evaporation of low-solubility contaminants from water bodies to atmosphere," *Environ. Sci. Technol.*, **7**, 611–614 (1973).

111. Mackay, D. and W.-Y. Shiu, "Aqueous solubility of polynuclear aromatic hydrocarbons," *J. Chem Eng. Data*, **22**, 524–528 (1977).

112. Mackey, D., *Volatilization of Organic Pollutants from Water*, EPA-600/3-82-019, U.S. Environmental Protection Agency, Athens, GA, 1982.

113. Martens, R., "Degradation of [8,9-^{14}C]Endosulfan by soil microorganisms," *Appl. Environ. Microbiol.*, **31**, 853–858 (1975).

114. Martin, H. (Ed), *Pesticide Manual*, 3rd Edition, British Crop Protection Council (1972).

115. May, W. E., S. P. Wasik, and D. H. Freeman, "Determination of the solubility behaviour of some polycyclic aromatic hydrocarbons in water," *Anal. Chem.*, **50**, 997–1000 (1978).

116. McAuliffe, C., "Solubility in water of parraffin, cycloparaffin, olefin, acetylene, cycloolefin, and aromatic compounds," *J. Phys. Chem.*, **70**, 1267–1275 (1966).

117. McCall, P. J., "Estimation of environmental of organic chemicals in model ecosystems," *Residue Rev.*, **85**, 231–244 (1983).

118. McConnell, G., D. M. Ferguson, and C. R. Pearson, "Chlorinated hydrocarbons and the environment," *Endeavour (oxf)*, **34**, 13–18 (1975).

119. McVeety, B. D. and R. A. Hites, "Atmospheric deposition of polycyclic aromatic hydrocarbons to water surfaces: A mass balance approach," *Atmos. Environ.*, **22**, 511–536 (1988).

120. Melnikov, N. N., *Chemistry of Pesticides*, Springer-Verlag, New York (1971).

121. Mercer, J. W., D. C. Skipp, and D. Giffin, *Basics of Pump-and-Treat Ground-Water Remediation Technology*, 600/8-90-003, U.S. Environmental Protection Agency, 1990.

122. Miller, M. M., S. Ghodbane, S. P. Wasik, Y. B. Tewari, and D. E. Martire, "Aqueous solubilities, octanol/water partition coefficients, and entropies of melting of chlorinated benzenes and biphenyls," *J. Chem. Eng. Data*, **29**, 184–190 (1984).

123. Mills, W. B., D. B. Porcella, M. J. Ungs, S. A. Gherini, K. V. Summers, L. Mok, G. L. Rupp, and G. L. Bowie, *Water Quality Assessment: A Screening Procedure for Toxic and Conventional Pollutants in Surface and Groundwater—Part I*, EPA-600/6-85-002a, U.S. Environmental Protection Agency, Washington, DC, 1985.

124. Mirvish, S. S., P. Issenberg, and H. C. Sornson, "Air-water and ether-water distribution of *n*-Nitroso compounds: Implications for laboratory safety, analytic methodology, and carcinogenicity for the rat esophagus, nose, and liver," *J. Natl. Cancer Inst.*, **56**, 1125–1129 (1976).

125. "PCB's-Aroclors," *Technical Bulletin O/PL 306A*, Monsanto Industrial Chemicals Co., St. Louis, MO (1974).

126. Montgomery, J. H., *Groundwater Chemicals Desk Reference. Volume 2*, Lewis Publishers, Chelsea, MI, 1991.

127. Montgomery, J. H., L. M. Welkom, *Groundwater Chemicals Desk Reference*, Lewis Publishers, Chelsea, MI, 1990.

128. Moreale, A., and R. Van Bladel, "Soil interactions of herbicide derived aniline residues: A thermodynamic approach," *Soil Sci.*, **127**, 1–9 (1979).

129. Moriguchi, I., "Quantitative structure—activity studies on parameters related to hydrophobicity," *Chem. Pharm. Bull. (Tokyo)*, **23**, 247–257 (1975).

130. Morrison, R. T., and R. N. Boyd, *Organic Chemistry*, Allyn & Bacon, Boston, MA, 1971.

131. Munz, C., and P. V. Roberts, "Air-water phase equilibria of volatile organic solutes," *J. Am. Water Works Assoc.*, **79**, 62–69 (1987).

132. Neely, W. B., D. R. Branson, and G. E. Blau, "Partition coefficient to measure bioconcentration potential of organic pesticides in fish," *Environ. Sci. Technol.*, **6**, 629–632 (1974).

133. *NIOSH Pocket Guide to Chemical Hazards*, U.S. Department of Health and Human Services, U.S. Government Printing Office, 1987.

134. Ofstad, E. B., and T. Sletten, "Composition and water solubility determination of a commercial tricresylphosphate," *Sci. Total Environ.*, **43**, 233–241 (1985).

135. Oliver, B. G., "Desorption of chlorinated hydrocarbons from spiked and anthropogenically contaminated sediments," *Chemosphere*, **14**, 1087–1106 (1985).

136. Onitsuka, S., Y. Kasai, and K. Yokishimura, "Quantitative structure-toxicity activity relationships of fatty acids and the sodium salts to aquatic organisms," *Chemosphere*, **18**, 1621–1631 (1989).

137. Osborn, A. G. and D. R. Douslin, "Vapor-pressure relationships for 15 hydrocarbons," *J. Chem. Eng. Data*, **19**, 114–117 (1974).

138. Owens, J. W., S. P. Wsaik, and H. DeVoe, "Aqueous solubilities and enthalpies of solution of *n*-alkylbenzenes," *J. Chem. Eng. Data*, **31**, 47–51 (1986).

139. Palit, S. R., "Electronic interpretations of organic chemistry. II. Interpretation of the solubility of organic compounds," *J. Phys. Chem.*, **51**, 837–857 (1947).

140. Pankow, J. F., and M. E. Rosen, "Determination of volatile compounds in water by purging directly to a capillary column with whole column cryotrapping," *Environ. Sci. Technol.*, **22**, 398–405 (1988).

141. Paris, D. F., D. L. Lewis, and J. T. Barnett, "Bioconcentration of toxaphene by microorganisms," *Bull. Environ. Contam. Toxicol.*, **17**, 564–578 (1977).

142. Petrasek, A. C., I. J. Kugelman, B. M. Austern, T. A. Pressley, L. S. Winslow, and R. H. Wise, "Fate of toxic organic compounds in wastewater treatment plants," *J. Water Pollut. Control Fed.*, **55**, 1286–1296 (1983).

143. Polak, J. and B. C.-Y. Lu, "Mutual solubilities of hydrocarbons and water at 0 and 25°C," *Can. J. Chem.*, **51**, 4018–4023 (1973).

144. Price, K. S., Waggy, G. T., and R. A. Conway, "Brine shrimp bioassay and seawater BOD of petrochemicals," *J. Water Pollut. Control Fed.*, **46**, 63–77 (1989).

145. Price, L. C., "Aqueous solubility of petroleum as applied to its origin and primary migration," *Am. Assoc. Pet. Geol. Bull.*, **60**, 213–244 (1976).

146. Radding, S. B., T. Mill, C. W. Gould, D. H. Lia, H. L. Johnson, D. S. Bomberger, and C. V. Fojo, *The Environmental Fate of Selected Polynuclear Aromatic Hydrocarbons*, EPA-560/5-75-009, U.S. Environmental Protection Agency, Washington, DC, 1976.

147. Rappe, C., G. Choudhary, and L. H. Keith, *Chlorinated Dioxins and Dibenzofurans in Perspective*, Lewis Publishers, Chelsea, MI, 1987.

148. Riederer, M., "Estimating partitioning and transport of organic chemicals in the foliage/atmosphere system: Discussion of a fugacity based model," *Environ. Sci. Technol.*, **24**, 829–837 (1990).

149. Rogers, R. D., and J. C. McFarlane, "Sorption of carbon tetrachloride, ethylene dibromide, and tetrachloroethylene on soil and clay," *Environ. Monit. Assess.*, **1**, 155–162 (1991).

150. Ruepert, C., A. Grinwis, and H. Govers, "Prediction of partition coefficients of unsubstituted PAHs from C_{18} chromatographic and structural properties," *Chemosphere*, **14**, 279–291 (1988).

151. Saeger, V. W., O. Hicks, R. G. Kaley, P. R. Michael, J. P. Mieyre, and E. S. Tucker, "Environmental fate of selected phosphate esters," *Environ. Sci. & Technol.*, **13**, 840–844 (1979).

152. Sanemasa, I., M. Araki, T. Deguchi, and H. Nagai, "Solubility measurements of benzene and the akylbenzenes in water making use of solute vapor," *Bull. Chem Soc. Jpn.*, **55**, 1054–1062 (1982).

153. Sax, N. I., (Ed.), *Dangerous Properties of Industrial Materials Report*, Van Nostrand Reinhold, New York, 1986.

154. Sax, N. I., and R. J. Lewis, Sr., *Hazardous Chemicals Desk Reference*, Van Nostrand Reinhold, New York, 1987.

155. Schroy, J. M., F. D. Hileman, and S. C. Cheng, "Physical/Chemical properties of 2,3,7,8-TCDD," *Chemosphere*, **14**, 877–880 (1985).

156. Schwarz, F. P., "Measurement of the solubilities of slightly soluble organic liqiuds in water by elution chromatography," *Anal. Chem.*, **52**, 10–15 (1980).

157. Schwarz, F. P., and S. P. Wasik, "Fluorescence measurements of benzene, naphthalene, anthracene, pyrene, fluoranthene, and benzo[*e*]pyrene in water," *Anal. Chem.*, **48**, 524–528 (1976).

158. Schwarzenbach, R. P., R. Stierli, B. R. Folsom, and J. Zeyer, "Compound properties relevant for assessing environmental partitioning of nitrophenols," *Environ. Sci. & Technol.*, **22**, 83–92 (1988).

159. Schwille, F. *Dense Chlorinated Solvents*, Lewis Publishers, Chelsea, MI, 1988.

160. Sims, R. C., W. C. Doucette, J. E. McLean, W. J. Grenney, and R. R. Dupont, *Treatment Potential for 56 EPA Listed Hazardous Chemicals in Soil*, EPA-600/6-88-001, U.S. Environmental Protection Agency, Washington, DC, 1988.

161. Smith, J. H., D. Mackay, and C. W. K. Ng, "Volatilization of pesticides from water," *Residue Rev.*, **85**, 57–71 (1983).

162. Southworth, G. R., "The role of volatilization in removing polycyclic aromatic hydrocarbons from aquatic environments," *Bull. Environ. Contam. Toxicol.*, **21**, 507–514 (1979).

163. Spencer, W. F. and M. M. Cliath, "Volatility of DDT and related compounds," *J. Agric. Food Chem.*, **20**, 645–649 (1972).

164. Standen, A. (Ed.), *Kirk-Othmer Encyclopedia of Chemical Toxicology. Volume 4*, 2nd Edition, John Wiley & Sons, New York, 1964.

165. Stephen, H. and T. Stephen, *Solubilities of Inorganic and Organic Compounds. Part I, Volume 1*, Pergamon Printing and Art Services, Ltd., London, 1963.

166. Stretiweiser, A., Jr. and L. L. Nebenzahl, "Carbon acidity equilibrium acidity of cyclohexylamine," *J. Am. Chem. Society*, **98**, 2188–2190 (1976).

167. Sutton, C. and J. A. Calder, "Solubility of alkylbenzenes in distilled water and seawater at 25°C," *J. Chem. Eng. Data*, **20**, 320–322 (1975).

168. Swann, R. L., "A rapid method for the estimation of the environmental parameters octanol/water partition coefficient, soil sorption constant, water to air ratio, and water solubility," *Residue Rev.*, **85**, 17–28 (1983).

169. Tewari, Y. B., M. M. Miller, S. P. Wasik, and D. E. Martire, "Aqueous solubility and octanol-water partition coefficients of organic compounds at 25°C," *J. Chem Eng. Data*, **27**, 451–454 (1982).

170. Thomas, R. G., "Volatilization from water," in Lyman, W. J., W. F. Rheel, and D. H. Rosenblatt (Eds.), *Handbook of Chemical Property Estimation Methods*, McGraw-Hill, New York, 1982.

171. Travis, C. C. and A. D. Arms, "Bioconcentration of organics in beef, milk, and vegetation," *Environ. Sci. Technol.*, **22**, 271–274 (1988).

172. *Treatability Manual. Volume 1. Treatability Data,* EPA-600/8-80-042a, U.S. Environmental Protection Agency, Washington, DC, 1980.

173. Ungnade, H. E. and E. T. McBee, "The chemistry of perchlorocyclopentadienes and cyclopentadienes," *Chem. Rev.,* **58,** 249–320 (1957).

174. Valsaraj, K. T., "On the physio-chemical aspects of partitioning of nonpolar hydrophobic organics at the air-water interface," *Chemosphere,* **17,** 875–887 (1988).

175. Valvani, S. C., S. H. Yalkowsky, and T. J. Roseman, "Solubility and partitioning. IV. Aqueous solubility and octanol-water partition coefficients of liquid nonelectrolytes," *J. Pharm. Sci.,* **70,** 502–506 (1981).

176. Veith, G. D., K. J. Macek, S. R. Petrocelli, and J. Carroll, *An Evaluation of Using Partition Coefficients and Water Solubility to Estimate Bioconcentration Factors for Organic Chemicals in Fish,* ASTM STP 707, American Society for Testing and Materials, Philadelphia, PA, 1980.

177. Verschueren, K., *Handbook of Environmental Data on Organic Compounds,* Van Nostrand Reinhold, New York, 1983.

178. Walton, W. C., *Practical Aspects of Ground Water Modeling,* National Water Well Association, Worthington, OH, 1985.

179. Warner, H. P., J. M. Cohen, and J. C. Ireland, *Determination of Henry's Law Constants of Selected Priority Pollutants,* EPA-600/D-87/229, U.S. Environmental Protection Agency, Washington, DC, 1987.

180. Wasik, S. P., "Determination of the vapor pressure, aqueous solubility, and octanol/water partition coefficient of hydrophobic substances by coupled generator column/liquid chromatographic methods," *Residue Rev.,* **85,** 29–42 (1983).

181. Watarai, H., M. Tanaka, and N. Suzuki, "Determination of partition coefficients of halobenzenes in heptane/water and 1-octanol/water systems and comparison with scaled particle calculation," *Anal. Chem.,* **54,** 702–705 (1982).

182. Weil, L., G. Dure, and K. E. Quentin, "Solubility in water of insecticide chlorinated hydrocarbons and polychlorinated biphenyls in view of water pollution," *Z. Wasser Forsch.* **7;** 169–175 (1974).

183. Weast, R., *Handbook of Chemistry and Physics,* CRC Press, Boca Raton, FL, 1989.

184. Weiss, G., *Hazardous Chemicals Data Book,* Noyes Data Corp., Park Ridge, NJ, 1986.

185. Wilhoit, R. C. and B. J. Zwolinski, *Handbook of Vapor Pressure and Heats of Vaporization of Hydrocarbons and Related Compounds,* Publication 101, Thermodynamics Research Station, College Station, TX, 1971.

186. Windholz, M., *The Merck Index,* Merck and Co., Rahway, NJ, 1976.

187. Wolf, N. L., "Methoxychlor and DDT degradation in water: rates and products," *Environ. Sci. & Technol.,* **11,** 1077–1081 (1977).

188. Yalkowsky, S. H., R. J. Orr, and S. C. Valvani, "Solubility and partitioning. 3. The solubility of halobenzenes in water," *I & EC Fundam.,* **18,** 351–353 (1979).

189. Yalkowsky, S. H., and S. C. Valvani, "Solubilities and partitioning. 2. Relationships between aqueous solubilities, partition coefficients, and molecular surface areas of rigid aromatic hydrocarbons," *J. Chem. Eng. Data,* **24,** 127–129 (1979).

190. Yalkowsky, S. H., S. C. Valvani, and D. Mackay, "Estimation of the aqueous solubility of some aromatic compounds," *Residue Rev.,* **85,** 43–55 (1983).

191. Yoshida, K., T. Shigeoka, and F. Yamauchi, "Non–steady state equilibrium model for the preliminary prediction of the fate of chemicals in the environment," *Ecotoxicol. Environ. Saf.,* **7,** 179–190 (1983).

192. Yoshida, K., Tadayoshi, S., and F. Yamauchi, "Relationship between molar refraction and *n*-octanol/water partition coefficient," *Ecotoxicol. Environ. Saf.,* **7,** 558–565 (1983).

193. Zepp, R. G., N. L. Wolfe, G. L. Baughman, P. F. Schlotzhauer, and J. N. MacAllister, "Dynamics of processes influencing the behavior of hexachlorocyclopentadiene in the aquatic environment," *178th Meeting of the American Chemical Society,* Washington, DC, 1979.

Index